Mechanisms in Heterogeneous Catalysis

CATALYTIC SCIENCE SERIES

ISSN 1793-1398 (Print)
ISSN 2399-4495 (Online)

Catalysis is at the forefront of the chemical industry and is essential to many fields in the chemical sciences. This series explores all aspects of catalysis in authored and edited volumes drawing on expertise from around the globe in a focussed manner. Volumes are accessible by postgraduate students and professionals in academia and industry.

Published

Vol. 22 *Mechanisms in Heterogeneous Catalysis*
by Rutger A. van Santen

Vol. 21 *Applications of X-Ray Photoelectron Spectroscopy to Catalytic Studies: From Routine Analysis to Cutting Edge Surface Characterization*
by Spyridon Zafeiratos

Vol. 20 *Noble-Metal-Free Electrocatalysts for Hydrogen Energy*
edited by Qingsheng Gao and Lichun Yang

Vol. 19 *Iron Catalysis: Design and Applications*
edited by Jose M. Palomo

Vol. 18 *Photoorganocatalysis in Organic Synthesis*
edited by Maurizio Fagnoni, Stefano Protti and Davide Ravelli

Vol. 17 *Hydroprocessing Catalysts and Processes: The Challenges for Biofuels Production*
edited by Bo Zhang and Duncan Seddon

Vol. 16 *Electro-Catalysis at Chemically Modified Solid Surfaces*
by Jacques Simonet

Vol. 15 *Noble Metal Noble Value: Ru-, Rh-, Pd-catalyzed Heterocycle Synthesis*
edited by Xiao-Feng Wu

Vol. 14 *Enantioselective Titanium-Catalysed Transformations*
by Hélène Pellissier

Vol. 13 *Gold Catalysis: An Homogeneous Approach*
edited by F. Dean Toste and Véronique Michelet

More information on this series can be found at http://www.worldscientific.com/series/css

(Continued at end of book)

CATALYTIC SCIENCE SERIES — VOL. 22

Series Editor: Chris Hardacre

Mechanisms in Heterogeneous Catalysis

Rutger A. van Santen

Eindhoven University of Technology, The Netherlands

World Scientific

NEW JERSEY • LONDON • SINGAPORE • BEIJING • SHANGHAI • HONG KONG • TAIPEI • CHENNAI • TOKYO

Published by

World Scientific Publishing Europe Ltd.
57 Shelton Street, Covent Garden, London WC2H 9HE
Head office: 5 Toh Tuck Link, Singapore 596224
USA office: 27 Warren Street, Suite 401-402, Hackensack, NJ 07601

Library of Congress Control Number: 2023006354

British Library Cataloguing-in-Publication Data
A catalogue record for this book is available from the British Library.

Catalytic Science Series — Vol. 22
MECHANISMS IN HETEROGENEOUS CATALYSIS

ISBN 978-1-80061-400-0 (hardcover)
ISBN 978-1-80061-401-7 (ebook for institutions)
ISBN 978-1-80061-402-4 (ebook for individuals)

For any available supplementary material, please visit
https://www.worldscientific.com/worldscibooks/10.1142/Q0412#t=suppl

Desk Editor: Shaun Tan Yi Jie

Typeset by Stallion Press
Email: enquiries@stallionpress.com

Preface

Catalysis science uncovers the rules that determine activity and selectivity of reactions, and relates this to composition and structure of the catalyst. In empirical practice, such rules develop by correlation of catalyst reactivity with its physical and chemical properties. The history of the transformation of catalysis based mainly on empirical correlation to a more deterministic science, based on molecular theory, is one of the leading themes in the book that follows. The book describes the mechanisms of catalytic reactions and their intimate relation with the inorganic chemistry of the catalyst. It also provides an exposition of the rich variety of known heterogeneous catalytic systems. Reaction mechanism is the network of elementary reactions that connects surface chemical reactivity with catalyst performance. It is part of the catalytic reaction cycles that define reaction kinetics. One of the reasons of the complexity of catalytic kinetics is that it integrates processes at different time and length scales. The relation between short-scale processes of molecular surface chemistry and longer-scale processes of catalytic reactor performance is central to catalytic kinetics.

Reaction mechanism in heterogeneous catalysis has only recently obtained a firm foundation. In science, discoveries are made within the context of state of knowledge and available tools. For the most part of the previous century, mechanistic theories of heterogenous catalytic reactions remained largely speculative because of limitations to probe the catalyst and reaction at the molecular level. Nonetheless, with time these theories contributed largely to increasing sophistication of kinetic simulation and chemical reactor engineering science. Only the past 30 years, due to advances in spectroscopies, computational science and design of molecular heterogeneous catalysts, it became possible to access catalytic reactivity at the molecular level. Early ideas on reaction mechanism as well as modern insights are described.

Catalysis is a science as well as a technology. Most of large-scale chemical processes employ catalysts. Also, smaller conversion devices for automotive exhaust treatment or electricity-generating fuel cells are based on catalysis. In the chapters on reaction mechanism, this relation with technology is made explicit in historic sketches of developing scientific understanding motivated by the discovery of new catalytic reactions. A main driver to new catalyst and reaction discovery is the search for a chemical process solution to the production of desirable chemical products. This is driven by opportunities created by new material or energy resources, need for alternatives to scarce natural products, and reduction of environmental waste. In the course of the previous century, raw material resources have changed from coal to oil and natural gas. At present, biomass conversion, renewable energy access and electricity storage are drivers for change.

Catalysis science developed at the interphase of technological invention and scientific exploration. It provides an interesting historic case study of the relation of industrial and academic research. This is another theme of the book. For respective reactions the mechanistic chapters contain a description of the evolution of main understanding of catalytic chemistry that developed in the course of time, which is complemented with the modern view that is often due to recent computational investigation. The chapters are organized by main reaction types and classes of catalyst systems. The mechanistic chapters are preceded by two additional chapters that deal with physical chemical aspects and introduces theories of surface chemistry and kinetics. In an introductory chapter the evolution of heterogeneous catalytic processes is sketched. This is preceded with a description of the founding catalytic discoveries and understanding of catalysis in the 19th century. In a final chapter the catalytic enterprise is reviewed. For catalysis, the dynamic process of scientific discovery, catalyst design and process innovation is described with examples from processes taken from the book. Scientific debate is important and some case stories are mentioned. Scientific growth relates to the cross-fertilization of ideas and techniques from many players. The outcome is often unexpected but valuable. The satisfactory stage of present understanding on heterogenous catalytic reaction mechanisms is that, at the molecular level, it is chemically described in similar terms as inorganic chemistry or physical organic chemistry. For instance, the surface complex of reaction intermediate and catalyst reactive site is an embedded organometal or coordination complex. Scientific questions that remain a challenge to resolve are also discussed. Amongst others, we look at the response of surface inorganic chemistry to catalytic reaction.

To write this comprehensive book asks also for a selection principle. Which topics to select and in what depth should reaction mechanism be discussed? I decided to bring together the mechanistic presentations in four chapters. Three chapters are organized along the reaction categories of hydrogenation, selective oxidation and solid acid catalysis. A fourth separate chapter deals with reactions which are mechanistically distinct and are catalyzed by molecular heterogenous catalytic systems. The difference in catalysts is whether the reactive surface is essentially part of a truncated solid, or is an organometal or coordination complex attached to a high surface area support. Each chapter presents major reactions and their mechanism. For many reactions, the context of their invention is also presented. The chapter presentations contain detailed chemical information for readers to learn the molecular aspects of reaction mechanism, its relation with kinetics, and how it relates to the inorganic chemistry of the catalyst. Most of the chapter sections are headed by a short summarizing motive, which highlights the content of a particular section. The focus of the discussion is on principles and major ideas. Conflicting and unresolved interpretations, and pros and cons, are mentioned. Results of experiment and simulation are presented but not the details. For the interested reader, an extensive list of references is added to each chapter.

The book aims at an advanced readership with a graduate-level knowledge of chemistry. It complements introductory catalysis books as the recent book by R. Prins, A. Wang, X. Li, and F. Sapountzi, *Introduction to Heterogeneous Catalysis*, World Scientific (2022) or the book by G. Rothenberg, *Catalysis: Concepts and Green Applications*, Wiley (2017).

It gave me great pleasure to work on this book for the past three years. The book is quite different from previous books I wrote on catalysis. For the past 40 years I have written a monograph in each decade. One together with Hans Niemantsverdriet, another with Matthew Neurock. This book would not have appeared without writing these previous books. They reflect the thinking in each time period and ideas limited by the status of experiment and theory from that decade. Writing this book helped me to rethink many of the catalytic issues I previously discussed. I never had the courage to focus singly on mechanism, since the molecular basis of surface reactions was not yet firmly formulated. Fortunately, mainly due to advances of the past 20 years, this situation has been altered. The uniqueness of this book on mechanism is that it is comprehensive and integrates classical fundamental concepts with modern molecular understanding.

Working on this book gave me an opportunity to also think back on my own wading through the maze of catalysis of the past 50 years. Through reading their papers or books in the writing process, I was virtually meeting again many friends and colleagues of the catalysis community I have met over the years. It made me remember pleasant meetings and our friendships, and also their lectures and the many discussions where statements and questions sometimes became hotly argued. I have been very fortunate to collaborate with great colleagues in many different locations, and meetings with great scientists on many occasions. I also enjoyed my contacts with many of the talented students with whom I have the privilege to undertake exciting adventures in the unknown. I treasure all these contacts, and the book is also a contribution to the catalysis community where I found my place in the past decade. Additionally, in the process of writing the book I was sometimes struck by the closeness in time of major recent industrial developments and my then presence in an industrial research environment, but without awareness of these great catalytic innovations and the chemical inventions that gave rise to them.

How I would have benefitted from expositions or discussions from experienced scientists that would have told me of such recent and exciting catalysis. I hope that this book will also serve such a role to the younger generation of catalytic chemists and engineers.

The book would not have appeared without the great help of Mustafa Doğan, who assisted me with the editing of text and drawing many of the figures. Figures where a reference number is indicated at the end of the caption are reproduced from published works, with kind permission from the respective publishers. Several colleagues gave me invaluable help by critical reading of parts of the manuscript. Specifically, I am very grateful for initial editorial advice by Bram Vermeer and the helpful suggestions and discussions with Rob (J.A.R) van Veen and Roel Prins. Due to their critical reading, many original mistakes and errors in the manuscript could be corrected for. The ones that are left are fully to my account.

Rutger A. van Santen
Amstelveen 2023

About the Author

Dr. Rutger van Santen is Emeritus Professor at the Institute for Complex Molecular Systems and Faculty of Chemistry and Chemical Engineering of the Eindhoven University of Technology, The Netherlands. He graduated in 1972 from the University of Leiden. After a postdoctoral stay at Stanford Research Institute, he joined Shell Research in 1973 and was appointed to the Chair of catalysis at the Eindhoven University of Technology in 1988, where he was Rector-Magnificus from 2001 to 2005. He is an Elected Member of the Royal Dutch Academy of Arts and Sciences (KNAW) and a Foreign Member of the U.S. National Academy of Engineering (NAE). He is considered one of the pioneers in the use of quantum chemical methods in computational heterogeneous catalysis. He has published over 800 papers, written and edited 17 books, and owns 22 patents, with a h-index of 102. His research achievements has been encapsulated in a Festschrift, *40 years of Catalysis Research: Rutger van Santen's Journey Through Chemical Complexity* (2012).

Professor van Santen is also the Founding Director of the Netherlands Institute for Catalysis Research (NIOK) and the Dutch National Research School Combination-Catalysis (NRSC-C). He has received many prestigious awards, such as the 1981 Gold Medal from the Royal Dutch Chemical Society, the 1992 Ciapetta Award from the North American Catalysis Society, the 1997 Bourke

Award from the U.K. Royal Society of Chemistry, the 1997 Spinoza Award from the Dutch Research Council, the 2000 Karl Ziegler Prize from the Max Planck Institut für Kohlenforschung, the 2001 Alwin Mittasch Medal from the German Catalysis Society, the 2009 Holst Award from Eindhoven University of Technology, and the 2010 Francois Gault Award from the European Federation of Catalysis Societies. He also received an Honorary Doctorate from the National Technical University of Ukraine and is a Knight of the Order of the Dutch Lion.

Contents

Chapter 1

Heterogenous Catalysis: History and Processes

1.1 Introduction

Reaction mechanism connects catalytic performance with catalyst chemistry.

Heterogeneous catalysis is of great practical interest. Catalytic chemical processes are the backbone of the chemical industry as we know it today. Catalysis science developed largely within the context of discovery and subsequent improvement of industrial processes. The large body of scientific catalytic knowledge which became discovered is at the core of this book.

The science of heterogeneous catalysis explains the chemical reactivity principles that cause the remarkable power of a catalyst to convert reactant to desired product. The heterogeneous catalytic reaction is a dynamic process wherein reactants are activated by an inorganic material, which has a unique feature in that it is not consumed by reaction and can be readily separated from product. The attraction of catalytic reactions is that they produce minimum waste and are energy efficient. A small amount of catalyst material can produce much product material, that is a multitude of the catalyst added. The catalyst functions as a product multiplier or rate accelerator. The inorganic catalyst is complex in composition and structure, which is tuned so as to achieve high reactant conversion rate and selective product formation. Catalysis is accomplished by a network of chemical reactions and reaction intermediates that constitutes the reaction mechanism. Catalyst performance is connected with the chemistry of the catalytic material through this reaction network. In the book, reaction mechanisms are discussed with a focus on this functional relation of catalytic reactivity and catalyst surface chemistry.

The evolution of reaction mechanistic ideas from their origin at the beginning of the 20th century to the present is an important part of the reaction mechanistic expositions in succeeding chapters. The chapters on reaction mechanism can be seen as a walk through the forest of different catalytic reactions, guided by concepts and general principles of catalytic reactivity. Reactions are discussed that are part of major chemical processes or devices, as well as reactions that may be of future interest.

The working of catalytic systems is scientifically fascinating due to its chemical richness and kinetic complexity. Macroscopic catalyst performance as measured in a reactor depends in a complex way on molecular events that take place on the catalyst surface. The role of reaction mechanism, which connects catalyst performance with surface reactivity, is visualized in Figure 1.1. In macroscopic kinetics, reaction rate is measured as a function of reaction mixture composition and reaction conditions. Reaction rate constants of this global kinetics are parameters that depend on reaction mechanism and are complex functions of elementary rate constants of reaction at surfaces. On the molecular level, surface chemical reactivity defines the stability and reactivity of reaction intermediates, which are formed on and react at the catalyst surface. This provides microkinetic expressions of reaction rate as a function of surface intermediate concentrations that contain the elementary surface reaction rate constants as kinetic reaction rate parameters. The reaction mechanism is the kinetic network that connects global kinetics with microkinetics. It is vital to be cognizant of the relation between catalytic reactivity and chemistry of the catalyst surface.

Global kinetics is essential to the chemical engineer for the design of catalytic reactor and process. Microkinetics provides the catalytic chemist with the relation of catalyst performance with composition and structure of the catalyst that is essential to catalyst

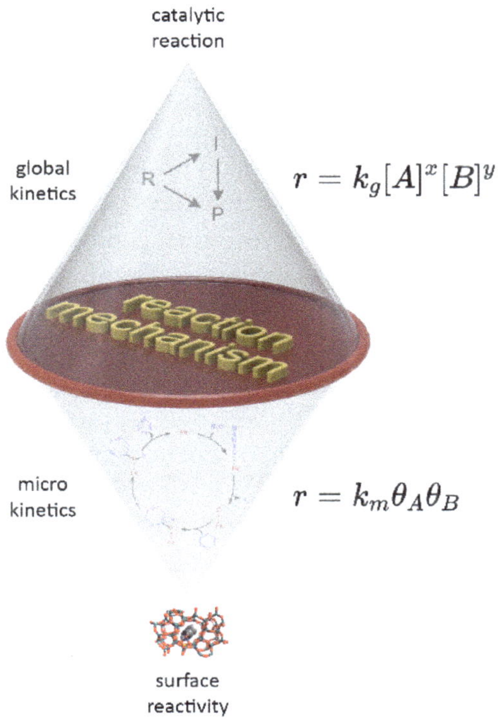

Figure 1.1 Reaction mechanism connects molecular surface chemistry with global catalytic reactivity. Global kinetics expresses reaction rate r as a function of reactant medium concentration, while microkinetics formulates reaction rate as a function of surface concentration θ_i of reaction intermediates [1].

design. The chemical reactions that are part of the reaction mechanism provide the connection with surface chemical reactivity. The chapters on reaction mechanism describe these chemical reactions and their relationship with structure and composition of the catalytically reactive surface.

The understanding of catalytic action took a long time to come. Initial scientific ideas and concepts were refined or had to be corrected with time. This is especially relevant for theories of reaction mechanisms, since for most of the previous century reaction mechanisms could not be directly studied at a molecular level. Only at the end of the previous century was a molecular foundation formulated.

Notwithstanding the speculative nature of early mechanistic theories, the mechanistic chapters illustrate the importance of these early theories to later molecular mechanistic understanding. This is worthwhile since many of the earlier ideas provide the context for later discoveries that are basic to modern catalytic science.

The large empirical body of inorganic chemistry and catalytic reactivity discovered in the 19th century and physical chemical understanding of catalytic activity provide the scientific basis of many of the important catalytic chemical processes that was invented in the 20th century. In the first part of this chapter a short history is presented of catalysis in the 19th century, when it became recognized as an independent chemical phenomenon. The three catalytic principles, that still are the founding axioms of catalysis, are introduced in Section 1.2.

A unique aspect of catalysis is the close interwovenness of fundamental scientific discovery and industrial practice (see also [2], [3]). These technical solutions became realized in the construction of large industries that make major impact to our society. As background to the later mechanistic chapters, the development of catalytic processes of the 20th century is described in Section 1.3. The 20th century can be viewed as the golden age of catalysis. Numerous catalytic processes were invented that became implemented in large-scale chemical industries. The first golden age episode started at the beginning of the 20th century with the invention of the iconic ammonia synthesis process. It is the episode of the discovery of major catalytic heterogenous hydrogenation processes. Fundamental to these developments are discoveries of new catalytic materials. Catalysts are inorganic solids that catalytic chemists adapt to desired catalytic reactivity. The then developed catalysts were bulk solids promoted with additives or they consisted of catalytically reactive components distributed on high surface area supports. Catalysis science developed with the founding of catalytic kinetics and proposal of early reaction mechanistic models.

In the second half of the previous century, the general expansion of the chemical and petrochemical industry provided a major new incentive to discovery and implementation of new catalytic processes. It gave rise to a second golden age episode of process innovation. Catalytic chemistry changed profoundly due to discoveries of molecular inorganic complex chemistry. Coordination and organometallic complexes also became explored for the synthesis of heterogenous catalysts. This new chemistry and the discovery of new solid-state catalysts had a major impact on catalytic material design. Catalysts became manipulated at the molecular level. It was the age of the molecularization of heterogenous catalysis. Also due to advances in surface spectroscopies and computational science, reaction mechanisms became molecularly founded. These physical chemical advances are described in Chapter 2, which deals also with the kinetics of heterogeneous catalysis. The subject matter of the reaction mechanistic chapters that are the main part of this book is introduced in Section 1.4 of this chapter.

1.2 The Definition of Catalysis

For catalysis science, the definitions of Berzelius, Ostwald and the Sabatier principle are paradigmatic truths.

Towards the end of the 18th century and at the beginning of the 19th century, modern chemistry made its entry via the law of conservation of mass by Antoine Lavoisier and the law of multiple and definite proportions of Joseph Proust and John Dalton [4]. According to the latter, elements combine in well-defined mass ratios.

Quantification became an essential tool to understand the chemistry of natural phenomena. In the 19th century, the chemical composition could be measured by a range of analytical tools. The balance that measures weight became complemented with electrochemical techniques and spectral measurements of compounds in a flame.

Jöns Jacob Berzelius, who defined the phenomenon of catalysis, was a great analytical chemist who, in addition to determining the atomic weights of many elements, also discovered at least three new elements at a time when only 45 elements were known [5].

The 19th century was the age of discovery of the elements. Its crown is the Mendeleev periodic table.

Joseph Priestley and Carl Wilhelm Scheele had discovered oxygen in the second half of the 18th century. Their work was fundamental to Lavoisier's discoveries. A rich inorganic chemistry thus developed around oxygen. Catalytic oxidation reactions became widely explored and catalysis became an important topic in chemistry in the 19th century [6]. The insight developed early that the catalytic effect depends specifically on the composition of the catalyst material.

Scientists such as Humphry Davy and Johann Wolfgang Döbereiner discovered Pt metal as an active oxidation catalyst of hydrogen and methane. Whereas the oxidation of sulfur to produce sulfuric acid was known since the Middle Ages, the discovery of its catalytic oxidation was new. A process with also platinum as catalyst became used in the catalytic oxidation of SO_2. Another large-scale catalytic oxidation process of the 19th century is the Deacon process that produces chlorine from HCl. It is catalyzed by a $CuCl_2/ZnCl_2$ catalyst. Some of this early history of oxidation catalysis is told in the introduction of the chapter on oxidation catalysis (Section 4.1).

Science of the 19th century provided three definitions of catalysis, which are still the cornerstones of present catalytic understanding. At the beginning of the 19th century, Berzelius proposed his famous definition of catalysis: The catalyst influences the rate of a reaction, but catalyst material is not consumed by reaction. In line with 19th century quantitative chemistry, the catalyst is a conserved quantity. Berzelius also coined the word catalysis (*kata* is Greek for down and *lysis* means loosen). The catalyst decomposes a reacting substance. The other definitions by Wilhelm Ostwald and Paul Sabatier had to wait for the discovery of chemical thermodynamics. This happened at the end of the 19th century and was the second major development in chemistry of that century. Its founding fathers are Ostwald and Jacobus Henricus van 't Hoff [7]. Chemical thermodynamics defines the equilibrium concentration of a chemical reaction. For a reacting system, equilibrium is defined by the respective free energies of reactants and products. The relation of equilibrium theory and kinetics is fundamental to catalysis. For catalysis, equilibrium theory is highly useful. It predicts the temperature and pressure where conversion to reactant is possible. In a time where catalytic action could not be predicted this was invaluable knowledge, since it provided a method to predict test conditions for an unknown reaction.

At the end of the 19th century Ostwald gave a second definition of catalysis, which exploits the then just formulated chemical thermodynamics. Catalysis became recognized as a kinetic phenomenon. The third catalysis principle was formulated by Sabatier in molecular terms. Catalytic action becomes understood as a chemical reaction at the catalyst surface. These three catalytic principles will be presented in more detail in the following subsections.

1.2.1 *The Berzelius Definition*

Catalyst material is not consumed by reaction.

The 1835 definition by Berzelius of the catalytic force was based on the observation of a wide variety of reactions, such as acid catalysis of starch as well as the then just discovered oxidation reactions catalyzed by transition metals.

Berzelius views catalysis as activation of a reactant by a catalytic force that derives from the catalytic active body. The catalyst does not take part in the reaction and remains unaltered after reaction [5]. He distinguishes the catalytic force from chemical affinity, which he understood as the interaction between substances that make them recombine or decompose. This derives from their chemical properties. The origin of the catalytic force was a mystery to him. The answer to this question is in essence the topic of this book. The first part of Berzelius' definition of the catalyst turns out to be a misconception, which is clarified by the Sabatier principle of Section 1.2.4. Sabatier made clear that catalytic action is based on the very participation of the catalyst in the reaction. This does not contradict the second part of Berzelius' definition that the catalyst is not consumed by reaction. That is still the generally accepted definition of the catalyst.

The term affinity has two different interpretations. The difference becomes clear when formulated within chemical thermodynamics. In modern terms chemical affinity, which relates to properties of chemical substances that make their recombination possible, is an equilibrium property. It relates to the free energy of product formation, which is a measure of chemical reactivity. A modern probe of surface chemical reactivity is the free energy of adsorption. Berzelius refers to chemical affinity in this sense.

Affinity is also used to indicate the driving force of the reaction. This can be regarded as kinetic affinity. Within kinetics, affinity measures the degree to which a reaction is outside its equilibrium conditions. It is only non-zero when reaction has a finite rate and becomes zero when reactant and product are at equilibrium. This affinity definition that is relevant to kinetics will be discussed in the next section, that deals with the chemical thermodynamic definition of catalysis.

1.2.2 *Chemical Thermodynamics and Catalytic Kinetics: The Ostwald Definition*

Catalysis is a kinetic phenomenon and catalyst influences only reaction rates. Equilibrium is not affected.

The equilibrium constant of chemical thermodynamics is also a relation of reaction rate constants. This can be used to give a thermodynamic definition of kinetic affinity that is the driving force of a chemical reaction. It is also the basis of Ostwald's definition of catalysis as a kinetic phenomenon.

In 1864 Cato Maximilian Guldberg and Peter Waage had defined the law of mass action for a reaction between A and B:

$$r = k[A]^x[B]^y \tag{1.1}$$

The rate of reaction r (which they called affinity, but which is not the driving force of the reaction) is proportional to the rate constant k

and the product of the concentrations of reactants [A] and [B]. For the ester formation from alcohol and acid that Guldberg and Waage studied, the orders of reaction *x* or *y* are equal to one. This suggests the intuitive idea that the reaction rate constant *k* is proportional to the probabilities that two reactants collide.

The reaction that Guldberg and Waage studied was a homogenous non-catalytic reaction. For catalytic reactions their interpretation is questionable, since generally the orders of *x* and *y* of catalytic reactions are non-natural numbers. The reason for this is discussed in detail in Section 2.4. It is shown that actually the reaction orders relate to surface concentrations of reaction intermediates.

Guldberg and Waage realized that for reversible reactions at equilibrium, the rates of the forward (r_f) and backward (r_b) reactions should be the same [8]. This gives the equilibrium relation:

$$\frac{k_f}{k_b} = \frac{[C^o]^{x'}[D^o]^{x'}}{[A^o]^x[B^o]^x} \tag{1.2}$$

In Eq. (1.2), $[X_i^o]$ are equilibrium concentrations.

According to van't Hoff, the reaction ratio $\frac{k_f}{k_b}$ of the rate constants is equal to the thermodynamic equilibrium constant K_{eq}, which results in Eq. (1.3a–b).

$$K_{eq} = \frac{k_f}{k_b} \tag{1.3a}$$

$$K_{eq} = \frac{[C^o]^{x'}[D^o]^{y'}}{[A^o]^x[B^o]^y} \tag{1.3b}$$

$$K_{eq} = e^{\frac{-\Delta rG^o}{R_gT}} \tag{1.3c}$$

Eq. (1.3c) relates the equilibrium constant with thermodynamic parameters. $\Delta_r G^\circ$ is the standard Gibbs free energy difference of reactants and products, R_g the gas constant, and T the temperature.

The modern age of heterogenous catalysis made its start once expressions such as Eq. (1.3) became established and thermodynamic properties were measured. Essential is the determination of free energies of the reactants and products, defined by their respective standard entropies and enthalpies. These define $\Delta_r G°$, as in Eq. (1.4):

$$\Delta_r G° = \Delta_r H° - T\Delta_r S° \tag{1.4}$$

As recounted in Section 1.3, the identification of the proper values of thermodynamic constants played an important role in the discovery of the ammonia synthesis reaction.

The kinetic affinity (A_f) follows when the overall rate of reaction R is calculated as the difference between the forward and backward reaction rates:

$$R = r_f - r_b = \left(1 - \frac{r_b}{r_f}\right) \tag{1.5a}$$

$$= rf\left(1 - e^{\frac{-A_f}{R_g T}}\right) \tag{1.5b}$$

$$A_f = R_g T \, ln\left\{\left(\frac{[C^o]^{x'}[D^o]^{y'}}{[A^o]^x[B^o]^y}\right)\left(\frac{[C]^{x'}[D]^{y'}}{[A]^x[B]^y}\right)^{-1}\right\} \tag{1.5c}$$

The kinetic affinity A_f is the driving force of the reaction. It is non-zero only as long as the reaction is not at equilibrium. It depends on the difference between equilibrium and non-equilibrium reaction concentrations [9]. The presence of the catalyst will not affect the value of A_f, which only depends on relative concentrations. The equilibrium concentrations on which it depends is defined by the equilibrium constant K_{eq}. However, the reaction rate constant k_f that defines r_f will be altered due to the presence of the catalyst. This is the insight of Ostwald.

Ostwald's fundamental law of catalysis can be formulated as: the catalyst will not change reaction equilibrium. Catalysis is a kinetic phenomenon and will influence only reaction rate constants [10].

Svante Arrhenius and van't Hoff realized the intimate relation between the temperature dependence of the equilibrium constant

(Eq. (1.3a)) and that of the reaction rate. They discovered that the temperature dependence of many reaction rates behaves according to what is now known as the Arrhenius equation of the reaction rate constant:

$$k = Ae^{\frac{-E_{act}}{R_g T}} \tag{1.6}$$

A is a constant and E_{act} is the activation energy. Usually, the presence of catalyst increases reaction rate and activation energy decreases. The understanding of the chemical cause of this reaction rate change and decrease in activation energy had to await development of kinetic theory as well as insight into the interplay of reactant molecules with surfaces. This is detailed in Chapter 2 (Sections 2.3 and 2.4). Importantly it will be seen for catalytic reactions that an Arrhenius-type rate equation does not generally apply and is only valid in a limited temperature regime. Essential is the understanding that a catalytic reaction is not a single elementary reaction step but is composed of a cyclic combination of several elementary reaction steps. This became clear after the formulation of the principle of Sabatier that is discussed next.

Their groundbreaking discoveries and formulation of chemical thermodynamics and kinetics are recognized by the respective Nobel prizes to van't Hoff (1901, the first Nobel prize in chemistry), Arrhenius (1903), Ostwald (1909), and Sabatier (1912).

1.2.3 *The Sabatier Principle*

Sabatier proposed formation of a temporary transient intermediate. The Sabatier principle states that for an optimum catalyst, elementary reaction rate constants of reactant activation and product formation are equal.

At the end of the 19th century, Sabatier explored catalytic hydrogenation. Together with his collaborator Jean-Baptiste Senderens, he discovered that apart from the then known catalysis by Pt other metals such as Ni, Co, and Fe, after reduction of the corresponding oxides, are also active hydrogenation catalysts [11]. At a temperature of 150–200°C, the hydrogenation of ethene and benzene with hydrogen to the saturated alkanes is readily catalyzed by finely dispersed Ni. This was followed by the discovery of many organic hydrogenation reactions and their catalysts [12], [13].

Sabatier was awarded the Nobel Prize in 1912 for the discovery of hydrogenation by finely divided metals.

Sabatier suggested a molecular mechanism of the catalytic hydrogenation reaction. The hydrogen molecule dissociates, which gives an intermediate surface hydride compound. This is unstable and hydrogenates the reactant molecule (he mentions that NiH_2 is a known compound) [14]. In the process it regenerates the metal. The concept of the temporary unstable intermediate is the fundamental idea of Sabatier on which later theories of the Sabatier principle build. The stability of these intermediates depends on catalyst composition and determines the rate and course of the reaction [15].

The Sabatier principle is based on the idea that the catalytic event consists of several reaction steps, where a surface site initially is consumed and after reaction is liberated (see Figure 1.2a).

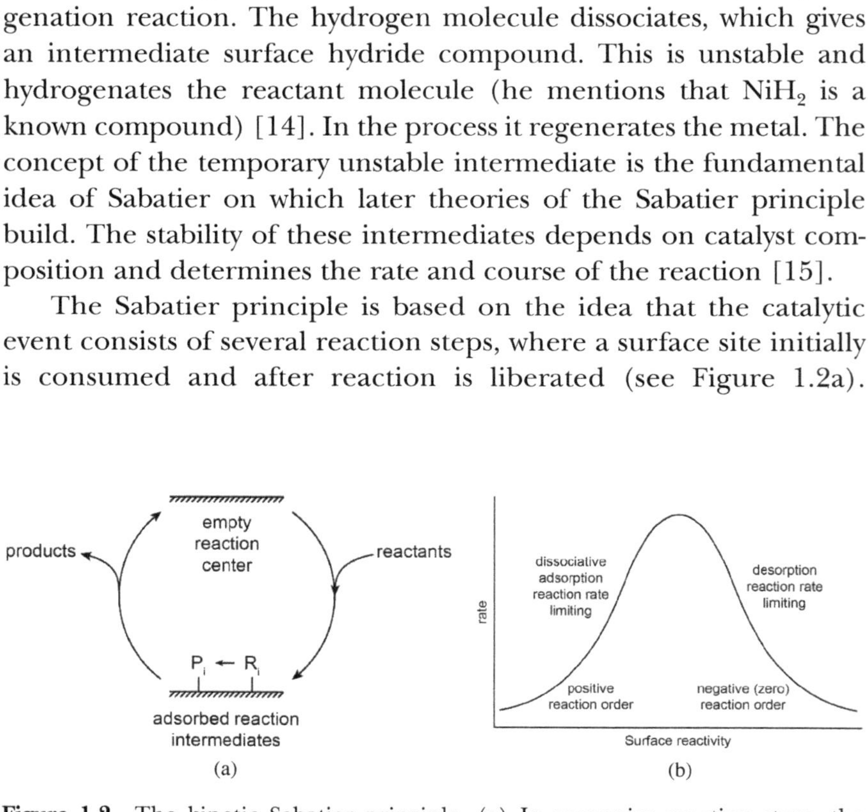

Figure 1.2 The kinetic Sabatier principle. (a) In successive reaction steps, the reactive catalytic site becomes consumed by reaction with reactant. The reaction intermediate R_i rearranges to product reaction intermediate P_i, which desorbs from the catalyst. This regenerates the reactive catalyst site. Adsorbed species on the surface should be stable enough to adsorb and react, but interaction with the surface should be weak enough that the product can desorb. (b) The volcano curve plot of overall reaction rate against surface reactivity. At the optimum rate, elementary reaction rate constant of formation of surface intermediate and that of product desorption balance.

Michel Boudart, one of the leading catalytic chemists of the middle of the previous century, phrased this in the following way: "A catalyst is a substance that transforms reactants into products, through an uninterrupted and repeated cycle of elementary steps in which the catalyst is changed through a sequence of reactive intermediates, until the last step in the cycle regenerates the catalyst in its original form [16]. The catalytic reaction cycle minimizes the temperature of a reaction and converts materials with minimum waste; thus, it saves energy and materials. The catalyst material is recycled in the process."

In the mid-1950s, catalytic kinetic theories were developed that reformulated the catalytic event as a two-step reaction of the reaction intermediate [17]. Under influence of the catalyst, the adsorbed surface intermediate becomes activated and transforms into a product intermediate that in a subsequent reaction desorbs. For the more reactive surface the elementary reaction rate of the surface intermediate will increase, but a stronger interaction with the surface will decrease the elementary reaction rate of desorption. The overall rate can be limited by the elementary reaction rate constant of reactant activation or product desorption (for details see Chapter 2). When the two are approximately equal the catalytic reaction rate is maximum. Figure 1.2b shows schematically the bell-shaped curve that results when reaction rate is plotted against the strength of adsorption of reaction intermediate. A measure of surface reactivity is the adsorption energy of reaction intermediate molecules or atoms (the chemical affinity).

Boudart related the two-step reaction model with the temporary unstable intermediate hypothesis of Sabatier. The Sabatier hypothesis is reformulated as the Sabatier principle: In a catalytic reaction, surface intermediates have to be stable enough to be formed but should not be too stable so as not to desorb.

The Russian scientists Aleksei Balandin and Mikhail Temkin [18] did show that when reaction rate is plotted as a function of interaction strength of adsorbed reaction intermediates, a bell-shaped dependence is found. This is called the volcano plot or Balandin plot. The volcano plot is an expression of the kinetic Sabatier principle of maximum catalytic reaction rate for optimum reactivity of the catalyst.

Previously electrochemists had shown that hydrogen evolution rates, when plotted as a function of the interaction energy of hydrogen with electrode material, also show volcano dependence [19], [20].

Kinetically there is an intimate relation between the order of reaction rate and catalyst reactivity. When surface reactivity is to the left of the volcano plot, the reaction rate should be positive order in reactant. Surface coverage is low and increases with surface reactivity. When surface reactivity is to the right of the volcano plot, the maximum reaction rate becomes suppressed by increase of reactant concentration. Surface concentration of reaction intermediate is high. Information on the relative position of the catalyst with respect to the volcano maximum helps to decide whether catalyst formulation should be changed to alter surface reactivity to stabilize or destabilize proposed reaction intermediates.

At the end of the previous century, due to advances in computational physics, first-principle micromolecular kinetics developed that enabled quantitative models of volcano rate plots as a function of reaction intermediate stability [21]. In Section 2.4.4 a detailed introduction is given.

In summary, the three axiomatic principles of catalysis are:

— The catalyst is not consumed by reaction.
— The catalyst changes reaction rate and not reaction equilibrium.
— Catalytic action is due to formation of temporary surface complexes.

1.3 The Golden Age of Heterogenous Catalysis

Catalysis is a chemical science as well as technology. Nobel Prize-winning scientific contributions largely influenced its early founding as well as modern catalytic science. Major discoveries by industrial scientists provided the catalysts of present large-scale chemical industrial processes.

1.3.1 *Introduction*

Major catalytic processes and systems discovered in the course of the 20th century will be presented in this section. As mentioned in Section 1.1, in the past century the development of catalysis science and technology went through two periods of exponential growth and innovation.

The first golden age of catalysis started around 1910 with the invention of ammonia synthesis. This was followed by a great period of innovation with many new catalytic processes; they are presented in Section 1.3.2. As with ammonia synthesis, most of these processes are catalytic hydrogenation processes. They can be considered as follow-ups to the discovery of catalytic hydrogenation by Sabatier. Additionally important was the successful contribution of the products of catalytic processes to satisfy the changing technical demands of evolving society.

The second golden age of catalysis started around the middle of the last century. Catalytic systems changed due to the impact of molecular inorganic chemistry such as coordination chemistry and organometallic chemistry that became mature. This is discussed in Section 1.3.3.

The impact of the new processes on society, agriculture, transport, and materials has been tremendous. Many of the materials that are currently common were previously unknown and are the product of catalytic processes.

In Table 1.1 a schematic overview is given of the main processes and their catalysts discovered since the beginning of the 20th century. Change or demand from society gave a large impetus to process innovation. Major societal episodes are indicated in the time column. The second column lists respective catalysts and the third column the corresponding catalytic reactions. The latter are discussed in following sections.

Table 1.1 illustrates that invention of a new reaction is a joined process with the discovery of a new catalytic material. The invention of new reactions requires exploration of new catalyst materials.

Table 1.1 Catalytic processes and their catalysts.

Year	Catalyst	Process
1900 Coal; 1910 World War I;	Nickel; Promoted iron; Copper;	Hydrogenation; Synthesis gas to methane, methanol and gasoline; Nitrogen to ammonia; Fat hardening;
1920	Metal sulfides;	Hydrodesulfurization; Hydrodenitrogenation;
1930 Oil; Automobiles;	Solid acids (clays);	Catalytic cracking;
1940 World War II;	Sulfuric acid; Hydrogen fluoride;	Synthetic kerosene;
1950 Major oil reserves; Discoveries in chemical/polymer industry;	Synthetic zeolites; Bifunctional acid catalysts; Complex reducible oxides; Supported coordination complexes; Solid base catalysts;	Refinery processes, hydrocracking, catalytic cracking (zeolites); Selective oxidation; Polymerization (polypropylene, metathesis); Detergents;
1970 Natural gas; Energy crisis (oil);	Novel synthetic zeolitic materials;	Methanol to gasoline/alkenes; Renaissance: synthesis gas to chemicals/gasoline;
1980 Environment;	Noble metals/reducible oxides; Zeolitic oxidation catalysis;	Exhaust catalysis (NO_x); Stackgas treating (SO_2); Fine chemical catalysis;
2000 Sustainable energy; Biomass; Electrification;	Lewis acidic zeolites; Au catalysis; Alloys; Complex oxides;	Lignin/cellulose/sugar conversion; Selective oxidation; Hydrogen evolution/fuel cell; Photocatalysis;

Sometimes they become modifications of known catalysts, but major breakthroughs are usually related to new material discovery and with it deepening insight into the inorganic chemistry of catalysts as well as their synthesis.

Catalyst synthesis and preparation became increasingly sophisticated, helped by parallel development of characterization tools. Serendipity, the unexpected discovery of new catalysis, is often part of the story of the invention. Insights in relation between catalyst performance and chemical composition of a catalyst, the subject of this book, took long to evolve and is still mainly empirical. For catalyst synthesis we refer to [22], [23].

There are several driving forces for process discovery. Similar to ammonia synthesis, catalytic process discovery was directed by engineering opportunities for product materials that are scarcely available. The fat hardening process is an example.

The change in feedstock (often energy related) provides another driving force. The large-scale chemical industry was initially based on coal as feedstock. In the course of the 20th century, ample crude oil reservoirs were discovered that made crude oil readily available. This, in combination with the growing automotive industry, generated a new and large demand for transportation fuels. This gave rise to the expansion of a large petrochemical industry and new refining-related processes.

Also, a large chemical industry came into existence based on alkene feedstock derived from refineries. Through new catalytic inventions a wide range of products was manufactured, for instance polymers, detergents, and antifreeze. The valuable polymer industry that became established generated a large demand for catalytic processes to produce necessary monomer intermediates.

After the oil crisis of the 1970s, interest in natural gas as feedstock became important as environmental concerns became a major societal issue. Amongst others it stimulated refineries to develop processes for fuels low or without sulfur. Also, automotive exhaust treatment catalysts were invented and, as a consequence, exhaust catalyst production became an important industry. A major effort has started to substitute stoichiometric organic reactions that produce much harmful waste by catalytic processes.

Around 2000, a demand arose for energy conversion reactions that are renewable and have low CO_2 emission. Biomass conversion has become of interest as well, as have electrocatalytic processes for the production of hydrogen. In biomass conversion the major challenge is to convert non-food-related biomass to fuel and chemicals.

Hydrogenation catalysis, oxidation catalysis, as well as solid acid catalysis are crucial to these processes and are introduced in mechanistic Chapters 3–6.

1.3.2 *The First Golden Age of Heterogenous Catalysis*

Major catalytic hydrogenation and hydrocarbon conversion processes became initially invented.

Since ammonia synthesis has an iconic position in catalysis, this section starts with an extensive presentation of the context and history of its invention [24]. Then a short review of the other processes of the first golden age period of heterogenous catalysis is given.

The invention of ammonia synthesis was the outcome of a great public debate. At the beginning of the 20th century, which was an imperialistic period, Western European countries realized their vulnerability with respect to the supply of Chile saltpeter, whose main ingredient is $NaNO_3$. The supply of Chile saltpeter is sensitive to overseas transport, as it can be readily disrupted in case of war.

In the 19th century, Chile saltpeter had become the main source of fertilizer and dynamite. Initially it came from Chile natural guano, that is, the droppings of birds. In the second part of the 19th century this was overtaken by Chile saltpeter, which is harvested in large mines in the Atacama Desert of Chile.

To substitute Chile saltpeter by local supply was a major driver to the invention of a chemical nitrate production process and the ammonia synthesis reaction, which gives the desired nitrate by subsequent oxidation.

The ammonia synthesis reaction was invented as part of the overall process to oxidize nitrogen, that is separated from air, to nitrate. The idea to combine a process that reduces nitrogen in a hydrogenation process with hydrogen to ammonia and then to oxidize ammonia to nitrate is remarkable. The necessary hydrogen could be derived via electrolysis of water or by gasification of coal as in the first industrial Haber-Bosch ammonia synthesis process [25].

In the overall process the Haber-Bosch ammonia synthesis is combined with the Ostwald process for ammonia oxidation, named after Ostwald who succeeded its technical realization [26]. IG Farben, the present BASF, took the ammonia synthesis reaction over from Haber to develop the process industrially. Ammonia synthesis became applied for the first time at large scale by IG Farben as part of the nitrogen to nitrate process in Oppau, Germany in 1913. The process became a huge success. It is operated at many locations in the world and is the major current producer of fertilizer.

Three important scientific and technical endeavors made the realization of ammonia synthesis possible. First, the conclusive determination of the free energy of the ammonia synthesis reaction. This was successfully done by Fritz Haber as the outcome of a scientific debate with Walther Nernst. Haber, then at Karlsruhe Technical University, demonstrated the process experimentally in 1909 (Haber obtained the Nobel Prize in 1918 for the invention of the ammonia synthesis reaction, and Nernst in 1920 for thermochemistry).

Second, the successful development of the practical catalyst by Alwin Mittasch and his colleagues at IG Farben. To the catalytic chemists, the main guideline to the search for catalyst materials was the Sabatier suggestion of formation of a temporary unstable surface intermediate.

For the ammonia synthesis reaction, Mittasch had to select catalysts that would not form intermediates with nitrogen that are too stable. In his search for new catalysts, the rich inheritance of the inorganic chemistry of the 19th century was available to him. Already in 1813, the French scientist Louis Jacques Thénard had discovered the reverse reaction of ammonia decomposition. Subsequent studies had shown that the activity of iron, copper, silver, gold, and platinum for decomposing ammonia decreased in the order given. According to a historic paper by A. J. B. Robertson, this is one of the earliest recorded examples of a pattern of catalytic activity [27].

The original catalyst discovered by Haber consisted of osmium. Modern chemistry teaches that when the nitrogen is part of a coordination complex with osmium, the nitride intermediates can decompose to nitrogen [28].

Mittasch discovered the present catalyst based on Fe, which is less poisonous, cheaper, and more readily available than osmium. Iron is also known to form nitrides. In order to promote its activity, additives have to be added. At conditions prescribed by chemical thermodynamics, about 6,500 experiments on about 2,500 different formulations were performed to arrive at the complex technical Fe catalyst promoted with potassium and alumina. It is noteworthy that Fe is located in the same column of the periodic system as Os. Ruthenium metal that is located in between is also an active ammonia synthesis catalyst and used today in an industrial process that uses Ru on a carbon support.

Third, the process engineering contribution by Carl Bosch (Nobel Prize 1931 jointly with Friedrich Bergius for chemical high-pressure methods. Bergius is the inventor of the coal liquefaction process by exposure of coal to hydrogen. Bosch and Bergius were the first engineers to obtain this honor). Because the reaction is exothermic, thermodynamics demands low reaction temperature. However, for finite reaction rate an activation barrier has to be overcome. The process operates with 15% product yield at a temperature of 700 K. At this temperature it requires the high pressure of 200 atmosphere [29]. An additional complexity is the low yield of ammonia. The ammonia has to be separated from the process stream and recycled. The solution of these high-pressure engineering challenges initiated the academic discipline of Chemical Reaction Engineering. Catalysis science became an important branch with focus on catalytic kinetics.

This short history of the invention of the ammonia synthesis process provides insight into the various aspects of catalytic innovation. It is the technological answer to challenges posed by society. It often relates to a change in feedstock that requires new conversion technologies or the need for new products. New catalyst discovery generates new catalytic conversion reactions. In addition to chemical thermodynamics, chemical reactor engineering also needs kinetic information for process design. The chemical kinetics of catalytic reactions that was developed as a result is introduced in Chapter 2.

As stated in Table 1.1, in addition to ammonia synthesis an early application of hydrogenation catalysis was the fat hardening process. As in Sabatier's hydrogenation process, Ni was the preferred catalyst. It produces margarine, that replaces butter, from vegetable or fish oils. In the process unsaturated C=C bonds are converted into saturated bonds.

Since in Western Europe there are no oil resources, coal conversion to liquid fuel became sought for. An important process is the earlier mentioned Bergius process that liquifies coal by exposure to high-pressure hydrogen. It produces a fuel that requires purification, because it contains, amongst others, sulfur-containing compounds. Suitable catalysts for desulfurization are inorganic transition metal sulfide compounds.

The Fischer-Tropsch process provides an alternative way for fuel production by conversion of synthesis gas produced by gasification of coal. As in the ammonia synthesis process, the initial Fischer-Tropsch reaction used a catalyst based on iron.

Due to the expanding oil discoveries before the middle of the previous century, crude oil replaced coal as feedstock. The increase of automotive transportation generated a demand for new processes to produce adequate fuel. Especially after the Second World War, with the discovery of huge cheap oil reservoirs, oil refineries and the petrochemical industry rapidly expanded.

Initial developments were mainly in the USA. One of the first processes was the Houdry catalytic cracking process that cracks the large molecules in crude oil to smaller molecules. The initial catalysts used were acidified clays.

Many additional processes were invented for the oil industry that are currently used at large scale in refineries. To supply fuel to bombers in the Second World War a process was formulated to produce alkylate, which consists of branched hydrocarbons, catalyzed by neat acids. It developed into a major refinery process. Currently solid acid catalysts are available for this reaction, which will be discussed in Chapter 5.

Shortage of crude oil from the Middle East, because of the oil crisis in the 1970s, provided the incentive to replace oil as feedstock with natural gas. Synthesis gas can be readily made from natural gas. This gave rise to a renaissance of the synthesis gas conversion processes that had been explored in the 1930s in Germany. An example is the renewed interest in the Fischer-Tropsch process that produces a mixture of aromatics and alkanes. Plants using natural gas were constructed in Malaysia, South Africa, and Qatar. Important was the Mobil discovery that methanol (readily made from synthesis gas and produced worldwide on large scale using mainly Cu/ZnO_x catalysts (see Chapter 3)) can be dehydrated on a synthetic zeolite material (ZSM-5) and converted into short olefins and aromatics (see Chapter 5).

In the 1970s as well, environmental concerns became a major societal issue. Amongst others it stimulated refineries to develop processes for fuel low in sulfur. New hydrocarbon conversion processes became in demand. Aromatics as the main component of gasoline were replaced by branched hydrocarbons.

Catalysts were also inverted to reduce the emission of NO_x in automotive exhaust gas and the production of exhaust catalysts became an important industry. A major effort also started to substitute stoichiometric organic reactions in fine chemical industry in order to reduce harmful waste by catalytic oxidation processes.

Catalytic hydrogenation is the dominant catalytic technology of the processes introduced so far. Catalysts are bulk inorganic materials of catalytically active particles dispersed on high surface area supports.

1.3.3 *The Second Golden Age of Heterogeneous Catalysis*

Molecular inorganic chemistry revolutionized heterogeneous catalysis.

The second golden age of heterogenous catalysis was in the second part of the 20th century. Not only scientific innovation but also economic changes of society contributed. Heterogenous catalysts changed into supported molecular systems. New classes of catalytic reactions were discovered.

One development was the rapid expansion of the oil-refining industry, which stimulated inventions of many new petrochemical processes. New solid acid catalysts for catalytic cracking of heavy oil were invented that had comparable impact as the ammonia synthesis catalyst discovery 50 years earlier.

Other major discoveries with major impact were new polymeric materials. Heterogeneous catalytic polymerization reactions of ethylene and propylene became discovered, which added to the growing polymer industry. The need for monomer intermediates led to innovation of catalytic oxidation processes. Table 1.1 illustrates some of these innovations.

As mentioned in Section 1.2, the inorganic chemistry that provides the material base for catalyst discovery underwent a major change in the middle of the 20th century. The earlier discovered catalysts are bulk materials or catalytically active particles distributed over a high surface area support. In the latter, dispersion increases the effective surface area of the catalytic active material. An important function of the support is to stabilize particles as small as a few nanometers. Because of its strength, usually an alumina support is used. The surfaces of these catalytically reactive particles are terminated bulk structures. This is very different from the molecular heterogenous catalysts that are introduced below.

In the 1940s, catalytically active molecular coordination complexes and metal-organic compounds were discovered. For instance, Co-carbonyl complexes discovered by Otto Roelen in 1938 are active in the homogenous hydroformulation reaction that produces aldehyde by insertion of CO into an alkene [30].

Use of such molecular complexes in supported form or as part of a heterogenous high surface area catalyst results in an important new class of catalysts. The production of polypropylene with a $TiCl_3/MgCl_2$ catalyst was discovered, where Ti is present as part of a coordination complex that is supported by $MgCl_2$. Single-site Cr cations attached on a silica support were discovered that in the USA became used as catalysts for polyethylene production. Metathesis catalysis, which was originally discovered as disproportionation catalysis with catalysts prepared with Co-carbonyl complexes immobilized on a high surface area support, also became important for polymer production. In the metathesis reaction, alkenes interchange part of their molecules and form other alkene molecules. Transition metal catalysts used in automotive exhaust catalysis to reduce NO_x to N_2 are single-site metal atoms that are partially dissolved in a reducible oxide as CeO_2.

The mechanism of these and related reactions, and some of the history of the discovery of these unique heterogeneous catalytic systems, are discussed in Chapter 6.

Also, scientists that contributed to this new episode of flourishing catalysis became recognized with Nobel prizes. In 1963 Karl Ziegler and Giulio Natta were awarded the Nobel Prize for the stereoselective polymerization of propylene, and in 2005 Yves Chauvin, Robert Grubbs, and Richard Schrock were awarded the Nobel Prize for the discovery of the metathesis reaction and its mechanism. The discovery of the disproportionation reaction, which is related to the metathesis reaction, was made in the 1960s by researchers from Phillips Petroleum Company.

Process innovation also took place by combination of homogenous and heterogenous catalytic reactions in complex chemical processes. An interesting example that illustrates the integration of different reactions is the Shell Higher Olefin Process (SHOP) for the production of biodegradable detergents [31], [32]. The SHOP process was invented in 1968 and commercialized in 1972. The linear higher olefins that are used to make detergents are produced by oligomerization of ethylene and are terminated with a polar group.

The size distribution of the short alkene oligomer mixture is shifted to longer molecules by the metathesis reaction that is catalyzed by a molecular heterogeneous catalyst with supported Mo or W coordination complexes. To enable the metathesis reaction to take place, the ϖ bonds in the individual alkene oligomers have to shift to different bond positions. This bond shift reaction is catalyzed by a heterogenous alkali-doped alumina catalyst [33]. Homogenous catalysis is used for two reactions: The oligomerization of ethylene, catalyzed by an organometallic Ni complex, and the hydroformulation reaction of Roelen, catalyzed by Co or Rh that adds the molecular polar ends by CO insertion (Chapter 3, Figure 3.20).

Very different catalysis contributed to innovation of refinery processes. The Houdry process for catalytic cracking of crude oil to transportation fuel was known since the 1930s. The process is catalyzed by solid acidic materials. Initially acidified clays were used. Discoveries of nanoporous synthetic zeolitic materials and their use in the catalytic cracking process had a major beneficial economic impact. The use of zeolite catalysts improved the yield of the catalytic cracking process largely by reduction of crude oil loss as gas or carbon residues. In 1979 the Mobil researchers Charles Plank and Edward Rosinski were inducted into the USA National Inventors Hall of Fame for their invention of 1966.

The mechanism of solid acid catalysis and solid acidic zeolitic materials is the subject of Chapter 5. In Figure 5.18 some structures of zeolite materials are shown. The unique feature of the nanoporous zeolite structure is that the reactive surface is inside the zeolite material and is part of its crystallographic structure. No chemical bonds are broken of the internal zeolite surface. This is very different from conventional catalyst surfaces, which can be thought as formed by the cleavage of bulk chemical bonds and are reactive because of the coordinative unsaturation of the surface atoms. In the case of the zeolites, their acidity is due to protons attached to the internal zeolite surface in the nanopores of the zeolites. The structure of the catalytically reactive proton is well defined at the atomic level. The high selectivity of the zeolites applied in the catalytic cracking process is due to the inhibition of the formation of large coke-forming molecules in the small nanopores.

Zeolite synthesis became possible due to the discoveries of Richard Barrer at the University of Aberdeen and Imperial College in London in the 1950s [34] and Union Carbide researchers in the USA. Major inventions of new zeolitic materials by Union Carbide and Mobil researchers resulted in important new processes. An example is the application of zeolites in conversion processes of natural gas to liquid fuels or chemicals. As is discussed in detail in Section 5.4.5, zeolitic catalysts are applied in the conversion of methanol to gasoline, or to ethylene and propylene. Methanol is produced from synthesis gas in large-scale industrial processes (see Section 3.2.2.2), while synthesis gas is in turn can be made from natural gas by catalytic oxidation or the steam-reforming reaction (see Section 4.2.2.3). The natural gas conversion route to gasoline or olefins via methanol is an alternative to the Fischer-Tropsch process, which is less selective.

A third major innovation of the 1950s was the discovery of selective oxidation of propene by complex reducible solid-state oxidic materials by Sohio research company in Cleveland. Previously the main path to oxygenate alkenes was the homogeneous catalytic hydroformylation reaction to aldehyde or ketone, in which CO is inserted in an alkene. Another reducible mixed metal oxide catalyst used in selective methanol oxidation to formaldehyde was invented at approximately the same time by British Imperial Chemical Industries (ICI) researchers (see Section 4.2.2.4). These inventions are part of the renaissance of oxidation catalysis that came after the period of oxidation catalyst innovation that had taken place earlier in the 19th century.

An important product molecule is acrylonitrile, produced by catalytic reaction of propylene and ammonia with oxygen. As with formaldehyde, it is an intermediate for the production of polymers. Other related important chemicals that are synthesized by catalysis with complex metal oxides are acrolein and acrylic acid. Catalysts are solid-state mixed metal oxides that have a rich composition variability. The mechanism of heterogeneous catalytic oxidation reactions is discussed in Chapter 4.

1.4 Summary; Introduction to the Book Chapters

The discovery of catalytic reaction and catalyst are joint processes.

The three axioms of catalytic action of Berzelius, Ostwald, and Sabatier were formulated in the course of the 19th century. In the first part of the 20th century, chemical thermodynamics became the scientific fundament for catalytic exploration. With the discovery of the continuous heterogenous catalytic process, this led to the many industrial processes of the current chemical and petrochemical industry.

New reactions require discoveries of new catalytic materials. No scientific principle comparable to the strictness of chemical thermodynamics exists for the selection of a catalyst. For this, extensive empirical knowledge of inorganic chemistry and also known catalytic reactivity was exploited. The latter became greatly extended in the 19th century after the discoveries of Davy and others. In addition, the Sabatier hypothesis, that a transient intermediate is formed at the surface, which is not too stable or unstable, became an important guiding principle.

What remained lacking for the most part of the 20th century was chemical understanding of the relation between catalytic reactivity and catalyst structure. The relevant surface chemistry has to be known at the molecular level in order to select the catalyst material. The tools to make this information available were only developed in the second half of that century. In the course of time, insights in the relation between catalyst reactivity and its material properties improved and provided the foundation of modern heterogeneous catalysis science.

The major drive for new process discovery is the ever-changing relation between available feedstock and desirable product. The latter can be the replacement of a scarce natural substance or sometimes the production of a newly discovered useful chemical. In time passed, the major feedstock change was from coal to oil, followed by transitions to natural gas, biomass, and electricity from renewable sources. Major examples are transportation fuels or base chemicals related to the polymer industry. In the latter part of the 20th century, environmental demands led to deep sulfur reduction in fuels and exhaust emission catalysis to reduce NO_x to N_2.

Within the context of catalytic process research, catalysis is a design activity. Chemical knowledge of catalytic reactivity is not sufficient. Hence, by necessity, empirical exploration is the main route to innovative progress.

Once useful reactions and catalysts became known, the search for the chemical cause of catalytic reactivity became a highly useful and also profitable activity. It led to adaptation and sometimes replacement of the original catalysts with often major improvements in process yields and stability.

The inorganic catalytic materials discovered in the first golden age of heterogenous catalysis had complex compositions of dispersed bulk compounds. Later particles of reactive materials became deposited on usually oxidic high surface area supports.

A major change in catalyst design happened around the middle of the 20th century. By that time molecular inorganic complex chemistry had come to fruition. It became a second molecular branch next to organic chemistry, which already from its early beginning in the 19th century was a molecular science. Molecular coordination complexes of metal atoms or cations and metal-organic complexes became known as useful catalysts. Such catalytically reactive molecular complexes can be immobilized by attachment to high surface area supports. As Chapter 6 describes, this new molecular class of catalysts led to the discovery of many new catalytic reactions and discovery of important new materials, for instance a variety of polymers.

According to the Berzelius definition, a catalyst is not consumed by reaction. However, in practice a catalytic reaction proceeds through reactivity stages: initiation, steady state, and deactivation. The three stages imply chemical (and sometimes physical) changes of the catalyst when exposed to reactants. Formation of surface intermediates may alter the reactive catalyst surface into a quasi-stable state in which, compared to its original state, it can be more or less reactive. When reaction proceeds continuously, deposition of side products or intrinsic instability of catalyst material will lead to catalyst deactivation. The additional challenge to catalyst design is to incorporate knowledge of these surface changes in the inorganic chemistry and material properties of the catalyst.

The relation between catalyst performance and material properties of the catalyst remained poorly understood for a long time. To a large extent this was due to lack of knowledge of the structure of the inorganic solids and their exposed surfaces. Instrumental techniques to study catalysis at the nano level still had to be developed. Still today the identification of the reactive surface state is experimentally not trivial. Catalyst testing in relation to catalyst synthesis is the major vehicle for creating the heterogeneous catalytic knowledge base.

The large body of new catalytic materials and catalytic reactions that were discovered in the course of the 20th century, and the introduction of increasingly refined physical characterization techniques of the catalysts, provide the foundation of modern heterogenous catalysis science. The relation between catalyst reactivity and its material properties obtained a molecular foundation. These advances are mainly due to academic research on molecularly well-defined model systems of known catalytic reactions [35].

In Chapter 2 the physical chemistry of heterogenous catalysis is introduced. Tools for the characterization of catalyst surface and morphology, and kinetic models based on chemical thermodynamics, are introduced. Molecular theories of chemical reactivity from surface science and computational catalysis are presented.

Kinetic theories are discussed that relate surface reactivity trends with trends in catalytic reactivity. This establishes a molecular foundation for the Sabatier hypothesis of the transient reaction intermediate. The Sabatier principle only indirectly relates to surface reactivity but is fundamentally a kinetic principle.

To devise a kinetic model of a catalytic reaction, the mechanism of a reaction must be known. The mechanism of a catalytic reaction is the network of elementary reactions that constitute the catalytic reaction cycle. The major advance of today is that elementary reaction rate constants can be deduced from the chemistry that occurs at the catalyst surface.

The kinetics of a reaction can be studied at two different levels: the molecular level and the global level. On the molecular level the bond-breaking reactions have to be described as they occur on the catalyst surface. This relates the chemistry of reaction intermediate transformation with surface structure and composition. For most of the reactions discussed in this book, this information has become available in the course of the past 208 years. Previously molecular transformation theories had often been formulated based on indirect experiments, which will appear to be very useful or even essential to the quantum-chemical computational modelling results on which modern theories are based.

The other level is the global kinetic level, in which catalyst performance is characterized in terms of conversion and selectivity. The full mechanistic reaction network has to be incorporated into a kinetic model. Through the introduction of hypotheses, such as the rate-controlling step and the surface composition, the global relation between catalyst action and catalyst composition can be identified. The identification of the rate-controlling step is helped by *in situ* experiments for the determination of the concentration of major reaction intermediates.

Agreement only between measured and simulated reaction kinetics based on a hypothetical mechanistic model will not validate the correctness of the model. Several different mechanistic models may give the same kinetics. Validation requires molecular information on elementary reaction steps and reaction intermediates, and agreement between simulated and measured surface compositions.

The mechanistic chapters describe the reaction mechanism on these two levels. The relation between catalyst performance and the structure and composition of the catalytically reactive surface site is of primary interest. Changes of surface phases and surface state under influence of contact with reaction mixture are observed in many catalytic systems.

Chapter 3 deals with the mechanism of hydrogenation reactions, Chapter 4 with selective oxidation reactions, Chapter 5 with solid acid catalysts, and Chapter 6 with molecular (single-site) heterogeneous catalysts.

In Chapter 3 the initial focus is on transition metal catalysts. The main early mechanistic models, developed with the introduction of isotope-labelled molecules, are presented. As in the other chapters, these models are compared with the molecular reaction models from modern spectroscopy and computation. The first part of the chapter deals with hydrocarbon conversion, and nitrogen and carbon monoxide hydrogenation catalysis. In the second part of this chapter, heteroatom removal for substituted hydrocarbons is discussed.

Hydrodesulfurization and hydrodenitrogenation catalysis by sulfide catalysts, and hydrodeoxygenation reactions of biomass molecules derived for lignin or carbohydrates, are presented. The latter catalytic systems can be complex bifunctional systems with a hydrogenation-active particle on a Brønsted or Lewis acidic support.

There is a rich variety of reaction mechanisms of oxidation catalytic reactions, which are the subject of Chapter 4. In this chapter selective oxidation reactions with molecular oxygen are presented. Catalysts are transition metals or reducible metal oxides. The latter may have a very complex composition.

Oxidation by molecular oxygen is complex, since selective reaction may involve surface-activated molecular oxygen as well as atomic oxygen. The reactivity of these oxygen species may differ widely and depend on catalyst composition. Especially for the reducible mixed metal oxides that have multi-component composition, heterogenous oxidation catalysis is complex since the overall reaction is due to a network of elementary reactions that are catalyzed by different reaction centers. Autocatalytic radical gas- or liquid-phase reactions with molecular oxygen, and selective surface reactions with molecular or atomic oxygen, are to be distinguished. Selective oxidation reactions of ammonia to nitrogen or NO_x, of methanol to formaldehyde and aromatics, and of alkenes or alkanes to oxygenated products are the main reactions presented in this chapter. In addition to these reactions the NO_x reduction with ammonia, important in exhaust catalysis, is presented. The final section that collects and compares the different elementary reaction steps also contains a subsection on reactivity descriptors. In this subsection the mechanism of selective oxidation is compared with that of electrocatalytic hydrolysis.

In Chapter 6 related catalysis by single-site molecular catalysts is presented for selective oxidation with activated oxygen as in H_2O_2 or N_2O. A comprehensive review of elementary reaction steps of heterogenous catalytic oxidation reactions of Chapters 4 and 6 is present at the end of Chapter 4.

A major question in solid Brønsted acid catalysis in Chapter 5 is how to distinguish the intrinsic reactivity of surface protons from the additional physical interactions of reactants and protonated intermediates with the solid. The first part of the chapter is concerned with the inorganic chemistry of solid acids and the relation of proton bond strength with the composition of the catalyst material. The second part deals mainly with the mechanism of zeolite-catalyzed reactions. The nanoporous zeolites are the workhorse of solid acid-catalyzed reactions. They are useful because their shape makes them selective catalysts. The relation between reaction mechanism and zeolite cavity structure, that is the cause of shape selectivity, is extensively discussed.

The molecular mechanism of solid acid-catalyzed reactions is related to catalysis in strong acids. The physical organic chemistry of acid-catalyzed reactions became well understood in the middle of the previous century. The hydrocarbon conversion reactions or dehydration reactions that are discussed proceed by cationic intermediates that result from proton activation. Different from reactions in the liquid phase, in solid acid catalysis carbocationic intermediates are activated intermediates and part of transition states.

Molecular heterogeneous catalytic systems are introduced in Chapter 6. Such catalysts consist of single-site (single reactive metal atom) or small molecule cluster catalytic sites. Catalysis is via Lewis acid non-redox and redox chemistry or metal-organic (coordination) chemistry. When distributed on non-reducible supports the single-site catalytic systems will mainly catalyze insertion reactions, as in selective alkene oxidation or polymerization catalysis. The Lewis acid catalysts also induce molecular rearrangement reactions. This is discussed for carbohydrate isomerization where hydride transfer bond transformation steps are key reaction steps, analogous to the Brønsted acid-catalyzed reactions of Chapter 5.

When distributed on reducible supports, dual sites are created, and molecular bond dissociation reactions become possible. Such catalysts find practical use in automotive exhaust catalysis. The reducible support provides reactive Lewis base oxygen sites as well as oxygen vacant Lewis acidic sites that in a specific reaction provide synergy to the reactivity of single transition metal atom sites.

Shape-selective catalysis is generated by incorporation of reactive catalyst centers in the nanopore framework of a zeolite. Examples are selective alkane oxidation and methane to methanol oxidation. The chapter is concluded with an introduction to Au catalysis. This metal, which is generally regarded as non-reactive, is promoted by dispersion on a reducible support and remarkably also by water. It is an interesting catalyst for low-temperature CO oxidation and selective oxidation of biomass-related alcohols.

In the final summarizing Chapter 7, catalysis science will be revisited from a different perspective than in the previous chapters.

Catalysis is a science as well as a technology. The history of its development provides an opportunity to ask about the interaction of fundamental understanding and application-oriented scientific investigation with that of the catalytic design approach, which is directed towards the invention of a new desirable reaction or the need to improve catalyst stability or other functionality. This will be discussed with reference to scientific philosophical literature that refers to this question.

To advance science, critical discussion is essential. Catalysis science provides an interesting case study of the role of scientific debate and its resolution. Over the years, catalytic reaction mechanisms have created substantial debate, due to the difficulty to relate the molecular base of catalyst reactivity with global catalyst performance. In the mechanistic chapters, the evolution of this understanding due to increased instrumental sophistication can be followed. The debate on mechanism was essential to catalysis science progress. As part of the summarizing review of reaction mechanisms, some of the unresolved issues will be revisited and also the challenge this poses to catalyst science.

In the past decades the science of heterogeneous catalysis has integrated the molecular sciences. The molecular view has become the major lens through which the chemistry of the catalyst surface is interpreted. This is a great accomplishment and is positioned far from the origin of heterogeneous catalysis in the first part of the previous century.

Current catalytic modelling has obtained a well-founded molecular base, and also the tools are available to connect with global catalyst performance. Notwithstanding this successful state of affairs, there still is a limit to the predictability of catalyst performance. Is this because of the intrinsic complexity of the catalytic reaction and still a lag in understanding of the dynamics of the catalytic reaction? How does this relate to the accuracy of simulations based on quantum-chemical and molecular dynamics data?

Modern catalysis theory of the 21st century is based on our understanding of reaction mechanism in relation to the inorganic chemistry of the catalyst. Catalyst reactivity modelling has become an important tool for catalyst selection, but catalysis is still not a predictive science. The practice of catalysis is still catalyst design and adaptation based on feedback from experiment.

References

[1] G. Li, E. A. Pidko, R. A. van Santen, Z. Feng, C. Li, and E. J. M. Hensen, “Stability and reactivity of active sites for direct benzene oxidation to phenol in Fe/ZSM-5: A comprehensive periodic DFT study,” *J. Catal.*, vol. 284, no. 2, pp. 194–206, 2011, doi: 10.1016/j.jcat.2011.07.008.

[2] A. Zecchina and S. Califano, *The Development of Catalysis: A History of Key Processes and Personas in Catalytic Science and Technology*. John Wiley & Sons, 2017.

[3] R. A. van Santen, “Catalysis in Perspective,” in *Catalysis: From Principles to Applications*, M. Beller, A. Renken, and R. A. van Santen, Eds. John Wiley & Sons, 2012, pp. 3–19.

[4] M. DeLanda, *Philosophical Chemistry*. Bloomsbury Publishing, 2015.

[5] J. L. Marshall, “Berzelius' World,” *CHEM 13 NEWS Mag.*, Sep. 2019, Accessed: Dec. 02, 2022. [Online]. Available: https://uwaterloo.ca/chem13-news-magazine/september-2019/feature/berzelius-world.

[6] J. Wisniak, “The history of catalysis. From the beginning to nobel prizes,” *Educ. Química*, vol. 21, no. 1, pp. 60–69, Jan. 2010, doi: 10.1016/S0187-893X(18)30074-0.

[7] K. J. Laidler, *The World of Physical Chemistry*. Oxford University Press, 1993.

[8] C. M. Guldberg and P. Waage, “Experiments concerning Chemical Affinity,” in *Ostwald's Klassiker der exakten Wissenschaften*, vol. 104. Wilhelm Engleman, 1899, pp. 10–125.

[9] I. Prigogine, *From Being to Becoming*. W.H. Freeman and Co., 1980.
[10] W. Ostwald, "Ostwald on catalysis," *Z. Phys. Chem.*, vol. 15, pp. 705–706, 1894, Accessed: Mar. 17, 2021. [Online]. Available: https://www.chemteam.info/Chem-History/Ostwald-1894.html.
[11] P. Sabatier, "Hydrogénations et déshydrogénations par catalyse," *Berichte der Dtsch. Chem. Gesellschaft*, vol. 44, no. 3, pp. 1984–2001, Jul. 1911, doi: 10.1002/cber.19110440303.
[12] M. Che, "Nobel Prize in chemistry 1912 to Sabatier: Organic chemistry or catalysis?," *Catal. Today*, vol. 218–219, pp. 162–171, Dec. 2013, doi: 10.1016/j.cattod.2013.07.006.
[13] P. Sabatier and E. E. Reid, *Catalysis in Organic Chemistry*, 2nd ed. Nostrand Company, 1923.
[14] P. Sabatier, "The method of direct hydrogenation by catalysis," *Nobel Lecture*, Dec. 1912, Accessed: Dec. 03, 2022. [Online]. Available: https://www.nobelprize.org/prizes/chemistry/1912/sabatier/lecture/.
[15] P. Sabatier, *La Catalyse en Chimie Organique*. Librarie Polytechnique, 1913.
[16] J. M. Thomas, K. I. Zamaraev, I. I. Moiseev, and M. N. Vargaftik, "Perspectives in Catalysis," in *Perspectives in Catalysis*, J. M. Thomas and K. I. Zamaraev, Eds. Blackwell Scientific, 1992, p. 91.
[17] M. Boudart, *Kinetics of Chemical Processes*. Prentice-Hall, 1968.
[18] A. A. Balandin, "The multiplet theory of catalysis — energy factors in catalysis," *Russ. Chem. Rev.*, vol. 33, no. 5, p. 275, May 1964, doi: 10.1070/rc1964v033n05abeh001407.
[19] H. Gerischer, "Mechanismus der Elektrolytischen Wasserstoffabscheidung und Adsorptionsenergie von Atomarem Wasserstoff," *Bull. des Sociétés Chim. Belges*, vol. 67, no. 7–8, pp. 506–527, Jan. 1958, doi: 10.1002/bscb.19580670714.
[20] R. Parsons, "The rate of electrolytic hydrogen evolution and the heat of adsorption of hydrogen," *Trans. Faraday Soc.*, vol. 54, no. 0, pp. 1053–1063, Jan. 1958, doi: 10.1039/TF9585401053.
[21] J. K. Nørskov, F. Studt, F. Abild-Pedersen, and T. Bligaard, *Fundamental Concepts in Heterogeneous Catalysis*. John Wiley & Sons, 2014.
[22] J. Regalbuto, Ed., *Catalyst Preparation*. CRC Press, 2006.
[23] K. P. de Jong, *Synthesis of Solid Catalysts*. John Wiley & Sons, 2009.
[24] K. Tamaru, "The History of the Development of Ammonia Synthesis," in *Catalytic Ammonia Synthesis: Fundamentals and Practice*, J. R. Jennings, Ed. Springer, 1991, pp. 1–18.
[25] M. Jones, K. H. R. Rouwenhorst, A. S. Travis, and L. Lefferts, "1921–2021: A century of renewable ammonia synthesis," *Sustain. Chem.*, vol. 3, no. 2, pp. 149–171, Apr. 2022, doi: 10.3390/suschem3020011.
[26] L. B. Hunt, "The ammonia oxidation process for nitric acid manufacture," *Platin. Met. Rev.*, vol. 2, no. 4, pp. 129–134, 1958.

[27] A. J. B. Robertson, "The early history of catalysis," *Platin. Met. Rev.*, vol. 19, no. 2, pp. 64–69, 1975.

[28] D. C. Ware and H. Taube, "Substitution-induced N-N coupling for nitride coordinated to osmium(VI)," *Inorg. Chem.*, vol. 30, no. 24, pp. 4605–4610, Nov. 1991, doi: 10.1021/ic00024a029.

[29] M. Appl, "The Haber-Bosch Process and the Development of Chemical Engineering," in *A Century of Chemical Engineering*, W. Furter, Ed. Springer, 1982, pp. 29–53.

[30] B. Cornils, W. A. Herrmann, and M. Rasch, "Otto Roelen, pioneer in industrial homogeneous catalysis," *Angew. Chem. Int. Ed.*, vol. 33, no. 21, pp. 2144–2163, Nov. 1994, doi: 10.1002/anie.199421441.

[31] E. F. Lutz, "Shell higher olefins process," *J. Chem. Educ.*, vol. 63, no. 3, pp. 202–203, 1986, doi: 10.1021/ed063p202.

[32] W. Keim, "Oligomerization of ethylene to $\alpha{\$}-olefins: discovery and development of the shell higher olefin process (SHOP)," *Angew. Chem. Int. Ed.*, vol. 52, no. 48, pp. 12492–12496, Nov. 2013, doi: 10.1002/anie.201305308.

[33] H. Pines, *The Chemistry of Catalytic Hydrocarbon Conversions*. Academic Press, 1981.

[34] D. E. W. Vaughan, "Contributions of R. M. Barrer to Zeolite Synthesis," in *From Zeolites to Porous MOF Materials — The 40th Anniversary of International Zeolite Conference*, R. Xu, Z. Gao, J. Chen, and W. Yan, Eds. Elsevier, 2007, pp. 87–95.

[35] G. A. Somorjai and Y Li, *Introduction to Surface Chemistry and Catalysis*, 2nd ed. John Wiley & Sons, 2010.

Chapter 2

Founding Principles of Heterogeneous Catalysis Science

2.1 Introduction

The catalytic event is a complex phenomenon that integrates chemistry at different length and time scales.

The question of how catalytic reactivity and catalyst properties relate gave rise to the development of tools to characterize the catalyst and measure catalyst performance. It led to the definition of the catalytic reaction site and knowledge of its structure and composition. Reactivity models were proposed as to how the catalyst surface activates reactants and how surface reactivity connects with the reaction network that transforms reactants into products. This chapter introduces the physical chemical principles that are basic to catalytic kinetics and surface reactivity.

Chemical thermodynamics provided the foundation to early catalysis science for formulation of reaction kinetics and characterization of catalytic materials. In the course of time, spectroscopies developed and the structure and composition of the catalyst surface became increasingly better understood. Ultimately it led to the discovery of the chemical laws that relate kinetics with the molecular chemistry of the catalyst. The function of reaction mechanism changes from a kinetic model with parameters to be fitted by experimental rate data to a chemical model with parameters derived from elementary reaction rate constants which can be deduced from surface chemistry.

The 1930s were a highly creative and productive period of physical chemical exploration of catalyst kinetics. The fundamental catalytic kinetic equations were formulated. The catalyst structure, composition, and surface became important subjects of study and first theories developed of their relationship with catalyst reactivity. The physical chemistry to define surface structure and catalyst morphology was based on chemical thermodynamics. Kinetics were formulated as proportional to reactive surface area. The proper definition of reactive surface area became an important question. Through the early development of vacuum techniques by Irving Langmuir, physical adsorption became understood and with it the methods to measure surface area and catalyst pore structure. Already in 1916 Langmuir devised the atomistic surface reactivity model now named after him, which became fundamental to catalytic kinetics [1].

Gradually with improvement of technical capabilities able to probe catalyst properties with increasing resolution, insights into the chemistry of catalytic reactivity developed. The concept of the catalytic reaction site became defined. Mainly within the context of chemical reactor engineering a rich theory of catalyst kinetics developed, which came to fruition around 1960.

Initially kinetics developed without an explicit relation to surface reactivity. We would do well to remember that in the first two decades of the previous century, many physical chemists still did not accept the physical existence of molecules. Ostwald, one of the founding fathers of physical chemistry, never accepted the idea of a molecule other than as a working hypothesis. Molecular theories of catalytic reactivity developed through catalyst model studies on metal films or filaments, exploitation of isotope-labelling techniques, and improved analytical product analysis techniques. These developments were part of the more general scientific advances that took place in the early part of the 20th century, which changed physical chemistry into a science founded on molecular theories and experiments to study matter [2].

Quantum mechanics gave an explanation of the periodic table that became the basis of theories of chemical bonding in molecules

Table 2.1 The development of the physical chemistry of heterogenous catalysis.

Year	Catalysis science discipline	Function
1910	Chemical thermodynamics	Predict temperature, pressure of reaction (composition catalyst-empirical testing)
1940	Chemical kinetics, isotopes, high vacuum systems	Reactor design, surface reaction mechanism (morphology, surface area and porosity, chemisorption, molecular dissociation)
1970	Spectroscopy for catalyst characterization	Catalyst particle size and composition (nm)
1980	Model catalysis/surface science ultra-high vacuum	Relation of catalyst surface structure and composition (atomic) and chemical reactivity
2000	Molecular heterogenous catalysis/computational catalysis	Calculate elementary reaction rate transition state energies for catalyst site models with quantum chemistry Microkinetic simulations
2020	Artificial intelligence	Multiscale simulation of inorganic surface chemistry/catalyst synthesis

and crystals [3]. Founding molecular theories of catalytic reactivity were introduced that relate with the electronic structure of the catalyst. Reaction rate expressions became available to predict reaction rate constants from the interaction free energies of reactants.

There is a succession of catalysis science episodes as illustrated in Table 2.1 and is discussed below.

At the beginning of the previous century, the formulation of chemical thermodynamics provided the early scientific basis for discoveries in heterogeneous catalysis. Chemical thermodynamics also played a major role in the immediate period that followed since it is basic to material surface and morphology characterization methods. These methods rely on measurements of physical adsorption of molecules and are discussed in the first sections of this chapter.

The first model of molecular adsorption on a surface was the Langmuir model. This is a surface checkerboard model where reaction sites are equivalent. Importantly it is based on the definition of the finite surface, which contains a finite number of surface sites. Surface titration methods were introduced that discriminate between surface sites that are not reactive and only physically adsorb and surface sites that are reactive. Whereas in the first case molecules may already desorb below room temperature, in the latter case molecular desorption only happens at higher temperature. A molecule is considered to be chemisorbed when strong surface chemical bonds are formed. This may induce molecular bond cleavage (dissociation). It is to be distinguished from physical adsorption, where molecules interact weakly with the surface by physical dispersion forces (the van der Waals interaction energy). This model does not distinguish between sites of different composition. Use of surface titration techniques led to the model of a catalytic site as a uniquely configured surface site. This Taylor reaction site model is an alternative to the uniform Langmuir checkerboard model [4].

Chemical kinetics were formulated in terms of surface reactions of adsorbates mainly based on the Langmuir adsorption model [5]. The kinetic equations became known as the Langmuir-Hinshelwood equations [6], [7]. The reaction mechanisms used in kinetics were initially hypothetical and selected to be as simple as possible. The two-step catalytic reaction model [8] became the standard approach to formulate kinetic equations of catalysis to the chemical engineer [9]. Kinetic parameters had to be fitted with experiment.

Molecular models of the chemistry of surface reactions were proposed once isotope labelling of molecules allowed discrimination between different reaction models. In the decades that followed this became widely explored due to developments in mass spectrometry and gas chromatography, which made accurate product analysis possible. This happened after the 1931 discovery by Harold Urey (Nobel Prize 1935) of deuterium [10]. Mechanistic experiments with deuterated molecules were pioneered by Eric Rideal of Great Britain and Hugh Taylor from Princeton University [4], [11]. These early mechanistic studies led to the first models of catalytic hydrogenation that are presented in the next chapter. Founding molecular mechanistic concepts mainly in hydrogenation catalysis, oxidation catalysis, and acid catalysis were formulated. These concepts by necessity were often speculative but were very fruitful and provided useful context for subsequent molecular mechanistic approaches.

The spectroscopic discoveries of the 1950s and 1960s became rapidly exploited to characterize the catalysts. Developments in electron microscopy made particle morphology, particle size and composition visible at the micro to nano level. Critically important are surface-sensitive techniques that enable discrimination between surface composition and bulk composition. Nuclear magnetic resonance and X-ray absorption or emission techniques allow determination of the local structure of surface clusters or agglomerates. Vibrational spectroscopies of adsorbates enabled us to probe the state of adsorbates and reaction intermediates on surfaces. Results obtained with these techniques will be extensively used in this and following chapters. For an introduction to spectroscopies for catalysis, the textbooks by Thomas *et al.* and Niemantsverdriet are excellent references [12], [13]. Because of the deepened insights in the structure of the catalyst, major progress was made in the correlation of catalytic reactivity with catalyst material properties. The catalytically reactive surface site became within direct reach of the experiment. However, catalysis still remained far from being a predictive science.

Further advances had to await high-resolution techniques and surface science approaches of the 1970s [14], [15]. Then single-crystal surfaces were studied as models of the catalytic reactive surface [14], [15]. Atomic information became available by discoveries of High-Resolution Electron Microscopy [16], Scanning Tunneling Microscopy [17] and related techniques. The relation between chemisorption and surface structure became understood on a molecular level. An important discovery is that surfaces will reconstruct to new surface phases when highly covered with adsorbate [18].

Twenty years later the increasingly powerful hardware capabilities of computation made quantum-chemical simulation of reactive surfaces possible. Atomic models of structures of the catalytic reactive site close to that experimentally observed became accessible to computation. Computational catalysis was thus applied as a tool for experimentation. Quantification of key physical chemical concepts of catalytic reactivity, that in some cases had already been formulated over half a century before, became possible [19].

Catalysis is a kinetic phenomenon and ultimately relates to the elementary rate constants of reaction intermediates adsorbed on the catalyst surface [5]. Progress in computational science not only enabled the calculation of ground state (free) energies but also of transition state (free) energies. The ability to calculate elementary reaction rate constants is a great advance in catalytic kinetics [20], [21]. As discussed in Section 2.4 these calculations are based on the Eyring reaction rate equation, which is based on the concept of activated complexes [22], [23]. This equation is already known since the 1930s, but so far had not been used in first-principle calculations for complex molecules or surface reactions.

First-principle prediction of a reaction rate constant implies that kinetics can be related to the surface structure and composition of the catalytic center. The kinetic model became formulated as microkinetics [24], [25] that explicitly includes all elementary steps of the reaction network. A kinetic simulation is possible without parameter fitting to experiment. Kinetics has become a tool to predict catalysts function-structure/composition relationships.

2.2 The Catalytic Reactive Site

The Langmuir adsorption isotherm is based on a checkerboard model of the surface. According to Taylor, the catalytic reaction depends on coordinative unsaturation of surface atoms as well as their topological arrangement.

Often new science is generated by technique development in a different scientific discipline. This has repeatedly happened in catalysis. An early example is the investigations by Langmuir that began in 1911 when he joined General Electric in Schenectady as the first academically schooled scientist. One of his major contributions is the substantially improved lifetime of the lightbulb. Research that led to this can be considered the early founding of surface science. He received the Nobel Prize in 1932. In the course of the succeeding century, it became a sophisticated discipline. The Langmuir unit (L) for gas adsorption carries his memory. It is the amount of gas at a pressure of 10^{-6} Torr that is adsorbed in one second on a surface.

The key technical advance on which his work was based is the use of vacuum techniques and accurate pressure measurements. In addition to vacuum technology the use of metal filaments, which can be heated by an electrical current, provided the possibility to study "clean" surfaces with quantitative adsorption measurements.

Langmuir's theoretical interpretation of his early experiments led to the formulation of what is now called the Langmuir adsorption isotherm (1913). It is fundamental to catalytic kinetics as well as to catalyst characterization. In 1938 the BET (Brunauer, Emmett, Teller, Section 2.2.3) equation was proposed to determine the surface area of the catalyst surface.

2.2.1 *The Langmuir Adsorption Isotherm*

An essential assumption is the conservation of the number of sites.

Langmuir's basic and then revolutionary idea was his proposal that gas phase molecules adsorb on the surface of a monolayer [6], [26]. He proposed that the interaction between adsorbate molecule and surface atom is of chemical nature and of short range. Reactivity of the surface atom is related to its available valence that is higher than that of bulk atoms due to coordinative unsaturation of the former.

Physical adsorption is to be distinguished from chemisorption. A physically adsorbed molecule interacts with the surface through the dispersive van der Waals force. Per interacting molecule the physical interaction energy is of the order of 5 kJ/mol. Molecules that are physically adsorbed do not form chemical bonds with the surface and do not discriminate between different surface atoms. The chemisorption bond is substantially stronger and can be of the order of 100 to 200 kJ/mol. Chemisorption discriminates between surface atoms. The strong chemical interaction with the surface is the source of surface chemical transformations.

The Langmuir adsorption isotherm is based on a model of an idealized surface with a checkerboard structure. Adsorption site positions are regularly distributed. For this Langmuir based himself on the then just published X-ray diffraction experiments of William Bragg from 1913 [27] that demonstrate that atoms in a solid have a regular and periodic arrangement.

The adsorption isotherm gives the surface coverage θ of adsorbed molecules as a function of gas phase concentration; θ = number of adsorbed molecules/total number of surface sites. It is defined by the adsorption equilibrium constant $K_{ads} = \frac{k_{ads}}{k_{des}}$, k_{ads} and k_{des} being respectively the elementary reaction rate constants of adsorption and desorption. Eq. (2.1) is the Langmuir adsorption equation written in current notation.

$$\frac{\theta}{1-\theta} = K_{ads} \cdot \frac{p}{p_0} \tag{2.1}$$

Figure 2.1 illustrates the coverage dependence as a function of partial pressure according to the Langmuir equation.

The three assumptions that lead to this equation are as follows.

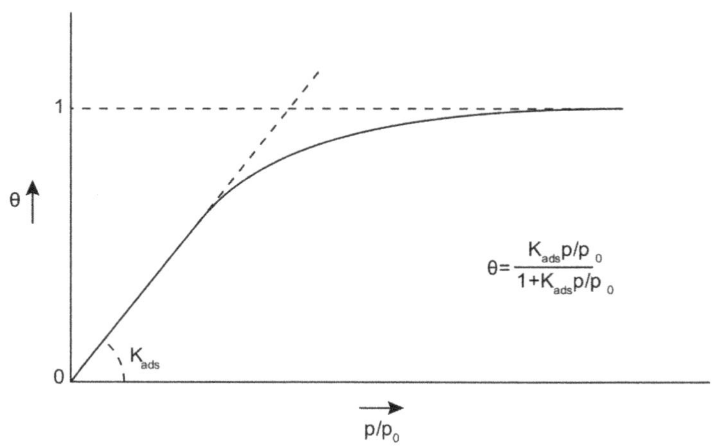

Figure 2.1 The Langmuir adsorption isotherm.

Postulate 1: The total number of adsorption sites is constant. Adsorbing molecules compete for vacant surface sites. This is expressed by Eq. (2.2):

$$\sum_i (\theta_i + \theta_v) = 1 \tag{2.2}$$

θ_i is the surface concentration of molecule i and θ_v is the concentration of vacant surface sites.

Postulate 2: Molecules adsorb individually. They only experience a site-blocking effect of other adsorbed molecules. There is no lateral interaction.

Langmuir [5] demonstrated for CO oxidation catalyzed by a platinum filament that surface coverage of reaction intermediates is an important kinetic variable. At low temperature, oxygen adsorption becomes blocked by adsorbed CO and reaction is negative order in CO pressure (the surface vacancy concentration decreases with CO pressure), whereas at high temperature CO concentration is low and the surface coverage is dominated by oxygen. Then reaction rate is proportional to CO pressure. The reaction mechanism of the oxygen and carbon monoxide reaction could not yet be unraveled.

Langmuir was well aware of the approximate nature of postulate 2. Generally, beyond a particular surface coverage the adsorption energy will decrease. This can be due to repulsive steric interaction when molecules become adsorbed at close distance or due to changes in surface atom chemical bonding induced by interaction with the adsorbed molecules.

Postulate 3: The substrate atoms do not reconstruct when molecules adsorb.

Especially when surface coverage becomes high, the surface atoms will rearrange. Chemisorption substantially weakens neighbor surface atom bonds. When a surface becomes highly covered with adsorbates, this may lead to surface structure reconstruction.

Notwithstanding the limited validity of postulates 2 and 3, the impact of the Langmuir isotherm formulation to catalytic kinetics has been enormous. This is mainly because it explicitly accounts for the finiteness of the number of adsorption sites.

2.2.2 *The Taylor Reaction Site*

The reactivity of catalytic reaction centers relates to coordinative unsaturation of surface atoms as well as their topological arrangement.

The checkerboard surface model of Langmuir became questioned in 1925 by Taylor. He suggested that catalyst surfaces are non-uniform and that reaction sites will have different reactivity, because surface atoms may differ in their number of neighboring surface atoms [4].

Taylor supported the idea of uniquely reactive surface sites by selective adsorption experiments. In practical catalysts, a large decrease in catalytic reaction rate is often found when an extremely small concentration of molecules is chemisorbed compared to total surface area as measured by physical adsorption experiments. The small number of molecules poison the reaction sites.

Langmuir suggested that differences in the strength of the adsorbate bond as a function of reaction center relate to differences in surface atom interatomic distances, which are a consequence of differences in the number of neighboring surface atoms. This idea has been taken over in the 1950s by chemists such as Aleksei Balandin and Otto Beeck [28], [29]. Taylor made the alternative and correct proposal that surface atom reactivity relates to differences in surface atom coordinative unsaturation. He even predicted that catalysis may occur on a single surface atom. The later discoveries in coordination chemistry and organometallic chemistry show that such single surface metal atom catalysis is indeed possible.

Differences in surface atom coordinative unsaturation as well as site topology play a role. In Section 2.3.2, surface science experiments by Gábor Somorjai are presented that demonstrate the large difference in reactivity of Fe single crystal model catalysts in the

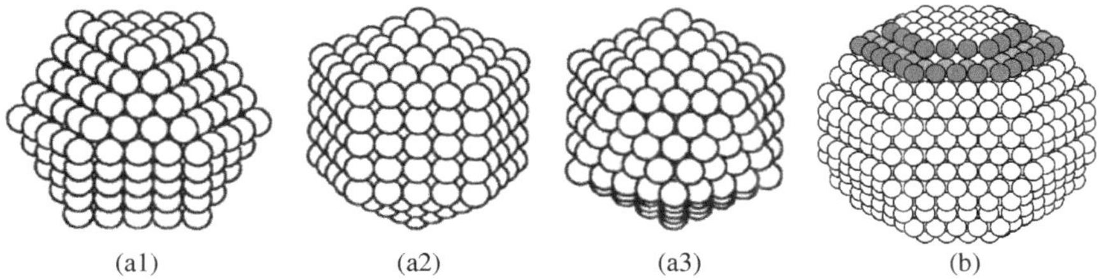

Figure 2.2 The differences in coordinative unsaturation of surface atoms of transition metal particles. (a) Comparison of particles of different shape. (b) Particle with step-edge (B5) sites (darkly colored) [32].

ammonia synthesis reaction [30]. Surfaces with atoms of larger degree of coordinative unsaturation show higher reactivity. A later experiment by Chockendorf *et al.* [31] on CO conversion with Ni single crystal surfaces that contain step-edge imperfections demonstrate that such step-edge sites are the catalytic reaction sites. They are selectively poisoned by sulfur, which decreases the catalyst reactivity by orders of magnitude. The step-edge sites are substantially more active than ideal smooth terrace surfaces.

Figure 2.2 illustrates the different kinds of surface sites one distinguishes on a transition metal particle.

One distinguishes corner and edge versus terrace sites. The larger the particles the smaller the relative concentration of corner and edge sites. Particle shape not only affects the relative concentration of corner and edge sites, but also the relative ratio of particular terraces. The corner site is coordinatively most unsaturated, to be followed by that of the edge surface atom. Surface atoms on terraces are the least coordinatively unsaturated and hence the least reactive. A surface site atom that has the highest degree of coordinative unsaturation is most unstable. Thermodynamics would predict that such surfaces should be in the minority.

According to the Taylor argument, these surface atoms should have higher reactivity. So generally, one expects reactive sites to be present in a minority concentration. In addition to differences in coordinative unsaturation of surface atoms, their relative arrangement also matters. This relates to surface atom steric demands for activation of molecular bonds in reactant molecules. This initiates the catalytic reaction.

When reactant molecular bonds cleave, at least two reaction intermediates form. This implies that several surface atoms need to be available to stabilize the fragments and make the bond dissociation thermodynamically possible. The step-edge configuration near the top of the particle of Figure 2.2b illustrates the configuration of such sites. They are called B5 sites as they consist of five combinations of triangular or square configuration arranged in steps. Such step-edge sites cannot be stabilized on particles that are too small. In Chapter 3, high-resolution electron microscopy experiments [33] are discussed that validate the presence and structure of these uniquely reactive sites. The relative concentration of reactive catalytic sites can be probed with transient kinetic experiments as discussed in Section 2.2.6. The explanation of the differences in reactivity as a function of catalyst site atomic structure had to await the end of the 20th century and is discussed in Section 2.3.

2.2.3 *The BET Adsorption Isotherm: Determination of Surface Area*

The BET adsorption isotherm determines the surface area based on multilayer physical adsorption.

Here the BET (Brunauer-Emmett-Teller) equation is presented, which determines the total surface area of a catalytic material from physical adsorption measurements. Since the method is based on physical adsorption, it is not sensitive to potential differences in surface composition. The BET equation is an extension of the Langmuir adsorption equation. Different from the Langmuir adsorption isotherm, which applies to monolayer adsorption, the BET adsorption isotherm describes multilayer adsorption.

The equation was proposed in 1938 by the chemists Stephen Brunauer and Paul Emmett and theoretical physicist Edward Teller [34]. Emmett was a leading catalytic chemist. In the Second World War he was engaged with separation of uranium isotopes. Teller became later known as the father of the hydrogen atomic bomb (he was made immortal by the Dr. Strangelove movie from 1964 starring actor Peter Sellers). Brunauer was a former student of Teller.

Practical catalysts are usually porous materials, so they have a high surface area. When the pore size distribution is in the 1–100 nm range the surface area can be of the order of 200 m^2/g or even larger. Often used supports are aluminum and silicon oxides. The rate of diffusion of molecules in or out of the pores of such supports may compete with the reaction rate of the surface-catalyzed reaction, as is discussed in Section 2.2.5. The pore dimensions are relevant; they can be deduced from so-called t-plots, discussed in Section 2.2.4. The BET method as well as the t-plot method are based on physical adsorption of inert gas molecules at or near conditions where gases condense.

The BET adsorption isotherm calculates the volume of gas adsorbed onto the surface of a material as a function of gas pressure. This pressure is chosen below the condensation pressure. The molecules in the condensing film are assumed to have the same density and interaction energies as in the corresponding liquid phase. The BET equation is:

$$v = \frac{v_m c p}{(p_0 - p)\left(1 + (c-1)(p/p_0)\right)} \tag{2.3}$$

In Eq. (2.3) v is the volume adsorbed, v_m the volume of the monolayer, p_0 the condensation pressure, and $c = e^{\frac{E_1 - E_l}{RT}} >> 1$. E_1 is the heat of adsorption of the first layer and E_l is that of the adsorbed layer. Eq. (2.3) is valid in the limiting case where there is an infinite number of layers. Related expressions can be deduced when the layer thickness is finite. Eq. (2.3) becomes equal to the Langmuir adsorption isotherm when the gas pressure $p << p_0$.

The parameters in Eq. (2.3) can be deduced from volume-pressure adsorption experiments. Such a curve is shown in Figure 2.3. Unlike the Langmuir isotherm, the adsorption volume of the BET isotherm does not have a plateau as a function of pressure but tilts upwards. Comparison of the experimental curves with the BET isotherm expression for finite layer thickness leads to determination of multilayer thickness.

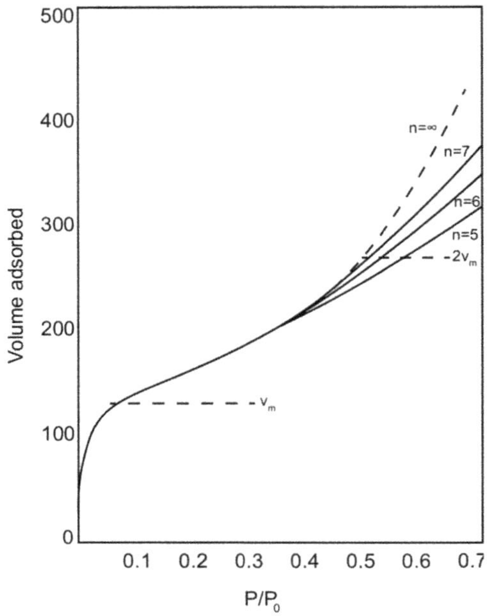

Figure 2.3 BET adsorption isotherm of nitrogen for different adsorbed layer thicknesses n [34].

The experimental Figure 2.3 shows that in this particular physisorption experiment, the layer thickness is 5 or 6 when $p/p_0 = 0.6$. Also, in the figure v_m is indicated. This is the volume of gas that covers the surface with a monolayer. Using the density of a nitrogen film, a surface area of approximate 500 m^2/gram can be deduced for the material used. Its precise value depends on the assumption made for the monolayer film density.

The BET adsorption isotherm equations are currently routinely used for surface area measurements with physical adsorption. Nitrogen is the gas that is most used.

2.2.4 *Pore Size Distribution: The T-plot*

Pore size distribution of materials with pore sizes between 1 and 100 nm can be determined using capillary condensation of inert gases. For larger pores, mercury porosimetry is of use. The latter derives pore size dimension from the pressure needed to force the mercury into the pores.

Capillary condensation as well as porosimetry rely on the Kelvin expression. The dimension of a pore relates to the curvature of the meniscus. For a cylindrical pore this equation can be written as [35]:

$$\ln p/p_0 = 2\frac{\sigma \cdot v}{RT \cdot r} \tag{2.4}$$

In Eq. (2.4) σ is the surface tension of the liquid, v is the molecular volume of the adsorbed gas, T is the temperature, R is the gas constant and r is the so-called Kelvin radius. For a convex meniscus curvature, r is positive and for a concave curve, r is negative.

When pore filling is studied with increase of pressure, the pore may gradually fill. Upon reversal of the pressure, a delayed pore depletion may show. As illustrated in Figure 2.4, this hysteresis depends on pore shape, size, and connectivity.

Pore filling starts with the smaller pores and the larger pores become filled at higher pressures. Pore depletion starts at a lower pressure, which is determined by the pressure the gas can pass the smaller pores.

In 1963, the t-plot method was developed by the Dutch scientist Jan Hendrik de Boer. This is an elegant tool, and currently a routine method, to extract pore size distribution from adsorption data [37].

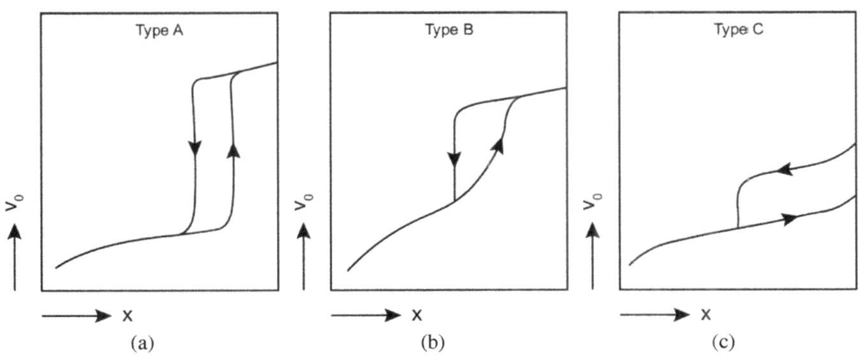

Figure 2.4 Three classes of hysteresis loops. V_0, the volume of gas adsorbed, is plotted as a function of pressure x. (a) Tubular or ink bottle-shaped pores; (b) non-uniform tubular pores; (c) slit-type pores [36].

In the t-plot, one plots the volume of adsorbed gas as a function of layer thickness *t* of condensed liquid gas molecules. This is equal to the volume of gas adsorbed divided by the volume of a monolayer of adsorbed gas. In the absence of pore condensation, layer thickness can be deduced from the surface area according to the BET equation. When, as a function of increasing pressure, a larger amount of gas volume is adsorbed than expected from the extrapolated BET isotherm, the value of *t* determines the pore size dimension. The extra volume of gas is equal to the pore volume. At present, electron microcopy can also be used to determine the distribution of pore size and shape.

2.2.5 *Pore Diffusion and Reaction Rate*

When reaction rate become diffusion limited, its dependence on surface site concentration changes from proportional to square root. The apparent activation energy becomes half the intrinsic activation energy.

Langmuir's monolayer adsorption and surface reaction theories were in opposition with ideas developed by the then famous physical chemist Max Bodenstein at Hannover. Bodenstein explained the catalytic role of a surface as a physical effect. He suggested the presence of a physically adsorbed film as in capillary condensation. Reactants adsorb in the film and reaction rate is enhanced due to the increase in local concentration [38]. This view was shared by Ostwald. (Interestingly, the Ostwald reaction is a fast reaction at high temperature and is limited by film diffusion [39]).

However, for reactions such as hydrogen or carbon monoxide oxidation, the temperature of reaction is too high for capillary condensation to be relevant. Instead, catalysis is a chemical phenomenon that relates to the reactivity of the catalyst surface. At mild temperature, pore diffusion may affect reaction rate.

High surface area catalysts are microporous. As long as their pores are at least a few nanometers wide, molecules move in and out of the pores through Knudsen diffusion. Knudsen diffusion physics is valid as long as the pressure of gas is high enough and pore sizes large enough so that the number of molecular collisions between molecules is large compared to the number of molecular collisions with the pore surface. The validity vanishes when the pore size becomes so small that it is comparable to the dimension of molecules. Pores of nanosize dimension dominate in zeolitic materials as used in solid acid catalysts (Chapter 5). Here the discussion of pore diffusion is limited to pore size dimensions where Knudsen diffusion is valid.

According to Knudsen diffusion physics, the diffusion rate of molecules is driven by a concentration gradient [40]. In a diffusion-limited catalytic reaction (Figure 2.5) a concentration gradient develops between pore mouth and molecules within the micropore. Along the length of the pore, concentration of reactant decreases because of catalytic reaction with the micropore wall. Reaction rate is suppressed compared to the case where diffusion is fast. Ernest Thiele deduced in 1939 [41] the rate equations that describe the altered reaction rate due to pore diffusion. For a first-order kinetic reaction and catalyst with cylindrical pores, the diffusion-limited reaction rate expression is [42]:

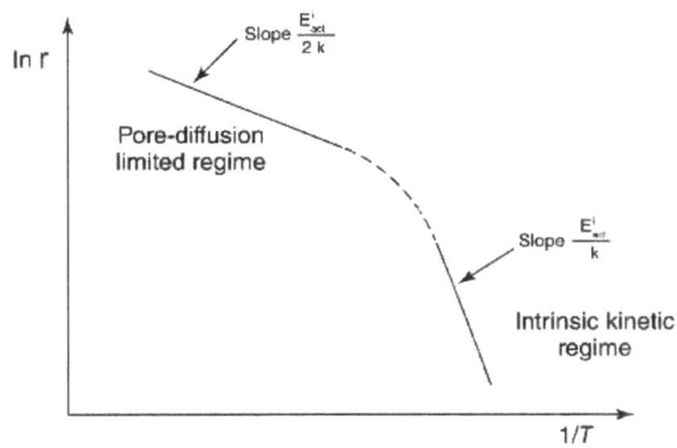

Figure 2.5 The reaction rate *r* of a diffusion-limited reaction as a function of temperature. At low temperature the reaction rate is small compared to the diffusion rate, and at high temperature the reaction rate becomes fast compared to the diffusion rate.

$$r_D = \sqrt{\frac{2D \cdot S.r_i}{L^2}} (L^2 / 2D(S.r_i)^{-1}) \quad (2.5)$$

In Eq. (2.5) r_D is the diffusion-limited reaction rate, D the diffusion constant, L the length of the micropore, S the reactive surface area, and r_i the intrinsic catalytic reaction rate per unit surface area. This is the reaction rate in the absence of pore diffusion.

In the diffusion-limited high temperature regime, where reaction rate r_i is fast and diffusion is reaction rate limiting, the reaction rate r is equal to r_d and proportional to $\sqrt{S}$, the reactive surface area, and the activation energy is half of the intrinsic activation energy: $E_{act} = \frac{1}{2} \cdot E^i_{act}$. The activation energy of the intrinsic reaction rate r_i is given by E^i_{act}. In the low temperature regime, where the diffusion is relatively fast and reaction rate r_i is slow and is reaction rate controlling, the activation energy of r is equal to the intrinsic activation energy: $E_{act} = E^i_{act}$. Diffusion limitation has the interesting consequence that reaction rate has non-Arrhenius dependence as a function of temperature.

Experimentally one can determine whether pore diffusion plays a role by comparative kinetic measurements by varying the particle size at constant contact time (linear gas velocity/mass catalyst). This will increase potential diffusional limitation. When no change in reaction rate is observed, reaction is not diffusion rate limited.

2.2.6 *The Concentration of Reactive Centers: Transient Kinetics*

The catalytic reactive site usually has a minority presence on the catalyst surface.

How to determine the concentration of the catalytic reaction centers has puzzled catalysis researchers, since Taylor suggested the presence of a unique reaction center. There is not only the question of how to determine reaction site concentration that is present at extremely low concentration, but also the question of how to cope with the change in surface state that occurs by reaction. That the catalytically reactive site is not independent of reaction had already been noted by Faraday [43], who established that the rate of reaction depends on catalyst pretreatment conditions and that the surface state is different after reaction. Characterization of the catalyst before or after reaction does not necessarily relate to its state during reaction.

Most ideal is identification of the surface state during reaction. Currently spectroscopic techniques such as Infrared (IR) and Raman spectroscopy are available to identify surface intermediates during reaction. The solid inorganic state of the surface can be investigated by X-ray photoelectron spectroscopy (XPS) or photoadsorption spectroscopic techniques such as extended X-ray absorption fine structure (EXAFS) (see [12], [13]). However, how do we know that the species that one observes is not a spectator, a non-reactive co-adsorbed species, but one that actually participates in the reaction?

In Taylor's days the only way to estimate the number of reactive sites was by using catalyst poisoning reagents that selectively interact with catalyst sites, and presumably strongest with sites of highest surface reactivity.

In situ identification of the concentration of reaction sites implies that the concentration of surface reaction intermediates that is part of the catalytic reaction cycle has to be identified. This requires a kinetic experiment. Here the use of transient kinetic experiments is discussed.

The kinetic reaction rate depends on the number of reaction sites and kinetic reaction rate constants. The issue is how to disentangle these two kinetic parameters. For simplicity the procedure is illustrated here for a monomolecular reaction that is not diffusion rate limited. The intrinsic reaction rate R of a monomolecular reaction is given by Eq. (2.6):

$$R = N_s k_i \theta_i \tag{2.6}$$

The reaction rate R is proportional to the total number of sites N_s that can be determined by the total surface area or by chemical titration. The fraction of catalytically active sites is defined by the ratio of sites occupied by reactive intermediates θ_i. k_i is a site-normalized elementary reaction rate constant, which is assumed to be reaction rate controlling (see Section 2.4).

At least two experiments are needed to determine the number of reactant sites $N_s\theta_i$. One experiment is needed to determine the total reaction rate R, the other to determine $N_s\theta_i$ or k_i.

When after reaction the surface is probed by temperature-programmed desorption (TPD) or temperature-programmed reaction (TPR), experiments that remove reaction intermediates, the amount of adsorbed reaction intermediate $N_s\theta_i$ can in principle be identified. Such a TPD or TPR experiment can be done for instance by switching to an inert nitrogen flow, or to a reactive hydrogen or oxygen flow. This postmortem analysis can determine the concentration and kind of surface reaction intermediate but assumes that the state of the surface remains the same after reaction. Connection of measured reaction intermediate with reactive intermediate is not proven.

With the inventions of mass spectrometry, gas chromatography and *in situ* IR spectroscopy, it is possible to determine the catalytic reaction center surface concentration *in situ* via transient experiments with use of isotope-labelled molecules. Once reaction is at steady state, the reactant molecule is switched to an isotope-labelled molecule or vice versa [44]–[46]. This ensures that chemically the reactant state of the surface does not change. When a non-labelled reactant is initially used, the concentration of adsorbed non-labelled reactant intermediates will decrease and become substituted by labelled reaction intermediates. The concentration of desorbing labelled product molecules will increase until a new surface steady state with isotope-labelled molecules is established. The rate of appearance of isotope-labelled product molecule is equal to the elementary surface reaction rate constant k_i. Substitution of k_i in the steady state reaction rate expression, Eq. (2.6), gives $N_s\theta_i$, the surface site concentration. The fraction of catalytic reaction sites follows once N_s is known from a determination of the total surface area. This is based on the reasonable assumption that because the reactive intermediate is reaction rate controlling, its catalyst site concentration is close to one. The following example indicates that this is not necessarily the case.

The power of the Steady State Isotopic Transient Kinetic Analysis (SSITKA) technique is illustrated by the experiment of Mirodatus *et al.* of the hydrogenation of benzene on a Ni/SiO_2 catalyst [47], [48]. They used the switch of C_6H_6 to C_6D_6 and identified surface site coverages with benzene and hydrogen as well as the relative concentration of reaction intermediates. At the low temperature of 400 K, their experiment showed that the surface coverage is dominated by adsorbed hydrogen atoms. This inhibits adsorption of benzene. The reaction rate-controlling step is hydrogenation of a partially hydrogenated adsorbed benzene species. The concentration of this reactive intermediate was determined by adsorption of hydrogen at only 1% of the Ni surface area. It illustrates that the fraction of reactive surface sites can depend on reaction conditions. In this case it depends on the hydrogen-blocking concentration. The topology of the Ni atoms is not important for this reaction, since the reaction is not surface structure sensitive (see Section 3.3). Another reaction that has been extensively investigated with transient kinetics is the hydrogenation of CO to give methane or higher hydrocarbons [49]. For the methanation reaction Ni is the preferred catalyst. Higher hydrocarbons are produced in the Fischer-Tropsch reaction, for which alkali-promoted Fe or Co are the preferred catalysts (see Section 3.2.2). The SSITKA experiments were able to distinguish a change in the number of reactive sites when the particle size changed. For methanation, the relative concentration of reaction centers decreased when the transition metal nanoparticle size decreased. In the case of Fischer-Tropsch catalyst promotion, the relative concentration of reactive intermediates "CH_x" that give longer hydrocarbons increased whereas the intrinsic reaction rate was unaffected.

Kinetic experiments only provide information on the relative concentration of proposed surface reaction intermediates. Correlation with surface atom arrangement and coordination requires atomistic information. Due to advances in spectroscopy and computation, the molecular basis of catalytic reaction dependence on catalyst structure and composition was founded at the end of the previous century and the beginning of this century. In the next section molecular theories of surface reactivity are presented.

2.3 Fundamentals of Surface Reactivity

The reactivity of the surface of a transition metal relates only indirectly to bulk electronic structure but relates directly to the quantum chemistry of a surface molecular complex.

Here we discuss how to characterize the chemical reactivity of the surface and what its relation is with the electronic structure of the catalyst material.

The chemisorption energy of molecules or atoms adsorbed on surfaces provides a measure of the surface reactivity. It is sensitive to composition and structure of the surface. An often-used probe for this energy is the temperature at which adsorbate desorbs. When desorption is a non-activated process, the activation free energy of the desorption rate can be related to the adsorption free energy. Other methods are calorimetric measurements or determination of adsorption equilibrium constants.

In Section 2.3.1 early ideas are discussed for the transition metals on the relation between bulk physical and chemical properties and adsorption energies. The relation between the heat of formation of bulk compounds and the adsorption energies provides the important insight that the surface chemical bond shares common features with chemical bonding within compounds. In order to make progress the structure of the adsorbed state has to be known. This became possible with advances in surface science. The structures of surfaces covered with adsorbates can be identified with atomic precision.

In addition to measurement of adsorption energies, determination of elementary reaction rate constants is also essential to probe surface reactivity. Quantum computational techniques have become available that predict adsorption energies as well as activation energies of elementary reaction rate constants on representative surface models. In Section 2.3.3, the nature of the surface chemical bond is discussed. Structure reactivity descriptors of surface electronic structure are introduced that relate to surface chemical reactivity.

The history of the exploration regarding the nature of the surface chemical bond was a slow process of recognition that chemisorption only indirectly relates to bulk material electronic properties. It is not a physical but rather a chemical phenomenon in which surface atoms react with the adsorbate. The chemisorbed state is a surface molecular complex that causes major surface bond weakening and sometimes disruption of surface metal atom-atom bonds. The spectroscopy of chemisorbed molecules indicates a close relation with bonding in coordination or organometallic complexes [50], [51].

2.3.1 *Physical Chemistry of Chemisorption*

The nature of the surface chemical bond and that of bulk compounds are correlated.

In the 1940s and 1950s, chemisorption was probed by adsorption studies on metal films or filaments in vacuum. Changes in physical properties of these materials due to exposure to reactive gases could be readily measured. It was observed that adsorption of gases correlates with changes in electronic conductivity, magnetism, or metal work function [29]. These properties relate chemisorption to bulk electronic variables and specifically to the electron density of states at the Fermi level. The Fermi level of a metal is defined as the electron energy where electrons are most weakly bound. It is related with the lowest photoexcitation energy where a metal emits electrons. Especially, investigations with alloys suggested that the electron density at the Fermi level is a reactivity descriptor of transition metals [52]. In the early 1960s, this view became supported by physical investigations of semiconducting compounds active as catalysts [53], [54]. Theories that relate chemisorption properties to the electron density at the Fermi level became known as the "electronic factor theory". However, due to improved insights into the nature of the surface chemical bond, the "electronic factor theory" is to be rejected. The physical justification of "electronic factor theory" is the assumption that the chemisorption interaction is weak compared to the interaction energies between the metal atoms. In Section 2.3.3, it is discussed that this view of the surface chemical bond is incorrect.

Modern theory as well as experiment demonstrate that the chemisorbed state is a molecular surface complex, reminiscent of a coordination complex that consists of a few metal atoms. In the adsorption complex, bonding interactions may vary between 100 and 1000 kJ/mol. The molecular adsorption complex is embedded in the surrounding surface. The dependence on surface structure relates to differences in the embedding energy of molecular cluster into the surrounding solid.

The quantum-chemical argument for the independence of the surface chemical bond on electron density at the Fermi level is that adsorption localizes metal electrons that otherwise are delocalized [55]–[59]. However, the position of the Fermi level with respect to adsorbate orbital energies is relevant since it determines the total number of electrons that is part of the surface chemical bond. Nonetheless, up to the 1980s the surface chemical bond remained often interpreted in terms of the "electronic factor theory".

Experimentally, a first dent in "electronic factor theory" came from work function measurements. The work function is a surface-sensitive property, which can be used to indirectly measure surface composition. It relates to the photoelectric effect discovered by Einstein [60]. When an electron is excited and emitted from its bulk Fermi level it has to pass through the surface, which has a small dipole moment. This dipole moment is sensitive to variation in surface composition. Before the advance of X-ray spectroscopic techniques, which measure elemental surface composition directly, work function measurements were used to measure the surface state of transition metals. When applied to study alloys, it was found that the surface composition of alloys often differs from that of the bulk. In this case correlation of reactivity with bulk physical properties as the electron density at the Fermi level became circumspect [61]. Later this became extensively validated when surface-sensitive techniques such as Auger electron spectroscopy as well as X-ray and UV photoemission techniques were introduced. The chemical nature of the adsorbate-surface bond became apparent. Reactivity relates with the reactivity of surface atoms and their configuration. Chemisorption is sensitive to the kinds of atoms present at the surface [62], [63].

Interestingly for the transition metals, a relation exists between adsorption energy and a bulk property known as the heat of formation. In 1963 the Japanese scientists Tanaka and Tamaru discovered this important relation [64], which is illustrated in Figure 2.6.

Figure 2.6 compares adsorption energies of different gases for a variety of transition metals. The striking feature of this figure is that different adsorbates correlate similarly with the bulk heat of formation of the corresponding metal oxide. The O_2 molecule adsorbs dissociatively. The plot of Figure 2.6 actually compares the energy of adsorption of the O adatom with the oxygen atom binding energy in the bulk oxide. The correlation of Figure 2.6 is another indication that chemisorption does not correlate with the electron density of

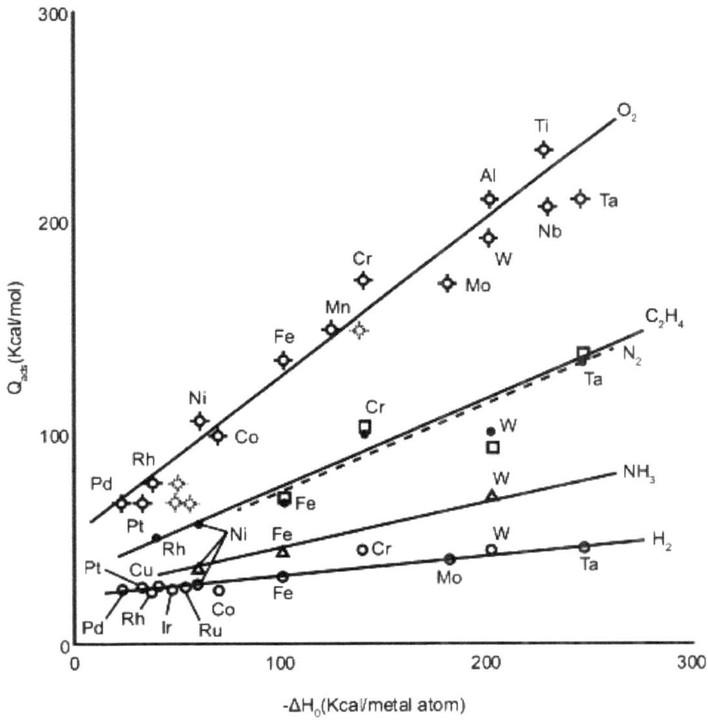

Figure 2.6 The correlation between the initial heat of adsorption of various gases (Q_{ads}) and the standard heat of formation of the highest oxide per metal atom ($-\Delta H_0$). Energies are in kcal/mol [64].

states at the Fermi level of the transition metal. Instead, chemical bonding features that determine the heat of formation of the bulk oxide relate to chemical bonding features of the chemisorption bond.

From Figure 2.6, a second interesting observation can be made by comparison of the heat of adsorption of O_2 with that of the bulk oxides. The chemical bond of the adsorbate is stronger than in the oxide. This is in line with the experimental observation that the surface of transition metals remains oxidized at a temperature where the bulk oxide would decompose. In the adsorption event, the bulk phase remains metallic whereas the surface can be covered with oxygen atoms.

An intuitive way to understand this phenomenon is the realization that oxide formation requires rupture of metal-metal bonds. In contrast, the chemisorbing atom adsorbs to surface atoms and no metal-metal atom bonds have to be disrupted. Surface metal-metal atom bonds have already been disrupted to create the surface. The adatom chemical bond adsorbs to a state activated with respect to the bulk. It leads to the correct expectation that the surface chemical bond is sensitive to the coordination of the surface metal atoms. Generally, one observes that the adatom bond energy will increase the lower the surface atom coordination with neighbor metal atoms is. A more open surface is less stable and more reactive because the surface metal atoms are less coordinatively saturated. The structure dependence of the surface chemical bond is a major topic of the next section.

2.3.2 *The Surface Science View*

2.3.2.1 *Surface Structure and Reactivity*

Chemical bond dissociation reactions are structure sensitive. Coordinative unsaturation of surface metal atoms as well as surface topology matter.

Predictive chemical science becomes possible with knowledge of the structure of chemical complexes. The breakthrough contribution of surface science to catalysis is that it provides the structure of the adsorbate covered surface with the same detail as crystallography for bulk structures. In this introductory subsection, the single crystal approach is introduced along with the discovery of surface reactivity dependence on structure. Dissociative adsorption of diatomic molecules is used to discuss surface structure dependence. Chemical bonding aspects of trends in chemisorption energy as a function of surface metal atom and its coordination number are presented in a following section. This is followed by the introduction of kinetic expressions that relate reaction rate with surface reactivity.

Surface science technology is based on the use of well-defined single crystals as catalyst models in combination with ultra-high vacuum and advanced spectroscopic tools to characterize the surface. The books by Somorjai [65], Ertl [15], and Masel [66] provide excellent introductions to surface science.

Surfaces are prepared by cutting single crystals of a chosen material along different crystallographic directions. The reactivity of such a surface can be determined by measuring the rate of adsorption of a gas.

In Figure 2.7 an early example shows the rate of dissociative adsorption of N_2 on Fe surfaces of different structure by Ertl *et al.*

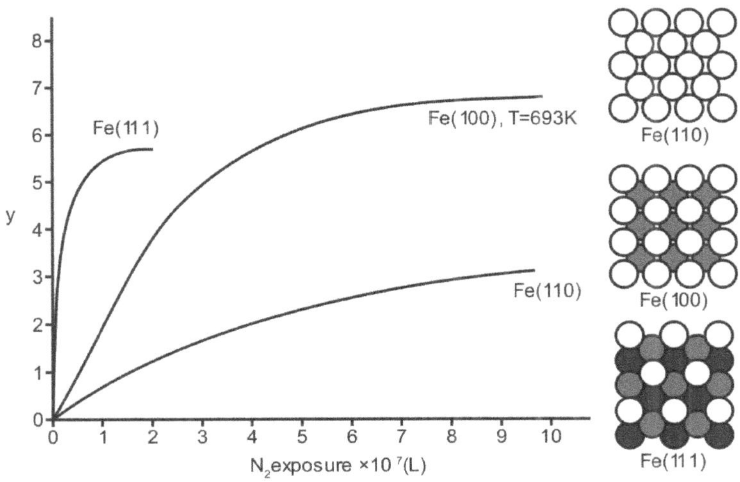

Figure 2.7 Variation of the relative coverage with N atoms (y) on different Fe single crystal surfaces with N_2 exposure at 693 K. The structures of respective surfaces are shown on the right [15].

[67], [68]. The reaction rate is followed from the increase in time of the surface coverage with N adatoms. The latter is determined by Auger electron spectroscopy. A striking difference in adsorption rate is observed.

Three Fe surfaces were studied: the least reactive Fe(110) surface, the less reactive Fe(100) surface, and the most reactive Fe(111) surface. The index notation indicates surface crystallographic orientation. All three surfaces are terrace surfaces but they differ by ordered surface atom arrangements. Since the sites at a particular surface (that consist of several metal atoms) are identical, the surfaces correspond to the Langmuir checkerboard surface model. Bulk Fe metal has a body-centered cubic structure. The Fe atom can be considered to be located in the center of a cube and is coordinated with eight neighbor Fe atoms located at the corners of the cube. A surface is created by metal atom bond cleavage. This decreases coordination of the surface metal atoms compared to the bulk. On the most stable Fe(110) surface, the coordination number of a surface atom is six. This determines its reactivity. The Fe atom at the Fe(100) surface has four nearest neighbors that are located in the subsurface layer. On the Fe(111) surface two coordinatively unsaturated atoms are exposed, one with coordination number three and the second with coordination number seven. In the next section, the relationship between surface atom coordination and reactivity is discussed. The most reactive surface has exposed atoms with the lowest coordination number. In Section 3.2.1.1, ammonia synthesis experiments catalyzed by the same Fe surfaces that also show orders of magnitude differences in reaction rate are discussed (Figure 3.16).

In addition to the coordination number, the connectivity of surface atoms is important. For this, one needs to consider the next nearest atom configurations. Corrugated surfaces contain step-edge sites as previously introduced for small particles (Figure 2.2). Such step-edge sites, which can have triangular or square tiling surface topology motives, are also called B5 sites. Their different surface topologies may lead to different surface reactivities.

For NO dissociation on different Pt surface sites, Figure 2.8 and Table 2.2 illustrates how the combination of coordinative unsaturation of surface atoms and site topology motive affects reactivity.

DFT-calculated adsorption and transition state energies and reprective intermediate structures for a stepped and non-stepped Pt

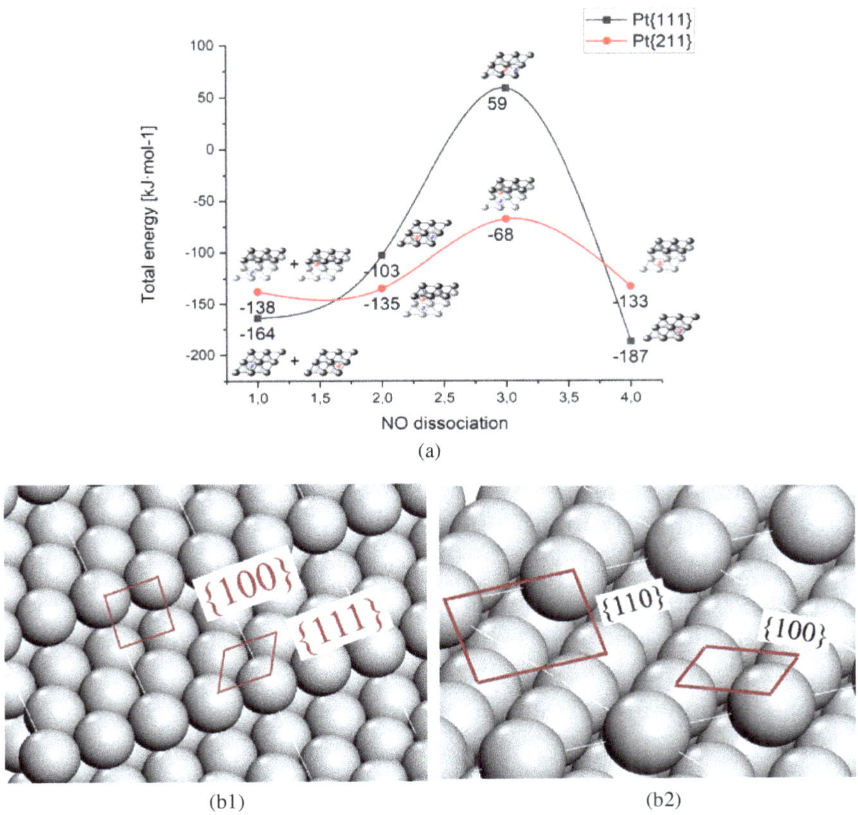

Figure 2.8 Dissociation of NO on stepped and non-stepped single crystal surfaces of Pt. (a) DFT potential energies of NO dissociation on Pt(111) surface and Pt(111) surface with (100) directed steps [69]. (b1) The corrugated Pt(211) surface. The step-edge site consists of square and triangular atom arrangements. Crystallographic orientations are indicated. The coordination numbers of the surface atoms are respectively 6, 9, and 11 [70]. (b2) The corrugated Pt(410) surface. The B5 sites are kinked sites. Coordination numbers of surface atoms are 6, 8, 9, and 11 [70].

Table 2.2 DFT-calculated activation energies of NO dissociation on Pt surfaces [70].

Surface	(111)	(100)	(211)	(410)
E_{act}(kJ/mol)	246	90	65	30

surface are shown in Figure 2.8a [69]. The surface step on this surface is similar as on the Pt(211) surface (Figure 2.8b1). This surface is often studied experimentally as a model of step-edge sites. Whereas the gas-phase dissociation energy of NO is 607 kJ/mol, the activation energy of dissociation of the adsorbed NO chemical bond can be as low as 65 kJ/mol. The decrease in bond dissociation energy barrier between gas phase and surface relates to the strong chemical interaction of reactant and surface. This reduction of bond cleavage energy reflects the beneficial action of the catalyst. A reduction of the activation energy of bond dissociation by interaction with the surface to 10% of the molecular bond energy is common for surface reactions.

Table 2.2 compares activation energies for several surface sites. The activation energy of dissociation of NO decreases by more than 150 kJ/mol from its high value on the Pt(111) surface to the low values of the reactive surface. A low activation energy barrier is also found for dissociation on the Pt(100) terrace. Step-edge sites have lower activation energies. The stepped Pt(410) surface shown in Figure 2.8b2 has the lower barrier.

The dominant reason for the decrease in activation energy relates to the difference in reaction path with that on the Pt(111) surface. Inspection of Figure 2.8a shows that on the Pt(111) surface, the NO molecule is perpendicularly adsorbed to the surface. In order to dissociate, the NO bond axis has to bend into a position parallel to the surface. When adsorbed next to the surface step, bending of the NO axis is not necessary as the oxygen atom of NO is close to the step-edge.

Figure 2.9 illustrates that this favors orbital overlap between surface electron density and the molecular chemical bond electron density. The NO π bond is weakened due to electron backdonation

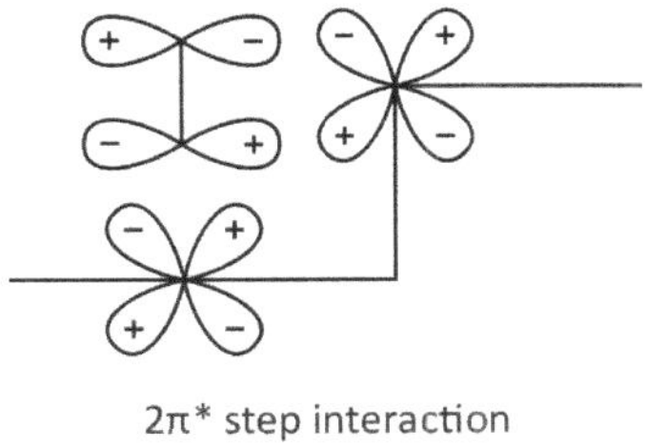

Figure 2.9 Schematic illustration of orbital interaction of step-edge atom d orbitals with the anti-bonding 2π* NO orbital.

from the step-edge atoms into the corresponding anti-bonding molecular orbitals (see Section 2.3.3). When the molecule dissociates, the separated atoms do not share a bond with the same metal atom. The oxygen atom binds twofold to the step-edge, and the nitrogen atom at a threefold site at the bottom of the step. Also on the [100] surface, the NO molecule dissociates with low activation energy barrier (see Table 2.2); the separated atoms of the NO molecule do not share a bond with the same metal atom. In the square configuration of the four surface atoms, in the transition state the O atom of NO binds to two surface metal atoms and the N atom of the NO molecule to the two Pt atoms at the opposite position. This is a major difference with the Pt(111) surface, where in the transition state the dissociating fragments share a chemical bond with the same surface metal atom. According to the Bond Order Principle discussed in the next section, this has considerable additional energy cost. On the Pt(111) surface, this causes the elementary reaction step of NO dissociation to be endothermic, whereas on the stepped surface this dissociation is thermodynamically neutral.

The N_2 molecule adsorbs side-on to the transition metal surface, but in contrast to NO it has a high barrier on the Pt(100) surface. Dissociation of the N_2 molecule with low transition state energy by late transition metals requires activation by step-edge B5 sites [71]. The step-edge site has the lower activation energy when one of the nitrogen atoms adsorbs in threefold coordination and the second nitrogen atom interacts with surface metal atoms that do not share a bond with the other N atom. The Pt(211) step-edge site has the topology that satisfies this requirement [32], [72]. The high rate of N_2 dissociation on the Fe(111) of Figure 2.7 has the same cause. Reaction sites have proper topology and contain transition metal atoms of low degree of coordination. The difference in reactivity of NO versus N_2 relates to preferred coordination of the N atom. This is determined by the valency of the adatoms. On late transition metals the preferred coordination of the trivalent N atom is threefold at a triangular surface atom site [73]. For divalent O a twofold coordination is optimum, and fourfold surface coordination for tetravalent C.

Diatomic molecules with π bonds require surface sites that consist of an ensemble of five or six atoms. This ensemble of several surface metal atoms is needed to accommodate the adatoms that require multiple surface metal atom coordination for strong binding. In addition, the activation barrier is sensitive to coordinative unsaturation of the surface atoms. Differently, chemical bond dissociation of σ chemical bonds such as the C–H bond of an alkane is not sensitive to the surface atom ensemble size, but only depends on the coordinative unsaturation of the surface atoms. In Section 2.3.3.4 (Figure 2.23), where chemical bonding aspects of dissociation reactions are discussed, it is shown that cleavage of a C–H bond proceeds through a transition state that involves only a single transition metal surface atom.

2.3.2.2 *Trends in Adsorption Energies*

The Bond Order Principle predicts that adsorbate chemical bonds become weaker the higher the coordination number of a surface atom is. Adsorbate bond energies increase for metal atoms of decreasing d-valence electron count along a row of the periodic table.

The structure dependence of an adatom relates to two topological properties of surface metal atoms. One has to distinguish between the adatom coordination number with surface metal atoms (c_{ad}) and the surface metal atom coordination number (c_m). Experimentally the contribution of these topological properties to adsorption energy differences is usually difficult to disentangle, because information on adsorbate structure at the atomistic level is necessary. DFT calculations are available that distinguish the energy contributions of the two determinants of the adatom energy. Table 2.3 and Figure 2.10 illustrate this for the interaction energy of a carbon atom with different Ru surfaces [74], [75]. A useful comparison of the bond energy can be made when one of the two coordination numbers is kept constant.

Table 2.3 The dependence of adsorbate adsorption energy as a function of surface atom coordination. (a) DFT-calculated bond energies of C attached to Ru surfaces as a function of adatom coordination number c_{ad} and surface metal atom coordination number c_m [74], [75].

Surface	Siting	Adsorption energy (kJ/mol)	C atom coordination number, c_{ad} (Ru atom coordination number, c_m)
Ru(0001)	top	497	1(9)
	bridge	631	2(9)
	hollow hcp	688	3(9)
	hollow fcc	648	3(9)
Ru$(10\bar{2}0)$	top up	549	1(7)
	top down	675	3(7) + 1(11)
	bridge short	666	2(7)
	bridge long	579	2(7) + 2(11)
Ru$(10\bar{1}0)$	hollow	678	2(11) + 1(7)
Ru(1015)	hcp	714	2(7) + 1(9)

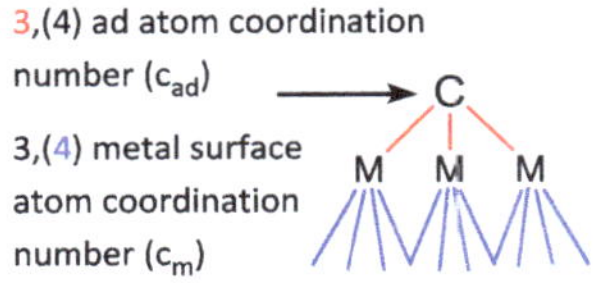

Figure 2.10 Schematic explanation of notation (c_{ad}, c_m) in Table 2.3.

Table 2.3 gives DFT-computed adsorption energies of a C atom adsorbed onefold, twofold, or threefold to surface metal atoms of different surfaces. Generally, the interaction energy increases the larger the adatom coordination number c_{ad} is. The table illustrates that adatoms tend to prefer adsorption in high-coordination sites. The table also shows that the adsorption energy increases when the surface metal atom coordination number c_m decreases. The surface metal atom coordination number is a structural reactivity descriptor.

Shustorovich [76] made an interesting suggestion on how to understand the surface metal atom coordination dependence. He postulates that the valency of a metal atom is a constant and that this valency has to be distributed over the bonds that direct to the atom. This proposition is valid as long as metal atomic orbital hybridization is minor. The Shustorovich suggestion is called Bond Order Conservation (BOC). The more bonds are directed towards a metal atom, the weaker each individual bond will be. This implies that an adatom that attaches to a metal atom that has a higher number of neighbors will have the weaker adatom bond strength.

When all atoms are equal Shustorovich derived the following nonlinear relation for the bond energy as a function of neighbors n of the central atom:

$$E(n+1)-E(n)=E(n=1)/n(n+1) \tag{2.7}$$

When coordination is large, the bond energy decreases as n^{-2}. The BOC principle applies generally. It also can be used to understand differences of the chemisorption bond when substitution to the molecular fragment atom that connects with the surface alters, e.g., CH_3 versus CH_2, CH, or C. The BOC principle predicts that the M–CH_3 bond is weaker than the M–CH_2 adsorbate bond. In M–CH_x the adsorbate-surface metal bond strength increases when x decreases. Such relations as a function of x are known as scaling laws and have been extensively documented for different transition metals and adsorbates [77].

The BOC principle also provides insight into how the adsorbate bond strength varies when metals become converted into metal

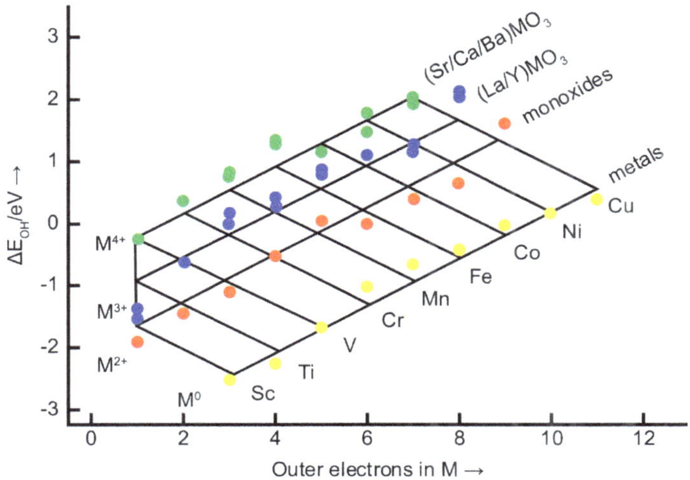

Figure 2.11 Adsorption energy grid of adsorbed OH on metals (yellow), monoxides (red), La/Y perovskites (blue), and Sr/Ca/Ba perovskites (green), at constant formal valence as a function of the number of valence (outer) electrons [78].

oxides. In Figure 2.11 the bond strength of OH is compared when adsorbed to transition metal atom as a function of number of transition metal d-valence electrons that have a different formal metal cation charge. The latter ranks the metals as in the 3d transition metal row of the periodic table.

The interaction energy increases when the position of the transition metal atom moves towards the left in its row of the periodic table. This increases the polarity of the M–O bond. The metal cation with the lower charge has the stronger interaction. The reactivity of cations is reduced compared to that of the neutral metal atoms. As for the transition metals, the higher the M–O coordination number (that increases with cation charge) the weaker the adsorbate OH bond.

The adsorption energies of O, C, and H atoms on different transition metals are compared schematically in Figure 2.12a. DFT-calculated energies are plotted as a function of valence electron count of the second transition metal row of the periodic table [73]. The dense [111] surfaces of the transition metals of face-centered cubic structure are compared. The adatoms are located in high-coordination sites (coordinated to three surface metal atoms).

The respective adsorption energies decrease gradually when adsorption moves to a transition metal to the right in the row of the periodic table. When the d-valence electron count of the metal becomes 10 (total number of valence electrons 11 as in Cu) the adsorption energy drops sharply.

The adsorption energies of C and O are higher than that of the H atom. The reason for this difference is that a H atom only binds with its 1s atomic orbital, but the surface chemical bond with C and O is mainly with their 2p atomic orbitals. There is a large difference in the respective trends of the binding energies of O versus C. Compared to the C adatom, the O adatom binds more weakly to the late transition metals with high d-valence electron count, but this inverts when compared with those adsorbed to transition metals with low valence electron count.

The difference in adatom bonding of C versus O relates to the difference in electron affinity of O and C atoms. The larger electron affinity of O increases the polarity of the adsorbate bond. This is beneficial when O is attached to an electronegative metal. However, when attached to a late transition metal with high work function, the polarity contribution to the adatom chemical bond decreases. Then covalent bonding interactions dominate.

This change in relative contributions of covalent and ionic bonds is schematically illustrated in Figure 2.12b [79]. The relative reduction in adatom bond energy of O compared to C is due to the higher electron occupation of anti-bonding electron density of the M–O bond (see Section 2.3.3).

The variation in bond energies is significantly larger for adatoms than for admolecules. In Figure 2.13 this is illustrated for the adsorption energy of CO and the adsorption energies of C and O adatoms (referred to the bond energy of CO). The bond energy change of the adatoms is dominated by the change in bond energies of the O adatoms. Variation in reaction energies of the dissociation reaction of adsorbed CO are dominated by variation in the bond energies of the adatoms. This is a general feature of surface reaction energies.

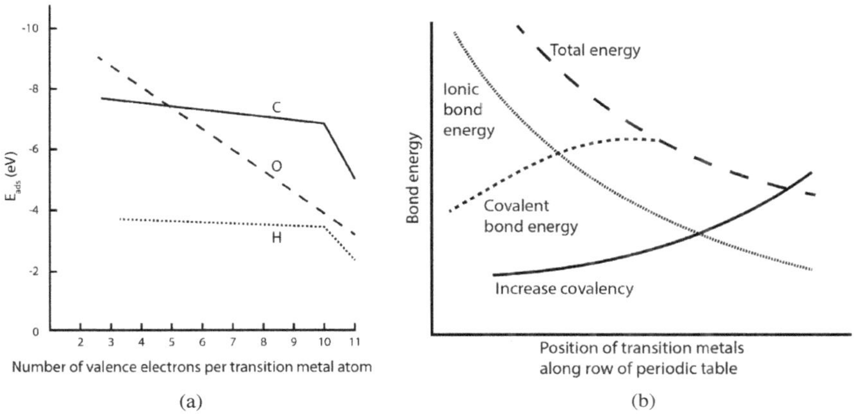

Figure 2.12 Trends in adatom bonding and bond energies as a function of transition metal. (a) Schematic trends of adsorption energies of O, C, and H in threefold coordination surface sites of 4d transition metals as a function of d-valence electron number. Energies are with respect to gas-phase atoms (eV). (b) Schematic presentation of the relative contributions of covalent and ionic bonding to the surface adatom chemical bond as a function of position of transition metal along a row of the periodic table [73].

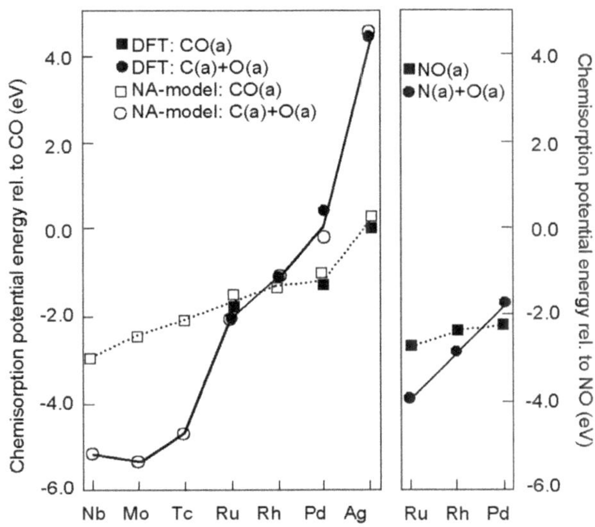

Figure 2.13 Energy differences of adatoms and admolecules as a function of transition metal position along a row of the periodic table. Admolecular adsorption energies change less than adatom bond energies. Bond energies of dissociation products of CO and NO are compared [80].

Some molecules, such as methane and nitrogen, adsorb weakly and mainly through the van der Waals dispersive interactions that vary little with surface composition. Other molecules such as CO and NO adsorb with higher adsorption energies. The reason that molecules with strong chemisorption bonds show less variation in bond energy than the corresponding adatoms is that for the adsorbed molecule the changes in orbital interactions counteract, whereas for adatoms this counter action is absent. For adatoms such as C, O, and N, chemical bonding is dominated by the interaction with the adatom 2p atomic orbitals.

The Chatt-Dewar-Duncanson chemical bonding model [81], [82] provides an illustration of the counteracting bonding changes of the admolecule. For chemisorption of CO this is schematically illustrated in Figure 2.14.

The Highest Occupied Molecular Orbital (HOMO) of CO is its lone pair 5σ orbital, while the Lowest Unoccupied Molecular Orbital (LUMO) is the $2\pi^*$ CO orbital. The CO molecule attaches through its C atom to the transition metal atom. When the transition metal varies, the number of metal valence electrons that interacts with CO changes. An increase in metal valence electrons or

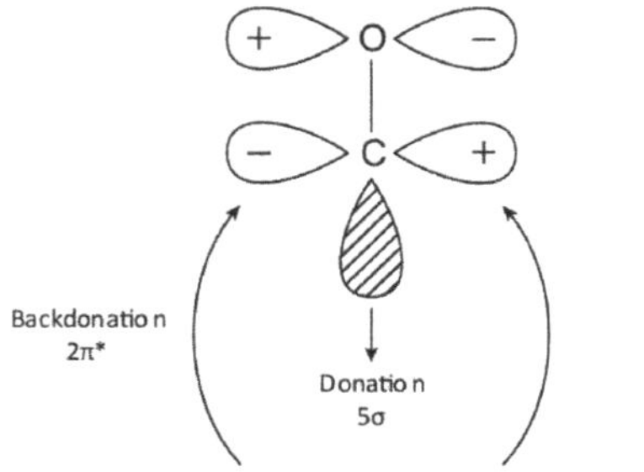

Figure 2.14 The Chatt-Dewar-Duncanson donation-backdonation chemical bonding scheme of CO attached to a surface transition metal atom. The highest occupied molecular orbital is the 5σ orbital that binds through electron donation to the surface, while the lowest unoccupied molecular orbitals are the two unoccupied $2\pi^*$ orbitals.

bond polarity will increase backdonation of electrons into the $2\pi^*$ CO orbital. This will increase the strength of the M–C bond.

In contrast, the increase in d-valence electrons decreases the donation probability of electrons from the CO 5σ orbital since empty surface valence electron density decreases. This will weaken the M–C bond. For adatoms as C, O, and N, chemical bonding to the transition metal is dominated by the interaction only with the adatom 2p atomic orbitals and compensating orbital interactions as for the admolecule are absent.

2.3.2.3 *Transition States and the Brønsted-Evans-Polanyi Relation*

Differences in surface reaction rate activation energies relate with the adsorption strength of adsorbed dissociation fragments. The activation entropy of surface reactions is small. The activation entropy of the desorbing molecule is orders of magnitude larger.

The Eyring reaction rate expression of elementary reaction rate constants is the fundamental expression on which microkinetics is based. The expression is presented in the first part of this section. This is followed by an introduction of the Brønsted-Evans-Polanyi (BEP) relation, which relates the transition state energy of the activated complex as defined by the Eyring expression with the reaction energy of the corresponding elementary reaction. The BEP relation is fundamental to the prediction of catalytic reactivity trends from surface reactivities.

In the 1930s, Eyring developed the reaction rate equation in close collaboration with Polanyi [22], [83], [84]. Different from Arrhenius, they formulated their reaction rate expression as an equation for elementary reaction rate constants. Essential to the Eyring expression is the idea of an activated complex.

In Figure 2.15 the change in potential energy for an elementary bond cleavage reaction is sketched. When the reacting molecular bond stretches, potential energy increases until it reaches an energy maximum. Beyond this maximum, further stretching of the bond

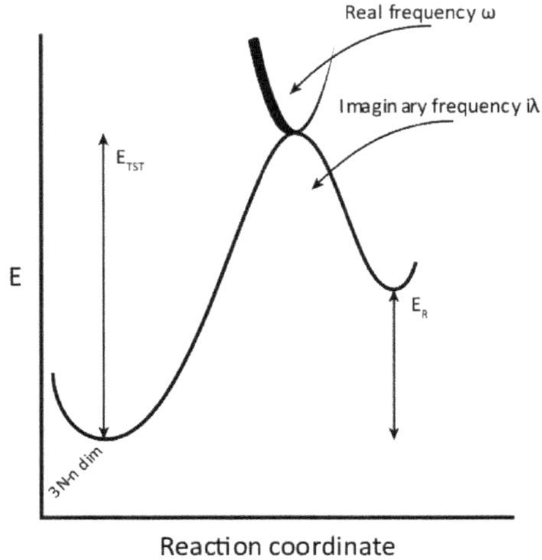

Figure 2.15 The relation between activated complex energy and transition state. Schematic potential energy of reacting system as a function of reaction coordinate [85].

lowers the energy until the product ground state is reached. Progress is measured by the reaction coordinate. The structure of the reacting system at its maximum energy is called the activated complex and the energy maximum the transition state energy.

The Eyring expression of the elementary reaction rate constant is given by Eq. (2.8a):

$$k_{TST} = \kappa \frac{ek_B T}{h} e^{\frac{\Delta S^{\#}}{k_B} - \frac{\Delta H^{\#}}{k_B T}} \tag{2.8a}$$

$$= \nu_p e^{\frac{-E_{act}}{k_B T}} \tag{2.8b}$$

In Eq. (2.8a) $\Delta S^{\#}$ and $\Delta H^{\#}$ are respectively the activation entropy and activation enthalpy of the activated complex, k_B is the Boltzmann constant, and h is the Planck constant. k is a transmission factor that is usually chosen to be equal to ½. Eq. (2.8a) can be interpreted as the rate of transformation $\kappa \frac{ek_B T}{h}$ of activated complex multiplied by its quasi-equilibrium concentration, which is defined by the free energy $\Delta G^{\#}$.

In Eq. (2.8b) the Eyring reaction rate expression is rewritten in the Arrhenius form. The rate expression has the dimension s^{-1}. It is proportional to ν_p, the pre-exponent, and the temperature-dependent exponential term that contains the activation energy E_{act} that relates to activation enthalpy $\Delta H^{\#}$. It has similar temperature dependence as the classical Arrhenius Eq. (1.6).

Whereas equilibrium thermodynamics has a precise founding in statistical mechanics, this is not the case with kinetics. The analogue of equilibrium statistical mechanics is irreversible statistical mechanics. Kinetic expressions are approximate and only valid with some restrictions. The condition of validity of the Eyring equation is that all vibrational and rotational degrees of freedom of the reacting system are in equilibrium except for the reaction coordinate that describes progress of the reaction. Exchange of energy between reaction complex and environment is to be fast compared to the rate of reaction [86].

An example of such a reaction coordinate for dissociation of a molecular chemical bond is the stretching of a C–H bond when methane dissociates or the bond distance of C–O when it cleaves. The ground state energies of Figure 2.15 are defined as the respective energy minima of the system. The first derivative of the energy with respect to degrees of freedom is zero and the second derivative is positive. The normal frequencies of a system are proportional to the square root of the second derivative of the energy. This gives positive frequency values for normal modes in the ground state. The transition state energy E_{TST} is defined as the energy of a molecular structure of maximum potential energy. Then for one normal mode the second derivative of the energy is negative and its frequency is imaginary. This defines the reaction coordinate. Since the reaction coordinate is treated separately, the activated complex contains one degree of freedom less than the ground state. The contribution of the activation entropy to the pre-exponent ν_p follows from equilibrium statistical mechanics [87]. According to statistical mechanics, equilibrium constants relate to the partition functions Q_i of respective components *i*. The partition function is the product of the partition functions of the normal modes of the system. In terms of partition functions Eq. (2.8a) can be rewritten as in Eq. (2.9a):

$$k_{TST} = \kappa \frac{k_B T}{h} \frac{Q^{\#}}{Q_0} e^{\frac{-E_{\text{act}}}{k_B T}} \tag{2.9a}$$

$$E_{act} = E_{\text{TST}} + \frac{1}{2} h \left(\sum_i \nu_i^{\#} - \sum_j \nu_j^{0} \right) + k_B T \tag{2.9b}$$

$$Q_{\nu_i} = \frac{e^{\frac{-h\nu_i}{2k_B T}}}{1 - e^{\frac{-h\nu_i}{k_B T}}} \tag{2.9c}$$

In Eq. (2.9a) $Q^{\#}$ is the partition function of the vibrational and rotational degrees of freedom of the activated complex. Q_0 is the ground state partition function. Within the approximation of harmonic potentials Eq. (2.9b) gives the relation between E_{act} and E_{TST}. The expression for the vibrational partition function of one normal mode is presented in Eq. (2.9c). The activation energy E_{act} is equal to E_{TST} when corrected for the zero-point vibrational frequencies. The activation entropy relates to the difference of mobility of the activation complex and that of the ground state.

A fundamental assumption for the validity of the Eyring activation complex reaction rate expression is that exchange of energy between reaction complex and environment is fast compared to the rate of reaction.

Usually E_{TST} is calculated from quantum-chemical calculations at zero temperature. Then the temperature dependence of $\Delta S^{\#}$ and $\Delta H^{\#}$ is ignored. Non-zero temperature calculations of transition states are possible by extending the calculations with molecular dynamics. Then the harmonic approximation for the frequencies can also be circumvented (see [21], [88], [89]).

The activation entropy relates to the difference in mobility of the activated complex and the ground states. On a surface the mobility of adsorbed species is low. Then the pre-exponential factor ν_p reduces to $ek_bT/h \approx 10^{13}$ s^{-1} and the entropy contribution is ignored. This pre-exponent frequency corresponds to the time scale of a vibration. It is sometimes referred to as the frequency of the movement of the reaction coordinate over the transition state barrier in the activated complex. κ is the friction coefficient for movement over the barriers. It is usually chosen 1 or ½. The proposition that the entropy of the transition state of a surface reaction is small is not unreasonable because a strong interaction between surface and molecule is needed to cleave molecular chemical bonds. This is in contrast to the activation entropy of a molecule that desorbs from the surface. Then there is a gain in rotational entropy. The non-mobile transition state is called a tight transition state. The desorption transition state is a state of high transition state mobility. This is called a loose transition state. The pre-exponent of the reaction rate constant of desorption is 10^4 to 10^6 larger than that of the surface reaction. It is the reason of the so-called pressure gap problem between catalysis and surface science. In a surface science experiment, reactivity is usually studied by heating an adsorbate-covered surface in vacuum. When the adsorption energy of adsorbate is of the same order of magnitude as the activation energy of reaction, the molecule desorbs from the surface before reaction happens. This is because of the larger pre-exponent of the desorption rate constant. On the other hand, in the catalytic experiment a high pressure of reacting gas is used. This maintains enough surface concentration that reaction happens with finite rate. Due to the absence of partial pressure, the surface composition in surface science experiments is generally very different from that in the catalytic experiment.

Whereas the concepts of activated complex and transition state had been discovered in the 1930s, their use in kinetics remained limited. The transition state is very difficult to deduce experimentally and quantitatively reliable energies for electronic structure calculations became only available in the 1990s. At present, the physical chemistry of transition states and their relation with surface reactivity have become revealed mainly from computational studies. Once such calculations became available, a remarkable discovery was made for chemical bond dissociation reactions. A linear dependence between calculated activation energy and energy of adsorption, not of reactant molecules, but of the atoms generated by dissociation was observed. This is illustrated for the activation energies of N_2 dissociation for different transition metals collected in Figure 2.16 by Nørskov and colleagues in 2001 [90].

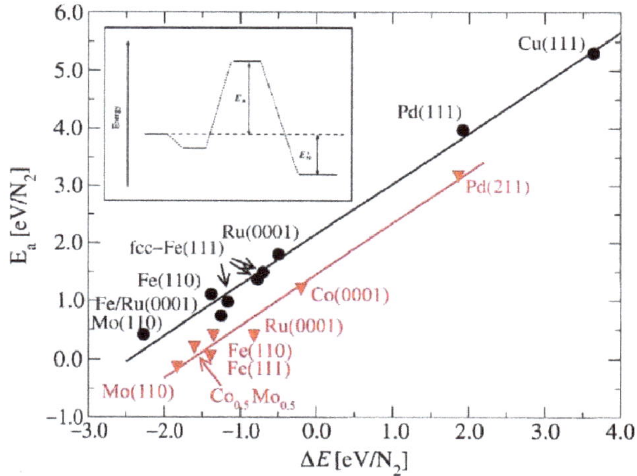

Figure 2.16 The calculated activation energies (transition state potential energies) E_a for N_2 dissociation ($E_a = E_{TST} - E_{ads}$) on a range of metal surfaces plotted as a function of the adsorption energy for two nitrogen atoms ($\Delta E(N_2)$). Results for both close-packed terrace surfaces (black circles) and more open surfaces (red triangles) are shown. The inset shows a sketch of the energetics for the N_2 dissociation reaction [90].

The calculated activation energy of the rate of N_2 dissociation is plotted against the adsorption energies of N adatoms. Both are calculated with respect to the energy of the free N_2 molecule. A linear dependence of activation energy versus dissociation energy is observed. This is a general result, which is found for many dissociation reactions. Similar linear dependencies of the activation energy of the rate of dissociation versus adsorption energies of adatoms or reaction fragments have been reported for other molecules, e.g., CO, NO, as well as methane and ammonia [91]–[93].

The activation energy relation with reaction energy is only linear as long as surfaces with similar structures are compared. On the stepped surfaces the activation energy is lower, but again a linear dependence is found.

This linear activation energy-reaction energy relationship is known in kinetics as the BEP relation. This relation was first formulated in the 1930s [94], [95]. Related linear free energy relationships are well known in physical chemistry and are found for many different, also non-catalytic reactions [96]. They are called Hammett relations in physical chemistry [97].

The BEP relation can be expressed as in Eq. (2.10):

$$\Delta E_{act} = \alpha_{BEP} \Delta E_{react} \left(0 < \alpha_{BEP} < 1\right) \quad (2.10)$$

This compares for an elementary reaction step the change in activation energy ΔE_{act} with the change in reaction energy ΔE_{react} when the same reaction is compared for a different material. a_{BEP} is the BEP proportionality constant that has a value between zero and one. The relation only applies to elementary reactions and is valid as long as the reaction path and reaction coordinates for the different materials are the same. It implies that the topology of the reaction center also has to be the same. The relation is extensively used in electrochemistry [98] as well as catalytic kinetics [99]. The value of the proportionality constant α_{BEP} remains unknown. It is usually chosen as ½.

The important contribution of computational catalysis is the determination of the value of α_{BEP}. The DFT computational data of Figure 2.16 and other calculations find for the dissociation reaction of a diatomic molecule $\alpha_{BEP} \approx 1$ [100].

This discovery that $\alpha_{BEP} \approx 1$ has a major implication for the structure and dynamics of the transition state. The similarity of change in reaction energy and transition state implies that the molecule in the transition state and dissociated state has a comparable strong interaction with the surface. Compared to the molecular state before dissociation, the transition state is late. The transition state structure is close to that of the dissociated state.

It is consistent with the low mobility and small activation entropy of the surface reaction transition state. Figure 2.8a illustrates for the dissociation of adsorbed NO the difference between the adsorbed state and transition state structures. In the transition state the N–O bond is already broken and the N and O atoms are adsorbed in configurations close to that of the respective dissociated atoms.

Because of microscopic reversibility (the reaction equilibrium condition), the activation energy change of the recombination of surface atoms has proportionality constant $-(1-\alpha_{BEP})$. It is the reverse reaction of molecular bond dissociation.

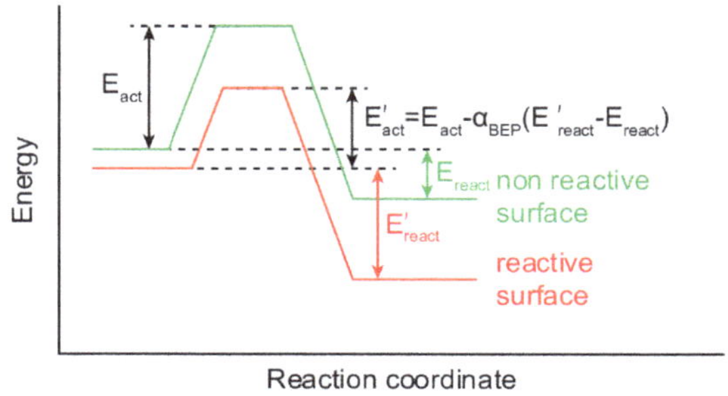

Figure 2.17 Schematic representation of the relation between change in activation energy E_{act} and reaction energy for a dissociating molecule E_{react}. The BEP relation is the linear relation $\Delta E_{act} = \alpha_{BEP}\,\Delta E_{react}$. This is valid as long as the reaction path of dissociation does not change. Because of microscopic reversibility, for the reverse recombination reaction $\Delta E_{act,rev} = -(1-\alpha_{BEP})\ \Delta E_{react}$.

Figure 2.17 illustrates schematically the BEP relation.

The late nature of the transition state of the surface dissociation reaction implies that in the transition state, information on the molecular adsorbed state is lost. This is reflected in the plot of Figure 2.16, which is not a plot of change in activation energy E_{act} versus reaction energy, but instead relates $E_a = E_{act} - E_{ads}$ (mol) with E_{react} . E_a is the apparent activation energy E_{app} of the dissociation reaction measured with respect to the gas-phase energy. The plot of Figure 2.16 has no reference anymore to the state of the adsorbed molecule [101].

Instead of the BEP expression Eq. (2.10), the modified BEP relation Eq. (2.11) is to be used.

$$\Delta E_{app} = \alpha' \cdot \Delta E_{ads}\left(atoms\right)\left(\alpha' \approx 1\right) \tag{2.11}$$

The reason for this difference is that Eq. (2.10) only applies to one-dimensional reaction coordinates. The relation is valid for electron transfer [102] or proton transfer reactions [103]. Eq. (2.10) also applies to C–H bond cleavage reactions. Then the initial state is the intact molecule, and the final state the dissociated alkyl fragment and hydrogen atom. The reaction coordinate is the stretching C–H bond (see Section 2.3.3.3).

For molecules that strongly adsorb on the surface, the late nature of the transitions state defines the stretch mode of the molecular bond that cleaves as the reaction coordinate. In the transition state, there is no memory of the initial adsorbed state of the molecule. Even for diatomic molecules that strongly adsorb, this implies that the energy of the free molecule is the proper energy reference.

A second major contribution of computational science to the kinetics of catalysis is the postulate that not the adsorption energy of the reactant molecules (see, however, [104]), but the adsorption energy of dissociated reaction intermediates, determines surface reactivity [101]. The latter provides a practical guide to select a catalyst for a particular reaction. Once the rate-controlling step is understood, surface reaction thermodynamics can be used to propose a particular transition metal as catalyst. As an example, the selection of the steam-reforming catalyst is shortly discussed. In this reaction, methane reacts with steam to produce carbon monoxide and hydrogen. Water has to decompose to O_{ad} and H_{ad} and methane into C_{ad} and H_{ad}. The catalyst surface has to be reactive enough that both decomposition reactions are possible. CO and methane are formed in subsequent adatom recombination reactions. Figure 2.12a shows that on the late transition metals, the bond energy of C is substantially stronger than that of O. Because of the low adsorption energy of atomic oxygen to the Pt surface, water will not decompose on the surface of this metal. In contrast, dissociation of methane on Pt is possible due to the stronger M–C bond energy. For water decomposition, more reactive metals are required. For this reason, the steam-reforming reaction is catalyzed by more oxophilic transition metals as Rh, Ru, or Ni [105]. In contrast, Pt is an active metal for the methane to synthesis gas oxidation reaction [106]. Rh is more reactive for methane decomposition, but has lower catalytic reactivity due to its higher oxygen atom affinity; the rate of formation by recombination of H_{ad} with O_{ad} is slow compared to that on Pt, which has the weaker M–O bond [107]. In Section 2.4.3, the kinetic expressions are presented that deduce reaction rate trends by use of the BEP relation.

2.3.3 *The Nature of the Surface Chemical Bond: Chemical Reactivity Descriptors*

The adsorbed state is a molecular cluster embedded on the surface of a material. This embedding gives constraints to the structure of the chemisorption site that affects the bond energies.

2.3.3.1 *Introduction*

The first modern formulation of the surface chemical bond is the molecular orbital theoretical view that stems from the chemist Roald Hoffmann (Nobel Prize 1981) from Cornell University [108] in 1988. This electronic theory is based on semi-empirical Extended-Hückel calculations. It provides a useful framework to interprete the later first-principle DFT-calculated results on the adsorbed state and its reactivity. A major advance of first-principle DFT-based calculations compared to the methods used by Hoffmann is that electrostatics is treated properly. The relative contributions of covalent and ionic bonding can be properly determined.

The molecular orbital theory of chemisorption refines earlier suggestions on the nature of the chemisorption bond to transition metals, which were based on Pauling's hybridization model of the transition metal. Pauling's valence bond model ascribes variation in transition metal bond energy to a change in the d-character percentage of the valence atomic orbitals of the transition metal [75], [109], [110]. The relative d-character of the metal bond was deduced from structure and physical properties of the transition metals. In the 1950s catalytic chemists used the Pauling d-character model to understand trends in reactivity as a function of the transtion metal [28]. The Pauling metal bonding theory became strongly criticized because it very poorly explains metal alloy properties [111].

These proposals contrast with alternative suggestions by Beeck [29] that correlate differences in surface reactivity with metal structural parameters such as the atom-atom distances between the surface atoms. Related is the multiplet theory of Balandin from the

Russian school of catalysis, which correlates reactivity of surface with surface atom topology and reactant molecular dimensions [112]. Alternative suggestions are the correlation of adsorbate energy with the electron density of states at the Fermi level (see Section 2.3.1).

These suggestions were rejected due to increased knowledge of the structure of the catalytic reaction center from surface science [15], [65], and through the use of first-principle quantum-chemical computational tools, a wealth of molecular information on the relation between surface reactivity and the nature of the surface chemical bond has become available [19]. Importantly, proper understanding arose once it became understood that the adsorbed state is a molecular surface complex (see Section 2.3.2).

2.3.3.2 *The Electronic Structure and Bond Energy of Transition Metals*

Trends in the bond energy of the transition metals follow from differences in electron distribution over bonding and anti-bonding electron densities.

The valence electron structure of the transition metal is formed by the interaction of its d-valence with s-, p-valence atomic orbitals. This creates delocalized orbitals with energies that are distributed over a continuous energy band. The partial density of state (PDOS) of the d-valence electron structure for the transition metals Mo, Ru, and Pd is illustrated in Figure 2.18a (red). The number of d-valence electrons varies between zero and ten per metal atom.

The s-, p-atomic orbitals of higher quantum numbers are spatially more extended than the d-atomic orbitals and therefore overlap stronger with neighboring atoms. This results in the broader s-, p-valence electron density band. The s-, p-valence electron PDOS is shown in blue in Figure 2.18a. In the s-, p-valence electron band for different transition metals, the number of valence electrons per atom is constant and approximately equal to one. The respective electron densities of Figure 2.18 have been calculated by DFT [113], [114].

In Figure 2.18a, next to the PDOS the bond order overlap densities are indicated. The sign of the bond order overlap densities

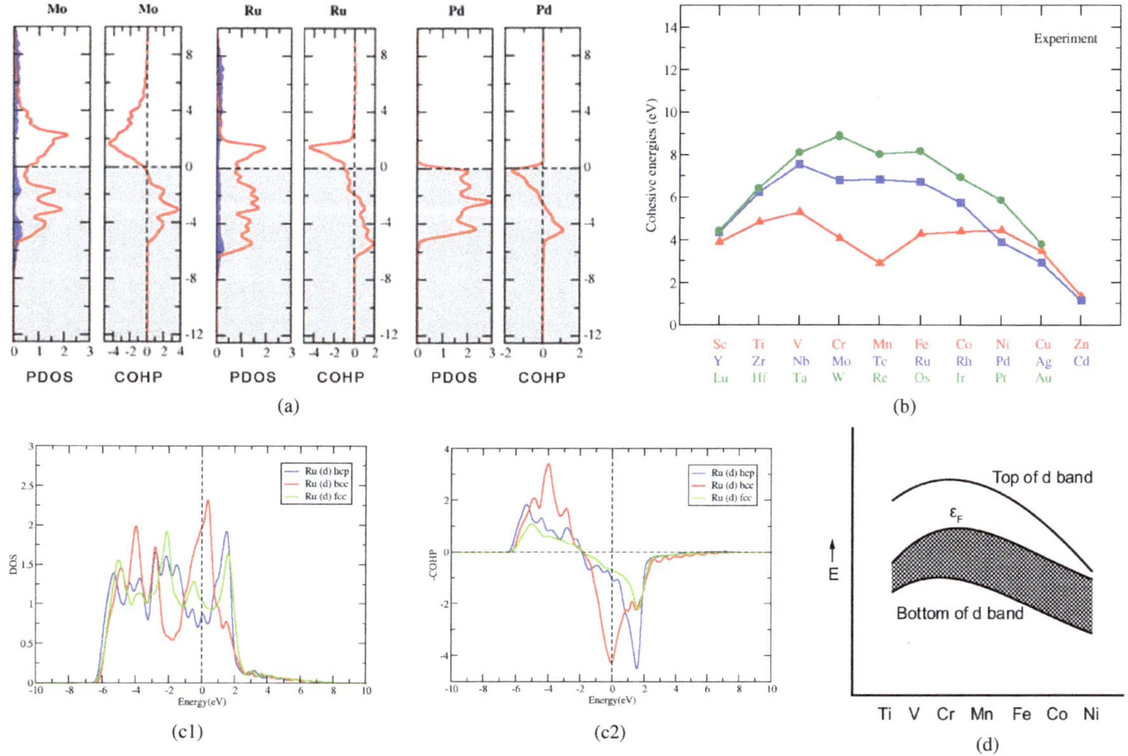

Figure 2.18 The electronic structure and cohesive energy of the transition metals. (a) Partial Densities of States (PDOS) and Crystal Orbital Hamiltonian Overlap (COHP) densities of Mo, Ru, and Pd; blue s-, p-valence PDOS, red d-valence PDOS, COHP. (b) Cohesive energies as a function of transition metal position in the periodic table. (c) Electron energy densities for the hcp, bcc, and fcc structures of Ru; c1 PDOS, c2 COHP [113], [114]. (d) Schematic presentation of the d-valence band energy as a function of transition metal position in a row of the periodic table according to Hoffmann. In grey is the d-valence electron occupation [108].

(COHP) indicates whether a particular PDOS at that electron energy is bonding or anti-bonding. When its sign is positive, the electron population of the corresponding orbital gives an attractive bonding contribution to the energy. For the negative (COHP) sign, orbital interactions are anti-bonding. Electron occupation of electron density in this energy regime weakens the chemical bond energy.

Figure 2.18b shows the computed cohesive energies of transition metals as a function of their position in the periodic table. For the second and third row transition metals, the cohesive energies of Figure 2.18b show a maximum when the d-valence electron count per atom is five. Then all bonding orbitals are occupied, and the anti-bonding orbitals are unoccupied. When electron count decreases, the cohesive energy decreases because of electron depletion of bonding orbital electron density. On the other hand, when electron occupation increases, anti-bonding orbitals become occupied and the bond energy decreases. The dip in covalent energy of the transition metals of the first row of the periodic table is due to magnetic interactions, which are uniquely strong for these metals. Calculations include magnetic spin polarization. Calculations without spin polarization show a maximum in cohesive energy when the number of d-valence electrons per atom is equal to five. This is similar as for the cohesive energies of the transition metals of the second and third rows of the periodic table.

In Figure 2.18c the d-valence electron densities are compared for different structures of the same transition metal. This comparison of electron densities is used to demonstrate that bond strength has no relation with electron density at the Fermi level. The electron densities of state of respectively the hexagonal close-packed (hcp), face-centered cubic (fcc), and body-centered cubic (bcc) structures are compared. The hcp structure is the most stable. Figure 2.18c illustrates that the electron density of states at the Fermi level (chosen as the zero energy) of this structure has the lower value. The higher value of the density of states at the Fermi level is found for the less stable bcc structure. The low stability of the bcc structure is because substantial anti-bonding electron density is occupied. In contrast, in the stable hcp structure, nearly only bonding electron density is occupied. This example illustrates that electron density of states at the Fermi level has no relation with strength of chemical bonds. The Fermi level defines the HOMO energy and the total electron occupation per atom. The bond strength relates to the distribution of electrons over bonding and anti-bonding orbitals.

The accurate results of calculated electron densities Figures 2.18a and 2.18c are different from the trends in electron energy density as deduced from early estimates by Hoffmann [108] that is shown in Figure 2.18d. These trends were extrapolated from computed electron density data of the late transition metals. This was based on the assumption that electron energy distributions in transition metals are independent of electron occupation. It leads to a very different conclusion on the relation between bond energy and electron density of states at the Fermi level. Comparison of Figure 2.18d with Figure 2.18b suggests a relation between density of states at the Fermi level and cohesive energy. This is invalidated by the accurate calculations of Figure 2.18c. The error also demonstrates the poor accuracy of solid state electronic structure calculations in the 1980s [115]. The misunderstandings that this gave is one of the reasons reliable chemisorption theory had to await until the beginning of the 20th century.

2.3.3.3 *Quantum Chemistry of the Surface Chemical Bond*

There is one-to one relation between electronic reactivity descriptors and the structure of the adsorption site.

Here the electronic structure of the molecular complex of adatom and transition metal surface is discussed. Electronic reactivity descriptors of the transition metal surface such as the width of the d-valence electron band, the average d-valence electron band energy, and the total d-valence electron occupation are introduced. To introduce these concepts, initially a schematic model of the surface chemical bond is discussed using Figure 2.19. It illustrates the electronic interaction of an adatom with a single atomic valence orbital of a transition metal surface atom. This provides a basis for

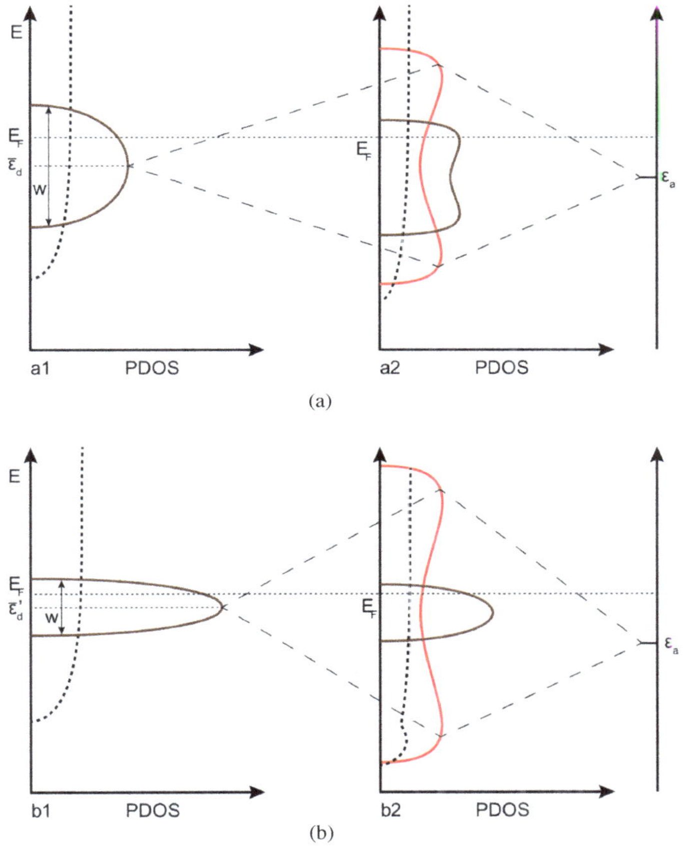

Figure 2.19 Schematic representation of the interaction of the adsorbate atomic orbital ε_a with the d- and s-, p-valence electron bands of the transition metal surface. The bandwidth of the d-valence electron band (unbroken lines) is denoted by *W*. The s-, p-valence electron energy band is indicated with broken lines. The Fermi level E_F and average d-valence electron energy $\bar{\varepsilon}_d$ are indicated. Comparison is made between surface atoms with (a) high coordination and (b) low coordination. (a1) and (b1) show the electronic structure of the undisturbed surface. (a2) and (b2) show the transition metal electron structure altered due to interaction with adsorbate.

the analysis of DFT-calculated adatom energies and electron densities of states of the H atom adsorbed on different adsorption sites of the transition metal surface. The quantum-chemical founding of electronic reactivity descriptor relations with adatom bond energies is presented. It will be seen that the reactivity descriptors primarily relate to coordinative unsaturation of the surface metal atoms.

Figure 2.19 schematically shows the interaction of adsorbate atomic orbital ε_a with the transition surface metal valence electrons.

Two situations are compared, one in which the adatom interacts with a less reactive surface (a) and one in which the adatom interacts with a more reactive surface (b). The more reactive surface atom has a lower number of neighboring atoms. In Figures 2.19a1 and 2.19b1 the undisturbed valence electron band energies are sketched. In Figures 2.19a2 and 2.19b2 the electronic structures of the interacting systems are shown. The electronic properties that relate to difference in coordination number of the surface metal atoms are the d-valence electron density band width $W$and the average d-valence electron band energy $\bar{\varepsilon}_d$. The relation with $\bar{\varepsilon}_d$ is explained below when Figure 2.20 is discussed. It is indicated in the respective Figure 2.19 that the width $W$is larger when the coordination number is higher. When the surface atom has more neighboring atoms, electrons have a larger probability to hop between atoms. It causes the electrons to be more delocalized. The band width $W$is a measure for the degree of delocalization.

A main feature of the electronic structure of the interacting system is the splitting of the d-valence electron energy band into upper and lower parts. These parts correspond to the bonding and anti-bonding electron densities between adsorbate and surface atoms. The interaction with the adsorbate alters the surface electronic structure strongly. Part of the d-valence electrons interact with the adatom, which leaves it unchanged. The other part of the d-valence electron energy splits into an upper anti-bonding and lower bonding electron density part (indicated in red in Figures 2.19a2 and 2.19b2). This splitting of the d-valence electron energy density is larger for the system with the narrower bandwidth and is a measure of stronger binding. The adatom bond energy follows from the distribution of electrons over bonding and anti-bonding orbitals. The Fermi level E_F that is also indicated in Figure 2.19 determines the total number of electrons involved in the adatom surface chemical bond. Its position with respect to ε_a defines also the distribution of electrons over the adatom and surface atoms. The ionicity of the surface chemical bond depends on this difference that defines the relative amount of electron transfer from surface to adatom or vice versa.

For a surface atom to bind with the adatom, surface electrons have to localize. The larger the electron density band width *W*, the more the surface electrons are delocalized. The larger delocalization energy requires a higher energy cost of electron localization, which results in a weaker adatom surface chemical bond. The localization of electron density on the surface atom involved in chemisorption weakens the metal-metal atom bond strengths with neighbor metal atoms. This surface metal atom bond energy weakens substantially because of the generally stronger interaction of the chemical bond between adatom and surface atom (order 150 kJ/mol) compared to that between two surface transition metal atoms (order 80 kJ/mol).

The adsorbed state can be considered a molecular coordination compound that is embedded in the surrounding environment of surface atoms. Compared to binding in the corresponding free molecular complex, bonding in the surface complex is weaker. This embedding in the solid comes at the cost of the weakening of surface metal atom interaction.

Figure 2.19 can also be used to illustrate how the covalent adatom bond energy changes with variation in d-valence electron count. In Figure 2.19 the Fermi level E_F is chosen such that part of the anti-bonding orbital fragments between surface metal atoms and hydrogen atom are occupied. When the d-valence electron band electron occupation decreases, the occupation of anti-bonding orbitals depletes and adsorbate bond energy increases. The covalent bond energy is maximum when anti-bonding orbitals become empty and only bonding orbitals are occupied (see Figure 2.12b, the trend in covalent bond energy).

In DFT calculations it was observed that increased reactivity of the surface not only relates to *W* but also to $\bar{\varepsilon}_d$. This correlation is only found in first-principle calculations such as DFT since such

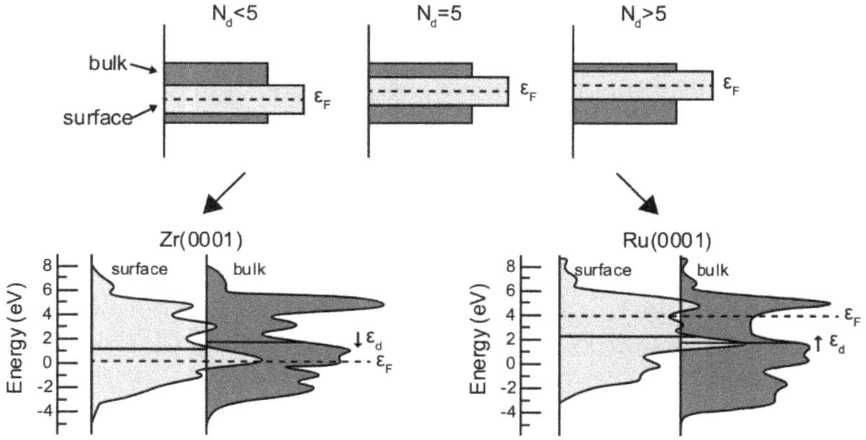

Figure 2.20 The electronic reactivity descriptor $\bar{\varepsilon}_d$. DFT d-valence electron energy band distributions, average d-valence electron band energy $\bar{\varepsilon}_d$ with varying positions of E_F are presented. Comparison is made of Zr(0001) and Ru(0001) surface PDOS with respective bulk electronic structures [116].

calculations include repulsive electron-electron interactions. Scheffer *et al.* [116] from the Max Planck Institute in Berlin discovered that $\bar{\varepsilon}_d$ as well as W relate to the surface atom coordination number. Figure 2.20 illustrates the shift of $\bar{\varepsilon}_d$ and that the relation with coordination number of the surface metal atom depends on d-valence electron count.

The calculated d-valence electron band structures of Zr and Ru illustrate also the decrease of band width W when the surface electronic structure (low coordination number) is compared with bulk electronic structure (high coordination number). $\bar{\varepsilon}_d$ shifts upward for a transition metal when the d-valence band electron occupation is more than half the total possible electron occupation, and vice versa. Larger coordinative unsaturation of a surface atom decreases W. When the d-valence electron band is more than half filled, this leads to electron transfer from the bulk d-valence electrons into the surface d-valence electron band. This increase in d-valence orbital electron occupation causes stronger electronic repulsive interactions and hence $\bar{\varepsilon}_d$ will move upward. Hence the upward shift of $\bar{\varepsilon}_d$ relates to increased surface atom reactivity. The reverse happens when the d-valence band is half or less filled with electrons. Then the increase in reactivity relates to a decrease of $\bar{\varepsilon}_d$.

$\bar{\varepsilon}_d$ as well as W are electronic descriptors of the surface reactivity. Since $\bar{\varepsilon}_d$ is easier to extract from theory or experiment, it is more often used. Both descriptors relate change of coordinative unsaturation of the surface metal atoms.

In the final part of this section the results of DFT calculations of the energies and electronic structure of a H atom adsorbed atop and threefold to the (111) surface terraces of fourth row transition metals are presented [114].

Figure 2.21 shows computed bond energies of H as a function of electron occupation of the d-valence electrons of transition metals. The H atom that has threefold coordination shows the expected stronger bond energy compared to that of the H atom adsorbed in onefold coordination. The bond energy of H adsorbed atop shows a maximum for Pd when valence electron occupation decreases. The bond energy of hydrogen increases more uniformly from Ag with decrease of d-valence electrons. The difference relates to the lower energy of the adatom bonding orbitals in threefold

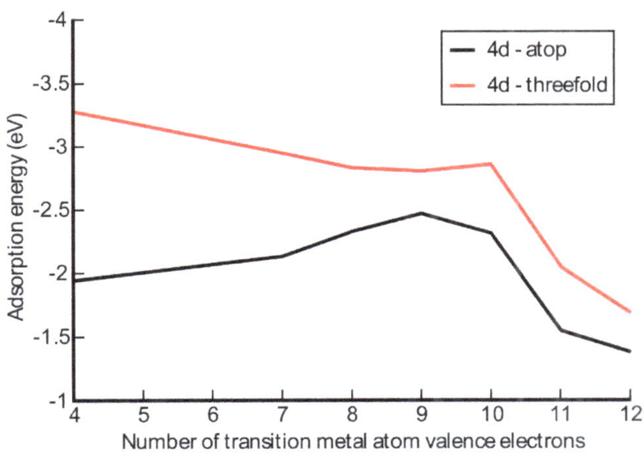

Figure 2.21 Adatom energy of H atop (black) and threefold adsorbed (red) on the [111] surface of the fcc crystal structure of transition metals as a function of the transition metal d-valence electron occupation [114].

coordination that leads to an increase of negative charge on the hydrogen atom.

Figures 2.22a and 2.22b show the respective PDOS for H adsorbed to Rh. The contribution of the five d-valence atomic orbitals energy distribution (blue), the lower intensity of the Rh s-, p-atomic orbitals (red and green, respectively) as well as the PDOS of the 1s atomic orbital of H (grey) are shown. The figure shows that for threefold-coordinated H, the stronger interaction with the surface shifts the PDOS intensity of the H 1s atomic orbital to lower energy. The maxima of the H 1s atomic orbital PDOS are located near or below the d-valence electron band of the Rh surface atoms. H 1s atomic orbital PDOS tails upwards to even above the Fermi level (in the figures, E_F is the zero-electron energy level (unit eV)). Whereas occupation of the low-energy PDOS has a bonding contribution to the bond energy, occupation of the PDOS near and above the Fermi level gives an anti-bonding contribution.

Figures 2.22a2 and 2.22b2 are difference plots of the transition metal d-valence electron PDOS in the presence and absence of the H adatom. For H adsorbed atop, Figure 2.22a2 shows the downward and upward shifts of d-valence electron band density. This comes at the cost of d-valence electron density in the center of the d-valence electron energy band of the free surface. In Figure 2.22b2 similar changes are observed for threefold-adsorbed H. Confirmation that the new densities belong respectively to bonding and anti-bonding electron energy regimes follows from the bond order overlap densities (COHP) plots of Figures 2.22a3 and 2.22b3. A positive sign of COHP indicates bonding PDOS, a negative sign anti-bonding PDOS.

Figure 2.22 of the electronic structure of the H atom to the Rh(111) surface can also be used to understand the respective trends in adsorption energy as a function of d-valence electron count. With a decrease in total electron density the Fermi level, which is the energy of the HOMO, shifts to a lower value. Since for the H atom adsorbed to the Rh(111) surface the Fermi level is located in the anti-bonding PDOS regime, the H adsorption energy will increase.

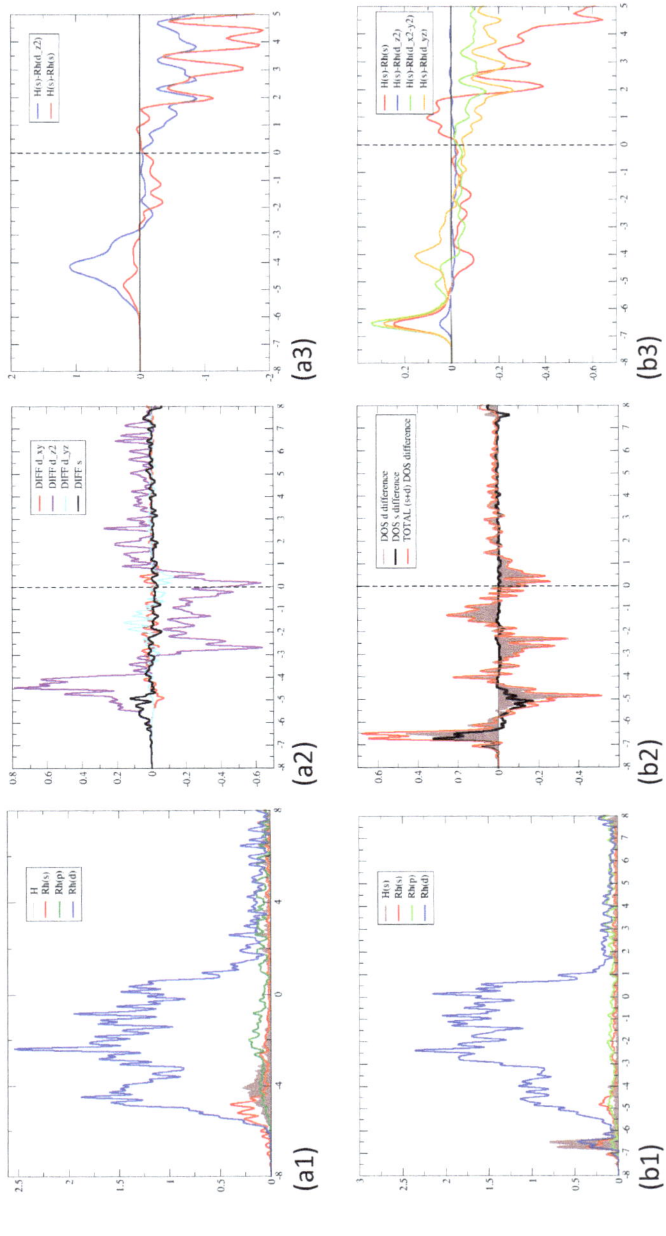

Figure 2.22 Electronic structure of H adsorbed on Rh(111) surface. (1) Partial densities of states of H atom, (2) Partial density of states difference with free surface, and (3) bond order overlap densities. (a) H adsorbed atop, and (b) H adsorbed threefold (E_F is zero level, unit eV) [114].

The computed electron densities can be used to estimate adatom bond energy changes when transition metal electron occupation varies. This is approximate since it assumes that electron density distribution does not change with electron occupation.

For threefold-adsorbed H, the bond energy initially increases when the transition metal d-valence electron occupation decreases. The electron occupation of the anti-bonding electron energy density decreases. With further decrease of transition metal electron occupation, an electron energy density regime is reached where the PDOS electron energy density is non-bonding. Only at lower electron occupation bonding orbitals deplete.

For atop-adsorbed H, the Fermi level is located in the non-bonding regime. Initially there is only a small change of the adatom, which has the maximum bond energy. When the d-valence electron occupation decreases, the bonding orbital energy density depletes and the bond energy decreases.

For threefold-adsorbed H the bond energy initially increases because the anti-bonding density depletes, but then shows a plateau when the valence electron density continues to decrease. First an energy density region is reached where the PDOS is non-bonding. Only at lower electron density bonding orbitals become depleted.

When transition metals are compared with decreasing valence electron count, their electron affinity also decreases. This decreased electron affinity will favor electron donation from transition metal to adsorbate. The increase in bond energy of threefold-adsorbed H with d-valence electron occupation is due to increased polarity of the M_3–H chemical bond. Its bonding orbital density has lower density than that of atop-adsorbed H. Whereas the change in covalent bond energy dominates the trend in the adsorption energy of atop-adsorbed H, the change in ionic energy dominates the bond energy trend for threefold-adsorbed H (see Figure 2.12b).

In summary, for chemisorption on the transition metal surface, four reactivity descriptors have been discussed: coordinative unsaturation of the surface transition metal atom, the transition metal electron affinity (work function), d-valence electron occupation, and d-valence electron band width. The electronic signatures of surface atom coordination are the average energy of the d-valence electrons, $\bar{\varepsilon}_d$, as well as the d-valence electron band width *W*. This relation between electronic features and surface topology is useful for the interpretation of experimental spectroscopic data of surfaces when information on surface structure or adsorption site structure is lacking.

2.3.3.4 *Molecular Bond Activation*

A low activation energy requires strong interaction with the surface. Electron backdonation into anti-bonding molecular orbitals weakens the molecular bond. In the transition state the molecular bond is broken. The energy is minimized by strong interaction of molecule fragments with the surface.

Here the quantum chemistry of the adsorption energy of admolecules and its relationship with activation of the molecular bond are discussed. The word dissociation is used differently in coordination chemistry and surface science. In coordination chemistry, dissociation means the disruption of a metal-ligand bond and the desorption of the ligand molecule. In surface science, on the other hand, the analogous process of metal-adsorbate bond disruption is called desorption and the intramolecular bond cleavage in an adsorbate is called dissociation. In Section 2.3.2.3 the transition state of the molecular bond cleavage reaction is described as late and tight. The transition state geometry of the activated complex is close to the dissociated state and interaction with surface is strong. Quantum-chemically this can be analyzed by a calculation of the bond orders and an analysis of the electronic structure of the molecule in the transition state. In this section such analyses are presented for the activation of methane and CO.

The DFT-calculated transition state of the C–H bond cleavage of methane is shown in Figure 2.23 and Table 2.4 [117]. On the Rh(111) surface, C–H bond cleavage over a single metal atom is assisted by interaction of the H atom with a second Rh surface atom.

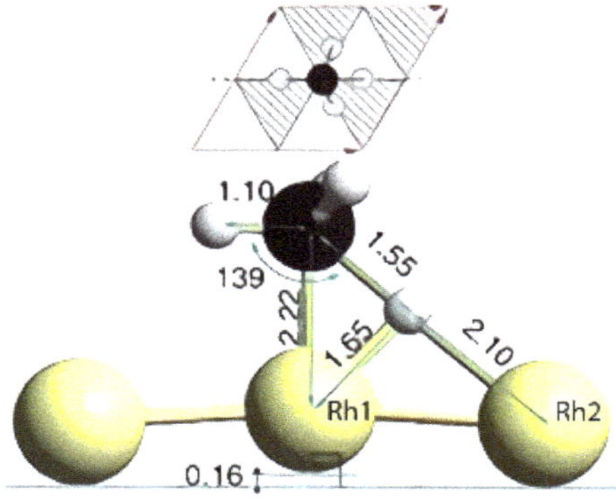

Figure 2.23 The transition state of C–H bond cleavage of CH_4 by the Rh(111) surface. Structure and bond distances (Å) are shown [117].

Table 2.4 Bond orders of reactant, transition state, and product state of methane dissociation on Rh(111) surface [101].

Bond	Reactant State	Transition State	Product State
C–H	0.99	0.29	< 0.2
C–Rh1		0.43	0.54
H–Rh1		0.38	0.42
H–Rh2		< 0.2	0.49

In the transition state, the C–H bond is stretched by approximately 35% and interaction between M–C and M–H atoms develops. The weakening or increase of the bond strength can be quantified by calculation of the respective bond orders, which are shown in Table 2.4. Whereas in adsorbed methane the bond order of C–H is close to one, in the transition state this decreases by 60%. Importantly there is an increase in bond orders of the M–C and M–H bonds. The latter slightly increase in the final state. The H atom then adsorbs in bridging position between two metal atoms, and the CH_3 fragment adsorbs atop. The weakened C–H bond and the well-developed M–C and M–H bond interactions are consistent with a late transition state.

In the following the quantum chemistry of CO chemisorption and chemical bonding features of the dissociating molecule are presented. The molecular orbitals of CO are shown in Figure 2.24a. The C–O bond consists of molecular orbitals that are σ and π symmetric versus the C–O axis. Relevant to the interaction with the surface are the HOMO, the 5σ orbital that is mainly located on the carbon atom, and two LUMO 2π* orbitals that are delocalized over the two CO atoms.

The 5σ as well as the 2π* orbitals are anti-bonding orbitals. This follows from the negative sign of the COHP in Figure 2.24a3. Electron depopulation of the doubly electron-occupied 5σ orbital strengthens the C–O bond, and electron population of the 2π* orbitals will weaken the C–O bond strength.

According to the Dewar-Chatt-Duncanson chemical bonding model (Figure 2.14), the electronic interaction with CO consists of an electron-donating and electron-backdonating interaction. A schematic representation of the molecular orbital scheme of CO chemisorption is given in Figure 2.24b for interaction with a single Cu metal atom. This bonding scheme relates to the early 1964 proposal of the surface adsorbate bond by George Blyholder [118]. In the adsorbed state, the interaction with surface also creates bonding and anti-bonding molecular orbital fragments between surface atoms and interacting molecules.

The CO 5σ orbital gives bonding and anti-bonding orbitals by interaction with a transition metal σ-symmetric (d_{z^2}) or s atomic orbital. In Figure 2.24b the respective bonding and anti-bonding orbitals that are formed by the interaction of CO 5σ and Cu d_{z^2} are schematically shown. The 5σ contributes two electrons as does the Cu d_{z^2}. Both bonding and anti-bonding orbitals are electron occupied. This causes the contribution to the adsorbate bonding energy to be repulsive. The embedding of the adsorption complex into the solid broadens the energy levels. When the molecular orbital in the free molecule is occupied and anti-bonding electron density is higher than the Fermi level, a fraction of electrons is donated into the metal. Generally, for the late transition metals which have high d-valence electron occupation, the interaction with doubly occupied

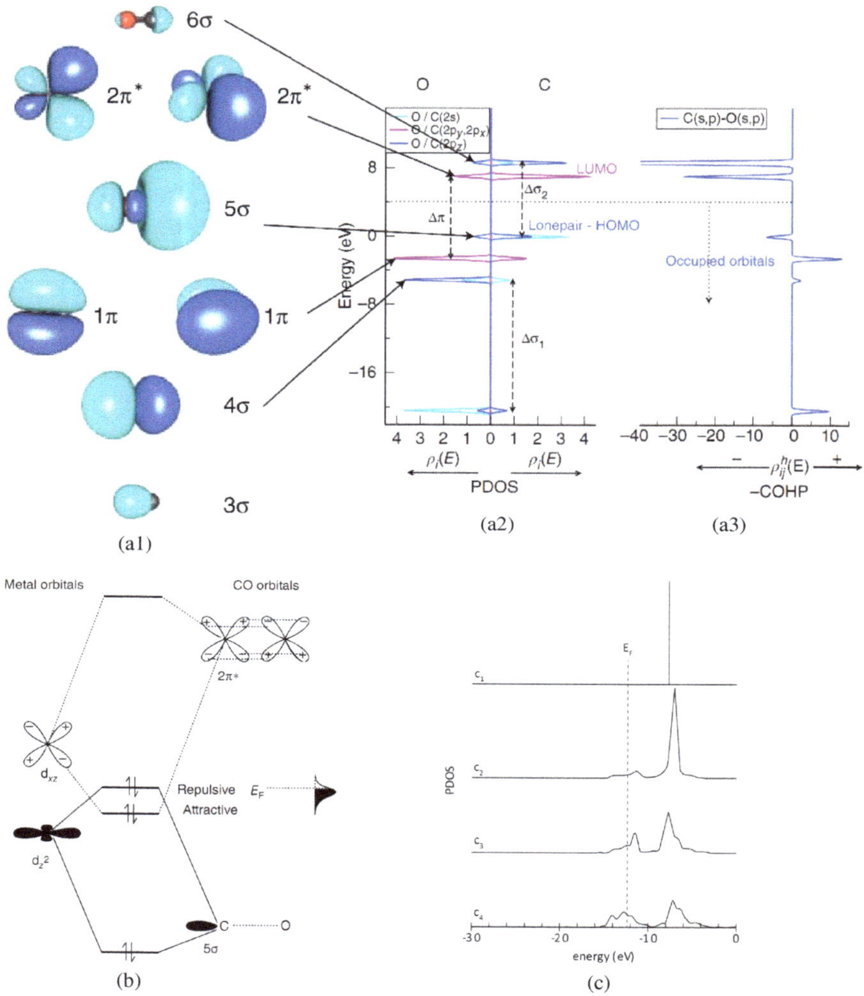

Figure 2.24 The electronic structure of adsorbed CO. (a) The electronic structure of the CO molecule: (a1) visualization of the molecular orbitals; (a2) partial density of states (PDOS) on C and O respectively; (a3) bond order overlap population (COHP) [119]. (b) Schematic bonding and anti-bonding orbital scheme of Dewar-Chatt-Duncanson chemical bonding model [118]. (c) The CO 2π* PDOS of differently coordinated CO to non-stepped and stepped Rh(111) surface as derived from Extended Hückel calculations. The Fermi level is indicated: (c1) free molecule, (c2) CO adsorbed atop, (c3) CO adsorbed threefold, (c4) CO adsorbed at step-edge [120].

molecular orbitals such as the σ-symmetric orbital of CO is repulsive. Bonding as well as anti-bonding surface fragment orbitals are electron occupied. When transition metal electron occupation is reduced, the repulsive interaction decreases. Different from the interaction with the CO 5σ orbital, the interaction with the empty CO 2π* orbitals is attractive since only bonding surface fragment orbitals become electron occupied. Bonding and anti-bonding orbitals are formed between 2π* and metal atom d_{xz} and d_{yz} atomic orbitals. The transition metal d-atomic orbitals contribute electrons to the π-symmetric orbital interaction. The 2π* orbital is initially empty and metal electrons are donated into the 2π* CO orbitals. In this case when electron density of the transition metal valence band decreases, the attractive interaction will be diminished. Change in transition metal valence electron occupation leads to counteracting bond strength contributions of the σ- and π-symmetric orbitals. This is the reason for the small variation of CO adsorption energies compared to that of C or O adatoms (see Figure 2.13).

The interaction with the metal surface weakens the C–O bond strength. This bond weakening is dominated by the π-symmetric orbital interaction of metal surface and CO. The C–O bond weakens because electrons donated by the surface occupy the anti-bonding CO 2π* orbital.

Figure 2.24c compares the changes in 2π* PDOS distribution for CO adsorbed to different adsorption sites of the non-stepped and stepped Rh(111) surface. The molecule is adsorbed perpendicular to the surface. The Fermi level E_F is indicated. The figure shows that broadened 2π* bonding and anti-bonding density evolve below and above, respectively, the discrete level of free CO. When CO adsorbed atop a single surface atom (c2) is compared with CO adsorbed threefold to the surface (c3), the difference between the 2π* maxima of the anti-bonding and bonding PDOS distributions increases. It indicates the stronger 2π* interaction with increased coordination of the C atom of CO with surface metal atoms. This interaction is even larger when the CO molecule is adsorbed near the step-edge site. The increased interaction is due to additional contact of the CO oxygen atom with step-edge metal atoms (see

Figure 2.9). The bonding PDOS that develops below the Fermi level is occupied by electrons. This confirms that interaction with the 2π* orbital gives an attractive contribution to the CO adsorption energy. Electron occupation of the 2π* CO orbital weakens the C–O bond, because of its anti-bonding electron density with respect to the C–O bond strength. The electronic distributions show that the more the C–O bond weakens, the better CO contacts with the surface.

Experiments confirm the sensitivity of the C–O bond strength with respect to adsorption site. Two experimental results are presented in Figure 2.25. Figure 2.25a shows IR absorption spectra of CO adsorbed on Pd-Ag alloys and Figure 2.25b depicts high-resolution energy electron loss spectroscopy (HREELS) data of CO adsorbed on the reactive Ru($11\bar{2}1$) surface.

Figure 2.25a shows that the vibrational frequency of CO adsorbed on Pd is shifted downward with respect to the gas-phase frequency of 2170 cm^{-1}. This agrees with the predicted decrease of C–O bond strength when CO adsorbs on the transition metal surface. Several adsorption peaks are present that indicate CO is adsorbed on different adsorption sites. Since Ag will weakly interact with CO, alloying reduces the Pd surface atom ensemble size accessible to CO. The decrease in intensity of the lower vibrational frequency peaks when Pd is alloyed with increasing concentration of Ag allows us to conclude that the adsorption peak at 2060 cm^{-1} corresponds to CO adsorbed to a single metal atom, and that the lower frequencies correspond to CO adsorbed with higher coordination to Pd transition metal atoms. The higher coordination of the C atom of CO to the metal surface weakens the C–O bond strength more.

The HREELS spectra of Figure 2.25b on the reactive stepped Ru($11\bar{2}1$) surface demonstrate that the C–O bond strength of CO adsorbed side-on to a step-edge site is weakened and readily dissociates. This reactive CO molecule has the extremely low vibrational frequency of 166 meV (= 1328 cm^{-1}) that disappears by bond dissociation near room temperature. This follows from the appearance of a peak at 70 meV of adsorbed O. The vibrational frequencies of less reactive atop- and twofold-adsorbed CO are found at 247 and

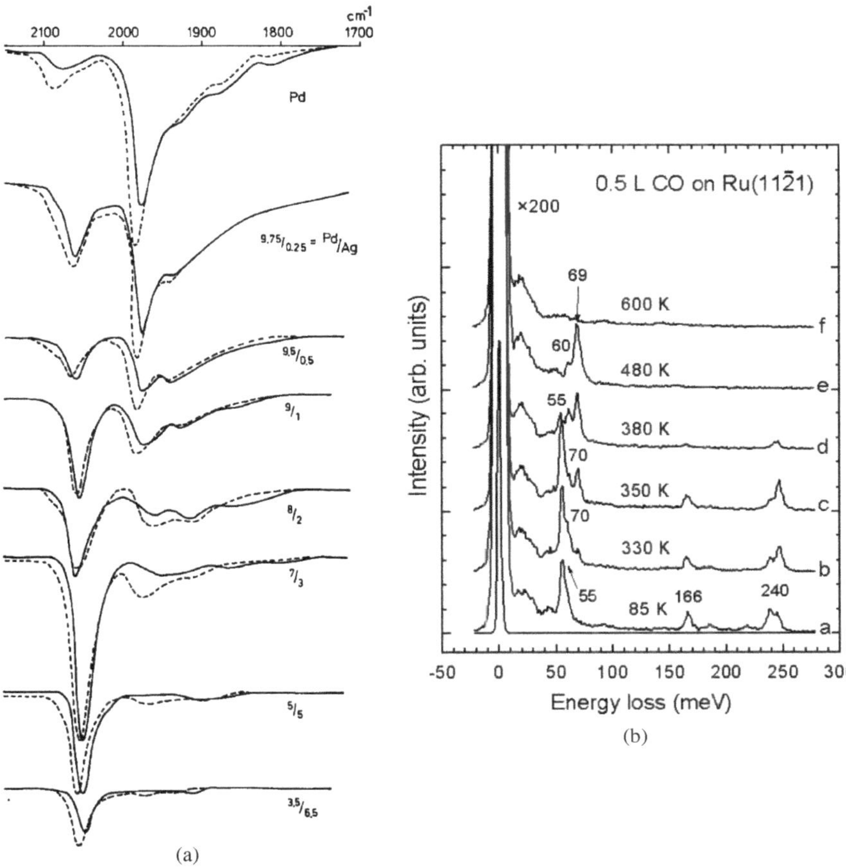

Figure 2.25 Vibrational spectra of adsorbed CO. (a) IR vibrational spectra of CO adsorbed on Pd/Ag alloy particles of varying concentration, P_{CO} = 0.01 Torr; vibrational frequency of atop-adsorbed CO equals 2060 cm^{-1}, while that of bridge-adsorbed CO equals 1960 cm^{-1} (frequency of free CO is 2170 cm^{-1}) [121]. (b) HREELS spectra of CO on the Ru(11$\bar{2}$1) surface after an exposure of 0.5 L CO at 85 K (1 meV = 8.066 cm^{-1}) [122].

238 meV and the related M–C frequencies of adsorbed CO at 55 and 62 meV, respectively. These adsorption states are less reactive and only disappear at higher temperature.

The lower vibrational frequencies of reactive CO agree with expectation from theory. DFT calculations of the electronic

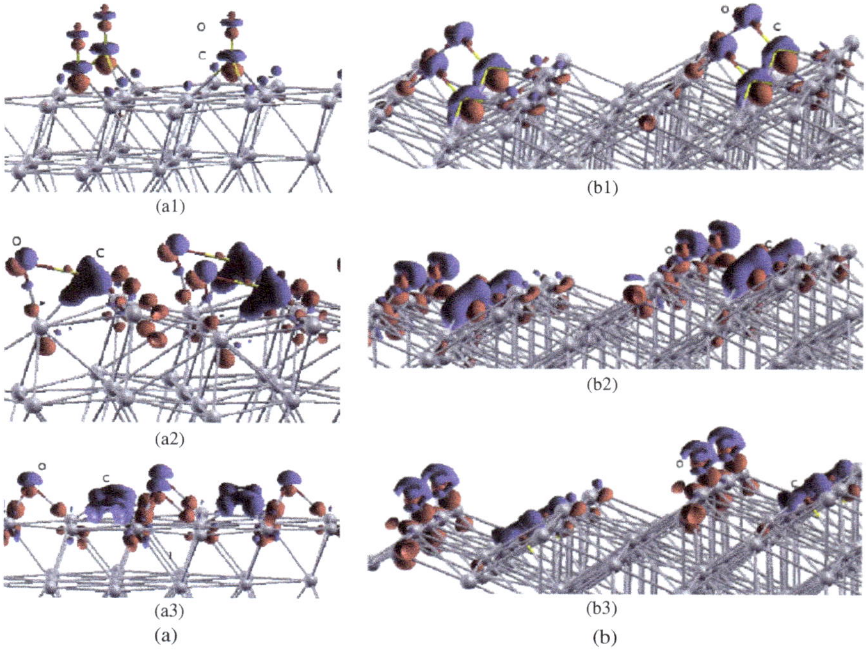

Figure 2.26 Electronic features of CO dissociation. (a) Charge difference plots of CO in the (a1) initial, (a2) transition, and (a3) final state on Ru(0001). (b) Charge difference plots of CO in the (b1) initial, (b2) transition, and (b3) final state on Ru($11\bar{2}1$). A blue region has gained electron charge and a red region has lost electron charge. Charge difference is measured between occupied electron densities of the same structure free from surface and in contact with surface [101], [123], [124].

structure of CO that dissociates on the Ru(0001) terrace and Ru($11\bar{2}1$) surface are shown in Figure 2.26. From top, calculated charge differences are shown of CO in the ground state, transition state and dissociated state. The charge differences are calculated as the difference of charge distribution of CO molecules in contact with the surface and charge distribution of a CO molecule stretched similarly but in gas phase.

Figure 2.26a shows dissociation of the CO molecule on the Ru(0001) terrace. Differences in charge density distribution for adsorbed CO indicate the increase in electron density of the π-symmetric states and decrease in electron density of the σ-symmetric states. In the transition state the CO molecule is side-on coordinated to three metal atoms. A large charge density accumulates in the carbon atom. Charge is removed from the surface metal atoms. In the final state the 2p atomic orbital of O accumulates charge similar as the carbon atom.

Figure 2.26b shows for CO adsorbed on the corrugated surface that in the initial state, the chemical bond of the side-on adsorbed CO molecule is already substantially weakened and its structure is close to that of the transition structure of Figure 2.26a. The transition structure of Figure 2.26b is closer to its final state. This is in line with the very low calculated transition state energy of 65 kJ/mol for the Ru(11$\bar{2}$1) surface compared to a barrier of 216 kJ/mol for dissociation of CO on the Ru(0001) terrace [125].

The principle that low activation energy of chemical bond dissociation requires strong interaction with the transition metal surface and substantial population of adsorbate anti-bonding orbitals is general. Activation of N_2 is discussed in [126] and O_2 dissociation in [127]. The transition state tends to be late, and its structure is close to that of the dissociated final state.

In the next section trends in catalytic reactivity are related to the trends of surface reactivity discussed in this section. Whereas surface reactivity is probed by a single elementary surface reaction, catalytic reactivity derives from the catalytic reaction network.

2.4 Catalytic Reactivity and Kinetics

Kinetics relates catalytic reaction rate and selectivity with surface reactivity. Reaction mechanism determines the kinetic network.

2.4.1 *Introduction*

Early catalytic kinetics began with the formulation of the catalytic reaction rate as a function of proposed surface reaction intermediates (Section 2.2.1.1). In the 1920s, inspired by the Langmuir checkerboard model of adsorption, the Langmuir-Hinshelwood kinetic equations were formulated. The essential feature is that there is a fixed total number of reactive surface sites. A reaction mechanism was postulated, and kinetic reaction rate parameters were obtained by fit with experiment.

Initially kinetics developed mainly within a reactor engineering context in order to determine optimum reaction conditions for design of catalytic reactors. Mechanistic models remained largely speculative until the end of the previous century. Kinetic concepts such as reaction rate-controlling step and major abundant reactive intermediate (MARI) that were subsequently developed are discussed below.

A major advance in kinetics happened due to increasingly better understanding of the molecular basis of surface chemical reactions. It changed the concept of surface reaction intermediate from postulated assumption to observable molecular entity. Elementary surface reaction constants became available and the state of the catalytic site during reaction was investigated. At the end of the previous century this led to the formulation of microkinetics, which is defined by molecular reaction networks with elementary surface reaction constants deduced from experiment or computational data.

In Section 2.4.2 the Langmuir-Hinshelwood-Hougen-Watson kinetic equations are introduced. They relate reaction rate with the elementary reaction rate constants of surface intermediate conversion steps, their concentration and reaction affinity. These equations are central to catalytic kinetics. The relation between global kinetics and microkinetics is discussed in Section 2.4.3. Global kinetic expressions are reaction rate expressions with mass law dependence on reactant or product concentration in the reaction mixture.

Catalytic global kinetics is essentially different from the Guldberg-Waage mass law kinetics (see Section 1.2.2) since the reaction order is fractional instead of a natural number. It will be shown that these reaction orders relate to the concentration of surface intermediates. The temperature dependence of the mass law rate expressions of a catalytic reaction is non-Arrhenius. This is due to

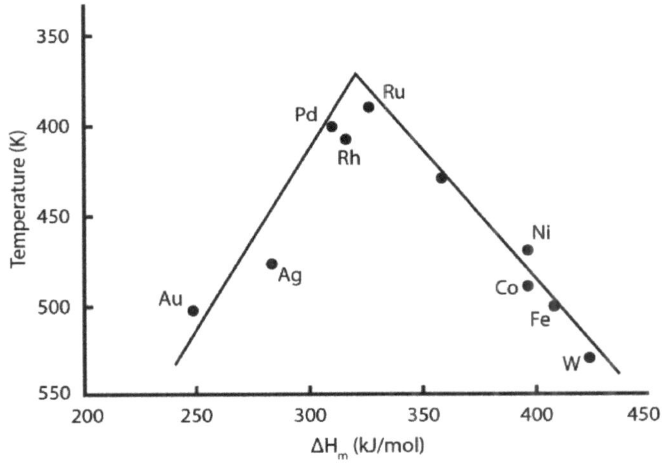

Figure 2.27 Volcano curve of the temperature of formic acid decomposition as a function of the heat of formation of the metal formate [128].

the change in concentration of surface reaction intermediates with temperature. The final Section 2.4.4 deals with the kinetic formulation of the Sabatier principle.

The major difference between surface reactivity and catalytic reactivity is that surface reactivity is measured by the chemisorption energy of a single reaction intermediate, whereas catalytic reactivity is a kinetic phenomenon that depends on rate of reactant activation and product desorption. An early illustration of this difference is the measurement of catalytic reaction rate as a function of heat of formation of formic acid decomposition by Rootsaert and Sachtler in 1960 [128]. Figure 2.27 plots for different metals the temperature of reaction (that is a measure of catalytic reactivity) as a function of heat of formation. It shows a maximum in reaction at an optimum heat of formation.

This volcano-type dependence of the catalytic reaction rate as a function of heat of formation is very different from the Tanaka-Tamaru plot of adsorption energies of different transition metals, which is shown in Figure 2.6. This figure shows a linear dependence of surface reactivity (as measured by heat of adsorption) and bulk heat of formation.

The maximum in reaction rate of Figure 2.27 is due to the Sabatier principle that has been introduced in Section 1.2.3. The reaction rate of substrate activation competes with the reaction rate of product desorption. An increase in surface reactivity will enhance the reaction rate of adsorbate activation but decrease the rate of product desorption. A maximum is found when rate of reactant activation becomes comparable to the rate of product desorption. Several such volcano curves are currently known for different catalytic reactions. Some of them will be presented in succeeding chapters.

Kinetic theories of volcano curve dependence cannot be based on the assumption of an invariant reaction rate-controlling step of the reaction. The reaction rate-controlling step changes with change of surface reactivity, when the reaction rate passes through the volcano curve maximum. In Section 2.4.4 the BEP equations will be used to relate activation energy with surface energy for elementary surface reaction rate constants. They will be used to model Sabatier principle kinetics.

2.4.2 *The Langmuir-Hinshelwood-Hougen-Watson Equations*

The rate of the catalytic reaction is proportional to elementary reaction rate constant, reaction intermediate concentration and kinetic affinity.

A useful kinetic model is the two-step Temkin-Boudart model [8], [129] that assumes a limited number of surface reaction intermediates. Kinetic equations contain reaction rate parameters that originally were empirically fitted. Here for this model the Langmuir-Hinshelwood kinetic equations are introduced. These are used to derive the Langmuir-Hinshelwood-Hougen-Watson kinetic expressions. This provides an opportunity to introduce key kinetic concepts that are fundamental to catalytic kinetics [5].

The Temkin-Boudart model is based on the following reaction scheme:

$$A_1 + A_2 \leftrightharpoons B_1 + B_2 \tag{2.12}$$

A_1 and A_2 are reactants and B_1 and B_2 are products. The catalyst surface has vacant catalyst sites θ_v and occupied sites θ_A. The corresponding reaction equations are:

$$A_1 + \theta_v \underset{r_1}{\overset{r_{-1}}{\leftrightharpoons}} \theta_A + B_1 \tag{2.13a}$$

$$A_2 + \theta_A \underset{r_2}{\overset{r_{-2}}{\leftrightharpoons}} \theta_v + B_2 \tag{2.13b}$$

An important simplification in this two-step kinetic model is that only one surface intermediate is proposed. As in the Langmuir adsorption isotherm the number of surface sites is a constant, so that:

$$\theta_v + \theta_A = 1 \tag{2.14}$$

The Langmuir-Hinshelwood kinetic equations for the rates of change of products C_{B1}, C_{B2} and θ_A are:

$$\frac{1}{N_S}\frac{dC_{B2}}{dt} = r_2\theta_A C_{A2} - r_{-2}\theta_v C_{B_2} \tag{2.15a}$$

$$\frac{1}{N_S}\frac{dC_{B1}}{dt} = r_1\theta_v C_{A1} - r_{-1}\theta_A C_{B1} \tag{2.15b}$$

$$\frac{d\theta_A}{dt} = r_1\theta_v C_{A1} + r_{-2}\theta_v C_{B2} - r_{-1}\theta_A C_{B1} - r_2\theta_A C_{A2} \tag{2.15c}$$

The rate of reactant consumption or production is proportional to the number of reaction sites N_S.

In Eqs (2.15) reaction intermediates do not interact and their surface concentration is assumed to be uniformly distributed over the catalyst surface. This is called the mean field approximation. The uniformity of surface intermediate distribution implies that surface diffusion processes are assumed to be fast.

Eqs. (2.15) can be solved by use of the Quasi-Steady State assumption of the surface intermediate concentration. This is also called the Bodenstein (Germany, 1871–1942) assumption:

$$\frac{d\theta_A}{dt} = 0 \tag{2.16}$$

This steady state assumption commonly applies. The quasi-steady state assumption can become invalid at very specific conditions where reactions are non-stationary and show oscillatory or more complex behavior [15]. These cases will not be considered here.

The solution of Eqs (2.15) has an interesting physical interpretation as has been elucidated by Hougen and Watson [130]. This is the reason that their name is sometimes added to the Langmuir-Hinshelwood-type expressions of Eqs. (2.17).

$$\frac{1}{N_s} R = \frac{r_1 r_2 C_{A_1} C_{A_2} \left(1 - e^{\frac{-A_f}{RT}}\right)}{r_1 C_{A_1} + r_2 C_{A_2} + r_{-1} C_{B_1} + r_{-2} C_{B2}} \tag{2.17a}$$

$$A_f = -RT \ln \frac{r_{-1} r_{-2}}{r_1 r_2} \frac{C_{B_1} C_{B_2}}{C_{A_1} C_{A_2}}\ ; K_{eq} = \frac{r_{-1} r_{-2}}{r_1 r_2} \tag{2.17b}$$

$$\theta_A = \frac{r_1 C_{A_1} + r_{-2} C_{B_2}}{r_1 C_{A_1} + r_2 C_{A_2} + r_{-1} C_{B_1} + r_{-2} C_{B2}} \tag{2.17c}$$

Eqs. (2.17) show the solution for the reaction rate R and surface coverage θ_A. One can distinguish in the expression of R three terms:

- The numerator $r_1 r_2 C_{A_1} C_{A_2}$ is the Guldberg-Waage mass law reaction rate expression as one would obtain in the homogenous phase (ignoring restriction of total number of sites).
- The driving force $(1 - e^{\frac{-A_f}{RT}})$. According to Ostwald the driving force of a chemical reaction is the difference in concentration of reaction mixture with equilibrium concentration (see Section 1.2.3). It can be calculated from the kinetic affinity A_f, defined in Eq. (2.17b) [131], [132]. The kinetic affinity is zero when products and reactants are equilibrated. Then reaction rate is also zero. Near equilibrium the reaction rate is proportional to A_{aff}.

– The denominator $\frac{1}{r_1 C_{A_1} + r_2 C_{A_2} + r_{-1} C_{B_1} + r_{-2} C_{B2}}$. This component reduces the reaction rate. The same denominator occurs in the expression for the surface coverage θ_A of Eq. (2.17c). It relates to the blocking of the reaction center by surface-adsorbed reaction intermediates. It is a consequence of the Langmuir assumption of the presence of a finite number of sites.

The concepts of reaction rate-controlling step and MARI [133], [134] can be illustrated with a simplified expression by ignoring the reverse reactions of products B_1 and B_2. Then Eqs. (2.17) reduce to:

$$\frac{1}{N_S} R = r_2 C_{A2} \theta_A \tag{2.18a}$$

$$\theta_A = \left(1 + \frac{r_2 C_{A2}}{r_1 C_{A1}}\right)^{-1} \tag{2.18b}$$

When elementary reaction rate r_2 of reactant $A2$ is fast, surface coverage θ_A is close to zero and the surface state is dominated by surface vacancies. Then adsorption and activation rate of $A1$ is rate controlling. In the opposite case when $r_1 C_{A1} > r_2 C_{A2}$, θ_A becomes close to one and the MARI is the surface reaction intermediate A. Now the elementary rate of product formation r_2 becomes reaction rate controlling. Whereas in the first case the reaction rate is first order in C_{A1} and zero order in C_{A2}, this inverts in the second case. This example illustrates that reaction order, reaction rate-controlling step, and MARI are related.

The term Langmuir-Hinshelwood mechanism is reserved for catalysis where reaction occurs between intermediates adsorbed on the surface. This is the dominant mechanism of surface reactions. It is different from that whereby a molecule directly reacts from the gas phase with adsorbed intermediates. This reaction is named the Eley-Rideal mechanism [135], [136], suggested in 1940 (see also comments from Prins [137]). Earlier in 1922 Langmuir proposed such a mechanism for the high-temperature CO oxidation reaction catalyzed by Pt [6]. The surface then is saturated with oxygen adatoms and CO reacts from the gas phase by impacting with an oxygen adatom. The kinetic expressions Eqs. (2.13) describe reaction events in accord with the Eley-Rideal mechanism. The next section explores the relation between the microkinetic Langmuir-Hinshelwood formulation of surface kinetics with global kinetics power law expressions.

2.4.3 *The Reaction Order and Apparent Activation Energy (from Micro to Global Kinetics)*

The reaction order of global kinetics relates to surface intermediate concentrations defined by microkinetics. There is a maximum in reaction rate when adsorption energy of reactant increases. There is also a maximum in reaction rate when temperature increases. Both relate to change in surface coverage and reaction rate-controlling step. The heterogenous catalytic reaction shows non-Arrhenius-type temperature dependence.

The interplay of adsorption and surface reactions determines the catalytic reaction rate. The apparent reaction rate constant r_{app} and reaction orders x and y of global power law kinetic expression Eq. (2.19) depend in a complex way on these elementary reactions.

$$R = N_S \, r_{app} \, C_{A_1}^{x} C_{A_2}^{y} \tag{2.19}$$

Reaction rate parameters in the power law rate expression are only constant in a limited reaction condition regime. Here for a monomolecular reaction, analytical expressions for apparent reaction rate constant r_{app} and reaction orders x or y that explicitly depend on the state of the reacting surface are presented. In the first part of this section, catalytic reaction rate expressions are presented that illustrate the dependence of catalytic reaction rate on adsorption equilibrium and elementary surface reaction rates. In the second part of this section, the relation between power law reaction rate expression and surface coverage of reaction intermediates is discussed.

In Figure 2.28 the rate of hydrogenation of unsaturated hydrocarbons catalyzed by Pd is compared as a function of hydrocarbon

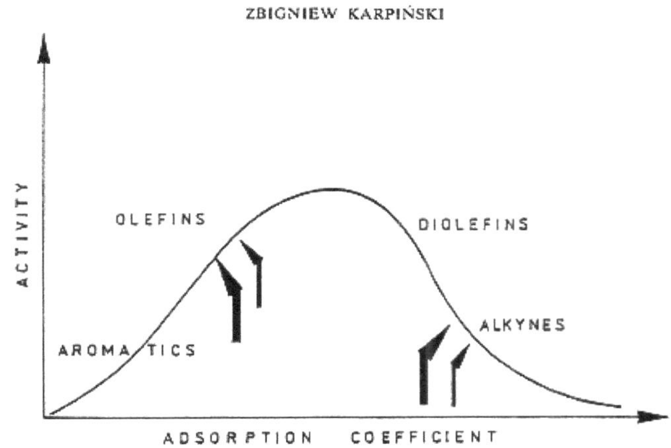

Figure 2.28 The rate of hydrogenation of unsaturated hydrocarbons (activity) catalyzed by Pd plotted as a function of hydrocarbon adsorption coefficients [138].

adsorption strength. A bell-shaped dependence of reaction rate as a function of increasing value of adsorption coefficients is observed [138]. The adsorption energies vary because of the difference in size and unsaturation of reactant molecules. At the left of Figure 2.28 the adsorption free energy is low. The reaction rate increases with increase of adsorption free energies because of the increase of reactant intermediate surface coverage. There are two possible explanations for the maximum in reaction rate of Figure 2.28 as a function of adsorption coefficient. Beyond the maximum the reaction becomes rate limited by product desorption or there is lack of supply of hydrogen when surface becomes covered with adsorbate. At low temperature in the case of benzene hydrogenation the surface is mainly covered with hydrogen, but strongly adsorbed hydrocarbons such as ethyne will suppress hydrogen adsorption [139]–[142].

The kinetics of catalysis with and without product poisoning are discussed next. This is followed by a kinetic model of suppressed reaction rate by limited hydrogen adsorption.

Reaction equations for a two-step model with product inhibition are given by:

$$A \underset{k_{des}}{\overset{k_{ads}}{\rightleftharpoons}} A_{ads}$$

$$A_{ads} \underset{r_r^-}{\overset{r_r^+}{\rightleftharpoons}} B_{ads}$$

$$B_{ads} \xrightarrow{k_{des}} B \tag{2.20}$$

Eq. (2.20) represents a lumped kinetic scheme of the hydrogenation reaction. Reaction with hydrogen is implicit in lumped reaction rate constants r_r^+ and r_r^-. Reactant A equilibrates with surface intermediate A_{ads}. In the next step, A_{ads} is converted with rate constant r_r^+ to intermediate B_{ads}. This surface hydrogenation is assumed reversible. B_{ads} desorbs as product B and is assumed to not readsorb. This is consistent with the saturated hydrocarbons as products since they only weakly adsorb.

The solution of the kinetic rate equations that correspond to Eq. (2.20) is summarized in the legend of Figure 2.29. The surface area normalized reaction rate R' is calculated as a function of product desorption rate constant k_{des} and elementary reaction rate constant r_r^+ of the conversion of A_{ads} into B_{ads}. k_{dess}varies, but reaction rate constants r_r^+ and r_r^- are assumed to be constant. Figures 2.29 compare the potential energies of reaction for the case of (a) relatively weak and (b) strong adsorption. Figure 2.29c gives a schematic representation of reaction rate R' versus decrease of k_{des}. The adsorption constant increases.

Figures 2.29a and 2.29b show the activation energy E_r of the elementary reaction constant r_r of the conversion of A_{ads} into B_{ads}, and the respective adsorption energies $E_{ads,A}$ and $E_{ads,B}$. E_r is defined with respect to the adsorbed state. For convenience $E_{ads,A}$ and $E_{ads,B}$ are chosen to be equal and so are the surface reaction rate constants: $r_r^+ = r_r^-$. The solution of the Langmuir-Hinshelwood kinetic equations that is illustrated in Figure 2.29c shows a bell-shaped dependence of reaction rate R' on the elementary reaction rate k_{des}. Also, the change in surface concentrations A_{ads} and B_{ads} are indicated in the figure. Whereas when k_{des} is fast surface concentration is dominated by A_{ads}, when k_{des} decreases beyond the reaction rate maximum $R'_{max} = \frac{1}{\sqrt{2}} r_r$ surface coverages of A_{ads} and B_{ads} are equal.

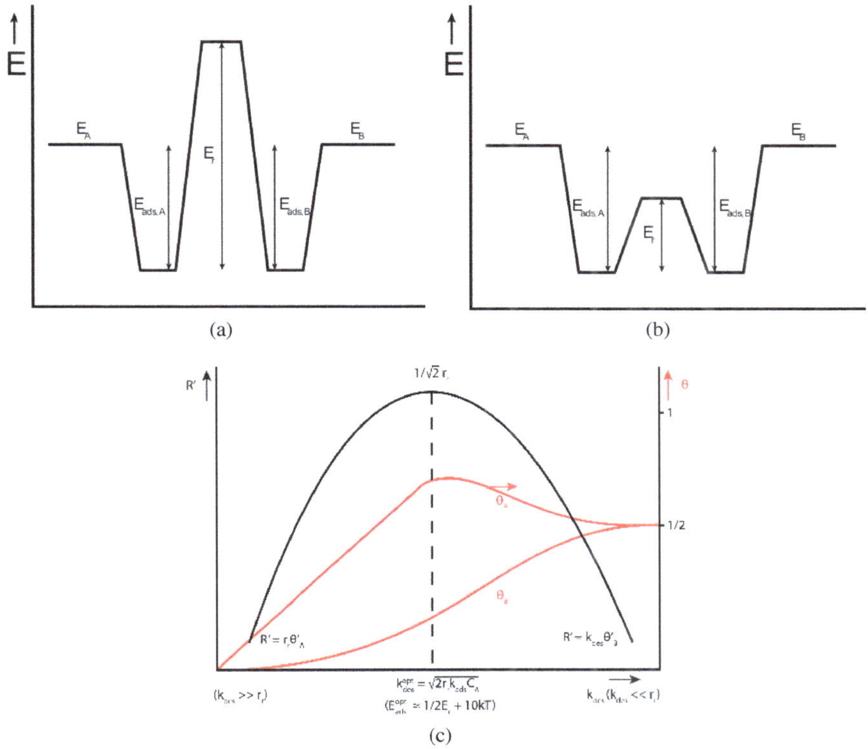

Figure 2.29 Schematic representation of the kinetics of a monomolecular reaction where adsorption energy varies. (a) Potential energy diagram when surface reaction rate r_r is limiting ($E_r > E_{ads,A}$). (b) Potential energies when k_{des} is reaction rate limiting ($E_r > E_{ads,B}$). The respective adsorption energies ($E_{ads,A} = E_{ads,B}$) and activation energy of surface reaction E_r are indicated. (c) Schematic representation of the solution of monomolecular kinetic model (Eq. (2.20)): dependence of catalytic reaction rate on product desorption reaction rate k_{des}. Normalized reaction rate $R' = \frac{R}{N_S}$ and surface coverage θ plotted as a function of k_{des}. Relevant expressions are $\theta'_A = \frac{K_{ads}c_A}{1+K_{ads}C_A}$ ($k_{des} >> r_r$); $\theta'_B = \frac{1}{2(1+1/K_{ads}C_A)}$ ($k_{des} >> r_r$); $\theta_B = \frac{r_r}{r_r + k_{des}}\theta_A$; $k_{des}^{opt} = \sqrt{2r_r k_{ads} C_A}$; $E_{ads}^{opt} = \frac{1}{2}E_r - k_B T \ln\sqrt{\frac{2r_r k_{ads} C_A}{\upsilon_{des}^2}} \approx \frac{1}{2}E_r + 10k_B$.

When E_{ads} increases towards E_r, the surface coverages of A_{ads} and B_{ads} increase and a kinetic transition occurs. The reaction rate R' approaches its maximum and decreases when reaction rate constant k_{ads} further decreases. The optimum value of E_{ads} where R' is

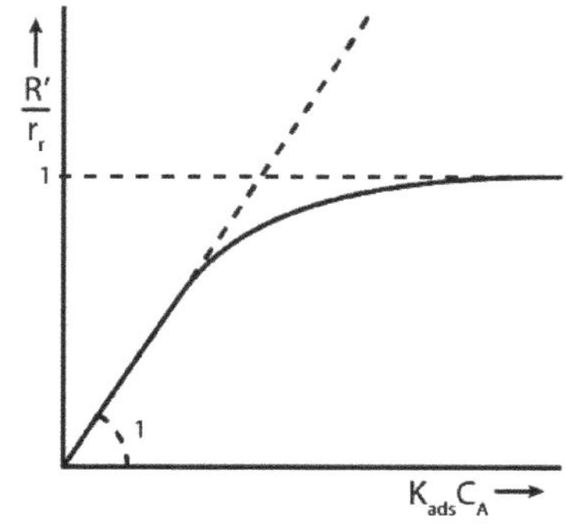

Figure 2.30 Schematic representation according to Eq. (2.21a) of the energetics of a monomolecular catalytic reaction when the rate of product desorption is fast.

maximum is given by $E_{ads}^{opt} \approx \frac{1}{2} E_r + 10 k_B T$ (see legend of Figure 2.29). The term proportional to $k_B T$ is due to differences in the pre-exponents of the reaction rate constants. The pre-exponent of the elementary reaction rate of desorption is orders of magnitude larger than that of the surface reaction rate constant r_r.

As long as k_{des} is large compared to r_r, the reaction rate is given by Eq. (2.21a). The reaction rate-controlling step is elementary reaction constant r_r, and reaction rate is proportional to θ_A. With increase of k_{ads}, the surface becomes covered with adsorbate A and surface coverage with adsorbate B remains low. It is illustrated in Figure 2.30 that, as long as E_r remains substantially larger than E_{des}, the reaction rate saturates at a maximum value and becomes equal to r_r with increase of $k_{ads}CA$.

In the extreme limit of very low k_{des}, the surface becomes equally covered by A_{ads} and B_{ads}, since these concentrations equilibrate. Then the reaction rate is rate limited by the elementary desorption reaction and given by Eq. (2.21b).

$$\lim_{k_{des} \gg r_r} \frac{R}{N_s} = r_r \theta_A = r_r \frac{K_{ads} C_A}{1 + K_{ads} C_A} \tag{2.21a}$$

$$\lim_{k_{des} \ll r_r} \frac{R}{N_s} = k_{des} \theta_B = k_{des} \frac{1}{2\left(1 + \frac{1}{K_{ads} C_A}\right)} \tag{2.21b}$$

A catalytic reaction will only have a finite rate when the temperature of reaction is high enough that surface vacancies are available to adsorb reactant. At lower temperatures the surface can become blocked by adsorbed reactant molecules. Catalyst poisoning by adsorption of product molecules is also not uncommon. To prevent this, a minimum temperature of reaction has to be selected with $k_BT > \frac{\Delta H_{ads}}{\Delta S_{ads}}$.

For ethyne, which has an adsorption coefficient that locates the hydrogenation rate at the right of the reaction rate maximum of Figure 2.28, the reaction product is ethene. It is a desirable reaction and is selective (with respect to ethane formation). When this reaction is catalyzed by Pd, product ethene desorption is fast compared to the desorption rate of ethyne [143]. This makes it unlikely that the maximum in reaction rate of Figure 2.28 is due to increase of product poisoning. It indicates that the reason for the maximum in reaction rate of Figure 2.28 is not blockage of surface by product poisoning but due to inhibition of hydrogen adsorption. At the temperature of benzene hydrogenation, the surface is mainly covered with hydrogen, which inhibits benzene adsorption [139]. In contrast, when ethyne is hydrogenated, at least 40 percent of the surface is covered with ethyne [142].

When product desorption is fast and the elementary reaction of hydrogen addition to adsorbate is reaction rate controlling, the reaction rate R' is given by:

$$R' = k_h \theta_A \theta_H^x \tag{2.22a}$$

$$= k_h \theta_A \left(1 - y\theta_A\right)^x \quad \left(\theta_v = 0\right) \tag{2.22b}$$

In Eq. (2.22), θ_A and θ_H are respective surface coverages of reactant intermediate A and H, y is a stoichiometric number, and x is the reaction order in surface hydrogen. Eq. (2.22a) can be rewritten into Eq. (2.22b) when surface vacancy θ_V coverage is low. Reaction intermediate coverage A increases with increase of the adsorption equilibrium constant. According to Eq. (2.22b), R' is maximum when $\theta_A = (y(1 + x))^{-1}$. The maximum in reaction rate of Figure 2.28 is due to the competition of hydrogen and adsorbate

A for vacant surface sites. At the maximum the MARI changes from being dominated by adsorbed hydrogen atoms to being dominated by reactant hydrocarbon A. The reaction rate limiting step changes from hydrogen atom addition to dissociative adsorption of H_2. The similarity with Eq. (2.20) is that the maximum in reaction rate corresponds to a change in surface state and reaction rate-controlling step.

Eq. (2.21a) is a generally useful kinetic expression in catalysis. It is similar to the Michaelis-Menten kinetic expression of the rate of the enzyme-catalyzed reaction [144]. This expression also implicitly assumes that total number of reaction sites is a constant, reaction is monomolecular and product desorption is fast. In the Michaelis-Menten expression, different symbols from those in Eq. (2.21a) are used. K_{ads} can be identified with K_m^{-1}, where K_m is the Michaelis constant, and r_r is the equivalent of V_{max}, the maximum enzyme rate. Differences in adsorption free energies that determine K_m^{-1} relate to the match of reactant shape with enzyme cavity dimension. This match tends to dominate differences in reactivity. This is the lock and key principle proposed by Emil Fischer in 1895 [145], [146]. In Chapter 5, reactions that are catalyzed by zeolitic materials are discussed. These materials contain nanopores with dimensions comparable to organic molecules. Also for these systems, a match of molecular dimension and that of nanocavity determines adsorption coefficients. When physical adsorption is modest, reaction rate differences are determined by elementary reaction rates of reactant activation and change in adsorption energies. Reaction rate expressions are similar to Eq. (2.21a). However, when physical adsorption is strong, surface intermediates equilibrate and reaction rate is limited by product desorption. Then reaction rate expressions relate to Eq. (2.21b).

An important discovery in kinetics is the observation that there is a close connection between reaction order in the power law rate relation of Eq. (2.19) and the surface intermediate concentrations as defined in the Langmuir-Hinshelwood microkinetic expressions Eqs. (2.15) [5], [147], [148].

It relates to the important difference between apparent activation energy E_{app} of the catalytic reaction and intrinsic activation energy E_{int} of elementary reaction rate constants. Reaction rate has a non-Arrhenius temperature dependence because E_{app} and reaction order x depend on temperature.

For a monomolecular reaction, the apparent activation energy E_{app} can be calculated from the temperature derivative $d/d\left(\frac{1}{T}\right)$ of Eq. (2.21a). Reaction order x follows from the expression of surface coverage (see also [5]). This gives the reaction rate power law expression:

$$R' = r_{app}C_A^x = r_{app}C_A^{1-\theta_A} \tag{2.23a}$$

$$r_{app} = r_r K_{ads}^{1-\theta_A} \tag{2.23b}$$

$$E_{app} = E_r - E_{ads}\left(1-\theta_A\right) \tag{2.23c}$$

The reaction order $x = 1 - \theta_A$ is a fractional number less than 1 that depends on surface coverage. It illustrates the general feature of catalytic kinetics that reaction order relates with surface coverage. Reaction order becomes less than 1 when intermediate coverage increases. When surface coverage θ_A is small, the apparent reaction rate r_{app} is proportional to the adsorption equilibrium constant $r_{app} = r_r K_{ads}$ and $E_{app} = E_r - E_{ads}$. When coverage is low, the apparent activation energy of the catalytic reaction is equal to the intrinsic activation energy of the elementary surface reaction measured with respect to reactant in the gas phase. The apparent activation energy increases linearly with increase of θ_A when coverage increases. Reaction order x decreases with increase in surface intermediate concentration. As long as K_{des} is fast, the maximum value is $E_{app} = E_r$. Then reaction order is zero since $\theta = 1$. At high coverage, the activation energy of reaction is measured with respect to the adsorbed surface state. In Section 3.2.2.3, analogous relations are presented for the CO dissociation reaction.

The relation between reaction order x and θ of Eq. (2.23a) was discovered first by the Russian Mikhail Temkin. The equality for a monomolecular reaction order $x = 1 - \theta_A$ is called the Temkin relation [149]. With increase of temperature, the surface coverage decreases and hence the apparent activation energy E_{app} decreases. This leads to non-Arrhenius-type temperature dependence. With increasing temperature, the apparent activation energy remains positive as long as $E_{ads} < E_r$ and reaction rate increases. However, when $E_{ads} < E_r$ beyond the maximum temperature T_{max}, the apparent activation energy becomes negative. Then beyond T_{max}, the rate of reaction decreases with temperature. At the reaction rate maximum, $E_{app} = 0$ and $\theta_A = \frac{E_{ads} - E_r}{E_{ads}}$.

A maximum in reaction rate as a function of temperature is found for most heterogenous catalytic reactions. More than a hundred years ago this was reported by Langmuir for the CO oxidation reaction [26] and by Rideal [150] for the hydrogenation of ethene with Ni. When an increase in catalyst deactivation rate can be excluded, the origin is kinetic: at high temperatures the surface coverage becomes low because of increasing rate of desorption. Beyond the reaction rate maximum, the decrease in surface concentration of reaction intermediates is not compensated by the increase in elementary surface reaction rate constants.

In the CO oxidation reaction at high temperatures, the CO adsorption equilibrium limits reaction rate; in the case of ethene hydrogenation, at high temperatures the adsorption of hydrogen is limiting [151], [152]. For the hydrogenation reaction of benzene catalyzed by Pd, the temperature of maximum reaction rate is 480 K [139]. For the latter hydrogenation reaction, the adsorption energy of hydrogen is higher than the activation energy of the elementary reaction rate of hydrogenation. The temperature where reaction rate is maximum is called the zur Strassen temperature [152].

In summary, kinetics has been discussed where reaction rate varies due to change in reactant adsorption energy. When reaction rate is plotted as a function of adsorption energy, a bell-shaped dependence is found. It relates to a change in MARI. When the rate of desorption becomes too slow, reaction rate is inhibited by product desorption. In the case where two reactants compete for the same adsorption site, bell-shaped dependence of reaction rate relates to the takeover of one of the reactants as MARI.

When the catalytic reaction rate is plotted as a function of temperature, a maximum in reaction rate can also be found. This is due to the decrease in surface concentration of reaction intermediates at high temperatures. Beyond the reaction rate maximum, the apparent activation energy can become negative when the adsorption energy of the reactant is larger than the intrinsic activation energy of the surface elementary reaction.

In the next section, kinetic changes are discussed when the catalyst, but not the reactant, is altered. Then due to variation of catalyst, both the elementary surface and product desorption reaction rate constants change.

2.4.4 *Sabatier Principle Kinetic Equations*

The rate of a catalytic reaction is maximum at an optimum adsorption energy of reaction intermediates.

Here it is discussed how trends in catalytic reactivity relate to surface reactivity change. This changes the activation free energies of the elementary surface reaction rate constants. The BEP relations of Section 2.3.2.3 provide a relation between activation energies of elementary reaction rate constants and adsorption energies of surface fragments. Since adsorption energies are a probe of surface reactivity, the BEP relations connect activation energies with surface reactivity. BEP relations apply as long as molecular bond dissociation paths are similar. It means that the structure of the catalytic reaction center has to be the same. Within these constraints, kinetic equations can be formulated that predict catalytic rate as a function of surface reactivity. The solution of such kinetic equations shows Sabatier principle kinetic behavior.

According to the Sabatier principle, a maximum in catalytic rate exists that is due to the opposite dependencies of the respective activation energies of bond dissociation and adsorbate desorption energy on surface reactivity. In the 1960s, kinetic equations of the Sabatier principle based on the application of BEP relations were first presented by the Russian scientists Georgii Boreskov [99] and

Aleksei Balandin [153]. Initial understanding of the origin of volcano curve-type dependence in catalytic kinetics is due to Balandin. Their results are presented in the first part of this section. A next major evolution was microkinetic simulations of Sabatier volcano curves with first-principle elementary reaction rate constants [154]–[156]. Computational quantum chemistry made this possible in the first decade of this century. This will be demonstrated in a final paragraph for the ammonia synthesis reaction.

A heuristic kinetic argument can be given of the generality of the bell-shaped rate dependence on surface reactivity. This follows from a consideration of the microkinetic expressions for a reaction that has bond cleavage as the rate-controlling step:

$$R' = k\,\theta_A\theta_v\ (\theta_A+\theta_p+\theta_v = 1) \tag{2.24a}$$

$$= k\,\theta_A\,(1-\theta_A-\theta_p) \tag{2.24b}$$

$$\approx k\,\theta_A\,(1-3\theta_A)\ (\theta_A \approx 2\theta_A) \tag{2.24c}$$

The normalized reaction rate R' is proportional to the elementary reaction rate constant k and the product of the surface concentration of adsorbed reaction intermediate θ_A and vacant site concentration θ_v. Reaction intermediate A dissociates into two product fragments with surface occupation θ_p. This concentration increases with increase of surface reactivity.

Because total number of surface sites is conserved (the Langmuir condition), θ_A and θ_p are not independent. This gives Eq. (2.24b). When product desorption is slow, $\theta_p \approx 2\theta_A$. Substitution of this relation into Eq. (2.24a) gives Eq. (2.24c). Reaction rate R' has an inverse parabolic dependence of surface coverage θ_A and hence on surface reactivity. For Eq. (2.23), R' is maximum when $\theta_A = \frac{1}{6}$. Beyond this surface coverage, the rate of product desorption becomes reaction rate controlling. In the previous section, a related argument for a bell-shaped maximum in reaction rate versus coverage was given for a bimolecular surface reaction (Eq. (2.22)).

The elementary reaction rate constant k increases with increase of surface reactivity. This shifts the maximum in reaction rate of Eq. (2.24) to higher surface reactivity compared to predicted value based on surface composition change only.

Here, for a two-step model, the kinetic expression for the reaction rate maximum R'_{max} is derived as a function of a surface reactivity descriptor. Using the BEP relations, catalytic rate expressions are deduced that relate to energy change of adsorbed bond dissociation fragments. The reaction model is represented by Eq. (2.25):

$$\begin{aligned} & A_2 \underset{k_{des,A2}}{\overset{k_{ads,A2}}{\rightleftharpoons}} A_{2,ads} \\ & A_{2,ads} \xrightarrow{k_r} 2A_{ads} \\ & BA_{ads} \xrightarrow{r_{des,B}} B \end{aligned} \tag{2.25}$$

Reactant A_2 adsorbs and dissociates into adatoms A, which desorb as product B in a final reaction step. Adsorption of A_2 depends on adsorption equilibrium constant K_{ads,A_2}. The elementary reaction rate constant of dissociation of $A_{2,ads}$ is k_r, which is considered to be irreversible. Adsorbed adatoms A_{ads} desorb as product B. In the kinetic model, the rate constant of this reaction is represented by the lumped reaction rate constant $r_{des,B}$. Figures 2.31a and 2.31b illustrate the corresponding potential energies of the reaction, and Figure 2.31c gives a schematic presentation of the calculated bell-shaped dependence of reaction rate on the adsorption energy of adatoms A_{ads}.

Figures 2.31a and 2.31b illustrate the change in activation energy E_r of the reaction rate constant of bond dissociation when surface reactivity changes. The activation energy of bond dissociation changes with adsorption energy of the adatoms according to the BEP relation. The figures also illustrate the small dependence of molecular adsorption on surface reactivity compared to that of the adatom adsorbates. In the kinetic model, the change in molecular adsorption energy has been ignored. The change in activation energy of the bond dissociation activation is proportional to that of $E_{ads,2A}$: $\delta E_r = -\alpha_{BEP}\delta E_{ads,2A} = -\alpha_{BEP}\Delta$. A related expression holds for the change in product desorption energy. This is assumed equal to the adsorption energy change: $\delta E_{des,B} = \Delta$ (the desorption is considered to be non-activated). Note that the sign of the changes δE_r and $\delta E_{des,B}$ are opposite. In the legend of Figure 2.31, it is indicated that this leads to opposite exponential dependencies on Δ of respectively k_r

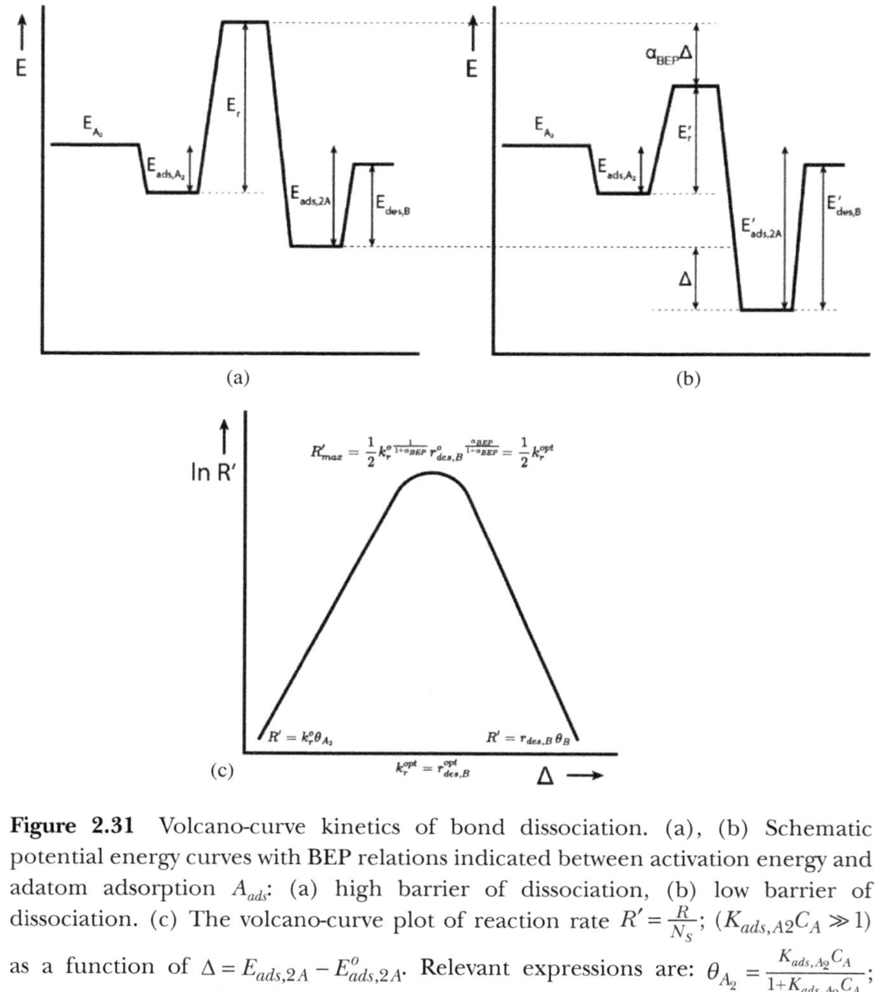

Figure 2.31 Volcano-curve kinetics of bond dissociation. (a), (b) Schematic potential energy curves with BEP relations indicated between activation energy and adatom adsorption A_{ads}: (a) high barrier of dissociation, (b) low barrier of dissociation. (c) The volcano-curve plot of reaction rate $R' = \frac{R}{N_S}$; $(K_{ads,A2}C_A \gg 1)$ as a function of $\Delta = E_{ads,2A} - E^o_{ads,2A}$. Relevant expressions are: $\theta_{A_2} = \frac{K_{ads,A_2}C_A}{1+K_{ads,A_2}C_A}$; $\left(r^o_{des,B} > k^o_r\right)$; $\theta_{A_2} = \frac{K_{ads,A_2}C_A}{1+K_{ads,A_2}C_A}$; $\left(r^o_{des,B} > k^o_r\right)$; $k_r = k^o_r e^{\alpha_{BEP}\frac{\Delta}{k_BT}}$; $r_{des,B} = r^o_{des,B} e^{\frac{-\Delta}{k_BT}}$; $\Delta^{opt} = -\frac{k_BT}{1+\alpha_{BEP}} ln \frac{k^o_r K_{ads}C_A}{r^o_{des,B}\left(1+K_{ads}C_A\right)}$.

and $k_{des,B}$. It is explained in Section 2.3.2.2 that the change in adsorption energy of the A_2 molecule can be assumed to be small compared to the adsorption energy change of the adatoms A and hence also compared to change in activation energy of the A_2 bond energy cleavage.

Figure 2.31c shows the volcano-type dependence of reaction rate R' as a function of Δ. The reaction rate maximum $R'_{max} = \frac{1}{2} k_r^{opt}$ occurs when $k_r^{opt} = r_{des,B}^{opt}$. At the left of R'_{max}, the reaction rate constant k_r is reaction rate controlling and the MARI is the concentration of adsorbed molecule A_2. At the right of R'_{max}, the reaction rate constant $r_{des,B}$ becomes rate controlling and the MARI is the reactant intermediate concentration B.

First-principle kinetics deduces trends in catalytic reaction rate as a function of surface reactivity with essentially the same procedure. For one catalyst, a complete quantum-chemical simulation of all elementary reaction rate constants is to be done. Then the elementary reaction rate constants of other surfaces can be deduced from calculations of ground state energies only by application of the BEP relation.

In Figure 2.32, first-principle simulated rates of the ammonia synthesis reaction are shown for catalysts with different composition but with the same stepped surface structure [157]. The optimal metal composition is predicted from the simulated optimum adatom bond energy.

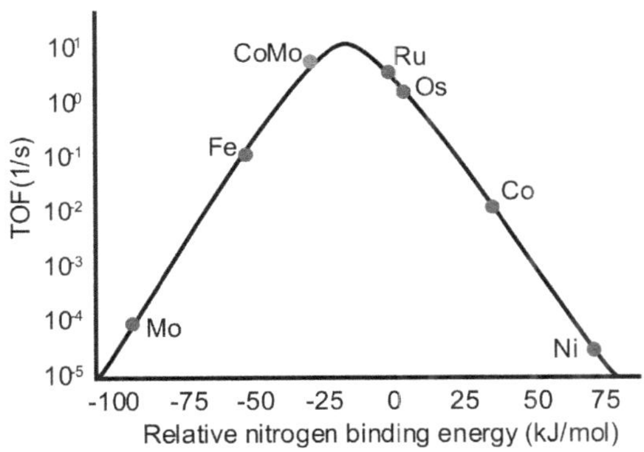

Figure 2.32 The normalized reaction rate of ammonia synthesis (TOF/s) as a function of the adsorption energy of a N atom. Adatom energy is selected equal to zero on Ru surface from which quantum chemical results derive. The synthesis conditions are 400°C, 50 bar, gas composition H_2:N_2 = 3:1 containing 5% NH_3. The numbers are obtained by combining a microkinetic model describing ammonia synthesis rates with the linear BEP relation existing between the potential energy and the activation energy for N_2 dissociation. Stepped surfaces are compared (see also Figure 2.16) [157].

The kinetic simulations on which the results of Figure 2.32 are based use kinetic equations where formally there is no change in rate-controlling step. This is implicit due to the dependence of adsorption equilibrium constants on elementary reaction rate constants of adsorption and desorption [158].

The rate-limiting step of the ammonia synthesis reaction is experimentally identified with the dissociative adsorption step of N_2 (see Section 3.2.1). This generates adsorbed N atoms. In subsequent steps, hydrogen atoms are added that give rise to adsorbed NH_x intermediates.

The reaction rate R can be approximated as:

$$R \approx k_{diss} K_{ads} P_{N_2} \left(1 - \theta_N\right)^2 \tag{2.26}$$

since the N_2 molecule requires two site positions in order to dissociate. K_{ads} is the adsorption constant of N_2 adsorption.

According to the BEP relation, with increasing interaction energy of adsorbed nitrogen adatoms, the elementary reaction rate constant k_{diss} increases and the rate of desorption decreases. The vacant surface concentration decreases due to the increase of θ_N. The opposing dependencies of $(1 - \theta_N)$ and k_{diss} give the maximum in reaction rate. Beyond this maximum, nitrogen removal is reaction rate limiting.

The simulated volcano curve of Figure 2.32 shows the reaction rate R as a function of nitrogen atom adsorption energies for different materials. The positions of respective catalyst materials are denoted. Figure 2.32 predicts that the most active catalyst material is located in between the CoMo alloy and Ru metal. It suggests that the conventional Fe-based ammonia catalyst can be substantially improved. Transition metal catalysts can be modified by comparing different metals, but also by use of catalyst promoters or alloys. The BEP method combined with microkinetics makes identification possible of suitable catalysts for reactor design and process conditions [159], [160].

The use of plots of reaction rate versus surface reactivity (adsorption energy of dissociated reaction intermediates) teaches us that whether for a particular system reaction rate can be increased by increase or decrease of catalyst site reactivity. This depends on whether experiment or simulation positions the catalytic system left or right of the volcano maximum.

One of the additional lessons of this section is that reaction mechanism and kinetics are closely intertwined. However, the same reaction mechanism can lead to very different kinetics. Catalyst performance depends strongly on the kinetic regime where the catalyst operates. Hence one cannot deduce the reaction mechanism solely on the basis of fit of modelled reaction kinetics with experiment. Additional information of the surface state and adsorbed reaction intermediates is required.

2.5 Summary

This chapter highlights the physical chemical concepts that are fundamental to heterogenous catalysis science. It provides the scientific background for the mechanistic Chapters 3–6.

The scientific evolution is sketched from its chemical thermodynamics foundation at the beginning of the previous century to catalytic kinetics based on a molecular science of chemical reactivity. It is a discovery process of increasingly better understanding of the relation of catalytic reactivity with catalyst structure and composition. It translates into the three physical chemical issues that are addressed in this chapter:

- what is the nature of the catalyst center
- what determines the reactivity of the catalytic surface
- what is the relation between catalytic kinetics and surface reactivity.

Chemical thermodynamics provided the means to determine reaction conditions and early context to develop tools for the determination of surface area and catalyst particle morphology. It also guided the formulation of the Langmuir-Hinshelwood kinetic equations that are still fundamental to catalytic kinetics. Transient kinetic isotope exchange experiments are useful to establish catalytic reaction site concentration. It was soon discovered that in practical catalytic systems, only a fraction of the surface is present as the catalytic reactive center. Catalytic studies of particle size dependence of transition metals that relate with variation in surface structure demonstrate that uniquely reactive surface sites exist. Their atomic structure is different from surface terraces that usually have low reactivity. The reactive sites are often an ensemble of surface atoms that have a step-edge structure. Such sites will only be present on larger particles. These are the preferred sites especially for dissociation of molecular π bonds.

C–H bonds have σ symmetry and are activated by contact with only a single surface atom. Small transition metal particles have high reactivity for reactions with C–H bond activation since they contain a high fraction of reactive edge atoms.

With the advance of spectroscopy that probes material and surfaces at the molecular level, and the development of chemical bonding theories of solid materials, a molecular theory of surface reactivity emerged. The chemisorption energy is an important probe of surface reactivity. Computational models of surfaces and catalytic sites that closely simulate experimental systems became available. This creates a wealth of information on the relation of surface structure and composition and the energetics of chemisorption. The coordination number of adatom or admolecular fragment adsorption to surface atoms, and the coordination number of surface atoms with surrounding surface or bulk atoms, became understood as descriptors of surface reactivity. Concepts such as surface atom coordinative unsaturation and adsorbate complex embedding were developed. The surface electronic structure that relates to these structural descriptors became unraveled. This developed into a tool for spectroscopies to characterize surfaces. It also created a theoretical basis of surface chemical reactivity. The adsorbate chemical bond is described by its polarity as well as covalent character. Polarity relates to electron affinity of the surface. Covalency relates to the distribution of electrons over bonding and anti-bonding orbitals.

The program to relate surface reactivity with catalytic reactivity made a significant step forward with the computational capability to also calculate transition states of surface chemical reactions. Elementary surface reaction rate constants could be studied in relation to the surface chemistry of the catalytic reactive site. The significant discovery was made that rate differences between catalysts are dominated by differences in the adsorption energies of surface fragments (or adatoms) from molecular bond dissociation rather than differences in molecular adsorption energies. Adsorption energies of adsorbed reaction intermediates may determine reaction temperature. The BEP relation is known since the 1930s. It is a linear relation between activation energy and reaction energy of elementary reactions. Due to computational access to the transition state, trends in surface reactivity can be quantitatively determined. Transition states of reaction intermediates that are part of elementary surface reaction steps have low mobility. The low activation barriers require strong interaction with the surface that causes a near frozen position of reactant bond atoms. Therefore the activation entropy of surface reaction steps tends to be low.

These developments made it possible to formulate first-principle kinetic simulations for the prediction of catalytic reaction rate as a function of catalyst site structure and composition. Quantum-chemical calculations provide elementary reaction rate constants. Langmuir-Hinshelwood microkinetic equations can be solved to predict kinetics. In the simulations, relevant temperatures and reaction conditions can be chosen. In addition to conversion and selectivity, the state of the surface is predicted, from which the reaction rate-controlling step is deduced. Structure-function relations between catalyst reactivity and material are derived.

Useful trends in catalytic reactivity can be observed by plotting simulated reaction rates as a function of calculated adsorption energies of dissociated reaction fragments. This leads to volcano-type dependencies that can be interpreted by the Sabatier principle. It makes it possible to identify the optimum catalytic material for a particular reaction. The contribution of computational simulations has been indispensable for modern theories of reaction mechanisms presented in the following chapters. Kinetic simulations can also assist identification of the proper reaction mechanism out of several options. Accuracy of quantum-chemical simulations is adequate to distinguish between different mechanisms.

Surface reaction mechanistic concepts introduced in this chapter are:

- Reaction occurs through adsorption and activation of reactant molecules by the catalyst surface. This generates adsorbed surface intermediates.
- Product formation occurs in a sequence of surface intermediate bond-breaking or bond formation steps and reaction intermediate recombination steps. In a final step, the product molecule desorbs.
- The relative stability of reaction intermediates relates not only strongly to the chemical composition of the reaction center, but also to its structure.
- Elementary rate constants as well as reaction conditions determine the relative concentration of surface intermediates. Often one intermediate dominates. This is the MARI.

References

[1] I. Langmuir, "The constitution and fundamental properties of solids and liquids. Part I. Solids," *J. Am. Chem. Soc.*, vol. 38, no. 11, pp. 2221–2295, Nov. 1916, doi: 10.1021/ja02268a002.

[2] K. J. Laidler, *The World of Physical Chemistry*. Oxford University Press, 1993.

[3] L. Pauling, *The Nature of the Chemical Bond: An Introduction to Modern Structural Chemistry*. Cornell University Press, 1960.

[4] H. S. Taylor, "A theory of the catalytic surface," *Proc. R. Soc. London. Ser. A*, vol. 108, no. 745, pp. 105–111, May 1925, doi: 10.1098/rspa.1925.0061.

[5] G. B. Marin, G. S. Yablonsky, and D. Constales, *Kinetics of Chemical Reactions: Decoding Complexity*, 2nd ed. John Wiley & Sons, 2019.

[6] I. Langmuir, "The mechanism of the catalytic action of platinum in the reactions 2CO + O2 = 2CO2 and 2H2 + O2 = 2H2O," *Trans. Faraday Soc.*, vol. 17, no. 0, pp. 621–654, Jan. 1922, doi: 10.1039/tf9221700621.

[7] C. N. Hinshelwood, *The Kinetics of Chemical Change.* Clarendon Press, 1926.

[8] M. Boudart, "Two-step catalytic reactions," *AIChE J.*, vol. 18, no. 3, pp. 465–478, May 1972, doi: 10.1002/AIC.690180303.

[9] K. Marshall Watson and O. A. Hougen, *Chemical Process Principles, Part Three: Kinetics and Catalysis.* Chapman and Hall, 1948.

[10] H. C. Urey, G. M. Murphy, and F. G. Brickwedde, "A name and symbol for H2," *J. Chem. Phys.*, vol. 1, no. 7, p. 512, Nov. 2004, doi: 10.1063/1.1749326.

[11] A. Farkas, L. Farkas, and E. K. Rideal, "Experiments on heavy hydrogen IV—The hydrogenation and exchange reaction of ethylene with heavy hydrogen," *Proc. R. Soc. London. Ser. A*, vol. 146, no. 858, pp. 630–639, Oct. 1934, doi: 10.1098/RSPA.1934.0177.

[12] J. M. Thomas and W. J. Thomas, *Principles and Practice of Heterogeneous Catalysis*, 2nd ed. John Wiley & Sons, 1997.

[13] J. W. Niemantsverdriet, *Spectroscopy in Catalysis: An Introduction*, 3rd ed. John Wiley & Sons, 1993.

[14] G. A. Somorjai, *Chemistry in Two Dimensions: Surfaces.* Cornell University Press, 1981.

[15] G. Ertl, *Reactions at Solid Surfaces.* John Wiley & Sons, 2009.

[16] J. C. H. Spence, *High-Resolution Electron Microscopy*, 4th ed. Oxford University Press, 2013.

[17] C. J. Chen, *Introduction to Scanning Tunneling Spectroscopy*, 2nd ed. Oxford University Press, 2008.

[18] G. A. Somorjai, "Surface reconstruction and catalysis," *Annu. Rev. Phys. Chem.*, vol. 45, no. 1, pp. 721–751, Nov. 1994, doi: 10.1146/annurev.pc.45.100194.003445.

[19] J. K. Nørskov, F. Studt, F. Abild-Pedersen, and T. Bligaard, *Fundamental Concepts in Heterogeneous Catalysis.* John Wiley & Sons, 2014.

[20] D. G. Truhlar, B. C. Garrett, and S. J. Klippenstein, "Current status of transition-state theory," *J. Phys. Chem.*, vol. 100, no. 31, pp. 12771–12800, Aug. 1996, doi: 10.1021/JP953748Q.

[21] K. De Wispelaere, L. Vanduyfhuys, and V. Van Speybroeck, "Entropy Contributions to Transition State Modeling," in *Modelling and Simulation in the Science of Micro- and Meso-Porous Materials*, C. R. A. Catlow, V. van Speybroeck, and R. A. van Santen, Eds. Elsevier, 2018, pp. 189–228.

[22] H. Eyring, "The activated complex in chemical reactions," *J. Chem. Phys.*, vol. 3, no. 2, pp. 107–115, Feb. 1935, doi: 10.1063/1.1749604.

[23] M. G. Evans and M. Polanyi, "Some applications of the transition state method to the calculation of reaction velocities, especially in

solution," *Trans. Faraday Soc.*, vol. 31, no. 0, pp. 875–894, Jan. 1935, doi: 10.1039/TF9353100875.

[24] A. Dumesic, D. F. Rudd, L. M. Aparicio, J. E. Rekoske, and A. A. Trevino, *The Microkinetics of Heterogeneous Catalysis.* American Chemical Society, 1993.

[25] R. A. van Santen and C. Liu, "Theory of Zeolite Catalysis: An Introductory Account," in *Modelling and Simulation in the Science of Micro- and Meso-Porous Materials*, C. R. A. Catlow, V. van Speybroeck, and R. A. van Santen, Eds. Elsevier, 2018, pp. 151–188.

[26] I. Langmuir, "The adsorption of gases on plane surfaces of glass, mica and platinum.," *J. Am. Chem. Soc.*, vol. 40, no. 9, pp. 1361–1403, Sep. 1918, doi: 10.1021/ja02242a004.

[27] W. H. Bragg, "X-rays and crystals," *Nature*, vol. 90, no. 2256, pp. 572–572, 1913, doi: 10.1038/090572c0.

[28] K. Ogasawara, F. Alluqmani, H. Nagoshi, U. Khimii, and A. A. Balandin, "The multiplet theory of catalysis. Structural factors in catalysis," *Russ. Chem. Rev.*, vol. 31, no. 11, p. 589, Nov. 1962, doi: 10.1070/RC1962V031N11ABEH001323.

[29] O. Beeck, "A challenge to the physicist (as exemplified by the hydrogenation of ethylene over metal catalysts)," *Rev. Mod. Phys.*, vol. 17, no. 1, p. 61, Jan. 1945, doi: 10.1103/RevModPhys.17.61.

[30] G. A. Somorjal and J. Carrazza, "Structure sensitivity of catalytic reactions," *Ind. Eng. Chem. Fundam.*, vol. 25, no. 1, pp. 63–69, 2002, doi: 10.1021/I100021A009.

[31] J. K. Nørskov, T. Bligaard, B. Hvolbaek, F. Abild-Pedersen, I. Chorkendorff, and C. H. Christensen, "The nature of the active site in heterogeneous metal catalysis," *Chem. Soc. Rev.*, vol. 37, pp. 2163–2171, 2008, doi: 10.1039/b800260f.

[32] K. Honkala, A. Hellman, I. N. Remediakis, A. Logadottir, A. Carlsson, S. Dahl, C. H. Christensen, and J. K. Nørskov, "Ammonia synthesis from first-principles calculations.," *Science*, vol. 307, no. 5709, pp. 555–558, Jan. 2005, doi: 10.1126/science.1106435.

[33] D. S. Su, B. Zhang, and R. Schlögl, "Electron microscopy of solid catalysts - Transforming from a challenge to a toolbox," *Chem. Rev.*, vol. 115, no. 8, pp. 2818–2882, Apr. 2015, doi: 10.1021/cr500084c.

[34] S. Brunauer, P. H. Emmett, and E. Teller, "Adsorption of gases in multimolecular layers," *J. Am. Chem. Soc.*, vol. 60, no. 2, pp. 309–319, Feb. 1938, doi: 10.1021/ja01269a023.

[35] R. von Helmholtz, "Untersuchungen über Dämpfe und Nebel, besonders über solche von Lösungen," *Ann. Phys.*, vol. 263, no. 4, pp. 508–543, Jan. 1886, doi: 10.1002/ANDP.18862630403.

[36] J. H. de Boer and B. C. Lippens, "Studies on pore systems in catalysts II. The shapes of pores in aluminum oxide systems," *J. Catal.*, vol. 3, no. 1, pp. 38–43, Feb. 1964, doi: 10.1016/0021-9517(64)90090-9.

[37] J. H. de Boer, B. C. Lippens, B. G. Linsen, J. C. P. Broekhoff, A. van den Heuvel, and T. J. Osinga, "Thet-curve of multimolecular N2-adsorption," *J. Colloid Interface Sci.*, vol. 21, no. 4, pp. 405–414, Apr. 1966, doi: 10.1016/0095-8522(66)90006-7.

[38] E. K. Rideal, "Concepts in catalysis. The contributions of Paul Sabatier and of Max Bodenstein," *J. Chem. Soc.*, no. 0, pp. 1640–1647, Jan. 1951, doi: 10.1039/JR9510001640.

[39] M. A. Salam, E. Nassef, E. Elkheriany, and Y. El Tawel, "Modelling and simulation of gauze reactor of ammonia oxidation," *Am. J. Chem. Eng.*, vol. 4, no. 1, p. 16, 2016, doi: 10.11648/J.AJCHE.20160401.13.

[40] C. N. Satterfield, *Mass Transfer in Heterogeneous Catalysis.* M.I.T. Press, 1969.

[41] E. W. Thiele, "Relation between catalytic activity and size of particle," *Ind. Eng. Chem.*, vol. 31, no. 7, pp. 916–920, 1939, doi: 10.1021/ie50355a027.

[42] M. Boudart and G. Djega-Mariadassou, *Kinetics of Heterogeneous Catalytic Reactions*, Course Book. Princeton University Press, 1984.

[43] M. Faraday, "V. Experimental researches in electricity. — Sixth series," *Philos. Trans. R. Soc. London*, vol. 124, pp. 55–76, Dec. 1834, doi: 10.1098/RSTL.1834.0007.

[44] P. Biloen, "Transient kinetic methods," *J. Mol. Catal.*, vol. 21, no. 1–3, pp. 17–24, Oct. 1983, doi: 10.1016/0304-5102(93)80108-7.

[45] C. Mirodatos, "Use of isotopic transient kinetics in heterogeneous catalysis," *Catal. Today*, vol. 9, no. 1–2, pp. 83–95, Mar. 1991, doi: 10.1016/0920-5861(91)85011-V.

[46] S. L. Shannon and J. G. Goodwin, "Characterization of catalytic surfaces by isotopic-transient kinetics during steady-state reaction," *Chem. Rev.*, vol. 95, no. 3, pp. 677–695, 1995, doi: 10.1021/cr00035a011.

[47] C. Mirodatos, "Benzene hydrogenation: An isotopic transient study," *J. Phys. Chem.*, vol. 90, no. 3, pp. 481–487, 1986, doi: 10.1021/j100275a026.

[48] C. Mirodatos, J. A. Dalmon, and G. A. Martin, "Steady-state and isotopic transient kinetics of benzene hydrogenation on nickel catalysts," *J. Catal.*, vol. 105, no. 2, pp. 405–415, Jun. 1987, doi: 10.1016/0021-9517(87)90069-8.

[49] C. Ledesma, J. Yang, D. Chen, and A. Holmen, "Recent approaches in mechanistic and kinetic studies of catalytic reactions using

SSITKA technique," *ACS Catal.*, vol. 4, no. 12, pp. 4527–4547, Dec. 2014, doi: 10.1021/CS501264F.
[50] E. L. Muetterties, "A coordination chemist's view of surface science," *Angew. Chem. Int. Ed.*, vol. 17, no. 8, pp. 545–558, Aug. 1978, doi: 10.1002/ANIE.197805453.
[51] J. Robert, M. Gavin, J. Reutt, and E. L. Muetterties, "Metal-hydrogen bridge bonding of hydrocarbons on metal surfaces," *Proc. Natl. Acad. Sci.*, vol. 78, no. 7, pp. 3981–3985, Jul. 1981, doi: 10.1073/PNAS.78.7.3981.
[52] D. D. Eley, "Catalysis by palladium gold and related alloys," *J. Res. st. Catal. Hokkaido Univ*, vol. 16, no. 1, pp. 101–114, 1968.
[53] T. Wolkenstein, "The electron theory of catalysis on semiconductors," *Adv. Catal.*, vol. 12, no. C, pp. 189–264, Jan. 1960, doi: 10.1016/S0360-0564(08)60603-3.
[54] K. Hauffe, "Über den Mechanismus von Gasreaktionen an Oberflächen halbleitender Katalysatoren," *Adv. Catal.*, vol. 9, no. C, pp. 187–203, Jan. 1957, doi: 10.1016/S0360-0564(08)60167-4.
[55] J. Koutecký, "A contribution to the molecular-orbital theory of chemisorption," *Trans. Faraday Soc.*, vol. 54, no. 0, pp. 1038–1052, Jan. 1958, doi: 10.1039/TF9585401038.
[56] D. M. Newns, "Self-consistent model of hydrogen chemisorption," *Phys. Rev.*, vol. 178, no. 3, p. 1135, Feb. 1969, doi: 10.1103/PhysRev.178.1123.
[57] T. B. Grimley and C. Pisani, "Chemisorption theory in the Hartree-Fock approximation," *J. Phys. C Solid State Phys.*, vol. 7, no. 16, p. 2831, Aug. 1974, doi: 10.1088/0022-3719/7/16/014.
[58] R. A. van Santen and M. Neurock, "Theory of surface chemical reactivity," *Russ. J. Phys. Chem. B*, vol. 1, no. 4, pp. 261–291, Aug. 2007, doi: 10.1134/S1990793107040021.
[59] R. A. van Santen, *Theoretical Heterogeneous Catalysis.* World Scientific, 1991.
[60] A. Einstein, "Über einen die Erzeugung und Verwandlung des Lichtes betreffenden heuristischen Gesichtspunkt," *Ann. Phys.*, vol. 322, no. 6, pp. 132–148, Jan. 1905, doi: 10.1002/ANDP.19053220607.
[61] W. M. H. Sachtler and G. J. H. Dorgelo, "The surface of copper-nickel alloy films: I. Work function and phase composition," *J. Catal.*, vol. 4, no. 6, pp. 654–664, Dec. 1965, doi: 10.1016/0021-9517(65)90265-4.
[62] W. M. H. Sachtler and P. Van Der Plank, "The role of individual surface atoms in chemisorption and catalysis by nickel-copper alloys," *Surf. Sci.*, vol. 18, no. 1, pp. 62–79, Nov. 1969, doi: 10.1016/0039-6028(69)90267-2.

[63] W. M. H. Sachtler, "Surface composition of alloys," *Appl. Surf. Sci.*, vol. 19, no. 1–4, pp. 167–180, Nov. 1984, doi: 10.1016/0378-5963(84)90060-6.

[64] K. I. Tanaka and K. Tamaru, "A general rule in chemisorption of gases on metals," *J. Catal.*, vol. 2, no. 5, pp. 366–370, Oct. 1963, doi: 10.1016/0021-9517(63)90101-5.

[65] Y. L. Gabor, A. Somorjai, *Introduction to Surface Chemistry and Catalysis,* 2nd ed. John Wiley & Sons, 2010.

[66] R. I. Masel, *Principles of Adsorption and Reaction on Solid Surfaces.* John Wiley & Sons, 1996.

[67] F. Bozso, G. Ertl, M. Grunze, and M. Weiss, "Interaction of nitrogen with iron surfaces. I. Fe(100) and Fe(111)," *J. Catal.*, vol. 49, no. 1, pp. 18–41, Jul. 1977, doi: 10.1016/0021-9517(77)90237-8.

[68] F. Bozso, G. Ertl, and M. Weiss, "Interaction of nitrogen with iron surfaces: II. Fe(110)," *J. Catal.*, vol. 50, no. 3, pp. 519–529, Dec. 1977, doi: 10.1016/0021-9517(77)90063-X.

[69] R. A. van Santen, "Complementary structure sensitive and insensitive catalytic relationships.," *Acc. Chem. Res.*, vol. 42, no. 1, pp. 57–66, Jan. 2009, doi: 10.1021/ar800022m.

[70] Q. Ge and M. Neurock, "Structure dependence of NO adsorption and dissociation on platinum surfaces," *J. Am. Chem. Soc.*, vol. 126, no. 5, pp. 1551–1559, Feb. 2004, doi: 10.1021/JA036575O.

[71] Á. Logadóttir and J. K. Nørskov, "Ammonia synthesis over a Ru(0001) surface studied by density functional calculations," *J. Catal.*, vol. 220, no. 2, pp. 273–279, Dec. 2003, doi: 10.1016/S0021-9517(03)00156-8.

[72] T. R. Munter, T. Bligaard, C. H. Christensen, and J. K. Nørskov, "BEP relations for N2 dissociation over stepped transition metal and alloy surfaces," *Phys. Chem. Chem. Phys.*, vol. 10, no. 34, pp. 5202–5206, Aug. 2008, doi: 10.1039/B720021H.

[73] R. A. van Santen and I. Tranca, "How molecular is the chemisorptive bond?," *Phys. Chem. Chem. Phys.*, vol. 18, no. 31, pp. 20868–20894, Aug. 2016, doi: 10.1039/C6CP01394E.

[74] I. M. Ciobica and R. A. van Santen, "A DFT study of CHx chemisorption and transition states for C–H activation on the Ru($11\bar{2}0$) surface," *J. Phys. Chem. B*, vol. 106, no. 24, pp. 6200–6205, Jun. 2002, doi: 10.1021/jp013210m.

[75] I. M. Ciobîcă, F. Frechard, R. A. van Santen, A. W. Kleyn, and J. Hafner, "A DFT study of transition states for C–H activation on the Ru(0001) surface," *J. Phys. Chem. B*, vol. 104, no. 14, pp. 3364–3369, Apr. 2000, doi: 10.1021/jp993314l.

[76] E. Shustorovich, "Chemisorption phenomena: Analytic modeling based on perturbation theory and bond-order conservation,"

Surf. Sci. Rep., vol. 6, no. 1, pp. 1–63, Jul. 1986, doi: 10.1016/0167-5729(86)90003-8.

[77] F. Abild-Pedersen, J. Greeley, F. Studt, J. Rossmeisl, T. R. Munter, P. G. Moses, E. Skúlason, T. Bligaard, and J. K. Nørskov, "Scaling properties of adsorption energies for hydrogen-containing molecules on transition-metal surfaces," *Phys. Rev. Lett.*, vol. 99, no. 1, pp. 16105–16109, Jul. 2007, doi: 10.1103/PHYSREVLETT.99.016105.

[78] F. Calle-Vallejo, N. G. Inoglu, H.-Y. Su, J. I. Martínez, I. C. Man, M. T. M. M. Koper, J. R. Kitchin, and J. Rossmeisl, "Number of outer electrons as descriptor for adsorption processes on transition metals and their oxides," *Chem. Sci.*, vol. 4, no. 3, p. 1245, Feb. 2013, doi: 10.1039/C2SC21601A.

[79] W. Gao, Y. Chen, B. Li, S. P. Liu, X. Liu, and Q. Jiang, "Determining the adsorption energies of small molecules with the intrinsic properties of adsorbates and substrates," *Nat. Commun.*, vol. 11, no. 1, pp. 1–11, Mar. 2020, doi: 10.1038/s41467-020-14969-8.

[80] B. Hammer and J. K. Nørskov, "Theoretical surface science and catalysis—calculations and concepts," *Adv. Catal.*, vol. 45, no. C, pp. 71–129, 2000, doi: 10.1016/S0360-0564(02)45013-4.

[81] J. Chatt and L. A. Duncanson, "586. Olefin co-ordination compounds. Part III. Infra-red spectra and structure: attempted preparation of acetylene complexes," *J. Chem. Soc.*, no. 0, pp. 2939–2947, Jan. 1953, doi: 10.1039/JR9530002939.

[82] C. Wentrup, "Zeise, Liebig, Jensen, Hückel, Dewar, and the Olefin π-Complex Bonds," *Angew. Chem. Int. Ed.*, vol. 59, no. 22, pp. 8332–8342, May 2019, doi: 10.1002/ANIE.201913840.

[83] K. J. Laidler and M. C. King, "Development of transition-state theory," *J. Phys. Chem.*, vol. 87, no. 15, pp. 2657–2664, 1983, doi: 10.1021/J100238A002.

[84] H. Eyring and M. Polanyi, "On simple gas reactions," *Zeitschrift fur Phys. Chemie,* vol. 227, no. 11, pp. 1221–1245, Nov. 2013, doi: 10.1524/ZPCH.2013.9023.

[85] R. A. van Santen and J. W. Niemantsverdriet, *Chemical Kinetics and Catalysis.* Plenum Press, 1995.

[86] H. A. Kramers, "Brownian motion in a field of force and the diffusion model of chemical reactions," *Physica,* vol. 7, no. 4, pp. 284–304, Apr. 1940, doi: 10.1016/S0031-8914(40)90098-2.

[87] O. K. Rice, *Statistical Mechanics, Thermodynamics and Kinetics.* W.H. Freeman and Company, 1967.

[88] T. Bučko, L. Benco, J. Hafner, and J. G. Ángyán, "Monomolecular cracking of propane over acidic chabazite: An ab initio molecular

dynamics and transition path sampling study," *J. Catal.*, vol. 279, no. 1, pp. 220–228, Apr. 2011, doi: 10.1016/J.JCAT.2011.01.022.

[89] P. M. Zimmerman, D. C. Tranca, J. Gomes, D. S. Lambrecht, M. Head-Gordon, and A. T. Bell, "Ab initio simulations reveal that reaction dynamics strongly affect product selectivity for the cracking of alkanes over H-MFI," *J. Am. Chem. Soc.*, vol. 134, no. 47, pp. 19468–19476, Nov. 2012, doi: 10.1021/JA3089372.

[90] S. Dahl, A. Logadottir, C. J. H. Jacobsen, and J. K. Nørskov, "Electronic factors in catalysis: the volcano curve and the effect of promotion in catalytic ammonia synthesis," *Appl. Catal. A Gen.*, vol. 222, no. 1–2, pp. 19–29, Dec. 2001, doi: 10.1016/S0926-860X(01)00826-2.

[91] V. Pallassana and M. Neurock, "Electronic factors governing ethylene hydrogenation and dehydrogenation activity of pseudomorphic PdML/Re(0001), PdML/Ru(0001), Pd(111), and PdML/Au(111) surfaces," *J. Catal.*, vol. 191, no. 2, pp. 301–317, Apr. 2000, doi: 10.1006/JCAT.1999.2724.

[92] Z. P. Liu and P. Hu, "General trends in CO dissociation on transition metal surfaces," *J. Chem. Phys.*, vol. 114, no. 19, p. 8244, May 2001, doi: 10.1063/1.1372512.

[93] A. Michaelides, Z.-P. Liu, C. J. Zhang, A. Alavi, D. A. King, and P. Hu, "Identification of general linear relationships between activation energies and enthalpy changes for dissociation reactions at surfaces.," *J. Am. Chem. Soc.*, vol. 125, no. 13, pp. 3704–3705, Apr. 2003, doi: 10.1021/ja027366r.

[94] J. N. Brönsted, "Acid and basic catalysis," *Chem. Rev.*, vol. 5, no. 3, pp. 231–338, Oct. 1928, doi: 10.1021/CR60019A001.

[95] M. G. Evans and M. Polanyi, "Further considerations on the thermodynamics of chemical equilibria and reaction rates," *Trans. Faraday Soc.*, vol. 32, no. 0, pp. 1333–1360, Jan. 1936, doi: 10.1039/TF9363201333.

[96] J. Shorter, *Correlation Analysis in Organic Chemistry: An Introduction to Linear Free-Energy Relationships.* Clarendon Press, 1973.

[97] L. P. Hammett, *Physical Organic Chemistry.* John Wiley & Sons, 1940.

[98] D. Pletcher and F. Walsh, *Industrial Electrochemistry.* Chapman and Hall, 1990.

[99] G. K. Boreskov, *Heterogenous Catalysis.* Boreskov Instutute of Catalysis, 1988.

[100] J. K. Nørskov, T. Bligaard, A. Logadottir, S. Bahn, L. B. Hansen, M. Bollinger, H. Bengaard, B. Hammer, Z. Sljivancanin, M. Mavrikakis, Y. Xu, S. Dahl, and C. J. H. Jacobsen, "Universality in

heterogeneous catalysis," *J. Catal.*, vol. 209, no. 2, pp. 275–278, Jul. 2002, doi: 10.1006/JCAT.2002.3615.

[101] R. A. van Santen, M. Neurock, and S. G. Shetty, "Reactivity theory of transition-metal surfaces: A Brønsted-Evans-Polanyi linear activation energy-free-energy analysis," *Chem. Rev.*, vol. 110, no. 4, pp. 2005–2048, Apr. 2010, doi: 10.1021/CR9001808.

[102] R. A. Marcus, "Chemical and electrochemical electron-transfer theory," *Annu. Rev. Phys. Chem.*, vol. 15, no. 1, pp. 155–196, Nov. 1964, doi: 10.1146/ANNUREV.PC.15.100164.001103.

[103] R. P. Bell and O. M. Lidwell, "Potential energy curves in proton transfer reactions," *Proc. R. Soc. Lond. A. Math. Phys. Sci.*, vol. 176, no. 964, pp. 114–121, 1940.

[104] M. A. Vannice, "The catalytic synthesis of hydrocarbons from H2CO mixtures over the Group VIII metals: V. The catalytic behavior of silica-supported metals," *J. Catal.*, vol. 50, no. 2, pp. 228–236, Nov. 1977, doi: 10.1016/0021-9517(77)90031-8.

[105] G. Jones, J. Jakobsen, S. Shim, J. Kleis, M. Anderson, J. Rossmeisl, F. Abild-Pedersen, T. Ligaard, S. Helveg, and B. Hinnemann, "First principles calculations and experimental insight into methane steam reforming over transition metal catalysts," *J. Catal.*, vol. 259, no. 1, pp. 147–160, 2008, doi: 10.1016/j.jcat.2008.08.003.

[106] Y. H. Chin, C. Buda, M. Neurock, and E. Iglesia, "Selectivity of chemisorbed oxygen in C–H bond activation and CO oxidation and kinetic consequences for CH4–O2 catalysis on Pt and Rh clusters," *J. Catal.*, vol. 283, no. 1, pp. 10–24, Oct. 2011, doi: 10.1016/J.JCAT.2011.06.011.

[107] D. A. Hickman and L. D. Schmidt, "Synthesis gas formation by direct oxidation of methane over Pt monoliths," *J. Catal.*, vol. 138, no. 1, pp. 267–282, Nov. 1992, doi: 10.1016/0021-9517(92)90022-A.

[108] R. Hoffmann, *Solids and Surfaces : A Chemist's View of Bonding in Extended Structures.* John Wiley & Sons, 1988.

[109] L. Pauling, "The nature of the interatomic forces in metals," *Phys. Rev.*, vol. 54, no. 11, p. 899, Dec. 1938, doi: 10.1103/PhysRev.54.899.

[110] L. Pauling and F. J. Ewing, "The ratio of valence electrons to atoms in metals and intermetallic compounds," *Rev. Mod. Phys.*, vol. 20, no. 1, p. 112, Jan. 1948, doi: 10.1103/RevModPhys.20.112.

[111] W. Hume-Rothery, H. M. Irving, and R. J. P. Williams, "The valencies of the transition elements in the metallic state," *Proc. R. Soc. Lond. Ser. A.*, vol. 208, no. 1095, pp. 431–443, Sep. 1951, doi: 10.1098/RSPA.1951.0172.

[112] A. A. Balandin, "The multiplet theory of catalysis - energy factors in catalysis," *Russ. Chem. Rev.*, vol. 33, no. 5, p. 275, May 1964, doi: 10.1070/RC1964V033N05ABEH001407.

[113] I. Tranca and R. A. van Santen, "Basic Quantum-Chemical Concepts, The Chemical Bond Revisited," in *Modern Heterogeneous Catalysis: An Introduction*, R. A. van Santen, Ed. John Wiley & Sons, 2017, pp. 173–207.

[114] R. A. van Santen, "Chemical Bonding and Reactivity of Transition Metal Surfaces," in *Modern Heterogeneous Catalysis: An Introduction*, R. A. van Santen, Ed. John Wiley & Sons, 2017, pp. 209–291.

[115] C. M. Varma and A. J. Wilson, "Systematics of the binding energy of oxygen and hydrogen on transition-metal surfaces. I.," *Phys. Rev. B*, vol. 22, no. 8, p. 3795, Oct. 1980, doi: 10.1103/PhysRevB.22.3795.

[116] A. Michaelides and M. Scheffler, "An introduction to the theory of metal surfaces," pp. 1–40, 2010, Accessed: Oct. 28, 2021. [Online]. Available: http://th.fhi-berlin.mpg.de/th/publications/Michaelides_Scheffler_Chapter_Submission.pdf.

[117] B. S. Bunnik, G. J. Kramer, and R. A. van Santen, "Symmetric transition state analysis: An analysis of dissociative methane adsorption on Rh{111} using quantum chemical calculations," *Top. Catal.*, vol. 53, no. 5–6, pp. 403–416, May 2010, doi: 10.1007/S11244-010-9453-4.

[118] G. Blyholder, "Molecular orbital view of chemisorbed carbon monoxide," *J. Phys. Chem.*, vol. 68, no. 10, pp. 2772–2777, 1964, doi: 10.1021/j100792a006.

[119] R. A. van Santen, *Modern Heterogeneous Catalysis: An Introduction*. John Wiley & Sons, 2017.

[120] A. De Koster, A. P. J. Jansen, R. A. van Santen, and J. J. C. Geerlings, "Reactivity of CO on stepped and non-stepped surfaces of transition metals," *Faraday Discuss. Chem. Soc.*, vol. 87, no. 0, pp. 263–273, Jan. 1989, doi: 10.1039/DC9898700263.

[121] Y. Soma-Noto and W. M. H. Sachtler, "Infrared spectra of carbon monoxide adsorbed on supported palladium and palladium-silver alloys," *J. Catal.*, vol. 32, no. 2, pp. 315–324, Feb. 1974, doi: 10.1016/0021-9517(74)90081-5.

[122] C. Y. Fan, H. P. Bonzel, and K. Jacobi, "CO adsorption on the multiple-site Ru($11\bar{2}1$) surface: The role of bonding competition," *J. Chem. Phys.*, vol. 118, no. 21, p. 9773, May 2003, doi: 10.1063/1.1570392.

[123] S. Shetty and R. A. van Santen, "Site structure sensitivity differences for dissociation of diatomic molecules," *Top. Catal.*, vol. 53, no. 13–14, pp. 969–975, Aug. 2010, doi: 10.1007/S11244-010-9516-6.

[124] S. Shetty, A. P. J. Jansen, and R. A. van Santen, "CO dissociation on the Ru(1121) surface," *J. Phys. Chem. C*, vol. 112, no. 36, pp. 14027–14033, Sep. 2008, doi: 10.1021/JP801754N.

[125] S. Shetty, A. P. J. J. Jansen, and R. A. van Santen, "Direct versus hydrogen-assisted CO dissociation," *J. Am. Chem. Soc.*, vol. 131, no. 36, pp. 12874–12875, Sep. 2009, doi: 10.1021/ja9044482.

[126] G. Kour, X. Mao, and A. Du, "Computational screening of single-atom alloys TM@Ru(0001) for enhanced electrochemical nitrogen reduction reaction," *J. Mater. Chem. A*, vol. 10, no. 11, pp. 6204–6215, Mar. 2022, doi: 10.1039/D1TA08246A.

[127] M. M. Montemore, M. A. Van Spronsen, R. J. Madix, and C. M. Friend, "O2 Activation by metal surfaces: implications for bonding and reactivity on heterogeneous catalysts," *Chem. Rev.*, vol. 118, no. 5, pp. 2816–2862, Mar. 2018, doi: 10.1021/ACS.CHEMREV.7B00217.

[128] W. J. M. Rootsaert and W. M. H. Sachtler, "Interaction of formic acid vapour with tungsten," *Zeitschrift fur Phys. Chemie*, vol. 26, no. 1–2, pp. 16–26, Oct. 1960, doi: 10.1524/ZPCH.1960.26.1_2.016.

[129] M. I. Temkin, "Relaxation of the rate of two-step catalytic reaction," *Kinet. Catal.*, vol. 17, p. 1095–1099, 1976.

[130] O. A. Hougen and K. M. Watson, *Chemical Process Principles Part One: Material and Energy Balances.* John Wiley & Sons, 1947.

[131] I. Prigogine, P. Outer, and C. Herbo, "Affinity and reaction rate close to equilibrium," *J. Phys. Colloid Chem.*, vol. 52, no. 2, pp. 321–331, 2002, doi: 10.1021/J150458A004.

[132] T. de Donder, *L'affinité.* Gauthier-Villars, 1927.

[133] M. Boudart, *Kinetics of Chemical Processes.* Elsevier, 1991.

[134] J. A. Dumesic, G. W. Huber, and M. Boudart, "Principles of Heterogeneous Catalysis," in *Handbook of Heterogeneous Catalysis.* John Wiley & Sons, 2008, doi: 10.1002/9783527610044.HETCAT0001.

[135] D. D. Eley and E. K. Rideal, "Parahydrogen conversion on tungsten," *Nature*, vol. 146, no. 3699, pp. 401–402, 1940, doi: 10.1038/146401d0.

[136] D. D. Eley, "The interchange of hydrogen in the adsorbed film on tungsten," *Proc. R. Soc. Lond. Ser. A. Math. Phys. Sci.*, vol. 178, no. 975, pp. 452–464, Aug. 1941, doi: 10.1098/RSPA.1941.0067.

[137] R. Prins, "Eley–Rideal, the other mechanism," *Top. Catal.*, vol. 61, no. 9–11, pp. 714–721, Jun. 2018, doi: 10.1007/S11244-018-0948-8.

[138] Z. Karpiński, "Catalysis by supported, unsupported, and electron-deficient palladium," *Adv. Catal.*, vol. 37, no. C, pp. 45–100, Jan. 1990, doi: 10.1016/S0360-0564(08)60363-6.

[139] P. Chou and M. A. Vannice, "Benzene hydrogenation over supported and unsupported palladium: I. Kinetic behavior," *J. Catal.*, vol. 107, no. 1, pp. 129–139, Sep. 1987, doi: 10.1016/0021-9517(87)90278-8.

[140] X. Zhao, Y. Chang, W. J. Chen, Q. Wu, X. Pan, K. Chen, and B. Weng, "Recent progress in Pd-based nanocatalysts for selective hydrogenation," *ACS Omega*, vol. 7, no. 1, pp. 17–31, Jan. 2022, doi: 10.1021/ACSOMEGA.1C06244.

[141] D. Teschner, J. Borsodi, Z. Kis, L. Szentmiklósi, Z. Révay, A. Knop-Gericke, R. Schlögl, D. Torres, and P. Sautet, "Role of hydrogen species in palladium-catalyzed alkyne hydrogenation," *J. Phys. Chem. C*, vol. 114, no. 5, pp. 2293–2299, Feb. 2010, doi: 10.1021/JP9103799.

[142] D. Mei, P. A. Sheth, M. Neurock, and C. M. Smith, "First-principles-based kinetic Monte Carlo simulation of the selective hydrogenation of acetylene over Pd(111)," *J. Catal.*, vol. 242, no. 1, pp. 1–15, Aug. 2006, doi: 10.1016/J.JCAT.2006.05.009.

[143] D. Teschner, J. Borsodi, A. Wootsch, Z. Révay, M. Hävecker, A. Knop-Gericke, S. D. Jackson, and R. Schlögl, "The roles of subsurface carbon and hydrogen in palladium-catalyzed alkyne hydrogenation.," *Science*, vol. 320, no. 5872, pp. 86–89, Apr. 2008, doi: 10.1126/science.1155200.

[144] K. A. Johnson and R. S. Goody, "The original Michaelis constant: translation of the 1913 Michaelis–Menten paper," *Biochemistry*, vol. 50, no. 39, pp. 8264–8269, Oct. 2011, doi: 10.1021/BI201284U.

[145] E. Fischer, "Ueber den Einfluss der Konfiguration auf die Wirkung der Enzyme III," *Berichte der Dtsch. Chem. Gesellschaft*, vol. 28, no. 2, pp. 1429–1438, May 1895, doi: 10.1002/CBER.18950280243.

[146] E. Fischer, "Ueber ein neues dem Amygdalin ähnliches Glucosid," *Berichte der Dtsch. Chem. Gesellschaft*, vol. 28, no. 2, pp. 1508–1511, May 1895, doi: 10.1002/CBER.18950280261.

[147] L. L. van Reijen and G. C. A. Schuit, "The power rate law in heterogeneous catalysis and absolute rates of reactions," *Bull. des Sociétés Chim. Belges*, vol. 67, no. 7–8, pp. 489–505, Jan. 1958, doi: 10.1002/BSCB.19580670713.

[148] I. A. W. Filot, *Introduction to Microkinetic Modeling.* Technische Universiteit Eindhoven, 2018.

[149] G. C. Bond, "The use of kinetics in evaluating mechanisms in heterogeneous catalysis," *Catal. Rev. Sci. Eng.*, vol. 50, no. 4, pp. 532–567, 2009, doi: 10.1080/01614940802480338.

[150] E. K. Rideal, "XXXIX.—The hydrogenation of ethylene in contact with nickel," *J. Chem. Soc. Trans.*, vol. 121, no. 0, pp. 309–318, Jan. 1922, doi: 10.1039/CT9222100309.

[151] G. I. Jenkins and E. Rideal, "The catalytic hydrogenation of ethylene at a nickel surface. Part II. The reaction mechanism," *J. Chem. Soc.*, no. 0, pp. 2496–2500, Jan. 1955, doi: 10.1039/JR9550002496.

[152] H. zur Strassen, "Zur Kinetik der katalytisclien Äthylenhydrierung," *Zeitschrift für Phys. Chemie*, vol. 169A, no. 1, pp. 81–90, Feb. 1934, doi: 10.1515/ZPCH-1934-16908.

[153] A. A. Balandin, "Modern state of the multiplet theory of heterogeneous catalysis," *Adv. Catal.*, vol. 19, no. C, pp. 1–210, Jan. 1969, doi: 10.1016/S0360-0564(08)60029-2.

[154] T. Bligaard, J. K. Nørskov, S. Dahl, J. Matthiesen, C. H. Christensen, and J. Sehested, "The Brønsted–Evans–Polanyi relation and the volcano curve in heterogeneous catalysis," *J. Catal.*, vol. 224, no. 1, pp. 206–217, May 2004, doi: 10.1016/j.jcat.2004.02.034.

[155] A. Hellman *et al.*, "Predicting catalysis: Understanding ammonia synthesis from first-principles calculations," *J. Phys. Chem. B*, vol. 110, no. 36, pp. 17719–17735, Sep. 2006, doi: 10.1021/JP056982H.

[156] A. Logadottir, T. H. Rod, J. K. Nørskov, B. Hammer, S. Dahl, and C. J. H. Jacobsen, "The Brønsted–Evans–Polanyi Relation and the volcano plot for ammonia synthesis over transition metal catalysts," *J. Catal.*, vol. 197, no. 2, pp. 229–231, Jan. 2001, doi: 10.1006/JCAT.2000.3087.

[157] C. J. H. Jacobsen, S. Dahl, B. G. S. Clausen, S. Bahn, A. Logadottir, and J. K. Nørskov, "Catalyst design by interpolation in the periodic table: bimetallic ammonia synthesis catalysts," *J. Am. Chem. Soc.*, vol. 123, no. 34, pp. 8404–8405, 2001, doi: 10.1021/JA010963D.

[158] P. Stoltze and J. K. Nørskov, "Bridging the 'pressure gap' between ultrahigh-vacuum surface physics and high-pressure catalysis," *Phys. Rev. Lett.*, vol. 55, no. 22, p. 2502, Nov. 1985, doi: 10.1103/PhysRevLett.55.2502.

[159] C. J. H. Jacobsen, S. Dahl, A. Boisen, B. S. Clausen, H. Topsøe, A. Logadottir, and J. K. Nørskov, "Optimal catalyst curves: connecting density functional theory calculations with industrial reactor design and catalyst selection," *J. Catal.*, vol. 205, no. 2, pp. 382–387, Jan. 2002, doi: 10.1006/jcat.2001.3442.

[160] J. K. Nørskov, T. Bligaard, J. Rossmeisl, and C. H. Christensen, "Towards the computational design of solid catalysts," *Nat. Chem.*, vol. 1, no. 1, pp. 37–46, Apr. 2009, doi: 10.1038/nchem.121.

Chapter 3

Catalytic Hydrogenation Reactions

Founding mechanistic concepts in heterogenous catalysis developed initially mainly with the exploration of hydrogenation reactions.

Catalytic hydrogenation is the chemical heart of several major catalytic processes that were invented in the first part of the 20th century. It began with the discovery by Sabatier in 1897 of the hydrogenation of ethene to ethane by highly dispersed nickel. Soon after Sabatier's discoveries, the ammonia synthesis process was born. The history of the invention and introduction of these catalytic hydrogenation processes is described in Section 1.3. The mechanisms of catalytic hydrogenation reactions are the subject of this chapter. Emphasis is on the molecular chemistry that defines catalytic reactivity in relation to catalyst structure and composition. Initial mechanistic discoveries and concepts are compared with recent molecular insights.

Early research on the mechanism of catalytic hydrogenation reactions was mainly based on kinetic studies. Major progress was made once the deuterium isotope was discovered and isotope-labelled studies became possible. In this early episode, mechanistic study of the hydrogenation of ethene, nitrogen, and carbon monoxide led to the first formulations of their molecular mechanism. Some of these early mechanistic concepts became iconic to the science of heterogeneous catalysis. In the course of time, molecular and atomic aspects of catalytic reactions became increasingly better understood. It relates to the increasingly better instrumentation to access the structure of the catalyst surface at the molecular level. The molecular view of the relation between catalyst performance and structure and composition of surfaces owes significantly to contributions of surface science as well as to quantum-chemical simulation.

The first three sections of this chapter deal with the conversion of hydrocarbons and the diatomic molecules carbon monoxide and nitrogen. These reactions are for the main part catalyzed by transition metals and share several common reactivity features. Therefore, the first part of the chapter can also be considered an introduction to transition metal catalysis. An additional section formulates general principles within the context of structure-function relations of the respective reactions.

Hydrocarbon conversion consists of dehydrogenation or hydrogenation reactions as well as skeleton isomerization and hydrogenolysis reactions. In the case of N_2 hydrogenation, the main product is ammonia. In the case of catalytic hydrogenation of CO, main products are methane or longer hydrocarbons and methanol.

An important subject is the selectivity of a particular reaction. Hydrocarbon conversion as well as CO hydrogenation provide an opportunity to discuss this in detail. Why do some metals give selective hydrocarbon isomerization instead of mainly methane? Why do some catalysts show high selectivity of CO conversion to long hydrocarbon chains versus methane or alcohol? A complexity of transition metal catalysts is that the structure of transition metal particle and surface state is altered by the catalytic reaction. It is a significant feature of hydrocarbon conversion as well as CO conversion.

In the second part of the chapter the mechanisms of the removal of heteroatoms from hydrocarbons that contain heteroatoms are presented. Hydrogenation reactions such as hydrodesulfurization (HDS) and hydrodenitrogenation (HDN) are important for the purification of oil-derived liquids. Hydrodeoxygenation reactions of molecules derived from biomass are important to the treatment of bio-oils.

Whereas the mechanism of conversion reactions of hydrocarbons, N_2, and CO mainly concern catalysis by transition metals, metal sulfides are catalysts of HDS and HDN reactions. Metal sulfides such as MoS_2 and WS_2 promoted with Ni or Co are used.

Study of the role of these sulfide promoters has importantly contributed to an understanding of the reaction mechanism of HDS and HDN.

Hydrodeoxygenation catalysts are often bifunctional. Catalysts contain a hydrogenation active component that is distributed on Brønsted- or Lewis-acidic supports. The reaction mechanism is complex because of the presence of the different functionalities of the catalyst. Ring opening, dihydroxylation, and decarboxylation reactions are part of catalytic hydrodeoxygenation catalysis.

3.1 Mechanism of Hydrocarbon Activation

3.1.1 *Ethene Hydrogenation by Transition Metals*

The correspondence between the chemistry of surface reactions and organometallic complexes is discovered.

Here the mechanism of the hydrogenation of ethene is presented. The classical Horiuti-Polanyi model of catalytic alkene hydrogenation that was formulated in 1934 is introduced. This mechanistic model has been very influential for the formulation of many of the later molecular reaction mechanistic models that derive from surface science and computational studies.

The early debate on ethene hydrogenation that is catalyzed by a transition metal dealt with the question of whether this is a direct reaction with the hydrogen molecule or that the hydrogen molecule has to dissociate and hydrogenation occurs in a succession of hydrogen addition reaction steps [1].

Early experiments by Farkas, Bonhoeffer [2], and Rideal [3] in Great Britain established that the H_2-D_2 exchange reaction to give HD or the ortho-para nuclear spin state change of H_2 proceeds through initial dissociation of the molecule. Hydrogen atoms become adsorbed on the transition metal surface. The product molecules are formed by consecutive atom-atom recombination reactions.

In the presence of ethene the hydrogen-deuterium exchange reaction competes with reaction with ethene. Horiuti and Polanyi [4]

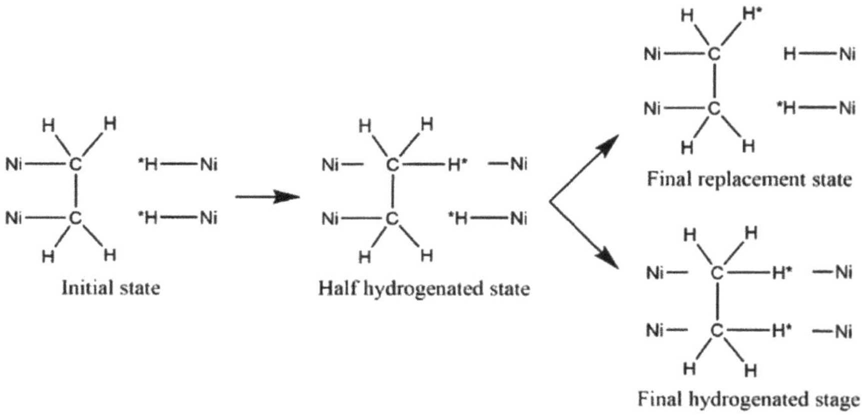

Figure 3.1 The original model of ethene hydrogenation by Horiuti and Polanyi. The hydrogen atom exchange reaction is in competition with ethene hydrogenation [4].

demonstrated that hydrogenation of ethene proceeds through subsequent addition of a H or D atom to adsorbed ethene. Their proposed reaction mechanism of ethene hydrogenation is shown in Figure 3.1.

Horiuti and Polanyi formulated the following postulate concerning the elementary steps that lead to hydrogenation of ethene:

— By contact with the catalyst, the hydrogen molecule dissociates into two adsorbed hydrogen atoms.
— The ethene molecule adsorbs on the catalyst. The hydrogen atom breaks its chemical bond with the surface and forms a new C–H bond by its addition to the ethene molecule.
— Ethane formation follows after a second H atom addition to the ethyl intermediate. The reaction from ethene to ethane proceeds through two successive hydrogen addition steps. Coordinatively saturated ethane desorbs from the surface. In modern language, the half-hydrogenated state is an adsorbed ethyl intermediate.

This molecular model of the mechanism of a catalytic reaction formulated in terms of a surface reaction is iconic to fundamental catalysis.

Horiuti is one of the giants of the kinetics of catalysis from the first half of the previous century. He is the founder of the renowned Japanese Catalysis Institute in Hokkaido, Japan. Polanyi is an important physical chemist. His name is also connected with the Brønsted-Evans-Polanyi (BEP) relation between activation energy and reaction energy of an elementary surface reaction (see Section 2.3.2.3) that was formulated around the same time as the Horiuti-Polanyi proposal.

In the period that followed from 1940 to 1970, the hydrogenation reaction of ethene continued to be extensively explored. An important question that was addressed is how hydrogen dissociation and surface reactivity relate to catalyst surface composition. In particular, the transition metals and their alloys were investigated. Alloy catalysis is discussed in Section 3.1.4.2.

The next major insight after Horiuti and Polanyi came from breakthroughs in organometallic chemistry and the development of surface science. In 1973 Geoffrey Wilkinson obtained the Nobel prize for his 1966 discovery of an active ethene hydrogenation molecular inorganic coordination [5].

The mechanism of the Wilkinson hydrogenation reaction as formulated by Halpern [6] is sketched in Figure 3.2. Its importance is that intermediates of catalytic hydrogenation are chemically identified, and the reaction cycle is closed.

Molecular structural information available from crystallographic determination of the structures of related compounds, NMR identification of reaction intermediates, and kinetic experiments established the Halpern reaction mechanism.

Figure 3.2 illustrates that the Wilkinson catalyst is a molecular coordination complex in which a Rh^+ cation has four ligands that are located in a plane. The ligands are three PPh_3 molecules and a Cl^- ion. Reaction is initiated when one PPh_3 ligand is substituted by a dissociatively adsorbed H_2 molecule.

The hydrogen molecule dissociates into two hydrogen atoms. One hydrogen atom occupies the ligand position in the ligand plane, the other adsorbs in apical position.

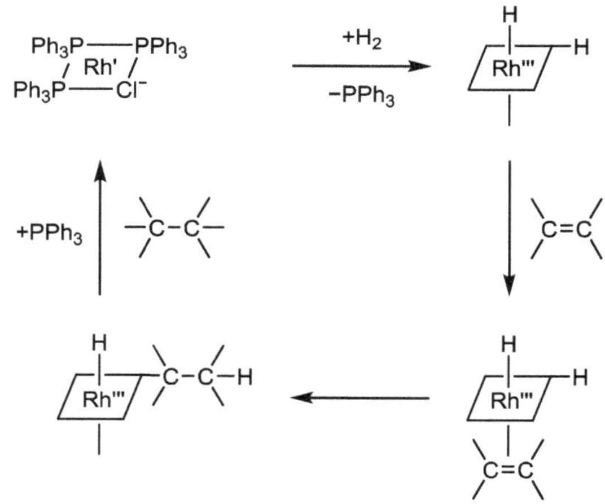

Figure 3.2 The Halpern mechanism of the Wilkinson Rh phosphine-catalyzed ethene hydrogenation reaction. Initially three phosphine (PPh_3) ligands and one Cl^- surround the Rh^+ cation in a square arrangement. A phosphine ligand becomes replaced by H_2 with a valency change of the metal from Rh^+ to Rh^{3+}. Ethene adsorbs and the complex becomes an approximate octahedron. By a first hydrogen atom addition step, the alkyl intermediate is formed. In this step a phosphine ligand reabsorbs. After a second hydrogen addition step, the ethane molecule is formed and the Rh cation reduces [6].

In this oxidative addition reaction, the Rh^+ is formally oxidized to Rh^{3+} and the adsorbed hydrogen atoms are hydride ions.

The ethene molecule adsorbs through a π bond and coordinates sideways opposite to the apical hydride ion.

In the next step, the ethene molecule reacts with the in-plane hydride ion to give an alkyl anion. In a final step the ethane molecule is formed by a second hydride addition.

When ethane desorbs, the Rh^{3+} reduces back to Rh^+. This is called a reductive elimination step [7].

Similar to the Horiuti-Polanyi surface reaction, hydrogen addition occurs in two consecutive reaction steps of hydrogen atoms and adsorbed alkyl is reaction intermediate.

According to Wilkinson, the formation of the alkyl intermediate is key to the metal organic reaction he discovered [8].

Twenty years after the Halpern coordination complex reaction mechanism proposal, it was confirmed by first-principle quantum-chemical calculations of Morokuma *et al.* [9].

Geometrically there is a difference between the octahedral reaction complex of Wilkinson and the structure of surface intermediates of the hydrogen addition reaction.

Later surface science and computational studies did demonstrate the analogy between binding in the organometallic complex and coordination of ethene on the surface, and change in local electron densities in the hydrogen addition steps.

Important is the discovery that also on the surface, when ethene reacts with the hydrogen atoms, it is π coordinated to a transition metal surface atom.

The contribution of surface science is the third major step towards establishing the hydrogenation mechanism.

It came from surface science experiments in the 1990s. Surface crystallography of well-defined single crystal surfaces provide structural data on an atomistic level of chemisorption complexes of molecules on surfaces [10]. This in combination with first-principle quantum-chemical calculations gave a detailed atomistic and energetic model of the bond breaking and bond formation steps of the Horiuti-Polanyi mechanism.

On a reactive transition metal, ethene can adsorb in two modes. It can strongly coordinate to a surface with two bonds directed to the carbon atom (di-σ) or weakly adsorb to a single surface metal atom (the π-adsorbed mode).

These two adsorption modes are schematically drawn in the inserts of Figure 3.3a, which presents an infrared–visible sum frequency generation spectrum. It shows the presence of these two adsorption modes on the Pt(111) surface. When exposed to hydrogen, Somorjai *et al.* from UC Berkeley [11] identified three intermediates: the two ethene adsorbed states and adsorbed ethylidyne (see Figure 3.3b).

The concentration of π-adsorbed ethene is small, but di-σ-adsorbed ethene remains present in significant concentration.

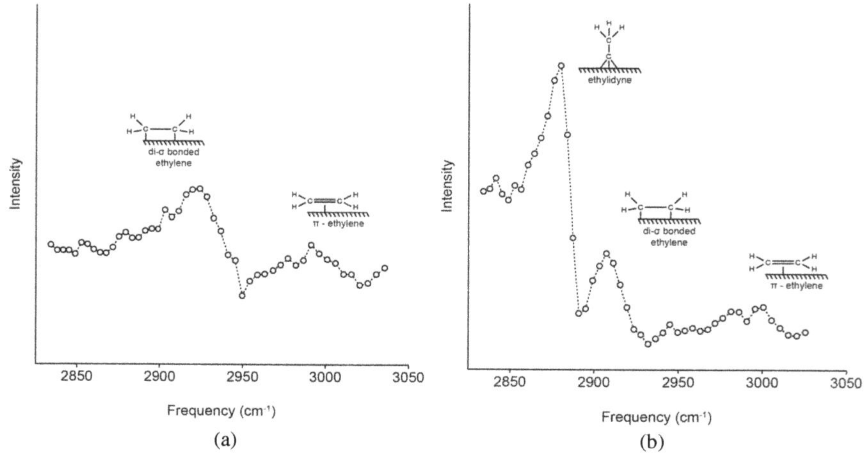

Figure 3.3 Infrared-visible sum frequency generation spectra of reaction intermediates of ethene hydrogenation on Pt(111) surface [11]. (a) A mixture of di-σ- and π-bonded ethene at 120 K on Pt(111). (b) The Pt(111) surface during ethene hydrogenation with 100 Torr of H_2, 35 Torr of C_2H_4, and 615 Torr of He at 295 K.

Only π-adsorbed ethene gives reaction to the reactive ethyl intermediate and consecutive ethane formation.

Adsorbed ethylidyne (see insert of Figure 3.3b) only slowly reacts with hydrogen and is therefore not an intermediate of reaction. It is a so-called spectator species similar to di-σ-adsorbed ethene. Such spectator species are quite common in catalysis and may block the surface. A high temperature will remove such species.

Shortly after the Somorjai experiment, Neurock *et al.* [12] provided the first DFT-calculated potential energies of respective transition states and reaction intermediates of ethene hydrogenation.

In Figure 3.4 the results of ethene hydrogenation calculations by quantum-chemical DFT calculations are shown for the reaction on Pt(111) and Rh(111) surfaces [13]. Potential energy changes are shown as a function of progress of the hydrogenation reaction. The successive energies of adsorbed intermediates as well as transition states of successive bond formation steps are given.

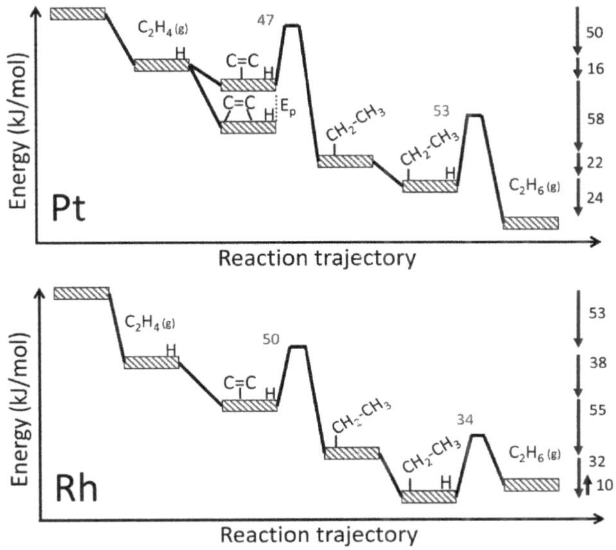

Figure 3.4 Potential energy diagram for ethene hydrogenation according to the Horiuti-Polanyi mechanism over Pt and Rh. Energy differences for each step and the activation energies for the hydrogenation steps are given in kJ/mol. E_p corresponds to the promotion energy required to convert the di-σ-ethene species into the π-ethene species [13].

The calculations confirm the two-step hydrogen addition mechanism. The first major activation barrier to overcome is reaction of ethene to give the adsorbed ethyl intermediate. In the next step the ethyl intermediate is converted into ethane.

For different metals, the relation between respective activation energy barriers may alter. In the comparison of Pt and Rh, the slower step changes from ethyl hydrogenation in Pt to adsorption of ethene in Rh.

The basic idea of the Horiuti-Polanyi mechanism is that surface reactions are initiated by bond dissociation reactions of adsorbed molecules that recombine to yield product molecules in consecutive steps. This mechanistic idea is fundamental to the Langmuir-Hinshelwood kinetics (Section 2.4.2) and is a cornerstone of surface kinetics.

The kinetics of the ethene hydrogenation reaction is positive order in hydrogen reactant concentration and changes from zero order in ethene at low temperature (surface is covered with ethene) to positive order in ethene at higher temperature [14].

The reaction order of hydrogen reactant concentration depends on which of the two consecutive hydrogen addition steps is reaction rate controlling.

At low hydrogen pressure, when hydrogen atom addition to adsorbed ethene is reaction rate controlling, reaction rate will be half order in H_2. When instead addition of the hydrogen atom to adsorbed ethyl is reaction rate controlling, reaction is first order in H_2.

Alkane dehydrogenation is the reverse reaction of alkene hydrogenation. The reaction enthalpy of ethane to ethene dehydrogenation is $\Delta H = 83$ kJ/mol [15]. It is a high-temperature reaction. The rate of reaction is first order in alkane and negative order in hydrogen.

The reaction mechanism is the reverse of the Horiuti-Polanyi alkene hydrogenation reaction. The β C–H bond cleavage of adsorbed intermediate alkyl is usually reaction rate controlling, and is formed after the first C–H bond splitting of the alkane.

Transition metal catalysts rapidly deactivate because dehydrogenation also leads to deactivating carbonaceous residue. This is discussed in some detail in Section 3.3.

The preferred catalysts for dehydrogenation of short alkanes are reducible oxide catalysts such as Cr_2O_3 [16]. These catalysts are more robust with respect to deactivation and catalyst regeneration.

3.1.2 *Hydrogenation and Dehydrogenation Catalyzed by Lewis Acid Oxides*

Bond dissociation on a non-reducible oxide surface is a heterolytic bond cleavage reaction.

Cr_2O_3 and ZnO are used in a variety of hydrogenation and dehydrogenation reactions. Alcohol dehydrogenation and alkene hydrogenation catalyzed by Cr_2O_3 have been discovered by Sabatier at the beginning of the last century [17]. Lazier and Adkins studied in 1925 competitive dehydration of ethanol and dehydrogenation by ZnO [18]. Catalytic dehydrogenation of alkanes became investigated in the 1930s by Frey and Hupke [19]. ZnO/Cr_2O_3 is an active catalyst for high-temperature conversion of CO or CO_2 to methanol [20]. However, it has been replaced since the 1960s by the more active Cu/ZnO catalyst [21][20]. Methanol synthesis reactions are discussed in detail in Section 3.2.2.4.

Related catalysts are used for the water-gas shift reaction that produces H_2 from the reaction of CO with H_2O [22].

Chemical bond dissociation of the H_2 molecule or a hydrocarbon C–H bond by a Lewis acid oxide is a heterolytic reaction. In heterolytic bond activation, the chemical bond formally splits into a positive and negative fragment. The alkane C–H bond dissociates into a negatively charged alkyl anion that binds to the oxide cation. The hydrogen atom reacts as a proton with the oxide oxygen atom. This is different from homolytic bond dissociation by a transition metal surface, where the electrons of the chemical bond that cleaves are equally distributed between the dissociated surface adsorbate fragments.

The mechanism of the heterolytic hydrogenation reaction has been proposed by Burwell from Northwestern University in 1960 [23]. He formulated the elementary reaction scheme of Eq. (3.1) for ethene hydrogenation by a surface site of Cr_2O_3 [24]:

$$\begin{aligned} &Cr^{3+}O^{2-} + H_2 \rightarrow [Cr^{3+}H^-]OH^- \\ &[Cr^{3+}H^-]OH^- + C_2H_4 \underset{k_d}{\overset{k_{ins}}{\rightleftharpoons}} [Cr^{3+}C_2H_5^-]OH^- \\ &[Cr^{3+}C_2H_5^-]OH^- \rightarrow C_2H_6 + Cr^{3+}O^{2-} \end{aligned} \tag{3.1}$$

Dissociation of the hydrogen molecule gives a Cr-hydride intermediate that is charge compensated by protonation of the oxygen anion. In the second step, the hydride ion inserts into adsorbed ethene (k_{ins}) and intermediate Cr-alkyl is formed. The negatively charged alkyl anion recombines with the proton adsorbed to lattice oxygen in a final step. The charge of Cr^{3+} is maintained in this reaction. The reverse step of decomposition of adsorbed alkyl to hydride anion and ethene (k_{cl}) is the β C–H bond cleavage reaction.

As a general rule for hydrogenation or dehydrogenation, the catalyst cations should act as soft Lewis acids [25]. Cations are relatively large, have low charge and are polarizable. The cation does not change valence by reaction. ZnO or Ga_2O_3 can also dehydrogenate alkanes. However, reduction of the oxides may limit reaction lifetime [26], [27].

Heterolytic bond dissociation reactions are not limited to soft Lewis acids. Hard Lewis acid oxides (cations are small and non-polarizable) such as Al_2O_3 are able to dissociate polar molecules such as H_2O.

This generates a Brønsted acidic proton that attaches to the oxygen anions which bridge cations and Brønsted basic site when hydroxyl is attached to a single Al^{3+} cation.

Such hydroxylated surfaces are discussed in Section 5.2.4. Hard Lewis acid oxides will activate polar bonds but will not activate covalent C–H bonds.

Figure 3.5 provides a schematic illustration of heterolytic propane dehydrogenation catalyzed by ZnO. The intermediate alkyl fragment adsorbs on the Zn^{2+} cation. Alkene is formed by β C–H bond cleavage. Hydrogen evolves by recombination of the hydrogen atoms. The competitive oxide reduction reaction is indicated.

Heterolytic bond cleavage reactions are ubiquitous and not limited to metal oxides. In Section 3.4.1, HDS and HDN reactions are presented. These reactions are catalyzed by metal-sulfide catalysts.

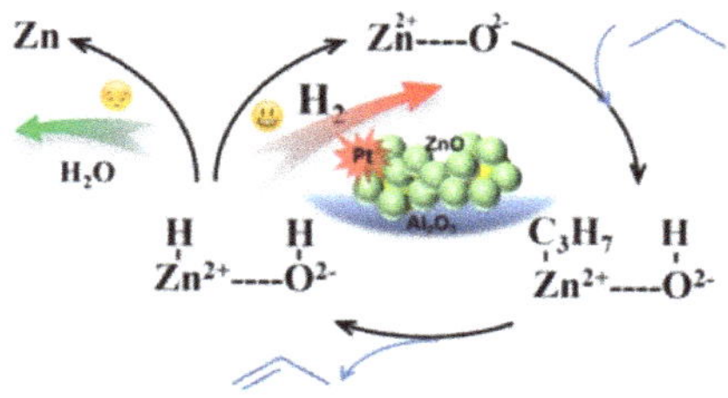

Figure 3.5 Heterolytic bond splitting by propane on ZnO [27].

On the edges of these catalysts, S–H sulfhydryl groups are formed by heterolytic dissociation of H_2. Then sulfide accepts a proton that gives sulfhydryl. The other hydrogen atom adsorbs as anion to the sulfide metal cation. H_2S will heterolytically dissociate on a vacant metal sulfur site with formation of two sulfhydryls.

3.1.3 *Hydrogenolysis of n-Alkane Molecules*

C=C double bond cleavage of surface intermediates is the reaction rate-controlling step of hydrogenolysis.

Hydrogenolysis is a hydrogenation reaction catalyzed by transition metals by which alkanes are converted into shorter alkanes and methane. The key reaction is cleavage of the alkane C–C bond. This is an indirect reaction and happens as part of a sequence of several elementary steps.

The alkane molecule initially interacts by physical adsorption. Contact with the metal surface is through the hydrogen atoms. The interaction energy of a C–H bond contact with the surface is of the order of 5 kJ/mol.

Direct contact of a metal atom with a carbon atom or C–C bond is screened by hydrogen atoms. This is called the umbrella effect [28]. Whereas the σ C–C bond energy is much weaker than the C–H bond energy ($E_{C\text{–}H}$ = 410 kJ/mol, $E_{C\text{–}C}$ = 350 kJ/mol), reaction is initiated by C–H bond cleavage that has a relatively low activation energy (≈ 50 kJ/mol).

Early kinetic evidence in 1948 of the relative ease of C–H bond cleavage versus C–C bond cleavage was provided by Kemball and Taylor [29]. Hydrogen-deuterium exchange experiments with alkanes catalyzed by Ni demonstrated that the temperature of this exchange reaction is substantially lower than that of reactions involving C–C bond cleavage.

Flaherty and Iglesia from UC Berkeley deduced in 2013 from detailed kinetic experiments [30], [31] the three main steps of the reaction schematically illustrated in Figure 3.6. Once a C–H bond is broken, direct chemical bonding contact of a carbon atom of alkyl

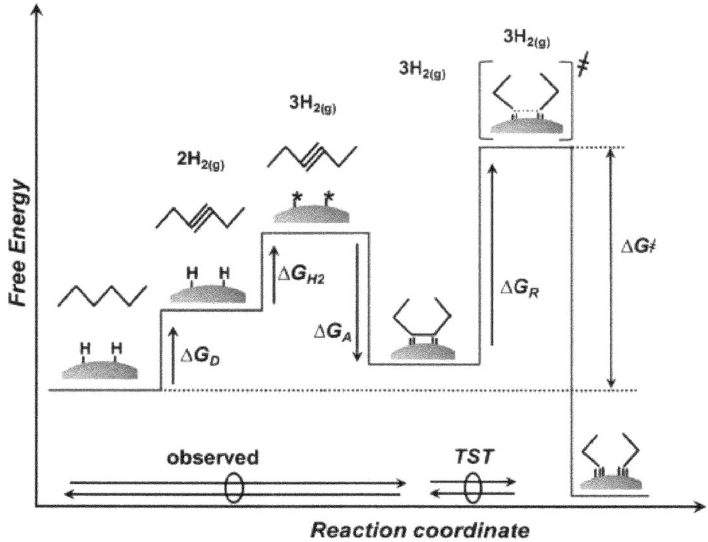

Figure 3.6 Schematic representation of the reaction mechanism of C–C bond cleavage reaction of hexane [31].

is possible with the transition metal surface atoms. The nature of the initial σ C–C is changed and the C–C bond next to the carbon atom that loses a hydrogen atom gets π bond character.

The Kemball and Taylor experiments also demonstrated that cleavage of this C=C bond is the reaction rate-controlling step. This counterintuitive kinetic observation is that C–C bond cleavage is the rate-limiting step of the reaction. This initially was a puzzling result, since the alkane σ C–C bond energy (348 kJ/mol) is weaker than the C–H bond energy (410 kJ/mol).

As is sketched in Figure 3.6, the Major Abundant Reactive Intermediate (MARI) is generated after several subsequent C–H bond cleavage steps. The C=C or C=-C bond of this reaction intermediate cleaves and $M{=}CRH_x$ intermediates form. After hydrogenation steps, short alkanes and methane desorb.

The reaction rate has a high negative H_2 reaction order (−2.3). This supports the proposal that formation of the MARI occurs after loss of several hydrogen atoms.

The hydrogenolysis reaction proceeds through intermediates that are also precursors to deactivation of transition metal catalysts.

The reaction competes with other hydrocarbon conversion reactions and is usually undesirable. The selectivity and deactivation of hydrocarbon conversion reactions is the subject of the next section.

3.1.4 *Selectivity of n-Alkane Conversion Catalyzed by Transition Metals; Isomerization*

There is a large difference in the reactive surface atom ensemble requirement for different hydrocarbon conversion modes.

The selectivity of the n-alkane hydroconversion reactions catalyzed by transition metals is the subject of this section. It will be seen that selectivity is a strong function of transition metals and of reaction site structure as well.

The n-alkane conversion reactions are respectively dehydrogenation, isomerization, and hydrogenolysis, as illustrated in Figure 3.7.

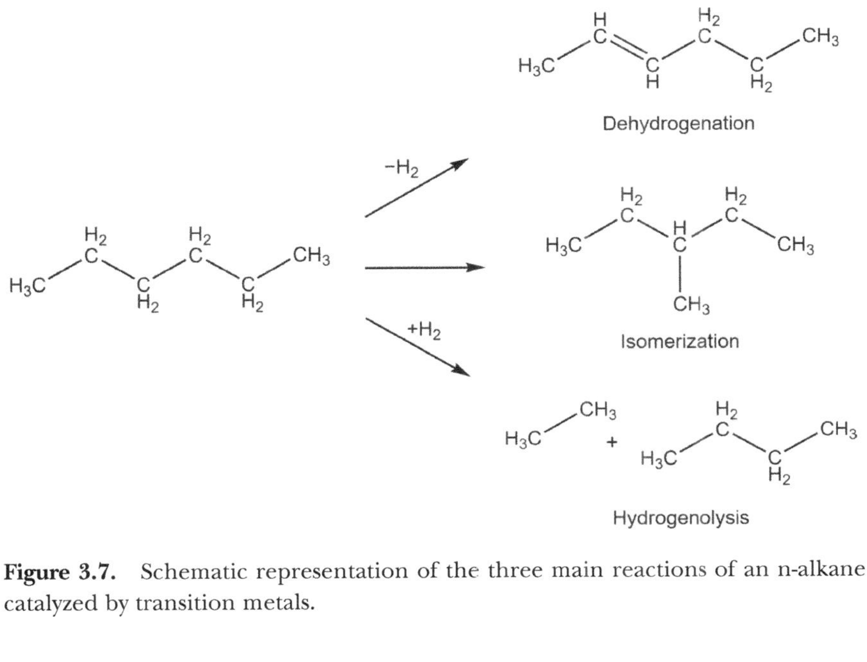

Figure 3.7. Schematic representation of the three main reactions of an n-alkane catalyzed by transition metals.

The reaction mechanism of dehydrogenation and hydrogenolysis was discussed in previous sections. After introducing the mechanism of the isomerization reaction, catalyst selectivity as a function of metal alloy composition is discussed. Alloying is an important catalytic tool to suppress the hydrogenolysis reaction.

An important reaction site structure concept that is used in this and following sections is the number of transition metal surface atoms that are part of the catalytic site. This is called the surface atom ensemble size (see also Sections 2.3.2.1, 2.3.3.3).

Whereas for the hydrogenolysis reaction the reaction center consists of an ensemble size of five or six transition metal surface atoms, the isomerization or dehydrogenation reactions are catalyzed by reaction sites that consist of an ensemble of two or only one surface atom. Alloying affects surface ensemble size of reaction sites.

The dependence of the selectivity of n-hexane conversion on transition metals is shown in Figure 3.8. The selectivity of the isomerization reaction versus that of hydrogenolysis is compared as a function of transition metal reactivity (measured by the strength of chemisorption, see Figure 2.12). The reaction was executed in excess hydrogen and alkane dehydrogenation is minimal.

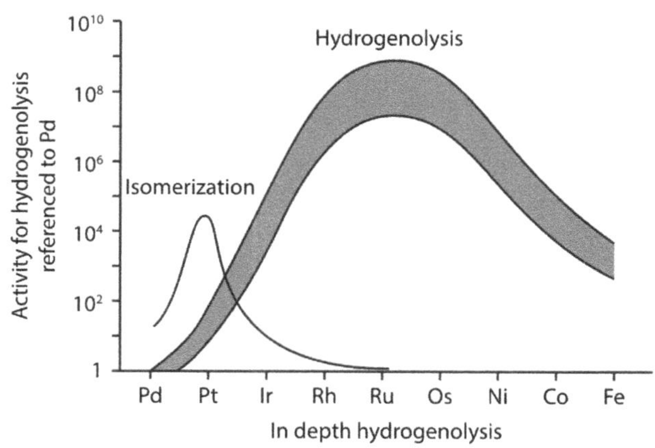

Figure 3.8 Schematic dependence of selectivity of the isomerization and hydrogenolysis reactions as a function of transition metal. Left line indicates isomerization activity. Curves with grey in between indicate hydrogenolysis activity. Temperature of reaction ≈ 600 K and high excess hydrogen [36].

Both the hydrogenolysis and isomerization reaction are exothermic. For hydrogenolysis $\Delta H_{C_2H_6+H_2 \to CH_4} = -66$ kJ/mol, and for isomerization $\Delta H_{n\text{-alkane} \to i\text{-alkane}} = -6$ kJ/mol. Thermodynamically a low temperature is beneficial, but reaction temperature is determined by kinetics.

The discovery that transition metals will not only catalyze hydrogenation, dehydrogenation, and hydrogenolysis reactions but also isomerization was made in the 1960s [32]–[35]. The mechanism of alkane isomerization is discussed below.

A high selectivity is only found for transition metals at the lower left corner of the transition metal block in the periodic table. These metals are the least reactive.

The activation energy of the hydrogenolysis reaction on Pt (E_{act} = 230 kJ/mol) is high compared to the that of the isomerization reaction (E_{act} = 180 kJ/mol), but decreases strongly when the reactivity of transition metal increases.

This is due to the BEP relation (Section 2.3.2.3). The increase in surface reactivity stabilizes the strongly bound products of the C–C bond cleavage. The decrease in activation energy is less for the isomerization reaction because, as discussed below, no such strongly adsorbed intermediates are part of the reaction mechanism.

The maximum in reaction rate of isomerization is due to increased negative competition with hydrogenolysis. The maximum in hydrogenolysis reaction rate is due to increasing deactivation of the catalyst.

The deposition of carbonaceous residue, which results in partial carbiding of the transition metal when exposed to hydrocarbon conversion reactions, is a general phenomenon [12], [37]. For instance, alkene hydrogenation reactions ($\Delta H_{C_2H_4+H_2 \to C_2H_6} = -130$ kJ/mol) also develop carbonaceous surface-deactivating intermediates as well as deactivating surface carbides.

In Section 3.1 (Figure 3.3), for the hydrogenation of ethene by Pt, the presence of spectator species ethylidyne was discussed. Because it suppresses hydrogenolysis conversion of bare metal surface to partially carbided surface, ethylidyne may even enhance stability of reaction as is argued in Section 3.3.

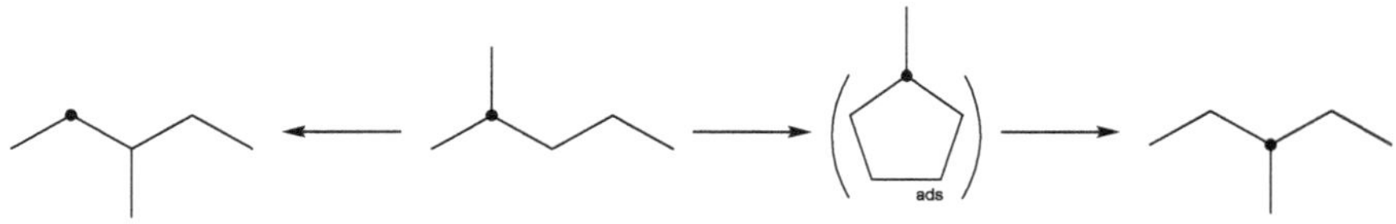

Figure 3.9 Bond shift (E_{act} = 180 kJ/mol), left arrow, versus cyclic mechanism (E_{act} = 280 kJ/mol), right arrow [39].

The mechanism of the isomerization reaction catalyzed by highly dispersed Pt or Pd particles on an alumina support has been elucidated in the 1970s by Gault *et al.* from the University of Strasbourg [38], [39]. They developed a sophisticated mass spectrometric analysis technique so as to follow the carbon atom relocation of ^{13}C labels in reacting alkanes.

As for the hydrogenolysis reaction, the isomerization is initiated by C–H bond cleavage. There are two isomerization pathways of the alkane [34], [35]: bond shift and cyclic mechanisms. The latter mechanism is only possible for alkanes that contain at least five carbon atoms (see Figure 3.9).

Use of ^{13}C-labelled molecules enable them to determine that the bond shift mechanism has the lower activation energy (E_{act} = 180 kJ/mol, pentane, Pt) compared to that of the cyclic mechanism (E_{act} = 280 kJ/mol, pentane, Pt). The cyclic mechanism explains the often-observed presence of cyclopentanes as product.

The Gault proposal of the mechanism of C–C and C–H bond rearrangement of reaction intermediates adsorbed to the metal surface is illustrated in Figure 3.10.

Partially dehydrogenated intermediates rearrange and desorb after hydrogen addition steps. Figure 3.10a illustrates their mechanistic suggestion. A reaction site that consists of one or two surface atoms is required for C–C bond cleavage and formation steps [40]. It proceeds by formation of a cyclopropyl intermediate. Adsorption of the partially dehydrogenated alkane intermediate is possibly analogous to the $C_3H_6PtCl_2$ complex [41] that is formed by reaction with cyclopropane (see Figure 2.10b).The small surface ensemble requirement of alkane isomerization relates to that of alkene

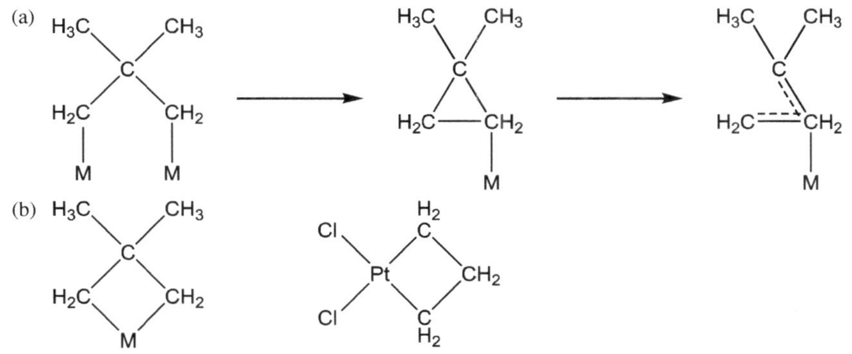

Figure 3.10 The neopentane isomerization of reaction intermediates as proposed by Garin and Gault. (a) Schematic representation of neopentane adsorption and isomerization. The reaction intermediates require only a single bond with the transition metal. (b) The initial adsorbed state of neopentane does not necessarily require independent bonding with the surface. Molecular complexes are known that show a shared bond with a single atom [39].

hydrogenation in Section 3.4.1. Also, this reaction has a low surface atom ensemble demand.

The ring-opening reaction of methyl cyclopentane is also an isomerization reaction. The selectivity of this ring-opening reaction is surface structure demanding. The ring-opening reaction is known since 1933 [42]. Its mechanism has been extensively investigated in the 1960s by Gault *et al.* [43]. For catalysts with high Pt dispersion (Pt nanoparticle size ≈ 2–4 nm) or low dispersion (Pt nanoparticle size ≈ 10–20 nm), the selectivity pattern is very different. On small Pt particles there is close to equal probability for each C–C bond to cleave (see Figure 3.11a1). This is different for the larger particles where dissociation of the C–C bond next to the methyl substituent is suppressed (see Figure 3.11b2). DFT calculations by Rosch *et al.* in 2013 [44] have elucidated its relation with surface structure.

The small particle contains step-edge sites as indicated in Figure 3.11a2. Different from the planar surface, this creates space for the methyl group. The arrangement of adsorbed methyl cyclopentane at the step-edge site is not possible on the planar surface because of repulsive interactions with the surface metal atoms that

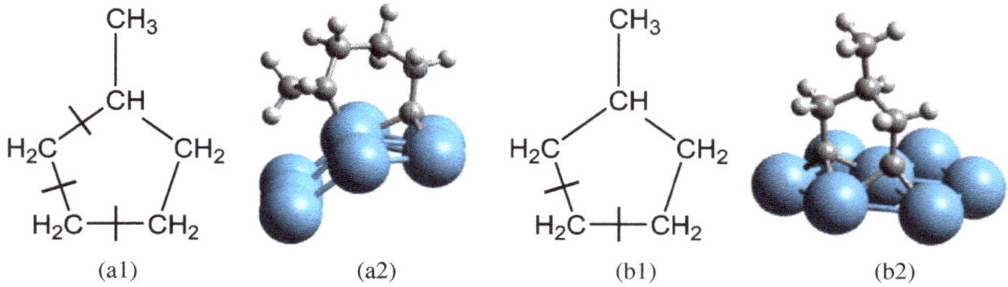

Figure 3.11 Ring-opening reaction of methyl cyclopentane. The difference in selectivity as a function of surface structure is illustrated. (a) Ring-opening on a step-edge site, (b) ring-opening on a surface terrace [44].

then arises. This is the reason the C–C bond next to the methyl substituent has low probability to be cleaved.

Similarly for the hydrogenolysis reaction, this reaction that proceeds by C–C bond cleavage requires a surface ensemble of at least 5 or 6 surface atoms. Coordinatively unsaturated C atoms are coordinated to several surface atoms.

The kinetic reaction rate expressions of all three reactions shown in Figure 3.7 are negative order in hydrogen. It relates to the need of C–H bond cleavage steps to produce the reaction rate-controlling intermediate. The dehydrogenation reaction has a reaction order in hydrogen between -0.5 and -1. Transition metal-catalyzed isomerization and hydrogenolysis have comparable and higher negative hydrogen reaction orders between -2 and -3.5 [45].

A major discovery of the 1950s is the superior activity and selectivity of transition metal catalysts for hydrocarbon conversion reactions when dispersed on an acidic support. In the presence of excess hydrogen, these catalysts provide stable performance of isomerization, aromatization, and selective cracking reactions. In particular, Pt metal-promoted solid acid catalysts have superior activity for hydrocarbon activation.

These reactions are due to a breakthrough discovery of a Pt supported on acidified alumina catalyst by Haensel [46], [47] in 1947. This catalyst has the highly desirable property of upgrading gasoline to high octane number gasoline. The latter makes efficient high pressure-compression automotive engines possible.

The catalyst is currently used in large refinery plants for the production of high-octane gasoline in the platforming process.

At high temperatures and excess hydrogen, such catalysts convert alkanes into aromatics (800 K) and at more modest temperatures (530-600 K), isomerization reactions are catalyzed [48]–[50].

The invention came as a large surprise because no one had imagined that the high cost of Pt would make such a process viable. However, the catalyst has a long life and needs only an extremely low concentration of Pt (0.01 percent weight). A great novelty was the combination of catalytic hydrogenation/dehydrogenation function with solid acid-catalyzed reactions.

For these catalysts it became an important issue to extend catalyst life. Deactivation of the transition metal contributes substantially to reduced lifetime of the bifunctional [51]–[53].

In the following section the use of transition metal alloying is discussed for the decrease of this transition metal deactivation. In the 1970s it was extensively explored by Sinfelt of Exxon research [52]–[54]. A significant part of the presentation on alloys is based on his discoveries.

As background to the discussion of alloy catalysis, here follows a short summary of why bifunctional catalysts with transition metal dispersed on a solid acid are superior aromatization and isomerization catalysts.

The transition metal dehydrogenates alkane to alkene. Isomerization and aromatization of alkene are catalyzed by the protons of the acidic support. The activation energy of proton-catalyzed alkene isomerization is $\approx$ 120 kJ/mol (the activation energy of alkane dehydrogenation is comparable). The reason for the superior activity of the hydroisomerization reaction is the substantially lower activation energy of proton-activated alkene isomerization compared to that of Pt-catalyzed alkane isomerization.

Also, aromatic formation is due to proton-catalyzed reactions of intermediated alkenes. Deactivation of the transition metal will suppress the reaction rate of dehydrogenation and hence the steady state alkene concentration necessary for the isomerization or aromatization reactions.

The reaction mechanism of these bifunctional catalysts is extensively discussed in the chapter on solid acid catalysts (see Section 5.4.3).

3.1.5 *Alloy Catalysis; the Ensemble Effect*

Dilution of reactive transition metal surface with non-reactive substituents changes the reactive size of surface atom ensemble. This reduces the rate of reactions through intermediates that demand large surface atom ensembles.

The selectivity of hydrocarbon isomerization or aromatization changes by alloying a reactive transition metal (defined as metal that dissociates H_2) with a non-reactive element (defined as element that does not dissociate H_2). This is due to suppression of the hydrogenolysis reaction that is the main contributor to transition metal deactivation.

Here the mechanistic role of alloys in hydrocarbon conversion catalysis is discussed. The study of these catalytic systems has contributed largely to clarify catalyst structure-catalyst function relations.

Since the 1950s, alloy catalysis has been part of fundamental studies that probe the relation with chemical bonding properties of the catalyst. As mentioned in Section 2.3.1, it was initially thought that surface reactivity relates to the electron density of states at the Fermi level of the metal bulk electronic structure [55]–[57].

This hypothesis became challenged with the observation [58]–[61] that the surface composition of an alloy may be very different from the average bulk composition. In addition, a view developed whereby chemisorption is selective with respect to the surface atoms to which it binds. This can be exploited to chemically titrate the surface to determine surface composition of particular atoms [59]. The modern chemical bonding interpretation of the surface bond is discussed in Section 2.3.3.

Figure 3.12a is an example of the titration of Ni atoms on the surface of a Ni/Cu alloy. It shows the amount of hydrogen adsorbed on a Ni/Cu alloy. One notes the steep decline in surface Ni atom concentration compared to average bulk Ni concentration with increase of bulk Cu concentration. The surface becomes enriched

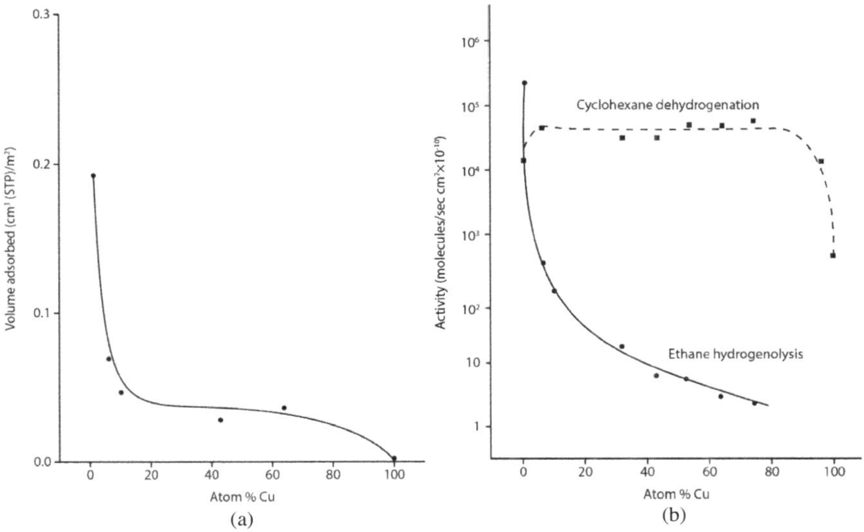

Figure 3.12 Catalysis by Ni-Cu alloys. (a) The chemisorption of hydrogen on copper-nickel catalyst at room temperature as a function of average bulk composition. (b) Activities of copper-nickel alloy catalysts for the hydrogenolysis of ethane to methane and the dehydrogenation of cyclohexane to benzene. The activities refer to reaction rates measured at 316°C in tenfold excess hydrogen [54].

with Cu metal atoms because it lowers the surface energy. Cu has a lower sublimation energy than Ni [62].

The measured changes in the reaction rates of ethane hydrogenolysis and cyclohexane dehydrogenation as a function of bulk Ni/Cu concentration are compared in Figure 3.12b. The reaction rate of the ethane hydrogenolysis reaction decreases by several orders of magnitude when Ni is alloyed with Cu. This happens already when the bulk alloy Cu composition is only 10% and the surface Ni atom concentration, as determined by hydrogen titration, has reached its nearly constant value of 25% of the Cu free surface. In contrast to the hydrogenolysis reaction, the cyclohexane dehydrogenation reaction rate is not much affected and even increases upon alloying with Cu.

The cause of these differences is the very different surface atom ensemble demand of the hydrogenation reaction versus that of hydrogenolysis.

According to the Horiuti-Polanyi and Gault mechanistic propositions, when the hydrogenation and isomerization reactions are catalyzed by a transition metal, they require a reaction center with only one or two surface metal atoms. In contrast, hydrogenolysis according to the Iglesia mechanistic proposal requires an ensemble of five or six surface metal atoms.

Dilution of the reactive Ni transition metal atoms with the non-reactive Cu atoms decreases the ensemble size of the reactive surface atoms [63].

Within the Langmuir checkerboard model of the surface for the hydrogenolysis reaction, this is consistent with the orders in magnitude decrease in reaction rate since the probability to have, for instance, five atoms as a surface ensemble is proportional to 0.25^5.

Comparable changes in the reaction rate of hydrogenolysis are found for alloys of non-hydrogen dissociation-active metals such as Au, Ag, or Sn [64].

The postulate that the ethane hydrogenolysis reaction rate decreases due to reduction of reactive surface atom ensemble size is called the surface ensemble effect.

One distinguishes the primary from secondary ensemble effects. The primary ensemble effect decreases the number of atoms of reactant molecule in contact with the metal surface.

In the example of neopentane isomerization, the alloy will suppress the multipoint contact of Figure 3.10a with a surface site and only an adsorption mode as in intermediate 3.10b is possible.

As indicated in Figure 3.6, two-point contact of the reaction intermediate with the surface is a precondition for hydrogenolysis. Generally, dissociation of a π molecular bond such as the C=C bond of ethene, or that of C=O or O=O, requires a two-point adsorbate contact with a large surface ensemble (see Sections 2.3.2.1, 3.3).

In the 1970s and 1980s, the details of the binding mode of unsaturated carbon atom and surface were not well known, and a surface site was often indicated with a [*] symbol or M. At present it is well understood that usually a single adsorbate atom coordinates to a surface site of several surface atoms. This secondary ensemble effect also changes adsorption energy when a reactive surface atom is substituted by a non-reactive surface atom.

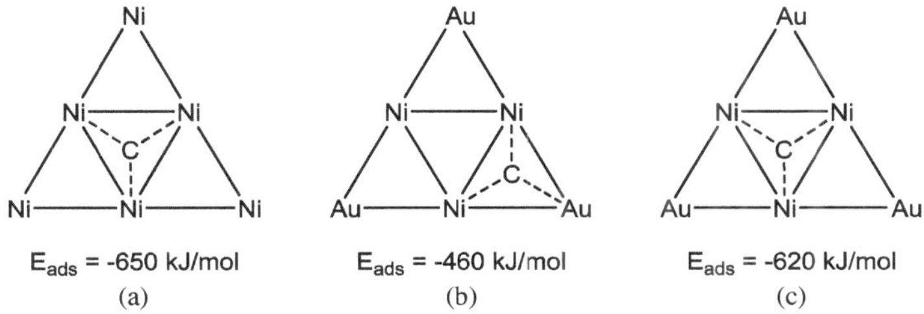

Figure 3.13 The calculated adsorption energy of a carbon atom on a Ni/Au(111) surface. Comparison of secondary ensemble and ligand effect. (a) C threefold-adsorbed to Ni(111) surface, (b) C adsorbed to site with a Au atom substituted for Ni in the coordination shell of the adsorbed carbon atom, (c) C adsorbed to the same site as (a) but surrounded by Au atoms [65].

Figure 3.13 illustrates the secondary ensemble effect. The DFT-computed adsorption energy of C adsorbed in threefold site of the Ni(111) surface (Figure 3.13a) is compared with adsorption of the C atom in a site in which one Ni atom is replaced by Au (Figure 3.13b) [65].

A 30% decrease in adsorption energy is found. It is the result of the substantially weaker M–C bond with Au than Ni. In the Ni_2Au site, the interaction of C with Au expresses itself as a loss of coordination with one atom.

As indicated in Figure 3.13c, the indirect effect of Au substitution around the reaction center gives only a decrease of 5% of the C adsorption energy. There is a small decrease in reactivity of Ni atoms. This effect can be considered a ligand effect analogous to that in a coordination complex [62].

Analysis of the electronic structure of the surface Ni atoms shows that this is due to an increase in Ni d-valence electron occupation. The d-valence electron band width decreases due to virtual decrease in coordination of the Ni atoms (see Figure 2.19, the Ni-Au interaction is weaker than the Ni-Ni interaction). There is a small electron flow from Au to Ni that results in an upwards shift of the average d-valence electron band [66]. The decrease in bond energy is due to the slight increase in the occupation of the anti-bonding orbitals of the Ni_3C complex.

Alloying a reactive transition metal with non-reactive metal changes selectivity mainly because of a site-blocking effect. Reaction intermediates that require several molecular atom contact points with the surface become most affected. The strength of the individual chemical bonds between hydrocarbon carbon atom and surface metal atoms decreases mainly because of the secondary ensemble effect. High coordination becomes replaced by low coordination. Reactions that proceed through intermediates with a single atom contact are hardly affected.

One expects that hydrogenation, dehydrogenation, and isomerization reaction rates that only require a single or dual surface atom ensemble decrease in proportion to the reactive surface atom concentration change. However, in contrast to this expectation, Figure 3.12b indicates that dehydrogenation reaction rate actually slightly increases.

The selectivity of the dehydrogenation rate of cyclohexane is negatively affected by hydrogenolysis. Alloying Ni with Cu suppresses the reaction rate of the hydrogenolysis reaction, that not only increases benzene yield, but also decreases deactivating carbonaceous intermediate deposition. On the alloyed surface, less deactivating carbon is deposited that compensates for the loss of Ni surface concentration due to the presence of surface Cu atoms.

Similar to the rate of cyclohexane dehydrogenation, the reaction rates of hydrogenation of ethene and cyclohexene also increase by a factor of 4 over PdAu catalysts compared to supported Pd [67]–[69].

Coadsorbed C or S have a similar beneficial effect on hydrocarbon conversion reactions as alloying. Carbides of Mo, W, or V are selective hydrogenation catalysts whereas these reactive transition metals rapidly deactivate when exposed to hydrocarbons [70]. Sulfur is a known promoter of selective hydrocarbon conversion catalysis by platinum [71].

An additional reason for the increase in rate of hydrogenation is the weakening of the adsorption energy of ethene by alloying. Due to the secondary ensemble effect (Figure 3.13b) on the alloyed surface or the partially carbon covered surface, the adsorption mode of ethene changes from di-σ to π adsorbed (see inserts, Figure 3.3).

This also happens when the surface contains a dense overlayer of ethene molecules [12]. Then lateral interactions between the adsorbed molecules inhibit di-σ coordination of ethene to multiple metal surface atoms. Instead ethene becomes π adsorbed to a single surface metal atom.

Quantum-chemical computed potential energy reaction diagrams of ethene hydrogenation by the Pd(111) surface for the low and high ethene surface coverage cases are shown respectively in Figures 3.14a and 3.14b [72].

One notes the difference in respective adsorption energies of ethene in Figure 3.14a (di-σ-adsorbed ethene) versus Figure 3.14b (π-adsorbed ethene).

For an isolated ethene molecule adsorbed on the Pd surface, the barriers for the first and second H addition steps are respectively 72 and 71 kJ/mol. Because of the weaker adsorption of ethene on the surface highly covered with ethene, the activation energy of the first hydrogenation step is reduced to 38 kJ/mol. This substantially increases the steady state concentration of the ethyl intermediate and with it the reaction rate of ethene hydrogenation.

Kinetic Monte Carlo calculations of ethene hydrogenation on Pd/Au alloys by Neurock *et al.* [73] demonstrate that the reaction rate of hydrogenation, normalized per Pd atom, only slightly increases due to a combination of the decreased activation energy for alkyl formation from ethene and lower concentration of adsorbed hydrogen atoms due to a smaller adsorption energy.

The shift from di-σ-adsorbed ethene to π-adsorbed ethene also explains the beneficial effect of promotion by metals such as Ag, Au, and Ga in selective hydrogenation of ethyne to ethene [74], [75].

Whereas ethene is a desirable product, its hydrogenation to alkane is to be suppressed. Alloying reduces the adsorption energy of ethene more than the activation energy of ethene hydrogenation (Figure 3.14b).

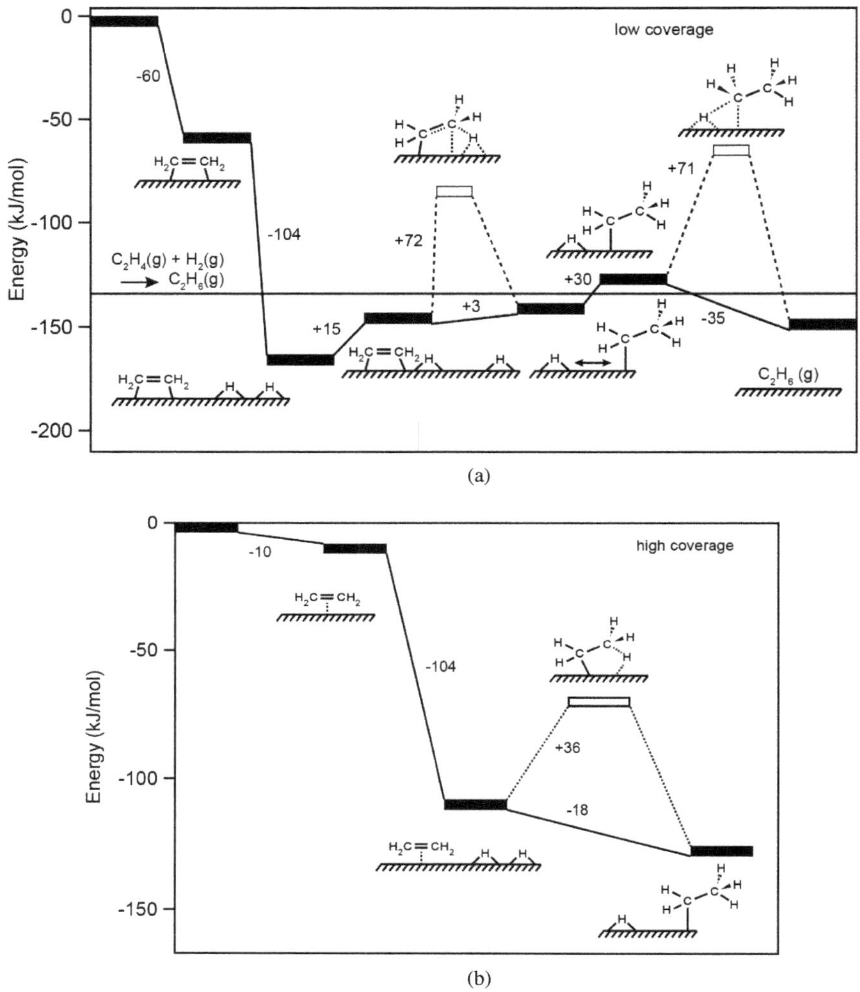

Figure 3.14 Comparison of potential energies of ethene hydrogenation on Pd(111) surface as a function of ethene coverage. (a) The overall potential energy diagram for ethene hydrogenation over Pd(111) at low surface coverage. (b) The potential energy diagram for ethene hydrogenation over Pd(111) at high surface coverage. The solid boxes refer to reactant, intermediate, and product states, whereas the two white boxes denote the transition state structures and energies for the surface reactions of ethene to ethyl and ethyl to ethane [72].

In summary, the dilution of the transition metal surface with non-reactive components reduces the relative rate of reactions that require a large surface ensemble of reactive transition metal atoms.

On the other hand, it promotes the reactivity of intermediates on small surface atom ensemble sites by their shift from a strongly adsorbed to less strongly adsorbed state.

So far it has not been mentioned that alloying also affects favorably a bimolecular reaction where one component adsorbs so strongly that it suppresses adsorption with a more weakly adsorbing coreactant. This results in a low reaction rate.

In the alloy, the reactant that strongly adsorbs to the transition metal is next to unoccupied Ag or Au neighbor atoms. This creates surface sites next to strongly adsorbed reactant for the weakly interacting coreactant and reaction probability increases.

An example not from hydrogenation catalysis but from selective oxidation catalysis is the formation of vinyl acetate from reaction of acetic acid with ethene. Intermediate acetate adsorbs strongly on the Pd atoms. The presence of Au creates vacant surface sites for ethene to adsorb and react with acetate [76].

Catalytic reactivity depends on surface composition and structure. In addition to surface topology, the degree of coordinative unsaturation and unique reaction site structure also play a role.

In this section the effect of alloying on catalysis was discussed within the Langmuir checkerboard model of the catalytic site. An alternative view is that hydrogenolysis is catalyzed by a uniquely reactive Taylor site that is selectively poisoned by the alloying metal. This is discussed in Section 3.3 in the context of metal particle size and shape dependence of catalytic reactions.

3.2 Mechanism of Hydrogenation Catalytic Reactions with N_2 and CO

Nitrogen and carbon monoxide are isoelectronic. While the chemistry of their bond activation has similarities, catalysis of the two molecules is quite different. Hydrogenation of nitrogen has a single product ammonia, whereas hydrogenation of carbon monoxide leads to a broad range of important products such as methane, longer alkanes and aromatics, as well as alcohols. The latter implies a complex reaction mechanism. Selectivity is directed by choice of transition metal catalyst.

Over the past hundred years the mechanisms of both reactions have been extensively investigated. On the molecular level, their reaction mechanisms have become well understood, as is described in the following sections.

The mechanism of the ammonia synthesis reaction is the archetype of a heterogenous catalytic reaction that is initiated by molecular dissociation. Research towards its mechanism has led to paradigmatic insights. The Nobel award to Gerhard Ertl in 2007 is recognition for these contributions.

An important general insight is the related site structure and composition sensitivity of the N_2 and CO bond cleavage reactions. The similarity of bond activation relates to the π bond character and their respective bond strengths ($E_{diss}(N_2)$ = 970 kJ/mol; $E_{diss}(CO)$ = 1110 kJ/mol; compare $E_{diss}(H_2)$ = 440 kJ/mol). As in the Horiuti-Polanyi mechanism, the reaction mechanism is a sequence of surface dissociations of reactant molecules and consecutive surface fragment recombinations that by desorption give the product molecules.

3.2.1 *Nitrogen Hydrogenation; Ammonia Synthesis*

In the ammonia synthesis reaction, N_2 dissociation is reaction rate controlling.

The ammonia synthesis reaction is exothermic ($\Delta H_{N_2+3H_2\rightarrow NH_3}$ = –46 kJ/mol). Reaction conditions are determined by the need for low temperature to have finite ammonia formation versus a higher temperature to overcome the activation energy of reaction. For this reason, reaction is executed at high pressure. For commercial ammonia synthesis catalysts, an apparent activation energy of 47 kJ/mol is reported; that, however, may vary largely with reaction conditions [77]. The ammonia catalyst as originally discovered and used at large scale in commercial synthesis processes consists of Fe promoted with potassium and alumina [78]. The ammonia synthesis reaction is a structure-sensitive reaction and the reaction rate of N_2 dissociation is reaction rate controlling. Adsorbed ammonia suppresses the reaction rate [79].

From N_2 adsorption measurements, Brunauer and Emmett concluded already in 1934 [80] that N_2 adsorbs in a dissociated state on the Fe ammonia catalyst surface. Extensive kinetic studies of the ammonia synthesis reaction by the Russian physical chemist Mikhail Temkin [81] established that reaction is first order in N_2 and proportional to the ratio $(H_2)^3/(NH_3)^2$. At the high pressure (200 atm) of reaction, the equilibrium of product ammonia and H_2 gas-phase concentration determines the concentration of surface-adsorbed NH_x intermediates. Later work by Taylor *et al.* [82] with deuterium-labelled ammonia and hydrogen confirmed that nitrogen dissociation is reaction rate controlling. The degree of reduction of the Fe catalysts determines whether N_{ads} or NH_{ads} dominates the concentration of reaction intermediates.

Twenty years later Ertl *et al.* studied dissociation kinetics of N_2 [83] on a reactive Fe(111) surface and concluded that the extremely low apparent rate of dissociative adsorption of N_2 is due to the low surface concentration of N_2 at reaction condition. The fast desorption rate of the adsorbed N_2 (E_{ads} = 30 kJ/mol) competes with the slow dissociation rate constant of N_2. Whereas N_2 adsorption energy and activation energy of dissociation are comparable, the elementary reaction rate constant of N_2 dissociation is orders of magnitude smaller than the desorption reaction rate constant.

The reason for this is the extremely low activation entropy of N_2 dissociation compared to that of N_2 desorption. Whereas in desorption the N_2 molecule gains rotational entropy, surface N_2 dissociation has low entropy.

A low activation energy of dissociation requires close contact with surface metal atoms. The loose transition state of desorption and the tight transition state of the surface reaction

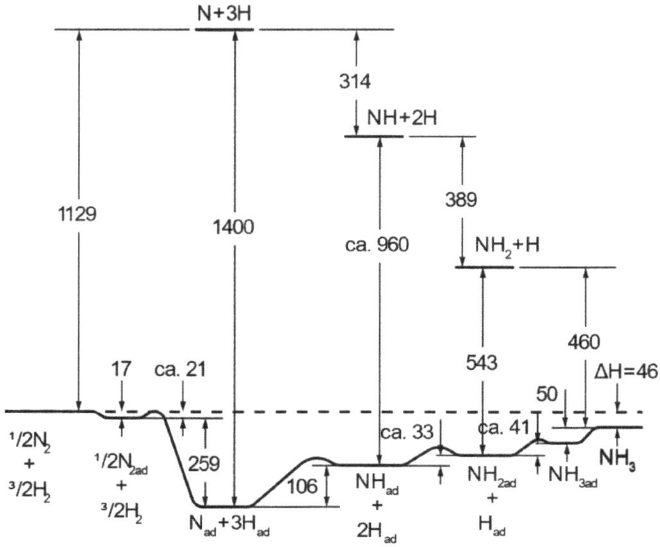

Figure 3.15 Mechanism and potential energy diagram of ammonia synthesis on potassium-promoted iron catalyst. The energies are given in kJ/mol [84].

(see Section 2.3.2.3) give a relative reduction of 10^{-4} to 10^{-6} to the reaction constant of dissociation.

The otherwise low surface coverage of N_2 also demands a high nitrogen pressure.

The reaction mechanism of the N_2 hydrogenation reaction is well understood. Energetics of reaction intermediates and reaction barrier of the ammonia synthesis reaction for the promoted Fe catalyst as experimentally determined by Ertl [84] is shown in Figure 3.15.

The low barrier of N_2 dissociation and consecutive steps of the strongly adsorbed N atoms that give adsorbed NH_x intermediates with increased value of x by hydrogen addition steps are shown. Once NH_3 is formed, it desorbs. At working condition the surface is partially covered also by coadsorbed ammonia (for a critical discussion see Bowker *et al.* [85]).

In the 1970s, Japanese [86] and British scientists at British Petroleum (BP) [87] realized that a process based on the more active alkali-promoted Ru catalyst is an attractive alternative to the Fe-based catalyst. In 1992 the Kellog company successfully started operation of ammonia synthesis plants based on the BP catalyst.

Thirty years after its discovery, surface science kinetic studies by Chorkendorff *et al.* from Technical University of Denmark [88], [89] demonstrated that the catalytically reactive site is a Taylor reaction site, which is only present at very low surface concentration. It is the step-edge, B5 site introduced in Section 2.2.1.2. The higher activity of the Ru catalyst versus that of Fe is ascribed to the stronger adsorption energy of nitrogen to the Fe surface. Product ammonia poisons Fe more severely than Ru.

3.2.1.1 *Structure Sensitivity of the Ammonia Synthesis Reaction*

The catalytic reactive site is a step-edge Taylor reaction center.

From X-ray studies of the crystal orientation of Fe crystallites of the ammonia synthesis catalyst, Brill *et al.* [90] deduced in 1967 that the (111) surface of Fe is substantially more reactive than other surfaces. Different from metals such as Pt or Pd, Fe has a body-centered cubic crystal structure. This stimulated Boudart *et al.* at Stanford [91] to study the Fe particle size dependence on a MgO-supported catalyst. The Fe particle size was varied between 1 and 30 nm. The Fe particles that are most active are the larger particles. This particle size dependence relates to reconstruction of the catalyst particles by contact with the reactants. The larger particles have the capacity to reconstruct by N adsorption to the more reactive phase [92]. As computationally shown in 1999 [93], the N atom preferentially adsorbs in a surface site where it coordinates to four Fe atoms. The formation of such stable Fe_4N sites provides the driving force for surface reconstruction.

The structure of the active site is closely related to the crystal structure of bulk Fe_4N. Interestingly this confirms intuitive ideas of Mittasch (see Section 1.3.2) on the necessity for the ammonia synthesis catalyst to be able to form intermediate nitrides. As we will discuss in the summarizing Section 3.3, surface reconstruction also

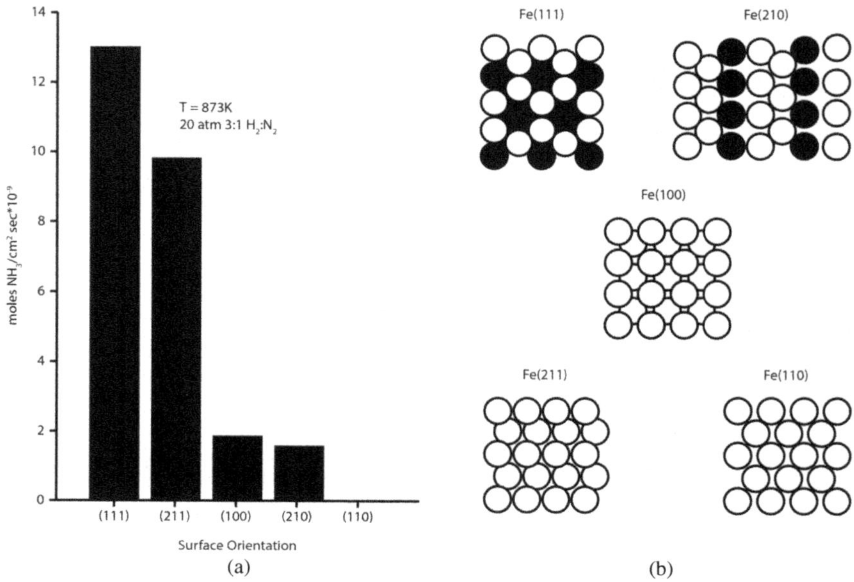

Figure 3.16 Surface structure sensitivity of the ammonia synthesis reaction according to low-pressure Fe single crystal measurements [97]. (a) The reaction rate of ammonia production normalized per surface Fe atoms is compared for different surfaces. (b) The crystal faces of the different Fe surfaces studied.

causes surface corrugation, which may create reactive Taylor-type sites [94].

In 1981 elegant surface science experiments on well-defined surfaces of Fe single crystals by Somorjai *et al.* [95], [96] confirmed the observed structure dependence of Brill *et al.* The measured reactivity of surfaces of different crystallographic orientation is compared in Figure 3.16.

At chosen reaction conditions, the normalized rates increase sharply when more reactive surfaces are compared. The reactivity of a surface increases when metal atoms of increasingly lower coordination in top and subsurface layers are exposed. The most reactive site has a close to step-edge configuration. The Fe(111) surface is most reactive.

The structure dependence can also be revealed by study of catalysts with different transition metal particle size. For Ru catalysts, a maximum reaction rate of ammonia synthesis occurs for Ru particles of 2 nm. This has been shown by scientists from the Topsøe Company Research and Technical University of Denmark [98]–[100]. This optimum in reaction rate is due to high activity of step-edge sites. Their concentration is maximum for the particles of 2 nm size. The structure of such particles is visualized in Figure 2.2b.

It confirms early suggestions in 1966 by van Hardeveld *et al.* at DSM in The Netherlands [101], [102].

The then just developed surface infrared spectroscopy was used to study the infrared spectra of N_2 adsorbed at low temperature to transition metal catalyst surfaces. Because N_2 has no dipole moment, it is infrared non-active. When adsorbed in asymmetric configuration, infrared activity is induced. Only infrared-active N_2 species were observed for metal particles of Ni, Pd and Pt particles in between 1.5 and 7 nm.

Van Hardeveld suggested that step-edge B5 sites as illustrated in Figure 3.17 provide the site for asymmetric adsorption of N_2.

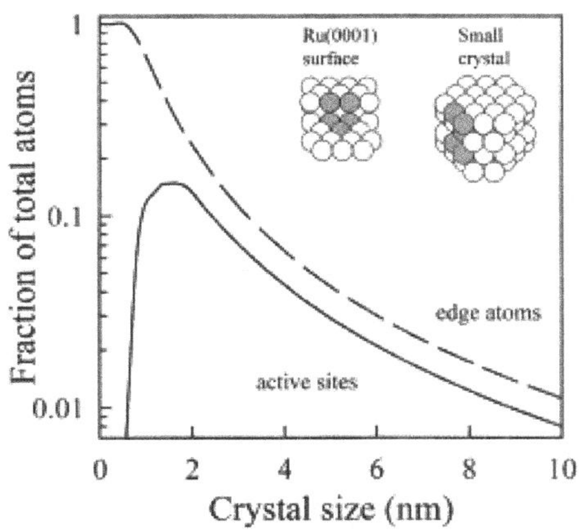

Figure 3.17 Fraction of edge atoms and active sites on small Ru crystals relative to the total number of atoms as a function of crystal size. The numbers are obtained from crystal models (see insert) exposing only (001) and (100) surface planes. The Ru crystal has hexagonal close-packed (hcp) structure. The active sites are present on the (100) surface. They consist of five Ru atoms exposing a threefold hollow hcp site and a bridge site close together (a B5-type site) where part of the atoms are edge atoms. This criterion is based on the structure of the active step site on the Ru(0001) surface also shown [99].

With Montfoort, a model was developed that predicts the maximum in surface concentration of such sites as a function of particle size.

This suggests that these B5 sites are the Taylor reaction centers for the ammonia synthesis reaction. Measurements by Nørskov *et al.* [100] in 2005 with High-Resolution Transmission Electron Microscopy demonstrated the presence of B5 sites on Ru particles of 2 nm size on a supported catalyst. With first-principle microkinetic simulations, they showed that the reaction rate predicted on the basis of the B5 sites agrees with experimental kinetic data.

The energetics and mechanism of the reaction are summarized in Figure 3.18, which is based on quantum-chemical DFT calculations using slab models of the Ru surface [103]. Potential energy diagrams of ammonia formation are compared for the non-stepped Ru(0001) surface and a stepped surface. The activation energy for N_2 dissociation is significantly reduced at the step-edge compared to that on the planar Ru(0001) surface. N_2 adsorbs weakly. On the non-stepped Ru(0001) surface, there is a high barrier for N_2 dissociation.

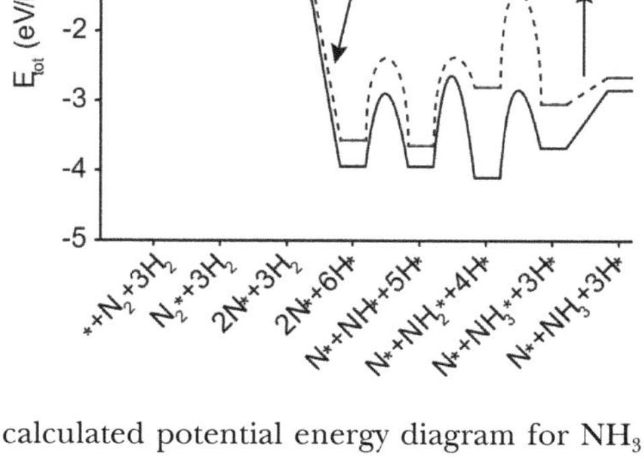

Figure 3.18 The calculated potential energy diagram for NH_3 synthesis from N_2 and H_2 over close-packed Ru(0001) (drawn line) and stepped Ru surface (dotted line). A* denotes an empty site and X* an adsorbed species [103].

This barrier of 200 kJ/mol reduces to 80 kJ/mol on the stepped surface. This reduction in activation energy of N_2 is the reason for the preference of the Taylor step-edge sites.

Adsorbed N, NH, and NH_2 are at the minimum of the potential energy diagram. Elementary reaction activation energies are of the same order of magnitude as the activation energy of N_2 dissociation on the stepped surface. The energy of adsorption of ammonia is 130 kJ/mol. This relatively high adsorption energy of ammonia is the reason that the presence of ammonia reduces the reaction rate.

Reconstruction of the catalyst surface by reaction is one of the reasons for the complex composition of practical ammonia synthesis catalysts. One of the roles of catalyst promoters is to maintain the catalyst surface at optimum site composition and structure [104]. The role of reconstruction in transition metal catalysis is discussed in Section 3.3.

3.2.1.2 *The Associative N_2 Activation Reaction*

In biological nitrogen fixation by the hydrogenase enzyme, the N_2 bond cleaves after addition of three hydrogen atoms.

In previous sections, nitrogen molecule activation by heterogeneous transition metal catalysts at high pressure and medium temperature was discussed. In these systems, reaction is initiated by dissociation of N_2. In biological nitrogen fixation, the ammonia-forming reaction mechanism is different: H atom addition to N_2 occurs before the N–N bond cleaves [105]. The biological enzymatic ammonia synthesis reaction is a low-temperature electrochemical reaction. In the hydrogenase enzyme, the N_2 reduction reaction is catalyzed by a sulfidic inorganic cluster of composition $MoFe_7NS_8$. This is the cofactor of the enzyme.

Hydrogen attachment happens in a sequence of proton addition and electron transfer steps. N–N bond cleavage occurs after formation of intermediate N_2H_3. In subsequent elementary reaction steps, H atoms are added to NH_x intermediates to finally give NH_3. This low-temperature route is driven by the electrochemical potential of the biological system.

Organometallic complexes have been explored for low-temperature nitrogen activation reactions. Systems have been discovered that activate N_2 by a reaction analogous to that of the hydrogenase enzyme. On organometallic catalysts with Mo or Fe cations as reaction center, the N_2 molecule adsorbs end-on and N–N bond rupture happens after protonation of one of the nitrogen atoms [106]. Catlow *et al.* have recently suggested that associative nitrogen activation, where the N–N bond cleaves after hydrogen addition, may play a role also in defected heterogeneous alloy catalysts of composition Mo_3Co_3N [107].

In the next sections on CO activation by heterogeneous catalysts, direct CO dissociation is also seen to compete with the associative activation mechanism.

3.2.2 *CO Hydrogenation Reactions*

An important kinetic difference between N_2 and CO activation is the stronger adsorption energy of CO.

The Sabatier-Senderens reaction that gives methane by hydrogenation of CO_2 and is catalyzed by finely dispersed Ni was discovered in France in 1897 [108]. It was part of Sabatier's Nobel award-winning investigations of catalytic hydrogenation. This system is a combination of two reactions catalyzed by the same catalyst:

1) The inverse water-gas shift reaction that is endothermic:

$$CO_2 + H_2 \rightarrow CO + H_2O, +41 \text{ kJ/mol} \tag{3.2}$$

and 2) the methanation reaction of CO that is exothermic:

$$CO + 3H_2 \rightarrow CH_4 + H_2O, -206 \text{ kJ/mol} \tag{3.3}$$

The CO_2 hydrogenation process currently attracts increasing interest due to efforts to reduce CO_2 emission as part of climate change abatement. The CO and CO_2 methanation reactions can be used to convert coal into methane. For this, coal is converted into synthesis gas by the endothermic steam gasification reaction that produces a mixture of H_2, CO_2, and CO. Hydrogen is derived from synthesis gas by application of the water-gas shift reaction that is the reverse of Eq. (3.2), i.e., converts CO to CO_2.

Closely related to the methanation reaction of CO is the Fischer-Tropsch reaction that converts synthesis gas into liquid fuel molecules. In 1923 Fischer and Tropsch [109] discovered the process named after them at the then Kaiser-Wilhelm Institute, which is now the Max Planck Institute in Mühlheim.

Another important hydrogenation reaction of CO_2 and CO is the production of methanol. This reaction was also discovered by Senderens and Sabatier in 1905 [110] but is catalyzed by Cu instead of Ni.

These reactions, that at present are part of large industrial plants at several locations around the world, will be the subject of the following sections on CO conversion catalysis. The main subject is the reaction mechanism of the respective reactions. This is complemented with a short exposition of their industrial invention background.

Whereas the methanation and Fischer-Tropsch reactions involve cleavage of the C–O bond, the bond of CO remains intact in the reaction that gives methanol. The mechanism of the inverse water-gas shift reaction and methanol synthesis are related.

As with the previous reactions, the key mechanistic question that is addressed is the nature of the surface intermediates that determine the selectivity of respective reactions. This is investigated in relation to structure and composition of the catalyst.

One of the major reasons for the differences in kinetics between ammonia synthesis and CO conversion is the difference in respective adsorption energies of N_2 and CO. On a transition metal such as Ru, they are –40 kJ/mol for N_2 and –180 kJ/mol for CO. Very differently from N_2, it implies that the catalyst surface is covered to a large extent by CO. Ammonia synthesis kinetics is first order in N_2, but CO conversion kinetics is negative order to zero order in CO, except for methanol formation that is catalyzed by weakly adsorbing Cu.

The role of promoters in the two reactions is also quite different. Potassium is an important promoter of the Fe catalyst that is used for ammonia synthesis. Promoted Fe catalysts are used as well in the Fischer-Tropsch reaction. Potassium addition to the Fe catalyst increases the N_2 adsorption energy [111]. This decreases the apparent activation energy of N_2 dissociation and enhances its rate.

In the Fischer-Tropsch reaction, potassium promotes selective formation of higher hydrocarbons. Potassium will not affect the chemistry of CO. Instead the main role of potassium is to suppress dehydrogenation of partially hydrogenated CH_x surface intermediates that lead to chain growth [82].

3.2.2.1 *Mechanism of the Fischer-Tropsch Reaction*

The reaction rate of CO dissociation versus that of chain growth determines Fischer-Tropsch selectivity.

The Fischer-Tropsch reaction converts synthesis gas into methane and liquid fuels. The last product is especially important to nations that do not have access to oil. In the first part of the previous century, this provided the main incentive to develop the Fischer-Tropsch process in Germany. In the second part of the century, large Fischer-Tropsch plants were constructed in South Africa because this country was excluded from oil imports.

After the oil crisis in 1973 that threatened oil access to the Western countries, the Fischer-Tropsch process gained interest again. It was explored and implemented between 1990 and 2010 in large plants in Malaysia and Qatar that convert natural gas into liquid fuels. In the past twenty years, conversion of synthesis gas produced from biomass is also explored.

The different processes are optimized with different catalysts because the H_2/CO ratio varies with carbon resource. The preferred catalyst for conversion of coal-derived synthesis gas is a potassium-promoted Fe catalyst. For conversion of natural gas-derived synthesis gas, Co is used.

Around 1975 an alternative to Fischer-Tropsch was discovered at Mobil Oil Company. This led to the methanol to olefin (MTO) or gasoline (MTG) processes [112], catalyzed by solid acid catalysts. These processes are based on natural gas as feedstock.

In New Zealand, Norway, and China these reactions have become the heart of large industrial scale developments. The mechanism of these reactions is discussed in Chapter 5.

The historic development of the discovery of the Fischer-Tropsch mechanism is a fascinating story. It shows how, since its discovery in 1923, a succession of mechanistic hypotheses became initially accepted and subsequently rejected. In a timespan of one or two decades, this happened at least four times. And still the dispute is not completely settled. Slowly a fruitful approach towards the prediction of the optimum site structure and composition has developed.

A major reason that mechanistic interpretation changes with time is instrumental technique improvement. Product analysis, catalyst characterization, as well as kinetics experimentation improved.

The Fischer-Tropsch reaction network is illustrated in Figure 3.19. It consists of a succession of three main reaction types.

— Activation of chemisorbed CO gives "C1" surface reaction intermediates. The nature of the "C1" species has been speculated for the most part of the previous century. A "CHOH"-type intermediate as well as "CH_x"-type intermediate have been suggested. C–O bond cleavage can be direct or happens after hydrogen atom addition to CO.

— The chain growth reaction, which couples the "C_1"surface intermediate monomers to long-chain hydrocarbons, is analogous to a polymerization process. It covers the surface with adsorbed

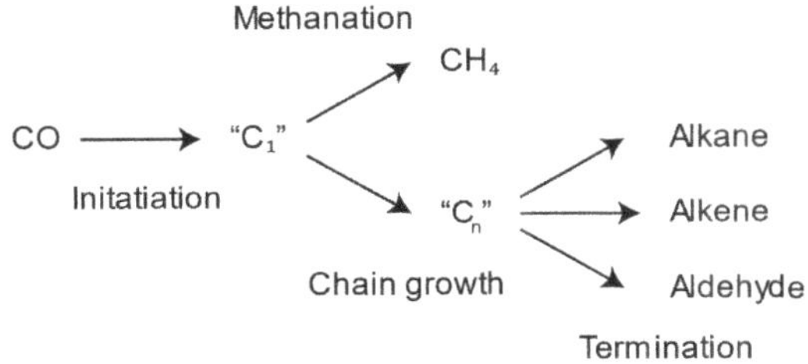

Figure 3.19 Scheme that illustrates the three mechanistic regimes of the Fischer-Tropsch reaction.

alkyl species of different chain length. It has been proposed that the chain growth proceeds through CO insertion into "C_1"or "C_n" intermediates of the growing hydrocarbon chain [113], [114], [115] or that chain growth happens by insertion of partially hydrogenated CH_x intermediates into the growing hydrocarbon chain [116], [117]. Alternatively, Emmett *et al.* [118] suggested chain growth by the partially hydrogenated CO intermediates. Of the different chain growth proposals, the mechanism with "CH_x" monomers is most probable.

— The termination reaction that removes the final intermediates as products from the surface. Hydrogenation gives alkanes, β C–H bond cleavage or other reaction give alkenes, and CO insertion can give aldehydes and alcohols or acids by consecutive reactions.

The relative rate of chain growth versus termination determines largely selectivity with respect to the methanation reaction.

Here the mechanisms of the elementary reactions that contribute to the respective regimes are presented.

Following the classical Mittasch idea that catalysis proceeds through intermediate compound formation, Fischer and Tropsch proposed that surface carbide intermediates formed by dissociation of CO polymerize in the chain growth reaction [109].

Emmett and colleagues, then at Gulf company [119], challenged this suggestion and devised an elegant experiment using radioactive ^{14}C-labelled molecules. They introduced a now classic approach to discriminate between reaction with surface intermediates versus gas-phase molecules. This experiment has since been followed by many researchers working on different catalytic systems.

Their idea was to carbide the surface by initial decomposition of ^{14}C-labelled CO and to expose the catalyst to synthesis gas at reactive conditions. They were interested in the ^{14}C distribution in the product molecules. They used the Fe catalyst for their studies.

Since no radioactivity is incorporated into the final product, they concluded that bulk iron carbide atoms are not reaction intermediates for chain growth. Instead Emmett [118] proposed in 1953 the partially hydrogenated CO intermediates, "CHOH", as monomers in the surface polymerization reaction. They based this on experiments with ^{14}C-labelled alcohols and noted that their radiochemical labels became incorporated into the product hydrocarbons.

Twenty years later, Pichler and Schulz of the University of Karlsruhe [120], [121] proposed a different intermediate for the Fischer-Tropsch reaction. Instead of insertion of the intermediate "CHOH" into the growing chain, they suggested that adsorbed CO inserts into the growing hydrocarbon. After CO insertion, the C–O bond cleaves and the growing chain lengthens. This insertion reaction of CO is thought to be analogous to the liquid-phase hydroformylation reaction that is catalyzed by the $Co_2(CO)_8$ carbonyl complex. The catalytic cycle of its reaction mechanism is illustrated in Figure 3.20.

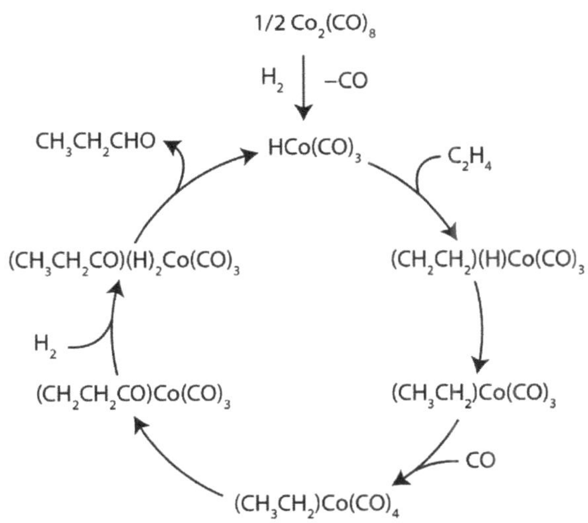

Figure 3.20 The catalytic cycle for the $HCo(CO)_4$-catalyzed hydroformylation of ethene. Hydroformylation is the reaction of an alkene with carbon monoxide and hydrogen to yield an aldehyde. The reaction was discovered by Otto Roelen in the laboratories of Ruhrchemie AG, Oberhausen-Holten in 1938 when he tried to recycle product olefins to the Fischer-Tropsch synthesis reactor. The reaction was initially named the "oxo reaction" [122].

The hydroformylation reaction produces an aldehyde or ketone from reaction of alkene (a product of the Fischer-Tropsch reaction) with CO and H_2. This remarkable homogeneous reaction was discovered in 1938 by Roelen at Ruhrchemie as a liquid-phase reaction in the liquid product fluid of the Fischer-Tropsch reaction in 1938 [123]. The active catalytic complex is molecular $HCo(CO)_4$ that is formed *in situ* by decomposition of the Co carbonyl dimer. Reaction is initiated by desorption of CO and subsequent alkene adsorption. The hydroformylation reaction is currently used at large scale to oxygenate hydrocarbons mainly in the detergent industry.

There is no direct evidence that the CO insertion reaction is also part of the chain growth reaction catalyzed by the heterogenous catalyst. The CO insertion of adsorbed surface alkyl intermediates is a reaction that terminates the chain growth and leads to oxygenated products.

In the 1970s at Shell, in transient CO hydrogenation experiments of Biloen and Sachtler, Co, Ni, and Ru catalysts were pretreated with ^{13}CO. Conditions were used where CO dissociates and adsorbs as C and O adatoms. When exposed to a flow of CO and hydrogen, ^{13}C-labelled hydrocarbons from pre-dissociated ^{13}CO were found in the product [124].

Co, Ni, and Ru are active Fischer-Tropsch or methanation catalysts but do not form stable carbide phases [125], [126]. These experiments led to the conclusion that reactive CH_x surface species are the "C_1" intermediates that are part of the Fischer-Tropsch chain growth reaction.

That CH_x intermediates indeed are able to oligomerize was proven in an elegant model experiment with CH_2N_2 and ^{13}CO-labelled intermediates. Brady and Pettit from the University of Austin [127], [128] decomposed CH_2N_2 on a Co catalyst that they then exposed to ^{13}CO and H_2. They demonstrated the presence of the ^{13}C label in product hydrocarbons.

Twenty years later, advanced NMR measurements in combination with gas chromatography / mass spectrometry analysis by Maitlis *et al.* from the University of Sheffield [116] confirmed, also from quantum-chemical calculations [117], [129], that "CH" is the

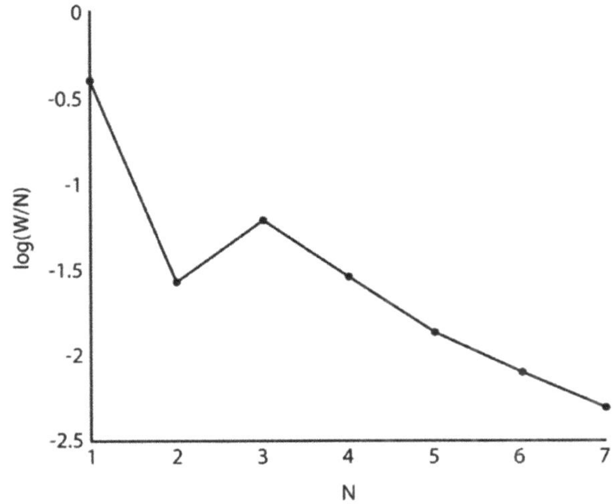

Figure 3.21 Anderson-Schulz-Flory plot of the formation of hydrocarbon products from CO hydrogenation. Plot of log(W/N) against N, where W is the weight fraction of products having carbon number N [130].

main "C_1" intermediate that inserts into surface-adsorbed alkylidene or alkenyl intermediates.

The product distribution as a function of hydrocarbon chain length is nearly logarithmic as illustrated in Figure 3.21. Methane and ethene or ethane are exceptions. This is not surprising since only for hydrocarbons with three or more carbon atoms "CH_x" did insert into a growing hydrocarbon chain.

The experimentally observed near-logarithmic dependence of the chain length of the product molecules is common in polymerization chemistry [131], and for the Fischer-Tropsch reaction it is called the Anderson-Schulz-Flory distribution [132], [133]. According to the Schulz-Flory polymerization model, chain growth happens by addition of the monomer to a growing polymer chain that is end-on adsorbed to the catalyst. For logarithmic chain length dependence, the insertion reaction rate has to be independent of growing polymer chain length. Then $\frac{C_{n+1}}{C_n} = \alpha$ is a constant. When the only product is a linear alkane, C_n is the surface concentration of the alkyl chain and α is the chain growth probability that varies between zero and one. It depends mainly on the ratio of the reaction rate of chain lengthening $r_{CC}^{f}\theta_1$ of intermediate C_n to C_{n+1} versus the reaction rate of surface intermediate "C_n" termination r_t that leads to product desorption. α is the solution of Eq. (3.4) [134].

$$\alpha = \frac{r_{CC}^{f}\theta_1 + r_{CC}^{b}\theta_V\alpha^2}{r_{CC}^{f}\theta_1 + r_t + r_{CC}^{b}\theta_V} \tag{3.4}$$

In Eq. (3.4), θ_1 is the surface concentration of the "C_1" monomer species that inserts into the growing hydrocarbon chain, θ_1 is the surface vacancy concentration, and r_{CC}^{f} and r_{CC}^{b} the respective forward "C_1" insertion and reverse alkyl C–C bond cleavage reaction rate constants.

In most kinetic modelling studies, r_{CC}^{b} is assumed to be zero. However, the chain growth reaction is reversible [135]. The "CH_x" insertion chain growth reaction mechanism is consistent with this reversibility.

A schematic representation of the chain growth reaction according to the Pichler-Schultz and Biloen-Sachtler reaction mechanisms is given in Figures 3.22a and 3.22b.

An important difference of the Fischer-Tropsch reaction with conventional polymerization is that in the Fischer-Tropsch reaction, the monomer "C_1"is to be generated *in situ.* This causes an important kinetic difference between the two mechanisms of Figure 3.22.

Within the Biloen-Sachtler mechanism, the reaction rate of "CH_x" formation from adsorbed CO, r_{CO-CHx}, has to be fast compared to the reaction rate of chain growth termination r_t. Otherwise there can be no buildup of "CH_x" concentration and it will be reacted away as methane. This implies that CO dissociation has to be fast and cannot be reaction rate controlling (in order to have a high α, r_t also has to be slow compared to the reaction rate of chain growth r_{CC}^{f}). As is discussed in the next section, this conclusion has an important implication for the structure dependence of the reaction.

Within the Biloen-Sachtler mechanism, the consumption rate of CO is maximum when $r_{CO-CH_x} = r_{cc}^{f}$ [136].

The Biloen-Sachtler mechanism is very different from the Pichler-Schulz chain growth kinetics of Figure 3.22b in which CO is

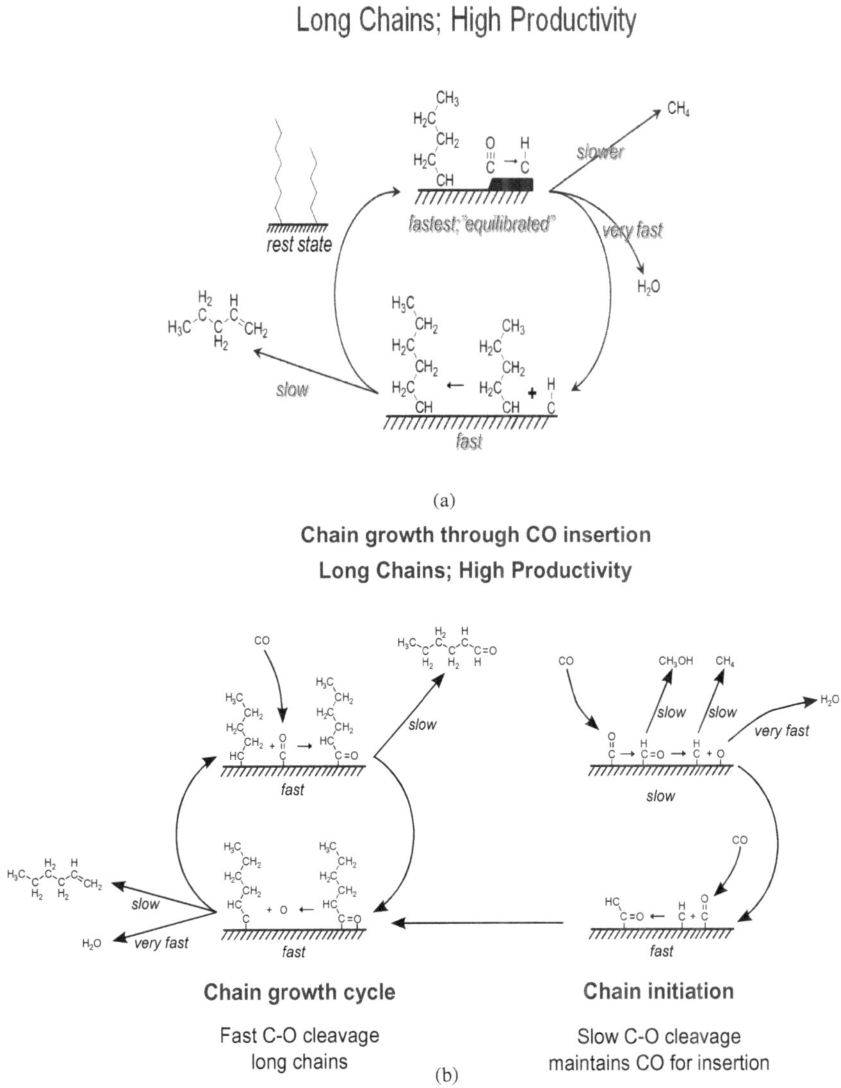

Figure 3.22 Comparison of the (a) Biloen-Sachtler and (b) Pichler-Schulz mechanistic proposals of the Fischer-Tropsch reaction towards linear alkanes. The relative rates of respective elementary steps for long alkane selectivity are indicated. For selective hydrocarbon chain growth versus methanation, the forward rate r_{CC}^{f} of chain growth has to be fast compared to reaction rate of termination r_t. Only alkyl chain intermediate termination by hydrogen atom addition or β C–H bond cleavage is indicated.

monomer, or the Emmett mechanism where "CHOH" is monomer. Here non-dissociated CO participates in the chain growth reactions. Its concentration should remain high. In the Pichler-Schulz mechanism, the reaction rate of CO dissociation has to be slow compared to the insertion rates of the "CHOH" or CO species into the growing hydrocarbon chain. In turn this is to be fast compared to the reaction rate of chain growth termination [137].

DFT calculations that compare the activation barriers of "CHO" insertion into the growing hydrocarbon chain indicate that the rate of this insertion reaction is incompatible with a high chain growth rate [117], [129], [138]. This is illustrated in Figure 3.23.

One notes the substantially higher barrier of the overall C–C bond formation reaction through CO insertion compared to the "CH_x" route.

The acceptance that the Biloen-Sachtler mechanism is the dominating reaction path for chain growth relates strongly to the increasing experimental evidence that CO dissociation has to precede chain growth.

With respect to the methanation reaction, Ponec at the University of Leiden [139] did also demonstrate with ^{13}C-labelled CO redeposition experiments on Ni that the surface "CH_x" species are the intermediates that give methane. Dissociation of CO must precede hydrogenation of the "C_1" intermediate. A similar conclusion was made by Rabo from Union Carbide [140]. He demonstrated that only reactive metals such as Ni, Co, and Ru that dissociate CO give methane or additional hydrocarbons, but a less reactive metal such as Pd or Cu gives methanol as product.

Importantly, as is discussed in the next section, the energies of the adatoms C_{ads} and O_{ads}, rather than the adsorption energy of CO, is a good measure of surface reactivity [141]. When reactivity of different metals is compared, trends in adsorption energies of the atoms instead of that of the molecules determine differences in activation barriers of the CO dissociation elementary reaction.

For selective chain growth, CO dissociation is to be fast compared to the elementary reaction rate constant of "C_1" insertion into

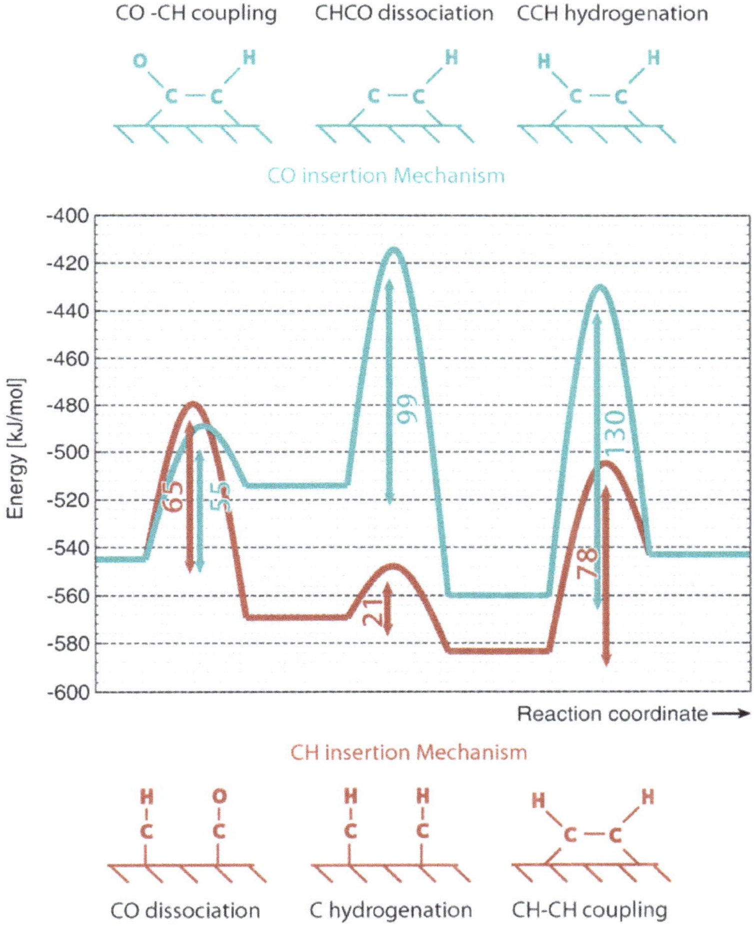

Figure 3.23 A comparison of activation barriers of elementary reaction steps for C–C coupling according to the "CH_x "mechanism (red) and the CO insertion chain growth mechanism (blue) on Ru(11$\bar{2}$1) surface according to DFT calculations [136]. Energy reference is CO and H_2 in gas phase.

the adsorbed hydrocarbon chain. Otherwise "C1" species are depleted and methanation dominates. This criterion determines catalyst composition as well as structure dependence.

Less reactive Ni with a relatively slow CO dissociation reaction rate is the preferred methanation catalyst, while more reactive Co and Ru with a high dissociation rate of CO are excellent Fischer-Tropsch catalysts.

The structure dependence often reveals itself by catalyst reactivity variation with change in transition metal particle size. The CO conversion rate increases with particle size and levels off for particle sizes in between 4–6 nm [142]–[145]. There is also a large change in selectivity as a function of particle size. Only the larger particles are selective for formation of higher hydrocarbons. The small particles mainly produce methane.

For carbon nanofiber-supported Co particles, this particle size dependence is shown in Figure 3.24. The levelling off of the increase in rate of normalized CO conversion occurs for the same particle size where methanation selectivity drops.

In contrast, surface science experiments with Ni single crystal catalysts by Goodman *et al.* from Texas A&M University [146] indicated that reaction rate of CO conversion is independent of surface structure. This apparent contradiction was resolved once the structure of the single crystal surfaces was studied by Scanning Tunnelling Spectroscopy [147].

Single crystal surfaces are never ideal, but always contain a small concentration of imperfections that are of the step-edge type. Similar to N_2, such step-edge sites catalyze CO bond cleavage significantly faster than the surface terraces (Section 2.3.3.4). This happens essentially because of dramatic reduction of the activation energy of the C–O bond-cleaving elementary step. The difference in reaction rate relates to the relative concentration of such step-edge sites in the respective surfaces. This appears to be very similar.

As discussed in Section 3.2.1.1 (Figure 3.17), the concentration of the step-edge sites depends on particle size. The concentration is maximum for particles with size of 2 nm.

For the methanation and Fischer-Tropsch reactions, the question of why reactivity becomes maximum at a particle size larger than predicted and then also becomes independent of particle size was answered once it was realized that large particles reconstruct. Surfaces highly covered with reaction intermediates or carbon atoms can reconstruct. This reconstruction releases stress

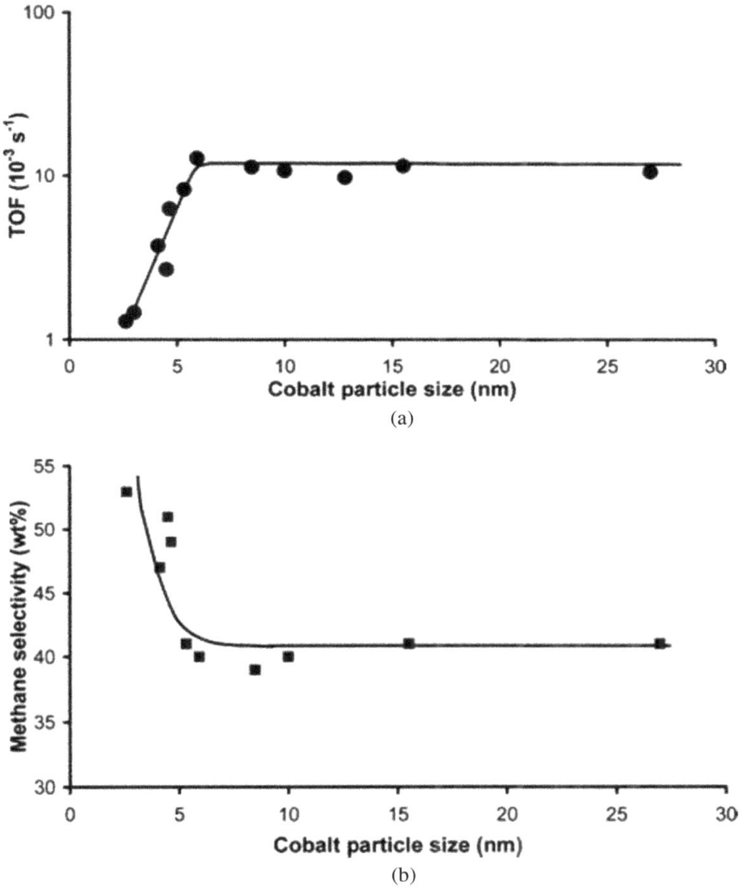

Figure 3.24 Dependence of Fischer-Tropsch reaction on particle size. (a) Turnover frequency (TOF) as a function of the Co particle size (220°C, H_2/CO, 2.1 bar). (b) The influence of cobalt particle size on methane selectivity (220°C, H_2/CO, 2.1 bar) [144].

by metal atom overlayer formation that creates step-edge sites [94], [148], [149].

In Figure 3.25 this is illustrated by simulations of the transformation of a Co surface terrace induced by high coverage adsorption with C atoms [149]. An overlayer of Co surface islands is formed with a size of the order of 2 nm. The edges of these islands contain step-edge B5 sites.

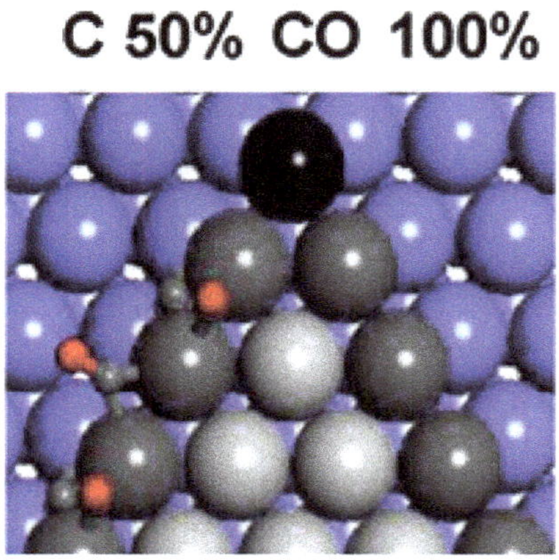

Figure 3.25 Reconstruction of Co(0001) surface by exposure to synthesis gas in the Fischer-Tropsch reaction. Simulated structure of C/CO-covered triangular islands with CO and C adsorbed on step-edge sites [149].

The formation of B5 sites by surface reconstruction also on terraces explains the particle size dependence of Figure 3.24.

The structure dependence of CO dissociation depends on the details of C–O bond activation. Low activation energy of the direct CO dissociation reaction requires step-edge sites. However, associative CO activation with C–O bond cleavage after initial H attachment to CO is less structure demanding. It has generated a debate [150] as to whether at synthesis condition on Co catalysts the reaction dominates on step-edges [151], [152] or on non-stepped terrace sites by an associative CO activation mechanism [153], [154] .

3.2.2.2 *Microkinetics of the Fischer-Tropsch Reaction*

The ratio of Taylor step-edge sites versus terrace sites determines the selectivity of chain growth versus methanation.

The reactivity difference of transition metal terraces versus stepped surfaces is generally large. Whereas bond dissociation reactions tend to prefer step-edge sites, recombination reactions as well as desorption reactions are expected to be faster for the less reactive terraces. The bond dissociation reaction requires strong interaction of their surface fragments with the catalyst surface, which is preferential for surface fragments to recombine or desorb a weak interaction with the surface.

This is relevant to the selectivity of the Fischer-Tropsch reaction.

As explained in the previous section, the Sachtler-Biloen "CH_x" mechanism implies that the selectivity for high chain growth versus methanation sensitively depends on the ratio of the lumped reaction rate constant of the C–O bond cleavage that generates CH_x intermediates and the reaction rate of CH_x insertion into the growing hydrocarbon chain. For low methanation selectivity, the reaction rate constant of the former has to be larger than the latter. However, high chain growth also requires that the reaction rate of hydrocarbon termination, which is responsible for product desorption, is slow compared to the reaction rate of chain growth.

Here quantum-chemical studies of elementary reaction rate constants and microkinetic simulations are discussed that provide insight into the surface structure and transition metal dependence. First the relation of chain growth rate and termination rate is discussed. This is followed by a discussion of CO dissociation and "CH_x" formation. The main focus is on structure sensitivity. The section is concluded with microkinetic simulations of selectivity and activity as a function of transition metal surface reactivity. A volcano-type relation between catalyst performance and surface reactivity is deduced (see also Section 2.4).

DFT calculations indicate that the activation energy for the CH_x to methane reaction varies between 100 kJ/mol and 200 kJ/mol when a terrace or step-edge site of a transition metal such as Co or Ru is compared. This is a measure for the activation energy of the termination reaction rate r_t. Activation energies for the chain growth reaction can be as low as 60 kJ/mol on surface terraces, as has been found for instance for the recombination of CH_{ads} and CH_2 on the Ru(0001) surface [117]. For these Fischer-Tropsch active metals, quantum chemistry is consistent with the requirement that the reaction rate of chain growth by insertion of CH_x into

Table 3.1 Activation energies (eV; 1 eV = 96 kJ/mol) for CH_x and CH_y recombination on flat terrace and stepped surface of Co(0001) [136].

	C+C	C+CH	C+CH_2	C+CH_3	CH+CH	CH+CH_2	CH+CH_3	CH_2+CH_2	CH_2+CH_3
Flat	1.22	0.91	0.74	0.94	0.86	0.76	1.05	0.70	1.11
Step	2.43	1.96	1.34	1.09	1.76	1.32	1.55	0.22	0.73

growing hydrocarbon chain is fast compared with the reaction rate of chain growth termination.

An impression of structure sensitivity of activation energies of CH_x–CH_y recombination reactions on the Co surface is shown in Table 3.1 [129].

For C–C bond formation, whether a terrace or step-edge site is preferred depends on the degree of hydrogenation x or y of the recombining "C_1"species. CH_x–CH_y bond formation has a lower barrier on a terrace, except for the recombination of $CH_{2,ads}$. The latter relates to the π bond character of product ethene (see Section 3.3).

For the non-stepped Co(0001) surface, the activation energies for the chain growth termination reaction are between 80 and 100 kJ/mol [129]. Surface terraces sustain the chain growth since the activation energy of the hydrocarbon chain termination is higher than that of most of the carbon-carbon bond-forming reactions. The CH_{ads} and $CH_{2,ads}$ recombination reaction is preferred. For insertion into the growing reaction chain, $CH_{2,ads}$ is to be substituted by CHR_{ads}.

The structure sensitivity of C–O bond dissociation depends on its mechanism. Is C–O bond dissociation a direct reaction or an associative reaction that is preceded by addition of an hydrogen atom? The direct C–O dissociation reaction path is highly structure sensitive and requires step-edge sites. In contrast, the associative reaction path is less structure demanding. Respective reaction intermediates are HCO, COH, or CHOH. Activation energies for the different reactions are collected in Table 3.2.

On the transition metal terrace, the activation energy of direct CO dissociation is substantially higher than that of the CH_x to methane hydrogenation reaction, which as mentioned earlier on transition metal terraces varies between 100 and 140 kJ/mol.

Table 3.2 Direct versus associative CO bond cleavage (activation energies in kJ/mol).

	Terrace surface			Stepped surface		
	Co(0001) [155]	Ru(0001) [156]	Ni(111) [147]	Co(211) [157]	Ru(11$\bar{2}$1) [158]	Ni(211) [147]
CO → C+O	367	205	285	142	65	194
CO+H → [HCO] → CH+O	130	146		123	125	133
CO+H → [COH] → C+OH	250			158	149	146
CO+2H → [CHOH] → CH+OH	125					

Therefore, on a surface terrace for the Fischer-Tropsch reaction that proceeds via direct CO bond dissociation, selective methanation is predicted.

This is different on the surface terraces for the associative C–O bond cleavage reaction path, because due to the lower C–O bond cleavage barrier the Fischer-Tropsch reaction is possible.

The first experimental observation of the associative C–O bond dissociation mechanism via hydrogen atom attachment to CO was a surface science experiment where H atoms react from the gas phase with an adsorbed CO overlayer. Using high-resolution electron energy loss spectroscopy, Mitchell *et al.* observed intermediate CHO formyl species as well as formaldehyde [159]. Advanced spectroscopic experiments twenty years later by the Salmeron group in UC Berkeley [160] on CO dissociation by supported Co particles corroborated the presence of the associative CO activation reaction and showed also an enhanced rate of dissociation on larger nanoparticles. In agreement with the quantum-chemical calculations of Table 3.2 it suggests that, also on Co step-edge sites, the associative mechanism has the lower barrier.

The differences in activation energies between direct and associative C–O bond dissociation are generally less on step-edge sites or corrugated surfaces but the respective active energies are sensitive to the specific structure of the step-edge sites. It can even happen on

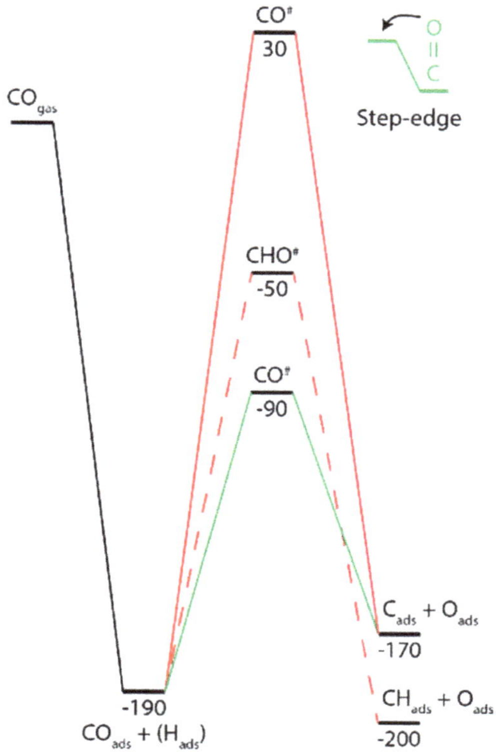

Figure 3.26 Comparison of potential energies for direct and associative CO activation on the Ru(1000) terrace (red) and direct CO dissociation on the reactive Ru($11\bar{2}1$) surface (green) [158].

step-edge sites that direct CO dissociation becomes more favorable than associative CO activation [147], [153], [156]–[158].

Complementary to Table 3.2, Figure 3.26 shows the extremely low barrier for direct CO dissociation on the Ru($11\bar{2}1$) step-edge. On the reactive step-edge sites, the associative C–O bond dissociation activation barrier is higher than that of direct C–O bond dissociation. This site is a favorable Taylor site for selective Fischer-Tropsch activity [176].

In the presence of hydrogen, CO dissociation towards CH_x formation can happen on stepped as well as non-stepped surfaces with activation energies that are comparable to CH_x hydrogenation to methane. On terrace sites, this will only be the case for C–O bond cleavage by the associative mechanism.

For transition metals of lower reactivity, the reaction rate of CH_x hydrogenation will increase, whereas that for the C–O cleavage reaction will decrease. Ni has low reactivity and the higher barrier for the C–O bond cleavage reaction. However, the activation energy of "CH_x" hydrogenation is lower and below the C–O bond cleavage barrier. This is different for the more reactive metals Co and Ru, which have stronger M–CH_x bonds. This is the reason that Ni is the preferred catalyst for the methanation reaction whereas Co, Ru, or Fe are preferred for the Fischer-Tropsch reaction (at reaction conditions the Fe catalyst consists of a reactive iron-carbide phase [125], [161], [162]).

Formation of methane, which is an undesirable co-product of Fischer-Tropsch active catalysts, is suppressed when the relative reaction rate of "CH_x" formation from CO exceeds that of the reaction rate of "CH_x" insertion into the growing hydrocarbon chain and with it the "CH_x" termination rate.

At step-edges, the low barriers of associative C–O bond dissociation are less than those of recombination reactions of "C" or "CH" with adsorbed hydrocarbon intermediates. Also on the step-edge sites, the activation energy of the hydrocarbon chain growth termination reaction is high compared to that of C–O bond cleavage as well as C–C bond formation.

In contrast, on terrace sites the activation energies of the C–O bond cleavage reactions are high compared to those of C–C recombination. This predicts that for a Fischer-Tropsch active catalyst, high selectivity of methane formation versus higher hydrocarbons formation is due to the presence of terrace sites.

In the second part of this section, microkinetic simulations are discussed, which probe selectivity as a function of the terrace site ratio versus that of step-edge B5 sites [151].

A question that arises is whether the reactive surface sites that give hydrocarbon chain growth will not become poisoned by strongly bound C atoms. This potential poisoning reaction is reduced by synergy of reaction at step-edge sites and terrace sites [163], [164].

The microkinetic simulations show that once a CH_x fragment is formed at the step-edge, it recombines with a growing hydrocarbon chain that is located on terrace sites next to the surface edge sites. Whereas CO bond activation occurs at the step-edge, the chain growth reaction happens on the surface terrace sites next to the dissociating CO molecule and CO dissociation is not suppressed by the presence of the growing hydrocarbon chain.

According to Iglesia *et al.* [154], [155], [165], the non-stepped Co surface is a selective Fischer-Tropsch catalyst. They suggest that the activation energy of the associative C–O bond cleavage reaction has lowered the activation energy such that it has become lower than that of chain growth termination. This disagrees with conclusions from the microkinetic simulations presented below. The explanation for this disagreement is that surface reconstruction has to be considered in the comparison with experiment. As explained in the previous sections, carbon adsorption induces surface reconstruction that generates reactive step-edge sites.

Microkinetic simulation with elementary reaction rate constants that are specific for a particular transition metal or reaction site structure can be used to determine the composition and structure of the reaction site for optimum catalytic performance. For a chosen surface structure, a plot of catalytic reactivity as a function of reactivity of transition metal has a bell-shaped dependence.

As discussed in Section 2.4, this dependence is predicted by the Sabatier principle and the curves are called kinetic volcano curves.

For the methanation reaction, Nørskov with Bligaard *et al.* [166] demonstrated in 2006 a volcano-type relation between the measured rate of methanation and the sum of the C and O adatom adsorption energies. The maximum in reaction rate was found for Co and Ru. In 2011 they accomplished a first-principle microkinetic simulation of the methanation reaction, and a two-dimensional plot of the rate of reaction as a function of C_{ads} and O_{ads} was constructed [167]. This simulation predicts that the Ni_3Fe alloy is most active.

The microkinetic simulations of the Fischer-Tropsch reaction by Filot *et al.* of 2014 [151] that are presented here illustrate the difference in selectivity of the terrace surface sites versus step-edge sites,

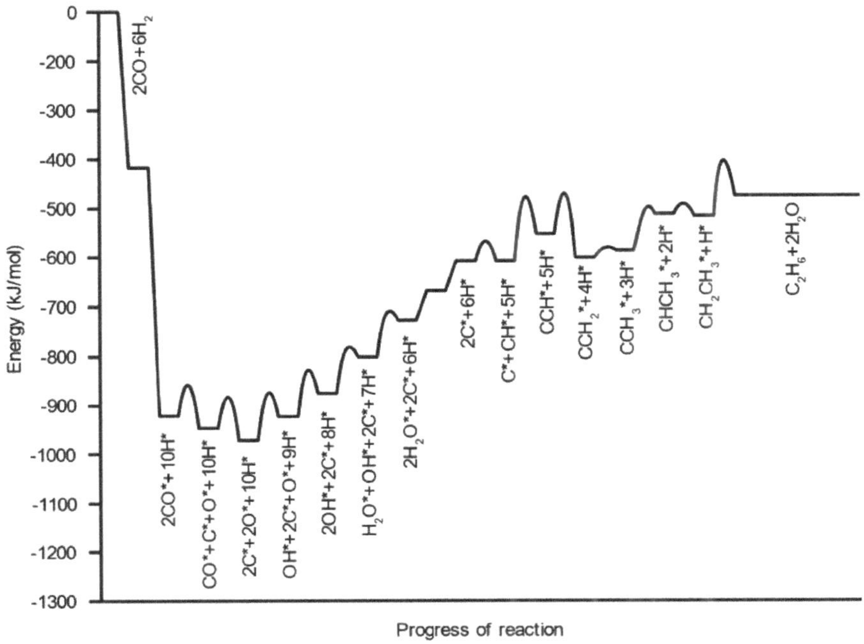

Figure 3.27 Potential energy diagram for ethane formation from carbon monoxide and hydrogen catalyzed by the Ru($11\bar{2}1$) surface [168].

and predict surface reactivity descriptor values for maximum selectivity.

From the DFT calculations, potential energy diagrams analogous to Figure 3.18 for the ammonia synthesis reaction are constructed. This is illustrated in Figure 3.27 for the conversion of CO and hydrogen to ethane.

In the simulations, all of the different competing reaction paths are considered, but Figure 3.27 shows only the potential energies of the C_{ads} and CH_{ads} recombination pathway. The potential energy initially shows a deep valley that upon removal of water moves up against the potential energy hill.

The reaction energies of hydrogen atom addition to C_{ads} and C–C bond formation show less variation. Differences in elementary reaction rate constants are mainly determined by differences in activation energies.

With the activation energies shown in Figure 3.27, elementary reaction rate constants are calculated according to the Eyring expression (Eq. (2.8a)) in Chapter 2. The activation entropies of surface reactions are considered small, but full rotational entropies are included for reactions between surface and gas phase [169] (see Section 2.3.2.3).

For the simulation results of Figures 3.28 of the Fischer-Tropsch reaction, 44 elementary reactions were included. They involve the many variations of elementary reactions and reaction intermediates that give C–O bond cleavage, C–C bond formation, H_2O formation, and hydrocarbon termination.

In order to include the surface polymerization reaction, the elementary rate constants of C_3 and higher alkanes are assumed to be independent of hydrocarbon chain length. Using such procedures, microkinetic calculations produce hydrocarbon product distribution curves very similar to those shown in Figure 3.21.

To compare the reactivity of different transition metals without having to redo the quantum-chemical calculations, an elegant procedure devised by Nørskov *et al.* [167] is to extrapolate data obtained for one transition metal to other metals. They suggest to use the BEP relation between the adatom adsorption energies and activation energies of elementary reactions. This only allows us to compare the reactivity of catalysts with the same surface structure, because the BEP relation only applies when the structure of transition states does not change. Rules analogous to the Bond Order Conservation rules (Section 2.3.2.2) can be used to relate the changes in adatom adsorption energies to those of partially hydrogenated or oxygenated surface fragments [170], [171].

Based on the elementary reaction rate constants of the Fischer-Tropsch reaction and its mechanism, microkinetic equations are formulated (Section 2.3.4). They determine the rate of surface fragment transformation and product formation as a function of temperature and reaction conditions.

For Ru($11\bar{2}1$), the results of microkinetic simulations are shown as two-dimensional volcano plots of reaction rate and chain growth probability α in Figure 3.28a [151]. In Figure 3.28a1 the rate of CO

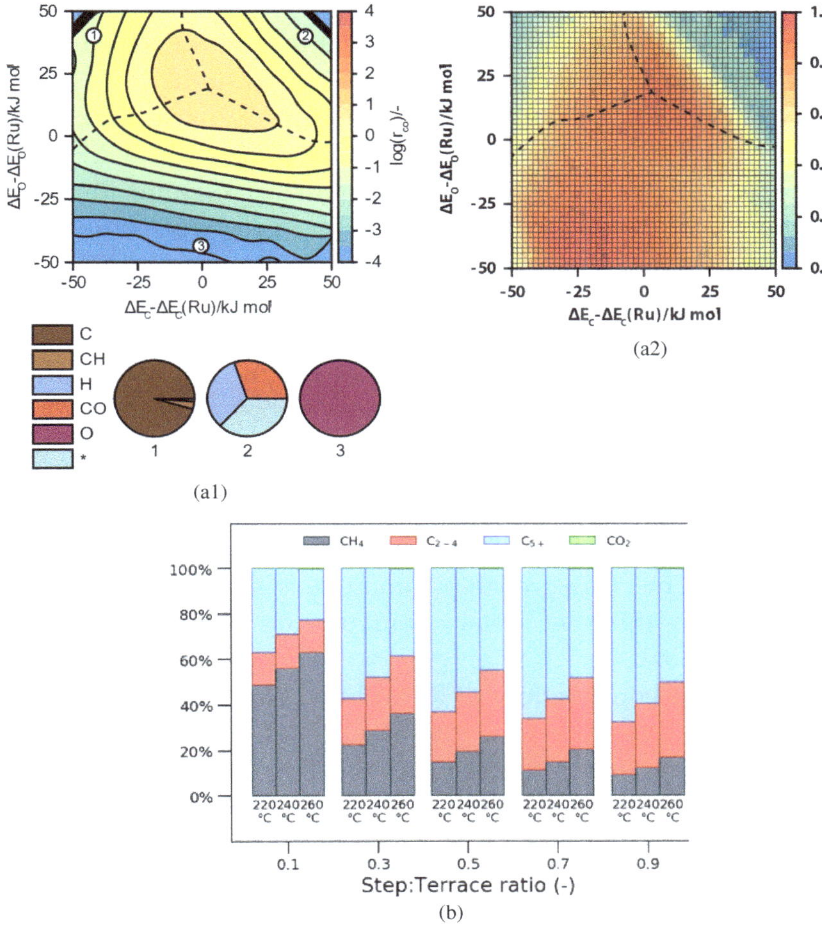

Figure 3.28 Microkinetic simulations of the Fischer-Tropsch reaction. (a1) Two-dimensional volcano plot of CO consumption rate (r_{CO}) as a function of C_{ads} and O_{ads} adsorption energies (T = 500 K; p = 20 bar; H_2/CO = 2) [151]. (a2) Two-dimensional volcano plot of chain growth probability α as a function of C_{ads} and O_{ads} adsorption energies (T = 500 K; p = 20 bar; H_2/CO = 2). Adsorption energies are indicated as difference compared to that on Ru (T = 500 K; p = 20 bar; H_2/CO = 2) [151]. (b) Hydrocarbon selectivity simulated on a combined surface consisting of Co(11$\bar{2}$1) and Co(0001) with varying Co(11$\bar{2}$1):Co(0001) ratio (T = 500, 520, 540 K, p_{H2} = 667 mbar, p_{CO} = 333 mbar) [172].

consumption is plotted as a function of the M–C and M–O adatom energies that are reactivity descriptors for a particular surface. Three different surface composition regions are indicated (1,2,3). In region 1 the surface is mainly covered with C_{ads}, in region 2 there are surface vacancies with CO_{ads} and H_{ads}, and in region 3 the main surface intermediate is O_{ads}. Region 2 dominates when the M–C and M–O bonds are not too strong. The other two regions realize when the respective M–C and M–O bonds increase in strength.

The presence of regions with different surface intermediate compositions implies that the rate-controlling step is different. The volcano maximum is at the triple point where the three surface composition regions meet. Here the rates of the respective rate-controlling steps are of the same order of magnitude. Figure 3.28a2 illustrates that the chain growth probability α is high in a substantial wide reactivity descriptor region. The maximum in CO reaction rate is close to the maximum value of the chain growth parameter α. The chain growth parameter α is also close to one over a wide range of different surface compositions where the CO conversion rate is low.

The surface compositions of Ru and Co (their reactivity is determined by the respective M–C and M–O bond energies) are quite different. On the more reactive Ru adsorbed O atoms dominate, whereas on Co there is a more equal distribution of surface intermediates. At conditions of the simulation, Co will have the higher rate.

The dramatic effect of relative step-edge site concentration on the selectivity of the Fischer-Tropsch reaction is illustrated by the selectivity of the synthesis gas conversion reaction derived from microkinetic simulations with Co surfaces that contain a mixture of the Co(0001) terraces and the step-edge sites of the Co(11$\bar{2}$1) surface of Figure 3.28b [172]. The chain growth probability is low at the Co(0001) terrace. When step-edge concentration increases, there is initially a large increase in long-chain hydrocarbon selectivity that levels off when the step/terrace ratio is half. In the mixed surface catalyst, main CO dissociation occurs on the step-edge sites. The chain growth factor α is high for the chain growth reactions that happen at the edge of the step-edge sites. The selectivity to undesirable methane increases when the relative concentration of the stepped surface decreases.

3.2.2.3 *Selectivity of the Fischer-Tropsch Reaction*

Incorrect adsorption energies of CO derived from calculations may give erroneous simulation results. The desorption energy of CO determines the temperature of reaction. Selective Fischer-Tropsch reaction requires low reaction temperature.

Insightful to the mechanism of the Fischer-Tropsch reaction is the understanding of the temperature dependence of the reaction. Figure 3.29 provides a schematic representation of the temperature dependence of the Fischer-Tropsch reaction and its surface reaction network.

In Figure 3.29 the temperature dependence of elementary reaction rate constants, chain growth parameter α, and surface coverage θ_{co} are compared.

The surface kinetic model of the Fischer-Tropsch reaction is given in Figure 3.29b [173], which is helpful for the interpretation of Figure 3.29a. It is also used here to summarize the Fischer-Tropsch reaction mechanism. Dependence on hydrogen pressure is implicit to the lumped elementary reaction rate constants. The scheme applies in the reactivity region where the rate of oxygen removal is fast.

For transition metals such as Co or Ru, when exposed to synthesis gas at low temperature the surface is covered with adsorbed CO. The minimum temperature necessary for the Fischer-Tropsch reaction to proceed relates to the temperature where CO desorbs. For CO to dissociate a surface site vacancy next to adsorbed CO has to be present, so that the C and O adatoms that are generated upon CO dissociation can be accommodated. As long as the surface is completely covered with CO, no surface vacancies are present and C–O bond cleavage remains inhibited. Therefore, the minimum temperature of reaction is determined by the temperature where CO begins to desorb.

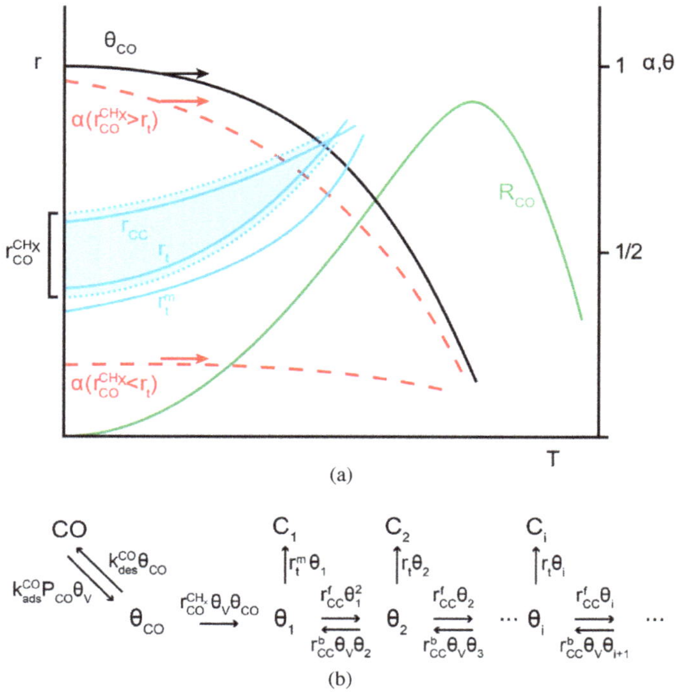

Figure 3.29 Temperature dependence of the selectivity of the Fischer-Tropsch reaction. (a) Schematic presentation of the temperature dependence of rate of CO consumption R_{CO}, elementary reaction rate r, chain growth probability α, and θ_{co} as a function of temperature. Left coordinate indicates reaction rate, right coordinate indicates α and surface coverage θ_{co}. (b) The lumped kinetic model of the Fischer-Tropsch reaction [136].

The decrease of ϑ_{CO} with temperature is indicated in Figure 3.29a. Once CO dissociates and a "CH_x" intermediate is formed by hydrogen atom addition, the chain growth reaction initiates and follow-up reactions that are part of the Fischer-Tropsch reaction can happen.

For high chain growth, $r_{CO}^{CH_x}$ (the lumped rate constant of "CH_x" formation from adsorbed CO) has to be fast compared to r_{cc}, the rate constant of chain growth (for simplicity this rate constant is considered unidirectional). Also, the rate constants of chain growth termination, r_t, and CH_x hydrogenation to methane, r_t^m, have to be smaller than r_{cc}.

In Figure 3.28a r_t^m is chosen to be smaller than r_t.

The rate constants of the respective elementary reactions are a strong function of the structure of the catalyst center. CO activation with low activation energy requires B5 step-edge sites. C–C bond formation is less sensitive to surface structure. There is an optimum in the reactivity of the transition metal. For high rate of C–O bond dissociation, a reactive transition metal is preferred. However, M–C or M–O bond energies should not be too large to prevent chain growth or surface blockage by carbiding or oxidation.

Fast formation rate of methane from "CH_x" or fast chain growth termination r_t happens on transition metals with the lower M–C bond energies such as Ni or surface terraces of more reactive metals. Surface terraces also have the weaker M–C bond energies compared to step-edges.

The higher activation energies for C–O bond cleavage of the surface terraces also favor methane formation.

Chain growth selectivity depends on how $r_{CO}^{CH_x}$ relates to r_{cc} and r_t. High chain growth requires r_{cc} to be larger than r_t. But an additional condition has to be satisfied.

When $r_{CO}^{CH_x}$ is smaller than r_{cc}, surface coverage will be dominated by adsorbed CO and reaction rate will be limited by "CH_x" generation. Selectivity to methane will be high.

In the opposite case, when $r_{CO}^{CH_x}$ is faster than r_{cc}, the surface coverage is dominated by growing hydrocarbons. Chain growth is also favored by a low surface vacancy concentration so that the reverse reaction of C–C bond cleavage of the growing chain is suppressed.

However, because of the low surface vacancy concentration, the rate of CO dissociation becomes suppressed.

Within the Biloen-Sachtler mechanism, the reaction rate is optimum when $r_{CO}^{CH_x} = r_{cc}$ and r_t is the slower rate constant [136].

The temperature dependence of the reaction rate constants depends on their activation energies. Because of their high activation energies, the reaction rates of termination increase and chain growth probability α decreases with temperature.

There is a maximum in the CO consumption rate R_{CO} as a function of temperature. At the temperature of the reaction rate maximum and beyond, the methanation rate dominates because CO coverage becomes low. The change in reaction with temperature is accompanied by a change in reaction order of CO from its negative value of −1 at low temperature to a positive value beyond the rate maximum. At low temperature, reaction is initially poisoned by CO and reaction rate increases because CO desorbs. Beyond the temperature of the reaction rate maximum, the decrease in surface coverage is not compensated anymore by the increased rate of CO dissociation.

A simplified kinetic model in which surface intermediate concentration is dominated by adsorbed CO (this implies that the rate of product desorption is fast) illustrates the relation between change in CO reaction order x and reaction rate maximum. This is given by Eq. (3.5):

$$R_{CO} = k_{diss}\,\theta_{co}\,(1 - \theta_{co}) \tag{3.5a}$$

$$= k_{diss}k_{ads}\,[CO]/(1 + k_{ads}[CO])^2 \tag{3.5b}$$

$$= k_{app}\left[CO\right]^{1-2\theta_{CO}} \tag{3.5c}$$

$$E_{app} = E_{diss} - E_{ads}(1 - 2\theta_{CO}) \tag{3.5d}$$

The adsorption energy of CO (appr. −180 kJ/mol) is larger than the activation energy of CO dissociation (appr. 130 kJ/mol). The condition for the existence of a temperature where R_{CO} is maximum is that the adsorption energy of the reactant intermediate is higher than the activation energy of the elementary reaction constant of bond dissociation.

Eqs. (3.5) can be used to determine θ_{CO}^{max} where R_{CO} is maximum: $\theta_{CO}^{max} = \left(E_{ads} - E_{diss}\right)/\left(2E_{ads}\right) <$ □. T_{max} follows from the adsorption isotherm expression of θ_{CO}. At T_{max} the order of reaction x is near zero and has already changed to slightly positive (see Section 2.4.2). The sign of E_{app} changes from positive to negative and $E_{app}^{max} = 0$. T_{max} shifts to higher temperature with increase of CO adsorption energy.

At the condition of high chain growth, simulations and experiment indicate that the CO consumption rate R_{CO} is close to zero order in CO [151], [174]. Then the surface is only partially covered with CO. Apart from the presence of surface vacancies, the surface is partially covered with growing hydrocarbon chains. High selectivity of chain growth happens at the lower temperature, but decrease in surface vacancies reduces the CO dissociation rate. There is a temperature optimum for maximum long hydrocarbon yield.

Quantum-chemical calculation of energies can have significant systematic errors. These partially cancel for reactions of surface intermediates that quasi-equilibrate.

This is different when reaction between surface and gas phase is considered. Then accurate prediction of adsorption energies is important. DFT-calculated adsorption energies tend to be higher than the experimental values. In kinetic simulations, this causes the predicted temperature of reaction to be too high, which has a large negative effect on simulated selectivity.

3.2.2.4 *Mechanism of Methanol Synthesis*

Surface formate is intermediate of methanol synthesis. It is formed by hydrogenation of CO_2.

In this section the mechanism of methanol formation is presented. Catalysts that are active in the formation of methanol from synthesis gas have a weak interaction with CO, since the C–O bond cleavage reaction rate must be relatively slow. The latter condition implies alternative mechanistic paths than direct hydrogen addition steps.

The history of the understanding on the reaction mechanism of methanol synthesis is intimately related to unravelling the nature of the catalytically reactive center. The catalyst that originally was used in industry is a mixed oxide and has a complex composition.

Methanol catalysis was discovered in 1913 and the first commercial production took place in 1923 in Germany by what is now the BASF company [21]. Matthias Pier developed the catalytic process that operates at the high pressure of 300 atm and at a temperature of 300–400°C. It uses a zinc chromite (ZnO-Cr_2O_3) catalyst that had been discovered by Alwin Mittasch. This catalyst is related to the

Fe_2O_3-Cr_2O_3 catalyst then also in use by BASF for the water-gas shift process that transforms a mixture of CO and H_2O into CO_2 and H_2. This process had been invented by the British industrialist Ludwig Mond in 1888.

Both the water-gas shift as well as the reverse reaction are important parallel reactions that play a role in the methanol reaction. The water-gas shift reaction catalyzed by the Fe_2O_3-Cr_2O_3 catalyst operates at a temperature comparable to the methanol synthesis reaction. The main role of Cr_2O_3 in this catalyst is to prevent sintering of Fe_2O_3 that is the catalytically active component [175]. Earlier (Section 3.1.2) Cr_2O_3 had been introduced as an active alkene hydrogenation or alkane dehydrogenation catalyst.

In the water-gas shift reaction, the redox properties of Fe_2O_3 facilitate decomposition of H_2O to hydrogen and adsorbed O. In the next step, CO_2 forms by the oxidation of CO with adsorbed oxygen atoms. This is called the regenerative mechanism [176]. Within the context of oxidation catalysis, related redox reactions are discussed in detail in Chapter 4.

Already in 1920 Armstong and Hilditch [176] had made an insightful mechanistic study of the water-gas shift reaction catalyzed by Cu-chromite. They concluded that CO and H_2O or reverse CO_2 and H_2 transform through formation of a formic acid intermediate. They deduced the formic acid intermediate by its trapping with ammonia. It is called the associative mechanism since hydrogen atom addition activates CO_2. These two mechanistic ideas, the regenerative mechanism versus the associative mechanism, are fundamental also to the mechanism of the methanol reaction.

In early methanol process development, Cu had been mainly rejected because of its rapid rate of deactivation.

This changed with the advance of synthesis gas that is produced from natural gas instead of coal. Because this synthesis gas is low in sulfur impurities, Cu, which is sensitive to sulfur poisoning, became an option as catalyst component. Industrially this is attractive because it makes replacement of the high-pressure process based on zinc chromite to a low-pressure process for methanol synthesis possible. Such a process was developed in 1963 by Davies and Snowdon and colleagues of British ICI Company [21], [177] that revolutionized the methanol synthesis industry. Their catalyst consists of Cu, Zn, and chromium, a formulation invented by Błasiak in 1947 [178].

Methanol is produced in a process that operates at lower temperature and pressure than the process catalyzed by the ZnO-Cr_2O_3 catalyst of Pier and Mittasch. Currently in many industrial plants, methanol is made from $CO/CO_2/H_2$ (10:10:80) mixtures over a $Cu/ZnO/Al_2O_3$ (60:30:10) catalyst at 530 K between 50–100 bar.

Mechanistic interest at that time shifted from the Cr- to Cu-based methanol or water-gas catalysts. However, it is useful to mention the early study on the Zn/Cr catalyst of methanol synthesis by Molstad *et al.* [20]. They established that ZnO as well as Cr_2O_3 individually catalyze conversion of synthesis gas to methanol. The rate of this reaction is enhanced for mixtures of these oxides.

They concluded that Cr_2O_3 stabilizes ZnO loss in the methanol catalyst. In this context, it is of interest to mention modern studies that indicate methanol formation by ZnO is a structure-sensitive reaction, catalyzed by its polar surface [179]. Cr_2O_3 may be a structural promoter that stabilizes this reactive ZnO phase. As discussed previously (Section 3.1.2), ZnO activates reactant bonds by a heterolytic mechanism.

The mechanism of the Cu/ZnO catalyst has been extensively investigated at ICI Research. In 1984 Waugh *et al.* discovered that the primary pathway to methanol formation is through conversion of CO_2 instead of CO. ^{14}C-labelled experiments demonstrated that CO_2 conversion to methanol is 100 times faster than CO [180], [181]. This reaction with CO_2 occurs together with the water-gas shift reaction that converts CO to CO_2 [182]. The water-gas shift is a fast reaction compared to methanol synthesis. Intermediate formate formation is identified as one of the key reaction intermediates for methanol from CO_2. When the formate decomposes, a methoxy intermediate is formed. Hydrogenation of methoxy to methanol is rate controlling.

Waugh *et al.* [182] proposed that in contrast to the methanol reaction, the water-gas shift reaction does not follow this associative pathway, but instead proceeds through the regenerative pathway where CO reacts with surface oxygen atoms or hydroxyl from decomposition of H_2O by the Cu catalyst. Surface OH_{ads} is readily formed by reaction of H_2O with oxygen adatoms. The water molecule then protonates the oxygen adatom and two OH_{ads} intermediates are generated.

The regenerative redox mechanism of the water-gas shift reaction is validated by single crystal experiments of Nakamura and Campbell [183]. CO reacts with OH through a COOH intermediate.

The discovery of the pathway to methanol by activation of CO_2 provides an alternative to earlier ideas on direct hydrogenation of CO by sequential H atom addition [184]. This reaction turns out to have a low rate. For this mechanism, Klier from Lehigh University suggested that the synergetic effect of Cu and Zn is the stabilization of Cu^{2+} cations dissolved in the ZnO matrix. This increases the reaction rate since CO adsorbs more strongly on Cu^{2+} than on reduced Cu. Twenty years later, high-resolution electron microscopy studies demonstrated instead that reduced copper crystallites present on the working catalyst are responsible for its catalytic action [185], [186].

The CO_2 pathway to methanol is consistent with Cu present as a metallic phase during methanol synthesis. Model experiments with commercial catalysts as well as single crystal surfaces probed with thermal desorption and infrared spectroscopy led to the formulation of the mechanism that consists of the following elementary steps [182], [187]:

— CO_2 decomposes into an adsorbed oxygen atom and CO.
— CO_2 reacts with adsorbed O to give intermediate carbonate.
— Consecutive hydrogen addition steps to carbonate give adsorbed formate. Alternatively, formate is directly formed from CO_2.
— Hydrogenation of CO is a slow reaction. The formation of methoxy intermediate is reaction rate controlling.
— Hydrogenation of formate gives intermediate methoxy that in consecutive reaction steps converts to methanol. The reaction

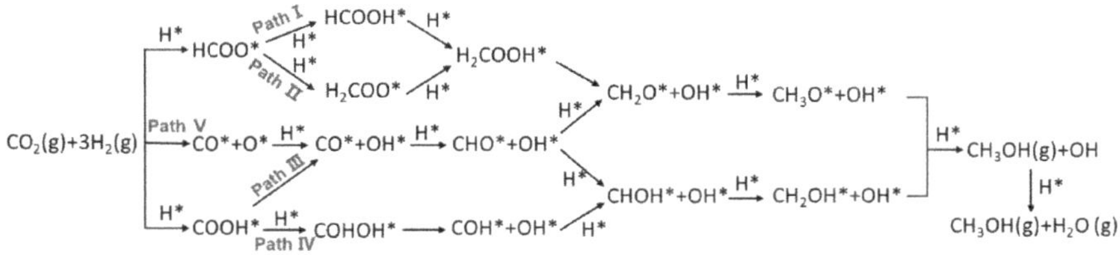

Figure 3.30 Reaction pathways considered for methanol formation from CO_2 hydrogenation [188].

rate of the hydrogenation of methoxy to methanol is rate controlling.

Figure 3.30 gives a schematic representation of the complex reaction network of the methanol synthesis reaction.

Methanol formation from CO_2 can in principle follow three pathways. The HCOO and CO+O paths are the main routes for methanol formation. The COOH path is the main route for the water-gas reaction. This is consistent with the regenerative route. CO then forms via initial reaction with adsorbed OH [187].

In the previous section, within the context of the Fischer-Tropsch reaction, it was mentioned that the minimum temperature of this reaction is determined by the temperature where CO desorbs so that vacancy sites become available for coadsorption of hydrogen and CO dissociation becomes possible. In methanol synthesis, the minimum temperature of reaction is determined by the temperature where formate decomposes and does not block surface sites.

The methanol synthesis reaction is structure sensitive [181], [189], [190]. Step-edge sites of corrugated Cu surfaces give increased reaction rates. The activity of Cu nanoparticles increases when particle size varies between 2 and 15 nm [191]. A plateau in activity is reached when particle size becomes 8 nm. This dependence on particle size is reminiscent of the particle size dependence of the earlier discussed Fischer-Tropsch reaction (Figure 3.23). It is consistent with step-edge sites as the reactive catalytic sites. The step-edge site concentration increases for larger particles and levels off

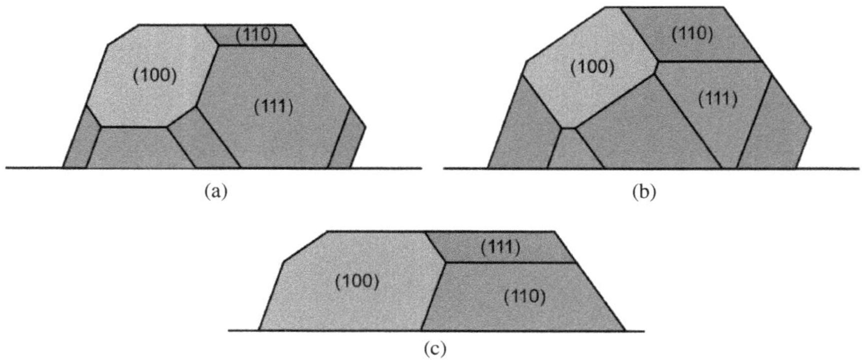

Figure 3.31 Structures of Cu particles on ZnO support as deduced from *in situ* TEM images of a Cu/ZnO catalyst in various gas environments [185]. (a) Cu particle shape at a pressure of 1.5 mbar of H_2 at 220°C. (b) Cu particle shape in a gas mixture of H_2 and H_2O, H_2:H_2O = 3:1 at a total pressure of 1.5 mbar at 220°C. (c) Cu particle shape in a gas mixture of H_2 (95%) and CO (5%) at a total pressure of 5 mbar at 220°C.

because surface terrace B5 sites are created by (oxygen-induced) surface reconstruction.

The promoting role of ZnO on the activity of the Cu particles has been elucidated by high-resolution electron microscopic studies in 2002 by research scientists of the Danish Topsøe company [185]. They demonstrated that the ZnO support affects the shape of the Cu particles. This is important because of the structure sensitivity of the methanol synthesis reaction.

Figure 3.31 shows changes in Cu particle shape when the catalyst is exposed to different compositions of gas. In the presence of CO, a high concentration of corrugated Cu(110) surface evolves [185].

In a later study with aberration-corrected high-resolution electron microscopy, Schlögl *et al.* demonstrated the presence of step-edge sites on the Cu particles of the Cu/ZnO catalyst [186]. They suggested that ZnO not only affects Cu particle shape but also composition.

Some surface ZnO reduces and alloys with the Cu surface atoms. The Zn atoms, which have lower surface energy, decorate the Cu particle step-edges. DFT calculations indicate that the more reactive Zn atoms strengthen M–O interaction between surface and reaction intermediates and Cu activates hydrogen.

The dominant effect of the presence of Cu step-edge sites and their alloyed counterparts is an increase in the stability of reaction intermediates, which affects the surface concentrations of the reactive systems and lowers the overall activation energy barriers.

This concludes the history of the elucidation of complex structure-function relation of the Cu/Zn methanol catalyst that happened over the past seventy years.

In time a remarkable reinterpretation of the synergetic effect of Cu and ZnO happened. Klier proposed in 1982 that Cu^{2+} cations are dissolved in a ZnO matrix. This is suggested to increase the concentration of reactant CO.

The other important new understanding is that the methanol synthesis reaction is not a CO hydrogenation reaction, but instead is dominated by hydrogenation of CO_2. An alternative view developed on the nature of the catalytically active centers. They consist of Taylor step-edge sites on reduced Cu particles. Zn atoms migrate to Cu and decorate the reactive Cu step-edge sites.

3.3 Structure-Function Relation of Transition Metal-catalyzed Hydrogenation Reactions

Structure-function differences relate to the type of chemical bond that is activated in the reaction rate-controlling step. This determines largely particle size dependence of transition metal particles. Metal particle deactivation and reconstruction may have large additional effects.

In the 1960s, Boudart from Stanford University was one of the first to draw attention to the differences in reactivity of catalytic systems that relate to particle size variation [192]–[194]. The dimensions of the nanoparticles vary between 2 and 20 nm.

Structural characterization of a catalyst at a molecular level was not yet possible, therefore he gave the following empirical definitions:

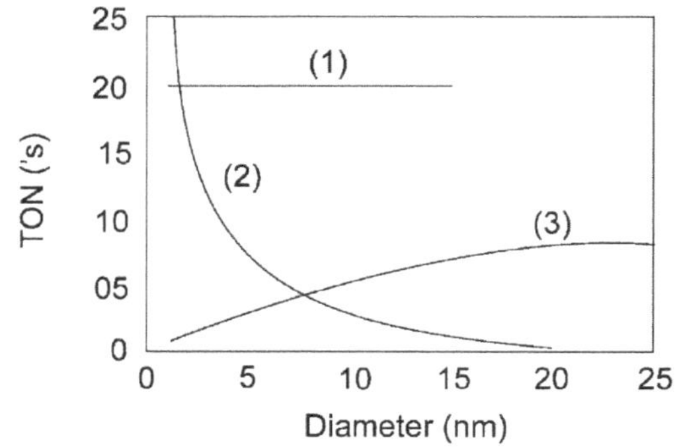

Figure 3.32 The three different kinds of structure sensitivity–particle size relationship plotted as turnover number (rate normalized per exposed surface atom) for selected reactions versus particle size: (1) benzene hydrogenation on Pt/SiO_2; (2) methane steam reforming [196] and ethane dehydrogenation; (3) CO hydrogenation on Ru/Al_2O_3 [197].

— a facile reaction on a given metal would be one for which the specific activity of the catalyst is practically independent of its mode of preparation

— a demanding reaction would require some special surface configurations that are produced on a metal surface only as a result of a special mode of preparation

The difference in particle size dependence for a variety of hydrogenation reactions has been summarized in 1997 by the French scientist Henry [195] and is illustrated in Figure 3.32.

A facile reaction is the hydrogenation of benzene, illustrated by reaction (1) in Figure 3.32. Demanding reactions are of different types. Figure 3.32 illustrates that the rate of reactions of type (2) declines with increase in particle size, and for reaction (3) activity shows a maximum or levels off.

Type (1) is a reaction where C–H bonds are formed. Type (2) reactions concern C–H or C–C bond cleavage. Broken chemical bonds are σ-symmetric molecular bonds. In reactions of type (3), π molecular bonds are activated as in N_2 or CO.

Demand for surface atom ensemble size to stabilize reaction intermediates varies for reactions. This is one of the reasons for differences in particle size-dependent catalysis.

The major difference in surface site distribution is the enhanced ratio of terrace versus edge sites with increase of particle size (Figure 2.2). Coordinative unsaturation of the edge atoms is larger than that of the terrace atoms. Therefore, reactivity of edge atoms is larger. Reactions which have a molecular bond cleavage reaction as rate-limiting step and are activated by contact with a single surface metal atom show type (2) behavior. An example is reactions with C–H bond cleavage as reaction rate-controlling step.

With increase in particle size, step-edge B5 sites are also generated. Their relative concentration shows an optimum as a function of particle size (Figure 3.17). CO or N_2 bond dissociation is sensitive to the presence of step-edge B5 sites. Reactions that contain these bond dissociation reactions as elementary reaction steps show type (3) behavior. Reaction rate should show a maximum.

A key question is the explanation of the actual value of the particle size at this rate maximum. Topology predicts a reaction rate maximum for 2 nm, but experimentally this maximum is often found for larger particle sizes. Also, in some cases there is no maximum but reaction rate levels off at large particle size. It is discussed below that these differences relate to surface reconstruction of metal particles and exposed surfaces.

There are several reasons for reactions of type (1) being independent of particle size. Different from type (2), these reactions all concern C–H bond formation.

Because most reactions have different reaction channels, the interpretation of particle size dependences is generally complex. It will not always be dominated by the reaction that is of primary interest. Particle size dependence can be dominated by suppression or activation of catalyst-deactivating intermediates. The surface can become covered with spectator adsorbates that affect structure of catalyst surface. Also, the surface and particle can reconstruct, which alters the state of the surface compared to that of the fresh catalyst.

Particle size dependence of hydrocarbon conversion reactions that illustrates this complexity is discussed in the following paragraphs. Structure sensitivity of CO conversion is presented next. Within this context, surface reconstruction is also revisited.

In the final part of this section, the intimate relationship is discussed between particle size dependence and the nature of the reactant chemical bond that is activated.

An example of a reaction where particle size dependence may relate to the presence of competitive reactions is the hydrocarbon conversion reaction of the alkanes.

Figure 3.7 illustrates the three different reaction channels of an alkane. The surface atom ensemble size requirement of the three reactions is quite different. A single or dual atom site catalyzes isomerization and alkane dehydrogenation. Differently, the hydrogenolysis reaction requires a larger ensemble of reactive surface atoms preferably arranged as a step-edge site.

In alkane conversion, the hydrogenolysis reaction is the main reaction that causes catalyst deactivation. It partially carbides the surface or forms non-reactive carbonaceous surface intermediates.

Experimentally the reaction rate of dehydrogenation is found to decrease with particle size (type (2) behavior). It can be due to reduced concentration of coordinatively unsaturated surface metal atoms but also due to increased deactivation of the particle with increase of particle size. Once the surface is converted into its partially deactivated state, alkane dehydrogenation and also alkene hydrogenation have stable and selective performance [198].

According to early data by Sinfelt *et al.* [199], hydrogenolysis of ethane catalyzed by Ni nanoparticles dispersed on silica-alumina support is a type (2) reaction. The reaction rate declines with increase in particle size. However, reports on particle size dependence of the hydrogenolysis reaction are contradictory. An earlier paper by Sinfelt *et al.* on Ni particles dispersed on silica reports independence on particle size [200] whereas for Rh particles a maximum in particle size dependence is reported [201].

A difficulty with the interpretation of these data is that no information on the surface state of the particles has been gathered. Reactive transition metal particles become, within seconds, covered with carbonaceous-deactivating compounds when exposed to hydrocarbons in a flow of hydrogen [202].

There is experimental evidence by Boudart *et al.* [203] that the presence of the competitive hydrogenolysis path causes the isomerization reaction to be "demanding" whereas this reaction is intrinsically structure insensitive. The experiments of Boudart *et al.* concern selective catalysis of neopentane catalyzed by Pt nanoparticles. Neopentane can isomerize to isopentane or hydrogenolyze to methane and isobutane (see Figure 3.10). The rate of the neopentane isomerization reaction sensitively depends on the pretreatment conditions of the Pt catalyst. Catalysts are compared with Pt particles in the same size range. A high temperature treatment increased the selectivity of the isomerization reaction. Surface analysis shows that this treatment tends to reduce surface imperfections. Imperfections such as step-edge sites have a high activity for the hydrogenolysis reaction. The reduction in concentration of the step-edge sites decreases the relative rate of hydrogenolysis and, with it, of the deactivation rates. The rate of isomerization is not affected.

The hydrogenolysis reaction is readily poisoned by sulfur adsorption. Sulfur adsorption has a beneficial effect on catalyst lifetime. This is due to site blocking that suppresses the hydrogenolysis reaction, which is the main cause of Pt catalyst deactivation [54]. The function of Pt metal dispersed on the solid acid bifunctional catalyst is to dehydrogenate alkane to alkene. Only very low concentrations of adsorbed sulfur are necessary. They are used to extend lifetime of the Pt metal in the catalytic platforming reaction that converts alkanes to aromatics (Section 3.1.4).

For a different reaction, surface science studies by Kiskinova and Goodman from Texas A&M University [204]–[207] demonstrated that sulfur indeed poisons reactive step-edge sites. CO conversion to methane (type (3)) catalyzed by Ni shows several orders of magnitude reaction rate decrease when a small amount of S is adsorbed. As discussed in Section 3.2.2.2, CO is activated by step-edge B5 sites.

Quantum-chemical calculations demonstrate the difference in ethane hydrogenolysis catalyzed by the Pt(111) terrace surface and the Pt(211) stepped surface [208], [209]. There is a complex interplay of dehydrogenation and C–C bond cleavage reactions.

Table 3.3 Comparison of activation energies (eV, 1 eV = 96 kJ/mol) of C–C bond cleavage reactions on Pt(111) and Pt(211) [209].

	Pt(111)	Pt(211)
CH_3CH_2	1.84	1.11
CH_2CH_2	2.22	1.67
CH_3C	1.95	1.70
CHCH	1.07	1.28
CH_3CH	1.18	1.34

C–H bond dissociation is a low barrier reaction with activation energies of the order 60 kJ/mol. The activation energies of C–C bond cleavage energies are 100 kJ/mol or higher. Most intermediates have higher reactivity at the Pt(211) surface.

Table 3.3 shows that different surfaces prefer different C–C bond cleavage steps. On the Pt(211) surface the CH_2CH_3 intermediate has the lower C–C bond cleavage energy, but on the Pt(111) surface this is the case for adsorbed CHCH or CH_2CH intermediates.

On both surfaces the CH_3C spectator surface intermediate observed by Somorjai *et al.* (see Figure 3.3) is the most stable.

The rate of type (1) reactions is independent of particle size. This not only implies that reaction is independent of site topology but also that reactivity does not relate with differences in coordinative unsaturation of surface atoms.

For Pt and Ni the hydrogenation of alkene is a facile reaction [210], [211]. It is consistent with the small surface atom ensemble size requirement of this reaction. Experiments with alloys also suggest a small surface atom ensemble size requirement for the hydrogenation of benzene [212].

Type (1) particle size dependence may be intrinsic to the C–H bond formation reactions but may also be due to catalyst deactivation that increases with particle size.

Exposure of a reduced transition metal to ethene and hydrogen initially deactivates the metal particles. A significant fraction of the catalyst surface becomes covered with unreactive carbonaceous residue. The surface is partially carbided [14], [213]–[217].

Catalyst deactivation by ethene hydrogenation as a potential cause of apparent particle size independence is demonstrated in an experiment of the CO_2 hydrogenation reaction by catalysts previously exposed to ethene. When, after use as ethene hydrogenation catalyst, Ni catalysts are used to hydrogenate CO_2 (that is a type (3) reaction when catalyzed by non-treated Ni catalysts), then the CO_2 hydrogenation reaction is found to be independent of Ni particle size [218]. It behaves as a type (1) reaction.

Taylor step-edge sites catalyze CO_2 hydrogenation. The Taylor step-edge sites have a higher presence in the larger particles. These are poisoned by deposition of carbon in the pretreatment reaction by deactivating side reactions of ethene hydrogenation. The rate of incorporation of carbon in surface layers of Ni particles increases with particle size [219].

Reactive metals such as Fe or Mo readily form metal carbides when exposed to hydrocarbons or CO. Whereas the non-carbided metal has a high selectivity for hydrogenolysis of hydrocarbons and the catalyst rapidly deactivates, once carbided it is a useful and stable catalyst for hydrocarbon or CO conversion reactions.

The reduced reactivity makes conversion of hydrocarbons selective for hydrogenation or isomerization because hydrogenolysis is suppressed.

In the Fischer-Tropsch reaction the initially reduced Fe catalyst converts to an iron carbide phase. Reduced Fe is too reactive and non-selective, the deactivated iron carbide phase is a selective Fischer-Tropsch catalyst [125], [220], [221].

Type (2) particle size dependence is found for reactions where C–H or C–C bonds cleave. The more reactive the surface atom, the lower the activation energy of bond activation. When particle size increases, the relative number of more active surface atoms decreases. Methane steam reforming [196] is a type (2) reaction. The reaction rate normalized per surface atom decreases with increase of particle size. The reaction transforms methane with water into carbon monoxide and hydrogen. In the reaction, C–H and O–H bonds are broken, generating H_{ads}, O_{ads}, and CH_{ads} intermediates that recombine as CO and H_2 product molecules. The mechanism of this oxidation reaction is discussed in Chapter 4 (Figure 4.3). The elementary reactions of C–H or O–H bond activation of methane and water require only an ensemble of a single surface metal atom [222]–[224].

The type (3) reactions of Figure 3.32 require step-edge catalyst centers. Cleavage of molecular π bonds is sensitive to the presence of step-edge B5 sites.

The rate-controlling step of the hydrocarbon hydrogenolysis reaction also involves cleavage of double bonds of surface intermediates.

Figure 3.17 illustrates that the probability for step-edge site configurations as a function of particle size has a maximum at 2 nm. For the ammonia synthesis reaction, which is rate limited by bond cleavage of N_2, the rate is indeed maximum for particles of this size [100].

Similar to N_2, CO activation is also step-edge site sensitive. But Figure 3.24 illustrates that the maximum in reaction rate is at substantially larger particle size. The rates of methanation and Fischer-Tropsch synthesis do not show a maximum as a function of increased particle size but level off. This is due to surface reconstruction that generates step-edge sites on terraced surfaces.

Surface science and catalytic model experiments have largely contributed to an understanding of this phenomenon of reconstruction of transition metal surfaces exposed to reacting gases [225], [226]. Relative surface energies may change when covered with an adsorbate overlayer. This may induce changes in particle morphology as shown in Figure 3.30 for Cu particles.

Incorporation of additional atoms such as C or O into the transition metal lattice causes strain because the space originally available only to surface metal atoms now also has to be shared with the incorporating atoms. Strain is released when metal atoms are pushed out of the surface layer. This is the process that leads to surface corrugation [94] and creates step-edge sites. Such transformations are particle size dependent and happen more readily for larger particles.

In the previously mentioned steam reforming reaction catalyzed by Ni particles, the strong interaction of carbon with step-edge sites provides a driving force for the creation of additional step-edge sites.

The Scanning Tunneling Microscopy (STM) study by Wilson *et al.* from Shell Research [148] visualizes the change in topology of a Co(0001) surface terrace when exposed to synthesis gas in the Fischer-Tropsch reaction. Figure 3.33 compares the structure of the Co surface in ultra-high vacuum before reaction with that after exposure to reactant gas.

Due to reaction with synthesis gas, the surface becomes puckered and covered with small Co particles of the order of a few nanometers. The cause of surface corrugation is incorporation of C atoms into the Co surface and strain release by atom rearrangement. The small Co particles contain the B5 sites necessary for the high rate and selectivity of the Fischer-Tropsch reaction. A simulation of the structure of this reconstructed surface at atomic scale is given in Figure 3.25.

The surface reconstruction phenomenon is general. Using sophisticated atomistic *in situ* surface scanning techniques, the formation of such terraced nanoparticles has also been observed in electrocatalytic experiments of the oxygen evolution reaction from water of a Pt(111) single crystal electrode [227], [228]. This is a reaction of interest, for water hydrolysis to produce hydrogen from electricity.

Ambient pressure STM experiments with a Cu(111) crystal of the water-gas shift reaction show that CO adsorbs preferentially on step-edges and that the strong interaction of CO with these sites induces the surface to create more reactive step-edges [227]. These step-edges in turn have the proper reactivity to activate H_2O and generate surface hydroxyls that can react with CO to form CO_2.

The relation between the nature of the reactant chemical bond that is activated and particle size dependence is addressed in the following concluding paragraphs [229].

Type (1) and type (2) reactions are chemically related. In reactions of type (1) C–H bonds are formed, whereas in reactions of type

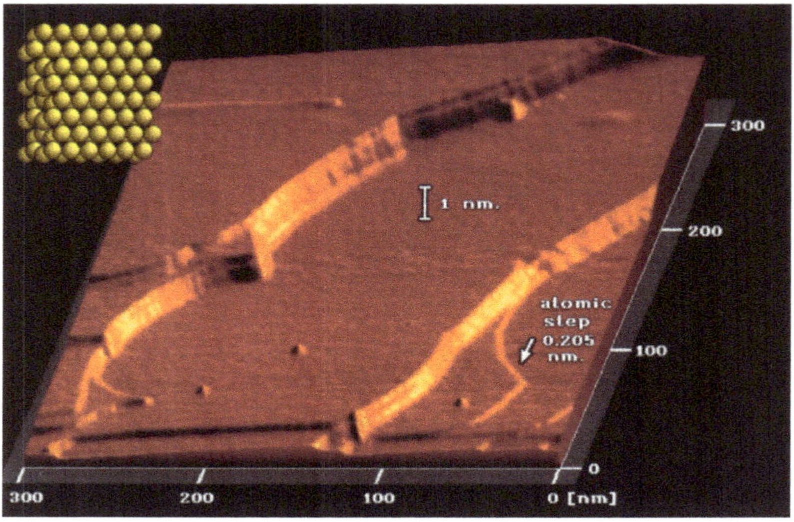

(a)

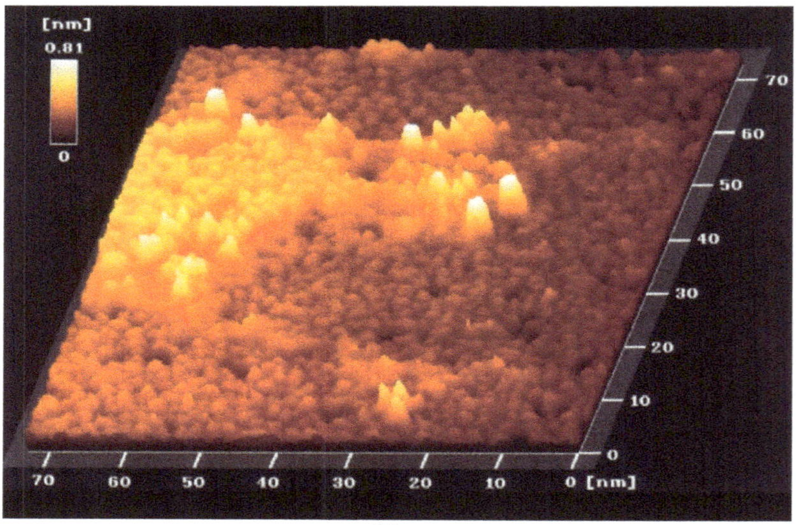

(b)

Figure 3.33 (a) STM image of the clean Co(0001) surface (prior to reaction) showing atomically flat terraces 150 nm (ca. 600 atoms) in width. The smallest step visible is monoatomic in height, 0.205 nm being the expected single atom step height on Co(0001). Inset: hard-sphere model of the bulk-terminated Co(0001). (b) STM image of the Co(0001) surface after 1 h exposure to high-pressure CO hydrogenation conditions [148].

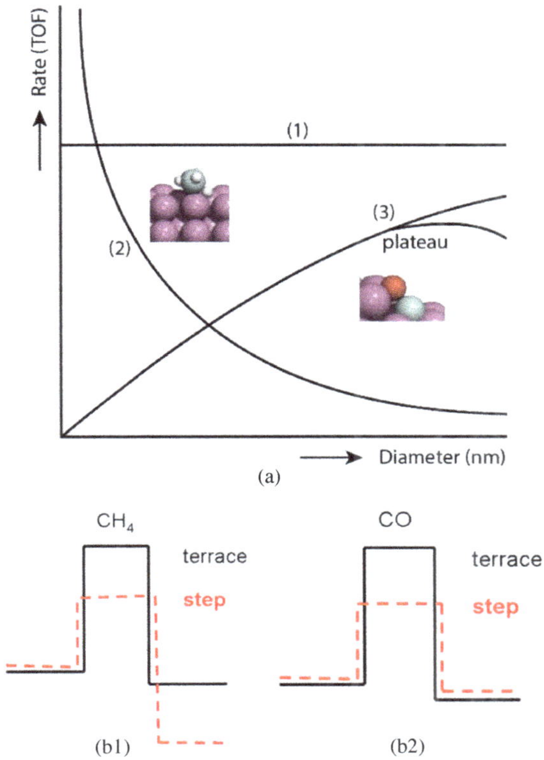

Figure 3.34 The relation between chemical bond activation and particle size dependence. (a) Schematic presentation of the relation between the variation of particle size dependence and transition state structure of reactant [229]. The inserts show representative transition states for C–H and C=O bond cleavage (white hydrogen atom, light blue carbon atom, mauve metal atom, red oxygen atom). (b1) Schematic comparison of potential energy change of C–H bond cleavage for a less reactive and more reactive surface. (b2) Schematic comparison of potential energy change of C=O bond cleavage for a non-stepped and stepped reaction site.

(2) C–H bonds are broken. The difference with reactions of type (3) is that in this case it concerns cleavage of molecular π bonds.

For C–H bond cleavage and formation, the difference between particle size dependence is a manifestation of the BEP relation. This is illustrated by Figure 3.34.

Figure 3.34a is a redrawing of Figure 3.32. Inserts show transition state structures of C–H and C–O bond activation. The structure of the transition state of methane C–H bond cleavage illustrates that activation occurs by contact with a single metal atom. This is relevant to reactions of type (1) and (2). The structure of the CO transition state for a stepped surface relates to type (3) behavior.

Particle size dependence of types (1) and (2) is related because the C–H bond formation and cleavage steps go through the same transition state. Particle size dependence relates to the relative increase in surface atoms that are less coordinatively unsaturated and hence are less reactive.

The difference in activation energies for C–H bond activation is illustrated in Figure 3.34b1. Reaction fragments H_{ads} and $CH_{3,ads}$ of methane are most strongly bound to a surface atom that is more coordinatively unsaturated.

Due to the BEP relation, the change in reaction energy relates to the difference in activation energies of C–H bond cleavage. The BEP relation applies only when elementary reactions are compared with similar reaction paths, a condition that is satisfied for C–H bond cleavage by a single metal atom.

Figure 3.34b1 indicates that this lowers the activation energy of C–H bond dissociation when in contact with a coordinatively more unsaturated surface atom. However, importantly the activation energy of C–H bond formation by recombination of H_{ads} and $CH_{3,ads}$ is unaffected.

Activation energy change difference of forward and reverse elementary steps are related with reaction energy by the microscopic reversibility relation:

$$\Delta E_{react} = \Delta E_{act,forw} - \Delta E_{act,rev} \tag{3.6}$$

In Eq. (3.6), $\Delta E_{act,forw}$ and $\Delta E_{act,rev}$ are respective changes in forward and reverse activation energies and ΔE_{react} is the change in reaction energy. A large change in forward activation energy comparable with ΔE_{react} limits activation energy of the reverse C–H bond formation reaction to a small value.

As discussed in Section 2.3.2.3, for surface reactions the change in activation energy of the forward bond dissociation and reaction energy is proportional and nearly equal ($\alpha_{BEP} \approx 1$). Then due to the microscopic reversibility, the activation energy of the reverse bond formation reactions is invariant. This is illustrated by Figure 3.34b1.

It explains insensitivity of reaction type (1) with respect to particle size change and the steep decrease of reaction type (2) when particle size increases.

Activation of a molecular bond by single metal atom is representative for cleavage of σ bonds. The C–H bond on methane is an example of a σ-symmetric bond.

When elementary reactions are compared on sites of different topology, the BEP relation does not apply. Then elementary reaction energies cannot be related with change in reaction energies. Activation energies can change whereas there is no change in reaction energy.

Figure 3.34b2 illustrates this for the activation energy difference of the C=O bond cleavage energy at B5 step-edge versus that on the terrace.

The activation energy is high on the surface terrace and largely reduced on the step-edge site. In contrast, differences in reaction energy of CO bond cleavage are small (see also Figure 3.26).

The independence of activation energy from reaction energy is representative for cleavage of π bonds.

The site dependence of π bond cleavage leads to particle size dependence of type (3) reactions. The rate of reaction increases due to the increased probability of formation of reactive step-edge B5 sites for larger particles.

Whether reaction rate goes through a maximum or levels off depends on additional surface reconstruction.

3.4 Activation of Hydrocarbons with Heteroatoms

The mechanisms of hydrodesulfurization (HDS), hydrodenitrogenation (HDN), and hydrodeoxygenation (HDO) reactions are presented. A major challenge is selective hydrogen consumption.

Catalytic chemistry of heteroatom removal from hydrocarbons by hydrogenation reactions is presented here. Sulfur- and nitrogen-containing hydrocarbons are present in coal liquids and crude oil, and oxygen is the major atom to be activated in biomass feedstock. In refinery processes heteroatom removal such as sulfur or nitrogen is needed to make possible upgrading processes, which often use transition metal-containing catalysts that otherwise are readily poisoned. For environmental reasons, only transportation fuels with a low content of sulfur and nitrogen are acceptable.

Oxygen-containing hydrocarbons dominate in pyrolyzed biomass oil from wood or plant waste or in diesel oils derived for plant seeds. They contain large fractions of oxygenates that originate from oxygenated molecules such as cellulose, lignin, or carboxylic acids. Reduction of oxygen content in biomass-derived oil is a necessity to improve liquid fuel quality. Biomass-derived oxygenates are attractive platform molecules for substitution of oil-derived chemicals.

Sulfide catalysts are the main catalysts used for the hydrotreating processes that remove heteroatoms. An important difference with fossil-derived oils is that biomass oils have a very low sulfur content. Hence transition metal-containing catalysts can also be used.

The chemistry of heteroatom removal relates to hydrocarbon hydrogenolysis. Instead of the C–C bond, C–S, C–N, and C–O bonds have to be cleaved. C–S and C–N bonds are weaker than the C–C bond, but the C–O bond is stronger (compare the bond energies: C–C = 346 kJ/mol, C–S = 272 kJ/mol, C–N = 305 kJ/mol, C–O = 358 kJ/mol). The heteroatoms of the respective molecules can be more readily activated than carbon-carbon bonds, because they can be more accessed by surface atoms. Unless molecular activation is sterically constrained due to bulkiness of heteroatomic molecules, the heteroatoms are not shielded from contact with surface atoms because of their lower coordination with hydrogen atoms.

A major catalytic challenge in HDS and HDN is to minimize hydrogen consumption and limit its use to that of removal of the heteroatoms, but not at the same time to hydrogenate unsaturated aromatics. In HDO catalysis the selectivity problem to solve is to obtain hydrocarbons with minimum loss of carbon from the initial reactant molecule. Carbon dioxide loss is to be suppressed at the cost of water loss.

Hydrotreating catalysis is applied in oil-refining processes since the middle of the previous century. The main catalysts used are MoS_2 or WS_2 promoted with Ni or Co. Such catalysts had originally been applied in the hydrogenation in coal liquefaction. In the presence of H_2S they are active and robust hydrogenation, dehydrogenation, and hydrogenolysis catalysts that cleave C–S, C–N, and C–O chemical bonds.

The reactivity of MoS_2 and related sulfides for hydrogen activation makes them also useful as electrodes for the electrocatalytic hydrogen evolution reaction. This is discussed in references [230]–[232].

WS_2 was originally used as bulk material. In the course of time the metal sulfide catalysts became supported on high surface area alumina or silica-alumina. The latter bifunctional catalyst is also widely explored as hydrocracking catalysts (see Section 5.4.3).

When promoted by Ni or Co, the composition of the catalysts is complex. To determine the distribution of the sulfide particles, their composition, and reactivity took the main part of the previous century. This was in combination with investigations of catalyst preparation chemistry to optimize catalyst performance. The history of hydrotreating catalysis is presented in references [233]–[235].

In the past decades, especially for the HDS reaction, molecular models of the relation between reactivity and structure of the sulfide catalytic reaction have been developed. These mechanistic models are the main subject of the next Section 3.4.1.

In the 1980s, biomass as feedstock for the chemical industry was intensively explored. New reactions were discovered and their catalytic chemistry explored [149]–[151]. Also for HDO the sulfides are useful. Because of the low sulfur content of the bio-oil, a rich variety of additional catalysts, most of them bifunctional and containing a transition metal, that are specific for the different oxygenates derived from biomass has been explored. The mechanism of HDO is often a complex combination of dehydroxylation, dehydration, decarboxylation, or decarbonylation reactions.

Reaction mechanism in relation with catalyst composition and structure is discussed in Section 3.4.2.

3.4.1 *The Hydrodesulfurization and Hydrodenitrogenation Reactions*

M–S bond energy of the M_xS_y catalyst is a reactivity descriptor.

Here the mechanisms of S and N atom removal from heteroatomic hydrocarbons, which are catalyzed by metal sulfide compounds, are introduced. After an introduction of HDS and HDN reaction networks, reactivity trends as a function of metal sulfide catalyst composition are presented. Such trends are used to identify reactivity descriptors. Molecular models of the catalytic reaction center and its promotion by Ni or Co are described.

The S or N atoms are often part of a cyclic hydrocarbon. Cleavage of chemical bonds happens in combination with H atom addition steps. Reaction networks have been formulated for HDS and HDN of a variety of molecules that are part of oil fractions [236]. Reaction networks representative for these systems are that of the HDS of dibenzothiophene (DBT) and thiophene, often used in model catalytic studies (Figures 3.35).

HDS of DBT follows two reaction pathways: direct desulfurization (DDS) with subsequent hydrogenation of the resulting sulfur atom, or indirect desulfurization initiated by hydrogenation of the hydrocarbon followed by C–S hydrogenolysis (HYD). In the latter case, one aromatic ring of DBT is first hydrogenated and C–S bonds cleave after initial SH formation. DDS is the preferred reaction route, since it minimizes hydrogen consumption and dominates.

When direct access of the catalyst cation to the sulfur atom of the reacting molecule becomes sterically constrained, for instance by the presence of alkyl substituents on the carbon atoms next to sulfur, the HYD route takes over. At commercial conditions, high hydrogen pressure is used and HYD tends to dominate.

Thiophene is often used as model molecule in mechanistic studies.

For this molecule subsequent hydrogen addition and C–S cleavage steps are schematically indicated in Figure 3.35b.

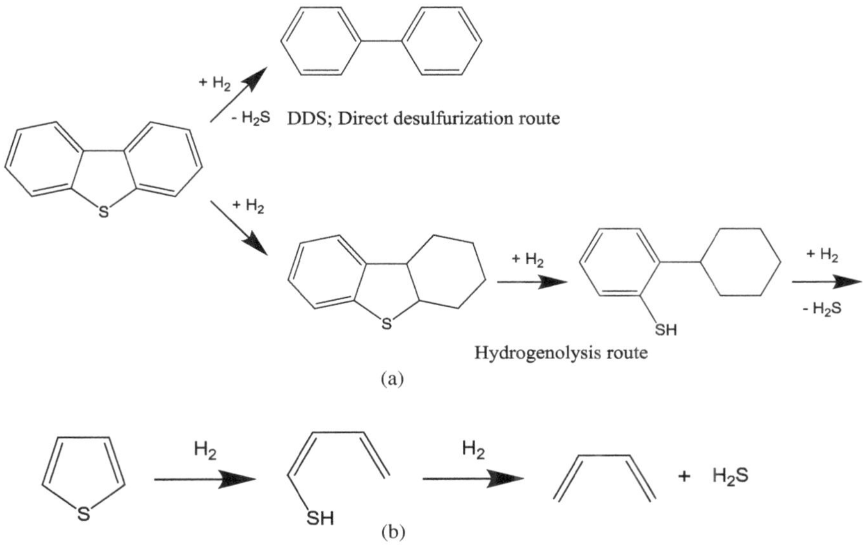

Figure 3.35 Hydrodesulfurization reaction paths of dibenzothiophene and thiophene. (a) Dibenzothiophene hydrodesulfurization pathways, direct desulfurization (DDS) versus hydrogenation followed by hydrogenolysis. (b) Thiophene desulfurization by the DDS mechanism [236].

HDN is a more demanding reaction than HDS. The C–N bonds will only cleave once the ring structure, which contains the nitrogen atom, is completely hydrogenated (see Figure 3.36a). After cleavage of one C–N bond, the aliphatic amine can be removed by two reaction routes: the Hoffmann elimination reaction that is acid-base pair catalyzed (Figure 3.36b1) or through an intermediate substitution reaction with H_2S (Figure 3.36b2). The latter reaction dominates in the presence of H_2S. The C–S bond hydrogenolysis reaction is faster than C–N bond hydrogenolysis.

Trends in catalytic reactivity as a function of metal sulfide composition are presented in Figures 3.37 for the HDS reaction. When catalytic reactivity is plotted as a function of the cohesive energy of a transition metal sulfide, the HDS reaction rate shows a bell-shaped dependence [237], [238]. Differently, a uniform decrease in

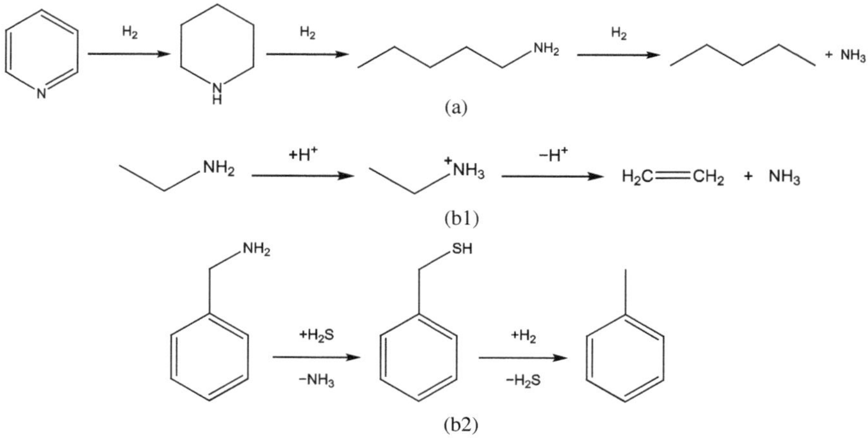

Figure 3.36 Hydrodenitrogenation (HDN) reaction networks [234]. (a) HDN of pyridine. (b) Nitrogen removal by (b1) Hoffman elimination through protonation and β C–H bond cleavage or (b2) nucleophilic substitution with H_2S.

reaction rate is found when this is plotted as a function of increasing M–S bond strength [239].

The interpretation of these trends led to a debate between a school of thought that suggests that maximum reactivity relates to a minimum of M–S bond energy [235], [240] and proponents of the idea that the volcano curve-type dependence reflects the Sabatier principle [241]. An optimum reactivity of the metal-sulfide compound defines the catalyst with the best performance. Both schools of thought based themselves on DFT-computed sulfide bonding energies. The dispute relates to how the M–S bond strength is determined. The Topsøe group calculated the M–S bond energy from the substitution energy of S atom in the transition metal bulk [244]. The group at the Institut Français du Pétrole considered the M–S bond energy as the energy per bond in the solid with reference to the energy of the free atoms [241], [242]. The latter definition corresponds to the chemical definition of the M–S bond energy as the bond energy of the individual bulk metal atom sulfide bond with the metal cation in its proper valence state. It relates to the M–S bond energy of sulfur adsorbed to a sulfide surface site.

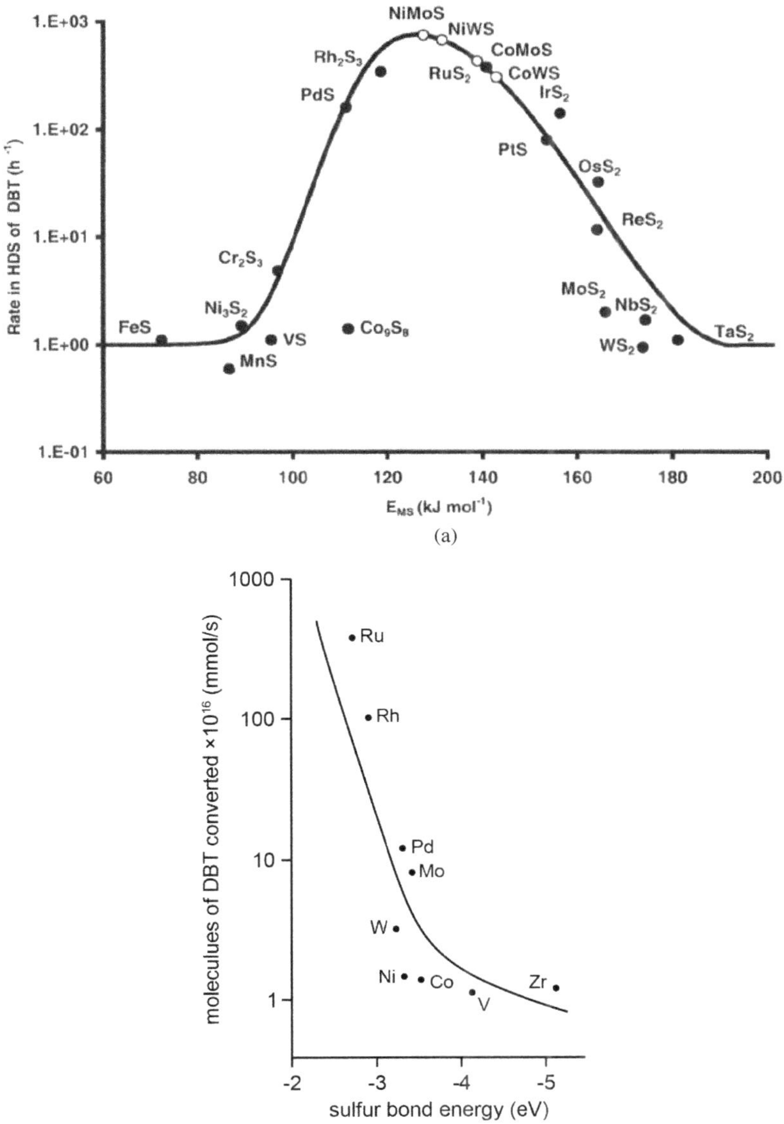

Figure 3.37 The reactivity of transition metal sulfides. (a) Variation of the HDS activity of DBT of the TMS catalysts as measured by Pecoraro and Chianelli [237] as a function of the metal-sulfur bond energy. This bond energy is derived from DFT calculations and is defined as the energy of the M–S bond with respect to the free atoms [243]. (b) HDS activity versus calculated metal-sulfur bond energy as deduced from the heat of formation according to Topsøe *et al.* [235].

The S substitution energy used by the Topsøe group is not a measure of the reactivity of the metal sulfide but rather that of the transition metal.

In view of the strong bonding of sulfur to the transition metal, the finding in Figure 3.37b that for minimum S substitution energy the HDS reaction rate is maximum can be understood, because then poisoning by sulfur is minimum.

The volcano curve dependence that is found for the activity plot against the M–S bond energy of the bulk metal-sulfide predicts that an optimum M–S bond energy exists where reaction rate is maximum. This is consistent with the Sabatier principle that there is an optimum in M–S surface bond energy when the rate of C–S bond activation and that of H_2S desorption are comparable (2.4.3). According to the Sabatier principle, the orders of reactants have different dependencies when reactivity is compared left or right of the volcano maximum.

Hensen *et al.* [244] measured the reaction rate of the HDS of thiophene and the reaction orders of thiophene, hydrogen, and H_2S for several metal sulfide catalysts dispersed on a carbon support.

Figure 3.38 shows a maximum in the thiophene HDS reaction rate when plotted as a function of transition metal d-valence electron count. The corresponding reaction orders are given in Table 3.4.

Figure 3.38 also includes the reaction rate of Co-promoted MoS_2. The HDS reaction rate increases by a factor of ten compared with the reactivity of the MoS_2 catalyst. Its reactivity is close to the maximum reactivity of the non-promoted Rh_2S_3 catalyst. The chemistry of catalyst promotion is discussed in the final paragraphs of this section.

The positive reaction order of the reactant thiophene shows a minimum when rate is maximum. The reaction order of hydrogen is also positive and increases continuously. The reaction order of H_2S is negative and decreases uniformly.

The positive reaction orders of hydrogen and thiophene imply that their activations are limiting. A decrease in reaction order implies that the surface concentration of the related intermediate

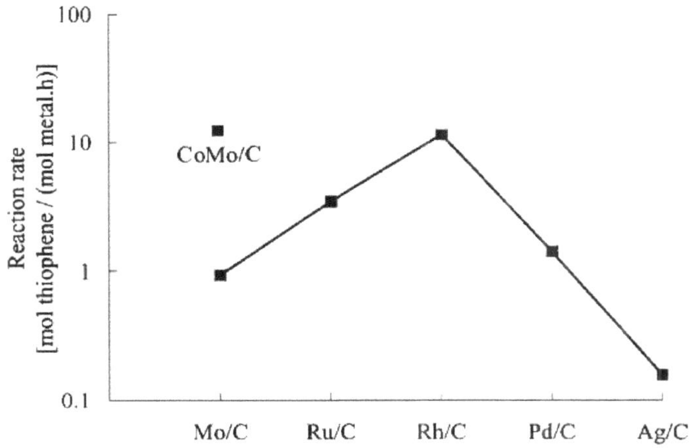

Figure 3.38 Thiophene HDS activity for the different carbon-supported transition metal sulfides at standard conditions (3.33 kPa thiophene, 1 kPa H_2S, 10 kPa H_2 and 573 K) [244].

Table 3.4 Reaction orders of thiophene (n_T), H_2S (n_S), and H_2 (n_H) under different conditions. Inlet: H_2S; partial pressure: 1 kPa; temperature: 573 K [244].

	n_T	n_S	n_H
Mo/C	0.50	−0.32	0.57
Ru/C	0.39	−0.25	0.53
Rh/C	0.31	−0.83	0.93
Pd/C	0.65	−1.04	0.99
CoMo/C	0.12	−0.46	0.78

increases and that the apparent elementary rate of activation increases.

The negative reaction order of H_2S indicates that reaction rate is inhibited by H_2S adsorption. The increase in positive order of hydrogen shows that it causes dissociative hydrogen adsorption to become more difficult.

Thiophene and hydrogen compete for adsorption. Their relative surface concentration varies. The rate optimum relates to the

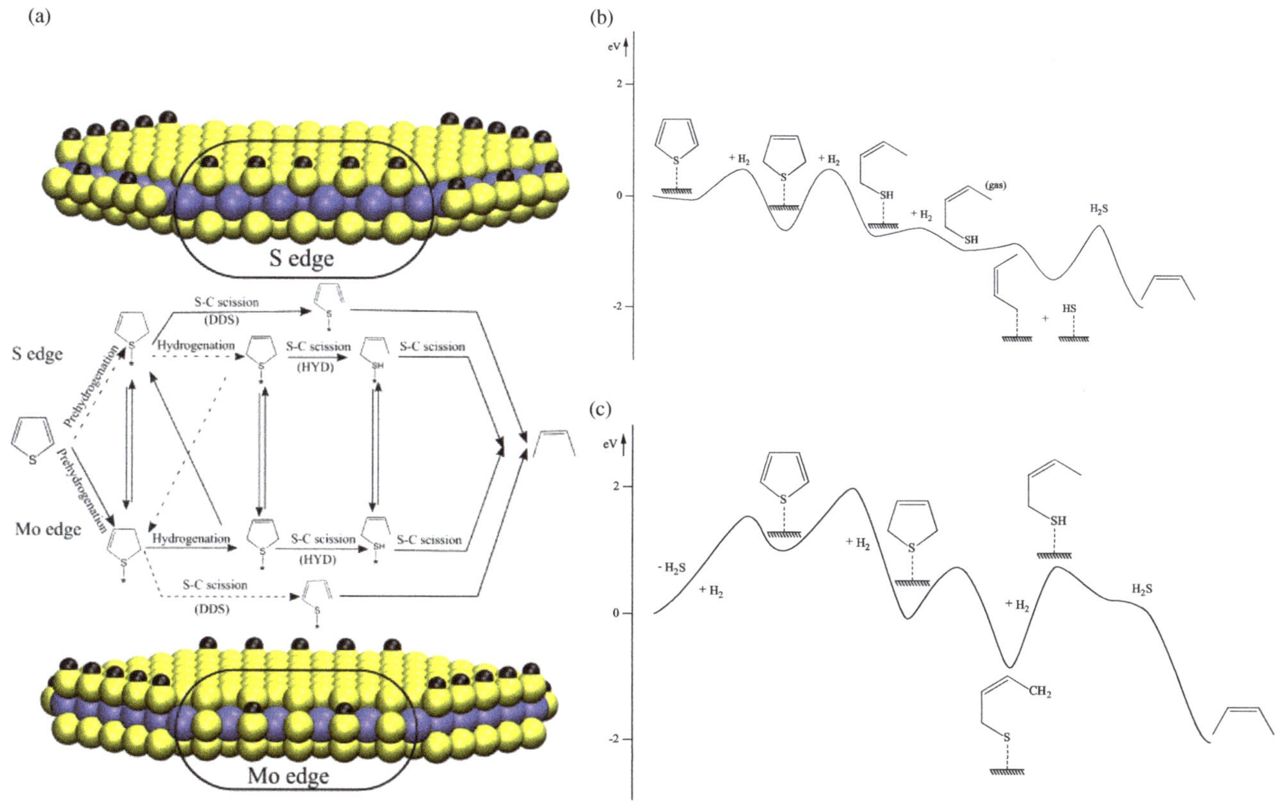

Figure 3.39 The catalytic reaction centers of MoS_2 and the mechanism of thiophene HDS [252]. (a) Schematic of the elementary reaction steps and edge structures of a single MoS_2 layer. The upper part is a side view of MoS_2 perpendicular to the sulfur-terminated S(1010) edge, with S and H coverages representative of HDS conditions. The middle part is a schematic

overview of the reactions involved in HDS of thiophene, including the possible interaction between the S(1010) edge and the terminated Mo(1010) edge. The dotted arrows denote reactions found to be slow. The lower part is a side view of MoS_2 perpendicular to the Mo(1010) edge, also with the S and H coverage present at HDS conditions. (b) Potential energies of elementary reaction steps of thiophene hydrogenation at the Mo edge of MoS_2 (schematic). The reference energy is the equilibrium edge configuration under HDS conditions (Mo edge with 50% S and 50% H) and thiophene in the gas phase. (c) Potential energies of elementary reaction steps of thiophene hydrogenation at the S edge of MoS_2 (schematic). The reference energy is the equilibrium edge configuration under HDS conditions (S edge with 100% S and 100% H) and thiophene in the gas phase.

surface condition where relative coverages are optimum and surface is least deactivated by adsorbed sulfide (see Section 2.4.3, Eq. (2.22)).

Reaction rate is maximum for optimum M–S bond energy (the uniform increase in negative order of H_2S indicates increase in M–S bond strength for the sulfide compounds) and not for minimum M–S bond energy. It indicates that there is a Sabatier principle relation between M–S bond energy and HDS reaction rate. This is in agreement with the suggestions of Chianelli and Toulhoat *et al.* [241], [242].

Figure 3.37a shows that M–S bond energy decreases for metal sulfides when the metal position in the second row of the periodic table moves to the right. In parallel the reaction order of hydrogen increases, which is an indication of reduced reactivity. This also disfavors C–S bond cleavage and decreases the relative surface concentration of the reaction intermediate.

At the reaction rate maximum, the rate of H_2S desorption and thiophene activation match. The low reaction order in thiophene indicates a maximum concentration of reaction intermediate.

In the Co-promoted MoS_2 system, H_2S inhibition is less and thiophene reaction rate is near zero order. This is consistent also with a relative increase in reaction intermediate concentration. Reduced H_2S inhibition agrees with the observed increase of the reduction rate of promoted sulfide systems [245].

Commonly used catalysts are promoted MoS_2 catalysts. This sulfide has the layered structure of Figure 3.39a. This figure shows that the basal plane atoms are coordinatively saturated sulfur atoms and that the reactive metal sites are located at the basal plane edges.

In the MoS_2 particles, the MoS_2 layers are stacked. The interaction energy of the layers is determined by weak van der Waals interactions between the polarizable sulfur atoms. Layers are readily separated. Catalyst sulfide particle size dependence relates to the diameter of the layer basal planes and their stacking.

In the case of Ni- or Co-promoted catalysts, a close contact of promoting metal sulfide and layered MoS_2 and WS_2 particles is essential.

Also, catalysts based on WS_2, which has a similar layered structure as MoS_2, can be used as hydrotreating catalysts. Hydrotreating is not only used to reduce the S or N content of oils, but also to hydrogenate aromatic compounds in a H_2S environment. WS_2 catalysts have a higher hydrogenation activity but lower HDS activity than MoS_2-based catalysts [246]. They have been much less investigated at the molecular level than the MoS_2 catalysts. Catalysis with WS_2 is discussed in [247], [248].

The promoted and supported sulfide catalyst is a complex system. The metal sulfide components can be present in different separate and mixed phases. Also, interaction with support affects metal-sulfide surface chemistry. The HDS and HDN reactions involve a complicated interplay of sulfide edge structures, adsorption energies of reactants and intermediates, and activation barriers of elementary reaction rates, which is a strong function of reaction conditions and especially H_2S concentration.

Important books by Topsøe, Clausen, and Massoth from 1996 [234] and Raybaud and Toulhoat from 2013 [249] describe the evolution of understanding of the structure and composition of practical hydrotreating catalysts.

Molecular understanding of HDS catalysis based on atomic knowledge of the structure of the catalytic reaction center is recent. It became possible once well-defined model sulfide particles could be studied at the atomic level with surface science techniques. Catalyst models could be simulated with quantum-chemical computation [250]–[255], with thiophene HDS as the preferred model reaction.

Here the detailed mechanistic model of the thiophene HDS reaction on MoS_2 due to Nørskov, Besenbacher, and Topsøe *et al.* [252], [256] is presented. It is based on experimental STM studies of model MoS_2 slabs deposited on Au substrate. Similar model systems have also been used to study the promoted systems, which are discussed in following paragraphs.

With reference to practical systems, the model MoS_2 slabs relate specifically to fully sulfided particles that are commonly present on the supported sulfide catalysts. In practical systems, due to interaction with the support, metal sulfide particles that are only partially sulfided are also present and metal cations also interact directly with oxygen atoms of the support.

Figure 3.39 shows simulated reaction site models of MoS_2 and the reaction paths of thiophene hydrogenation [252] based on the Nørskov, Besenbacher and Topsøe studies. In the MoS_2 layer, Mo^{4+} cations are sandwiched between an upper and lower S atom layer. The Mo cations are surrounded by six S atoms in trigonal prismatic coordination. Reaction happens at the edges of the MoS_2 particles. Two non-equivalent edge structures are formed. One surface is edge terminated with S atoms and the other surface is edge terminated with coordinatively unsaturated Mo atoms.

Quantum-chemical calculations combined with equilibrium thermodynamics predict composition and structure of these surface edges at reaction conditions.

Representative structures that derive from such simulations are those shown in Figure 3.39a. The S edge is 100% saturated with S, whereas at the Mo edge S occupation is 50%. At this site the S atom bridges two Mo cations. The H atom coverage at the Mo edge is also 50% whereas at the S edges this is 100%. H is adsorbed as S–H. Such a sulfhydryl group has weak Brønsted acidity. Whereas an adsorbate S atom can be added to the Mo edge, this is only possible on the S edge when first a vacancy of S is created.

Figure 3.39a also shows the reaction network of hydrogen addition steps that leads to product butene. Dominant reaction paths are different on the two edges. On the S edge DDS is the preferred HDS reaction path and on the Mo edge C–S bond hydrolysis is preferred. Reaction intermediates equilibrate by moving between the edges. A partial hydrogenated intermediate at the Mo edge is desulfurized at the S edge.

The reason for this mechanism is the presence of so-called brim sites at the Mo edge. Brim stands for bright rim features that are observed in the STM measurements [244]. These have been identified as the weakly acidic sulfhydryl groups on the Mo edge. The sulfhydryl groups adsorb thiophene that on this site has no direct contact with the Mo cations.

Partial hydrogenation of thiophene occurs and hydrogenated thiophene intermediates transfer between the Mo and S edge sites. The latter are more reactive with respect to C–S bond cleavage (compare Figures 3.39a and 3.38b).

After C–S bond cleavage the sulfur atom is accommodated by S vacancy site at the sulfur edge.

However, when at the S edge, due to the high activation barrier for sulfur atom removal, no S vacancy sites are available anymore and more fully hydrogenated thiophene intermediates will cleave C–S bonds on the Mo edge. Consistent with the earlier kinetic analysis it suggests that the slow regeneration rate of sulfur vacancy sites inhibits C–S bond cleavage.

Hydrogenation of aromatics is only slightly suppressed by H_2S. No surface vacancies are necessary when aromatics hydrogenation is catalyzed by the brim sites at the Mo edge [257]. On the other hand, hydrogenation is strongly suppressed by basic nitrogen compounds [258], [259]. These act as poisons since they adsorb strongly to the acidic surface sulfhydryl groups.

The model studies present important atomistic insights on the reactivity of MoS_2 slabs deposited on Au. In practical systems the sulfide particles are in contact with oxidic supports and additional chemistry will play a role.

Early suggestions on the action of sulfide catalyst promoters relate the promotional effect of Ni or Co to interfacial contact of single component sulfide particles such as Co_9S_8 and MoS_2 [260]. This was thought to be mechanistically due to hydrogen spillover effects. Hydrogen atoms from H_2 activated by Co or Ni sulfide particles would activate the molybdenum sulfide.

Instead, later hydrogen-deuterium exchange studies showed that reactivity of S atoms is changed. These S atoms are located on MoS_2 edges and share binding to the promoting cations of Co or Ni with Mo [261]. Initial atomistic models of sulfide promotion and successive models are discussed below.

In the 1970s researchers at Shell proposed the pseudo-intercalation model of the promoted reaction site as an alternative to the sulfide phase synergy model. For the hydrogenation reaction of benzene catalyzed by nickel-promoted WS_2, they demonstrated that promotion does not alter the reactivity of the W sites but affects only their concentration. According to their pseudo-intercalation model, electron donation to the sulfide corrugates the tungsten sulfide edge [262].

A decade later Mössbauer spectroscopic data of Topsøe scientists [263] showed that Co atoms decorate the edges of MoS_2 layers and form a so-called "Co–Mo–S" phase. Thirty years later catalyst model studies with surface science techniques in combination with computational modelling also by Topsøe researchers in collaboration with Danish academic groups demonstrated that Co^{2+} or Ni^{2+} cations substitute for Mo^{4+} cations at the MoS_2 edges [255], [264]–[266].

An electrostatic model of the MoS_2 edge predicts that charge-neutral substitution of a $[MoS]^{2+}$ unit by two-valent Co^{2+} or Ni^{2+} cations reduces S concentration at the MoS_2 edges [253].

The M–S bond energies are also weakened for chemical bonding reasons. This is due to the lower redox potential of the Ni^{2+} and Co^{2+} cations compared to that of Mo^{4+}.

Early quantum-chemical calculations by Harris and Chianelli in 1986 [267] suggested electron donation into bond-weakening antibonding M–S orbitals. The M–S bond weakening is larger for Ni cation substitution than for the Co cation. This change in chemical bonding is confirmed in later calculations by Byskov *et al.* [268] and Raybaud *et al.* [264].

The weakening of the M–S bond implies that Ni and Co substitution increases S vacancy formation on the surface of MoS_2 and is possibly the reason for the higher reactivity of the Ni- and Co-promoted catalysts. From calculated M-S bond energies of mixed sulfides, Raybaud and Toulhoat [243] predicted the following order of reactivity: NiMoS > NiWS > CoMoS > CoWS. In the four systems the M–S bond of NiMoS is the weakest.

In simulations of the HDS reaction of DBT, the kinetic volcano curve that results when reaction rate is plotted against M–S bond energy has its maximum for NiMoS.

In addition to differences in M–S bond energy, there are also subtle site structural changes that affect the difference in reactivity of the Co- versus Ni-promoted system. STM studies by Lauritsen *et al.*

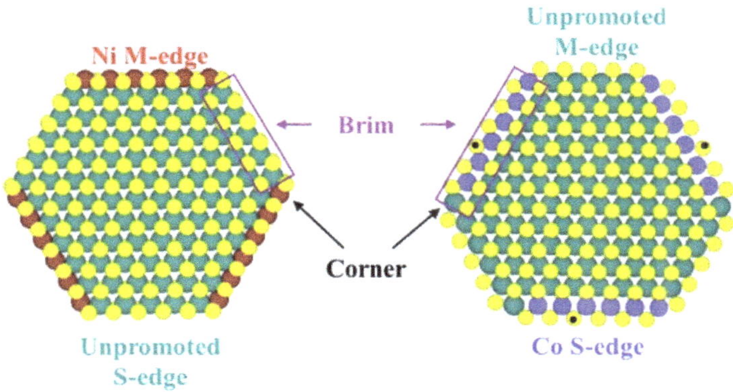

Figure 3.40 Particle models of NiMoS (left) with Ni atoms replacing Mo atoms on the molybdenum edge, and CoMoS (right) with Co atoms replacing Mo atoms on the sulfur edge of MoS_2. The three types of sites (edge, corner, and brim) are marked for reference. Color scheme: violet, cobalt; red, nickel; cyan, molybdenum; yellow, sulfur; black, hydrogen [269].

show that Co substitutes preferentially at the S edge of MoS_2. This is illustrated in Figure 3.40. Ni substitutes at the S edge on the larger MoS_2 particles, but on smaller particles Ni prefers decoration of the Mo edge [266].

This difference in promoter cation sites has an interesting consequence for HDS and HDN of crude oils. Compared to Co/MoS_2, the Ni/MoS_2 catalysts are more efficient aromatics hydrogenation catalysts. The nickel cation weakens the M–S bonds more than the Co cation. It creates additional sulfur vacant sites where nitrogen- or sulfur-containing molecules can be activated.

One difference between the Ni- and Co-promoted system is that the Ni substituted on the Mo edge adsorbs NH_3 stronger than H_2S, whereas for Co the reverse order is found [269]. These adsorption energies are a measure for their respective activities with respect to C–N versus C–S bond cleavage.

Co/MoS_2 is the preferred catalyst for HDS. Co-promoted MoS_2 activates the C–S bond more efficiently than the Ni-promoted system. Ni/MoS_2 catalysts are the preferred HDN catalysts [270]–[272].

The ammonia produced in HDN adsorbs more strongly than H_2S to the acidic sulfhydryl at the Mo edges. It is one of the causes of the reduction of HDS activity when HDN occurs jointly.

The brim site model of hydrogenation provides an explanation for the sulfide particle size-dependent selectivity of HDS as measured by Exxon researchers in 1994 [273]. Daage and Chianelli observed for DBT HDS that large particles remove S selectively without hydrogenation of aromatic rings (DDS), whereas HDS by the hydrogenolysis reaction path and also aromatics hydrogenation occur selectively when the MoS_2 particle is only a single layer. This is called the rim-edge selectivity model.

On supported MoS_2 catalysts, the particle size alters due to variation in slab layer diameter and the number of stacked MoS_2 layers of a MoS_2 particle. The layer at the top is similar to the single-layer MoS_2 model catalyst discussed above.

The earlier mentioned Topsøe studies (Figure 3.39) demonstrate that the basal surface of single-layer MoS_2 contains brim sites responsible for hydrogenation of the sulfur-containing organic molecules. When layers are stacked, these brim sites cannot be accessed by reactants anymore and reaction with sulfur vacant sites at the MoS_2 edges dominates.

The rim-edge model of Daage and Chianalli is consistent with the presence of active hydrogenation sites only in the top sulfur layer. These can be identified as the brim sites. It leads to dominance of the hydrogenolysis HDS reaction path for particles that consist of a single slab.

Since hydrogenation of the aromatic ring of DBT is less important for catalysts with a high stacking of MoS_2 layers, the alternative DDS reaction path is dominated by reaction of DBT with the sulfur vacancy sites at the edges.

Not only layer stacking affects selectivity of hydrogenation, but also the diameter of the MoS_2 slabs. When the diameter size increases, the ratio of corner over edge sites decreases [274]. The coordinatively unsaturated corner sites are active hydrogenation sites, whereas desulfurization by DDS is catalyzed preferentially by the edge sites.

The structure-function relation of the HDS reaction shows an increase of DDS versus aromatic ring hydrogenation when stacking of number of MoS_2 layers as well as diameter perpendicular to stacking direction increases.

3.4.2 *The Hydrodeoxygenation Reaction*

The deoxygenation reaction is a combination of reactions such as decarboxylation, decarbonylation, dehydration, and dihydroxylation.

In the 1980s research in HDO catalysis gained interest because of the need to also deoxygenate liquid fuels of high oxygen content derived from biomass by hydrotreating reactions [275], [276]. This is in contrast to the low oxygen content of oil-derived liquid fuels. Initially food biomass such as corn and plant oil were used as feedstock. Currently bio-oil that derives from non-edible biomass or renewable waste [277]–[279] is used. Its high oxygen content needs to be reduced so as to make it useful as transportation fuels or chemicals.

Here the mechanism of reactions that reduce oxygen content, mainly of biomass feedstock that consists of cellulose, hemicellulose, and lignin, by deoxygenation is discussed. A final section also considers the HDO mechanism of fatty acids. This reaction is relevant to the upgrading of waste cooking oils.

Plant biomass-derived feedstock consists of cellulose, hemicellulose, and lignin, that each requires specific catalysis for deoxygenation. Cellulose and hemicellulose are carbohydrates and consist of oligomers of C_5 or C_6 sugars. Lignin is a three-dimensional crosslinked polymer of propyl phenols and related molecules.

The pyrolysis reaction is a major process for conversion of biomass. It gives an oil of complex composition with a high content of oxygenates. For processing of pyrolysis oil to transportation fuel, deoxygenation is essential [277], [280], [281].

Figure 3.41 summarizes the main deoxygenation reactions that are part of the upgrading process.

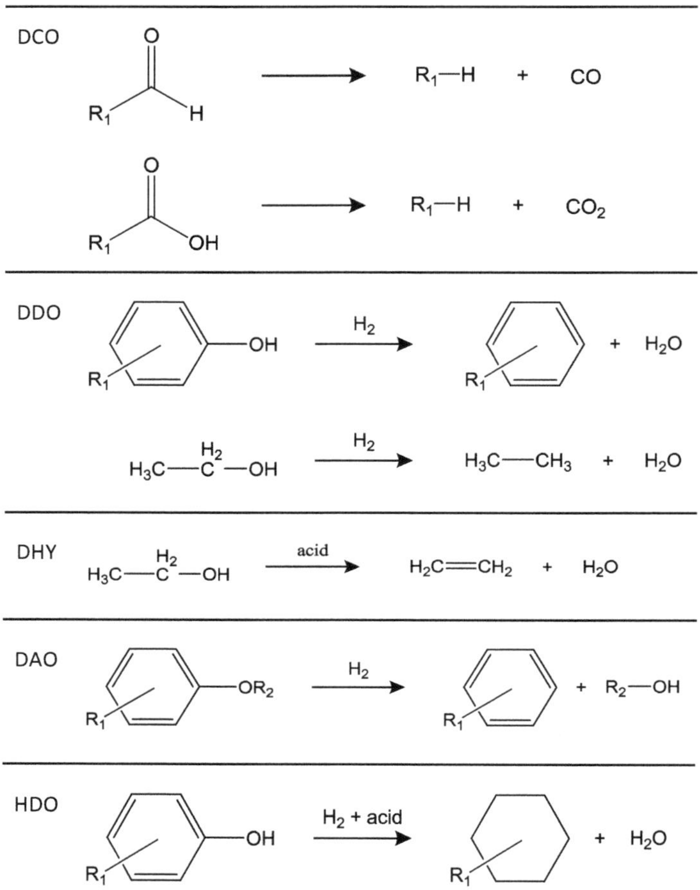

Figure 3.41 Deoxygenation reactions of bio-oil component molecules. Abbreviations are explained in the main text [277].

The removal of oxygen can occur via several main routes: decarbonylation or decarboxylation (DCO) of ketones or acids, direct deoxygenation (DDO) as dihydroxylation of phenol or alcohol, dehydration (DHY) of an alcohol to give alkene, dealkoxylation (DAO) of ethoxylated aromatics and HDO.

HDO consists of a HYD followed by a dehydration reaction. This reaction competes with DDO where phenol is dehydroxylated without saturation of the aromatic ring. The deoxygenation of phenolics is important for the upgrading of lignin-derived feedstock [282]. It is the subject of Section 3.4.2.1.

The other reactions are part of deoxygenation reactions of molecules such as furan and furan-related ring compounds, discussed in Section 3.4.2.2, that derive from cellulose [283]. Carboxylic acids derive also from cellulose or hemicellulose as well as from fatty esters. Their deoxygenation is discussed in the final Section 3.4.2.3.

A great variety of catalysts is used for these reactions. The catalysts are usually bifunctional, able to catalyze different combinations of the reactions shown in Figure 3.41.

3.4.2.1 *Hydrodeoxygenation of Phenolics*

The selectivity of direct deoxygenation without hydrogenation of unsaturated ring versus hydrogenolysis after ring hydrogenation is promoted by synergetic interaction with Lewis acid sites of the support.

The deoxygenation chemistry of guaiacol is representative for deoxygenation of phenolics. A schematic of the reaction network of guaiacol and some of the most important intermediates is shown in Figure 3.41 [284], [285].

The guaiacol molecule is a benzene ring substituted with a methoxy and hydroxyl group. The HDO reaction network is complex.

In Figure 3.42 three pathways are distinguished that give phenol: DDO of the hydroxyl group with hydrogen, hydrolysis of the methoxy group with water (DME), or by hydrogen (DMO). A related reaction deoxygenates phenol: the DDO reaction of phenol gives benzene. Phenol can also be deoxygenated by hydrogenation to cyclohexanol, followed by removal of the OH group to cyclohexane [284].

The selectivity of guaiacol deoxygenation depends sensitively on the catalyst. There is a large difference in selectivity between catalysts that only have a hydrogenation function versus bifunctional catalysts that in addition contain Lewis or Brønsted acid sites or oxophilic oxides. The reaction scheme of Figure 3.42 is representative for HDO of guaiacol catalyzed by bifunctional catalysts.

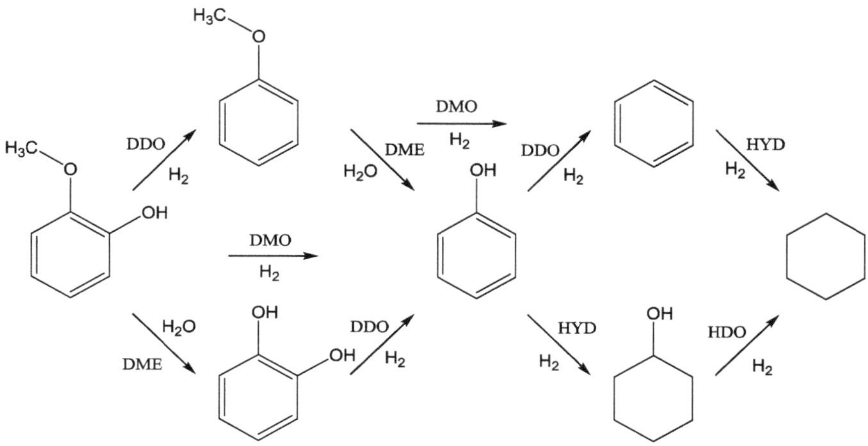

Figure 3.42 Reaction pathways of the guaiacol hydrodeoxygenation reaction [284]. Abbreviations are given in the text.

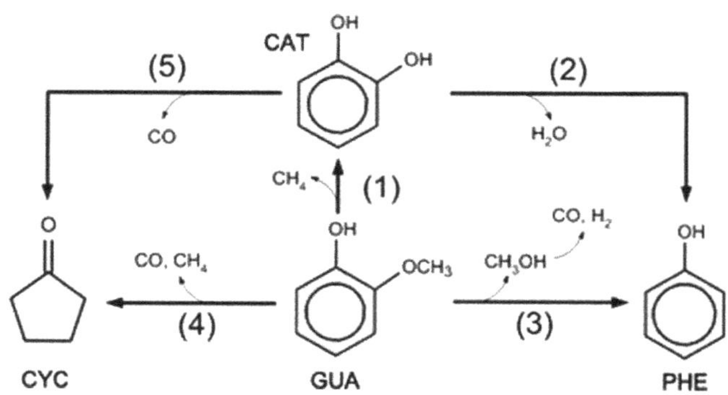

Figure 3.43 Reaction pathways of partial deoxygenation of guaiacol over a Pt/C catalyst [286].

Figure 3.43 illustrates the selectivity of guaiacol HDO when catalyzed by a transition metal on a neutral support such as carbon [286]. Platinum or ruthenium are amongst the most reactive catalysts [287]. Catechol is the main intermediate molecule. The important difference with the bifunctional reaction is that cyclopentanone is product in addition to phenol.

Cyclopentanone is formed from catechol by hydrogenation and successive decarbonylation of the diketone intermediate [286].

On a bifunctional acidic catalyst, the ketone further hydrodeoxygenates by a sequence of reactions as ketone hydrogenation to alcohol- and acid-catalyzed dehydration.

The transition metal demethylates by cleavage of the O–CH_3 bond and methane formation. Thermodynamics favors O–CH_3 (E_{diss} = 217 kJ/mol) over C–OCH_3 (E_{diss} = 356 kJ/mol) cleavage.

According to DFT calculations, cleavage of the phenolic C–OH bond (E_{act}(C–OH) ≈ 124 kJ/mol) as in the DDO reaction has a substantially higher barrier than that of the phenolic C–OCH_3 bond (E_{act}(C–OCH_3) ≈ 60-70 kJ/mol) [288]–[290]. This is in line with ready formation of catechol and phenol from guaiacol.

The reactivity of the metal sulfides is less than that of the transition metals, but the product pattern is comparable. Bulk MoS_2 or the more active $CoMoS_2$ give mainly phenol as product. Catechol is the main reaction intermediate [291]–[293]. As is also the case in HDS, $NiMoS_2$ is more active than $CoMoS_2$ [294]. Phenol is dehydroxylated and cyclohexane is the co-product. The reactivity of MoO_3 is comparable to that of the Mo sulfide [295].

Catalyst selectivity of bifunctional catalysts varies strongly with Brønsted or Lewis acidity.

Deoxygenation of guaiacol by a $CoMoS_2$ bifunctional catalyst has a different selectivity when catalyzed by Brønsted acidic sites of γ-Al_2O_3 than when supported on a Lewis acidic catalyst such as TiO_2 or ZrO_2. When supported by the Brønsted acidic support, the main products are catechol and methylated catechol [296]. When promoted by the Lewis acidic support, main products are catechol and phenol.

The Brønsted acidity of γ-Al_2O_3 catalyzes the bimolecular trans-alkylation reactions that give methyl-substituted catechol or cresol (for trans-alkylation reactions see Section 5.4.4).

Dispersion of transition metals on Lewis acidic supports increases selectivity of DDO without hydrogenation of the aromatic ring versus deoxygenation and hydrogenation of the aromatic ring (HYD) [280], [297]. This has been shown for demethoxylation of anisole and dihydroxylation of phenol catalyzed by Ru, Rh, and Ni supported on TiO_2, ZrO_2, and CeO_2 [298], [299].

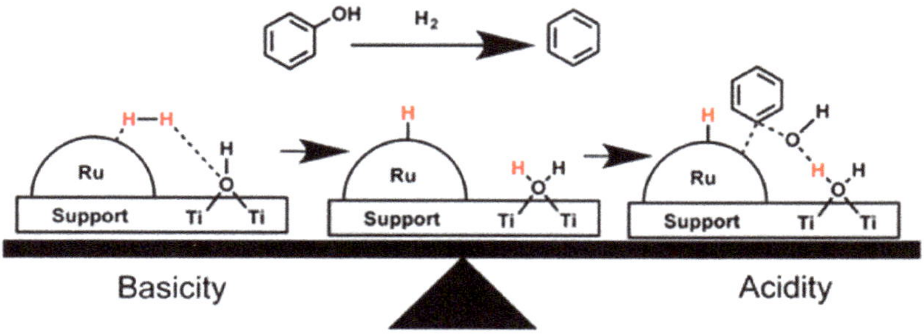

Figure 3.44 Direct deoxygenation of phenol at the interface of Ru and TiO_2 [300].

The mechanism of the DDO reaction has been elucidated for Ru dispersed on TiO_2 support by isotope exchange experiments and DFT quantum-chemical calculations by Grabow *et al.* [300]. They propose dual site synergy as schematically illustrated in Figure 3.44.

Reaction initiates by heterolytic H_2 dissociation at the interphase of the Ru particle and hydroxylated TiO_2. It creates an activated H_2O molecule and hydrogen atom on Ru. Phenol adsorbs with the phenyl ring on the Ru particle and the phenolic hydroxyl is activated by interaction with the hydrated Lewis acidic cation. The proton of water adsorbed to the Ti cation activates the oxygen atom of the phenol OH. This causes cleavage of C–OH and additional H_2O and benzene are generated.

When the experiment is performed with Ru dispersed on a non-hydroxylated TiO_2 surface (or without water addition) instead of DDO, the aromatic ring is hydrogenated and hexane is the main product.

The TiO_2 promotion effect is more general. Also, MoO_3 and Mo_2C, when dispersed on TiO_2, dehydroxylate phenol through DDO [301], [302].

Metal phosphide catalysts are also bifunctional HDO catalysts that hydrodeoxygenate mainly through the DDO mechanism to benzene, phenol, or anisole. The phosphorus in the reduced Ni_2P or Co_2P catalyst is mainly present as phosphide. The catalysts combine the Lewis acidic reactivity of $M^{\delta+}$ with oxophilic $P^{\delta-}$ [284], [285], [303].

The C–OH bond cleaves after heterolytic bond cleavage of the O–H bond. The H^+ atom is accommodated by phosphide and phenolate by Ni [304].

HDO of guaiacol catalyzed by Ni_2P/SiO_2 catalyst has as main reaction intermediates anisole and phenol, whereas the Co_2P/SiO_2 catalyst has as main intermediate products catechol (this is not a DDO product) and phenol.

In summary, HDO of guaiacol is a complex reaction that is sensitive to catalyst composition as well as reaction conditions [305]. There is an intricate interplay of Brønsted acidity, Lewis acidity, and hydrogenation activity. The DDO reaction is promoted via activation of phenolic OH by Lewis acidic or oxophilic sites. To suppress aromatic ring hydrogenation, the hydrogenation activity component of the bifunctional catalyst should not be too active.

3.4.2.2 *Hydrodeoxygenation of Furan and Furfural*

Ring opening requires initial hydrogenation of the furan ring to tetrahydrofuran. Product selectivity depends on competition between decarbonylation and hydrogenation.

Furan and furfural (see Figure 3.45) are important components of bio-oils. They are O atom-containing ring compounds that result from dehydration reactions of cellulose or hemicellulose [306]–[308].

They are also important intermediates in the overall conversion of hexose monosaccharides into linear alkanes for liquid fuels [283]. Furfural-derived oxygenates can be used as additives to fuel.

The ring-opening and deoxygenation reactions of furan and furfural are discussed here. The mechanistic question that

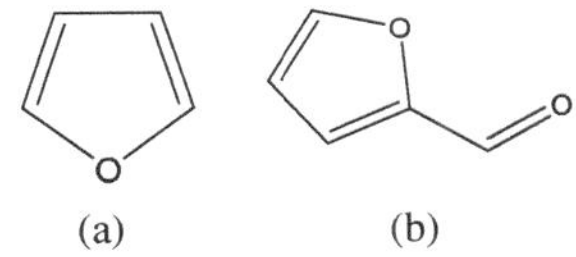

Figure 3.45 The oxygen-containing five-ring compounds: (a) furan, (b) furfural.

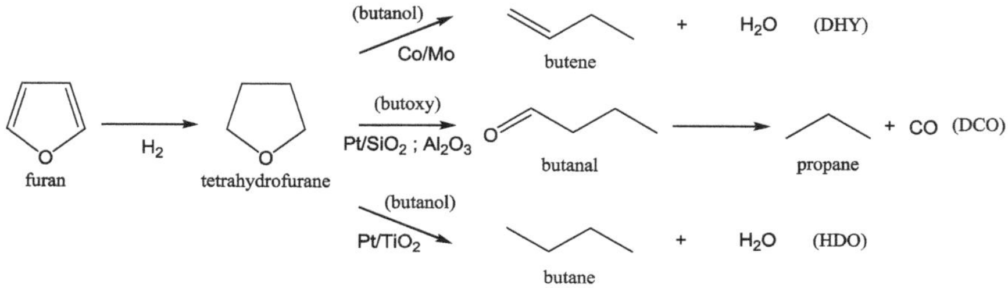

Figure 3.46 Ring-opening reactions of furan. Reaction intermediates are indicated in brackets.

determines selectivity of deoxygenation of furane and furfural is the relative contributions of hydrogenation and decarbonylation reactions. Tuning of bifunctional catalysts determines the difference. Deoxygenation catalysts are sulfides, transition metals, and metal phosphides dispersed on a variety of supports.

The ring-opening reaction of furan has some similarity with that of thiophene, but furan is less reactive than thiophene.

As is indicated in Figure 3.46, hydrogenation of the furan ring to tetrahydrofuran (THF) precedes the ring-opening reaction of furan. The subsequent ring-opening reaction of THF gives as initial products butanol, butanal, or butane, that in consecutive reactions dehydrate or decarbonylate to butene, propane, and CO [309].

The selectivity of the reaction depends sensitively on catalyst composition. When catalyzed by Co/MoS_2 or supported Pt, the HDO product distribution consists of C_4 hydrocarbons such as butane or butene. They are formed via butanol as reaction intermediate that dehydrates or dehydroxylates.

According to Bartok, in the absence of hydrogen, deoxygenation happens by the decarbonylation reaction that gives propane and carbon monoxide (the reaction path via butoxy) [310].

Deoxygenation produces selectively butane by hydrogenation of butanol when Pt is dispersed on Lewis acidic TiO_2.

With Ni_2P, hydrogenation of furan also has decarbonylation as a side reaction [311]. The oxophilicity of phosphorus contributes to this chemistry. The reaction is stereoselective. Substituted furan such as sterically hindered 2-methyltetrahydrofuran (MTHF) does not decarbonylate when catalyzed by Ni_2P [312], due to the repulsive interaction of the methyl substituent with the catalyst surface. In this case main products are pentane and some pentanone (the latter is an indication of adsorbed pentoxy as an intermediate).

Based on DFT calculations on Pd(111), Vlachos *et al.* [313] suggest that ring-opening of furan only has the lower barrier when hydrogen atom addition is limited to a single hydrogen atom to the α C atom next to the furan O atom. Continued hydrogenation of furane gives THF instead of the ring-opened product.

Surface science experiments indicate that decarbonylation of furan is suppressed by surface-adsorbed hydrogen. Decarbonylation is driven by the strong adsorption energy of CO to the transition metal [314].

The selective hydrogenation of furfural to 2-methylfuran is a desirable reaction, since 2-methylfuran is preferred as additive to gasoline over MTHF with its saturated ring system [315].

Pentanediol is attractive as chemical intermediate in polymer production (see Section 6.3.2).

As is illustrated in Figure 3.47 the chemistry of furfural catalysis is complex, because of its rich product pattern [316]. The ring-opening selectivity of furfural competes with decarbonylation of its CHO substituent.

Hydrogenation of the unsaturated ring system and CHO substituent to corresponding alcohol are also competitive reactions.

The selectivity of the reaction strongly depends on the transition metal catalyst. Selective formation of 2-methylfuran is catalyzed by Cu or Mo_2C that weakly interact with the furan ring and favor interaction with polar substituents. Hydrogenation by Cu gives preferentially furfuryl alcohol and has low selectivity for 2-methylfuran.

Catalysis by more reactive Mo_2C gives 2-methylfuran. After selective hydrogenation of the aldehyde to the alcohol, the C–OH is hydrogenated in a DDO reaction [317], [318].

The latter reaction is also found in surface science experiments with Pd(111) [319], [320]. However, because of the strong interaction of CO with Pd, catalysis by Pd has as main reaction furfural decarbonylation to furan and CO.

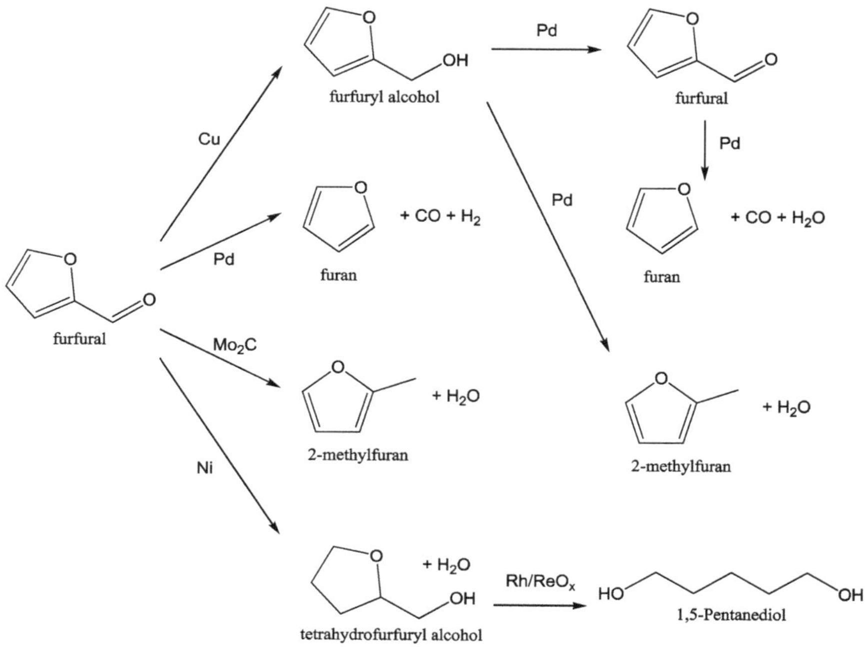

Figure 3.47 Ring-opening reactions and deoxygenation reactions of furfural.

Ni has a strong interaction with the furan ring and weaker interaction with CO, which favors selective formation of tetrahydrofurfuryl alcohol [321], [322].

The selectivity of the reaction is particle size dependent. Reactive small Pt particles promote C–CHO hydrogenolysis and give furan, whereas hydrogenation of furfural by the less reactive larger particles (that are dominated by terrace sites) give furfuryl alcohol [323], [324].

The ring-opening reaction of furfural to pentanediol is a two-step reaction. In an initial separate reaction, furfural is hydrogenated to tetrahydrofurfuryl alcohol.

In a second reaction, tetrahydrofurfuryl alcohol is converted to pentanediol catalyzed by Rh promoted with ReO_x. This reaction is bifunctional. Its reaction mechanism is discussed below, as has been resolved recently by experiment and DFT calculations [325]. The mechanism is illustrated in Figure 3.48.

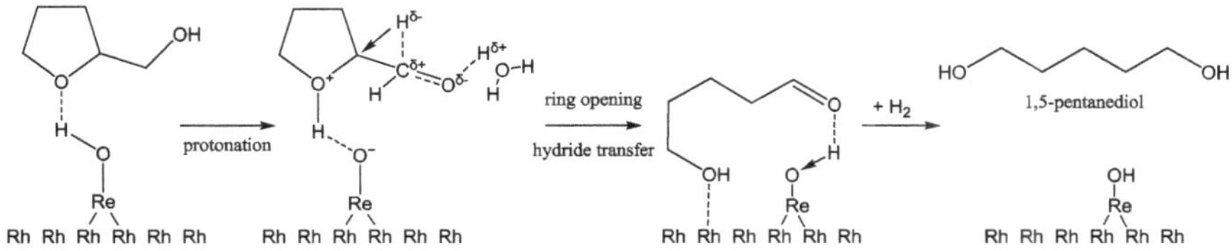

Figure 3.48 Ring-opening reaction of tetrahydrofurfuryl alcohol by ReO_x-promoted Rh in water [325].

The ReO_x dispersed over the Rh metal particle converts in water to a strong Brønsted acid. Rh functions as hydrogenation catalyst. The ring-opening reaction of tetrahydrofurfuryl alcohol is catalyzed by acidic ReO_x. Pentanediol is formed by subsequent hydrogenation.

In the ring-opening reaction, the ring oxygen atom of tetrahydrofurfuryl is activated by a proton and C–O bond cleavage is facilitated due to coactivation by the CH_2OH substituent [326]–[328]. Once the ring oxygen atom is protonated, a negatively charged hydrogen (hydride) atom moves from CH_2OH to the α carbon atom that connects with the ring oxygen atom.

The positive charge on $CHOH^+$ is stabilized by surrounding water. Proton backdonation to the negatively charged ReO_x^- generates the aldehyde on one end of the molecule and an alcohol group at the other end.

After the proton-catalyzed ring opening by the Rh/ReO_x catalyst, hydrogenation of the ring-opened intermediate gives product pentanediol.

In addition to the Rh/ReO_x system, Ir promoted by MoO_x can also be used for this reaction.

A key reaction step of the acid-catalyzed ring-opening reaction is the hydride transfer step of a hydrogen atom attached to the carbon atom of CH_2OH to the C atom of the ring system. Such a hydride transfer step is common to Brønsted or Lewis acid-catalyzed conversion reactions of carbohydrates (see Section 6.3.1.2, Figure 6.9) and combines usually with ring-opening or ring closure reactions. Hydride ion transfer happens jointly with proton transfer.

As in the tetrahydrofurfuryl alcohol reaction next to the ring opening, an alcohol group is converted to a ketone or the reverse hydrogenation reaction occurs.

3.4.2.3 *Carboxylic Acid Deoxygenation*

Decarboxylation of lactic acid by strong Brønsted acidic supports involves an internal hydride transfer reaction. Decarboxylation of aliphatic carboxylic acid is a metal- or metal sulfide-catalyzed reaction of dehydration followed by decarbonylation.

The biorefinery platform molecule lactic acid is produced by fermentation of biomass. It is a bifunctional carboxylic acid useful to the production of a variety of chemicals or liquid fuel [281].

The reaction network of the HDO of lactic acid is presented in the first part of this section.

The related mechanism of the deoxygenation of propionic acid is discussed in a final paragraph of this section. This is a prototype reaction of fatty acid deoxygenation. The latter reaction contributes to the conversion of fatty esters that are part of waste bio-oils that consist of fatty esters [327].

Propionic acid can be produced from lactic acid by dehydration catalysis.

The reaction network of the variety of deoxygenation reactions of lactic acid and their catalysts is shown in Figure 3.49.

A high caloric value of fuel requires high H/C ratio. The H/C ratio of product molecules increases by high reaction rates of decarboxylation or decarbonylation versus lactic acid oxygen reduction by dihydroxylation or dehydration.

The lactic acid molecule contains a carboxylic acid as well as OH substituent. Catalysis by a transition metal in the absence of hydrogen gives decarboxylation. Solid acids catalyze decarboxylation as well as dehydration. The relative rate of decarboxylation versus rate of dehydration depends on Brønsted acid strength. Strongly acidic materials such as Keggin-type polyacids have the higher reaction rate for decarboxylation [329]. Less acidic solid acids such as protonic zeolites or Nb_2O_5 catalyze dehydration selectively [330] (see Section 5.2.2 for details on solid acids).

Lactic acid is a potential platform molecule to produce acetaldehyde from lactic acid. This decarbonylation reaction is preferably done with strong acids. The solid acid-catalyzed decarboxylation is

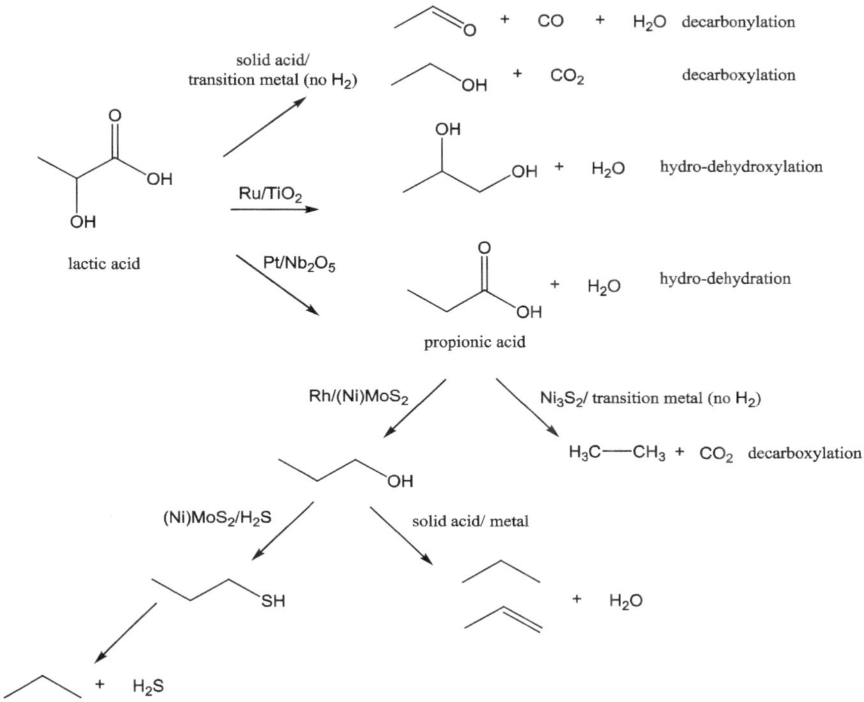

Figure 3.49 The reaction network of lactic acid (hydro)deoxygenation. All reactions are hydrogenation reactions except when differently indicated.

also a decarbonylation reaction. It is preceded by a dehydration step of the carboxyl group.

Dehydration of carboxyl is due to reaction of solid acid proton with carboxyl O–H [330]. In synchrony with proton backdonation to the solid from $CH_3CHO\text{-}H^+$, the C–CO bond cleaves to give carbonyl. In the overall reaction acetaldehyde is formed with CO and H_2O as co-products.

The Brønsted acid-catalyzed decarboxylation reaction of benzoic acid is different. This reaction is assisted by protonation of the benzene ring. This is followed by C–C bond cleavage of the protonated benzoate that gives benzene and CO_2 [331].

In the presence of hydrogen, dehydration and decarboxylation are suppressed in bifunctional catalysts promoted with hydrogenation-active catalytic material.

Dialcohols and carboxylic acid are produced by DDO from lactic acid. The selectivity of catalysis depends sensitively on whether the support is Lewis or Brønsted acidic. The HDO reaction of lactic acid catalyzed in water by Ru dispersed on a Lewis acid support such as TiO_2 gives 1,2-propanediol as main product [332]. Over Pt/Nb_2O_5, a transition metal supported by a Brønsted acid support, lactic acid hydrogenation does not have 1,2-propanediol but propionic acid as main product [281], [333]. The mildly acidic Nb_2O_5 support catalyzes selective dehydration of the "CH_3CHOH" part of the molecule. Pt hydrogenates intermediate acrylic acid to propionic acid.

A high H/C ratio in product molecule requires hydrogen addition to lactic acid.

Propionic acid deoxygenation reactions are shown in the bottom half of Figure 3.49. In the absence of hydrogen, catalysis by transition metal or metal sulfide gives ethane by a decarboxylation reaction from propionic acid.

Hydrogenation of propionic acid by Pd or Rh gives propanol or propenal [334]–[336].

Similar to the lactic acid to dialcohol transformation reaction, the propionic acid reaction is initiated by DDO of the carboxyl C–O, followed by hydrogenation. Catalysis by MoS_2 and Ni/MoS_2 is similar. Ni/MoS_2 is more active. In the presence of H_2S and H_2, the Ni/MoS_2 catalyst converts propanol to propane via intermediate propanethiol. With H_2S, initially intermediate propanethiol is formed from propanol. Subsequent hydrogenation regenerates H_2S and gives propane.

HDO catalysis of aliphatic carboxylic acids by metal sulfide catalysts has been experimentally [337] as well computationally investigated [338]–[340]. Computations provide a detailed molecular mechanism of the decarboxylation reaction of propionic acid catalyzed by $(Ni)MoS_2$ [340]. The reaction mechanism consists of a dehydration step that is followed by decarbonylation. The overall decarboxylation reaction is thought to happen through intermediate formation of ketene. The mechanism is schematically illustrated in Figure 3.50.

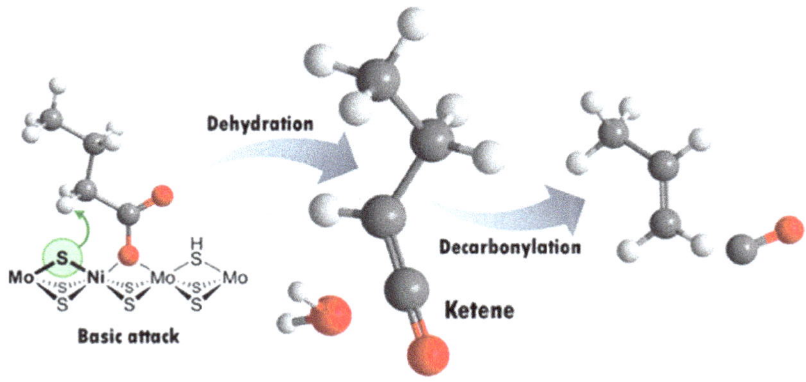

Figure 3.50 The decarboxylation reaction of propionic acid on Ni/MoS_2 [340].

The carboxylic acid adsorbs as carboxylate to a S vacant site in a reaction step in which the carboxyl OH bond cleaves heterolytically. The proton is donated to a surface S atom and carboxylate adsorbs to a cation. In a second step the reactive protonic α C–H bond cleaves and a second hydrogen atom is attached to a surface S atom.

One oxygen atom of the carboxylate reacts with the two hydrogen atoms to give water and intermediate ketene.

In a final step ketene decarbonylates and with an internal hydrogen step produces propene. This decarboxylation mechanism is possibly general for metal-catalyzed decarboxylation reactions [341].

3.5 Summary and List of Reactions

Reaction mechanisms involving hydrogen atom addition to adsorbed reaction intermediates are the central theme of this chapter. A major function of a hydrogenation catalyst is to dissociate the hydrogen molecule into hydrogen atoms that adsorb on the catalyst surface. A second important theme is the relation of reaction rate with the structure and composition of the catalytic center.

Elementary reactions share the common features of molecular bond dissociation and surface fragment recombination. Mechanistic differences relate to altered catalyst surface chemistry in response to different reactant reactivity.

For hydrocarbon conversion and hydrogenation of CO and N_2, transition metals are the most commonly used catalysts. These reactions are not only sensitive to transition metal choice but especially to the structure of the reaction center. The rate of reaction may vary with metal particle size since surface site structure varies with nanosize of transition metal particles.

Also, heteroatom removal from sulfur- and nitrogen-containing heteroatomic hydrocarbons is discussed. These reactions are catalyzed by metal sulfides. On the metal sulfide hydrogen dissociation is heterolytic. One hydrogen atom binds to a sulfur atom, the other to the metal cation.

This is different from homolytic dissociation of the hydrogen molecule on a transition metal surface. Then two equivalent hydrogen adatoms are generated.

Differences in catalyst reactivity relate to the sulfur vacancy surface concentration of the system.

HDO of oxygen-containing biomass molecules is the third class of reactions that is discussed. These oxygenated molecules are catalyzed by bifunctional catalysts with hydrogenation and acidic functions. This gives an additional complexity to the mechanism of these reactions.

Reaction sites for selective cleavage of C–O–C or C–OH bonds are often interfacial and consist of transition metal in contact with solid Lewis or Brønsted acid. In the reaction complex, metal function and acid site operate synergetically.

Tuning of reaction center reactivity by promotion with additional components is a common catalyst preparation strategy. Alloys have this function in the case of transition metals and metallic promoters for metal sulfides.

The unique reactivity of heterogeneous catalysts relates to the geometric extension of their reactive surface. It is the reason that transition metals are able to catalyze reactions that involve bond cleavage reactions of C–C, C=O, and N–N bonds. Such reactions require an ensemble of at least five or six transition metal atoms.

Part of this ensemble size requirement relates to the need for high coordination of the adatoms or surface fragments which are generated by bond dissociation to transition metal surface atoms. This is in contrast to C–H bond activation or H atom insertion reactions that only need contact with a single transition metal surface atom.

The latter reactions have their analogue in organometallic complex-catalyzed reactions that contain a single metal atom or cation.

A key reaction intermediate of the hydrocarbon conversion reactions catalyzed by heterogenous transition metal surfaces as well as organometallic complexes is the adsorbed alkyl intermediate.

It is readily formed by bond cleavage of C–H, hydrogen atom insertion into adsorbed alkene, or by "CH_2" insertion into another adsorbed alkyl intermediate. The adsorbed alkyl intermediate can undergo a variety of reactions that determines the selectivity of the reaction.

Progress of alkyl transformation is sensitive to the surface atom ensemble size of the reaction center. Additional C–H bond as well as C–C bond cleavage reactions can happen as well as intermediate hydrogen atom addition.

When additional C–H bond cleavage makes a bonding interaction of a second carbon atom of the hydrocarbon with the transition metal possible, this multipoint contact of the hydrocarbon fragment with the metal surface initiates the hydrogenolysis reaction. C–C bonds cleave and shorter surface hydrocarbon fragments are formed. Upon hydrogen addition this gives methane and short alkanes as product.

The most difficult elementary reaction step is the C–C bond cleavage reaction. This requires a large surface atom ensemble preferably arranged as a step-edge B5 site with coordinatively unsaturated surface atoms.

The selectivity of hydrocarbon product formation reactions that maintain the hydrocarbon skeleton, such as hydrogenation, dehydrogenation, and isomerization reactions requires suppression of the hydrogenolysis reaction. Selective reaction requires a low transition metal reactivity, such as that of platinum. It is reactive enough to activate C–H bonds, but not too reactive with respect to C–C bond cleavage.

Hydrogenation, dehydrogenation, and hydrocarbon isomerization reactions are possible on sites with small surface atom ensembles. Transition metal selectivity is promoted by its alloying with non-reactive transition metals as Cu, Ag, or Au. This reduces surface ensemble size and suppresses hydrogenolysis by reducing the reactivity of step-edge B5 sites.

Selectivity of hydrogenolysis versus that of other desired reactions is a general issue in hydrogenation catalysis.

In the HDS reaction of sulfur-containing aromatic molecules, the selectivity issue is to remove the sulfur atom without hydrogenation of the aromatic rings that are often part of the heteroatomic molecule.

Such DDS preferably happens on vacant surface sites of the metal sulfide catalyst. The sulfur atom of the heteroatomic molecule then is activated by contact with the metal cation.

Hydrogenation of the unsaturated sulfur-containing molecule, which leads to hydrogenolysis-type cleavage of the C–S bond, occurs on sites that involve the sulfhydryl. A metal sulfide particle of many stacked MoS_2 layers is selective to the DDS reaction, since the MoS_2 interlayer presence suppresses the sulfhydryl-catalyzed hydrogenolysis reaction.

Also in HDO, there is the selectivity question of DDO of the oxygenated molecules versus the hydrogenolysis reaction, where C–O bond cleavage only happens after hydrogenation of unsaturated hydrocarbon parts. The selective DDO site is a combination of a transition metal that activates H_2 and stabilizes reaction intermediate bonding, and a Lewis acid cation that activates the polar C–O bond.

In CO conversion catalysis a primary question is whether reaction is initiated by C=O bond cleavage or a reaction where the C–O bond remains intact. A reactive metal such as Ni cleaves the C=O bond, whereas Cu will leave the CO bond intact. Cu is the preferred catalyst for methanol formation (via intermediate formate), but Ni is the preferred catalyst for the methanation reaction. A catalyst where both reactions happen in parallel is Pd, which has reactivity in between Ni and Cu.

The Fischer-Tropsch reaction, which is catalyzed by more reactive metals such as Co and Ru, is an example where selectivity depends on the relation of consecutive surface reaction rates, namely, the elementary rates of C=O bond dissociation, C–C bond formation, and the termination reaction that removes growing alkyl chains as product from the surface.

Methane is the main product when hydrogenation of "CH_x" to methane is fast compared to the C=O bond cleavage reaction. The C–O bond of C=O can cleave by direct reaction and, after bond weakening, by addition of a hydrogen atom. In the presence of hydrogen atoms it generates "CH_x" intermediates.

The Fischer-Tropsch reaction has high selectivity when the elementary rate constant of the C=O bond cleavage reaction is fast compared to the "CH_x" to methane hydrogenation. For long hydrocarbon chain selectivity, the elementary rate constant of the insertion reaction of "CH_x" intermediates into the growing hydrocarbon alkyl chain intermediates is also slow or comparable to the elementary rate constant of C=O bond cleavage. The elementary rate of chain growth termination has to be slower.

For the same transition metal, differences in selectivity are due to different structure sensitivity of the respective elementary steps of the reaction. C=O dissociation (and also of other molecules with π bonds such as N_2) has only a low activation energy barrier when the structure of the reaction center consists of several metal atoms and has the structure of a step-edge site.

The termination reaction that requires M–C bond cleavage induced by hydrogen atom addition or CO insertion is fast on a surface terrace with surface atoms of low reactivity due to their high coordinative saturation.

The selectivity for methanation is high on the surface terrace where reaction rate constant of hydrocarbon chain termination is fast and reaction rate constant of C–O dissociation small. However, long hydrocarbon chain selectivity requires a catalyst surface with step-edge sites.

Such differences in site requirement are also fundamental to nanoparticle size dependence of hydrogenation reactions. Small particles have higher concentration of coordinatively unsatured surface metal atoms, which have a high rate for C–H bond activation reactions.

Step-edge B5-type sites are only stable on nanoparticles of intermediate size. It is the reason that the reaction rate normalized per surface atom of CO or N_2 conversion increases with nanoparticle size.

The state of the catalyst surface adapts to catalysis and reaction condition. High coverage with reaction intermediates may induce the catalyst surface to reconstruct. Such reconstruction can increase the concentration of step-edge B5 sites compared to the pristine surface and will also affect particle size dependence. In contrast to the expected low reactivity of the surface terrace, large particles will also contain selective B5 sites. This is the explanation of the observed high selectivity of the Fischer-Tropsch reaction for the larger particles as well.

In hydrocarbon conversion catalysis, the transition metal surface becomes partially deactivated by carbonaceous residue or carbide formation. This can be beneficial to hydrocarbon conversion selectivity since reactions that require small surface atom ensembles will remain functional. After a short initiation period, dehydrogenation or isomerization catalysis becomes selective and stable. Deactivating hydrogenolysis is suppressed.

This example indicates that trends in reactivity and selectivity can be significantly affected by catalyst deactivation that results from parasite reactions parallel to a desired reaction.

The Sabatier principle applies strictly to a reaction with a single product. The Sabatier maximum results from catalyst poisoning when catalyst reactivity is too large.

On the other hand, selectivity of a reaction that has several products may depend on selective poisoning of a particular reaction site. Sites of one reaction channel may deactivate whereas another reaction channel that uses different sites remains operational.

Also, when the non-selective reaction channel, which deactivates the catalyst, is a high activation reaction, the optimum catalyst can

Table 3.5 List of reactions of this chapter.

Type	Section
ammonia synthesis	3.2.1
acetic acid with ethene to vinyl acetate	3.1.3.1
anisole deoxygenation	3.4.2
alkane dehydrogenation/isomerization/hydrogenolysis	3.1.1–3.1.3
alkene hydroformylation	3.2.2.1
ethene hydrogenation	3.1.1
ethyl amine hydrodenitrogenation	3.4.1
ethyne to ethene hydrogenation	3.1.3.1
carbon monoxide hydrogenation (methanation, Fischer-Tropsch)	3.2.2
dibenzothiophene hydrodesulfurization	3.4.1
furan deoxygenation	3.4.2.2
furfural deoxygenation	3.4.2.2
guaiacol hydrodeoxygenation	3.4.2
lactic acid decarbonylation/hydrodeoxygenation	3.4.3.3
methanol synthesis	3.2.2.4
naphthene ring opening	3.1.4
phenol deoxygenation	3.4.2
pyridine hydrodenitrogenation	3.4.1
propionic acid to propanol	3.4.2.2
thiophene hydrodesulfurization	3.4.1
water-gas shift reaction	3.2.2.4

be the catalyst with lowest reactivity. The hydroisomerization reaction of n-alkanes catalyzed by Pt is such a reaction.

References

[1] G. H. Twigg and E. K. Rideal, "The exchange reaction between ethylene and deuterium on a nickel catalyst," *Proc. R. Soc. Lond. Ser. A.*, vol. 171, no. 944, pp. 55–69, May 1939, doi: 10.1098/RSPA.1939.0053.

[2] K. F. Bonhoeffer and A. Farkas, "On adsorption and reflection processes in the interaction of hydrogen and metals," *Trans. Faraday Soc.*, vol. 28, no. 0, pp. 242–247, Jan. 1932, doi: 10.1039/TF9322800242.

[3] A. Farkas, L. Farkas, and E. K. Rideal, "Experiments on heavy hydrogen IV—The hydrogenation and exchange reaction of ethylene with heavy hydrogen," *Proc. R. Soc. Lond. Ser. A*, vol. 146, no. 858, pp. 630–639, Oct. 1934, doi: 10.1098/RSPA.1934.0177.

[4] I. Horiuti and M. Polanyi, "Exchange reactions of hydrogen on metallic catalysts," *Trans. Faraday Soc.*, vol. 30, p. 1164, 1934, doi: 10.1039/tf9343001164.

[5] J. A. Osborn, F. H. Jardine, J. F. Young, and G. Wilkinson, "The preparation and properties of tris(triphenylphosphine) halogenorhodium(I) and some reactions thereof including catalytic homogeneous hydrogenation of olefins and acetylenes and their derivatives," *J. Chem. Soc. A*, vol. I, no. 0, pp. 1711–1732, Jan. 1966, doi: 10.1039/J19660001711.

[6] J. Halpern, "Mechanistic aspects of homogeneous catalytic hydrogenation and related processes," *Inorganica Chim. Acta*, vol. 50, no. C, pp. 11–19, Jan. 1981, doi: 10.1016/S0020-1693(00)83716-0.

[7] J. Halpern, T. Okamoto, and A. Zakhariev, "Mechanism of the chlorotris(triphenylphosphine) rhodium(I)-catalyzed hydrogenation of alkenes. The reaction of chlorodihydridotris(triphenylphosphine)rhodium(III) with cyclohexene," *J. Mol. Catal.*, vol. 2, no. 1, pp. 65–68, Jan. 1976, doi: 10.1016/0304-5102(77)80006-0.

[8] G. Wilkinson, "The Long Search for Stable Transition Metal Alkyls," *Nobel Lecture*, Dec. 11, 1973. https://www.nobelprize.org/prizes/chemistry/1973/wilkinson/lecture/ (accessed Dec. 23, 2022).

[9] C. Daniel, N. Koga, K. Morokuma, J. Han, and X. Y. Fu, "Ab initio Mo study of the full catalytic cycle of olefin hydrogenation by the Wilkinson catalyst RhCl(PR3)3," *J. Am. Chem. Soc.*, vol. 110, no. 12, pp. 3773–3787, 1988, doi: 10.1021/ja00220a010.

[10] Y. L. Gabor and A. Somorjai, *Introduction to Surface Chemistry and Catalysis*, 2nd ed. John Wiley & Sons, 2010.

[11] P. S. Cremer, X. Su, Y. R. Shen, and G. A. Somorjai, "Ethylene hydrogenation on Pt(111) monitored in situ at high pressures using sum frequency generation," *J. Am. Chem. Soc.*, vol. 118, no. 12, pp. 2942–2949, Mar. 1996, doi: 10.1021/ja952800t.

[12] M. Neurock, V. Pallassana, and R. A. van Santen, "The importance of transient states at higher coverages in catalytic reactions," *J. Am. Chem. Soc.*, vol. 122, no. 6, pp. 1150–1153, Feb. 2000, doi: 10.1021/ja992723s.

[13] C. J. Heard, C. Hu, M. Skoglundh, D. Creaser, and H. Grönbeck, "Kinetic regimes in ethylene hydrogenation over transition-metal surfaces," *ACS Catal.*, vol. 6, no. 5, pp. 3277–3286, May 2016, doi: 10.1021/acscatal.5b02708.

[14] H. Juro and M. Koshiro, "Hydrogenation of ethylene on metallic catalysts," 1968. https://srd.nist.gov/NSRDS/NSRDS-NBS-13.pdf (accessed Mar. 30, 2021).

[15] I. M. Gerzeliev, A. M. Gyul'Maliev, A. Y. Popov, and S. N. Khadzhiev, "Thermodynamic and quantum-chemical study of the oxidative dehydrogenation of ethane to ethylene," *Pet. Chem.*, vol. 55, no. 2, pp. 146–153, Mar. 2015, doi: 10.1134/S0965544115020097.

[16] B. M. Weckhuysen and R. A. Schoonheydt, "Alkane dehydrogenation over supported chromium oxide catalysts," *Catal. Today*, vol. 51, no. 2, pp. 223–232, Jun. 1999, doi: 10.1016/S0920-5861(99)00047-4.

[17] "Paul Sabatier, 1859-1941," *Obit. Not. Fellows R. Soc.*, vol. 4, no. 11, pp. 63–66, Nov. 1942, doi: 10.1098/RSBM.1942.0006.

[18] W. A. Lazier and H. Adkins, "Dehydrogenation and dehydration of alcohols over a zinc oxide catalyst," *J. Am. Chem. Soc.*, vol. 47, no. 6, pp. 1719–1722, Jun. 1925, doi: 10.1021/ja01683a033.

[19] F. E. Frey and W. F. Huppke, "Equilibrium dehydrogenation of ethane, propane, and the butanes," *Ind. Eng. Chem.*, vol. 25, no. 1, pp. 54–59, Jan. 1933, doi: 10.1021/IE50277A013.

[20] M. C. Molstad and B. F. Dodge, "Zinc oxide–chromium oxide catalysts for methanol synthesis," *Ind. Eng. Chem.*, vol. 27, no. 2, pp. 134–140, Feb. 1935, doi: 10.1021/ie50302a005.

[21] D. Sheldon, "Methanol production — A technical history," *Johnson Matthey Technol. Rev.*, vol. 61, no. 3., pp. 172–182, Jul. 01, 2017, doi: 10.1595/205651317X695622.

[22] E. Baraj, K. Ciahotný, and T. Hlinčík, "The water gas shift reaction: Catalysts and reaction mechanism," *Fuel*, vol. 288, p. 119817, Mar. 2021, doi: 10.1016/J.FUEL.2020.119817.

[23] R. L. Burwell, A. B. Littlewood, M. Cardew, G. Pass, and C. T. H. Stoddart, "Reactions between hydrocarbons and deuterium on chromium oxide gel. I. General," *J. Am. Chem. Soc.*, vol. 82, no. 24, pp. 6272–6280, Dec. 1960, doi: 10.1021/ja01509a019.

[24] P. P. M. M. Wittgen, C. Groeneveld, P. J. C. J. M. Zwaans, H. J. B. Morgenstern, A. H. van Heughten, C. J. M. van Heumen, and G. C. A. Schuit, "Hydrogenation of olefins and polymerization of ethene over chromium oxide/silica catalysts. III. Hydrogenation of olefins and H/D exchange," *J. Catal.*, vol. 77, no. 2, pp. 360–381, Oct. 1982, doi: 10.1016/0021-9517(82)90179-8.

[25] R. G. Pearson, "Hard and soft acids and bases," *J. Am. Chem. Soc.*, vol. 85, no. 22, pp. 3533–3539, Nov. 1963, doi: 10.1021/JA00905A001.

[26] E. J. M. Hensen, E. A. Pidko, N. Rane, and R. A. van Santen, "Water-promoted hydrocarbon activation catalyzed by binuclear gallium

sites in ZSM-5 zeolite," *Angew. Chem. Int. Ed.*, vol. 46, no. 38, pp. 7273–7276, 2007, doi: 10.1002/anie.200702463.

[27] G. Liu, L. Zeng, Z. J. Zhao, H. Tian, T. Wu, and J. Gong, "Platinum-modified ZnO/Al2O3 for Propane dehydrogenation: minimized platinum usage and improved catalytic stability," *ACS Catal.*, vol. 6, no. 4, pp. 2158–2162, Apr. 2016, doi: 10.1021/acscatal.5b02878.

[28] C. Zheng, Y. Apeloig, and R. Hoffmann, "Bonding and coupling of C1 fragments on metal surfaces," *J. Am. Chem. Soc.*, vol. 110, no. 3, pp. 749–774, Feb. 1988, doi: 10.1021/ja00211a015.

[29] C. Kemball and H. S. Taylor, "The catalytic decomposition of ethane and ethane-hydrogen mixtures," *J. Am. Chem. Soc.*, vol. 70, no. 1, pp. 345–351, Jan. 1948, doi: 10.1021/ja01181a108.

[30] R. D. Cortright, R. M. Watwe, B. E. Spiewak, and J. A. Dumesic, "Kinetics of ethane hydrogenolysis over supported platinum catalysts," *Catal. Today*, vol. 53, no. 3, pp. 395–406, Nov. 1999, doi: 10.1016/S0920-5861(99)00133-9.

[31] D. W. Flaherty and E. Iglesia, "Transition-state enthalpy and entropy effects on reactivity and selectivity in hydrogenolysis of n -alkanes," *J. Am. Chem. Soc.*, vol. 135, no. 49, pp. 18586–18599, Dec. 2013, doi: 10.1021/ja4093743.

[32] T. R. Ritliths, K. T. A. E. Lot, M. C. R. Symons, I. I. Zeitlin, A. Niimoto, J. J. Naughton, M. M. Frodyma, M. M. Frouyma, and K. Shibata, "Hydrocracking of neopentane and neohexane over evaporated metal films," *Nature*, vol. 187, no. 4741, pp. 937–938, 1960, doi: 10.1038/187937a0.

[33] J. R. Anderson and B. G. Baker, "The hydrocracking of saturated hydrocarbons over evaporated metal films," *Proc. R. Soc. Lond. Ser. A.*, vol. 271, no. 1346, pp. 402–423, Jan. 1963, doi: 10.1098/RSPA.1963.0026.

[34] Y. Barron, D. Cornet, G. Maire, and F. G. Gault, "The mechanism of isomerization of hexanes on platinum catalysts," *J. Catal.*, vol. 2, no. 2, pp. 152–155, Apr. 1963, doi: 10.1016/0021-9517(63)90023-X.

[35] Y. Barron, G. Maire, J. M. Muller, and F. G. Gault, "The mechanisms of hydrogenolysis and isomerization of hydrocarbons on metals: II. Mechanisms of isomerization of hexanes on platinum catalysts," *J. Catal.*, vol. 5, no. 3, pp. 428–445, Jun. 1966, doi: 10.1016/S0021-9517(66)80062-3.

[36] G. L. C. Maire and F. G. Garin, "Metal Catalysed Skeletal Reactions of Hydrocarbons," in *Catalysis: Science and Technology*, J. R. Anderson and M. Boudart, Eds. Springer, 1984, pp. 161–226.

[37] G. Webb, "The formation and role of carbonaceous residues in metal-catalysed reactions of hydrocarbons.," *Catal. Today*, vol. 7, no. 2, pp. 139–155, Apr. 1990, doi: 10.1016/0920-5861(90)85013-E.

[38] C. Corolleur, D. Tomanova, and F. G. Gault, "The mechanisms of hydrogenolysis and isomerization of hydrocarbons on metals: VII. Isomerization of labeled hexanes and hydrogenolysis of methyl (13C) cyclopentane on a 10% platinum-alumina catalyst," *J. Catal.*, vol. 24, no. 3, pp. 401–416, Mar. 1972, doi: 10.1016/0021-9517(72)90124-8.

[39] F. Garin and F. G. Gault, "The mechanisms of hydrogenolysis and isomerization of hydrocarbons on metals. VIII. Isomerization of 13C labeled pentanes on a 10% Pt-Al2O3 catalyst," *J. Am. Chem. Soc.*, vol. 97, no. 16, pp. 4466–4476, Aug. 1975, doi: 10.1021/ja00849a004.

[40] W. R. Patterson and J. J. Rooney, "Single atom sites and hydrocarbon reaction mechanisms," *Catal. Today*, vol. 12, no. 2–3, pp. 113–129, Feb. 1992, doi: 10.1016/0920-5861(92)85034-J.

[41] D. M. Adams, J. Chatt, R. G. Guy, and N. Sheppard, "149. The structure of 'cyclopropane platinous chloride,'" *J. Chem. Soc.*, vol. 45, no. 0, pp. 738–742, Jan. 1961, doi: 10.1039/JR9610000738.

[42] N. D. Zelinsky, B. A. Kazansky, and A. F. Plate, "Über 2-Methyl-bicyclo-[1.2.2]-heptan und dessen Verhalten bei der Dehydrogenisations- und Hydrogenisations-Katalyse. Aufspaltung Des Cyclopentan-Ringes durch katalytische Hydrierung," *Berichte der Dtsch. Chem. Gesellschaft (A B Ser.*, vol. 66, no. 9, pp. 1415–1419, Sep. 1933, doi: 10.1002/CBER.19330660950.

[43] G. Maire, G. Plouidy, J. Prudhomme, and F. Gault, "The mechanisms of hydrogenolysis and isomerization of hydrocarbons on metals I. Hydrogenolysis of cyclic hydrocarbons," *J. Catal.*, vol. 4, no. 5, pp. 556–569, 1965, doi: 10.1016/0021-9517(65)90160-0.

[44] Z. J. Zhao, L. V. Moskaleva, and N. Rösch, "Ring-opening reactions of methylcyclopentane over metal catalysts, M = Pt, Rh, Ir, and Pd: A mechanistic study from first-principles calculations," *ACS Catal.*, vol. 3, no. 2, pp. 196–205, Feb. 2013, doi: 10.1021/CS3005924.

[45] A. Cimino, M. Boudart, and H. Taylor, "Ethane hydrogenation-cracking on iron catalysts with and without alkali," *J. Phys. Chem.*, vol. 58, no. 9, pp. 796–800, 1954, doi: 10.1021/J150519A022.

[46] V. Haensel and G. R. Donaldson, "Platforming of pure hydrocarbons," *Ind. Eng. Chem.*, vol. 43, no. 9, pp. 2102–2104, Sep. 1951, doi: 10.1021/ie50501a036.

[47] "Vladimir Haensel," in *Biographical Memoirs*, vol. 88, National Academies Press, 2006, pp. 174–187.

[48] F. G. Ciapetta and J. Hunter, "Isomerization of saturated hydrocarbons in presence of hydrogenation-cracking catalysts: normal hexane," *Ind. Eng. Chem.*, vol. 45, no. 1, pp. 147–155, Jan. 1953, doi: 10.1021/IE50517A046.

[49] R. C. Archibald, B. S. Greensfelder, G. Holzman, and D. H. Rowe, “Catalytic hydrocracking of aliphatic hydrocarbons,” *Ind. Eng. Chem.*, vol. 52, no. 9, pp. 745–750, Sep. 1960, doi: 10.1021/IE50609A021.

[50] H. F. Schulz and J. H. Weitkamp, “Zeolite catalysts. Hydrocracking and hydroisomerization of n-dodecane,” *Ind. Eng. Chem. Prod. Res. Dev.*, vol. 11, no. 1, pp. 46–53, Mar. 1972, doi: 10.1021/I360041A007.

[51] J. Barbier, “Deactivation of reforming catalysts by coking — a review,” *Appl. Catal.*, vol. 23, no. 2, pp. 225–243, Jun. 1986, doi: 10.1016/S0166-9834(00)81294-4.

[52] N. Macleod, J. R. Fryer, D. Stirling, and G. Webb, “Deactivation of bi- and multimetallic reforming catalysts: influence of alloy formation on catalyst activity,” *Catal. Today*, vol. 46, no. 1, pp. 37–54, Nov. 1998, doi: 10.1016/S0920-5861(98)00349-6.

[53] J. L. Carter, G. B. McVinker, W. Weissman, M. S. Kmak, and J. H. Sinfelt, “Bimetallic catalysts; application in catalytic reforming,” *Appl. Catal.*, vol. 3, no. 4, pp. 327–346, Aug. 1982, doi: 10.1016/0166-9834(82)80267-4.

[54] J. H. Sinfelt, “Catalysis by alloys and bimetallic clusters,” *Acc. Chem. Res.*, vol. 10, no. 1, pp. 15–20, Jan. 1977, doi: 10.1021/ar50109a003.

[55] G. M. Schwab, “Alloy catalysts in dehydrogenation,” *Discuss. Faraday Soc.*, vol. 8, no. 0, pp. 166–171, Jan. 1950, doi: 10.1039/DF9500800166.

[56] D. D. Eley, “Catalysis by palladium-gold and related alloys,” *J. Res. Inst. Catal. Hokkaido Univ.*, vol. 16, no. 1, pp. 101–113, 1968.

[57] R. J. Best and W. W. Russell, “Nickel, copper and some of their alloys as catalysts for ethylene hydrogenation,” *J. Am. Chem. Soc.*, vol. 76, no. 3, pp. 838–842, 1954, doi: 10.1021/JA01632A060.

[58] P. van der Plank and W. M. H. Sachtler, “Surface composition of equilibrated copper-nickel alloy films,” *J. Catal.*, vol. 7, no. 3, pp. 300–303, Mar. 1967, doi: 10.1016/0021-9517(67)90115-7.

[59] W. M. H. Sachtler and P. van der Plank, “The role of individual surface atoms in chemisorption and catalysis by nickel-copper alloys,” *Surf. Sci.*, vol. 18, no. 1, pp. 62–79, Nov. 1969, doi: 10.1016/0039-6028(69)90267-2.

[60] J. H. Sinfelt, J. L. Carter, and D. J. C. Yates, “Catalytic hydrogenolysis and dehydrogenation over copper-nickel alloys,” *J. Catal.*, vol. 24, no. 2, pp. 283–296, Feb. 1972, doi: 10.1016/0021-9517(72)90072-3.

[61] W. M. H. Sachtler, “Surface composition of alloys in equilibrium,” *Le Vide*, vol. 164, p. 67, 1973.

[62] W. M. H. Sachtler and R. A. van Santen, “Surface composition and selectivity of alloy catalysts,” *Adv. Catal.*, vol. 26, no. C, pp. 69–119, Jan. 1977, doi: 10.1016/S0360-0564(08)60070-X.

[63] W. M. H. Sachtler, "Ensemble and Ligand Effects in Metal Catalysis," in *Handbook of Heterogeneous Catalysis*, G. Ertl, H. Knözinger, and J. Weitkamp, Eds. John Wiley & Sons, 2008, pp. 1585–1593.

[64] V. Ponec, "Alloy catalysts: the concepts," *Appl. Catal. A Gen.*, vol. 222, no. 1–2, pp. 31–45, Dec. 2001, doi: 10.1016/S0926-860X(01)00828-6.

[65] F. Besenbacher, I. Chorkendorff, B. S. Clausen, B. Hammer, A. M. Molenbroek, J. K. Nørskov, and I. Stensgaard, "Design of a surface alloy catalyst for steam reforming," *Science*, vol. 279, no. 5358, pp. 1913–1915, 1998, doi: 10.1126/science.279.5358.1913.

[66] F. Abild-Pedersen, J. Greeley, and J. K. Nørskov, "Understanding the effect of steps, strain, poisons, and alloying: methane activation on Ni surfaces," *Catal. Lett.*, vol. 105, no. 1, pp. 9–13, Nov. 2005, doi: 10.1007/S10562-005-7998-9.

[67] R. J. Davis and M. Boudart, "Ethylene hydrogenation by Pd-Au alloy," in *1st Tokyo Conference on Advanced Catalytic Science and Technology*, S. Yoshida, N. Takezawa, and T. Ono, Eds. Kodansha Weinheim VCH, 1991, p. 129.

[68] R. J. Davis and M. Boudart, "Structure of supported PdAu clusters determined by X-ray absorption spectroscopy," *J. Phys. Chem.*, vol. 98, no. 21, pp. 5471–5477, 1994, doi: 10.1021/J100072A012.

[69] M. Neurock and D. H. Mei, "Effects of alloying Pd and Au on the hydrogenation of ethylene: an ab initio-based dynamic monte carlo study," *Top. Catal.*, vol. 20, no. 1, p. 5, 2002, doi: 10.1023/A:1016339029623.

[70] J. G. Chen, "Carbide and nitride overlayers on early transition metal surfaces: Preparation, characterization, and reactivities," *Chem. Rev.*, vol. 96, no. 4, pp. 1477–1498, 1996, doi: 10.1021/cr950232u.

[71] C. R. Apesteguía and J. Barbier, "The role of catalyst presulfurization in some reactions of catalytic reforming and hydrogenolysis," *J. Catal.*, vol. 78, no. 2, pp. 352–359, Dec. 1982, doi: 10.1016/0021-9517(82)90319-0.

[72] M. Neurock and R. A. van Santen, "A first principles analysis of C–H bond formation in ethylene hydrogenation," *J. Phys. Chem. B*, vol. 104, no. 47, pp. 11127–11145, Nov. 2000, doi: 10.1021/jp994082t.

[73] D. Mei, E. W. Hansen, and M. Neurock, "Ethylene hydrogenation over bimetallic Pd/Au(111) surfaces: Application of quantum chemical results and dynamic Monte Carlo simulation," *J. Phys. Chem. B*, vol. 107, no. 3, pp. 798–810, Jan. 2003, doi: 10.1021/JP0139890.

[74] F. Studt, F. Abild-Pedersen, T. Bligaard, R. Z. Sørensen, C. H. Christensen, and J. K. Nørskov, "Identification of non-precious

metal alloy catalysts for selective hydrogenation of acetylene," *Science*, vol. 320, no. 5881, pp. 1320–1322, Jun. 2008, doi: 10.1126/SCIENCE.1156660.

[75] J. Osswald, K. Kovnir, M. Armbrüster, R. Giedigkeit, R. E. Jentoft, U. Wild, Y. Grin, and R. Schlögl, "Palladium–gallium intermetallic compounds for the selective hydrogenation of acetylene: Part II: Surface characterization and catalytic performance," *J. Catal.*, vol. 258, no. 1, pp. 219–227, Aug. 2008, doi: 10.1016/J.JCAT.2008.06.014.

[76] M. Neurock, "Perspectives on the first principles elucidation and the design of active sites," *J. Catal.*, vol. 216, no. 1–2, pp. 73–88, Jun. 2003, doi: 10.1016/S0021-9517(02)00115-X.

[77] P. Stoltze and J. K. Nørskov, "An interpretation of the high-pressure kinetics of ammonia synthesis based on a microscopic model," *J. Catal.*, vol. 110, no. 1, pp. 1–10, Mar. 1988, doi: 10.1016/0021-9517(88)90291-6.

[78] A. Mittasch and W. Frankenburg, "Early studies of multicomponent catalysts," *Adv. Catal.*, vol. 2, no. C, pp. 81–104, Jan. 1950, doi: 10.1016/S0360-0564(08)60375-2.

[79] J. R. Jennings, *Catalytic Ammonia Synthesis*. Springer, 1991, p. 451.

[80] P. H. Emmett and S. Brunauer, "The adsorption of nitrogen by iron synthetic ammonia catalysts," *J. Am. Chem. Soc.*, vol. 56, no. 1, pp. 35–41, Jan. 1934, doi: 10.1021/ja01316a011.

[81] M. Temkin and V. Pyzhev, "Kinetics of the synthesis of ammonia on promoted iron catalysts," *J. Phys. Chem. [USSR]*, vol. 13, p. 851, 1939.

[82] A. Ozaki, H. S. Taylor, and M. Boudart, "Kinetics and mechanism of the ammonia synthesis," *Proc. R. Soc. Lond. Ser. A.*, vol. 258, no. 1292, pp. 47–62, Oct. 1960, doi: 10.1098/rspa.1960.0174.

[83] G. Ertl, S. B. Lee, and M. Weiss, "Kinetics of nitrogen adsorption on Fe(111)," *Surf. Sci.*, vol. 114, no. 2–3, pp. 515–526, Feb. 1982, doi: 10.1016/0039-6028(82)90702-6.

[84] G. Ertl, "Surface science and catalysis—studies on the mechanism of ammonia synthesis: The P. H. Emmett award address," *Sci. Eng.*, vol. 21, no. 2, pp. 201–223, Jan. 1980, doi: 10.1080/03602458008067533.

[85] M. Bowker, I. B. Parker, and K. C. Waugh, "Extrapolation of the kinetics of model ammonia synthesis catalysts to industrially relevant temperatures and pressures," *Appl. Catal.*, vol. 14, no. C, pp. 101–118, Jan. 1985, doi: 10.1016/S0166-9834(00)84348-1.

[86] K. ichi Aika, H. Hori, and A. Ozaki, "Activation of nitrogen by alkali metal promoted transition metal I. Ammonia synthesis over ruthenium promoted by alkali metal," *J. Catal.*, vol. 27, no. 3, pp. 424–431, Dec. 1972, doi: 10.1016/0021-9517(72)90179-0.

[87] D. E. Brown, T. Edmonds, R. W. Joyner, J. J. McCarroll, and S. R. Tennison, "The genesis and development of the commercial BP doubly promoted catalyst for ammonia synthesis," *Catal. Lett.*, vol. 144, no. 4, pp. 545–552, Apr. 2014, doi: 10.1007/S10562-014-1226-4.

[88] S. Dahl, J. Sehested, C. J. H. Jacobsen, E. Törnqvist, and I. Chorkendorff, "Surface science based microkinetic analysis of ammonia synthesis over ruthenium catalysts," *J. Catal.*, vol. 192, no. 2, pp. 391–399, Jun. 2000, doi: 10.1006/jcat.2000.2857.

[89] S. Dahl, E. Törnqvist, and I. Chorkendorff, "Dissociative adsorption of N2 on Ru(0001): A surface reaction totally dominated by steps," *J. Catal.*, vol. 192, no. 2, pp. 381–390, Jun. 2000, doi: 10.1006/jcat.2000.2858.

[90] R. Brill, E.-L. Richter, and E. Ruch, "Adsorption of nitrogen on iron," *Angew. Chem. Int. Ed.*, vol. 6, no. 10, pp. 882–883, Oct. 1967, doi: 10.1002/anie.196708822.

[91] J. A. Dumesic, H. Topsøe, S. Khammouma, and M. Boudart, "Surface, catalytic and magnetic properties of small iron particles. II. Structure sensitivity of ammonia synthesis," *J. Catal.*, vol. 37, no. 3, pp. 503–512, Jun. 1975, doi: 10.1016/0021-9517(75)90185-2.

[92] F. Bozso, G. Ertl, M. Grunze, and M. Weiss, "Interaction of nitrogen with iron surfaces. I. Fe(100) and Fe(111)," *J. Catal.*, vol. 49, no. 1, pp. 18–41, Jul. 1977, doi: 10.1016/0021-9517(77)90237-8.

[93] J. J. Mortensen, M. V. Ganduglia-Pirovano, L. B. Hansen, B. Hammer, P. Stoltze, and J. K. Nørskov, "Nitrogen adsorption on Fe(111), (100), and (110) surfaces," *Surf. Sci.*, vol. 422, no. 1–3, pp. 8–16, Feb. 1999, doi: 10.1016/S0039-6028(98)00802-4.

[94] X.-Q. Q. Zhang, R. A. van Santen, and E. J. M. M. Hensen, "Carbon-induced surface transformations of cobalt," *ACS Catal.*, vol. 5, no. 2, pp. 596–601, Feb. 2015, doi: 10.1021/cs501484c.

[95] N. D. Spencer, R. C. Schoonmaker, and G. A. Somorjai, "Structure sensitivity in the iron single-crystal catalysed synthesis of ammonia," *Nature*, vol. 294, no. 5842, pp. 643–644, 1981, doi: 10.1038/294643a0.

[96] N. Spencer, "Iron single crystals as ammonia synthesis catalysts: Effect of surface structure on catalyst activity," *J. Catal.*, vol. 74, no. 1, pp. 129–135, 1982, doi: 10.1016/0021-9517(82)90016-1.

[97] N. Musselwhite and G. A. Somorjai, "Investigations of structure sensitivity in heterogeneous catalysis: From single crystals to monodisperse nanoparticles," *Top. Catal.*, vol. 56, no. 15–17, pp. 1277–1283, Aug. 2013, doi: 10.1007/S11244-013-0150-Y.

[98] S. Dahl, A. Logadottir, R. C. Egeberg, J. H. Larsen, I. Chorkendorff, E. Törnqvist, and J. K. NØrskov, "Role of steps in N2 activation on

Ru(0001)," *Phys. Rev. Lett.*, vol. 83, no. 9, pp. 1814–1817, Aug. 1999, doi: 10.1103/PhysRevLett.83.1814.

[99] C. J. H. Jacobsen, S. Dahl, P. L. Hansen, E. Törnqvist, L. Jensen, H. Topsøe, D. V. Prip, P. B. Møenshaug, and I. Chorkendorff, "Structure sensitivity of supported ruthenium catalysts for ammonia synthesis," *J. Mol. Catal. A Chem.*, vol. 163, no. 1–2, pp. 19–26, Dec. 2000, doi: 10.1016/S1381-1169(00)00396-4.

[100] K. Honkala, A. Hellman, I. N. Remediakis, A. Logadottir, A. Carlsson, S. Dahl, C. H. Christensen, and J. K. Nørskov, "Ammonia synthesis from first-principles calculations.," *Science*, vol. 307, no. 5709, pp. 555–558, Jan. 2005, doi: 10.1126/science.1106435.

[101] R. van Hardeveld and A. van Montfoort, "The influence of crystallite size on the adsorption of molecular nitrogen on nickel, palladium and platinum. An infrared and electron-microscopic study," *Surf. Sci.*, vol. 4, no. 4, pp. 396–430, Jul. 1966, doi: 10.1016/0039-6028(66)90016-1.

[102] R. Van Hardeveld and F. Hartog, "The statistics of surface atoms and surface sites on metal crystals," *Surf. Sci.*, vol. 15, no. 2, pp. 189–230, Jun. 1969, doi: 10.1016/0039-6028(69)90148-4.

[103] Á. Logadóttir and J. K. Nørskov, "Ammonia synthesis over a Ru(0001) surface studied by density functional calculations," *J. Catal.*, vol. 220, no. 2, pp. 273–279, Dec. 2003, doi: 10.1016/S0021-9517(03)00156-8.

[104] R. Schlögl, "Catalytic synthesis of ammonia — A 'never-ending story'?," *Angew. Chem. Int. Ed.*, vol. 42, no. 18, pp. 2004–2008, 2003, doi: 10.1002/anie.200301553.

[105] B. Hinnemann and J. K. Nørskov, "Catalysis by enzymes: the biological ammonia synthesis," *Top. Catal.*, vol. 37, no. 1, pp. 55–70, Mar. 2006, doi: 10.1007/s11244-006-0002-0.

[106] Y. Roux, C. Duboc, and M. Gennari, "Molecular catalysts for N_2 reduction: state of the art, mechanism, and challenges," *ChemPhysChem*, vol. 18, no. 19, pp. 2606–2617, Oct. 2017, doi: 10.1002/cphc.201700665.

[107] C. D. Zeinalipour-Yazdi, J. S. J. Hargreaves, and C. R. A. Catlow, "Low-T mechanisms of ammonia synthesis on Co3Mo3N," *J. Phys. Chem. C*, vol. 122, no. 11, pp. 6078–6082, Mar. 2018, doi: 10.1021/acs.jpcc.7b12364.

[108] P. Sabatier and J. B. Senderens, "New methane synthesis," *C. R. Acad. Sci. Paris*, no. 134, pp. 514–689, 1902.

[109] F. Fischer and H. Tropsch, "The synthesis of petroleum at atmospheric pressures from gasification products of coal," *Brennstoff-Chem 7*, no. 97, 1926.

[110] P. Sabatier and J. B. Senderens, “Nouvelles méthodes générales d’hydrogénation et de dédoublement moléculaire basées sur l’emploi des métaux divisé,” *Ann. Chim. Phys*, vol. 4, pp. 319–432, 1905.
[111] G. Ertl, M. Weiss, and S. B. Lee, “The role of potassium in the catalytic synthesis of ammonia,” *Chem. Phys. Lett.*, vol. 60, no. 3, pp. 391–394, Jan. 1979, doi: 10.1016/0009-2614(79)80595-3.
[112] C. D. Chang, “MTG revisited,” *Stud. Surf. Sci. Catal.*, vol. 61, no. C, pp. 393–404, Jan. 1991, doi: 10.1016/S0167-2991(08)60105-6.
[113] H. Pichler and H. Schulz, “Neuere Erkenntnisse auf dem Gebiet der Synthese von Kohlenwasserstoffen aus CO und H2 (Recent results in the synthesis of hydrocarbons from carbon monoxide and hydrogen),” *Chemie Ing. Tech.*, vol. 42, no. 18, pp. 1162–1174, 1970.
[114] R. F. Heck and D. S. Breslow, “The reaction of cobalt hydrotetracarbonyl with olefins,” *J. Am. Chem. Soc.*, vol. 83, no. 19, pp. 4023–4027, Oct. 1961, doi: 10.1021/ja01480a017.
[115] R. F. Heck, “Addition reactions of transition metal compounds,” *Acc. Chem. Res.*, vol. 2, no. 1, pp. 10–16, Jan. 1969, doi: 10.1021/ar50013a002.
[116] P. M. Maitlis and V. Zanotti, “The role of electrophilic species in the Fischer-Tropsch reaction,” *Chem. Commun.*, no. 13, pp. 1619–1634, Mar. 2009, doi: 10.1039/b822320n.
[117] I. M. Ciobîcă, G. J. Kramer, Q. Ge, M. Neurock, and R. A. van Santen, “Mechanisms for chain growth in Fischer-Tropsch synthesis over Ru(0001),” *J. Catal.*, vol. 212, no. 2, pp. 136–144, Dec. 2002, doi: 10.1006/jcat.2002.3742.
[118] J. T. Kummer and P. H. Emmett, “Fischer-Tropsch synthesis mechanism studies. The addition of radioactive alcohols to the synthesis gas,” *J. Am. Chem. Soc.*, vol. 75, no. 21, pp. 5177–5183, Nov. 1953, doi: 10.1021/ja01117a008.
[119] J. T. Kummer, T. W. DeWitt, and P. H. Emmett, “Some mechanism studies on the Fischer-Tropsch synthesis using C,” *J. Am. Chem. Soc.*, vol. 70, no. 11, pp. 3632–3643, Nov. 1948, doi: 10.1021/ja01191a029.
[120] G. Henrici-Olivé and S. Olive, “Mechanism of the Fischer-Tropsch synthesis: Origin of oxygenates,” *J. Mol. Catal.*, vol. 24, no. 1, pp. 7–13, Apr. 1984, doi: 10.1016/0304-5102(84)85034-8.
[121] H. Pichler and H. Schulz, “Neuere Erkenntnisse auf dem Gebiet der Synthese von Kohlenwasserstoffen aus CO und H2,” *Chemie Ing. Tech. — CIT*, vol. 42, no. 18, pp. 1162–1174, Sep. 1970, doi: 10.1002/cite.330421808.
[122] O. Roelen, “Präparative organische Chernie, Part 1,” *Naturforsch. und Medizin Deutschl.*, vol. 36, p. 157, 1948.

[123] G. D. Frey, "75 Years of oxo synthesis — The success story of a discovery at the OXEA Site Ruhrchemie," *J. Organomet. Chem.*, vol. 754. pp. 5–7, Mar. 2014, doi: 10.1016/j.jorganchem.2013.12.023.

[124] P. Biloen, J. N. Helle, F. G. A. van den Berg, and W. M. H. Sachtler, "On the activity of Fischer-Tropsch and methanation catalysts: A study utilizing isotopic transients," *J. Catal.*, vol. 81, no. 2, pp. 450–463, Jun. 1983, doi: 10.1016/0021-9517(83)90183-5.

[125] J. W. Niemantsverdriet, A. M. Van Der Kraan, W. L. Van Dijk, and H. S. Van Der Baan, "Behavior of metallic iron catalysts during Fischer-Tropsch synthesis studied with Mössbauer spectroscopy, X-ray diffraction, carbon content determination, and reaction kinetic measurements," *J. Phys. Chem.*, vol. 84, no. 25, pp. 3363–3370, 1980, doi: 10.1021/j100462a011.

[126] H. Schulz, "Short history and present trends of Fischer–Tropsch synthesis," *Appl. Catal. A Gen.*, vol. 186, no. 1–2, pp. 3–12, 1999, doi: 10.1016/S0926-860X(99)00160-X.

[127] R. C. Brady and R. Pettit, "Reactions of diazomethane on transition-metal surfaces and their relationship to the mechanism of the Fischer-Tropsch reaction," *J. Am. Chem. Soc.*, vol. 102, no. 19, pp. 6181–6182, 1980, doi: 10.1021/ja00539a053.

[128] R. C. Brady and R. Pettit, "On the mechanism of the Fischer-Tropsch reaction. The chain propagation step," *J. Am. Chem. Soc.*, vol. 103, no. 5, pp. 1287–1289, 1981, doi: 10.1021/ja00395a081.

[129] J. Cheng, X.-Q. X. Q. Gong, P. Hu, C. M. Lok, P. Ellis, and S. French, "A quantitative determination of reaction mechanisms from density functional theory calculations: Fischer–Tropsch synthesis on flat and stepped cobalt surfaces," *J. Catal.*, vol. 254, no. 2, pp. 285–295, Jan. 2008, doi: 10.1016/j.jcat.2008.01.009.

[130] M. L. Turner, H. C. Long, A. Shenton, P. K. Byers, and P. M. Maitlis, "The alkenyl mechanism for Fischer-Tropsch surface methylene polymerisation; the reactions of vinylic probes with CO/H2 over rhodium catalyst," *Chem. Eur. J.*, vol. 1, no. 8, pp. 549–556, Nov. 1995, doi: 10.1002/chem.19950010809.

[131] P. J. Flory, "Molecular size distribution in linear condensation polymers," *J. Am. Chem. Soc.*, vol. 58, no. 10, pp. 1877–1885, Oct. 1936, doi: 10.1021/ja01301a016.

[132] R. A. Friedel and R. B. Anderson, "Composition of synthetic liquid fuels. I. Product distribution and analysis of C5—C8 paraffin isomers from cobalt catalyst," *J. Am. Chem. Soc.*, vol. 72, no. 3, pp. 1212–1215, Mar. 1950, doi: 10.1021/ja01159a039.

[133] J. R. Platt, "Spectroscopic moment: A parameter of substituent groups determining aromatic ultraviolet intensities," *J. Chem. Phys.*, vol. 19, no. 3, pp. 263–271, Mar. 1951, doi: 10.1063/1.1748192.

[134] R. A. van Santen, M. M. Ghouri, A. J. Markvoort, and E. J. M. Hensen, "The Molecular Kinetics of the Fischer-Tropsch Reaction," in *Bridging Heterogeneous and Homogeneous Catalysis*, C. Li and Y. Liu, Eds. John Wiley & Sons, 2014, pp. 553–606.

[135] W. Chen, I. A. W. Filot, R. Pestman, and E. J. M. Hensen, "Mechanism of cobalt-catalyzed CO hydrogenation: 2. Fischer-Tropsch synthesis," *ACS Catal.*, vol. 7, no. 12, pp. 8061–8071, Dec. 2017, doi: 10.1021/acscatal.7b02758.

[136] R. A. van Santen, A. J. Markvoort, I. A. W. Filot, M. M. Ghouri, and E. J. M. Hensen, "Mechanism and microkinetics of the Fischer–Tropsch reaction," *Phys. Chem. Chem. Phys.*, vol. 15, no. 40, p. 17038, 2013, doi: 10.1039/c3cp52506f.

[137] R. A. van Santen and A. J. Markvoort, "Chain growth by CO insertion in the Fischer–Tropsch reaction," *ChemCatChem*, vol. 5, no. 11, pp. 3384–3397, Nov. 2013, doi: 10.1002/CCTC.201300173.

[138] M. Zhuo, K. F. Tan, A. Borgna, and M. Saeys, "Density functional theory study of the CO Insertion mechanism for Fischer–Tropsch synthesis over Co catalysts," *J. Phys. Chem. C*, vol. 113, no. 19, pp. 8357–8365, May 2009, doi: 10.1021/jp900281h.

[139] M. Araki and V. Ponec, "Methanation of carbon monoxide on nickel and nickel-copper alloys," *J. Catal.*, vol. 44, no. 3, pp. 439–448, Sep. 1976, doi: 10.1016/0021-9517(76)90421-8.

[140] J. A. Rabo, A. P. Risch, and M. L. Poutsma, "Reactions of carbon monoxide and hydrogen on Co, Ni, Ru, and Pd metals," *J. Catal.*, vol. 53, no. 3, pp. 295–311, Jul. 1978, doi: 10.1016/0021-9517(78)90102-1.

[141] M. A. Vannice, "The catalytic synthesis of hydrocarbons from H2CO mixtures over the Group VIII metals: V. The catalytic behavior of silica-supported metals," *J. Catal.*, vol. 50, no. 2, pp. 228–236, Nov. 1977, doi: 10.1016/0021-9517(77)90031-8.

[142] J. W. E. Coenen, W. M. T. M. Schats, and R. Z. C. van Meerten, "Structure effects in the interaction of hydrogen with benzene, cyclopropane and carbon monoxide over silica supported nickel catalysts," *Bull. des Sociétés Chim. Belges*, vol. 88, no. 7–8, pp. 435–452, Sep. 2010, doi: 10.1002/bscb.19790880701.

[143] M. A. Vannice, "The catalytic synthesis of hydrocarbons from carbon monoxide and hydrogen," *Catal. Rev.*, vol. 14, no. 1, pp. 153–191, Jan. 1976, doi: 10.1080/03602457608073410.

[144] G. L. Bezemer *et al.*, "Cobalt particle size effects in the Fischer-Tropsch reaction studied with carbon nanofiber supported catalysts," *J. Am. Chem. Soc.*, vol. 128, no. 12, pp. 3956–3964, Mar. 2006, doi: 10.1021/ja058282w.

[145] J. P. Den Breejen, P. B. Radstake, G. L. Bezemer, J. H. Bitter, V. Frøseth, A. Holmen, and K. P. De Jong, "On the origin of the cobalt particle size effects in Fischer-Tropsch catalysis," *J. Am. Chem. Soc.*, vol. 131, no. 20, pp. 7197–7203, May 2009, doi: 10.1021/JA901006X.

[146] D. W. Goodman, "Model catalytic studies over metal single crystal," *Acc. Chem. Res.*, vol. 17, no. 5, pp. 194–200, 1984, doi: 10.1021/ar00101a007.

[147] M. P. Andersson, F. Abild-Pedersen, I. N. Remediakis, T. Bligaard, G. Jones, J. Engbæk, O. Lytken, S. Horch, J. H. Nielsen, J. Sehested, J. R. Rostrup-Nielsen, J. K. Nørskov, and I. Chorkendorff, "Structure sensitivity of the methanation reaction: H2-induced CO dissociation on nickel surfaces," *J. Catal.*, vol. 255, no. 1, pp. 6–19, Apr. 2008, doi: 10.1016/j.jcat.2007.12.016.

[148] J. Wilson and C. De Groot, "Atomic-scale restructuring in high-pressure catalysis," *J. Phys. Chem.*, vol. 99, no. 20, pp. 7860–7866, 1995, doi: 10.1021/j100020a005.

[149] A. Banerjee, V. Navarro, J. W. M. Frenken, A. P. Van Bavel, H. P. C. E. Kuipers, and M. Saeys, "Shape and size of cobalt nanoislands formed spontaneously on cobalt terraces during Fischer-Tropsch synthesis," *J. Phys. Chem. Lett.*, vol. 7, no. 11, pp. 1996–2001, Jun. 2016, doi: 10.1021/ACS.JPCLETT.6B00555.

[150] H. Niemantsverdriet *et al.*, "Catalysis for fuels: General discussion," *Faraday Discussions*, vol. 197, no. 0., pp. 165–205, May 2017, doi: 10.1039/c7fd90010d.

[151] I. A. W. Filot, R. A. van Santen, and E. J. M. Hensen, "The optimally performing Fischer-Tropsch catalyst," *Angew. Chem. Int. Ed.*, vol. 53, no. 47, pp. 12746–12750, Aug. 2014, doi: 10.1002/anie.201406521.

[152] B. Zijlstra, R. J. P. Broos, W. Chen, H. Oosterbeek, I. A. W. Filot, and E. J. M. Hensen, "Coverage effects in CO dissociation on metallic cobalt nanoparticles," *ACS Catal.*, vol. 9, no. 8, pp. 7365–7372, Aug. 2019, doi: 10.1021/acscatal.9b01967.

[153] M. Ojeda, R. Nabar, A. U. Nilekar, A. Ishikawa, M. Mavrikakis, and E. Iglesia, "CO activation pathways and the mechanism of Fischer–Tropsch synthesis," *J. Catal.*, vol. 272, no. 2, pp. 287–297, 2010, doi: 10.1016/J.Jcat.2010.04.012.

[154] B. T. Loveless, C. Buda, M. Neurock, and E. Iglesia, "CO chemisorption and dissociation at high coverages during CO hydrogenation on Ru catalysts," *J. Am. Chem. Soc.*, vol. 135, no. 16, pp. 6107–6121, Apr. 2013, doi: 10.1021/ja311848e.

[155] M. Ojeda, R. Nabar, A. U. Nilekar, A. Ishikawa, M. Mavrikakis, and E. Iglesia, "CO activation pathways and the mechanism of Fischer-Tropsch synthesis," *J. Catal.*, vol. 272, no. 2, pp. 287–297, Jun. 2010, doi: 10.1016/j.jcat.2010.04.012.

[156] I. M. Ciobica and R. A. van Santen, "Carbon monoxide dissociation on planar and stepped Ru(0001) surfaces," *J. Phys. Chem. B*, vol. 107, no. 16, pp. 3808–3812, Apr. 2003, doi: 10.1021/jp030010x.

[157] P. Van Helden, J. A. Van Den Berg, and I. M. Ciobîcă, "Hydrogen-assisted CO dissociation on the Co(211) stepped surface," *Catal. Sci. Technol.*, vol. 2, no. 3, pp. 491–494, Mar. 2012, doi: 10.1039/c2cy00396a.

[158] S. Shetty, A. P. J. J. Jansen, and R. A. van Santen, "Direct versus hydrogen-assisted CO dissociation," *J. Am. Chem. Soc.*, vol. 131, no. 36, pp. 12874–12875, Sep. 2009, doi: 10.1021/ja9044482.

[159] W. J. Mitchell, J. Xie, T. A. Jachimowski, and W. H. Weinberg, "Carbon monoxide hydrogenation on the Ru (001) surface at low temperature using gas-phase atomic hydrogen: spectroscopic evidence for the carbonyl insertion mechanism on a transition metal surface," *J. Am. Chem. Soc.*, vol. 117, no. 9, pp. 2606–2617, 1995, doi: 10.1021/ja00114a025.

[160] A. Tuxen, S. Carenco, M. Chintapalli, C. H. Chuang, C. Escudero, E. Pach, P. Jiang, F. Borondics, B. Beberwyck, A. P. Alivisatos, G. Thornton, W. F. Pong, J. Guo, R. Perez, F. Besenbacher, and M. Salmeron, "Size-dependent dissociation of carbon monoxide on cobalt nanoparticles," *J. Am. Chem. Soc.*, vol. 135, no. 6, pp. 2273–2278, Feb. 2013, doi: 10.1021/JA3105889.

[161] E. De Smit and B. M. Weckhuysen, "The renaissance of iron-based Fischer–Tropsch synthesis: on the multifaceted catalyst deactivation behaviour," *Chem. Soc. Rev.*, vol. 37, no. 12, pp. 2758–2781, Dec. 2008, doi: 10.1039/B805427D.

[162] H. Zhao, J. X. Liu, C. Yang, S. Yao, H. Y. Su, Z. Gao, M. Dong, J. Wang, A. I. Rykov, J. Wang, Y. Hou, W. X. Li, and D. Ma, "Synthesis of iron-carbide nanoparticles: identification of the active phase and mechanism of Fe-based Fischer–Tropsch synthesis," *CCS Chem.*, vol. 3, no. 11, pp. 2712–2724, Nov. 2021, doi: 10.31635/CCSCHEM.020.202000555.

[163] S. G. Shetty, I. M. Ciobîcă, E. J. M. Hensen, and R. A. van Santen, "Site regeneration in the Fischer–Tropsch synthesis reaction: a synchronized CO dissociation and C–C coupling pathway," *Chem. Commun.*, vol. 47, no. 35, pp. 9822–9824, Aug. 2011, doi: 10.1039/C1CC11499A.

[164] A. J. Markvoort, R. A. van Santen, P. A. J. Hilbers, and E. J. M. Hensen, "Kinetics of the Fischer–Tropsch reaction," *Angew. Chem. Int. Ed.*, vol. 51, no. 36, pp. 9015–9019, Sep. 2012, doi: 10.1002/ANIE.201203282.

[165] D. D. Hibbitts, B. T. Loveless, M. Neurock, and E. Iglesia, "Mechanistic role of water on the rate and selectivity of Fischer-Tropsch synthesis

on ruthenium catalysts," *Angew. Chem. Int. Ed.*, vol. 52, no. 47, pp. 12273–12278, 2013, doi: 10.1002/anie.201304610.

[166] M. P. Andersson, T. Bligaard, A. Kustov, K. E. Larsen, J. Greeley, T. Johannessen, C. H. Christensen, and J. K. Nørskov, "Toward computational screening in heterogeneous catalysis: Pareto-optimal methanation catalysts," *J. Catal.*, vol. 239, no. 2, pp. 501–506, Apr. 2006, doi: 10.1016/J.JCAT.2006.02.016.

[167] J. K. Nørskov, F. Abild-Pedersen, F. Studt, and T. Bligaard, "Density functional theory in surface chemistry and catalysis.," *Proc. Natl. Acad. Sci.*, vol. 108, no. 3, pp. 937–943, Jan. 2011, doi: 10.1073/pnas.1006652108.

[168] I. A. W. Filot, R. A. van Santen, and E. J. M. Hensen, "Quantum chemistry of the Fischer–Tropsch reaction catalysed by a stepped ruthenium surface," *Catal. Sci. Technol.*, vol. 4, no. 9, pp. 3129–3140, Aug. 2014, doi: 10.1039/C4CY00483C.

[169] I. A. W. Filot, *Introduction to Microkinetic Modeling.* Technische Universiteit Eindhoven, 2018.

[170] F. Abild-Pedersen, J. Greeley, F. Studt, J. Rossmeisl, T. R. Munter, P. G. Moses, E. Skúlason, T. Bligaard, and J. K. Nørskov, "Scaling properties of adsorption energies for hydrogen-containing molecules on transition-metal surfaces," *Phys. Rev. Lett.*, vol. 99, no. 1, pp. 16105–16109, Jul. 2007, doi: 10.1103/PHYSREVLETT.99.016105.

[171] R. A. van Santen, M. Neurock, and S. G. Shetty, "Reactivity theory of transition-metal surfaces: A Brønsted-Evans-Polanyi linear activation energy-free-energy analysis," *Chem. Rev.*, vol. 110, no. 4, pp. 2005–2048, Apr. 2010, doi: 10.1021/CR9001808.

[172] B. Zijlstra, R. J. P. Broos, W. Chen, G. L. Bezemer, I. A. W. Filot, and E. J. M. Hensen, "The vital role of step-edge sites for both CO activation and chain growth on cobalt fischer-tropsch catalysts revealed through first-principles-based microkinetic modeling including lateral interactions," *ACS Catal.*, vol. 10, no. 16, pp. 9376–9400, Aug. 2020, doi: 10.1021/acscatal.0c02420.

[173] R. A. van Santen, A. J. Markvoort, M. M. Ghouri, P. A. J. Hilbers, and E. J. M. Hensen, "Monomer formation model versus chain growth model of the Fischer-Tropsch reaction," *J. Phys. Chem. C*, vol. 117, no. 9, pp. 4488–4504, Mar. 2013, doi: 10.1021/JP310245M.

[174] R. A. van Santen, I. M. Ciobîcâ, E. Van Steen, and M. M. Ghouri, "Mechanistic issues in Fischer–Tropsch catalysis," *Adv. Catal.*, vol. 54, pp. 127–187, Jan. 2011, doi: 10.1016/B978-0-12-387772-7.00003-4.

[175] C. Rhodes, G. J. Hutchings, and A. M. Ward, "Water-gas shift reaction: finding the mechanistic boundary," *Catal. Today*, vol. 23, no. 1, pp. 43–58, Jan. 1995, doi: 10.1016/0920-5861(94)00135-O.

[176] E. F. A. Rmstrong and T. P. H. Ilditch, "A study of catalytic actions at solid surfaces. —IV. The interaction of carbon monoxide and steam as conditioned by iron oxide and by copper," *Proc. R. Soc. Lond. Ser. A*, vol. 97, no. 684, pp. 265–273, May 1920, doi: 10.1098/rspa.1920.0032.

[177] P. Davies, F. F. Snowdon, G. W. Bridger, D. O. Hughes, and P. W. Young, "Water-gas conversion and catalysts therefor," GB1010871A, Jun. 08, 1962.

[178] E. Błasiak, "Sposób Wytwarzania Wysokoaktywnego Katalizatora do Syntezy Metanolu," Polish Patent 34,000, May 14, 1947.

[179] H. Wilmer, M. Kurtz, K. V. Klementiev, O. P. Tkachenko, W. Grünert, O. Hinrichsen, A. Birkner, S. Rabe, K. Merz, M. Driess, C. Wöll, and M. Muhler, "Methanol synthesis over ZnO: A structure-sensitive reaction?," *Phys. Chem. Chem. Phys.*, vol. 5, no. 20, pp. 4736–4742, Oct. 2003, doi: 10.1039/b304425d.

[180] G. C. Chinchen, P. J. Denny, D. G. Parker, G. D. Short, M. S. Spencer, K. C. Waugh, and K. A. Whan, "Mechanism of methanol synthesis from CO2/CO/H2 mixtures over copper/zinc oxide/alumina catalysts — use of C-14-labeled reactants," *Am. Chem. Soc. Div. Fuel Chem.*, vol. 30, no. 2, pp. 333–338, 1987.

[181] G. C. Chinchen, P. J. Denny, D. G. Parker, M. S. Spencer, and D. A. Whan, "Mechanism of methanol synthesis from CO2/CO/H2 mixtures over copper/zinc oxide/alumina catalysts: use of 14C-labelled reactants," *Appl. Catal.*, vol. 30, no. 2, pp. 333–338, Apr. 1987, doi: 10.1016/S0166-9834(00)84123-8.

[182] K. C. Waugh, "Methanol synthesis," *Catal. Today*, vol. 15, no. 1, pp. 51–75, May 1992, doi: 10.1016/0920-5861(92)80122-4.

[183] J. Nakamura, J. M. Campbell, and C. T. Campbell, "Kinetics and mechanism of the water-gas shift reaction catalysed by the clean and Cs-promoted Cu(110) surface: A comparison with Cu(111)," *J. Chem. Soc. Faraday Trans.*, vol. 86, no. 15, pp. 2725–2734, Jan. 1990, doi: 10.1039/FT9908602725.

[184] K. Klier, "Methanol synthesis," *Adv. Catal.*, vol. 31, no. C, pp. 243–313, Jan. 1982, doi: 10.1016/S0360-0564(08)60455-1.

[185] P. L. Hansen, J. B. Wagner, S. Helveg, J. R. Rostrup-Nielsen, B. S. Clausen, and H. Topsøe, "Atom-resolved imaging of dynamic shape changes in supported copper nanocrystals," *Science*, vol. 295, no. 5562, pp. 2053–2055, Mar. 2002, doi: 10.1126/science.1069325.

[186] M. Behrens, F. Studt, I. Kasatkin, S. Kühl, M. Hävecker, F. Abild-Pedersen, S. Zander, F. Girgsdies, P. Kurr, B. L. Kniep, M. Tovar, R. W. Fischer, J. K. Nørskov, and R. Schlögl, "The active site of methanol synthesis over Cu/ZnO/Al2O 3 industrial catalysts,"

Science, vol. 336, no. 6083, pp. 893–897, May 2012, doi: 10.1126/science.1219831.

[187] L. C. Grabow and M. Mavrikakis, "Mechanism of methanol synthesis on Cu through CO2 and CO hydrogenation," *ACS Catal.*, vol. 1, no. 4, pp. 365–384, Apr. 2011, doi: 10.1021/cs200055d.

[188] P. Wu, J. Zaffran, and B. Yang, "Role of surface species interactions in identifying the reaction mechanism of methanol synthesis from CO_2 hydrogenation over intermetallic PdIn(310) steps," *J. Phys. Chem. C*, vol. 123, no. 22, pp. 13615–13623, Jun. 2019, doi: 10.1021/ACS.JPCC.9B01847.

[189] C. V. Ovesen, B. S. Clausen, J. Schiøtz, P. Stoltze, H. Topsøe, and J. K. Nørskov, "Kinetic implications of dynamical changes in catalyst morphology during methanol synthesis over Cu/ZnO catalysts," *J. Catal.*, vol. 168, no. 2, pp. 133–142, Jun. 1997, doi: 10.1006/jcat.1997.1629.

[190] J. Yoshihara and C. T. Campbell, "Methanol synthesis and reverse water-gas shift kinetics over Cu(110) model catalysts: Structural sensitivity," *J. Catal.*, vol. 161, no. 2, pp. 776–782, Jul. 1996, doi: 10.1006/jcat.1996.0240.

[191] R. Van Den Berg, G. Prieto, G. Korpershoek, L. I. Van Der Wal, A. J. Van Bunningen, S. Lægsgaard-Jørgensen, P. E. De Jongh, and K. P. De Jong, "Structure sensitivity of Cu and CuZn catalysts relevant to industrial methanol synthesis," *Nat. Commun.*, vol. 7, no. 1, pp. 1–7, Oct. 2016, doi: 10.1038/ncomms13057.

[192] M. Boudart, "Catalysis by supported metals," *Adv. Catal.*, vol. 20, no. C, pp. 153–166, Jan. 1969, doi: 10.1016/S0360-0564(08)60271-0.

[193] M. Boudart and M. A. McDonald, "Structure sensitivity of hydrocarbon synthesis from CO and H2," *J. Phys. Chem.*, vol. 88, no. 11, pp. 2185–2195, 1984, doi: 10.1021/J150655A004.

[194] M. Boudart, A. Aldag, J. E. Benson, N. A. Dougharty, and C. Girvin Harkins, "On the specific activity of platinum catalysts," *J. Catal.*, vol. 6, no. 1, pp. 92–99, Aug. 1966, doi: 10.1016/0021-9517(66)90113-8.

[195] C. R. Henry, C. Chapon, S. Giorgio, and C. Goyhenex, "Size Effects in Heterogeneous Catalysis," in *Chemisorption and Reactivity on Supported Clusters and Thin Films*, R. M. Lambert and G. Pacchioni, Eds. Springer, 1997, pp. 117–152.

[196] G. Jones, J. Jakobsen, S. Shim, J. Kleis, M. Anderson, J. Rossmeisl, F. Abild-Pedersen, T. Ligaard, S. Helveg, and B. Hinnemann, "First principles calculations and experimental insight into methane steam reforming over transition metal catalysts," *J. Catal.*, vol. 259, no. 1, pp. 147–160, 2008, doi: 10.1016/j.jcat.2008.08.003.

[197] M. Che and C. O. Bennett, "The influence of particle size on the catalytic properties of supported metals," *Adv. Catal.*, vol. 36, no. C, pp. 55–172, Jan. 1989, doi: 10.1016/S0360-0564(08)60017-6.

[198] R. A. van Santen, B. G. Anderson, R. H. Cunningham, A. V. G. Mangnus, L. J. van Ijzendoorn, and M. J. A. De Voigt, "In situ observation of transient reaction phenomena occurring on zeolite catalysts with the aid of positron emission profiling," *Angew. Chem. Int. Ed.*, vol. 35, no. 2324, pp. 2785–2787, Dec. 1997, doi: 10.1002/anie.199627851.

[199] J. L. Carter, J. A. Cusumano, and J. H. Sinfelt, "Catalysis over supported metals. V. The effect of crystallite size on the catalytic activity of nickel," *J. Phys. Chem.*, vol. 70, no. 7, pp. 2257–2263, 1966, doi: 10.1021/j100879a029.

[200] D. J. C. Yates, W. F. Taylor, and J. H. Sinfelt, "Catalysis over supported metals. I. Kinetics of ethane hydrogenolysis over nickel surfaces of known area," *J. Am. Chem. Soc.*, vol. 86, no. 15, pp. 2996–3001, Aug. 1964, doi: 10.1021/JA01069A007.

[201] D. J. C. Yates and J. H. Sinfelt, "The catalytic activity of rhodium in relation to its state of dispersion," *J. Catal.*, vol. 8, no. 4, pp. 348–358, Aug. 1967, doi: 10.1016/0021-9517(67)90331-4.

[202] S. Fuentes and F. Figueras, "The influence of particle size on the catalytic properties of alumina-supported rhodium catalysts," *J. Catal.*, vol. 61, no. 2, pp. 443–453, Feb. 1980, doi: 10.1016/0021-9517(80)90391-7.

[203] M. Boudart, A. W. Aldag, L. D. Ptak, and J. E. Benson, "On the selectivity of platinum catalysts," *J. Catal.*, vol. 11, no. 1, pp. 35–45, May 1968, doi: 10.1016/0021-9517(68)90007-9.

[204] D. Wayne Goodman and M. Kiskinova, "Chemisorption and reactivity studies of H2 and CO on sulfided Ni(100)," *Surf. Sci.*, vol. 105, no. 2–3, pp. L265–L270, Apr. 1981, doi: 10.1016/0039-6028(81)90001-7.

[205] M. Kiskinova and D. W. Goodman, "Modification of chemisorption properties by electronegative adatoms: H2 and CO on chlorided, sulfided, and phosphided Ni(100)," *Surf. Sci.*, vol. 108, no. 1, pp. 64–76, Jun. 1981, doi: 10.1016/0039-6028(81)90358-7.

[206] D. W. Goodman, "Catalysis: from single crystals to the 'real world,'" *Surf. Sci.*, vol. 299–300, no. C, pp. 837–848, Jan. 1994, doi: 10.1016/0039-6028(94)90701-3.

[207] B. McAllister and P. Hu, "A density functional theory study of sulfur poisoning," *J. Chem. Phys.*, vol. 122, no. 8, p. 084709, Feb. 2005, doi: 10.1063/1.1854125.

[208] R. M. Watwe, R. D. Cortright, J. K. Nørskov, and J. A. Dumesic, "Theoretical studies of stability and reactivity of C2 hydrocarbon

species on Pt clusters, Pt(111), and Pt(211)," *J. Phys. Chem. B*, vol. 104, pp. 2299–2310, 2000, doi: 10.1021/jp993202u.

[209] Y. Chen and D. G. Vlachos, "Hydrogenation of ethylene and dehydrogenation and hydrogenolysis of ethane on Pt(111) and Pt(211): A density functional theory study," *J. Phys. Chem. C*, vol. 114, no. 11, pp. 4973–4982, Mar. 2010, doi: 10.1021/JP909163Z.

[210] E. Segal, R. J. Madon, and M. Boudart, "Catalytic hydrogenation of cyclohexene: I. Vapor-phase reaction on supported platinum," *J. Catal.*, vol. 52, no. 1, pp. 45–49, Mar. 1978, doi: 10.1016/0021-9517(78)90121-5.

[211] S. Davis and G. A. Somorjai, "Correlation of cyclohexene reactions on platinum crystal surfaces over a ten-order-of-magnitude pressure range. 1. Variations of structure sensitivity, rates, and reaction probabilities," *J. Catal.*, vol. 65, no. 1, pp. 78–83, Sep. 1980, doi: 10.1016/0021-9517(80)90279-1.

[212] W. K. Hall and P. H. Emmett, "The hydrogenation of benzene over copper-nickel alloys," *J. Phys. Chem.*, vol. 62, no. 7, pp. 816–821, 1958, doi: 10.1021/J150565A011.

[213] O. Beeck, "Hydrogenation catalysts," *Discuss. Faraday Soc.*, vol. 8, no. 0, pp. 118–128, Jan. 1950, doi: 10.1039/DF9500800118.

[214] E. K. Rideal, "XXXIX.—The hydrogenation of ethylene in contact with nickel," *J. Chem. Soc. Trans.*, vol. 121, no. 0, pp. 309–318, Jan. 1922, doi: 10.1039/CT9222100309.

[215] G. I. Jenkins and E. Rideal, "The catalytic hydrogenation of ethylene at a nickel surface. Part II. The reaction mechanism," *J. Chem. Soc.*, no. 0, pp. 2496–2500, Jan. 1955, doi: 10.1039/JR9550002496.

[216] H. zur Strassen, "Zur Kinetik der katalytisclien Äthylenhydrierung," *Zeitschrift für Phys. Chemie*, vol. 169A, no. 1, pp. 81–90, Feb. 1934, doi: 10.1515/ZPCH-1934-16908.

[217] G. A. Somorjal and F. Zaera, "Heterogeneous catalysis on the molecular scale," *J. Phys. Chem.*, vol. 86, no. 16, pp. 3070–3078, 1982, doi: 10.1021/J100213A007.

[218] C. Vogt, F. Meirer, M. Monai, E. Groeneveld, D. Ferri, R. A. van Santen, M. Nachtegaal, R. R. Unocic, A. I. Frenkel, and B. M. Weckhuysen, "Dynamic restructuring of supported metal nanoparticles and its implications for structure insensitive catalysis," *Nat. Commun.*, vol. 12, no. 1, pp. 1–10, Dec. 2021, doi: 10.1038/s41467-021-27474-3.

[219] S. Helveg, C. López-Cartes, J. Sehested, P. L. Hansen, B. S. Clausen, J. R. Rostrup-Nielsen, F. Abild-Pedersen, and J. K. Nørskov, "Atomic-scale imaging of carbon nanofibre growth," *Nature*, vol. 427, no. 6973, pp. 426–429, Jan. 2004, doi: 10.1038/nature02278.

[220] H. Schulz, "Short history and present trends of Fischer-Tropsch synthesis," *Appl. Catal. A Gen.*, vol. 186, no. 1–2, pp. 3–12, Oct. 1999, doi: 10.1016/S0926-860X(99)00160-X.

[221] J. Cheng, P. Hu, P. Ellis, S. French, G. Kelly, and C. M. Lok, "Density functional theory study of iron and cobalt carbides for fischer-tropsch synthesis," *J. Phys. Chem. C*, vol. 114, no. 2, pp. 1085–1093, Jan. 2010, doi: 10.1021/jp908482q.

[222] B. Bunnink and G. Kramer, "Energetics of methane dissociative adsorption on Rh{{}111{}} from DFT calculations," *J. Catal.*, vol. 242, no. 2, pp. 309–318, Sep. 2006, doi: 10.1016/j.jcat.2006.06.015.

[223] Z.-P. Liu and P. Hu, "General rules for predicting where a catalytic reaction should occur on metal surfaces: a density functional theory study of C-H and C-O bond breaking/making on flat, stepped, and kinked metal surfaces.," *J. Am. Chem. Soc.*, vol. 125, no. 7, pp. 1958–1967, Feb. 2003, doi: 10.1021/ja0207551.

[224] A. S. Bazhenov, L. Lefferts, and K. Honkala, "Adsorption and activation of water on cuboctahedral rhodium and platinum nanoparticles," *J. Phys. Chem. C*, vol. 121, no. 8, pp. 4324–4331, Mar. 2017, doi: 10.1021/ACS.JPCC.6B11953.

[225] H. B. Lyon and G. A. Somorjai, "Low-energy electron-diffraction study of the clean (100), (111), and (110) faces of platinum," *J. Chem. Phys.*, vol. 46, no. 7, p. 2539, May 2004, doi: 10.1063/1.1841082.

[226] G. A. Somorjai and M. A. Van Hove, "Adsorbate-induced restructuring of surfaces," *Prog. Surf. Sci.*, vol. 30, no. 3–4, pp. 201–231, Jan. 1989, doi: 10.1016/0079-6816(89)90009-9.

[227] L. Jacobse, Y. F. Huang, M. T. M. Koper, and M. J. Rost, "Correlation of surface site formation to nanoisland growth in the electrochemical roughening of Pt(111)," *Nat. Mater.*, vol. 17, no. 3, pp. 277–282, Mar. 2018, doi: 10.1038/s41563-017-0015-z.

[228] L. Jacobse, M. J. Rost, and M. T. M. Koper, "Atomic-scale identification of the electrochemical roughening of platinum," *ACS Cent. Sci.*, vol. 5, no. 12, pp. 1920–1928, Dec. 2019, doi: 10.1021/acscentsci.9b00782.

[229] R. A. van Santen, "Complementary structure sensitive and insensitive catalytic relationships.," *Acc. Chem. Res.*, vol. 42, no. 1, pp. 57–66, Jan. 2009, doi: 10.1021/ar800022m.

[230] J. D. Benck, T. R. Hellstern, J. Kibsgaard, P. Chakthranont, and T. F. Jaramillo, "Catalyzing the hydrogen evolution reaction (HER) with molybdenum sulfide nanomaterials," *ACS Catal.*, vol. 4, no. 11, pp. 3957–3971, Oct. 2014, doi: 10.1021/CS500923C.

[231] T. F. Jaramillo, K. P. Jørgensen, J. Bonde, J. H. Nielsen, S. Horch, and I. Chorkendorff, "Identification of active edge sites for electrochemical H2 evolution from MoS2 nanocatalysts," *Science*, vol. 317, no. 5834, pp. 100–102, Jul. 2007, doi: 10.1126/SCIENCE.1141483.

[232] B. Hinnemann, P. G. Moses, J. Bonde, K. P. Jørgensen, J. H. Nielsen, S. Horch, I. Chorkendorff, and J. K. Nørskov, "Biomimetic hydrogen evolution: MoS2 nanoparticles as catalyst for hydrogen evolution.," *J. Am. Chem. Soc.*, vol. 127, no. 15, pp. 5308–5309, Apr. 2005, doi: 10.1021/ja0504690.

[233] J. A. R. van Veen, "What's new? On the development of sulphidic HT catalysts before the molecular aspects," *Catal. Today*, vol. 292, pp. 2–25, Sep. 2017, doi: 10.1016/J.CATTOD.2016.09.027.

[234] S. S. Bello, C. Wang, M. Zhang, H. Gao, Z. Han, L. Shi, F. Su, and G. Xu, "A review on the reaction mechanism of hydrodesulfurization and hydrodenitrogenation in heavy oil upgrading," *Energy & Fuels*, vol. 35, no. 14, pp. 10998–11016, Jul. 2021, doi: 10.1021/ACS.ENERGYFUELS.1C01015.

[235] H. Topsøe, B. S. Clausen, and F. E. Massoth, *Hydrotreating Catalysis.* Springer, 1996.

[236] R. Prins, "Hydrodesulfurization," in *Catalysis: From Principles to Applications*, M. Beller, A. Renken, and R. A. van Santen, Eds. John Wiley & Sons, 2012, pp. 390–398.

[237] T. A. Pecoraro and R. R. Chianelli, "Hydrodesulfurization catalysis by transition metal sulfides," *J. Catal.*, vol. 67, no. 2, pp. 430–445, Feb. 1981, doi: 10.1016/0021-9517(81)90303-1.

[238] R. R. Chianelli, M. Daage, and M. J. Ledoux, "Fundamental studies of transition-metal sulfide catalytic materials," *Adv. Catal.*, vol. 40, no. C, pp. 177–232, Jan. 1994, doi: 10.1016/S0360-0564(08)60658-6.

[239] R. A. van Santen, *Modern Heterogeneous Catalysis: An Introduction.* John Wiley & Sons, 2017.

[240] J. K. Nørskov, B. S. Clausen, and H. Topsøe, "Understanding the trends in the hydrodesulfurization activity of the transition metal sulfides," *Catal. Lett.*, vol. 13, no. 1, pp. 1–8, Mar. 1992, doi: 10.1007/BF00770941.

[241] R. R. Chianelli, G. Berhault, P. Raybaud, S. Kasztelan, J. Hafner, and H. Toulhoat, "Periodic trends in hydrodesulfurization: in support of the Sabatier principle," *Appl. Catal. A Gen.*, vol. 227, no. 1–2, pp. 83–96, Mar. 2002, doi: 10.1016/S0926-860X(01)00924-3.

[242] P. Raybaud, G. Kresse, J. Hafner, and H. Toulhoat, "Ab initio density functional studies of transition-metal sulphides: I. Crystal structure and cohesive properties," *J. Phys. Condens. Matter*, vol. 9, no. 50, p. 11085, Dec. 1997, doi: 10.1088/0953-8984/9/50/013.

[243] H. Toulhoat and P. Raybaud, “Kinetic interpretation of catalytic activity patterns based on theoretical chemical descriptors,” *J. Catal.*, vol. 216, no. 1–2, pp. 63–72, May 2003, doi: 10.1016/S0021-9517(02)00118-5.

[244] E. J. M. Hensen, H. J. A. Brans, G. M. H. J. Lardinois, V. H. J. de Beer, J. a. Rob van Veen, and R. A. van Santen, “Periodic trends in hydrotreating catalysis: thiophene hydrodesulfurization over carbon-supported 4d transition metal sulfides,” *J. Catal.*, vol. 192, no. 1, pp. 98–107, 2000, doi: 10.1006/jcat.2000.2824.

[245] G. Berhault, M. Lacroix, M. Breysse, F. Maugé, J. C. Lavalley, H. Nie, and L. Qu, “Characterization of acidic sites of silica-supported transition metal sulfides by pyridine and 2,6 dimethylpyridine adsorption: relation to activity in CH3SH condensation,” *J. Catal.*, vol. 178, no. 2, pp. 555–565, Sep. 1998, doi: 10.1006/JCAT.1998.2196.

[246] V. H. J. de Beer, J. G. J. Dahlmans, and J. G. M. Smeets, “Hydrodesulfurization and hydrogenation properties of promoted MoS2 and WS2 catalysts under medium pressure conditions,” *J. Catal.*, vol. 42, no. 3, pp. 467–470, Jun. 1976, doi: 10.1016/0021-9517(76)90122-6.

[247] M. Dekker, *The Desulfurization of Heavy Oils and Residua*, 2nd ed. CRC Press, 1981.

[248] K. Saiprasad Rao and G. Murali Dhar, “Catalytic functionalities of supported tungsten sulfide catalysts,” *J. Catal.*, vol. 115, no. 1, pp. 277–281, Jan. 1989, doi: 10.1016/0021-9517(89)90029-8.

[249] P. Raybaud and H. Toulhoat, *Catalysis by Transition Metal Sulphides*. Editions Technip, 2013.

[250] J. V. Lauritsen, M. Nyberg, J. K. Nørskov, B. S. Clausen, H. Topsøe, E. Lægsgaard, and F. Besenbacher, “Hydrodesulfurization reaction pathways on MoS2 nanoclusters revealed by scanning tunneling microscopy,” *J. Catal.*, vol. 224, no. 1, pp. 94–106, May 2004, doi: 10.1016/J.JCAT.2004.02.009.

[251] J. V Lauritsen, M. V Bollinger, E. Lægsgaard, K. W. Jacobsen, J. K. Nørskov, B. S. Clausen, H. Topsøe, and F. Besenbacher, “Atomic-scale insight into structure and morphology changes of MoS 2 nanoclusters in hydrotreating catalysts,” *J. Catal.*, vol. 221, no. 2, pp. 510–522, 2004, doi: 10.1016/j.jcat.2003.09.015.

[252] P. G. Moses, B. Hinnemann, H. Topsøe, and J. K. Nørskov, “The hydrogenation and direct desulfurization reaction pathway in thiophene hydrodesulfurization over MoS2 catalysts at realistic conditions: A density functional study,” *J. Catal.*, vol. 248, no. 2, pp. 188–203, Jun. 2007, doi: 10.1016/J.JCAT.2007.02.028.

[253] A. Travert, H. Nakamura, R. A. van Santen, S. Cristol, J. F. Paul, and E. Payen, "Hydrogen activation on Mo-based sulfide catalysts, a periodic DFT study," *J. Am. Chem. Soc.*, vol. 124, no. 24, pp. 7084–7095, Jun. 2002, doi: 10.1021/JA011634O.

[254] M. Salah, E. Hadri, M. Hehn, G. Malinowski, S. Albrecht, J. N. Winn, J. A. Johnson, B. Hinnemann, P. G. Moses, and J. K. Nørskov, "Recent density functional studies of hydrodesulfurization catalysts: insight into structureand mechanism," *J. Phys. Condens. Matter*, vol. 20, no. 6, p. 064236, Jan. 2008, doi: 10.1088/0953-8984/20/6/064236.

[255] E. Krebs, B. Silvi, and P. Raybaud, "Mixed sites and promoter segregation: A DFT study of the manifestation of Le Chatelier's principle for the Co(Ni)MoS active phase in reaction conditions," *Catal. Today*, vol. 130, no. 1, pp. 160–169, Jan. 2008, doi: 10.1016/J.CATTOD.2007.06.081.

[256] S. Helveg, J. V. Lauritsen, E. Lægsgaard, I. Stensgaard, J. K. Nørskov, B. S. Clausen, H. Topsøe, and F. Besenbacher, "Atomic-scale structure of single-layer MoS2 nanoclusters," *Phys. Rev. Lett.*, vol. 84, no. 5, p. 954, Jan. 2000, doi: 10.1103/PhysRevLett.84.951.

[257] M. Egorova and R. Prins, "Hydrodesulfurization of dibenzothiophene and 4,6-dimethyldibenzothiophene over sulfided NiMo/γ-Al2O3, CoMo/γ-Al2O3, and Mo/γ-Al2O3 catalysts," *J. Catal.*, vol. 225, no. 2, pp. 417–427, Jul. 2004, doi: 10.1016/J.JCAT.2004.05.002.

[258] M. Nagai, T. Sato, and A. Aiba, "Poisoning effect of nitrogen compounds on dibenzothiophene hydrodesulfurization on sulfided NiMoAl2O3 catalysts and relation to gas-phase basicity," *J. Catal.*, vol. 97, no. 1, pp. 52–58, Jan. 1986, doi: 10.1016/0021-9517(86)90036-9.

[259] V. La Vopa and C. N. Satterfield, "Poisoning of thiophene hydrodesulfurization by nitrogen compounds," *J. Catal.*, vol. 110, no. 2, pp. 375–387, Apr. 1988, doi: 10.1016/0021-9517(88)90328-4.

[260] B. Delmon, "A new hypothesis explaining synergy between two phases in heterogeneous catalysis the case of hydrodesulfurization catalysts," *Bull. des Sociétés Chim. Belges*, vol. 88, no. 12, pp. 979–987, Jan. 1979, doi: 10.1002/BSCB.19790881204.

[261] E. J. M. Hensen, G. M. H. J. Lardinois, V. H. J. De Beer, J. A. R. Van Veen, and R. A. van Santen, "Hydrogen–deuterium equilibration over transition metal sulfide catalysts: on the synergetic effect in CoMo catalysts," *J. Catal.*, vol. 187, no. 1, pp. 95–108, Oct. 1999, doi: 10.1006/JCAT.1999.2587.

[262] A. L. Farragher and P. Cossee, "Intercaliation model of benzene hydrogenation," in *Proceedings of the 5th International Congress on Catalysis 1972*, J. W. Hightower, Ed. North-Holland, 1973, p. 1301.

[263] H. Topsøe, B. S. Clausen, R. Candia, C. Wivel, and S. Mørup, "In situ Mössbauer emission spectroscopy studies of unsupported and supported sulfided Co-Mo hydrodesulfurization catalysts: Evidence for and nature of a Co-Mo-S phase," *J. Catal.*, vol. 68, no. 2, pp. 433–452, Apr. 1981, doi: 10.1016/0021-9517(81)90114-7.

[264] P. Raybaud, J. Hafner, G. Kresse, S. Kasztelan, and H. Toulhoat, "Structure, energetics, and electronic properties of the surface of a promoted MoS2 catalyst: an ab initio local density functional study," *J. Catal.*, vol. 190, no. 1, pp. 128–143, Feb. 2000, doi: 10.1006/JCAT.1999.2743.

[265] P. G. Moses, B. Hinnemann, H. Topsøe, and J. K. Nørskov, "The effect of Co-promotion on MoS2 catalysts for hydrodesulfurization of thiophene: A density functional study," *J. Catal.*, vol. 268, no. 2, pp. 201–208, Dec. 2009, doi: 10.1016/J.JCAT.2009.09.016.

[266] J. V. Lauritsen, J. Kibsgaard, G. H. Olesen, P. G. Moses, B. Hinnemann, S. Helveg, J. K. Nørskov, B. S. Clausen, H. Topsøe, E. Lægsgaard, and F. Besenbacher, "Location and coordination of promoter atoms in Co- and Ni-promoted MoS2-based hydrotreating catalysts," *J. Catal.*, vol. 249, no. 2, pp. 220–233, Jul. 2007, doi: 10.1016/J.JCAT.2007.04.013.

[267] S. Harris and R. R. Chianelli, "Catalysis by transition metal sulfides: A theoretical and experimental study of the relation between the synergic systems and the binary transition metal sulfides," *J. Catal.*, vol. 98, no. 1, pp. 17–31, Mar. 1986, doi: 10.1016/0021-9517(86)90292-7.

[268] L. S. Byskov, J. K. Nørskov, B. S. Clausen, and H. Topsøe, "DFT calculations of unpromoted and promoted MoS2-based hydrodesulfurization catalysts," *J. Catal.*, vol. 187, no. 1, pp. 109–122, Oct. 1999, doi: 10.1006/JCAT.1999.2598.

[269] S. Rangarajan and M. Mavrikakis, "On the preferred active sites of promoted MoS2 for hydrodesulfurization with minimal organonitrogen inhibition," *ACS Catal.*, vol. 7, no. 1, pp. 501–509, Jan. 2017, doi: 10.1021/ACSCATAL.6B02735.

[270] I. Shafiq, S. Shafique, P. Akhter, W. Yang, and M. Hussain, "Recent developments in alumina supported hydrodesulfurization catalysts for the production of sulfur-free refinery products: A technical review," *Catal. Rev.*, vol. 64, no. 1, pp. 1–86, 2020, doi: 10.1080/01614940.2020.1780824.

[271] G. Marroquín-Sá Nchez, J. Ancheyta-Juá, A. Ramírez-Zú, and E. Farfá N-Torres, "Effect of crude oil properties on the hydrodesulfurization of middle distillates over NiMo and CoMo catalysts," 2001, doi: 10.1021/EF010054.

[272] R. Wandas and T. Chrapek, "Hydrotreating of middle distillates from destructive petroleum processing over high-activity catalysts to reduce nitrogen and improve the quality," *Fuel Process. Technol.*, vol. 85, no. 11, pp. 1333–1343, Aug. 2004, doi: 10.1016/J.FUPROC.2003.09.009.
[273] M. Daage and R. R. Chianelli, "Structure-function relations in molybdenum sulfide catalysts: the 'rim-edge' model," *J. Catal.*, vol. 149, no. 2, pp. 414–427, Oct. 1994, doi: 10.1006/JCAT.1994.1308.
[274] C. Zhang, P. Li, X. Liu, T. Liu, Z. Jiang, and C. Li, "Morphology-performance relation of (Co)MoS2 catalysts in the hydrodesulfurization of FCC gasoline," *Appl. Catal. A Gen.*, vol. 556, pp. 20–28, Apr. 2018, doi: 10.1016/J.APCATA.2018.02.026.
[275] E. Furimsky, "Chemistry of catalytic hydrodeoxygenation," *Catal. Rev.*, vol. 25, no. 3, pp. 421–458, Aug. 1983, doi: 0.1080/01614948308078052.
[276] D. C. Elliott, D. Beckman, A. V. Bridgwater, J. P. Diebold, S. B. Gevert, and Y. Solantausta, "Developments in direct thermochemical liquefaction of biomass: 1983-1990," *Energy Fuels*, vol. 5, no. 3, pp. 399–410, May 2002, doi: 10.1021/EF00027A008.
[277] D. A. Ruddy, J. A. Schaidle, J. R. Ferrell, J. Wang, L. Moens, and J. E. Hensley, "Recent advances in heterogeneous catalysts for bio-oil upgrading via 'ex situ catalytic fast pyrolysis': catalyst development through the study of model compounds," *Green Chem.*, vol. 16, no. 2, pp. 454–490, Jan. 2014, doi: 10.1039/C3GC41354C.
[278] E. Furimsky, "Catalytic hydrodeoxygenation," *Appl. Catal. A Gen.*, vol. 199, no. 2, pp. 147–190, Jun. 2000, doi: 10.1016/S0926-860X(99)00555-4.
[279] "Neste MY Renewable Diesel™," *Neste.* https://www.neste.com/products/all-products/renewable-road-transport/neste-my-renewable-diesel#7c977f2c (accessed Dec. 17, 2022).
[280] M. W. Nolte and B. H. Shanks, "A perspective on catalytic strategies for deoxygenation in biomass pyrolysis," *Energy Technol.*, vol. 5, no. 1, pp. 7–18, Jan. 2017, doi: 10.1002/ENTE.201600096.
[281] J. C. Serrano-Ruiz, R. M. West, and J. A. Dumesic, "Catalytic conversion of renewable biomass resources to fuels and chemicals," *Annu. Rev. Chem. Biomol. Eng.*, vol. 1, pp. 79–100, Jun. 2010, doi: 10.1146/ANNUREV-CHEMBIOENG-073009-100935.
[282] S. Kang, X. Li, J. Fan, and J. Chang, "Hydrothermal conversion of lignin: A review," *Renew. Sustain. Energy Rev.*, vol. 27, pp. 546–558, Nov. 2013, doi: 10.1016/J.RSER.2013.07.013.
[283] D. M. Alonso, J. Q. Bond, and J. A. Dumesic, "Catalytic conversion of biomass to biofuels," *Green Chem.*, vol. 12, no. 9, pp. 1493–1513, Sep. 2010, doi: 10.1039/C004654J.

[284] X. Lan, E. J. M. Hensen, and T. Weber, "Hydrodeoxygenation of guaiacol over Ni2P/SiO2–reaction mechanism and catalyst deactivation," *Appl. Catal. A Gen.*, vol. 550, pp. 57–66, Jan. 2018, doi: 10.1016/J.APCATA.2017.10.018.
[285] S. K. Wu, P. C. Lai, and Y. C. Lin, "Atmospheric hydrodeoxygenation of guaiacol over nickel phosphide catalysts: Effect of phosphorus composition," *Catal. Lett.*, vol. 144, no. 5, pp. 878–889, Mar. 2014, doi: 10.1007/S10562-014-1231-7/FIGURES/8.
[286] D. Gao, Y. Xiao, and A. Varma, "Guaiacol hydrodeoxygenation over platinum catalyst: reaction pathways and kinetics," *Ind. Eng. Chem. Res.*, vol. 54, no. 43, pp. 10638–10644, Oct. 2015, doi: 10.1021/ACS.IECR.5B02940.
[287] D. Gao, C. Schweitzer, H. T. Hwang, and A. Varma, "Conversion of guaiacol on noble metal catalysts: reaction performance and deactivation studies," *Ind. Eng. Chem. Res.*, vol. 53, no. 49, pp. 18658–18667, Dec. 2014, doi: 10.1021/IE500495Z.
[288] C. C. Chiu, A. Genest, A. Borgna, and N. Rösch, "Hydrodeoxygenation of guaiacol over Ru(0001): A DFT study," *ACS Catal.*, vol. 4, no. 11, pp. 4178–4188, Oct. 2014, doi: 10.1021/CS500911J.
[289] J. Lu, S. Behtash, O. Mamun, and A. Heyden, "Theoretical investigation of the reaction mechanism of the guaiacol hydrogenation over a Pt(111) catalyst," *ACS Catal.*, vol. 5, no. 4, pp. 2423–2435, Apr. 2015, doi: 10.1021/CS5016244.
[290] M. L. Honkela, J. Björk, and M. Persson, "Computational study of the adsorption and dissociation of phenol on Pt and Rh surfaces," *Phys. Chem. Chem. Phys.*, vol. 14, no. 16, pp. 5849–5854, Mar. 2012, doi: 10.1039/C2CP24064E.
[291] E. Laurent and B. Delmon, "Study of the hydrodeoxygenation of carbonyl, carboxylic and guaiacyl groups over sulfided CoMo/γ-Al2O3 and NiMo/γ-Al2O3 catalyst: II. Influence of water, ammonia and hydrogen sulfide," *Appl. Catal. A Gen.*, vol. 109, no. 1, pp. 97–115, Feb. 1994, doi: 10.1016/0926-860X(94)85005-4.
[292] S. Mukundan, M. Konarova, L. Atanda, Q. Ma, and J. Beltramini, "Guaiacol hydrodeoxygenation reaction catalyzed by highly dispersed, single layered MoS2/C," *Catal. Sci. Technol.*, vol. 5, no. 9, pp. 4422–4432, Aug. 2015, doi: 10.1039/C5CY00607D.
[293] V. N. Bui, D. Laurenti, P. Afanasiev, and C. Geantet, "Hydrodeoxygenation of guaiacol with CoMo catalysts. Part I: Promoting effect of cobalt on HDO selectivity and activity," *Appl. Catal. B Environ.*, vol. 101, no. 3–4, pp. 239–245, Jan. 2011, doi: 10.1016/J.APCATB.2010.10.025.
[294] S. Mukundan, M. A. Wahab, L. Atanda, M. Konarova, and J. Beltramini, "Highly active and robust Ni–MoS2 supported on

mesoporous carbon: a nanocatalyst for hydrodeoxygenation reactions," *RSC Adv.*, vol. 9, no. 30, pp. 17194–17202, May 2019, doi: 10.1039/C9RA02143D.

[295] S. Ansaloni, N. Russo, and R. Pirone, "Hydrodeoxygenation of guaiacol over molybdenum-based catalysts: The effect of support and the nature of the active site," *Can. J. Chem. Eng.*, vol. 95, no. 9, pp. 1730–1744, Sep. 2017, doi: 10.1002/CJCE.22819.

[296] V. N. Bui, D. Laurenti, P. Delichère, and C. Geantet, "Hydrodeoxygenation of guaiacol: Part II: Support effect for CoMoS catalysts on HDO activity and selectivity," *Appl. Catal. B Environ.*, vol. 101, no. 3–4, pp. 246–255, Jan. 2011, doi: 10.1016/J.APCATB.2010.10.031.

[297] A. M. Robinson, J. E. Hensley, and J. Will Medlin, "Bifunctional catalysts for upgrading of biomass-derived oxygenates: a review," *ACS Catal.*, vol. 6, no. 8, pp. 5026–5043, Aug. 2016, doi: 10.1021/ACSCATAL.6B00923.

[298] C. Newman, X. Zhou, B. Goundie, I. T. Ghampson, R. A. Pollock, Z. Ross, M. C. Wheeler, R. W. Meulenberg, R. N. Austin, and B. G. Frederick, "Effects of support identity and metal dispersion in supported ruthenium hydrodeoxygenation catalysts," *Appl. Catal. A Gen.*, vol. 477, pp. 64–74, May 2014, doi: 10.1016/J.APCATA.2014.02.030.

[299] V. A. Yakovlev, S. A. Khromova, O. V. Sherstyuk, V. O. Dundich, D. Y. Ermakov, V. M. Novopashina, M. Y. Lebedev, O. Bulavchenko, and V. N. Parmon, "Development of new catalytic systems for upgraded bio-fuels production from bio-crude-oil and biodiesel," *Catal. Today*, vol. 144, no. 3–4, pp. 362–366, Jun. 2009, doi: 10.1016/J.CATTOD.2009.03.002.

[300] R. C. Nelson, B. Baek, P. Ruiz, B. Goundie, A. Brooks, M. C. Wheeler, B. G. Frederick, L. C. Grabow, and R. N. Austin, "Experimental and theoretical insights into the hydrogen-efficient direct hydrodeoxygenation mechanism of phenol over Ru/TiO2," *ACS Catal.*, vol. 5, no. 11, pp. 6509–6523, Nov. 2015, doi: 10.1021/ACSCATAL.5B01554.

[301] S. Boullosa-Eiras, R. Lødeng, H. Bergem, M. Stöcker, L. Hannevold, and E. A. Blekkan, "Catalytic hydrodeoxygenation (HDO) of phenol over supported molybdenum carbide, nitride, phosphide and oxide catalysts," *Catal. Today*, vol. 223, pp. 44–53, Mar. 2014, doi: 10.1016/J.CATTOD.2013.09.044.

[302] T. Prasomsri, T. Nimmanwudipong, and Y. Román-Leshkov, "Effective hydrodeoxygenation of biomass-derived oxygenates into unsaturated hydrocarbons by MoO3 using low H2 pressures," *Energy*

Environ. Sci., vol. 6, no. 6, pp. 1732–1738, May 2013, doi: 10.1039/C3EE24360E.

[303] Y. Li, Y. Zhao, B. Chen, and W. Wang, "Synergetic catalysis of nickel oxides with oxygen vacancies and nickel phosphide for the highly efficient hydrodeoxygenation of phenolic compounds," *ChemCatChem*, vol. 10, no. 12, pp. 2612–2619, Jun. 2018, doi: 10.1002/CCTC.201800010.

[304] J. S. Moon, J. H. Jang, E. G. Kim, Y. H. Chung, S. J. Yoo, and Y. K. Lee, "The nature of active sites of Ni2P electrocatalyst for hydrogen evolution reaction," *J. Catal.*, vol. 326, pp. 92–99, Jun. 2015, doi: 10.1016/J.JCAT.2015.03.012.

[305] Y. Li, J. Fu, and B. Chen, "Highly selective hydrodeoxygenation of anisole, phenol and guaiacol to benzene over nickel phosphide," *RSC Adv.*, vol. 7, no. 25, pp. 15272–15277, Mar. 2017, doi: 10.1039/C7RA00989E.

[306] Y. Román-Leshkov, C. J. Barrett, Z. Y. Liu, and J. A. Dumesic, "Production of dimethylfuran for liquid fuels from biomass-derived carbohydrates," *Nature*, vol. 447, no. 7147, pp. 982–985, Jun. 2007, doi: 10.1038/nature05923.

[307] Y. Zhang, Q. Xiong, Y. Chen, M. Liu, P. Jin, Y. Yan, and J. Pan, "Synthesis of ceria and sulfated zirconia catalysts supported on mesoporous SBA-15 toward glucose conversion to 5-hydroxymethylfurfural in a green isopropanol-mediated system," *Ind. Eng. Chem. Res.*, vol. 57, no. 6, pp. 1968–1979, Feb. 2018, doi: 10.1021/ACS.IECR.7B04671.

[308] V. M. Chernyshev, O. A. Kravchenko, and V. P. Ananikov, "Conversion of plant biomass to furan derivatives and sustainable access to the new generation of polymers, functional materials and fuels," *Russ. Chem. Rev.*, vol. 86, no. 5, pp. 357–387, May 2017, doi: 10.1070/RCR4700.

[309] E. Furimsky, "Thermochemical and mechanistic aspects of removal of sulphur, nitrogen and oxygen from petroleum," *Erdoel Kohle, Erdgas, Petrochem. Brennst.-Chem.; (Germany, Fed. Repub. of)*, vol. 36, no. 11, 1983.

[310] M. Bartók, G. Szöllösi, and J. Apjok, "Mechanism of hydrogenolysis and isomerization of oxacycloalkanes on metals, XVI. Transformation of tetrahydrofuran on platinum catalysts," *React. Kinet. Catal. Lett.*, vol. 64, no. 1, pp. 21–28, 1998, doi: 10.1007/BF02475365.

[311] R. H. Bowker, M. C. Smith, M. L. Pease, K. M. Slenkamp, L. Kovarik, and M. E. Bussell, "Synthesis and hydrodeoxygenation properties of ruthenium phosphide catalysts," *ACS Catal.*, vol. 1, no. 8, pp. 917–922, Aug. 2011, doi: 10.1021/CS200238V.

[312] P. Bui, J. A. Cecilia, S. T. Oyama, A. Takagaki, A. Infantes-Molina, H. Zhao, D. Li, E. Rodríguez-Castellón, and A. Jiménez López, "Studies of the synthesis of transition metal phosphides and their activity in the hydrodeoxygenation of a biofuel model compound," *J. Catal.*, vol. 294, pp. 184–198, Oct. 2012, doi: 10.1016/J.JCAT.2012.07.021.

[313] S. Wang, V. Vorotnikov, and D. G. Vlachos, "A DFT study of furan hydrogenation and ring opening on Pd(111)," *Green Chem.*, vol. 16, no. 2, pp. 736–747, Jan. 2014, doi: 10.1039/C3GC41183D.

[314] Y. Xu, "Decomposition of furan on Pd(111)," *Top. Catal.*, vol. 55, no. 5–6, pp. 290–299, Jun. 2012, doi: 10.1007/S11244-012-9797-Z.

[315] J. P. Lange, E. Van Der Heide, J. Van Buijtenen, and R. Price, "Furfural — a promising platform for lignocellulosic biofuels," *ChemSusChem*, vol. 5, no. 1, pp. 150–166, Jan. 2012, doi: 10.1002/CSSC.201100648.

[316] S. Chen, R. Wojcieszak, F. Dumeignil, E. Marceau, and S. Royer, "How catalysts and experimental conditions determine the selective hydroconversion of furfural and 5-hydroxymethylfurfural," *Chem. Rev.*, vol. 118, no. 22, pp. 11023–11117, Nov. 2018, doi: 10.1021/ACS.CHEMREV.8B00134.

[317] W. S. Lee, Z. Wang, W. Zheng, D. G. Vlachos, and A. Bhan, "Vapor phase hydrodeoxygenation of furfural to 2-methylfuran on molybdenum carbide catalysts," *Catal. Sci. Technol.*, vol. 4, no. 8, pp. 2340–2352, Jul. 2014, doi: 10.1039/C4CY00286E.

[318] M. M. Sullivan and A. Bhan, "Acetone hydrodeoxygenation over bifunctional metallic-acidic molybdenum carbide catalysts," *ACS Catal.*, vol. 6, no. 2, pp. 1145–1152, Feb. 2016, doi: 10.1021/ACSCATAL.5B02656.

[319] S. H. Pang and J. W. Medlin, "Adsorption and reaction of furfural and furfuryl alcohol on Pd(111): Unique reaction pathways for multifunctional reagents," *ACS Catal.*, vol. 1, no. 10, pp. 1272–1283, Oct. 2011, doi: 10.1021/CS200226H.

[320] V. Vorotnikov, G. Mpourmpakis, and D. G. Vlachos, "DFT study of furfural conversion to furan, furfuryl alcohol, and 2-methylfuran on Pd(111)," *ACS Catal.*, vol. 2, no. 12, pp. 2496–2504, Dec. 2012, doi: 10.1021/CS300395A.

[321] S. Sitthisa and D. E. Resasco, "Hydrodeoxygenation of furfural over supported metal catalysts: A comparative study of Cu, Pd and Ni," *Catal. Lett.*, vol. 141, no. 6, pp. 784–791, Jun. 2011, doi: 10.1007/S10562-011-0581-7.

[322] S. Sitthisa, T. Sooknoi, Y. Ma, P. B. Balbuena, and D. E. Resasco, "Kinetics and mechanism of hydrogenation of furfural on Cu/SiO2

catalysts," *J. Catal.*, vol. 277, no. 1, pp. 1–13, Jan. 2011, doi: 10.1016/J.JCAT.2010.10.005.

[323] Q. X. Cai, J. G. Wang, Y. G. Wang, and D. Mei, "Mechanistic insights into the structure-dependent selectivity of catalytic furfural conversion on platinum catalysts," *AIChE J.*, vol. 61, no. 11, pp. 3812–3824, Nov. 2015, doi: 10.1002/AIC.14902.

[324] V. V. Pushkarev, N. Musselwhite, K. An, S. Alayoglu, and G. A. Somorjai, "High structure sensitivity of vapor-phase furfural decarbonylation/hydrogenation reaction network as a function of size and shape of Pt nanoparticles," *Nano Lett.*, vol. 12, no. 10, pp. 5196–5201, Oct. 2012, doi: 10.1021/NL3023127.

[325] M. Chia, Y. J. Pagán-Torres, D. Hibbitts, Q. Tan, H. N. Pham, A. K. Datye, M. Neurock, R. J. Davis, and J. A. Dumesic, "Selective hydrogenolysis of polyols and cyclic ethers over bifunctional surface sites on rhodium-rhenium catalysts," *J. Am. Chem. Soc.*, vol. 133, no. 32, pp. 12675–12689, Aug. 2011, doi: 10.1021/JA2038358.

[326] Y. Nakagawa, M. Tamura, and K. Tomishige, "Catalytic reduction of biomass-derived furanic compounds with hydrogen," *ACS Catal.*, vol. 3, no. 12, pp. 2655–2668, Dec. 2013, doi: 10.1021/CS400616P.

[327] R. Fang, H. Liu, R. Luque, and Y. Li, "Efficient and selective hydrogenation of biomass-derived furfural to cyclopentanone using Ru catalysts," *Green Chem.*, vol. 17, no. 8, pp. 4183–4188, Aug. 2015, doi: 10.1039/C5GC01462J.

[328] Y. Nakagawa and K. Tomishige, "Production of 1,5-pentanediol from biomass via furfural and tetrahydrofurfuryl alcohol," *Catal. Today*, vol. 195, no. 1, pp. 136–143, Nov. 2012, doi: 10.1016/J.CATTOD.2012.04.048.

[329] B. Katryniok, S. Paul, and F. Dumeignil, "Highly efficient catalyst for the decarbonylation of lactic acid to acetaldehyde," *Green Chem.*, vol. 12, no. 11, pp. 1910–1913, Nov. 2010, doi: 10.1039/C0GC00203H.

[330] Z. Zhai, X. Li, C. Tang, J. Peng, N. Jiang, W. Bai, H. Gao, and Y. Liao, "Decarbonylation of lactic acid to acetaldehyde over aluminum sulfate catalyst," *Ind. Eng. Chem. Res.*, vol. 53, no. 25, pp. 10318–10327, Jun. 2014, doi: 10.1021/IE500988Q.

[331] G. W. Howe, A. A. Vandersteen, and R. Kluger, "How acid-catalyzed decarboxylation of 2,4-dimethoxybenzoic acid avoids formation of protonated CO2," *J. Am. Chem. Soc.*, vol. 138, no. 24, pp. 7568–7573, Jun. 2016, doi: 10.1021/JACS.6B01770.

[332] K. Liu, X. Huang, E. A. Pidko, and E. J. M. Hensen, "Hydrogenation of lactic acid to 1,2-propanediol over Ru-based catalysts," *ChemCatChem*, vol. 10, no. 4, pp. 810–817, Feb. 2018, doi: 10.1002/CCTC.201701329.

[333] J. C. Serrano-Ruiz and J. A. Dumesic, "Catalytic processing of lactic acid over Pt/Nb2O5," *ChemSusChem*, vol. 2, no. 6, pp. 581–586, Jun. 2009, doi: 10.1002/CSSC.200900004.

[334] I. Simakova, O. Simakova, P. Mäki-Arvela, and D. Y. Murzin, "Decarboxylation of fatty acids over Pd supported on mesoporous carbon," *Catal. Today*, vol. 150, no. 1–2, pp. 28–31, Feb. 2010, doi: 10.1016/J.CATTOD.2009.07.064.

[335] J. Lu, S. Behtash, M. Faheem, and A. Heyden, "Microkinetic modeling of the decarboxylation and decarbonylation of propanoic acid over Pd(1 1 1) model surfaces based on parameters obtained from first principles," *J. Catal.*, vol. 305, pp. 56–66, Sep. 2013, doi: 10.1016/J.JCAT.2013.04.026.

[336] W. Yang, R. V. Solomon, O. Mamun, J. Q. Bond, and A. Heyden, "Investigation of the reaction mechanism of the hydrodeoxygenation of propionic acid over a Rh(1 1 1) surface: A first principles study," *J. Catal.*, vol. 391, pp. 98–110, Nov. 2020, doi: 10.1016/J.JCAT.2020.08.015.

[337] O. I. Şenol, T. R. Viljava, and A. O. I. Krause, "Hydrodeoxygenation of methyl esters on sulphided NiMo/γ-Al2O3 and CoMo/γ-Al2O3 catalysts," *Catal. Today*, vol. 100, no. 3–4, pp. 331–335, Feb. 2005, doi: 10.1016/J.CATTOD.2004.10.021.

[338] C. Dupont, R. Lemeur, A. Daudin, and P. Raybaud, "Hydrodeoxygenation pathways catalyzed by MoS2 and NiMoS active phases: A DFT study," *J. Catal.*, vol. 279, no. 2, pp. 276–286, Apr. 2011, doi: 10.1016/J.JCAT.2011.01.025.

[339] M. Ruinart De Brimont, C. Dupont, A. Daudin, C. Geantet, and P. Raybaud, "Deoxygenation mechanisms on Ni-promoted MoS2 bulk catalysts: A combined experimental and theoretical study," *J. Catal.*, vol. 286, pp. 153–164, Feb. 2012, doi: 10.1016/J.JCAT.2011.10.022.

[340] M. F. Wagenhofer, E. Baráth, O. Y. Gutiérrez, and J. A. Lercher, "Carbon-carbon bond scission pathways in the deoxygenation of fatty acids on transition-metal sulfides," *ACS Catal.*, vol. 7, no. 2, pp. 1068–1076, Feb. 2017, doi: 10.1021/ACSCATAL.6B02753.

[341] J. Wu, J. Shi, J. Fu, J. A. Leidl, Z. Hou, and X. Lu, "Catalytic decarboxylation of fatty acids to aviation fuels over nickel supported on activated carbon," *Sci. Rep.*, vol. 6, no. 1, pp. 1–8, Jun. 2016, doi: 10.1038/srep27820.

Chapter 4

Selective Catalytic Oxidation Reactions

4.1 Introduction

In this chapter, mechanistic principles of selective oxidation reactions catalyzed by transition metal catalysts and solid-state reducible metal oxides are introduced. Main reactions discussed are selective oxidation of alkenes and alkanes, ammonia oxidation to NO_x, and reactions with NO_x to give N_2.

Selective oxidation reactions have a central position in catalysis. Introduction of oxygen atoms into hydrocarbons converts them to alcohols, ketones or other oxygenated molecules of great use in many chemical applications. Oxidative dehydrogenation reactions complement the non-oxidative dehydrogenation systems. Because of their exothermicity they allow for reactions at milder conditions. Also, catalytic oxidation of non-carbon-containing molecules are important. In order to produce fertilizer, ammonia is to be oxidized to nitrate. Related reactions are essential to the treatment of combustion exhaust systems.

In 1817 Humphrey Davy, then at the British Royal Institution in London and later president of the Royal Society, discovered the phenomenon of oxidation catalysis. Whereas Pt or Pd oxidize methane to CO_2 and H_2O, other metals such as Cu or Ag do not activate methane. The principle was discovered that catalytic performance varies with catalyst composition.

The experimental device Davy used for his experiments was a flame contained in a small-wire gauze lamp. Using this device, he could safely study flames since the gauze cools the flame and prevents the flame from evolving beyond the lamp.

The discovery of catalysis was made from the remarkable observation that when fed by methane a small Pt wire within the lamp remained glowing even after the flame was extinguished. It indicates that heat continues to be released. It had to imply that contact with Pt causes a reaction between methane and oxygen [1]. Davy's lamp became widely used as a mine lamp to warn miners of the presence of flammable gases.

The German chemist Johann Wolfgang Döbereiner used the related combustion reaction of H_2 in his commercially successful lighter device (tinderbox) that can easily be switched on to produce a flame. This was useful in a time where matches had not been invented. In the Döbereiner device a flame develops as hydrogen is ignited in air when passed over a Pt sponge. Hydrogen is generated by dripping sulfuric acid on zinc. The Döbereiner lighter from 1823 is the first practical application of a heterogeneous catalytic oxidation reaction [2].

Electrochemistry was of major interest in the same period. The fuel cell was invented by William Robert Grove in 1838. It generates electricity via the oxidation of H_2 by O_2. This electrochemical device uses also Pt as electrode material [3].

Ammonia oxidation to NO_x is one of the first large-scale oxidation processes catalyzed by a transition metal. This high-temperature reaction is at present catalyzed by a Pt/Rh alloy. The process is named after Wilhelm Ostwald, which he patented in 1902. The reaction itself was already known for more than half a century since the work of Charles Frédéric Kuhlmann in 1839. Ostwald made the technical discovery that high gas velocity improves the selectivity of the NO or NO_2 production (see Section 4.2.2.3) [4].

The Ostwald process became technologically important in combination with the ammonia synthesis reaction invented by Fritz Haber (see Section 1.4.2). These two processes are essential for the overall process that converts N_2 from air to nitrate. The first step in the process is hydrogenation of N_2 to NH_3, the second step the oxidation of NH_3.

The homogeneous lead chamber process where nitric oxide catalyzes SO_2 oxidation to SO_3 produces diluted sulfuric acid. Around 1880 the first commercial Contact process with Pt as catalyst was introduced. This was able to produce fuming sulfuric acid. The Pt-catalyzed SO_2 oxidation reaction had already been discovered in 1831 by Peregrine Phillips. In 1928 the Pt-catalyzed process was succeeded by the current process with V_2O_5 as catalyst, which is less sensitive than Pt to deactivating poisons [5].

V_2O_5 is an example of a reducible metal oxide. It readily reacts also with hydrocarbons and becomes reduced. With oxygen it is oxidized back to the metal oxide. It was discovered as catalyst for the oxidation reaction of benzene to maleic anhydride in 1933 in the USA. At that time benzene was readily available from liquefied coal or oil. Maleic anhydride is important in the manufacture of polyester resin for the automobile industry. Reducible metal oxides became widely applied to the selective catalytic oxidation of organic molecules in the course of the 20th century.

An early example of redox catalysis is the Deacon reaction. This process, operational since 1874, oxidizes HCl to Cl_2 with a $CuCl_2$ catalyst. Cl_2 is important in the soda production industry. In this reaction 2 Cu^{2+} ions are reduced by 2 Cl^- to 2 Cu^+ and Cl_2. Oxygen oxidizes Cu^+ back to Cu^{2+} and co-produces water.

Oxygen insertion into the hydrocarbon framework is chemically important since it functionalizes hydrocarbons to useful chemicals. Selective oxidation of alkenes and alkanes became important in the middle of the previous century. Then there was a growing need for catalytic processes to synthesize monomer molecules for the expanding polymer industry.

This period can be considered as the renaissance of oxidation catalysis. It led to the discovery of many new reactions, such as acrylonitrile from propene and maleic anhydride from butane. Some of these reactions became implemented in large-scale chemical processes in the petrochemical industry. Major inventions were based on the discovery of complex reducible mixed metal oxide compounds that often contain as main component MoO_3 or V_2O_5. With the increasing exploitation of natural gas after 1980, in addition to selective oxidation of alkenes, selective oxidation of alkanes became of more interest.

In addition to selective oxidation reactions catalyzed by solid-state mixed metal oxides, two selective oxidation reactions catalyzed by Ag became large-scale processes. One is the epoxidation of ethene, important amongst others for the production of anti-freeze, and the other is the oxidation of methanol to formaldehyde, important for the polymer industry.

In the second part of the previous century environmental concerns became important. There was a need to reduce NO_x emissions from high-temperature combustion processes. Exhaust treatment catalysts based on reducible metal oxides were developed, important also for the treatment of diesel engine exhausts as well as stack gas from industrial combustion processes. The selectivity of the ammonia reaction with NO_x to N_2 is the scientific issue here.

After 2000 climate issues generated an interest in energy conversion processes that are environmentally sustainable. Conversion processes of biomass-related molecules such as carbohydrates or lignin to chemicals by catalytic hydrogenation is discussed in Chapter 3. Research on electrocatalysis for hydrogen production and fuel cells has deepened our understanding of reactivity descriptors of reducible metal oxides. This is summarized at the end of this chapter.

In parallel with the inventions of mid-century selective oxidation processes due to advances in molecular chemistry, single-site heterogeneous catalysts were developed. Different from the reducible metal oxides and transition metals described in this chapter, they consist of organometallic complexes or coordination compounds immobilized on high surface area supports or are solid solutions of reactive transition metal cations in reducible metal oxides and zeolite-related nanoporous materials. These catalysts are a separate class by themselves, which led to the discovery of important new reactions that include selective oxidation processes of biomass molecules.

Also, new selective oxidation processes such as propene epoxidation with hydrogen peroxide or benzene hydroxylation with N_2O were discovered.

In this chapter the mechanisms of selective catalytic oxidation reactions with O_2 as reactant are introduced. Main catalysts are solid-state reducible oxides and noble metals.

Catalytic oxidation with activated oxygen such as H_2O_2 or N_2O and other oxidation reactions catalyzed by non-reducible and reducible Lewis acid single-site catalytic systems are presented in Chapter 6. For completeness, the respective elementary reaction steps discovered with these systems are also included in the summarizing Section 4.5 of this chapter.

4.2 The Four Main Catalytic Oxidation Systems

4.2.1 *Introduction and Oxidation Fundamentals*

For selective oxidation, the reactivity of the final product should be less than that of the reactant.

In the following part of this section four general features of oxidation reactions are discussed:

- The relative stability of oxidation product versus reactant, which is an important condition of catalyst reaction selectivity.
- The relation between surface and gas-phase reactivity.
- The different states of surface oxygen and their relation with catalyst selectivity.
- Mechanistic consequences of the triplet state nature of the oxygen molecule. The ground state of the molecule is paramagnetic.

The key catalytic problem of hydrocarbon oxidation is selectivity. Selective oxidation competes with total oxidation of the reactant. Total oxidation of reactant to CO_2 and H_2O is thermodynamically most exothermic and, unless chemistry directs otherwise, also kinetically favored.

This is the reason that desirable reactions such as oxidation of methane to methanol (see Section 6.3.2.3.1) or high-temperature selective oxidation of methane to ethene are not possible. The reaction rate of methanol oxidation is faster that than of methane, because the methanol C–H bond has a lower (E_{cov}(H–CH_2OH) = 400 kJ/mol) bond energy than the C–H bond of methane (E_{cov}(H–CH_3) = 415 kJ/mol).

In later sections selective catalytic oxidation of butane to maleic anhydride (Section 4.4.4.1) and propene to acrolein (Section 4.4.3.1) are discussed. The C–H bond energies of maleic anhydride as well as acrolein are substantially larger than that of butane. This suggests the rule that for selective oxidation the reactivity of the final product should be less than that of the reactant. It is a reasonable suggestion when the non-selective reaction is initiated by the reaction of a C–H bond with surface oxygen [6].

In selective oxidation, not only the surface reaction with adsorbed oxygen but also gas-phase radical reactions with oxygen molecules can be important. An example is the high-temperature oxidation of methane to ethene.

The reaction was discovered by Keller and Bhasin [7] in 1982 at Union Carbide. Because of its potential importance the reaction has been extensively investigated by many scientists.

Figure 4.1 summarizes the selectivity of this reaction as measured for many different catalytic systems [8]. The selectivity-conversion relation (a relation that depends on the relative rate of consecutive reaction versus initial reactions) is shown. The experimentally measured dependence appears to be independent of catalyst, as indicated by the grey band in Figure 4.1.

There is a steep decline in selectivity when the conversion of methane increases. Then the relative weight of the contributions of consecutive ethene combustion reactions to ethene conversion increases compared to that of the initial methane activation. This selectivity decline is similar for many different catalysts tested. It is the larger reactivity of ethene versus methane and the dominance of gas-phase radical reactions (see Section 4.2.1.1) rather than the difference in catalyst reactivity that determines this selectivity-conversion dependence [9]. The main role of surface-adsorbed

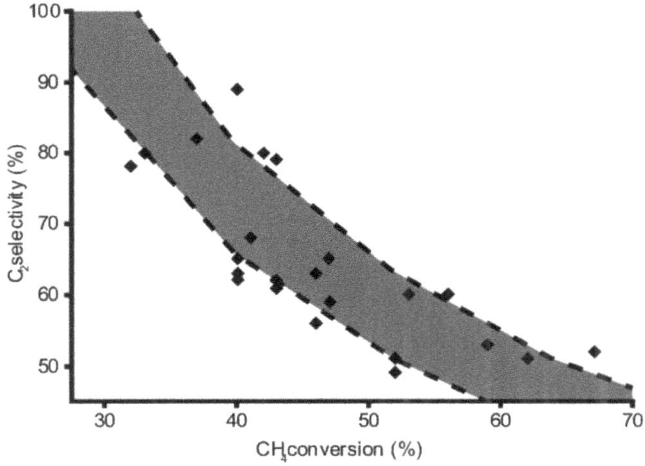

Figure 4.1 The selectivity of ethene formation as a function of methane conversion in the catalytic oxidation of methane: a comparison of catalysts with different composition. Reaction conditions: atmospheric pressure and temperatures from 943 to 1223 K [8].

oxygen is to activate the methane C–H bond. Ethene formation is due to consecutive radical reactions of the methyl radicals. Oxidation reactions with gas-phase oxygen interfere.

Differently from the oxidation of methane, benzene oxidation to maleic anhydride is a selective reaction even at high benzene conversion. Product stability is comparable to that of reactant. This catalytic reaction is a surface reaction. Activity and selectivity depend strongly on catalyst. V_2O_5 selectively catalyzes benzene oxidation with O_2 to maleic anhydride. Surface reactivity relates to the state of adsorbed oxygen.

Jerzy Haber [10], the first director of the Polish Institute of Catalysis in Krakow in the 1980s, distinguished two types of reactive oxygen:

- Oxygen intermediate radical ions such as O_2^- and O^- that adsorb on the surface of metal oxides. They mainly initiate non-selective radical-type oxidation reactions that lead to total oxidation. Such oxygen radicals are typical for low-temperature reactions.

Short lifetime radical oxygen intermediates are formed in between the O_2 adsorbed state and the lattice oxygen atom state. When this transformation is slow, the radical ions contribute to non-selective reactions.

- Lattice oxygen atoms with formal charge O^{2-} that give selective catalysis. They are the reactive oxygen atoms of reducible metal oxides at the temperature range of 600–700 K. Reaction of reactant with a lattice oxygen atom reduces the catalyst and decreases the cation valency. The dissociation reaction of molecular oxygen reoxidizes the reduced cation.

In addition to adsorption as O_2^-, O_2 can also adsorb as a neutral molecule or as O_2^{2-}. The difference between the radical state or non-radical oxygen state is essential. Molecular oxygen that is not adsorbed as a radical is proposed to react selectively in some reactions.

The difference in reactivity of non-dissociative molecular O_2, adsorbed as O_2^{2-}, and of lattice O^{2-} is illustrated by the difference in product selectivity of benzene.

V_2O_5 is a selective oxidation catalyst of benzene to maleic acid. This catalyst will not catalyze hydroxylation of benzene to produce phenol.

The reaction to produce phenol from benzene, when N_2O instead of O_2 is used as oxidant, was discovered by Panov *et al.* in Novosibirsk around 1990 [11]. This reaction is discussed in Section 6.2.2.2.

Mainly using $^{16}O_2$ and $^{18}O_2$ exchange experiments the mechanism of the benzene to maleic anhydride reaction has been unraveled by the Russian and Ukrainian scientists Boreskov [12] and Golodets [13], respectively. Early quantum-chemical studies were performed by the Polish Institute of Catalysis in Krakow [10], [14], [15]. In the benzene to maleic anhydride reaction, activated molecular oxygen (O_2^{2-}) gives initially intermediate hydroquinone from benzene. In following steps with atomic oxygen this converts into the maleic anhydride molecule. Phenol is not a reaction intermediate.

Differences in selectivity due to reaction with adsorbed molecular versus atomic oxygen is a recurring theme in oxidation catalysis. An example is the mechanism of the ethene epoxidation reaction in Section 4.3.3.

In the oxidation of benzene catalyzed by V_2O_5 the spin state of the oxygen molecule is changed from its free molecule triplet state into the singlet state of the O_2^{2-} anion. This change in spin state is essential because the spin state of chemical bonds in molecules is the singlet state of two electrons when they occupy the same orbital.

In a reaction the spin state of the reacting system is conserved unless external interactions interfere. For the O_2 molecules such an interaction is electron exchange with a reducible metal oxide or with transition metal electrons. Spin conservation can be overcome by reaction of the $^3O_2$ molecule with a paramagnetic cation. This became widely explored in the early 1950s [16]. The O_2^- species was identified by Haber on the surface of reducible oxides. Also, on metals or conductive metal oxides the triplet spin problem is resolved by electron exchange of molecule electrons with the electrons of opposite spin of the solid.

One way to circumvent the spin conservation rule is the free radical reaction Eq. (4.1):

$$RH + O_2 \rightarrow R\cdot + HO_2\cdot \tag{4.1}$$

This initiates radical chain reactions. In the liquid phase, radical chain reactions can be directed to produce hydroperoxides, a process called autocatalysis, by recombination of a radical intermediate R· with $HO_2\cdot$. A general feature of such processes is that these reactions are non-selective, and a mixture of alcohol, ketones and other oxidation products is obtained.

4.2.2 *Redox Systems*

4.2.2.1 *Autocatalytic Radical Reactions*

Hydrogen peroxide, hydroperoxide and OH radicals are intermediates of autocatalytic radical reactions.

Non-catalytic oxidation reactions in the gas phase are the prototype of radical reactions. Radical formation is initiated by reactions as shown in Eq. (4.1). Heterogenous catalytic oxidation reactions initiate radical formation by reaction of the reactant with surface oxygen atoms. Gas-phase oxygen reoxidizes the surface. Because of the presence of oxygen in the gas phase, reaction with gas-phase molecules will give total oxidation. Redox catalysis can also be used to tune the radical network. An example is the selective oxidation of toluene to its corresponding carboxylic acid.

This section begins with a return to the high-temperature methane to ethene oxidation reaction as an introduction to the gas-phase radical reaction network. In a second part the mechanism of redox radical reactions is introduced.

Previously in Section 4.2 the oxidation reaction of methane to ethene has been introduced [17]. The mechanism of this high-temperature reaction is a combination of elementary surface intermediate reaction steps and gas-phase radical reactions [17]. Reaction is initiated by generation of $CH_3\cdot$ radicals in the gas phase after abstraction of a hydrogen atom from methane by a surface oxygen atom.

Gas-phase radical reactions determine largely the selectivity of ethene formation versus combustion to CO or CO_2. Eqs. (4.2) provide a simplified version of the complete gas-phase radical network [16], [17]. Eqs. (4.2a) gives the elementary surface reactions that generate the $CH_3\cdot$ radical and reoxidizes the surface.

As indicated by Eqs. (4.2b), for the gas-phase reactions the $CH_3\cdot$ radicals can recombine with each other to give ethane. In a consecutive step, gas-phase radical reactions convert ethane to ethene. Hydrogen peroxide is formed as an intermediate that decomposes to $OH\cdot$ radicals. The $OH\cdot$ radical species in turn can also activate methane. This generates an autocatalytic gas-phase reaction chain. Non-selective reactions are initiated by reactions as shown in Eq. (4.2c) that generate formaldehyde by reaction of the $CH_3\cdot$ radical with oxygen radicals. At the high temperature of reaction, CH_2O is readily decomposed to CO and H_2. At high conversion of methane, the increased generation of $CH_3\cdot$ radicals by reaction of $OH\cdot$ with methane causes the decrease in ethene selectivity.

$$
\begin{aligned}
&O_2 + 2\square \rightarrow 2O_{ads} \\
&CH_4 + O_{ads} \rightarrow H_3C\cdot + OH_{ads} \\
&2OH_{ads} \rightarrow H_2O + O_{ads} + \square
\end{aligned}
$$

$$4CH_4 + O_2 \rightarrow 4H_3C\cdot + 2H_2O \tag{4.2a}$$

$$
\begin{aligned}
&2H_3C\cdot \rightarrow C_2H_6 \\
&C_2H_6 + O_2 \rightarrow HOO\cdot + H_5C_2\cdot \\
&HOO\cdot + H_5C_2\cdot \rightarrow C_2H_4 + H_2O_2 \\
&H_2O_2 \rightarrow 2OH\cdot \\
&2H_3C\cdot + O_2 \rightarrow C_2H_4 + 2OH\cdot
\end{aligned}
$$

$$CH_4 + OH\cdot \xrightarrow{\text{Auto catalysis}} H_3C\cdot + H_2O \tag{4.2b}$$

$$
\begin{aligned}
&H_3C\cdot + O_2 \rightarrow CH_2O + OH\cdot \\
&CH_2O \rightarrow H_2 + CO
\end{aligned}
\tag{4.2c}
$$

Liquid-phase oxidation reactions catalyzed by reducible cations had been discovered in the 1950s and 1960s [14]. Important reactions are acetic acid production from acetaldehyde or butane with Mn^{2+} or Co^{2+} ions and the production of terephthalic acid from paraxylene by solutions containing Mn^{2+} or Co^{2+} ions. Terephthalic acid is a monomer important for the production of the polymer PET (polyethene terephthalate).

With the discovery of zeolitic systems that can accommodate reducible cations, this chemistry has also been explored in analogous heterogenous catalytic reactions [18], [19] that are discussed in Section 6.3.2.1.

Hydroperoxide is a key reaction intermediate of these low-temperature oxidation reactions. Decomposition reactions of hydroperoxide are catalyzed by liquid-phase cations that lead to oxygenated product molecules. The reaction mechanism is illustrated by Eqs. (4.3).

$$
\begin{aligned}
&M^{(n-1)+} + RO_2H \rightarrow M^{n+} + RO\cdot + HO^- \\
&M^{n+} + RO_2H \rightarrow M^{(n-1)+} + RO_2\cdot + H^+ \\
&\text{Net: } RO_2H \rightarrow RO\cdot + RO_2\cdot + H_2O \\
&RO_2\cdot(RO\cdot) + RH \rightarrow R\cdot + RO_2H(ROH) \\
&R\cdot + O_2 \rightarrow RO_2\cdot \qquad (4.3)
\end{aligned}
$$

Abstraction of a hydrogen atom from a reagent by the ROO· radical generates hydroperoxide. Redox decomposition reactions of hydroperoxide regenerate ROO· and give RO· radicals.

With additional hydrogen abstraction, these low-temperature reactions can give acid, ketone or alcohol as product. With alkenes as reagent, hydroperoxide decomposition can also give epoxides [20]. In the redox reactions, the same amounts of OH^- anions and protons are formed. The reaction propagates by addition of O_2 to the intermediate hydrocarbon radical.

An important difference between high-temperature gas-phase radical oxidation and low-temperature redox radical reactions is the difference in propagation reactions. In the former OH· radicals dominate, in the latter peroxide radicals.

4.2.2.2 *Homogenous Selective Oxidation by Transition Metal Complexes*

The liquid-phase reaction is a two-step reaction. The oxygen atom that inserts into the alkene derives from water. The oxygen molecule reoxidizes the reduced catalyst.

The prototypical homogeneous non-radical oxidation catalytic system is the Wacker reaction. In 1957 at Wacker company the redox Pd/Cu catalytic system was discovered that selectively produces acetaldehyde from ethene by oxidation with molecular oxygen [21]. Acetaldehyde is important for acetic acid production. An industrial plant was constructed in 1960 in Köln.

The process is interesting since the single-site Pd^{2+} complex that reacts ethene to aldehyde cannot dissociate the oxygen molecule.

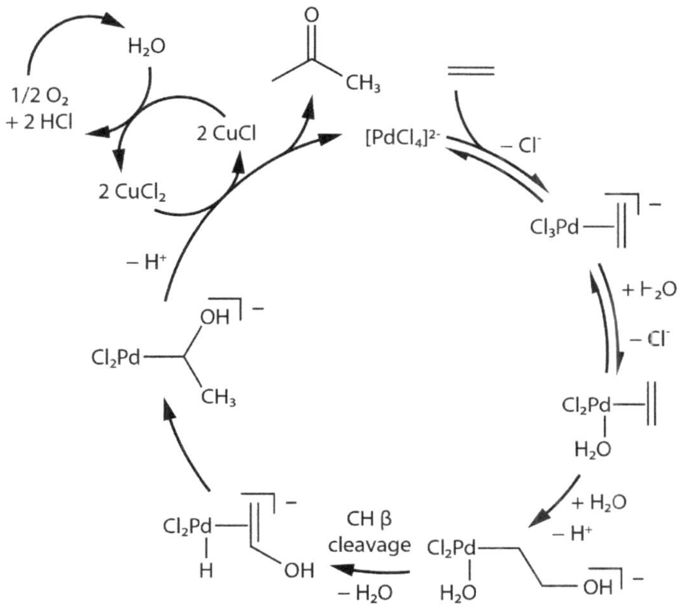

Figure 4.2 The two catalytic cycles of the Wacker reaction. The selective oxidation of ethene to aldehyde with Pd/Cu system [22].

Reoxidation happens by reaction with Cu^{2+} cations that are part of a copper chloride cluster complex that can be reoxidized by dissociation of the oxygen molecule.

As discussed in Section 4.4.2 this mechanistic principle of separate reaction centers for selective oxidation and oxygen dissociation is common also in heterogeneous catalysis with redox solid-state catalysts.

The liquid-phase catalytic cycles of the Wacker reaction [22] are illustrated in Figure 4.2. The reaction is catalyzed by a solution of $PdCl_2$ and $CuCl_2$ and proceeds through two cycles. A cycle that has similarity to the Wilkinson cycle of ethene hydrogenation (see Figure 3.2) oxidizes ethene. The catalyst precursor is $PdCl_4^{2-}$. Two of the Cl^- ligands become substituted by H_2O and ethene. The ethene ligand reacts with solvent H_2O and is converted into a negatively charged $CH_2CH_2OH^-$ ligand. The hydride transfer step of the H atom at the β position of $CH_2CH_2OH^-$ adsorbed to Pd^{2+} gives neutral CH_2CHOH.

In a consecutive step the metal-bonded hydrogen atom (H^-) attaches to CH_2CHOH, which upon deprotonation gives the acetaldehyde. After these elementary steps the Pd^{2+} cation is reduced to the Pd^0 atom, and two protons are released. The oxygen atom that inserted into ethene to give acetaldehyde derives from a water molecule. The zero-valent Pd atom is reoxidized by contact with Cu^{2+} cations that become reduced to Cu^+. The $CuCl_x$ complex is readily reoxidized by O_2 to $CuCl_2$ and H_2O. In this process the protons that were liberated in the Pd^{2+}-catalyzed cycle is reconsumed.

Heterogeneous catalysts for the Wacker reaction have been explored with Pd promoted by a variety of reducible oxides but so far no satisfactory system has been developed [23], [24].

An alternative to oxygen insertion into a hydrocarbon is the hydroformylation reaction. This homogeneous catalytic reaction is discussed in Chapter 3 (Figure 3.18). In this liquid-phase process CO is inserted into the π bond of an alkene of chain length n, which produces an aldehyde or ketone with n+1 carbon atoms. The hydroformylation reaction does not have the flexibility of selective oxidation reactions.

The important feature of oxygen insertion according to the Wacker mechanism is that the alkene molecule adsorbs as a ligand to the redox cation. This activates the molecule for reaction. Oxygen inserts by an outer ligand sphere reaction with water.

Similarly, in heterogenous catalytic systems reactant molecules often are activated by direct contact with reducible cations. However, especially for the reducible metal oxides (Sections 4.4.3, 4.4.4), the primary reaction can be hydrogen abstraction by surface oxygen as in Eq. (4.2a) and consecutive oxygen insertions without direct interaction with a surface cation. Only in a subsequent reaction after cation reduction the cation is reoxidized in a direct contact with molecular oxygen. The cation reduction and reoxidation steps are separate as in the Wacker reaction.

On the other hand, in many reactions molecular bond activation requires a surface vacant oxygen site and molecule activation requires activation by the reducible cation as in the Wacker reaction (see Section 6.4).

The mechanistic question of reactant activation by reducible cations or surface oxygen atom is another recurrent theme in selective oxidation catalysis.

4.2.2.3 *Heterogeneous Selective Transition and Noble Metal Catalysis*

The selectivity of oxidation is sensitive to the degree of oxidation of the transition metal surface. This defines the state of reactive oxygen. It depends sensitively on reaction conditions.

Ammonia and methanol oxidation are reactions where H atom transfer from reactant N–H or C–H bonds to surface oxygen or surface metal atoms is rate controlling. The ethene epoxidation reaction is different. In this reaction an oxygen inserts into the ethene π bond. In all three cases major questions relate to the chemistry that determines the selectivity of the reaction.

The oxidation of ammonia to NO_x catalyzed by Pt had already been discovered in 1838 by Kuhlmann [25]. This reaction is essential to the overall oxidation process for nitrate production from N_2 of air. An industrial process became possible with the invention of the ammonia synthesis process that was implemented in 1914. The commercially viable high-temperature oxidation of ammonia to NO was discovered by Ostwald [4].

The selectivity of ammonia oxidation varies largely with reaction temperature. A temperature above 1000 K is required for selective NO production. The catalyst is a gauze of a Pt/Rh alloy. The main issue is to prevent non-selective consecutive reactions of NO that give N_2O or N_2.

N_2O formation in these reactions is to be suppressed since N_2O is a greenhouse gas and also contributes to depletion of the atmospheric ozone layer. This is important in exhaust emission treatment systems as well as in the Ostwald process that produces nitric acid. Nitric acid plants are the largest source of N_2O emissions in the chemical industry [26].

An extremely low feed catalyst contact time of 10^{-3} to 10^{-4} s is used to prevent secondary reactions of NO. The then counterintuitive operation with high space velocity that suppresses consecutive reactions was the breakthrough discovery by Ostwald that made selective NO production possible.

In a modern interpretation, at high temperature surface reaction rates become fast compared to transport of reactant from gas phase to catalyst. Then reaction is film diffusion limited. Higher flow rate will not change conversion, but affects selectivity because products are more quickly removed.

Ostwald made his discovery in a time when the cause of catalytic action was not yet understood. Ostwald supported the now obsolete physical theory that catalysis relates to a unique condition of a reactant film in catalyst pores [27].

At low temperature, ammonia can be selectively oxidized to N_2. The ammonia reacts with intermediate surface nitrite or nitrate species. This reaction is exploited in automotive catalysis for diesel exhaust treatment. In this case the desirable reaction is the reduction of NO_x by reaction with NH_3 to produce N_2. Zeolite catalysts that contain Cu were discovered by König and co-workers around 1985 at Volkswagen, who then used a Cu-exchanged mordenite catalyst [28]. The mechanism of this reaction is discussed in Section 4.4.5.3.2.

A reaction related to the high-temperature ammonia oxidation is the steam reforming of methane to give synthesis gas [29]. Analogous to ammonia to NO_x oxidation, methane is oxidized to CO. It is catalyzed by transition metals such as Ni (see Figure 4.3). It provides the hydrogen for the ammonia synthesis reaction when derived from natural gas.

The original invention of the steam reforming reaction was in 1913 at BASF. In 1930 the first steam reforming plant was started up in the USA by Standard Oil company. After 1936 important additional catalyst innovations were introduced in industrial plants of ICI in Great Britain [30]. Steam reforming is catalyzed by transition metals such as Ni (see Figure 4.3). The steam reforming reaction is the reverse of the methanation reaction. The mechanism of this

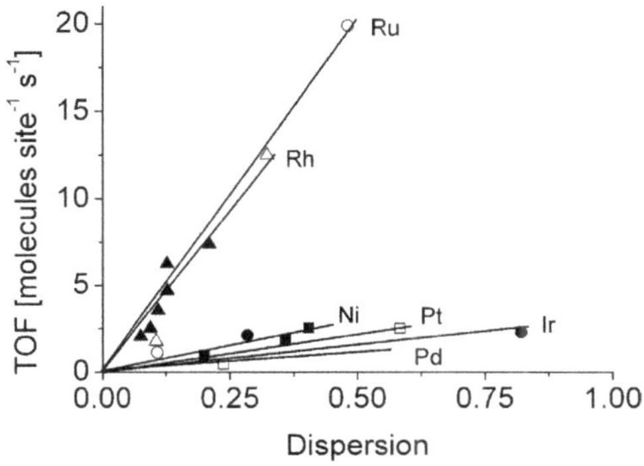

Figure 4.3 Reaction rate as a function of metal atom dispersion for CH_4–H_2O reforming reaction (773 K): Ru (○), 5 wt% Rh (▲), 1 wt% Rh (Δ), Ni (■), Pt (□), Ir (●) [31].

reaction is discussed in Section 3.2.2. It requires high temperature because the process has high endothermicity ΔH_{SR} = 206 kJ/mol. The preferred catalyst for the commercial process is supported Ni. The catalyst deactivates by carbon formation that is suppressed by addition of sulfur.

In the steam reforming reaction, water is the oxidant. The water molecule decomposes on the metal surface into O_{ads} and H_{ads}. CO is formed by recombination of adsorbed oxygen atoms with CH_{ads} intermediates generated by CH_4 decomposition on the transition metal surface. The transition metal will readily activate the C–H bond of methane. Hydrogen is formed by recombination of adsorbed hydrogen atoms. The advantage of the use of steam versus oxygen is high selectivity of synthesis gas. At lower temperature excess CO can be converted to hydrogen by the exothermic water-gas shift reaction that converts CO with H_2O to CO_2 and H_2. Since this reaction is mildly exothermic ($\Delta H = -9.84$ kJ/mol), the reaction conditions of the water-gas shift and steam reforming are different. The mixed metal oxide catalysts of the reaction are related to the methanol synthesis catalyst that hydrogenates a mixture of CO and CO_2 (Section 3.2.2.2).

In order for water to be used as oxidant, it has to decompose when in contact with the catalyst surface. The strength of the surface

M–O bond has to overcome the OH bond dissociation energy of the water molecule. The adsorption energy of the oxygen atom is too low for Pt, Ir or Pd to make water dissociation readily possible. Only for Ni or Rh water decomposition becomes thermodynamically possible.

This sensitivity to the surface adatom O bond energy is the reason for the large difference in reactivity of the transition metals shown in Figure 4.3. The transition metals with larger M–O bond energy are the most active. The strong dependence of the rate on transition metal particle size, normalized per surface atom, suggests that edge or corner sites of particles have substantially lower activation energies (see Section 2.2.1.2).

Whereas a reactive transition metal is needed for the steam reforming reactions, a less reactive metal has to be used in the oxidation of methanol to formaldehyde. Then molecular C–H bonds have to be partially maintained. For the methanol to formaldehyde reaction the preferred catalysts are the less reactive Cu or Ag metals. Two commercial processes are known for this reaction. In one commercial process methanol is oxidized by an Ag-containing catalyst. In an alternative process selective methanol oxidation is catalyzed by the reducible metal oxide $Fe_2(MO_4)_3$. Formaldehyde is an important reactive intermediate for many chemical products.

The methanol oxidation reaction had already been discovered initially in 1880 with Cu as catalyst. 30 years later in 1923 the more selective Ag catalyst of the industrial process was developed by BASF. The process oxidizes methanol at high temperature (900 K).

In 1959 scientists at ICI (now Johnson Matthey) in Great Britain developed the Formox process that uses the $Fe_2(MO_4)_3$ catalyst. This reaction is an early example of multifunctional redox catalysis. Catalysis by reducible metal oxides was already known since 1931 [32]. The $Fe_2(MO_4)_3$ makes practical use possible, because it suppresses MoO_3 evaporation. The temperature of reaction is 250 degrees lower than in the Ag-based process. This significantly increases selectivity of the reaction, due to the instability of formaldehyde at high temperature [33]. The mechanism of the silver- catalyzed reaction is presented in Section 4.3.2. Catalysis by reducible metal oxides is presented in Section 4.4.2.

Selective heterogenous oxidation of alkenes made an important start with the invention of selective oxidation of ethene to ethene epoxide by an Ag-containing heterogenous catalyst. Ethene epoxide can be readily converted to glycol, which is an important antifreeze chemical. It is also important in the production of polymers.

The discovery of the Ag-catalyzed reaction was made by Lefort in France in 1931. Soon after it became implemented as an industrial process by Union Carbide in the USA in 1936. It replaces the environmentally undesirable chlorohydrin process that uses epichlorohydrin as epoxidation agent. The process is currently executed in many plants distributed at many locations. The alumina-supported catalyst is promoted by Cl, alkali and rhenium. Understanding of the role of these promoters has substantially contributed to the mechanism of this reaction that is discussed in Section 4.3.3.

Epoxidation of propene by oxygen catalyzed by similar Ag or related Cu catalysts is not selective. Instead, especially on Cu, propene can be selectively oxidized to acrolein. This reaction is more efficient with reducible metal oxides as will be discussed in the mechanistic Section 4.4.2.1.

Whereas propene cannot be selectively oxidized to propene epoxide with oxygen, selective epoxidation is possible with hydrogen peroxide or a hydroperoxide. The catalyst contains non-redox single Lewis acidic sites that are part of single-site heterogeneous catalysts. The mechanism of this reaction is presented in Sections 4.5.3 and 6.3 on elementary oxidation reaction steps.

The discovery in 1982 by Haruta *et al.* in Japan that Au [34] can be an active catalyst came as a major surprise. Au is an inert noble metal that will not activate chemical bonds in molecules. As has been originally demonstrated for the oxidation reaction of CO to CO_2, it becomes a catalyst active at low temperature when distributed as single atoms or small metal particles on reducible metal oxide supports. With water it is also a low-temperature selective oxidation catalyst for alcohols when used in solutions of high pH. This can be usefully exploited to oxidize biomass molecules. The mechanism of Au catalyst systems is presented in Chapter 6 on single-site catalysis.

4.2.2.4 *Reducible Solid-state Metal Oxide Catalysts*

Molybdenum and vanadium oxides are essential ingredients of multicomponent reducible oxides for selective propene and propane oxidation.

One reason to call the 1960s a second golden age of heterogenous catalysis is the breakthrough discovery of solid-state reducible multicomponent metal oxides amongst others as selective alkene oxidation catalysts [5], [35]. It led to important discoveries of new reactions that are at the root of major chemical industries. Reactions catalyzed by reducible metal oxides that are discussed here are selective oxidation of propene and propane, the methanol to formaldehyde reaction and the reduction of NO_x by ammonia.

The chemical history of solid-state mixed metal oxide catalysis starts with the discovery of the reaction of propene to acrolein and acrylonitrile (the ammoxidation process) at Sohio in the USA (now BP) by Veatch and Callahan in 1957. The reaction is catalyzed by complex multicomponent catalysts. Bismuth-molybdate oxide is a representative example. The product acrylonitrile is an important base chemical for the polymer industry. The ammoxidation reaction became part of many industrial plants all over the world. The first plant was built in Lime, Ohio in 1960 [36].

The reaction mechanism of selective oxidation to acrolein and acrylonitrile by multicomponent metal oxide catalysts is discussed in Section 4.4.3.1. In the beginning of this century knowledge of the complex structures of catalysts and advances in computational chemistry made possible the modelling of elementary reactions activated by the metal oxide surfaces. The relation between selectivity of reaction and catalyst structure and composition has become understood in great detail.

In 1959, very soon after the discovery of selective oxidation of propene, the methanol oxidation process to formaldehyde that is catalyzed by $Fe_2(MO_4)_3$ was implemented by Johnson Matthey in Great Britain. Mechanistically relevant is the discovery of the very different surface composition of this catalyst compared to its bulk composition. The surface is covered with a monolayer of molybdate.

The mechanism of the methanol oxidation reaction is discussed in Section 4.4.3.2.

Because of the increasing importance in the 1970s of natural gas instead of oil as feedstock for the chemical industry, selective oxidation of alkanes instead of alkenes to base chemicals was explored. Use of alkane versus alkene is also economically advantageous. Catalysts containing vanadium oxide can activate alkane molecules.

Mo is a common component of selective alkene oxidation catalysts. V is a required component of catalysts that activate the less reactive alkanes. The lower reactivity of MoO_3 compared to that of V_2O_5 relates to the higher M–O bond energy (redox energies: $MoO_3/MoO_2 = 460$ kJ/mol; $V_2O_5/V_2O_4 = 374$ kJ/mol).

V_2O_5 is one of the oldest redox oxidation catalysts known. As mentioned in Section 4.1, initially V_2O_5 had been introduced as SO_2 oxidation catalyst. Since 1933 it was applied as catalyst for benzene oxidation to maleic acid. The butane to maleic anhydride reaction was discovered in 1966 by Bergman and Frisch at Monsanto (now Huntsman). The catalyst of this reaction has the composition $(VO)_2P_2O_7$. The first industrial plant became operational in 1975. As with the $Fe_2(MO_4)_3$ catalyst, this catalyst also reconstructs substantially during reaction. The mechanism of the reaction and its structure has been extensively investigated and is discussed in Section 4.4.4.1 [37].

Another desirable selective oxidation reaction of an alkane is the ammoxidation of propane instead of propene. Catalysts of complex composition, as for instance Mo/V/Nb/Te/O, have been introduced for this reaction. This system was discovered in 1988 by researchers from Mitsubishi.

The catalyst is a multisite system. As discussed in Section 4.4.3.1, vanadium activates the C–H bonds in the alkane, Te and Mo activate propene C–H bonds and induce N–H insertion reactions, and Nb is a structural promoter of the catalyst.

From the end of the 1980s, several companies have been actively involved in developing a selective propane ammoxidation process [6], [38]. One of the first was realized in 2007 by the Asahi Company in Japan.

Oxidative dehydrogenation of alkanes is a closely related reaction. The oxidation of short alkanes to alkenes is a radical reaction activated by V_2O_5. The activity of this catalyst is a strong function of its particle size [39]. The mechanism is presented in Section 4.4.4.2.

The other important function of vanadium oxide is as catalyst for the treatment of exhaust emission of diesel engines. Ammonia addition to the exhaust, to reduce NO_x to harmless N_2, is used since 1970 in stationary applications of power stations [40], [41]. This makes it possible to reduce NO_x at aerobic conditions. This first generation Selective Catalytic Reduction (SCR) is based on V_2O_5 as catalyst. A second generation of SCR catalysts for mobile diesel engine exhaust consists of zeolite-supported Ag and Cu catalysts [28], [42]. Since 2004 ammonia is added as so-called AdBlue liquid to the exhaust [43]. This liquid is a solution of urea in water. The V_2O_5 system is compared with noble metal zeolite catalysts in Section 4.4.4.3.

The exhaust emission of gasoline engine vehicles is treated at oxidative stoichiometric conditions. Gasoline exhaust catalysts became widely implemented in automobiles after 1981. The revolutionary catalyst invented for this process is discussed in Chapter 6. The catalyst consists of single metal atoms or small transition metal particles of Pt, Pd or Rh deposited on a reducible CeO_2 support.

4.3 The Mechanism of Selective Catalytic Oxidation by Transition and Noble Metals

4.3.1 *Ammonia Oxidation*

The selectivity of ammonia oxidation depends strongly on temperature. At high temperature NO_x is main product, at lower temperature N_2O and N_2 are products. The desorption temperature of NO is key to these differences.

Early molecular understanding of the ammonia oxidation reaction originates from the 1960s. Then mass spectrometry and molecular beam experiments were used to probe reaction intermediates. In 1969 Nutt and Kapur [44], [45] identified correctly the major reaction paths and intermediates which are discussed below.

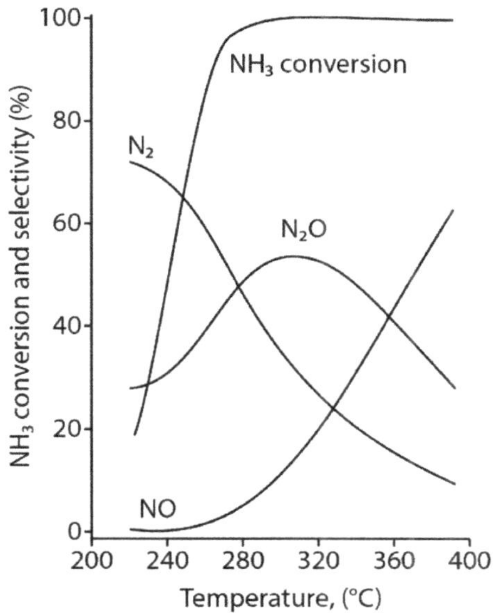

Figure 4.4 Ammonia conversion and selectivity as a function of temperature [47].

Later in the 1980s isotope-labelled experiments and *in situ* spectroscopic studies revealed the reaction network in molecular detail. Surface science data as well as computational chemistry after 2000 provided detailed molecular information on elementary surface reactions. This provides the basis for the mechanistic discussions of this section.

The selectivity of the ammonia oxidation reaction strongly depends on temperature [46], [47]. This is illustrated in Figure 4.4 by a Pt-containing catalyst.

It shows different product distributions for the oxidation of ammonia as a function of temperature. A maximum in N_2O production is found around 300°C. Below this temperature N_2 is the main product and at higher temperature NO is selectively produced.

The reaction network is schematically illustrated in Figure 4.5.

This scheme indicates the different molecules that participate in the ammonia oxidation reaction network. Their relative dominance will vary with metal catalyst and temperature.

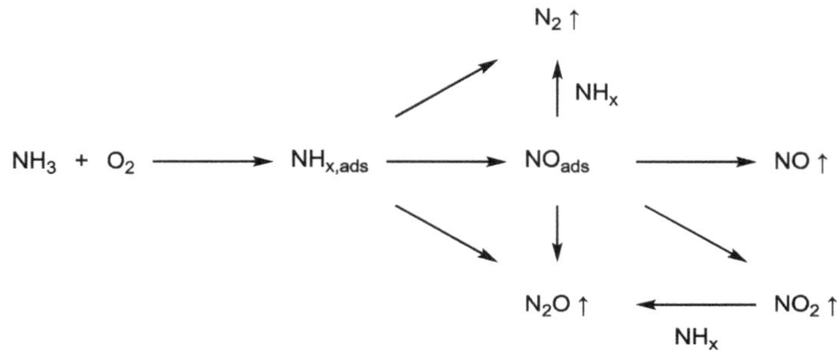

Figure 4.5 Schematic presentation of the reaction network of ammonia oxidation. Main products are N_2, N_2O and NO_x.

Similarly, as with the activation of hydrocarbon molecules of Chapter 3, the primary events are surface dissociation of the reacting molecules followed by recombination of adsorbed intermediates.

The primary reaction is activation of ammonia and O_2 with formation of adsorbed intermediates as NH_2, NH and N and adsorbed oxygen atoms. On a transition metal such as Pt, that has low reactivity, coadsorbed oxygen atoms promote dissociation of the ammonia. Then N–H bond dissociation is accompanied by formation of adsorbed OH. The surface OH hydroxyls recombine and desorb as water [48]–[50].

At low temperature the Pt catalyst surface is mainly covered with NH_x, OH and NO [51]. Recombination of NH_x with O gives NO and OH. This is a fast reaction. At low temperature NO remains adsorbed.

NO becomes the main product once at higher temperature its rate of desorption is not anymore reaction rate limited. The desorption temperature of NO is of the order of 230°C [52].

Low-temperature N_2 formation occurs mainly by the Fogel reaction of Eq. (4.4) that he proposed in 1964 [53]:

$$NO_{ads} + NH_{2,ads} \rightarrow [H_2N\text{-}NO]_{ads} \rightarrow N_2\ (g) + H_2O_{ads} \qquad (4.4)$$

The experiments by Nutt and Kapur [44], [45] confirmed Fogel's suggestion based on secondary ion mass spectrometry. NO is the intermediate that reacts with NH_2 to give N_2. The original Andrussow and Bodenstein proposal of 1926 with NH_2OH as main intermediate was thus rejected. Extensive isotope-labelling studies with ^{15}N- and ^{14}N-labelled NO and NH_3 [54]–[56] later demonstrated the validity of the low-temperature Fogel reaction for N_2 formation. In product N_2, one nitrogen atom stems from NO, the other from NH_3.

The Fogel mechanism is a generally important reaction. It is also the main reaction path of NO reduction with NH_3 to give N_2. This reaction reduces NO exhaust emission of combustion engines (see Section 4.4.4.3).

Selectivity of reaction changes when temperature increases and NO desorbs or dissociates [52]. Then the alternative path of N_2 formation by nitrogen adatom recombination becomes dominant. Also, N_2O formation happens by recombination of NO with N atoms or by recombination of two NO molecules [57]. When temperature increases further, NH_3 decomposes fast and the dominant ad-species on the Pt surface becomes atomic oxygen that recombines with N atoms to give NO.

N_2 formation by recombination of N adatoms or analogous NO formation from recombination of N_{ads} and O_{ads} are structure-sensitive reactions that readily occur on corrugated surfaces. The geometry of the Pt(100) surface is uniquely suitable for recombination and dissociation of NO [58]. Figure 4.6 shows DFT-calculated surface intermediates and reaction potential energies of some of the elementary steps mentioned for the Pt(100) surface [49].

NH_3 is activated on the Pt(100) surface by co-adsorbed oxygen atoms. NO formation by surface recombination of N_{ads} and O_{ads} is faster than the formation of N_2 by recombination of N_{ads} atoms. N_2O from recombination of adsorbed NO with N atom has a slightly higher activation energy. The selectivity to produce N_2O depends on the surface coverage with NO_{ads}.

The role of adsorbed oxygen can vary from promotion of ammonia N–H bond dissociation to poisoning of the reaction. Whether N–H bond cleavage is activated by the adsorbed oxygen atom depends on the relative bond strength of adsorbed O. On a surface

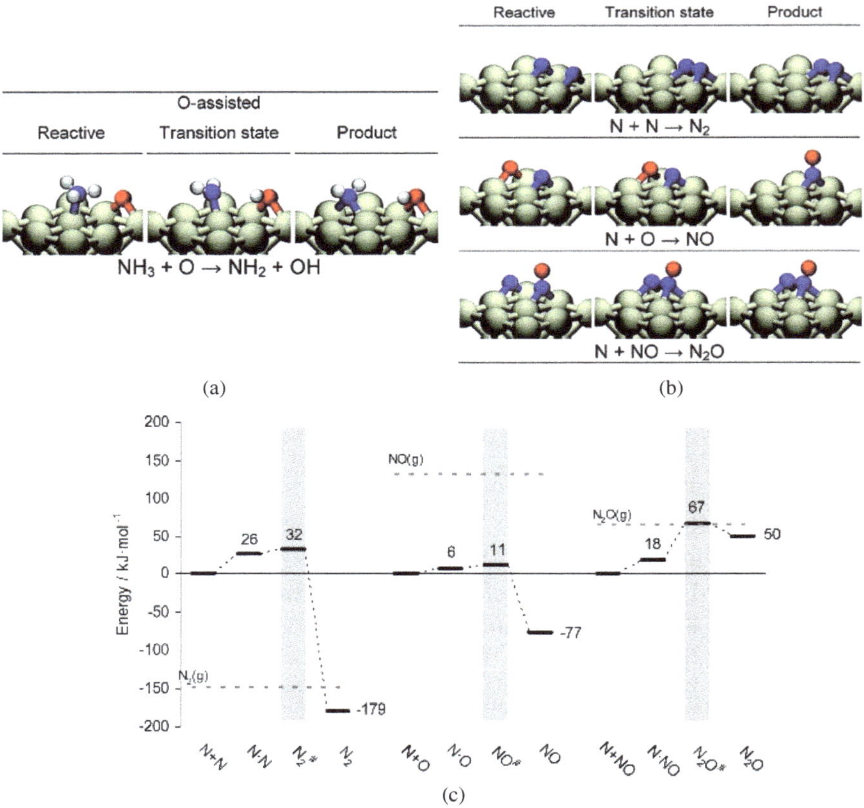

Figure 4.6 Reaction intermediates and potential energy diagrams of elementary reactions of ammonia oxidation on a Pt(100) surface [49]. (a) O adatom-assisted NH bond dissociation; (b) N_2, NO and N_2O formation; (c) potential energy diagrams of elementary reactions of Figure 4.6b.

of low reactivity, the reaction of N–H with O_{ads} gives OH_{ads} with a relatively strong O–H bond. The interaction of H_{ads} with metal is weaker than the O–H bond energy. When on a more reactive metal the O_{ads} bond energy is high, then as follows from the Bond Order Conservation rule (see Section 2.3.2.2) the O–H bond energy of the surface hydroxyl is decreased and activation of the N–H bond by the metal surface will be preferred. Then O_{ads} inhibits reaction because it blocks surface sites. Whereas on the less reactive Pt surface ammonia activation is assisted by co-reaction with adsorbed atomic oxygen, on the more reactive Rh surface NH bond activation is inhibited by co-adsorbed oxygen and needs a vacant Rh surface site to dissociate [59].

Microkinetic simulations [60] based on quantum-chemically computed rate constants of elementary reactions reproduce the temperature dependence of ammonia oxidation of Figure 4.4. A comparison of reactivities of Rh, Pd and Pt shows that Pt is most active but least selective to N_2, while Rh is most selective to N_2. This arises from the significantly lower barrier for N_2 formation than NO by recombination of the respective surface adatoms. The difference in selectivity relates to differences in the interaction of O atoms adsorbed to the metal surfaces. The larger O atom adsorption energy on the more reactive metal will reduce the rate of the N_{ads}, with O_{ads} recombination compared to the N_{ads} recombination reaction.

Ag and Cu are also active catalysts for low-temperature ammonia oxidation to N_2. The state of these catalytic materials at the reaction condition is not metallic, but oxidic. Raman spectroscopy identified a surface covered with nitrate and nitrite species [61]. NO readily oxidizes on these metals [62]. NH_3 reacts with surface oxygen atoms. With $NH_{2,ads}$ these nitrite and nitrate intermediates react to form N_2 in Fogel-type reactions. The NH_2NO intermediate decomposes to N_2, but the NH_2NO_2 intermediate decomposes to N_2O [63].

As discussed in Section 4.4.4.3, the mechanism of the reduction of NO_x with ammonia to nitrogen relates to the ammonia oxidation reaction. This reaction is important for diesel exhaust catalysts. V_2O_5 and Cu/zeolite are the most commonly used catalysts. Essential in the exhaust treatment reaction is also the prevention of non-selective N_2O formation.

The major difference between the transition metal catalysts of ammonia oxidation and the V_2O_5 and zeolite Cu catalysts for NO reduction is the presence of protons in the latter. With ammonia these react to form ammonium cations. Whereas on a Cu particle the Fogel reaction produces N_2O by decomposition of $[H_2N–NO_2]$, the bifunctional acidic catalyst decomposes NO_2 by the intermediate

[$H_4N–NO_2$] complex that contains additional hydrogen atoms and decomposes to N_2 and H_2O.

4.3.2 *Methanol Oxidation*

The selectivity of the methanol oxidation reaction depends on the state of oxygen-covered reactant sites. Metallic and oxidic states are distinguished.

The noble metal that selectively oxidizes methanol to formaldehyde is silver. The mechanism of selective oxidation by Ag catalysts is a complex topic and still not completely resolved. This is mainly because the state of adsorbed oxygen of the working catalytic system is difficult to ascertain. There are several oxygen atom bonding states. The relation between oxygen adatom bonding state and differences in reactivity is an important topic in this section as well as in the following Section 4.3.3 that deals with silver-catalyzed ethene epoxidation.

The methanol oxidation reaction is catalyzed by silver gauze and operates at a temperature of 920 K in excess methanol/air. As with the ammonia oxidation reaction, contact time has to be small to prevent consecutive decomposition and combustion of methanol. The temperature dependence of reaction is counterintuitive. As illustrated in Figure 4.7 the selectivity increases with temperature.

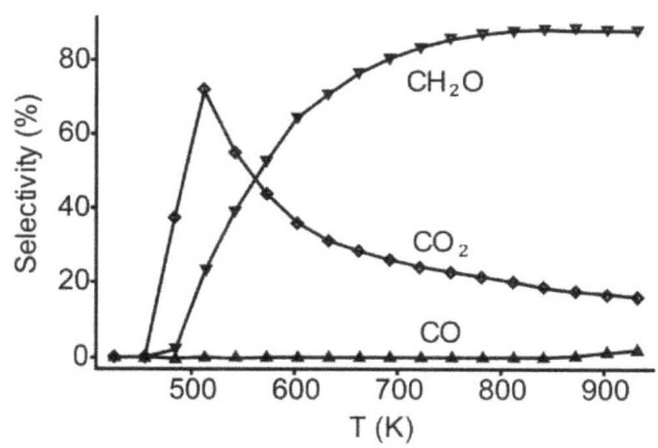

Figure 4.7 Partial oxidation of methanol to formaldehyde: selectivity to CH_2O (downward triangles), CO_2 (diamonds), and CO (upward triangles) as a function of reaction temperature [64].

The state of reactive oxygen strongly depends on temperature. This and differences in selectivity pattern of reaction at low or high temperature are exploited to identify the selective oxygen atom intermediate.

First, experiments are discussed on differences in reactivity of adsorbed oxygen. This is followed by the Wachs-Madix reaction mechanism as deduced from surface science experiments and concluded with an adapted version of this mechanism valid in the high-temperature regime.

Two different reactive oxygen species have been identified: an adsorbed oxygen state that only generates at high temperature, which is denoted as O_γ, and another species which is denoted as O_α that is only stable at lower temperatures. It is suggested that O_γ is the oxygen state that selectively activates methanol for formaldehyde formation. O_γ has only been observed on silver gauze-like material. In contrast to O_α, O_γ has not been identified by single crystal surface studies.

An important additional difference in reactivity of the two oxygen species is that methanol oxidation by O_α co-produces hydrogen and water, whereas methanol oxidation by O_γ gives only hydrogen co-product.

The temperature of reaction is far beyond the decomposition temperature of the bulk silver oxides. Figure 4.8a shows temperature-programmed desorption (TPD) data of an Ag particle exposed to oxygen at 573 K where O_α is stable and at 973 K, the temperature of the selective methanol oxidation reaction.

These TPD data indicate that at high temperature, oxygen dissolves in bulk Ag (the dissolved oxygen atoms are denoted as O_β) that leads to desorption. When oxygen treatment is at methanol oxidation condition, a large additional desorption peak is observed that is identified with O_γ. The latter results from the dissolved oxygen atoms that diffuse through the bulk Ag and desorb at high temperature. Note that this temperature is far beyond the decomposition temperature of Ag_2O of 560 K.

The Ag catalyst undergoes substantial morphology changes due to the oxygen dissolution processes. Bulk diffusion of O_β atoms leads to formation of a metastable silver sub-oxide [64].

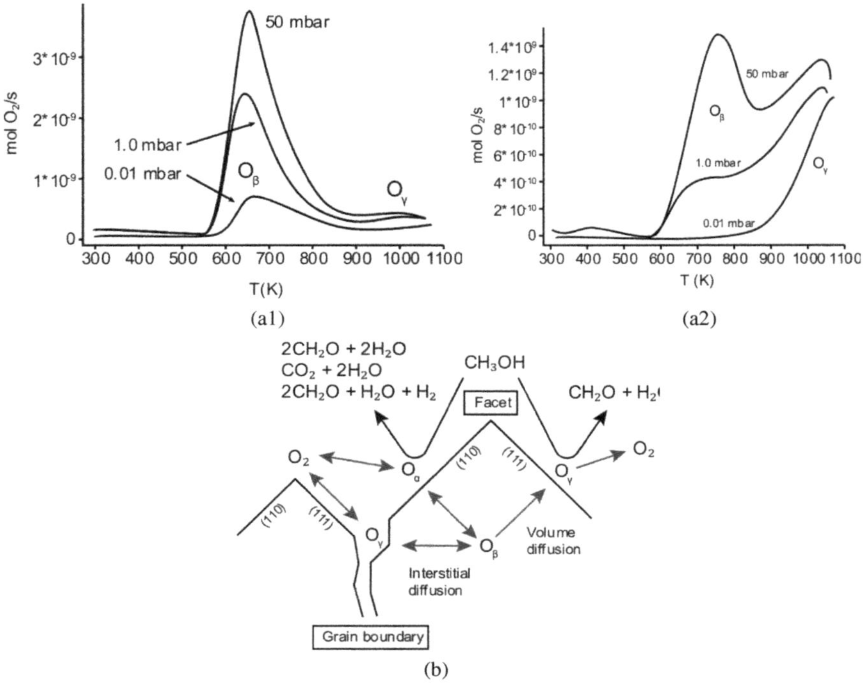

Figure 4.8 Oxygen adatom adsorption states on an Ag particle. Temperature-desorption (TPD) spectra and schematic representation of the state of the surface [64], [65]. (a1) shows TPD spectra for oxygen dosed at various pressures at 573 K. (a2) TPD spectra for desorption of oxygen dosed at various pressures at 973 K. The dosing condition of 973 K corresponds to the methanol oxidation condition. (b) Reaction scheme showing the formation of O_α, O_β and O_γ species from gas-phase O_2 at 923 K, and the initial product distributions obtained during methanol oxidation on the O_α and O_γ sites.

Figure 4.8b gives a schematic representation of the oxygen dissociation, transport processes and subsequent reaction of O_γ with methanol. In the high-temperature oxidation process, the site of selective oxidation (O_γ) is different from the location of O_2 dissociation. Rates of oxygen diffusion become rate controlling.

There is an important difference between reaction of methanol with O_α, the oxygen atom formed at lower temperature, and reaction with O_γ, formed at higher temperature.

The first molecular mechanism of the reaction catalyzed by silver came from the surface science studies on single crystal surfaces by Wachs and Madix at Stanford in 1978 [66], [67]. They supported their interpretation of the reaction mechanism with extensive isotope-labelling studies.

The Wachs-Madix methanol oxidation reaction scheme is given by Eqs. (4.5) [67]:

$$\begin{aligned}
&\text{a} \quad CH_3OH_{ads} + O_{ads} \rightarrow CH_3O_{ads} + OH_{ads}\\
&\text{b} \quad CH_3OH_{ads} + OH_{ads} \rightarrow CH_3O_{ads} + H_2O_{gas}\\
&\text{c} \quad 2CH_3O_{ads} \xrightarrow{\text{slow}} 2CH_2O_{gas} + 2H_{ads}\\
&\text{d} \quad 2\,H_{ads} \rightarrow H_{2,gas}\\
&\text{e} \quad CH_2O_{ads} + O_{ads} \rightarrow HCOO_{ads} + H_{ads}\\
&\text{f} \quad HCOO_{ads} \rightarrow CO_{2,gas} + H_{ads}\\
&\text{g} \quad 2\,H_{ads} \rightarrow H_{2,gas}
\end{aligned} \tag{4.5}$$

They made the major discovery that methanol is only activated once reacted with co-adsorbed O_{ad}. Reaction is initiated by heterolytic bond cleavage of the methanol hydroxyl. This is a kinetics phenomenon and not related to thermodynamics (note that $H_3CO–H$ bond energy is 427 kJ/mol and $H–CH_2OH$ bond energy is 400 kJ/mol). Deprotonated methanol adsorbs as a methoxy species on the silver surface. The C–H activation step of adsorbed methoxy is the slow step in the overall reaction to formaldehyde. Formaldehyde is formed by C–H bond cleavage on an oxygen vacant silver site. Water and hydrogen are co-products with a ratio of 2:1. The latter is the signature of reaction with O_α. This is the oxygen adatom state at the single crystal surface.

Non-selective oxidation of formaldehyde happens through intermediate formate formation. Also, in this reaction hydrogen is co-product.

This mechanism of the oxidation reaction refines earlier proposals that consider reaction to occur in two successive steps: dehydrogenation of methanol to give formaldehyde and hydrogen, and successive combustion of hydrogen by oxygen. The second exothermic reaction provides the heat for endothermic dehydrogenation.

Later in 2002 van Veen, Hinrichsen and Mühler from Ruhr-Universität Bochum [68] used transient kinetic experiments under high-temperature methanol oxidation condition, where the co-product of formaldehyde is only water. This is the signature of reaction with O_γ. They propose for the reaction with O_γ the following sequence:

$$\begin{aligned}
&CH_3OH_{gas} + O_\gamma \rightarrow CH_3O_{ads} + OH_{ads}\\
&CH_3O_{ads} + OH\ (O_{gas}) \rightarrow CH_2O_{gas} + H_2O_{gas}\ (OH_{ads})\\
&OH_{ads} + OH_{ads} \rightarrow H_2O_{gas} + O_\gamma
\end{aligned} \tag{4.6}$$

Similar to the Wachs-Madix proposal, reaction is initiated by formation of a methoxy intermediate. However, C–H bond cleavage is not activated by a vacant Ag site but happens with silver-adsorbed OH or another O_γ atom. Water is formed by recombination of adsorbed OH.

The important difference between the Wachs-Madix oxydehydrogenation path and the van Veen-Hinrichsen-Mühler mechanism is that the direct dehydrogenation path needs a surface oxygen vacancy. In the latter mechanism C–H bond cleavage occurs by reaction with adsorbed oxygen. This difference is consistent with the idea that O_γ is part of a metastable oxide cluster and O_α is an isolated adsorbed oxygen atom.

X-ray photoelectron spectroscopy (XPS) spectra of O_γ and O_α indicate that O_γ is more covalent (radical-like) than O_α. O_α is more polar [69]. The metallic environment of O_α more readily donates electrons to the adsorbed oxygen atom than the oxidized surface to O_γ. The oxygen atoms in a high oxygen atom environment have to compete for the silver electrons. This is consistent with an oxidic surface state of O_γ and metallic surface state of O_α.

The selectivity with O_γ is consistent with the increase in selectivity of the reaction with increasing temperature [70]. It is also consistent with C–H bond cleavage to be reaction rate controlling for methanol dehydrogenation to formaldehyde. Due to its high activation energy, its reaction rate will increase faster with temperature than that of formaldehyde oxidation that has lower activation energy. Overall conversion of methanol is rate limited by dissociative adsorption of oxygen [71].

The site of oxygen dissociation is different from that of selective methanol oxidation. Diffusion of oxygen through the bulk connects the two sites. As is discussed in Section 4.4.4.2 this decoupling of reaction sites is also a common feature of catalysis by reducible metal oxides.

The selectivity of O_γ adsorbed on Ag is similar to that of methanol oxidation catalyzed by $Fe_2(MO_4)_3$ (see Section 4.4.3). Also on this metal oxide catalyst only water is product and hydrogen is not a co-product.

In the next section the mechanism of the epoxidation reaction of Ag is discussed. Also, the selectivity of this oxygen insertion reaction will be seen to strongly depend on the state of the silver surface.

4.3.3 *Selective Oxidation of Ethene and Propene by Ag and Cu; Ethene Epoxidation*

The difference in reactivity of ethene versus propene is due to ready reactive intermediate formation of propene when activated by oxygen.

When catalyzed by Ag and Cu the selectivity of ethene and propene oxidation shows remarkable differences [72], [73]. Whereas Ag is uniquely selective for formation of ethene epoxide (EO), the selectivity to give propene epoxide (PO) from propene is quite low [74], [75]. Catalysis by Cu gives preferentially acetaldehyde from ethene [76]. The selectivity of propene oxidation to PO is higher on Cu, but the main product is acrolein [77]. Ag and Cu are the only metal catalysts of interest for these reactions, since transition metals that activate C–H bonds readily give total combustion. Au cannot be used since it will not dissociate O_2.

A main cause of the difference in selectivity between propene and ethene is the ready formation of the allyl intermediate from propene. It is the reason for non-selective total combustion of propene when catalyzed by Ag. Acrolein is dominantly formed when propene is catalyzed by cuprous oxide [78]–[80]. Important to the difference in reactivity of Ag and Cu is the state of the catalyst surface at reaction condition. Then the state of the Ag catalysts is mainly metallic, but the state of the Cu surface will be oxidic.

4.3.3.1 *Ethene Epoxidation*

The mechanistic question is whether epoxidation of ethene happens by the oxametallocycle versus oxirane surface reaction intermediate.

The main competitive product to the epoxide apart from total combustion is acetaldehyde. The silver catalyst for epoxidation of ethene consists of Ag particles of the order of 100 nm distributed over a low surface area α-Al_2O_3 support. Parts per million of chlorinated hydrocarbons are added to the ethene/oxygen feed. Its combustion leads to Cl deposition on the silver catalyst that significantly enhances the selectivity of the reaction. Cl deposited on the Ag catalyst is in dynamic equilibrium with ethene and chlorinated ethene [81]. Alkali promoters are essential for high selectivity of ethene conversion [82]. The understanding of the role of these promoters has significantly helped determine the mechanism of ethene epoxidation.

Figure 4.9a gives the selectivity of the reaction for different Ag-based epoxidation catalysts of different composition. There is a remarkable shift in dependence on ethene conversion for the promoted catalysts. Whereas on non-promoted silver selectivity decreases strongly with conversion, this is different for the promoted catalysts. In particular, catalysts promoted with Re not only increase initial selectivity but also suppress consecutive epoxide combustion reactions.

The global reaction network is given by Figure 4.9b. There are parallel reaction paths to EO (r_1) and total combustion (r_2) as well as a consecutive reaction (r_3) for total combustion of EO [72]. The parallel reactions are catalyzed by Ag, the consecutive reaction by a combination of support as well as silver metal component.

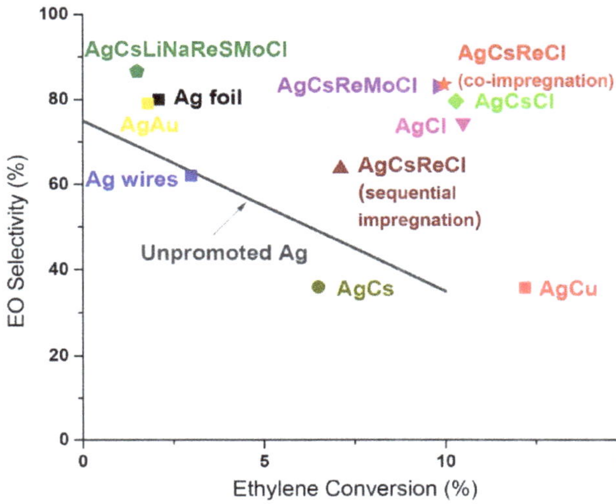

(a)

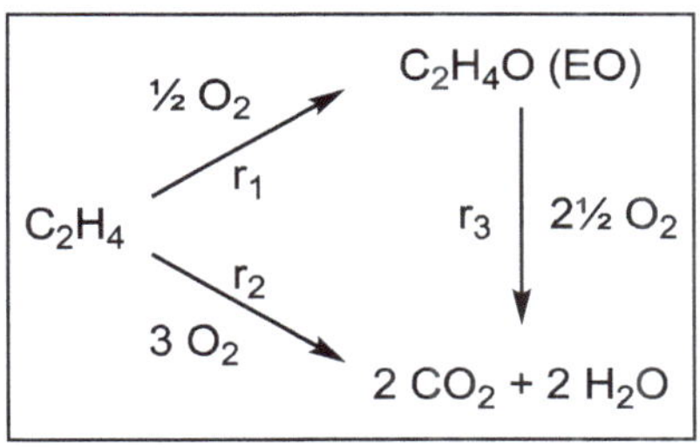

(b)

Figure 4.9 Ethene epoxidation reaction and selectivity. (a) Epoxidation selectivity of promoted catalysts (moderated with chlorinated hydrocarbon) (1–10 bar, 240°C, ethene/O_2 = 3) [82]. (b) The ethene epoxidation reaction network [72].

In 1946 Twigg *et al.* [83] proposed that consecutive total combustion (r_3) is due to a reaction where EO isomerizes to acetaldehyde. The isomerization of the epoxide to acetaldehyde is catalyzed by acidic protons of the α-Al_2O_3 support [84] and the aldehyde rapidly combusts by reaction catalyzed by Ag. The isomerization of EO is suppressed by addition of alkali cations that neutralize the support acidic site. Also, co-produced water will react with the epoxide. This gives reactive glycol that is rapidly oxidized by silver [85].

The reaction mechanism of ethene epoxidation has been intensively researched in the past fifty years. Even so, conflicting views still exist. A major issue is the structure of the selective reaction center on the silver particle. Below, the different views on the mechanism of the reaction are presented.

An initial mechanistic discussion evolved on the question of whether adsorbed O_2 or adsorbed atomic oxygen is the selective surface oxygen species. This question is important since it is relevant to the prediction of ultimate selectivity of the epoxidation reaction. The view that an atomically adsorbed oxygen species is responsible for selective catalysis is at present generally accepted.

The second debate that is still not completely settled is whether parallel reaction paths (r_1) and (r_2) are catalyzed by one type of reactive adsorbed oxygen atom or whether two types of reactive surface oxygen species are present as is the case for methanol oxidation.

In the following, first experiments are presented that support the mechanistic hypotheses. This is followed by computational studies that provide molecular models of suggested elementary reaction steps. This section concludes with a comparison of the selectivity of Ag and Cu.

Within the context of the question of whether molecular or atomic oxygen is the selective surface oxygen species, mechanistic studies in the 1970s focused on an understanding of the Cl-promoting effect [86]–[88].

Figure 4.10 gives experimental data for a supported Ag catalyst and an Ag crystal surface on epoxidation selectivity as a function of Cl coverage of the silver surface.

EO selectivity increases as a function of Cl coverage for both catalytic systems. Interaction of Cl with reactant site initially enhances the rate of EO formation, but beyond a certain Cl coverage the overall reaction rate becomes suppressed because site blocking by Cl takes over. In the surface science experiment the Cl coverage where r_1 decreases is lower than for the supported catalyst. The microscopic structure of the Ag particles on the supported catalysts is complex [91] and Cl becomes also located in subsurface Ag lattice sites.

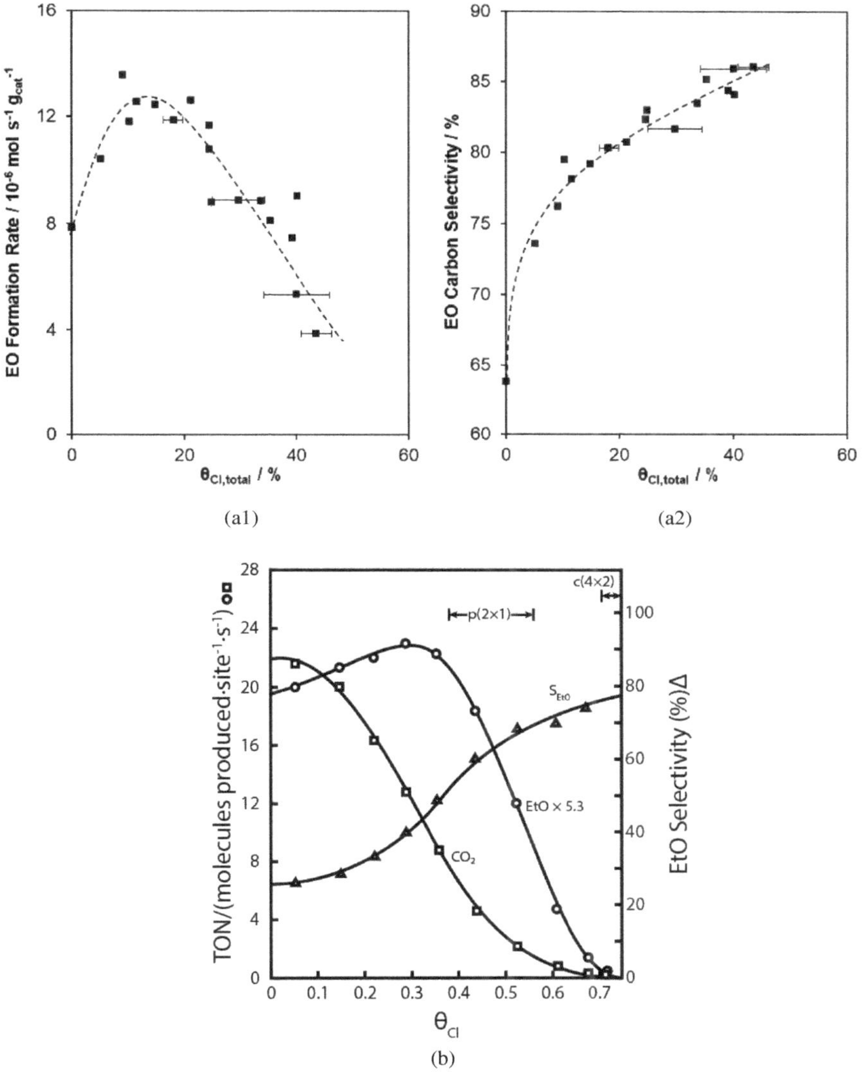

Figure 4.10 Comparison of the effect of Cl deposition on the ethene epoxidation reaction catalyzed by Ag catalyst and Ag(110) single crystal surface. (a) α alumina-supported Ag catalyst data, (a1) EO formation rate as a function of Cl coverage, (a2) EO selectivity as a function of Cl coverage [89]. (b) The dependence of rates of EO(EtO) and CO_2 production and selectivity upon Cl coverage conditions on Ag(110) [90].

In the 1970s, different from Twigg [83] who suggested that adsorbed atomic oxygen is responsible for EO formation, Kilty and Sachtler from Shell [87] suggested selective EO formation from adsorbed molecular oxygen. This was supported by infrared spectroscopic studies with isotope-labelled O_2 molecules that demonstrate the adsorption of molecular oxygen. The Kilty and Sachtler mechanism provides an explanation of the Cl-promoting effect. They suggest that Cl adsorption reduces the Ag surface atom ensemble size. This suppresses O_2 dissociation, which requires an ensemble of several Ag atoms. Atomically adsorbed oxygen is proposed to give total oxidation. The proposal is attractive since it predicts a maximum EO selectivity of 86%. At that time no experiments were available with higher selectivity. This maximum in selectivity is predicted with the following model: ethene reacts to CO_2 and H_2O by consuming six oxygen surface atoms. To generate these six adsorbed oxygen atoms, six O_2 molecules have to react with ethene to give epoxide. Therefore, for production of six EO molecules, one ethene molecule has to be totally oxidized.

It is important to establish the validity of this mechanism. It would imply that once selectivity of 86% is reached, further investment in new catalysts with higher selectivity is not useful.

As one can observe from Figure 4.9a the maximum selectivity reported is around this value. However, for the promoted catalysts of this figure, extrapolation of the high selectivity data to zero conversion indicates higher initial selectivity. Experiments with deuterated ethene also show initial selectivity higher than 86%. This suggests that higher selectivity is possible.

Deuterium substitution reduces the rate of C–H bond cleavage that initiates total combustion [92], [93]. This indicates that the C–H bond rupture is a limiting step for total combustion.

The competing proposal that atomic oxygen is responsible for ethene oxidation is also consistent with the Cl-promoting effect. It was proposed in 1976 by Force and Bell from UC Berkeley [86], [94]. They proposed that surface oxygen atoms only selectively react with ethene, when Cl atoms adsorbed around the oxygen atom block contact with the metal surface. This prevents activation of the ethene C–H bond and only insertion of the oxygen atom into the ethene π bond happens.

It also suggests two kinds of adsorbed oxygen atoms: non-selective oxygen atoms in a metallic surface environment and selective oxygen atoms in an inorganic chloride environment.

The proposal that atomic oxygen is responsible for selective oxidation agrees with the early suggestion of Twigg *et al.* [83].

Several successive experiments have demonstrated that surface oxygen atoms react selectively [95]–[97] with ethene to EO. Experiments with silver powder, alkali-promoted Ag surfaces, and Ag surfaces treated at high temperature with reactive gases such as NO_2 generate oxygen surface states that selectively epoxidize ethene. The Ag surfaces then have become highly reconstructed [98].

XPS experiments provided explicit proof of the presence of two oxygen species [99]. An electrophilic, electropositive and a nucleophilic, electronegative species are identified (see Figure 4.11).

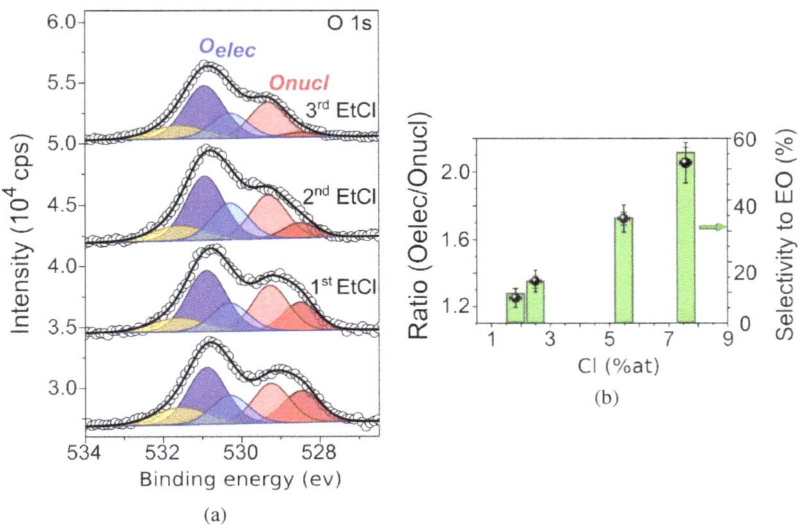

Figure 4.11 *In situ* stepwise chlorination of silver catalyst under ethene epoxidation condition at 0.3 mbar, 510 K, C_2H_4:O_2 = 1:2 [99]. (a) O1s XPS spectrum before (bottom) and after each chlorination step (1st, 2nd, 3rd EtCl pulses). (b) Ratio of electrophilic to nucleophilic oxygen species. O_{elec}/O_{nucl} (black dots) and selectivity (green bars) as a function of the Cl atomic concentration.

The relative concentration increase of electrophilic oxygen correlates with enhanced epoxidation selectivity due to increasing Cl surface coverage. The electrophilic oxygen inserts preferentially into the C=C π bond. The nucleophilic oxygen atom reacts preferentially with a C–H bond. The chemical interpretation of the different reactivities of the oxygen adatoms is explored in detail below.

In a paradigmatic paper Linic and Barteau [100], [101] proposed in 2002 that the surface intermediate for selective epoxidation is an oxametallocycle (OMC). It is shown in Figure 4.12a. The surface state is metallic. The alternative suggestion is shown in Figure 4.12b. EO formation occurs in an Eley-Rideal step by direct recombination of ethene with a surface oxygen atom. Ethene is not activated by contact with the silver atoms [100], [101]. This oxirane intermediate requires an oxidic or oxychlorinated state of the surface.

In the first case the effect of co-adsorbed promoters as Cl and alkali is mainly interpreted in terms of changes in electronic structure, while in the other case the local structure of the reaction center is emphasized. As schematically illustrated in Figure 4.12a, in the OMC the ethene molecule reacts in asymmetric coordination with the adsorbed O atom. One C–H bond interacts with a vacant site on the metal surface, while the other ethene carbon atom interacts with the oxygen atom. It is the analogue of the metalorganic cyclic intermediate discovered thirty years earlier in the metathesis reaction (Section 6.2).

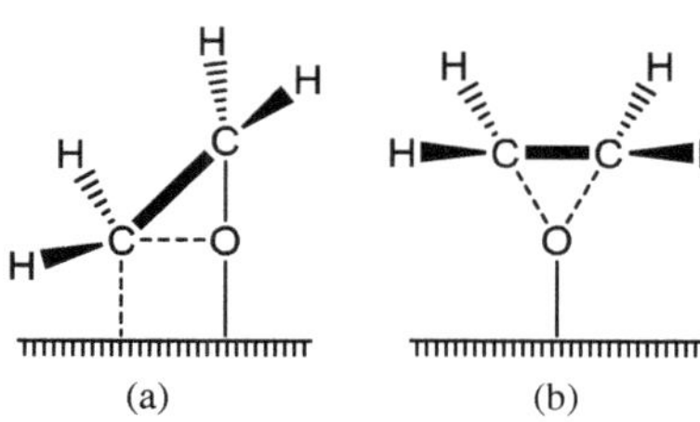

Figure 4.12 Schematic drawings of (a) oxametallocycle intermediate and (b) oxirane, the Eley-Rideal adsorbed EO intermediate.

Surface science spectroscopic experiments identified the structure of the OMC intermediate. In model experiments Barteau and Linic demonstrated that it is formed by adsorption of EO on the Ag surface. When the OMC intermediate decomposes, EO as well as acetaldehyde (AA) is formed.

Computational modelling of the OMC intermediate adsorbed on a metallic surface validated the interpretation of the surface science experiments [102]–[105]. The effect on activation energies of EO and AA formation by decomposition of OMC on co-adsorbed Cl or alkali has also been computationally investigated [102]–[105]. In these studies, there is no direct contact between the OMC and alkali or Cl atoms. Cl is assumed to be adsorbed subsurface as suggested by experiments of Waugh *et al.* [106].

The general conclusion of these studies is that the difference in activation energies for EO and AA relates to the strength of the M–O bond energy [107]. This is illustrated in Figure 4.13. The stronger the MO bond energy, the lower the relative activation energy for formation of EO versus AA (Figure 4.13c). On the Ag(111) surface, EO formation has the higher barrier. This becomes lower than the barrier for AA formation on the more reactive Ag(100) surface.

The M–O bond energy of the oxygen atom is larger when adsorbed on the Cu surface than on the Ag surface. These modelling studies of OMC decomposition predict a higher selectivity for EO formation on Cu than on Ag.

This generates a question since experimentally the selectivity of EO formation is higher when catalyzed by Ag compared to Cu. This issue will be addressed at the end of this section.

Recent quantum-chemical calculations indicate that selective formation of EO from the OMC intermediate is less probable than suggested by the abovementioned studies. Quantum-chemical and microkinetic simulation with use of machine learning techniques by Liu *et al.* at Fudan University in Shanghai [108] from 2021 identify a previously undiscovered deactivation mode of the OMC intermediate. The competing reaction path occurs by a C–H bond cleavage reaction of the α C–H bond next to the newly formed C–O bond.

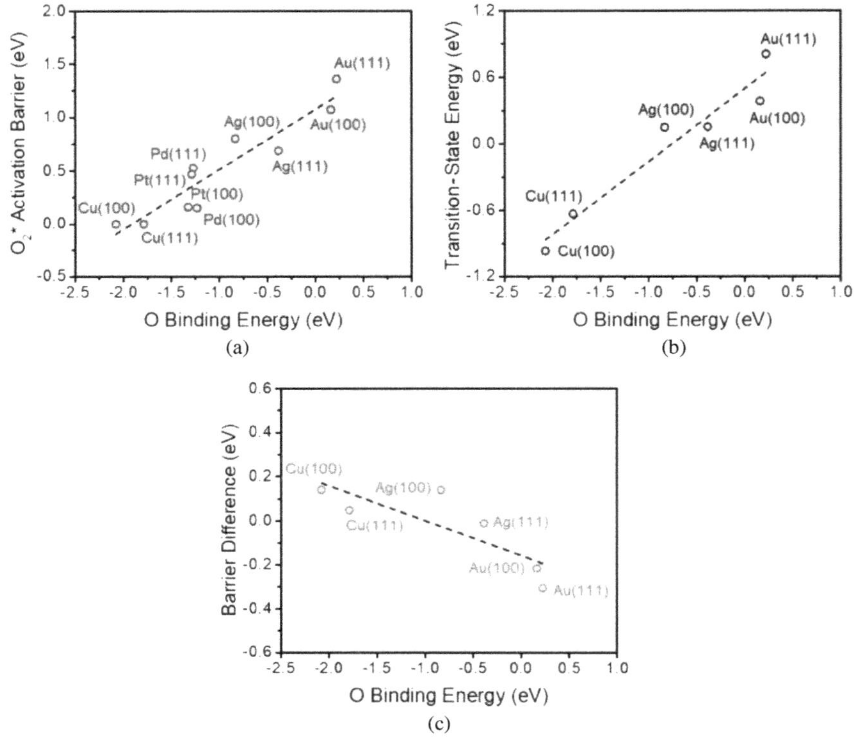

Figure 4.13 Scaling relations that relate ethene epoxidation activity and selectivity: a comparison of respective energies on different transition metal surfaces [107]. (a) O_2 activation barrier vs O binding energy. (b) Transition state energy of OMC formation (referring to gas-phase C_2H_4 and 1/2 O_2) vs O binding energy. (c) Barrier difference ($\Delta E = E_{AA} - E_{EO}$) of OMC decomposition for AA versus O binding energy.

This has a substantially lower activation energy for AA formation than the generally accepted reaction path with internal hydrogen transfer.

Results of these simulations are shown in Figure 4.14. They show the three decomposition modes of the OMC intermediate and microkinetic simulation results. The simulations give EO and AA as main products. However, due to low activation energy of the α C–H bond activation, the selectivity of EO formation is low.

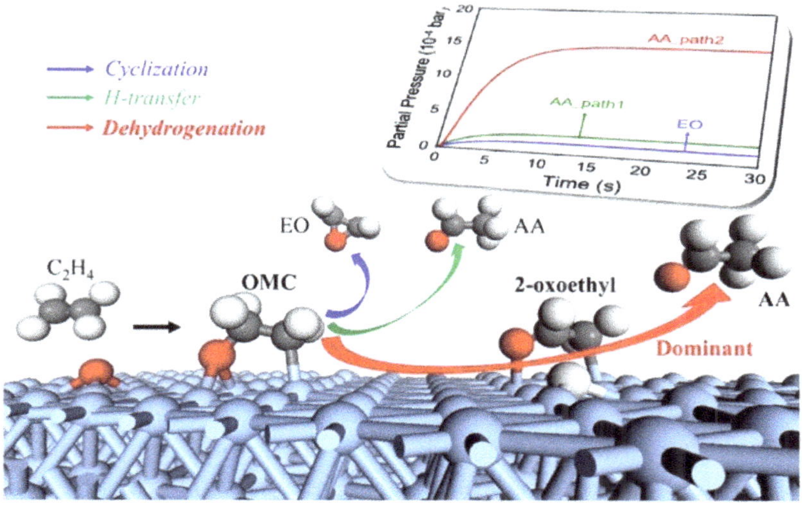

Figure 4.14 The reaction intermediates of the OMC reaction path. The two reaction pathways that form acetaldehyde are shown. The insert shows microkinetic simulations. The selectivity to AA formation is low due to the dehydrogenation pathway to AA [108]. So far there is no experimental confirmation of this pathway.

As will be seen in Section 4.4.2, this elementary step is an essential step in the oxidation reaction of propene to acrolein. These simulations are consistent with the possibility that another reaction intermediate is mainly responsible for EO formation.

The alternative proposition is that reaction proceeds through the adsorbed EO intermediate of Figure 4.12b. This path is representative for EO formation on the silver oxide surface without vacancies. The selectivity for EO versus AA formation depends on the presence of surface oxygen vacancies. This relates to the early Force and Bell suggestions that EO formation requires a reaction center where the O atoms have no nearest vacant site. It will be seen that OMC forms at the oxygen vacant site and decomposes nonselectively. Also, for this model the predicted activation energies match with experiment.

As a model of the oxidic surface reaction site, the reactivity of the $Ag_2O(100)$ surface has been quantum-chemically explored

[109]. DFT results of simulations on the non-oxygen vacant Ag_2O(100) (Figure 4.15a) and vacant oxygen surface (Figure 4.15b) show a dramatic difference in reactivity.

Only on the surface with an oxygen vacancy the OMC intermediate is formed. It has a low selectivity for EO formation, since it decomposes with a high barrier to EO and relatively low barrier to AA. This is different for decomposition of the OMC intermediate on the Ag(111) surface (also shown in Figure 4.15a). Then the barriers for formation of EO and AA are comparable (the direct α C–H bond activation path is not included in these simulations). No OMC intermediate is formed on the non-vacant Ag_2O(100) surface. At this surface there is an exothermic reaction between ethene and surface oxygen atom. EO desorbs without competitive formation of AA. The desorption energy of EO for the oxide surface is comparable with activation energy of EO on the Ag(111) surface. The oxidic surface without vacancy models selective EO formation.

The results of Figures 4.14 and 4.15 show that a low selectivity to EO is expected for reaction of ethene with an oxygen atom in a metallic environment, whereas this selectivity is high when oxygen is part of an oxidic environment, and no oxygen surface vacancy is present next to the reactive oxygen atom. This view is consistent with the experiments of Figure 4.11 that on the silver surface two kinds of oxygen atoms are present at reaction conditions, a nucleophilic oxygen atom of higher negative charge on the metallic surface and an electrophilic oxygen atom of lower charge (in a polar environment) that reacts with the olefinic π bond. The electrophilic oxygen can also be part of an OMC when a surface vacancy is present. Then it reacts non-selectively.

This model is consistent with the Ag surface atom ensemble size reduction effect of Cl as in the Force and Bell chlorine promotion model. Cl adsorbs strongly on silver and reduces surface vacant positions. The reactivity of oxygen adatoms then changes from OMC path to adsorbed EO intermediate pathway.

Bulk Ag_2O oxide is not stable at reaction conditions, but Cl adsorbs strongly. This determines its reduction of the size of the Ag

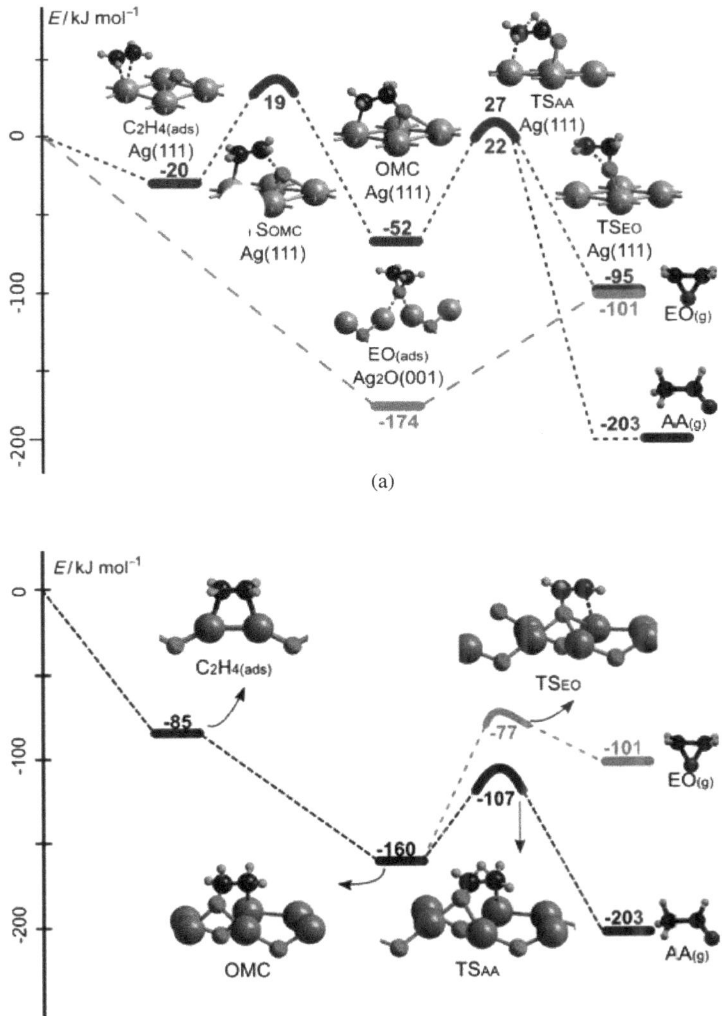

Figure 4.15 Comparison of energetics of OMC versus direct EO intermediate reaction paths [109]. (a) Potential energies and reaction intermediate structures of the EO and AA paths through OMC intermediate on Ag(111) surface and the direct EO formation path on the $Ag_2O(001)$ surface. (b) Potential energy diagram and intermediate structure for the non-selective reaction via OMC intermediate on $Ag_2O(001)$ surface with oxygen vacancy.

surface atom ensemble. In the presence of chlorine the effect of alkali on the reactivity of Ag is relatively small [110], [111]. It also helps to maintain a high surface concentration of Cl at high oxygen conversion. Alkali or Re promotion of the silver surface will assist local oxide formation. These promoters increase the rate of surface reoxidation since this reaction is rate controlling. The consumption rate of ethene is first order in oxygen and zero order in ethene [112].

Simulations and surface science experiments on the epoxidation reaction catalyzed by surfaces in the metallic state predict that Cu is more selective than Ag for the epoxidation reactions. However, in the catalytic experiment Ag is selective and Cu has a low selectivity. The reason for this difference is that under the epoxidation reaction conditions the Cu surface is oxidic [75]. Next to total oxidation of ethene the main product is acetaldehyde.

The difference in selectivity of silver oxide and copper oxide can be deduced from the computed potential energies of Figure 4.16.

The closed Ag_2O overlayer will selectively produce EO, but Cu_2O gives selectively acetaldehyde. This is due to the larger energy of desorption of EO from the Cu_2O surface than from the Ag_2O surface due to the stronger interaction of O with the Cu surface atoms. This strong M–O interaction in turn weakens the O–C bond strengths. It causes the ring opening of the adsorbed epoxide intermediate on Cu_2O to have a lower activation energy compared to that on Ag_2O (the activation energy is nearly zero). The acetaldehyde is readily oxidized, and the main reaction is total oxidation of ethene.

The chemistry of the copper oxide surface is complex. Structure and oxidation state of the Cu oxide system will affect selectivity of reaction. A unique reactive oxygen atom has been identified at the interphase of CuO or Cu_2O phases that selectively gives EO when reacted with ethene [114].

In summary, the mechanism of ethene epoxidation strongly depends on the state of the silver surface. Under reaction conditions bulk Ag_2O is not stable. Simulation as well as experiment suggest that oxygen atoms adsorbed on silver in its metallic state convert ethene with low selectivity to EO. An oxidic or oxychlorinated surface state is most likely the site of selective epoxidation.

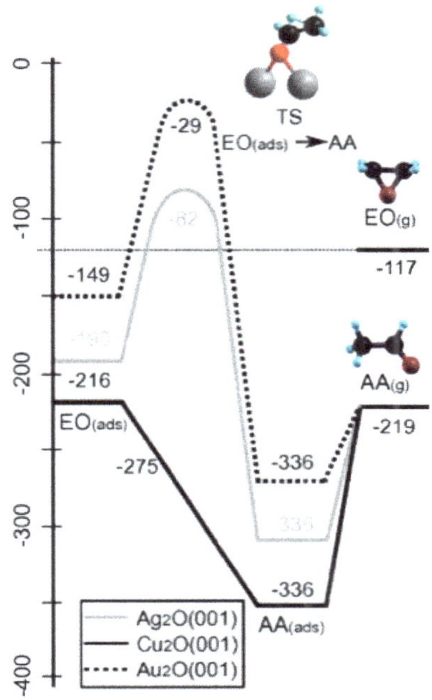

Figure 4.16 Comparison of the potential energy changes of the direct reaction path for EO formation on Ag_2O, Cu_2O and Au_2O(110) surfaces [113].

The role of Cl, alkali and Re is to maintain the oxychlorinated state of the reactive surface site.

Consecutive reactions are catalyzed by a combination of isomerizations or hydrations by acidic sites of the alumina support and total combustion of reaction intermediates by the silver surface. Alkali and possibly Re also play a role to suppress catalyst support reactivity.

4.3.3.2 *Selective Propene Oxidation*

The reactivity of propene allyl suppresses selective oxidation of propene to propene epoxide. Acrolein is the main product.

The desirable products of selective oxidation of propene by Ag or Cu with molecular oxygen are PO and acrolein (AC, CH_2=CH_2–CHO).

Both products are formed by catalysis with these metals but PO selectivity can be low. As discovered by Shell in 1964, copper is an attractive catalyst for acrolein production by propene oxidation.

The main reason for the unique reactivity of propene is its ready formation of the allyl intermediate.

There is nearly 90 kJ/mol difference between the C–H bond energy of the CH_3 part of propene compared to the C–H bond energy of ethene. Also, PO has a reactive CH_3 group that makes it liable to consecutive oxidation. This is different for acrolein, where this reactive C–H bond is absent.

Similar to oxidation catalysis of ethene described in the previous section, the chemistry of propene is very different when studied with a metallic surface or when catalyzed by the oxidized surface.

Because of its stronger M–O adatom oxygen bond strength, metallic copper is predicted to have a higher selectivity for PO than Ag [115], [116].

Experimentally, when reaction is executed below 530 K where Cu is in a metallic state, a PO selectivity of 50% at 45% conversion has been reported [116]. However, above 530 K the Cu surface is converted into the oxidic state. PO selectivity drops and the main product is acrolein with a selectivity of the order of 40% at 50% conversion. The increased selectivity of metallic Cu towards PO compared to Ag is consistent with the OMC intermediate. As shown in the previous section, for the case of ethene on oxidic Cu the oxygen insertion reaction is suppressed and C–H bond activation dominates [117], [118].

Selectivity to PO is sensitive not only to reaction conditions but will also vary with addition of promoters [77]. For Ag as well as Cu this affects the stability of a particular surface phase.

Surface science experiments indicate that selectivity is also sensitive to surface site structure. On the more open surface of Cu_2O [119] surface selectivity is higher for PO. This changes for the dense, more stable surfaces that favor acrolein.

Due to the lower Ag–O bond energy versus the Cu–O bond, the oxidic Ag site should have higher selectivity for propene oxide. This is only the case at low conversion. Due to the high Lewis basicity of

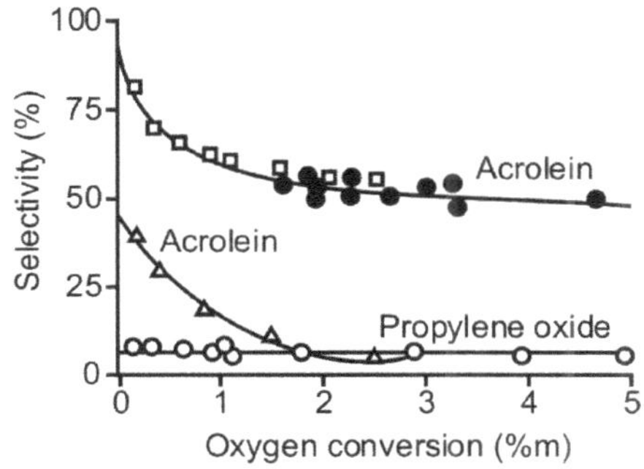

Figure 4.17 Oxygen conversion versus selectivity to acrolein and propene oxide at 480 K [122]. ○, 8%w Ag/α-Al_2O_3; △, 61%w Ag-39%w Au/α-Al_2O_3; •, 11%w Ag-89%w Au/α-Al_2O_3, □, 24%w Ag-76%w Au (pure alloy).

the oxygen adatoms, they activate the CH_3 substituent in propene or propene oxide. This will lead to consecutive combustion reactions. On Ag a representative selectivity is 5% PO at 20% propene conversion without acrolein as co-product [120]. Interestingly, in model experiments with Ag clusters of three atoms deposited at low temperature, a selectivity of 100% to PO is found. This is due to strong cluster Ag–O bonds that have reduced Lewis basicity [121].

Ag can be modified to produce acrolein with similar selectivity as Cu. An example is given in Figure 4.17 where Ag is alloyed with Au.

When the Au concentration increases, initial low selective formation of PO is overtaken by selective acrolein formation. Au will reduce the oxygen adatom concentration of the Ag/Au catalyst compared to its concentration on the Ag surface. On the metallic Ag surface allyl formation dominates, giving mainly total combustion. On oxidic surfaces, oxygen insertion competes with allyl formation and PO can be formed at low selectivity.

The change in selectivity of the Ag/Au alloys with increase of Au concentration is consistent with the suppression of PO formation and increasing rate of acrolein formation via intermediate allyl.

Similarly, experiments with Au deposited on SiO_2 catalysts with N_2O (O_2 will not dissociate on Au) as reactant give acrolein as main product [123]. Au will maintain a metallic surface state.

Isotope-labelling studies of propene labelled with ^{13}C or D by Adams and Sachtler at Shell in 1963 demonstrated allyl as the intermediate for acrolein formation [79], [80]. The probability for an oxygen atom to attach to one of the opposing primary carbon atoms of propene is found to be the same.

This is the case for acrolein formation by the Cu catalyst as well as the Bi/Mo oxide-related catalysts that are presented in Section 4.4.2.

The systems are mechanistically related. Kinetics is different; whereas acrolein production with the Cu system is rate limited by O_2 dissociation (first order in oxygen, zero order in propene concentration), the Bi/Mo system is rate limited by propene activation (first order in propene and zero order in oxygen concentration). The latter system is currently dominant in commercial selective oxidation of propene to acrolein.

At present, no catalyst has been found that gives a satisfactory yield of PO from propene when oxygen is used as oxidant. High selectivity for PO formation is found when hydrogen peroxide or hydroperoxides are used as oxidants. These reactions catalyzed by Lewis acidic cations are discussed in Chapter 6.

In summary, as sketched in Figure 4.18 there are three parallel elementary surface reactions that activate propene: activation by a surface oxygen atom on the metallic surface, intermediate epoxide formation on oxidic surface and reaction via OMC intermediate. Reactivity of Ag versus Cu is very different and as summarized below:

– Activation of allylic propene C–H happens on Ag by nucleophilic surface oxygen atoms. The state of the surface is metallic. This reaction is the analogue of the Wachs-Madix activation of O–H of methanol by co-adsorbed oxygen atoms (see Section 4.3.2). In the case of propene, it leads ultimately to acrolein or total combustion. An analogous reaction initiates combustion of PO by activation of the CH_3 substituent. On metallic surfaces, decomposition of OMC complex on Cu gives PO with higher selectivity than on Ag.

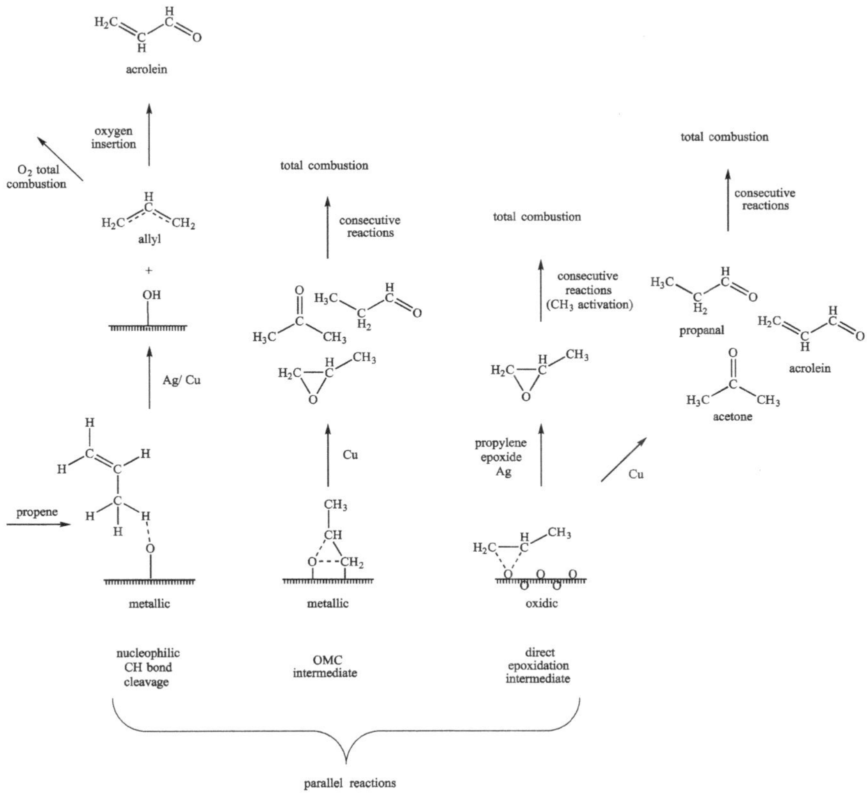

Figure 4.18 Propene oxidation reactions on Ag/Cu metal surfaces.

- On oxidic or oxychlorinated Ag, PO is formed in competition with total oxidation. On the oxidic Cu surface acrolein, propanal or acetone are products.

4.4 Reaction Mechanisms of Solid-state Redox Oxidation Reactions

4.4.1 *Introduction*

The history of the scientific understanding of these systems is a story of increasingly better characterization of the different phases in which these materials are composed and identification of the structure of the reactive surface.

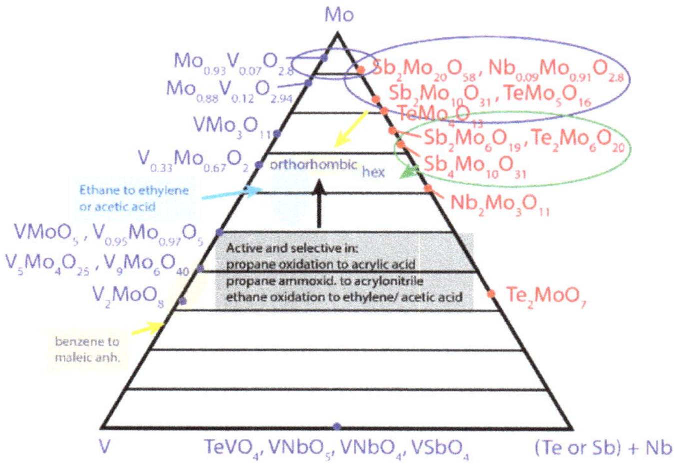

Figure 4.19 Diagram of complex metal oxide phase composition and relation with particular oxidation reactions [38].

An illustration of the complex composition of these systems is given in Figure 4.19. As the figure illustrates, compositions are mainly mixtures of molybdenum oxide and vanadium oxide. Other components are Te, Sb or Nb. The representative reactions of the materials in this figure are discussed in the following sections.

Advances in 1960–1990 mainly relate to the correlation of catalyst structure and composition data with catalyst performance. Important early contributions are due to the Polish Institute of Catalysis in Krakow and the French Institute de Catalysis in Villeurbanne, Lyon. As mentioned in Section 4.2.2.4 industrial research went through a period of large creativity with major discoveries of new reactions and materials. This also contributed largely to mechanistic understanding of these systems. An example is the work from Sohio (at present BP) that published important fundamental investigations of their complex mixed metal oxide systems.

The early insights are mainly from structural characterization, isotope kinetic measurements and surface spectroscopic characterization. After 2000 access became available to advanced spectroscopies and quantum-chemical calculations on model systems. Many of the hypotheses and theories of the previous period were revisited, which led to deepened molecular understanding of reaction mechanism. Chemical understanding of catalysis by solid-state metal oxides also were formulated in terms of inorganic complex and organometallic chemistry. A catalytic theory of structure-function developed.

4.4.2 *Kinetics and Reactivity Principles*

4.4.2.1 *Mars-van Krevelen Kinetics*

Fundamental to solid-state redox catalysis is the Mars-van Krevelen principle that the site of selective oxidation is not necessarily the same as the site of reoxidation.

Important to heterogeneous oxidation catalysis is the Mars-van Krevelen suggestion [124] that the reoxidation of a reduced catalyst after reaction and the reduction of the catalyst by reaction of surface oxygen with reactant are independent processes. This implies that the site of molecular oxygen activation and surface oxide consumption by reactant are not necessarily the same. This is not limited to selective oxidation by reducible metal oxides. An example of this separation of sites for the silver metal catalyst has been discussed in Section 4.3.2 for the methanol oxidation reaction.

In their early study of 1953 on the oxidation of aromatics catalyzed by V_2O_5, Mars and van Krevelen, then researchers at DSM, designed a simple two-step kinetic model. The corresponding kinetic expressions made an excellent fit to a large set of data of kinetic measurements of these reactions:

$$n_R^{\bullet} = k_1 P_R \theta \tag{4.7a}$$

$$n_{O_2}^{\bullet} = k_2 P_{O_2}^{n} (1-\theta) \tag{4.7b}$$

$$n_R^{\bullet} = \frac{1}{\beta} n_{O_2}^{\bullet} \tag{4.7c}$$

In Eq. (4.7) $n_R^{\bullet}$ is the rate of reagent consumption, while $n_{O_2}^{\bullet}$ is the rate of oxygen consumption. The reaction is assumed to be first order in reactant pressure P_R and of the order n in oxygen pressure P_{O_2}. The simplifying and reasonable assumptions are that reaction is proportional to the reactive surface oxygen concentration ϑ and that there is no oxygen desorption. Details of the mechanism are contained in global kinetic parameters k_1 and k_2. β is a stoichiometric number that relates to the amount of oxygen consumed per reactant.

The mechanism of the benzene to maleic acid oxidation reaction is discussed in Section 4.2.

Mars and van Krevelen interpreted their equation correctly that oxidation and reduction reactions are independent, but according to their formulation they do not necessarily occur on different sites.

At that time the Langmuir-Hinshelwood surface kinetics approach was not well known in the engineering community. The Mars-van Krevelen equations are actually an example of such kinetics, since they explicitly include a surface intermediate (the surface reactive oxygen). Their equation is similar as the Langmuir assumption that adsorbates do not interact other than by competition for adsorption sites. The linear dependence of the rate of oxygen consumption $\overset{\bullet}{n}_{O_2}$ on $(1-q)$ agrees with the then generally accepted postulate that molecular oxygen intermediates are the reactive oxygen intermediates (see Section 4.2).

The solution of Eqs. (4.7) gives the Mars-van Krevelen rate equation of redox oxidation catalysis:

$$\overset{\bullet}{n}_R = \frac{1}{\dfrac{\beta}{k_2 P_{O_2}^n} + \dfrac{1}{k_1 P_R}} \tag{4.8a}$$

$$\left(\overset{\bullet}{n}_R\right)^{-1} = \beta\left(k_2 P_{O_2}^n\right)^{-1} + \left(k_1 P_R\right)^{-1} \tag{4.8b}$$

Eq. (4.8a) has been derived without assuming a rate-determining step. The rate of reaction is determined by the balance of the rates of metal oxide reduction and cation oxidation (Eq. (4.7c)). These rates can be determined from the slope of $\left(\overset{\bullet}{n}_R\right)^{-1}$ of Eq. (4.8b). When reoxidation is fast, this slope will be independent of oxygen pressure. This is the desirable condition for selective oxidation that is often satisfied in reducible oxide catalytic systems.

The suggestion that oxidation and reduction reactions are independent is supported by oxygen isotope experiments [125]. The isotope label of oxygen in product then relates to the isotope composition of catalyst and not to that of gas-phase oxygen. This has been demonstrated in 1970 by Keulks [126] who measured the isotope distribution of a mixture of $^{18}O_2$ and propene catalyzed by a bismuth molybdate catalyst of Section 4.4.3. The catalyst initially only contained ^{16}O. For the bismuth molybdate catalyst system, the reactive intermediate is a surface lattice oxygen atom [126], [127].

Keulks noted that the initial acrolein product only contained ^{16}O. This confirms that propene reacts with oxygen atoms present in the metal oxide and not directly with the oxygen molecule. In the course of time oxide oxygen atoms are supplemented with oxygen atoms derived from gas-phase oxygen. Ultimately the acrolein molecule also becomes labelled with ^{18}O atoms. It also implies that oxygen insertion into reactant is the rate-controlling step and depletion of lattice oxygen atoms at the reaction site is fast.

A similar experiment with Ag-catalyzed epoxidation (see Section 4.3.3.1) gave an analogous result, but in this case for the labelling of the EO product [96]. It is one of the arguments that support atomic oxygen being the reactive surface intermediate for this reaction.

In summary, there are three fundamental kinetic postulates of reducible metal oxide catalysis:

- The only surface intermediate is activated oxygen.
- The elementary rate constant of O_2 activation, e.g., dissociation into adatoms, is large compared to the elementary rate constant of the surface oxidation reaction. The latter is fast compared to the rate of O_2 desorption that can be ignored.
- The diffusion of atomic oxygen or other reactive activated oxygen species through the bulk of the oxide is fast.

4.4.2.2 *Reactivity Determinants*

The seven pillars of selective solid-state oxidation catalysis of Graselli.
The additional dual site reaction motifs.

Robert Graselli, who was a research director at Sohio at the time that ammoxidation and related processes were developed, formulated requirements for high selectivity of solid-state reducible metal oxides. They can be considered inorganic chemical descriptors of the reactivity of surface M–O bonds in relation with reaction site topology.

Graselli defined seven key factors that he called the pillars of selective oxidation [35], [128]. They are based on the oxidation catalysis mechanism in which lattice oxygen atoms act as the reactive surface atoms. Later, two molecular structural motifs were discovered that has to be added to Graselli's definitions.

1. The chemical nature of lattice oxygen

Graselli defined the chemical nature of lattice oxygen anions as nucleophilic (selective) or electrophilic (non-selective, initiates total oxidation). This relates to the Haber proposal of Section 4.2 that transient O^- and O_2^- species induce non-selective radical reactions. However, whether the oxygen atom initiates a selective reaction or total oxidation also depends on the reactant.

Also, in metal oxidation catalysis reactive oxygen atoms can be distinguished into nucleophilic and electrophilic reactivity. Nucleophilic oxygen atoms activate C–H bonds and electrophilic oxygen atoms insert into the alkene π bond. In this case nucleophilic oxygen will initiate total oxidation of ethene, but selective oxidation of propene to acrolein.

A better difference between oxygen atoms is that they either react radical-like or covalent/polar. In redox catalytic systems C–H activation is a radical reaction. The oxygen insertion is a singlet spin state reaction. Synergy due to interaction between two cations may promote oxygen reactivity (see below, synergetic promotion).

2. Redox properties of the metal oxide

Dissociation of the oxygen molecule and adsorption of the oxygen atoms increase the valence state of the cation. Insertion of oxygen atoms into a reactant bond or water formation reduces cation valency. Oxygen reactivity relates to cation redox properties.

It explains the higher reactivity of V_2O_5 than of MoO_3 ($E(V_2O_5) - E(V_2O_4) = -29$ kcal/mol; $E(MoO_3) - E(MoO_2) = -50$ kcal/mol).

3. The M–O bond strength

This relates to cation redox properties, which can be complex for mixed metal oxides. The reactivity of the M–O bond relates to its polarity as well as covalent character (see Section 2.3.3). The cost of C–H bond cleavage has to be less than the strength of the newly formed O–H bond. The stronger the O–H bond strength, the weaker the M–O bond energy. The latter can be stronger than the M–O bond strength that has to be overcome in an oxygen insertion reaction. The M–O bond strength regulates selectivity. It should not be too weak (this gives total oxidation) or too strong (this gives low reactivity).

4. Host structure

The stability and adaptability of structure when it reduces or reoxidizes is important. Haber has suggested that metal oxides of group IV-VII are highly reactive since reduced structures stabilize by shearing of layers. This nucleates at the surface with release of oxygen [129].

5. Phase cooperation in multicomponent catalysts

Most materials consist of different phases. They can promote or stabilize each other. An example is the ammoxidation catalyst of propane. Whereas the pure phase $Mo_{7.5}V_{1.5}NbTeO_{29}$ has high yield, sometimes the presence of a second phase improves catalytic performance further. The $V_{1.5}NbTeO_{29}$ compound is activated by the presence of $Mo_6Te_2VO_{24}$. Graselli suggested that the presence of the other phases helps to form the desirable phase in the synthesis of the catalyst.

6. Multifunctionality

The presence of reaction centers of different reactivity due to the complex composition of the multicomponent metal oxide. For example, the different compositional elements of the propane ammoxidation catalyst $Mo_{7.5}V_{1.5}NbTeO_{29}$ have different functions. Vanadium oxide acts as propane dehydrogenation site to give propene, the Mo oxide site inserts O or N–H into the alkene, TeO activates C–H bonds, and Nb is a structural promoter. Sb can substitute for Mo in the propene ammoxidation reaction. Sometimes elements such as Fe or Ce are added to assist oxygen atom transport. The Mo and V cations also catalyze the reoxidation of reduced cations, which are generated when water desorbs or when oxygen is inserted into the hydrocarbon.

7. Active site isolation

This is the analogue of the surface ensemble effect [130]. A too high local concentration of reactive oxygen atoms should be prevented, because this will activate total oxidation (total oxidation requires many oxygen atoms to produce CO_2 and H_2O). Site isolation reduces local concentration of reactive oxygen atoms. It is suggested that the role of phosphorus in the $(VO)_2P_2O_7$ catalyst for butane oxidation to maleic anhydride (Section 4.4.4.1) is reduction of vanadium oxide cluster size. However, a dual oxide site can also be beneficial (see dual site effect below).

Computational chemistry has also identified synergetic cooperation between different cation components that changes the reactivity of oxygen compared to that of the mono cation oxide. Dual site reactivity has to be added to the Graselli pillars. Two molecular reactivity effects are to be distinguished: oxygen reactivity changes by dual cation synergy and dual site effect which requires at least two neighbor reactive oxygen sites.

– Synergetic promotion of oxygen reactivity

Lewis acid cations such as Bi^{3+}, Sb^{3+}, Te^{4+} and also P^{5+} promote the reactivity of oxygen atoms coordinated to reducible cations such as V^{5+} or Mo^{6+}. In the Bi–O–Mo bonding motif the lone pair of Bi promotes M–O radical-like reactivity for C–H bond cleavage of propene (Section 4.4.3.1). A synergy is proposed of Te with V [131]. In the ammoxidation catalyst the dual $O{=}Te^{4+}{-}O{-}V^{5+}$ activates C–H of propane to give $HO{-}Te^{4+}{-}O{-}V^{4+}$. In $(VO)_2PO_7$, which catalyzes the oxidation of butane to maleic anhydride, the reactivity of a O=P–O–V site promotes conversion of butane to butene (Section 4.4.4.1).

– Dual site effect

Many oxidation reactions require two intermediate hydrogen atom removals. This requires two reactive surface oxygen sites. The oxygen atoms are attached to different reducible cations. Catalysis is not single site but requires at least a cluster of two reducible cations. Examples are the Bi/Mo oxide catalyst of propene oxidation (Section 4.4.3.1), the Fe/Mo catalyst of methanol oxidation (Section 4.4.3.2) and the V oxide oxydehydrogenation catalyst (Section 4.4.3.3).

4.4.3 *Solid-state Multicomponent Mo Oxide Catalysts*

The inorganic chemistry of the noble metal systems and reducible metal oxides is very different, but the reaction mechanisms show many similarities. In both systems C–H bond activation is the rate-controlling step of reactant oxidation. The kinetics of reactions catalyzed by noble metals and reducible metal oxides will be seen to be quite different.

4.4.3.1 *Mechanism of Selective Propene and Propane Oxygenation*

Formation of allyl intermediate controls selectivity. Selective catalysis occurs by reaction of oxygen atoms of coordinatively saturated cations without direct contact with the cation. CH activation is a radical-type reaction. It requires dual cation complex sites where a non-reducible cation activates the reducible cation.

The mechanism of selective propene oxidation to acrolein catalyzed by reducible metal oxides as formulated in the 1980s is sketched in Figure 4.20.

This allylic mechanism is analogous to that of selective oxidation of propene to acrolein catalyzed by copper oxide in Section 4.3.3.2. Oxygen and propene isotope-labeling experiments on Bi/Mo

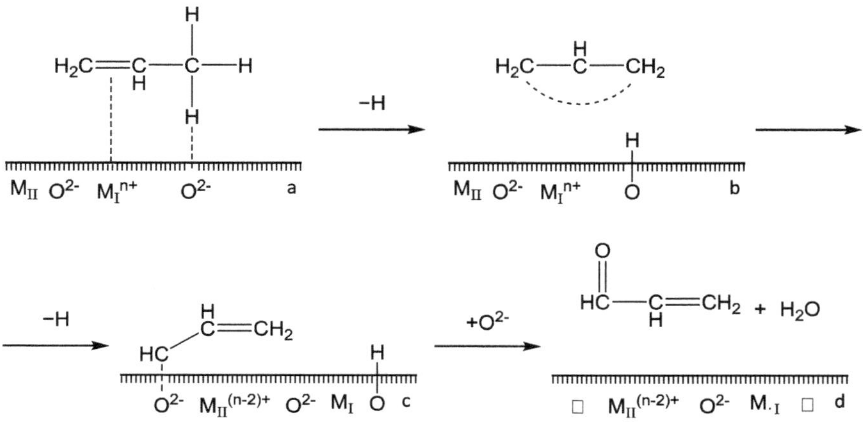

Figure 4.20 Schematic presentation of the mechanism of propene oxidation on mixed metal oxides [15]. (a) Activation of a propene molecule, (b) formation of allylic species, (c) formation of acrolein-like surface intermediate with participation of the catalyst oxygen and (d) desorption of products with formation of an oxygen vacancy.

oxide [132] established that the allylic intermediate of propene reacts with lattice oxygen to give initially a surface-bound alkoxy intermediate. After a second step where α hydrogen is transferred to the metal oxide, acrolein is formed.

In the acrolein molecule there is no memory of which of the end carbon atoms was originally part of the methyl group of propene. Two vacant oxygen atom positions and corresponding reduction of cations are generated by desorption of water via recombination of surface hydroxyls. The lattice oxygen atoms are replenished and cations reoxidized by oxygen atoms from dissociative adsorption of molecular oxygen at vacant oxygen atom sites.

At that time no direct information of the surface chemistry of reaction intermediates was available. Until the 1990s structural characterization and correlation with catalyst performance was the main approach to unravel the reaction mechanism.

Figure 4.21 illustrates the layer-like structure of MoO_3 and V_2O_5. The lateral surfaces are most stable, because coordinative saturation

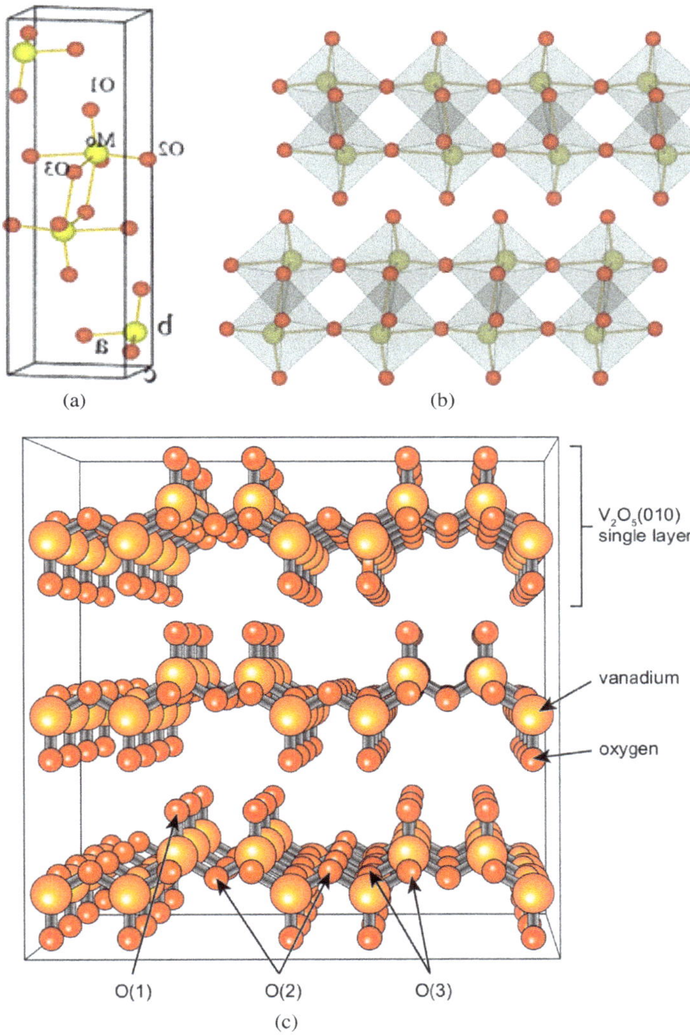

Figure 4.21 The crystal structures of MoO_3 and V_2O_5. (a) Unit cell of the crystal structure of MoO_3 [133]. (b) β corner-connected octahedra [133]. (c) Crystal structure of orthorhombic V_2O_5 with net plane stacking along the (010) direction. Vanadium and inequivalent oxygen centers, singly coordinated O(1), doubly coordinated O(2), and triply coordinated O(3), are marked. All atoms of the $V_2O_5(010)$ single-layer slab at the top are emphasized by darker shaded balls [134].

of the cations is maintained. In these systems Mo is close to octahedrally coordinated with six oxygen atoms. The vanadium cation is five-coordinated.

Two kinds of oxygen atoms have to be distinguished: an oxygen atom that is end-on coordinated with the cation (O(1) in Figure 4.21c) and oxygen atoms that share with two or three cations. The reactivity of the respective oxygen atoms is different. In vanadium oxide the end-on oxygen atom preferentially activates the alkane C–H bond. In the methanol oxidation reaction, the end-on oxygen atoms of Mo activate the C–H and O–H bonds. MoO_3 is not a reactive system for propene or propane.

The basal planes with coordinatively saturated cations are most selective. Surfaces different from the basal planes that expose coordinatively unsaturated cations are non-selective. Upon contact with molecular oxygen, they generate non-selective O^- and O_2^- intermediates. Since particular surfaces are more selective than others, the selectivity of these reducible metal oxide catalysts is particle shape dependent [135].

The structure of α-$Bi_2Mo_3O_{12}$ and several of the multicomponent catalysts derive from the Scheelite structure, with composition ABO_4. It is the structure of $CaWO_4$ (see Figure 4.22).

Different from the octahedral coordination of W in WO_3 or Mo in MoO_3, in this structure the reducible cations have tetrahedral coordination. In $CaWO_4$, the WO_4^{2-} tetrahedra are surrounded by Ca^{2+} cations. The latter cations have a cubic arrangement of eight oxygen atoms that stem from four surrounding WO_4^{2-} tetrahedra. Tetrahedral coordination is common in molybdates and tungstates, while vanadates are five-coordinate.

The α-$Bi_2Mo_3O_{12}$ structure, shown in Figure 4.23, is generated by replacement of WO_4^{2-} by MoO_4^{2-} and of 3 Ca^{2+} cations by 2 Bi^{3+} cations. This leads to A cation vacancies in the ABO_4 structure. These vacancies are ordered, and give rise to tetrahedral pairs of Mo_2O_8 dimers. The tetrahedral coordination of oxygen atoms around Mo becomes approximately fivefold. Coordination of the Bi^{3+} cation is decreased compared to the A site in Scheelite.

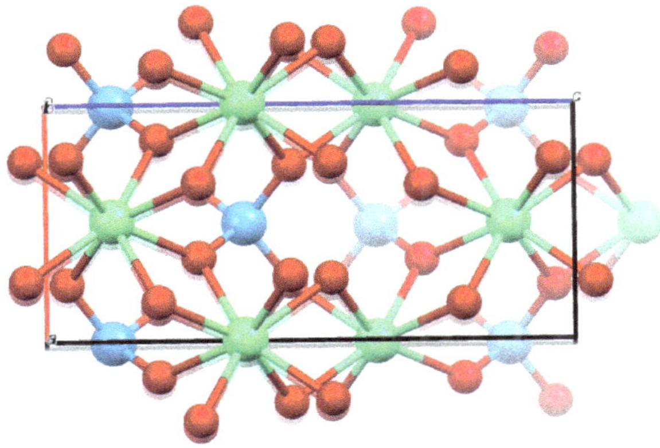

Figure 4.22 The Scheelite structure of $CaWO_4$: Ca, green; W, blue; O, red [136].

The structure of α-$Bi_2Mo_3O_{12}$ is shown in Figure 4.23. The importance of these structures is that the reactive Mo cation remains approximately tetrahedral, and that the oxygen anion bridges Mo and Bi cations.

The reaction mechanism of propene to acrolein and ammoxidation will be discussed next. The first question to address is the difference in reactivity of MoO_3 and $Bi_2Mo_3O_{12}$.

The Mo–O bond in MoO_3 is too strong to activate the allyl C–H of propene. Once the allyl intermediate is generated it will react with the MoO_3 lattice atoms to give acrolein.

Grzybowska *et al.* [138] demonstrated this in 1973 by reacting allyl iodine with MoO_3. The allyl iodine readily decomposes to the allyl intermediate and, at mild temperature, oxidation of the allyl to acrolein is 100% selective. The same experiment by exposure of allyl iodine to Bi_2O_3 only gave hexadiene and benzene. Hence oxygen insertion occurs selectively with the oxygen atom attached to Mo.

The promoting action of Bi on the reactivity of the oxygen atom in the Bi/Mo oxide system has for a long time not been properly understood [139]. Initially activation of the propene allyl C–H bond was ascribed to Bi [140], [141]. Quantum-chemical calculations, that include correlation energy corrections [142], show that actually the reactivity of oxygen atoms that bridge Bi^{3+} and Mo^{6+} changes.

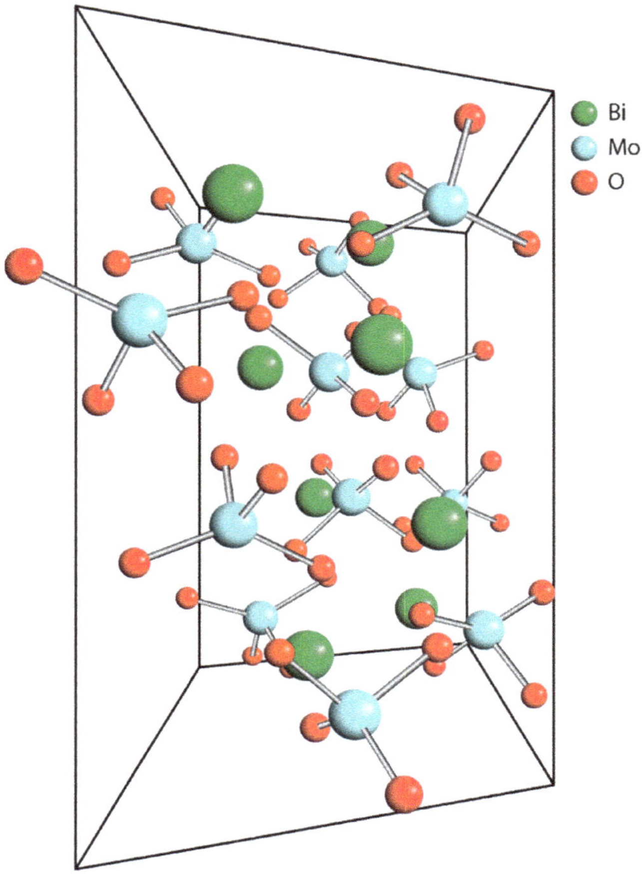

Figure 4.23 α-$Bi_2Mo_3O_{12}$ Scheelite-derived structure [137]. Molybdenum becomes part of $[Mo_2O_8]^{-4}$ dimers.

There is unique synergy of Bi^{3+} and Mo^{6+} chemical bonding features that induces radical-type reactivity to the oxygen atom. Bi^{3+} promotion of the M–O bond also weakens the Mo–O bond strength. The activated oxygen atom cleaves the propene allyl C–H bond. This activation is schematically illustrated in Figure 4.24.

The calculations that support Figure 4.24 are from Licht and Bell of the University of California [143], [144]. They studied the reaction mechanism of acrolein formation and ammoxidation of

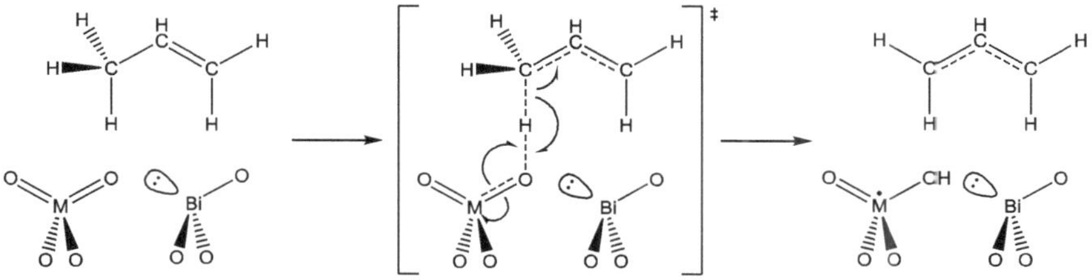

Figure 4.24 The lone pair electron on Bi activates the Mo–O bond of the molybdate tetrahedra that build the $Bi_2Mo_3O_{12}$ structure. It stabilizes electron spin exchange in the M–O bond so that a H radical atom can adsorb on the oxygen atom that bridges the Mo and Bi. The allyl radical that is generated forms an alkoxy intermediate with another oxygen atom attached to Mo [142].

propene on the (010) surface of reactive α-$Bi_2Mo_3O_{12}$ based on its known crystal structure [137], [145], [146] (see also Figure 4.23).

In Figure 4.25 the reaction mechanism for the oxidation of propene to acrolein is shown. When the allyl intermediate is formed only the Mo^{6+} cation becomes reduced and not the Bi^{3+} cation. The initial C–H bond cleavage reaction where the H atom attaches to the O atom that bridges Bi and Mo is the most difficult step.

After water formation and reoxidation of Mo, the allyl radical forms an alkoxy intermediate that attaches to the molybdate oxygen. Acrolein evolves after a second hydrogen atom is transferred to an additional molybdate oxygen atom.

The surface vacancies due to oxygen consumption become rapidly replenished with lattice oxygen atoms. The Licht and Bell proposal of the synergetic action of Bi and Mo resolves the long-standing issue of the molecular mechanism of CH bond activation of propene. Previously it had been thought that Bi–O activates the CH bond and molybdate stabilizes the allyl intermediate [140].

Acrylonitrile instead of acrolein is the product when ammonia is co-reactant. Figure 4.26 illustrates that, according to the Licht and Bell calculations, ammonia reacts with the allyl alkoxy to give the propyl amine. This dehydrogenates in consecutive N–H and C–H bond cleavage steps. The Licht and Bell ammoxidation mechanism is consistent with elementary steps that are only activated by oxygen atoms around the reducible cation. Interaction of ammonia with

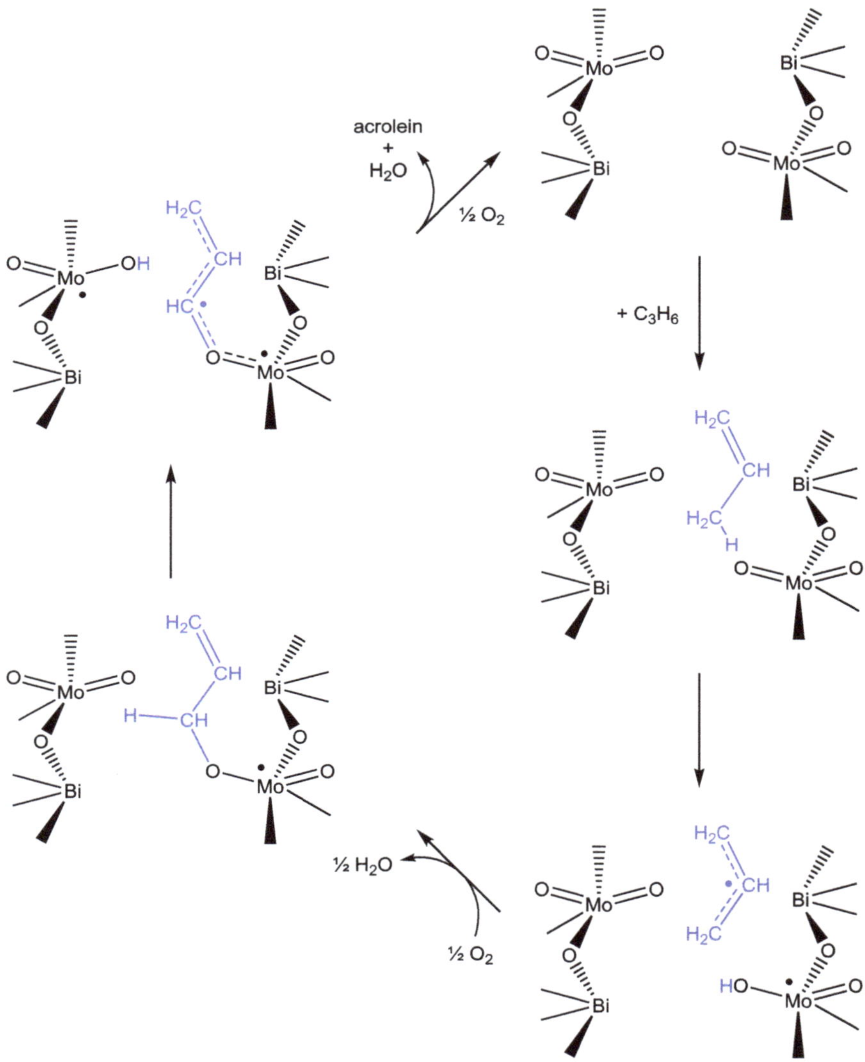

Figure 4.25 The reaction mechanism of selective acrolein formation by α-$Bi_2Mo_3O_{12}$ [143].

Lewis acid Bi^{3+} promotes N–H bond cleavage. It modifies previous suggestions that Mo–NH species are formed that insert into the allyl intermediate [147], [148].

The variation in composition of the multicomponent metal oxides optimizes yield and performance by optimizing the structure

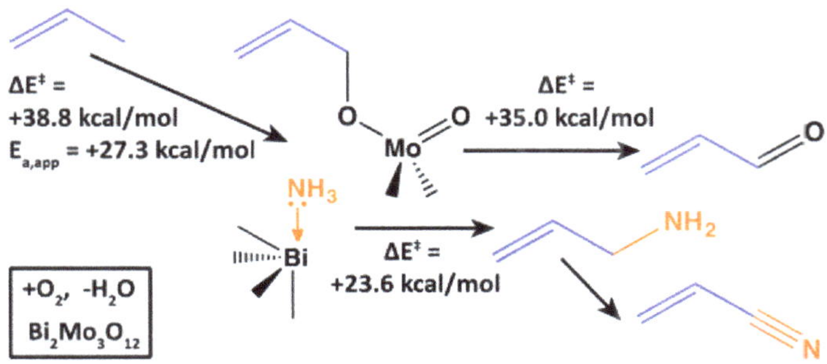

Figure 4.26 The elementary reaction steps and computed activation energies of acrylonitrile formation versus acrolein [143].

of the Bi–O–Mo bonding motif, where the lone pair of Bi promotes M–O reactivity for CH bond cleavage. It also creates the vacancy lattice structure for optimum O atom diffusion. Additional reducible cations such as Co^{2+} or Fe^{3+} are also added in order to create oxygen vacancies in the metal oxide that increase oxygen mobility. It may increase the reaction rate by thirtyfold [137], [146].

The best catalyst for ammoxidation of propane has the composition Mo–V–Nb–Te–O_x. It produces acrylonitrile with 60% yield from propane. An alternative catalyst with composition V–Sb–W–Te–Sn–O_x/Al_2O_3–SiO_2 gives 40% yield [149]. The oxidation catalysts have complex composition since they have to catalyze several elementary reactions under the same conditions. The crystallographic structure of $Mo_{7.5}V_{1.5}NbTeO_{29}$ is known [150]–[152]. In Section 4.4.2.2 the roles of the different cation components in the catalyst have been mentioned.

The specific role of vanadate is to activate the propane molecule by reaction with V=O to propene, which Te=O converts to the corresponding allyl intermediate [153], [154]. The structure has an isolated site with V=O that converts propane to propene, surrounded by a larger spatial region without V=O where propene reacts to acrylonitrile and propane is not activated. This minimizes total oxidation.

An important issue with multicomponent metal oxide catalysts is that the surface and bulk structures are different under reaction conditions. This is well known for the $Fe_2(MoO4)_2$ catalyst of methanol to formaldehyde oxidation that is discussed in the next section. For the Bi/Mo systems there are early papers that also suggest such differences in composition [155]–[157].

This section is concluded with a discussion of the surface composition of the $Mo_{7.5}V_{1.5}NbTeO_{29}$. Also here differences between surface and average bulk compositions are present [158].

Schlögl *et al.* from the Max Planck institute in Berlin studied in 2010 the system in the absence of ammonia. Propane oxidation then gives acrylic acid. They treated phase-pure materials at high temperature and in steam and discovered differences in reactivity notwithstanding that the bulk structure remained the same.

In particular, the Te concentration was higher at the catalyst surface. Selectivity of acrylic acid related positively to the Te/V ratio and anticorrelated to Mo concentration.

The bulk phase is a support for its surface and a packaging structure of Te_xO_y species as critical components that are liberated from its "container" only and possibly reversible under reaction conditions. Embedded in this Te_xO_y surface species are vanadium species. This is supported by the proposed synergy of Te^{4+} and V^{5+} that promotes bonding of hydrogen to V=O [131].

A selection criterion for the transition metal oxide catalysts is that they should not yield oxides that are too reactive and give total combustion or are non-reactive because they are too stable. This is the reason of the preference for oxides of Group 4–6 elements of the periodic table for selective oxidation.

Critical to the selective oxidation of propene and propane is the initial reaction step, which in the case of propene is activation of the allyl C–H bond. Vanadium oxide is too reactive and will give total oxidation. Molybdenum oxide is not reactive enough. The role of bismuth in molybdenum oxide is to activate the system for C–H bond cleavage and at the same time not cause total combustion. Selective oxidation of propane requires activation of C–H by more reactive vanadium oxide. Additional components have to be added to suppress total oxidation (site isolation effect) and selective conversion of the alkene intermediate generated from propane.

4.4.3.2 *Mechanism of Selective Methanol Oxidation*

Selective oxidation of methanol is optimum for an oxygen vacant basal plane of MoO_3. The adsorption energy of formaldehyde has to be low compared to the C–H bond activation energy of adsorbed methoxy.

The methanol oxidation catalyst to formaldehyde that is an alternative to Ag is $Fe_2(MO_3)_2$. For this catalyst the reaction mechanism is well understood since the 1980s [159]. It is closely related to the Wachs-Madix mechanism for the Ag-catalyzed reaction (see Section 4.3.2).

The structure of $Fe_2(MO_3)_2$ consists of molybdate MO_4^{2-} tetrahedra and Fe^{3+} cations octahedrally coordinated by its oxygen atoms. In methanol oxidation the Mo cations are reduced and reoxidized. The reaction cycle is schematically illustrated in Figure 4.27.

Reaction is initiated by heterolytic dissociation of the methanol O–H group. A methoxy species and a surface hydroxyl are formed. The methoxy species adsorbs on Mo and forms a methoxy molybdenum(VI) center. The rate-determining step is C–H activation of the methoxy to give formaldehyde. Lattice oxygen atoms are regenerated by oxygen dissociation. This does not necessarily occur at the same site.

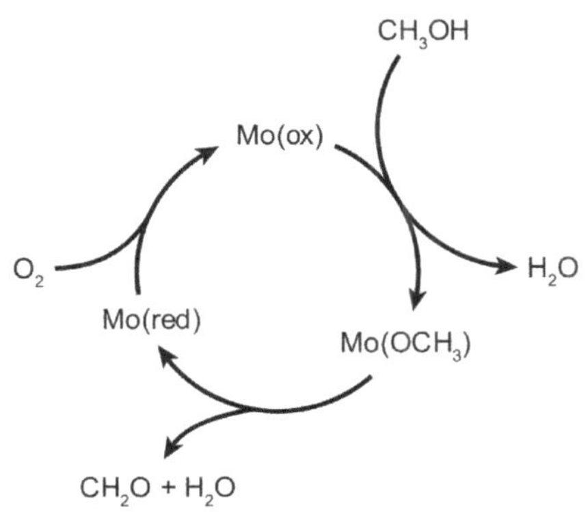

Figure 4.27 Catalytic reaction cycle of methanol oxidation to formaldehyde [160].

Non-selective formaldehyde oxidation proceeds via intermediate formate that decomposes to CO and H_2O [161], [162].

Here the surface chemistry of the methanol oxidation catalyst and its relation with the mechanism of the reaction is discussed.

The inorganic chemistry of the working catalyst is complex. Similar to the multicomponent catalytic systems of the previous section, the surface composition of the $Fe_2(MO_3)_2$ catalyst is different from the bulk.

With advanced surface-sensitive instrumentation such as low-energy ion surface scattering (LEISS) and high-resolution electron microscopy (HREM) it was demonstrated [161], [163], [164] that surface structure and composition of $Fe_2(MoO_3)_2$ change when the catalyst is exposed to reacting methanol and oxygen.

A thin monolayer of MoO_3 covers the Fe-molybdate system. The reactivity normalized to Mo site of this MoO_3 surface layer is substantially higher than that of $Fe_2(MoO_3)_2$.

The molecular basis for the interpretation of the reactivity of supported reducible overlayers owes significantly to the detailed investigations of supported MoO_3 [165] and V_2O_5 overlayers [166] in the Wachs laboratory in Bethlehem, USA in the 1990s.

The structures of supported Mo oxide particles vary from tetrahedral coordination for isolated metal cation species to octahedral coordination for oligomeric multi-Mo-oxo species.

Mo oxide [167] and V oxide particles [168], [169] have very similar chemical behavior. Kinetics conforms to the global methanol oxidation reaction mechanism of Figure 4.27.

Mechanistic and kinetic observations are:

- Equilibrium is established between gas-phase methanol and adsorbed methoxy. This is consistent with C–H activation of CH_3OH as rate-controlling step in the overall reaction to formaldehyde. The elementary rate constant of this bond cleavage reaction is independent of cluster oxide aggregate size.
- The elementary reaction that gives methoxy formation is reaction of the methanol proton with an oxygen atom that bridges a reducible Mo or V cation and a support cation (e.g., Al^{3+}, Ti^{4+}). The energy of this reaction determines the equilibrium

concentration of the methoxy species with respect to gas-phase methanol.

– This causes the equilibrium to be a sensitive function of the support. The variation of reaction rate normalized on reducible cation concentration is due to the altered equilibrium constant of the deprotonation reaction for different supports. Whereas for the vanadium oxide catalysts this gives a large difference in catalytic reactivity, this variation is much less for the molybdate catalysts. The probable reason is the lower Lewis basicity of the M–O–X bond for Mo^{6+} compared to V^{5+}.
– Molybdate is preferred over vanadate as methanol oxidation catalyst, because its lower reactivity slows down consecutive formaldehyde oxidation.

Whereas the Wachs studies conclude that the reactivity per metal cation of Mo oxide or V oxide is independent of cluster size, molecular studies demonstrate that this is only correct beyond a particular cluster size.

Gas-phase mass spectrometric investigations with molybdate clusters demonstrate that methanol oxidation requires at least a dimer Mo cluster. Reaction is not catalyzed by a single Mo reaction center [170]. Bowker *et al.* from Cardiff University demonstrated for molybdate supported with different concentration on Fe_2O_3 [171] that monomeric molybdate clusters are substantially less active and selective than Mo oxide oligomers.

In Chapter 6 we will return to the reactivity of single-site Mo complexes immobilized on high surface area supports. They are uniquely active in non-oxidation reactions such as alkene disproportionation (see Figure 6.4).

Comparison with other studies of supported reducible metal oxide clusters shows a relation with the M–O bond strength. The reactivity of W oxide clusters is substantially less. This is consistent with the stronger bond energy of the W–O chemical bond ($\Delta E(MoO_3–MoO_2)$ = 50 kJ/mol; $\Delta E(WO_3–WO_2)$ = 65 kJ/mol).

For the same reason the catalytic selectivity of ferric tungstate $Fe_2(WO_4)_3$ and ferric molybdate $Fe_2(MoO_4)_3$ for the oxidation of methanol are very different [172]. Although methanol conversion

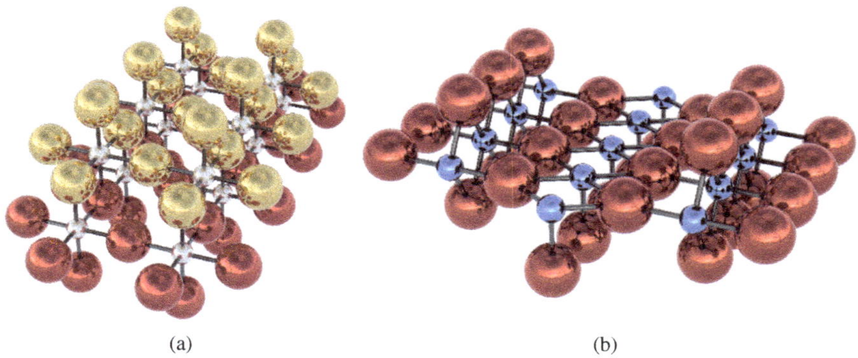

Figure 4.28 Structures of MoO_3 and MoO_2 surfaces. (a) MoO_3(010) surface: silver Mo^{6+}, yellow O^{2-} at (101) surface layer, red O^{2-} in second layer. (b) MoO_2(110) surface: blue Mo^{4+}, red O^{2-}.

rates are similar, the tungsten system produces mainly dimethyl ether, which is a solid acid-catalyzed reaction, whereas the molybdenum system gives selective formaldehyde production.

Model catalyst studies show that the reactivity of MoO_3 dispersed in a Fe_2O_3 catalyst is comparable to that of the MoO_3 layer on Fe-molybdate. The higher reactivity of the amorphous molybdate layer relates to a higher density of coordinatively unsaturated Mo cation sites [161], [173].

Very different from supported molybdate aggregates, methanol will not adsorb on the basal (010) surface of MoO_3 that only contains coordinatively saturated Mo cations (see Figure 4.28a). Heterolytic methanol O–H bond cleavage only occurs by an activated process that creates methoxy species. This low reactivity surface is most selective for formaldehyde formation.

Coordinatively unsaturated Mo cations are part of surfaces in crystallographic directions perpendicular to the basal layer. To create such a surface, cation-oxygen bonds have to be cleaved. Methanol readily adsorbs on non-basal plane surfaces [174]. However, selectivity of formaldehyde production is low and there is mainly total combustion of methanol.

When reaction is studied at low temperature and low conversion, the non-basal plane surfaces catalyze consecutive reactions.

Methanol reacts with formaldehyde to give methylal ($CH_2(CH_3O)_2$) and H_2O. Also dimethyl ether forms as co-product [175]. This indicates Lewis acid properties of the Mo cations.

The high reactivity and selectivity of the monolayer of MoO_3 dispersed on Fe-molybdate relates to the rapid diffusion of oxygen atoms through the Fe-molybdate lattice. The rate of diffusion of oxygen through the bulk MoO_3 lattice is substantially lower. This agrees with the Mars-van Krevelen oxidation kinetics (4.4.2.1) that requires lattice oxygen restoration to be fast compared to the elementary reaction rate of oxidation.

In the final part of this section the molecular mechanism of the methanol oxidation reaction is discussed in relation to the surface structure of the Mo oxide catalyst. This will be based on results of quantum-chemical calculations. One question that is addressed is the non-reactivity of monomeric molybdate with respect to the methanol oxidation reaction. The reason of the high methanol oxidation activity and selectivity of the oxygen vacant basal plane surface of MoO_3 versus low activity of the non-oxygen vacant surface and low selectivity of the surface with coordinatively unsaturated cations is the second question that will be addressed.

The quantum-chemical studies compare methanol reactivity of $MoO_3(010)$ and $MoO_2(110)$ surfaces. The two main differences between the two surfaces are the different Mo charge (6+ versus 4+) and the presence of coordinatively unsaturated Mo cations on the MoO_2 surface.

The respective structures are shown in Figure 4.28.

The $MoO_3(010)$ surface is the basal plane of the layered MoO_3 oxide. Its bulk structure is given in Figures 4.21a,b, while the (010) surface of MoO_3 is shown in Figure 4.28a. Also, at the surface the Mo^{6+} cations are coordinatively saturated by six oxygen anions. There are three kinds of oxygen atoms: oxygen atoms within the surface that connect two or three Mo-containing octahedra. In the latter case one of the Mo octahedra is below the surface plane. There is also a single coordinated oxygen atom, directed perpendicular to the surface. The excess charge on this oxygen atom is compensated by the shortage in charge of threefold coordinated surface oxygen. It explains the two-layer structure of bulk MoO_3.The end-on surface oxygen atom is most reactive.

The MoO_2 crystal has the rutile structure. Its most stable surface is the (110) surface, which is shown in Figure 4.28b. This surface has an equal number of coordinatively saturated and unsaturated Mo^{4+} cations. The latter are coordinated to four coordinatively saturated oxygen atoms in the plane and one oxygen atom below the plane. The in-plane surface oxygen atoms are three-coordinated with surface Mo^{4+} cations. Reactive oxygen anions are located perpendicular to the surface plane. They are part of the oxygen atom octahedra of coordinatively saturated Mo^{4+} cations and are attached to two Mo^{4+} cations.

Experiments suggest [170] that a cluster that contains at least two Mo cations is needed for selective oxidation of methanol. Twenty years before these experiments, in 1985 Goddard and Allison from the California Institute of Technology did quantum-chemical calculations [176] on Mo oxide cluster models and also made the suggestion that dual dioxo sites are needed for selective methanol oxidation.

Their proposal is shown in Figure 4.29.

Later DFT calculations, that are discussed below, which include the full geometry of the surface, confirm the need for the presence of two reactive oxygen atoms coordinated to a different Mo atom [177].

The dual dioxo site of Figure 4.29 consists of two molybdate tetrahedra. The discovery of Goddard and Allison is that the non-reacted Mo=O transforms to a stronger M≡O triple bond when in the first reaction step the methanol O–H bond cleaves and hydroxyl and methoxy are attached to molybdate. The M≡O triple bond is too strong to activate C–H cleavage of adsorbed methoxy. A second molybdate unit has to assist the reaction. Then C–H bond cleavage happens by a Mo=O of a second molybdate tetrahedron with formation of formaldehyde. The Mo–OH groups recombine to give water. The site is regenerated by lattice oxygen atom suppletion after dissociative O_2 adsorption.

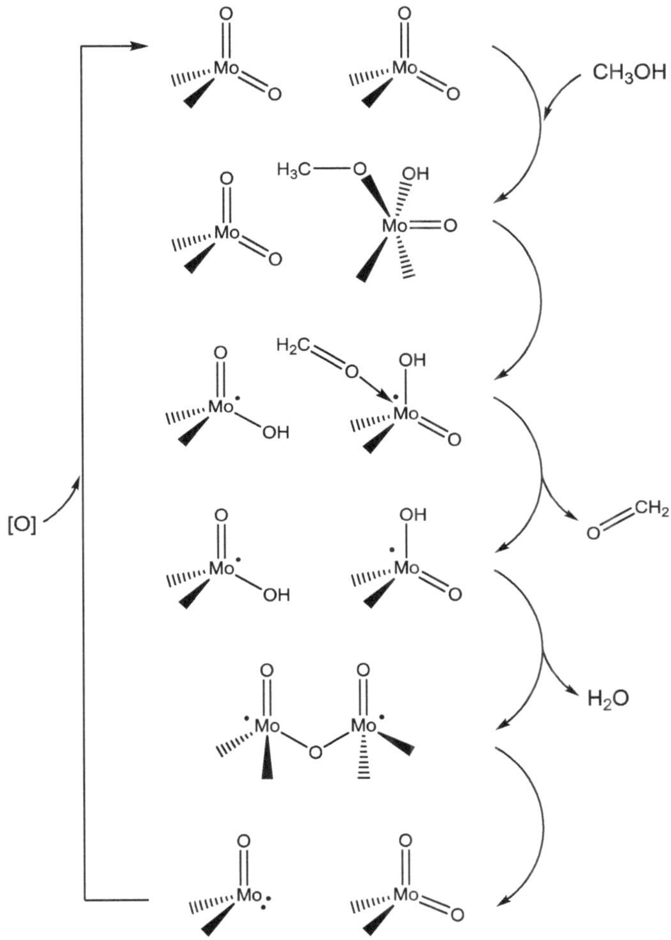

Figure 4.29 The mechanism for oxidative dehydrogenation on the dual dioxo site of MoO_3(010) [176].

DFT calculations by Rellán-Piñeiro and López [177] provide the explanation of why the basal MoO_3(010) plane with oxygen atom vacancies is the most selective for methanol oxidation. Calculated potential energies of methanol oxidation and formaldehyde decomposition are compared in Figure 4.30.

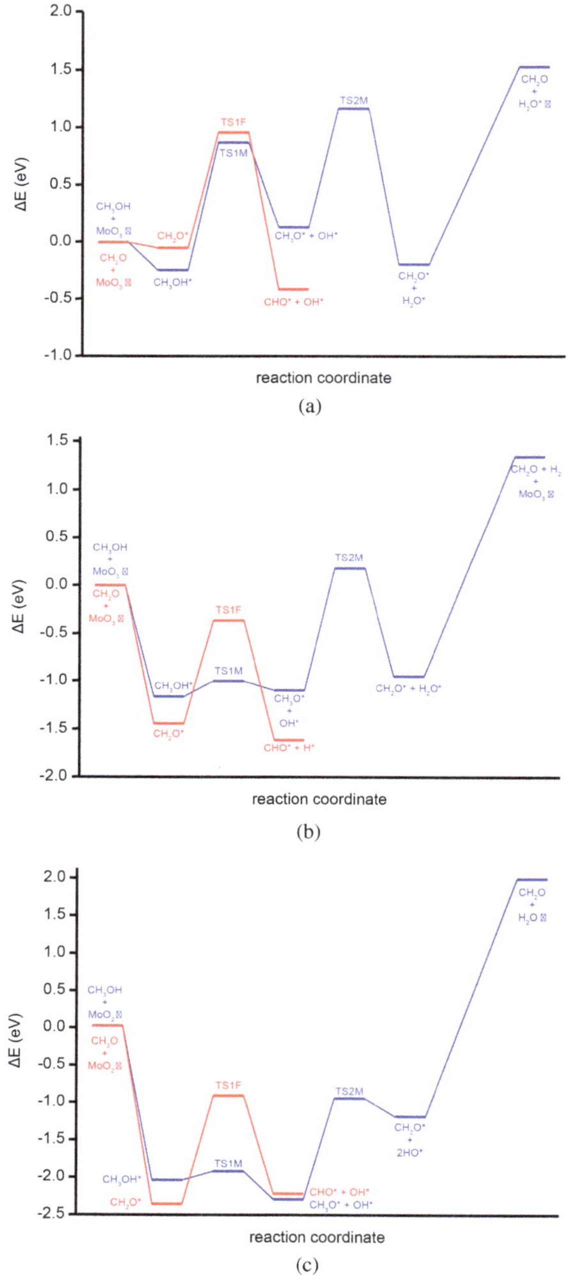

Figure 4.30 Potential energies of the methanol to formaldehyde oxidation reaction [177]. (a) Reaction on $MoO_3(010)$ surface; (b) reaction on vacant $MoO_3(010)$ surface; (c) reaction on $MoO_2(110)$ surface.

Selective oxidation of methanol to formaldehyde depends on the relative rates of formaldehyde formation versus consecutive combustion.

The reactivity of non-vacant $MoO_3(010)$ is low because of the coordinative saturation of the surface Mo cations. It is the reason for the low energy of methanol adsorption and the high barrier of methanol hydroxyl bond cleavage (see Figure 4.30a). The adsorption energy of formaldehyde, also shown in Figure 4.30a, is low compared to the activation energy of methanol activation. Under reaction conditions the formaldehyde coverage is low compared to that of methanol. Therefore, one expects that selectivity of methanol oxidation to formaldehyde will be high. But due to the high barrier of O–H bond cleavage on the $MoO_3(010)$ surface, the reaction rate of formaldehyde formation is predicted to be low.

In conflict with experiment this surface predicts O–H instead of C–H bond cleavage to be reaction rate controlling.

When formaldehyde is formed, as predicted by Goddard (his model with tetrahedra was based on the structure of $Fe_2(MoO_3)_2$) the two hydrogen atoms attach to the two coordinatively unsaturated end-on oxygen atoms of separate Mo octahedra.

On oxygen vacant surfaces more ready access to Mo is possible and selective methanol oxidation with high rate becomes possible. However, when the surface becomes too reactive as for the $MoO_2(110)$ catalyst surface, it is non-selective.

On the oxygen vacant $MoO_2(110)$ surface the energetics of methanol activation changes dramatically (see Figure 4.30b). This is because coordinatively unsaturated Mo sites are present to which methanol can bind.

The structure of the $MoO_2(110)$ surface is shown in Figure 4.28b. This surface is half-occupied by coordinatively saturated Mo cations and half-occupied by Mo cations with a vacant oxygen site. The valency of the cations is reduced to Mo^{4+}.

The oxygen vacant sites act as Lewis acidic sites that interact strongly with methanol and formaldehyde. The O–H bond of methanol is readily broken by heterolytic bond cleavage. C–H bond cleavage of methanol becomes the reaction rate-controlling step of methanol oxidation.

However, the adsorption energy of formaldehyde is larger than the activation energy of methanol conversion. As a consequence, formaldehyde will dominate surface coverage and cause the overall consecutive combustion reaction of formaldehyde to be fast compared to the oxidation rate of methanol to formaldehyde.

The oxygen vacant $MoO_3(010)$ surface provides a compromise between the two competing reactions of methanol and formaldehyde activation (see Figure 4.30c).

Due to the higher Mo^{6+} valence, the surface oxygen atoms are more Lewis basic than the surface oxygen atoms of the $MoO_2(110)$ surface that contains Mo^{4+} cations. Methanol and formaldehyde adsorb weaker on the surface vacant sites of $MoO_3(010)$ than the vacant sites of the $MoO_2(110)$ surface.

The adsorption energy of formaldehyde becomes less than the activation energy of the adsorbed methanol to formaldehyde reaction. The latter is dominated by the C–H bond cleavage activation energy. The low surface coverage of formaldehyde and the relatively high coverage with methanol is the reason that the overall rate of methanol conversion to formaldehyde is fast compared to the rate of formaldehyde combustion.

These model simulations explain the role of the oxygen vacant MoO_3 overlayer of the $Fe_2(MoO_3)_2$ methanol oxidation catalyst. The role of $Fe_2(MoO_3)_2$ as a support is that it stabilizes this layer and facilitates oxygen atom transport through its bulk to resupply oxygen atoms to the MoO_3 overlayer when it reduces in oxidative dehydrogenation of methanol.

4.4.4 *Vanadium Oxide and Related Catalyst Systems*

Vanadium oxide catalysts can be multicomponent bulk systems such as the $(VO)_2P_2O_7$ catalyst but are also used in the form of active oxide monolayers dispersed on a support. The reactivity of the vanadium oxide monolayer catalyst is smaller than that of bulk vanadium oxide. As with the Mo oxide monolayer catalysts of the previous sections, the oxygen atoms that bridge vanadium cations and support cations are most reactive.

For dispersed catalysts there is an important selectivity difference between single-center and multi-center catalyst systems. Different from the non-selective Mo single-site of the methanol oxidation reaction of Section 4.4.4.1, in oxydehydrogenation reactions of short alkanes single-site vanadium oxide catalysts are more selective.

The section is concluded with the mechanism of the NO_x reduction reaction by ammonia.

The reaction mechanisms of vanadium catalysts and Cu metal zeolite catalysts are closely related. Therefore, both catalyst systems are discussed in Section 4.4.4.3. In both catalytic systems protonic sites located at oxygen sites that bridge vanadium and support or are attached to the zeolite framework assist the metal oxide- or cation-catalyzed reaction.

4.4.4.1 *Oxidation of Butane to Maleic Anhydride*

Reaction of butane is initiated by a bond cleavage reaction between alkane methene C–H by vanadyl V=O or phosphate P=O. Vanadium oxide site isolation contributes to the selectivity of maleic anhydride production.

Selective oxidation of butane is possible since butane is less stable than the product maleic anhydride. According to Hodnett [178] a high selectivity of the oxidation reaction requires the weakest bond in the product to be at least 30–40 kJ/mol stronger than that of the reactant. For the butane to maleic anhydride system the difference is of the order of 60 kJ/mol.

The oxidation of butane to maleic anhydride is a complex reaction. It is catalyzed by a catalyst of approximate composition $(VO)_2P_2O_7$. Butane undergoes six C–H bond cleavage reactions and three oxygen insertion reactions. In a 14-electron transfer process and through a multitude of intermediate reaction steps butane, C_4H_{10}, is converted into maleic anhydride $C_4H_2O_3$. The two most important mechanistic proposals are presented.

At the reaction condition of 660–720 K and exposed to reactants, the $(VO)_2P_2O_7$ catalytic system consists of several oxide and phosphate phases. The difficulty in determining the surface state of the catalyst under reaction conditions is one of the reasons that there is an ongoing debate on the reaction mechanism. Also, the relation with the structure and composition of surface catalyst reaction sites cannot be considered conclusively settled [37].

The high reactivity of vanadium oxide is a necessary condition for the cleavage of the non-reactive C–H bonds of butane. However, the reactivity of vanadium oxide is too large, and the reaction is non-selective due to total combustion. Phosphorus is added to reduce the vanadium oxide reactivity. Phosphorus promotion as in the $(VO)_2P_2O_7$ catalyst also converts the V_2O_5 system from an oxydehydrogenation catalyst to a selective oxygenation catalyst.

The valence state of vanadium in $(VO)_2P_2O_7$ is +4. However, the active phase must possess an optimum V^{5+}/V^{4+} ratio [179], [180]. The presence of $VOPO_4$, with vanadium in valence state +5, is suggested [181], [182].

Phosphorus is thought to suppress total oxidation by site isolation of the vanadium cation. The reduced size of the vanadium oxide ensemble also reduces its reactivity. Goddard [183]–[185] proposed that phosphorus also acts as promoter of the reactivity of the reduced vanadium oxide ensemble.

In the next section the mechanism of the closely related oxydehydrogenation reaction of propane to propene is presented. Also, this reaction is sensitive to the cluster size of the vanadium oxide phase. In this case interaction of vanadium oxide monolayer oxygen atoms with support tunes its reactivity.

Kinetics is different from the reducible metal oxide systems of the previous section. The reaction order dependence on oxygen is non-zero. This implies that reoxidation of catalyst is not fast anymore compared to the oxidation of butane [188].

Oxygen isotope exchange experiments indicate that only the surface layer participates in the reoxidation reactions.

Not disputed is that the reaction rate-controlling step of butane activation is irreversible methene CH_2 bond rupture. This has been demonstrated by experiments using deuterium-labelled butane [186].

A primary question is whether the oxidation of butane is due to activation by surface lattice oxygen atoms or whether molecular oxygen is also active.

The latter cannot be excluded since activated O_2 is a known intermediate in the oxidation of benzene to maleic anhydride (see Section 4.2). Gleaves *et al.* suggest [187] that selective oxidation to maleic acid involves activation by atomic as well as molecular oxygen.

The complexity of the system follows from the detailed model of the relation between catalytic cycle and changes of the inorganic system induced by the reaction as deduced by the Schlögl group in Berlin [188]. Sophisticated *in situ* spectroscopic measurements have led to the coupled reaction phase model of Figure 4.31.

Different inorganic compositions and phases of the catalyst were identified as well as the dynamic changes that happen during reaction.

The chemistry is extremely complex, since many different crystal phases of inorganic compounds co-exist. Reaction conditions

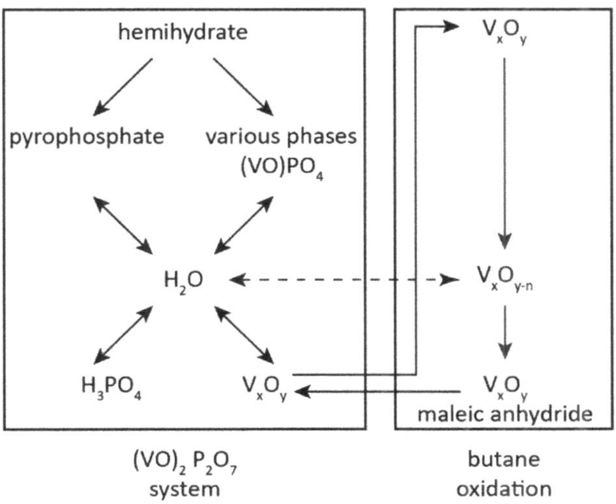

Figure 4.31 Schematic representation of the kinetic processes that determine the formation of the active catalyst phase from vanadium pyrophosphate [188].

determine the inorganic chemistry of the system and distribution of different phases.

$(VO)_2P_2O_7$ is metastable. The surface composition has less phosphorus than $(VO)_2P_2O_7$ and the average redox state of vanadium is V^{4+}. Disordered structures of the surface overlayer are proposed that form the catalytically active phase. This is built on the reduced V^{4+} phase [189] that only contributes to overall reactivity [181], [182]. Trifiro *et al.* from the University of Bologna suggested that selective V^{5+} has minority presence [188].

Water from the catalytic cycle drives the system in a surface phase-segregated state of a binary vanadium oxide that cannot crystallize into large crystals. Phosphate also plays a role in the inorganic chemistry of the catalyst system. Phosphate acts as a crystal growth inhibitor that maintains the surface layer in its two-dimensional amorphous oxidic state.

Excess oxygen oxidizes the V_xO_y compound into V_2O_5 that reacts with phosphoric acid to form $VOPO_4$.

In summary, a highly dynamic surface forms on the crystalline phases present in the butane to maleic anhydride catalyst under reaction conditions. This dynamic surface is thought to be composed of various vanadium phosphate species which can readily respond to changing operating conditions [190].

There are two major mechanistic proposals of selective maleic anhydride formation. According to the alkene intermediate pathway hypothesis of Centi and Trifiro [191], [192], [202] the reaction is a sequence of consecutive steps of gas-phase intermediates butene, butadiene and furane that adsorb, react and desorb. This is schematically shown in Figure 4.32a.

The alternative proposal by Vedrine *et al.* from Lyon [193], [194] is that once the butane methene C–H bond has been cleaved, the butyl radical intermediate adsorbs on a surface oxygen atom as alkoxy intermediate. This converts by successive C–H bond cleavage and oxygen insertion steps. The proposed reaction sequence is schematically illustrated in Figure 4.32b.

Elementary reaction steps relate to the hydrocarbon oxidation steps of Sections 4.4.3. C–H bond dissociation that is activated by lattice oxygen gives the alkyl radical and dehydrogenation

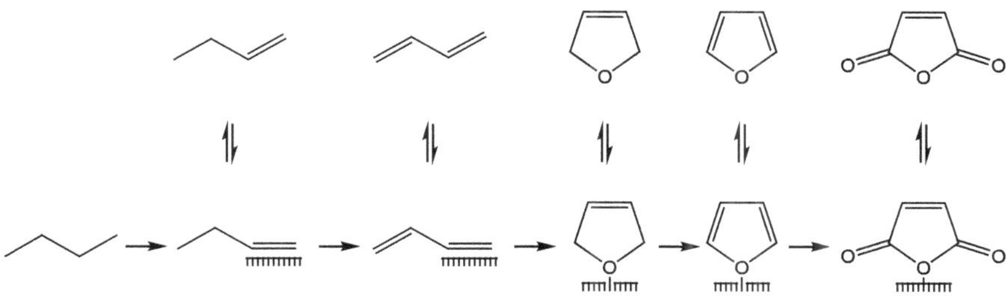
(a)

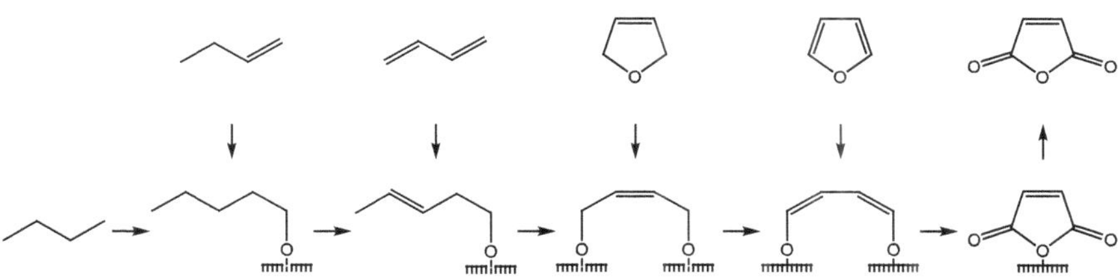
(b)

Figure 4.32 The butane to maleic anhydride reaction [193]. (a) The alkene intermediate pathway; (b) the alkoxide intermediate pathway.

intermediates. The alkyl radical adsorbs to surface oxygen that gives surface alkoxy intermediate.

Both mechanisms are supported by model kinetic experiments that only indirectly probe the surface intermediates. Known reactions that are part of the alkene pathway are the ring closure reaction of butadiene to furan that is catalyzed by Mo oxide [195] as well as the $(VO)_2P_2O_7$ catalyst [196], [197]. Maleic acid can be viewed as oxygenated furan. The experiments by Centi and Trifiro that produce significant amounts of gas-phase alkenes are executed in excess butane, but the surface is in a reduced state.

The major argument in favor of the alkoxy mechanism is that in model experiments the alkyl intermediate that forms after C–H bond cleavage irreversibly adsorbs. Also under butane reaction conditions, only butane gives maleic anhydride as product whereas oxidation of butene gives acetaldehyde, crotonaldehyde, and other partial oxidation products [198]. Only acrylic acid and acetic acid are minority species in the butane to maleic anhydride reaction.

Model experiments by Chen and Munson from Minnesota University [199], also with partially reduced catalysts and supported by detailed isotope-labelling experiments, propose the butyl intermediate. Their main argument is that in their experiments maleic anhydride formation competes with decomposition of the surface intermediate to ethene and 2 CO .This reaction pathway has some similarity with the concerted mechanism proposed by Ziolkowski *et al.* [200], where a butenyl intermediate binds with two alkoxy bonds to the catalyst surface.

Molecular models of the reactivity of vanadium phosphate surfaces are scarce. Here the results of quantum-chemical calculations of Goddard are presented that model the promoting role of phosphorus in $VOPO_4$ that contains five-valent V cations. The high concentration of phosphorus limits the vanadate connections to vanadium cation dimers. The oxygen atoms that coordinate vanadium are connected with P^{5+} cations that have tetrahedral coordination (see Figure 4.33). This topology agrees with the site isolation concept of Graselli (see Section 4.4.2.2).

Reduction of V_2O_5 particle size decreases its reactivity (see also next Section 4.4.4.2). It decreases the relative rate of non-selective oxidation reactions, but in the case of butane it suppresses its activation. According to Goddard *et al.* [183]–[185] phosphorus restores the reactivity for selective butane activation, but catalyst reactivity is

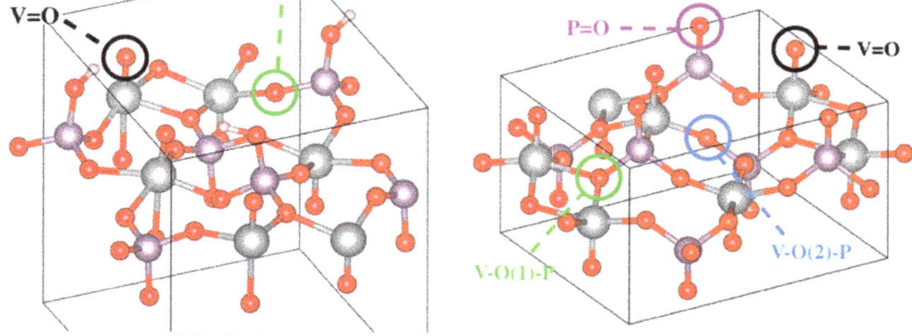

$(VO)_2P_2O_7$ X1-$VOPO_4$

Figure 4.33 Active surfaces of VPO with labeled surface motifs [184]. Oxygen atoms are red, vanadium atoms silver, and phosphorus atoms purple. For clarity the second layer of $(VO)_2P_2O_7$ is excluded.

low enough that it suppresses total oxidation. The Goddard model also explains the need of V^{5+} for this reaction.

Structures of $(VO)_2P_2O_7$ and the reactive X1-$VOPO_4$ phase are shown in Figure 4.33.

The $(VO)_2P_2O_7$ structure (that contains vanadium as V^{4+}) and the reactive X1-$VOPO_4$ are shown. The reduced and oxidized phases contain on their surface eight-membered rings of alternating vanadium and phosphorus atoms linked by oxygen atoms. The reactive V=O, V–O–P and P=O units are indicated. The X1-$VOPO_4$ phase is the only $VOPO_4$ phase that contains reactive P=O as well as V=O groups.

With quantum-chemical calculations corrected with electron correlation, Goddard *et al.* have been able to elucidate the reason for the low reactivity of $(VO)_2PO_7$ compared to $VOPO_4$. This is based on their discovery of the reduction-coupled oxo activation (ROA) mechanism of dual sites that contain a non-reducible and reducible cation. It is illustrated in Figure 4.34.

According to the calculations, reaction of C–H with bridging oxygen atoms that connect the cations is very endothermic. Reaction with V=O has also a substantial activation energy. C–H bond activation with P=O has the lower activation energy. In this reaction, when the H atom of C–H attaches to the P=O group and P–OH forms, the electron of H transfers to the vanadium cation that becomes reduced, but the V=O bond remains intact. The ROA mechanism counterintuitively proposes that promotion of oxidation is not due to activation of surface vanadyl V=O but by reaction of P=O that is assisted by reduction of V^{5+} to V^{4+}.

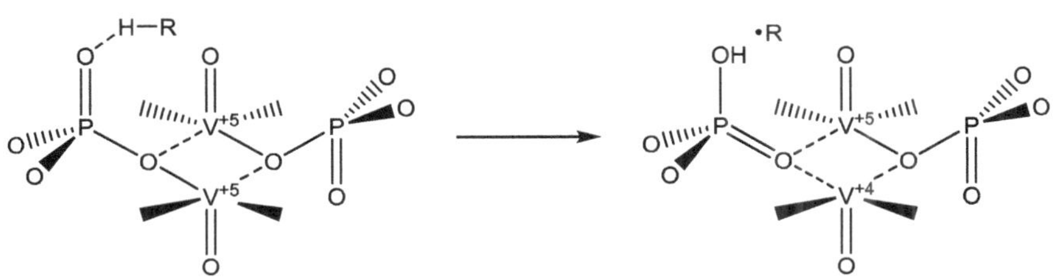

Figure 4.34 Valence bond description of the hydrogen abstraction process by P=O. This illustrates the ROA reaction mechanism [183].

It explains also the difference in reactivity of $(VO)_2P_2O_7$ and $VOPO_4$. The redox of the V^{4+}/V^{3+} couple in $(VO)_2P_2O_7$ is less favorable than that of the V^{5+}/V^{4+} couple of the $VOPO_4$ phase. Calculations show that the surface of $(VO)_2P_2O_7$ will not activate butane. Only the surface of $VOPO_4$ contains oxygen atoms reactive enough to cleave the C–H bond. The activation barrier for C–H bond cleavage is approximately 60 kJ/mol.

In favor of the alkoxy mechanism, bonding of reaction intermediates to the $VOPO_4$ phase is strong and reaction intermediates will not desorb, in contrast to the alkene intermediate pathway.

4.4.4.2 *Oxidative Dehydrogenation of Alkanes*

Oxidative dehydrogenation catalyzed by V_2O_5 is a radical reaction. Selective sites are preferentially single vanadium oxide complexes immobilized on non-reducible support. Activity correlates with bond energy of the V_xO_X oxygen bond.

The oxydehydrogenation (ODH) reaction that converts propane to propene by oxygen is in principle an attractive alternative to non-oxidative endothermic dehydrogenation because it is exothermic. In the non-oxidative reaction, deactivation of catalyst by coke deposition limits catalyst life (see Chapter 3).

However, in the ODH reaction non-selective oxidation reactions have to be overcome. Notwithstanding extensive research, no industrial process exists yet [35], [39], [201]. For butane the temperature of the ODH reaction is 100°C higher than that of the oxygenation reaction of butane to maleic acid. The temperature of propane dehydrogenation is higher than 900 K.

Here the mechanism of the ODH reaction of propane is presented for supported and unsupported vanadium oxide. Supported vanadium oxide catalysts are more selective. The main reactive descriptors that determine structure-function relationships are the coordination and bond energy of surface oxygen atoms.

Below a description of the reaction network and elementary steps is presented. The chemistry of C–H bond activation and oxygen atom insertion is discussed based on quantum-chemical calculations of propane activation by the basal plane of V_2O_5.

This provides background to the discussion of propane oxidation on supported catalysts that follows.

As for the other oxidation reactions, the global reaction network of the ODH of propane consists of different initial parallel steps and a consecutive non-selective reaction.

As illustrated in Figure 4.35, non-selective total oxidation can be due to consecutive reaction of propene (r_3) that is a function of conversion, or due to reaction (r_2) parallel to the propene-forming reaction (r_1).

In the case of propane ODH, the apparent activation energy of the reaction for consecutive propene combustion (r_3) is lower than that for alkene formation (r_1). The high rate of the consecutive propene oxidation (r_3) compared to that of propene formation (r_1) is due to the low activation energy of allyl formation from propene. The C–H bond of the methyl group of propene is weaker than that in propane or adsorbed intermediate isopropyl [203]. As discussed below, the parallel non-selective oxidation (r_2) originates from difference in the activation energies of decomposition of intermediate isopropoxide [204], [205].

Reaction is initiated by dissociation of C–H bonds (methene CH_2 in propane). This elementary step is reaction rate controlling. As expected for a reaction that is rate limited in C–H activation, the rate is first order in propane and zero order in oxygen. For bulk vanadium oxide, reoxidation of reduced catalyst is fast and follows the Mars-van Krevelen redox mechanism. The alkyl species generated by C–H bond cleavage reactions desorbs as olefin and the

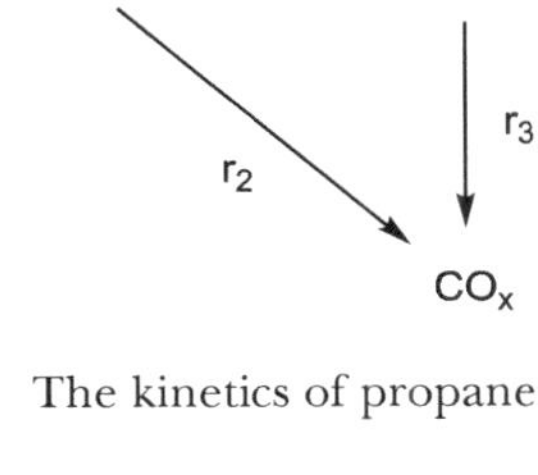

Figure 4.35 The kinetics of propane ODH [202].

remaining O–H groups recombine to form water and leave on the surface-reduced V centers; the latter are reoxidized by irreversible dissociative chemisorption of O_2.

Quantum-chemical calculations reveal the sensitivity and especially selectivity of reaction on the type of surface oxygen atom that activates the propane molecule [205]. Respective intermediates and potential energies of reaction are shown in Figure 4.36. The layered bulk structure of V_2O_5 is shown in Figure 4.21c. The vanadium cations are five-coordinated with oxygen atoms. There are three kinds of oxygen atoms: end-on coordinated vanadyl V=O and in the plane two- and three-coordinated oxygen atoms. The one- and

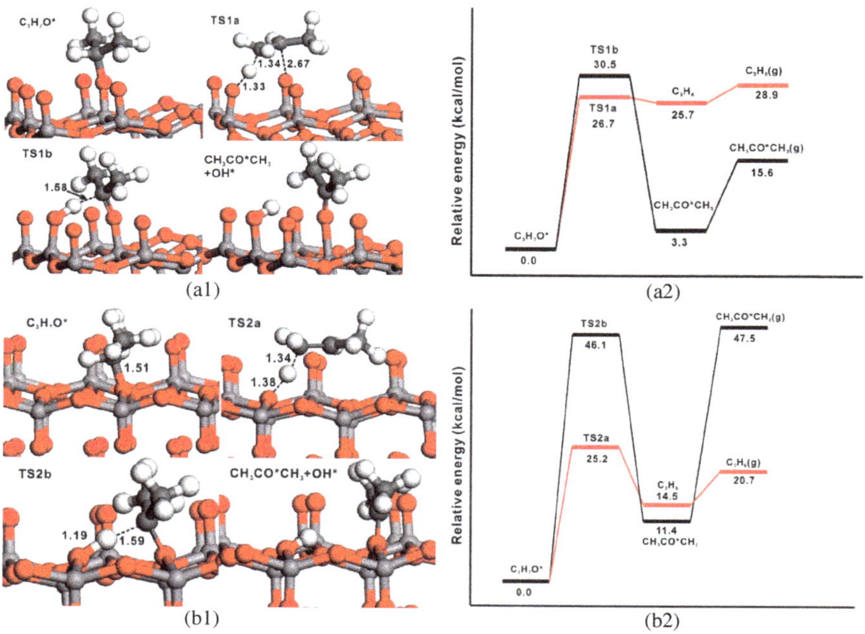

Figure 4.36 The competition of propene formation and oxygenation from the alkoxy state as calculated for $V_2O_5(001)$. Elementary bond cleavage steps for adsorbed propyl alkoxy to propene versus adsorbed acetone. (a) Reactivity of structures of vanadyl site, (a1) structures of reaction intermediates and transition states, (a2) potential energies. (b) Reactivity of two-coordinated oxygen surface site, (b1) structures of reaction intermediates and transition states, (b2) potential energies [205].

two-coordinated atoms are most reactive and are considered in the calculations.

The simulations start with adsorbed alkoxy species. C–H bond cleavage of propane by reaction with vanadyl oxygen gives a propyl intermediate that adsorbs as alkoxy on the oxide surface. Activation of the C–H bond occurs preferentially by the vanadyl V=O species. The computed activation energy of C–H bond cleavage is 105 kJ/mol. This is of the same order of magnitude as measured activation energies of the reaction [202].

The reactivity of the alkoxy intermediate adsorbed end-on on the vanadium cation (Figure 4.36a) has been compared with that of alkoxy adsorbed in the bridging site (Figure 4.36b) between two vanadium cations. The elementary step that gives propene involves a β C–H bond cleavage reaction, while the non-selective reaction by α C–H bond cleavage gives adsorbed acetone. Adsorbed acetone gives total combustion in consecutive reaction steps.

End-on adsorbed alkoxy decomposes into propene and acetone with comparable activation energies (Figure 4.36a1). The activation energy of alkoxy adsorbed twofold to vanadium cations to give propene is substantially lower than that for acetone (Figure 4.36b1).

This difference can be rationalized considering that in propene formation a surface M–O bond converts into MO–H and the M–O bond is only partially weakened, but for oxygenation the M–O has to be essentially broken. The bridging oxygen atom that has higher coordination will have the stronger M_x–O bond and will therefore be most sensitive to the difference in energy to form the O–H bond versus oxygen atom insertion. According to calculations by Hermann from Berlin [206], the difference in bond energy of the V=O and V–O–V bonds is 170 kJ/mol.

The calculations lead to the important conclusion that reaction with bulk vanadyl V=O gives non-selective propane conversion. In contrast, when oxygen insertion contributes to selectivity as in the conversion of butane to maleic acid, reaction may preferentially occur with vanadyl oxygen.

Reduction of V_2O_5 decreases the surface V=O concentration [207]. This is consistent with the correlation of V_2O_5 selectivity to butene versus maleic anhydride with relative concentration of V^{4+} or V^{3+} cationic sites [208].

The reactivity of V_2O_5 distributed as monolayers on a support is quite different than that of bulk vanadium oxide.

Due to refined catalyst synthesis techniques and advanced spectroscopies, the state of supported vanadium oxide as a function of concentration and support surface has become well understood [203], [209]–[213].

The relation between state of VO_x aggregate and propane conversion rate is schematically illustrated in Figure 4.37.

Isolated and polymeric surface VO_x can be present, and their reactivity is unaltered up to monolayer coverage. Beyond this coverage the normalized reaction rate (TOF) increases for nanoparticles in the 1–3 nm range [214]. The lower activity is found for the surface of bulk V_2O_5.

The reducibility of the respective vanadates is a function of support. A higher reducibility leads to higher ODH reactivity. Spectroscopic experiments indicate that for single vanadate sites the V=O groups do not disappear [215] but instead, as for instance for

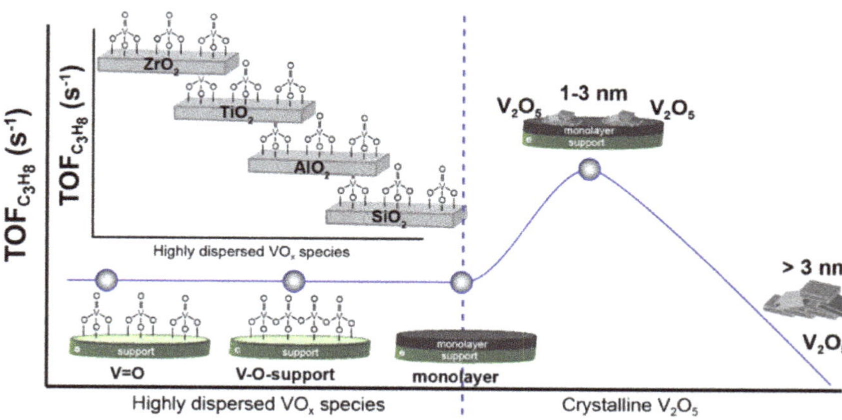

Figure 4.37 Schematic representation of the rate of oxidation of propane as a function of the state of supported VO_x species. Different supports are compared. The rate of reaction is normalized per exposed vanadium cation (TOF: turnover frequency) [203].

VO_x on TiO_2, the bridging oxygen atoms of the V–O–Ti bond are reduced.

This low reactivity of vanadyl V=O for monolayers of V_2O_5 versus that of a bridging oxygen atom is different from the high V=O reactivity of bulk V_2O_5. The bond energy of vanadyl in bulk V_2O_5 is substantially lower than that of the vanadyl in V_2O_5 monolayers [216]. In the bulk or when a subsurface vanadate layer is present, a vanadyl oxygen atom from the second layer completes the five-coordination of V^{5+} to approximate six-coordination by a Lewis acid-Lewis base interaction. When surface vanadyl reacts and the V=O bond weakens or breaks, increased interaction with the second layer oxygen atom compensates for part of the energy cost of V=O bond cleavage. This relaxation perpendicular to the basal layer directions is absent when vanadate is a monolayer. It causes the V=O bond energy of the monolayer to be more than 100 kJ/mol stronger than that of vanadyl on the bulk surface. It is consistent with the proposition that oxygen that is part of V–O–M activates the C–H bond of propane.

Bañares and Wachs [211] have demonstrated experimentally that similar to methanol oxidation by supported MO_x (see Section 4.3.2), the oxygen atoms that bridge vanadium cation and support cation take part in the catalytic cycle. The differences in bond energy of the V–O–M bond dominate the changes in propane activation energies. The bond energy of the V–O–Ti bond is lower than that of the V–O–V bond in bulk V_2O_5 and the vanadyl bond in the VO_x monolayer.

The sequence of apparent activation energies for sub-monolayer vanadium catalysts on different supports is $V/NbO_5 > V/SiO_2 > V/Al_2O_3 > V/ZrO_2 > V/CeO_2 > V/TiO_2$. The activation energies of propane activation vary from 45 to 150 kJ/mol and relate to the variation in the V–O–M bond energies [203].

For single-site VO_4 catalytic centers on a variety of supports, a linear relation exists between the activation energy of the propane to propene oxidation reaction and the computed energy of oxygen vacancy formation (see Figure 4.38a) [213].

The strength of the O–H bond that is formed by dissociation of reactant C–H with V–O–M relates to the activation energy of C–H

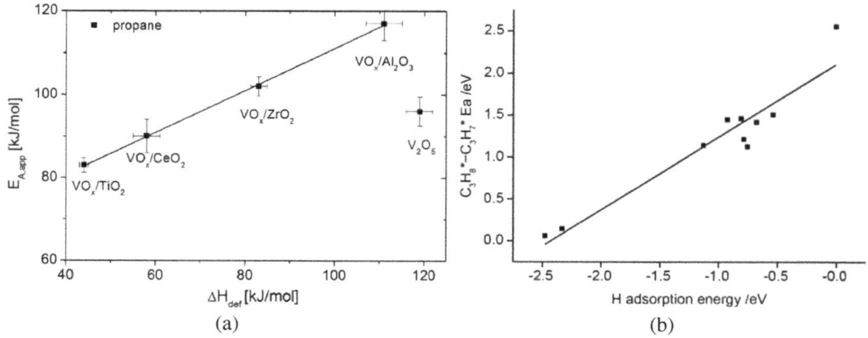

Figure 4.38 The activation of propane as a function of M–O or MO–H bond energy. (a) Correlation of defect formation enthalpy with apparent activation energy of propane oxidation [213]. (b) Correlation between the H adsorption energy and first-step dehydrogenation barrier of propane [207].

bond cleavage by the Brønsted-Evans-Polanyi (BEP) relation. According to the Bond Order Conservation rule this O–H bond energy in turn relates inversely with the strength of the V–O–M bond (see Section 2.3.2.3). Zhao and Gong from Tianjin University [207] demonstrated this BEP linear relation between the C–H bond activation energy and the O–H bond energy (Figure 4.38b).

The dependence of reaction rate on redox potential was also found when reactivity of different reducible metal oxides was compared by Berkeley scientists Bell and Iglesia [212]. They compared the activity of monolayers of VO_x, MoO_x and WO_x supported on ZrO_2. The activation energy for propane dehydrogenation increased in the order $VO_x/ZrO_2 < MoO_x/ZrO_2 < WO_x/ZrO_2$. Vanadium oxide is the most reactive because its redox energy is the lowest.

Non-selective parallel oxidation of propane (r_2) is strongly affected by the state of the vanadate support. Reyniers and Marin from the University of Ghent [204] found that the parallel non-selective oxidation of propane is lower for a monolayer of vanadate on TiO_2 than for bulk V_2O_5, which they ascribed to more difficult catalyst redox energy. The elementary step of vanadate reduction when propoxide converts to acetone generates three V^{4+} cations on bulk V_2O_5. In contrast, in the vanadate monolayer one of the three surface atoms must be reduced to V^{3+}. The lower redox potential of V^{3+} versus V^{4+} increases the activation energy of the oxygen insertion reaction.

The selectivity of propane to propene conversion due to the consecutive non-selective oxidation (r_3) increases with increase of VO_x surface density. The ratio of respective reaction rate constants is independent of support. When temperature increases the r_3/r_1 ratio decreases. The relative increase in apparent activation energy of consecutive combustion is dominated by the decrease in adsorption energy of propene when VO_x cluster size increases [202], [217], [218].

In summary, the topology of monomer, monolayer and nanoparticle on the support surface determines the relative reaction rate of propane activation versus that of parallel or consecutive combustion. The coordination and respective binding energies of vanadate oxygen atoms are important reactivity descriptors.

The most active ODH catalysts are supported by mildly reducible supports such as TiO_2. This relates to the reactivity of the V–O–M bond, which determines the activity of the system.

Supported vanadate catalysts with monolayer vanadate have the higher TOF. The reaction rate correlates inversely with surface reducibility. The oxygen atom that bridges vanadium and support cation activates C–H propane bonds. Whereas on the bulk V_2O_5 surface vanadyl V=O is more reactive than bridging V–O–V oxygen atoms, the V–O–M bond is more reactive than the vanadyl bond of VO_x monolayers. The latter V=O bond in the monolayer is substantially stronger than bulk V=O.

Consecutive reaction of propene (r_3) is fast because of the low activation barrier to give the reactive allyl. The ratio of r_3/r_1 does not vary with support. Variation in adsorption energy of propene compensates for difference in C–H bond activation energies.

Selectivity is a function of monolayer size. The relative rate of the non-selective consecutive reaction (r_3) is smaller for vanadate oligomers than for monomers. On the other hand, the relative rate of the parallel non-selective reaction (r_2) increases when the vanadium surface density increases. It relates to the elementary reaction rate of oxygen insertion.

In bulk V_2O_5 the reactivity of the oxygen atoms is too high for selective conversion of propane to propene. The non-selective parallel path (r_2) is an oxygen insertion reaction that is favored by weak M–O bonds.

4.4.4.3 *Aerobic Reduction of NO by Ammonia to N_2; Supported Vanadium Oxide and Metal-promoted Zeolite Catalysis*

The interplay of Brønsted acid, Lewis acid and redox catalysis. Single-site and dual-site systems. One of the nitrogen atoms of N_2 is from NO, the other from NH_3.

Selective catalytic reduction with NH_3 (NH_3–SCR) is a widely used technology to reduce the emission of nitrogen oxides (NO_x) in excess air. It was initially used to reduce emission of coal-fired power stations. At present it is also used to reduce standalone or automotive diesel emissions.

Two catalytic systems are dominant: V_2O_5 supported on TiO_2 (reminiscent of the system of the previous section) and zeolitic systems with chabazite structure promoted by Cu (see Chapter 5 for background on zeolites). One of the reasons these systems have been selected is their robustness with respect to SO_2 poisoning and sulfate formation.

The increasing requirements for a broad working temperature window, strong resistance against SO_2, alkali and heavy metals, and high hydrothermal stability have stimulated the investigation of improved NH_3–SCR catalysts with complex compositions. These include metal oxide catalysts ranging from VO_x, MnO_x, CeO_2, and Fe_2O_3 to CuO-based catalysts; acidic compound catalysts containing vanadate, phosphate and sulfate catalysts; and different zeolite catalysts ion-exchanged with Fe or Mn [42].

Here the reaction mechanisms of the NO–NH_3 reaction is presented as formulated for the supported V_2O_5/TiO_2 system [219]–[222] and compared with that of the Cu chabazite catalyst [28], [223]–[225].

The SCR reaction is a redox process in which NO and NH_3 participate in the reduction and O_2 is needed for the reoxidation. Typical operation temperatures are 580–600 K.

The stoichiometry of the standard SCR reaction is:

$$4\ NH_3 + 4\ NO + O_2 \rightarrow 4\ N_2 + 6\ H_2O \tag{4.9a}$$

Isotope-labeling experiments show that N_2 contains one N atom from NH_3 and one from NO_x. This suggests that the intermediate molecular complexes that give N_2 contain equal amounts of N atoms stemming from NH_3 and NO [226].

Whereas at low temperatures reoxidation of the catalyst is rate controlling, at high temperature reduction of the catalyst is rate controlling. This is due to the fast SCR reaction with participation of NO_2 [227], [228]:

$$2\ NH_3 + NO + NO_2 \rightarrow 2\ N_2 + 3\ H_2O \tag{4.9b}$$

NO_2 is produced in the diesel exhaust by oxidation of NO that is catalyzed by a Pt catalyst located in the exhaust pipe before the location where ammonia is added.

In the absence of NO there is also the slow SCR or NO_2 SCR reaction:

$$6\ NO_2 + 8\ NH_3 \rightarrow 7\ N_2 + 12\ H_2O \tag{4.9c}$$

A non-selective site reaction is the oxidation of ammonia, discussed in Section 4.3.1:

$$4\ NH_3 + 3\ O_2 \rightarrow 2\ N_2 + 6\ H_2O \tag{4.9d}$$

The chemistry of reactions involving NO and NO_2 can be quite complex. Dimers of NO_2 in presence of water decompose into adsorbed NO^+ and NO_3^-. NH_4^+ reaction with NO_3^- can give NH_4NO_3 that decomposes to N_2O. When reacted with NO, NH_4NO_3 reduces to NH_4NO_2 that in turn decomposes to N_2 [225].

The reactions catalyzed by the V_2O_5/TiO_2 catalyst and Cu-chabazite catalyst are related, but also show important differences. The mechanism of SCR of the respective catalysts is compared in Figure 4.39.

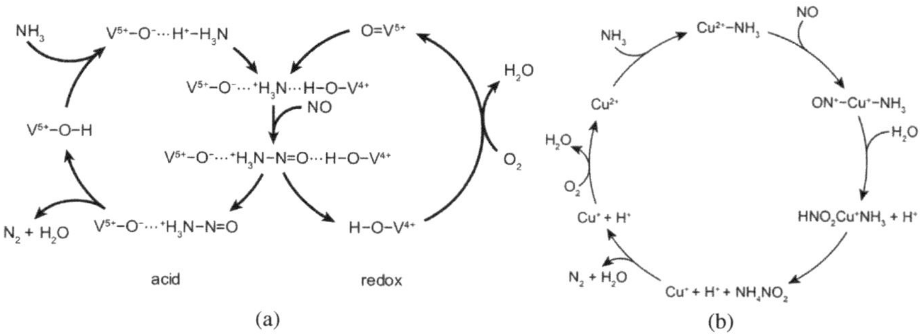

Figure 4.39 Catalytic cycles of the SCR reaction. (a) The reduction of NO by NH_3 catalyzed by vanadium oxide according to Topsøe and Dumesic [220], [229]. (b) Low-temperature standard NH_3-SCR reaction pathways catalyzed by isolated Cu ion monomers according to Peden *et al.* [225].

For the vanadium oxide catalyst, scientists from Topsøe company formulated in 1995 [220], [229], [233] the first mechanism of the SCR reaction by combined use of vibrational spectroscopy with transient kinetics. Twenty years later computational modelling studies also from Topsøe company in collaboration with Aarhus University [221] confirmed the early suggestion. The mechanism is shown in Figure 4.39a.

The ammonia molecule NH_3 and a surface proton equilibrate with NH_4^+. Then the NH_4^+ ion reacts with NO to give the NH_4^+NO intermediate. The proton is backdonated to the surface and N–H bond cleavage gives the Fogel intermediate H_2NNO (see Section 4.3.1). N–H bond cleavage is the rate-limiting step and V^{5+} reduces to V^{4+}. The Fogel intermediate rapidly decomposes to N_2 and H_2O, and V^{4+} reoxidizes in a consecutive reaction.

The detailed structure of the protonic sites is not known. A protonated vanadyl V=O–H [222] versus protonated V–OH–Ti site [221] is suggested.

In the previous section on the ODH of propane with the monomer, vanadate dispersed on TiO_2 with C–H does not occur with vanadyl V=O, but gives a proton attached to a bridging V–OH–Ti oxygen between reactive vanadate and TiO_2 support. A similar protonated site is proposed as the Brønsted acid site in the SCR reaction. Ammonium vanadate adsorbed on TiO_2 has been suggested as a model of the ammonium complex with vanadate [221]. This intermediate reacts in a consecutive step with NO.

O_2 dissociation has a high barrier on single-site vanadate, even when it is in the reduced state. Reoxidation occurs more readily when NO_2 is the oxidant. NO_2 decomposes into NO and an adsorbed oxygen atom. This oxidizes reduced V^{4+} and reacts with H. These steps are part of the fast SCR mechanism in Eq. (4.9b). On dual vanadate sites O_2 dissociation has a lower barrier [230]. Rate of reaction by vanadate oligomers is faster compared to that catalyzed by vanadate monomer [222].

Importantly Bell *et al.* [231] have shown that monomeric single-site vanadate has a unique high selectivity of N_2 production versus N_2O.

Whereas reactions of propane with the vanadium catalyst only happen with the oxygen atoms, without direct contact with the cation, this is different for SCR by the Cu zeolite catalyst. The reaction cycle for the Cu catalyst is shown in Figure 4.39b. The difference relates to direct access of ammonia and NO to the Cu^{2+} cations that are contained within the zeolite chabazite cage.

The Cu zeolite catalyst consists of Cu^+ and Cu^{2+}, cations located in the cavities of zeolites that have the chabazite structure. The site structure is shown in Figure 4.40.

The zeolite chabazite cavities are connected by narrow eight-ring openings formed by interconnected Si, Al or P tetrahedra (see Section 5.3). Their unique structure makes them stable with respect to hydrolysis and limits access of deactivating hydrocarbons that are also part of the exhaust emission.

The oxidation reaction of NH_3 by NO is catalyzed by a site consisting of two Cu^{2+} cations located in the chabazite cage [232], [233].

The Cu^{2+} cations are reduced by the N_2 evolving reaction. For reoxidation the oxygen molecules require at least two Cu^+ cations. In the reoxidation reaction two protons are consumed that stem from co-adsorbed ammonia (without co-reduction for the four-electron reaction at least 4 Cu^+ cations are needed). The rate of reaction is proportional to zeolite proton concentration (the protons are adsorbed on oxygen atoms of the zeolite lattice). An

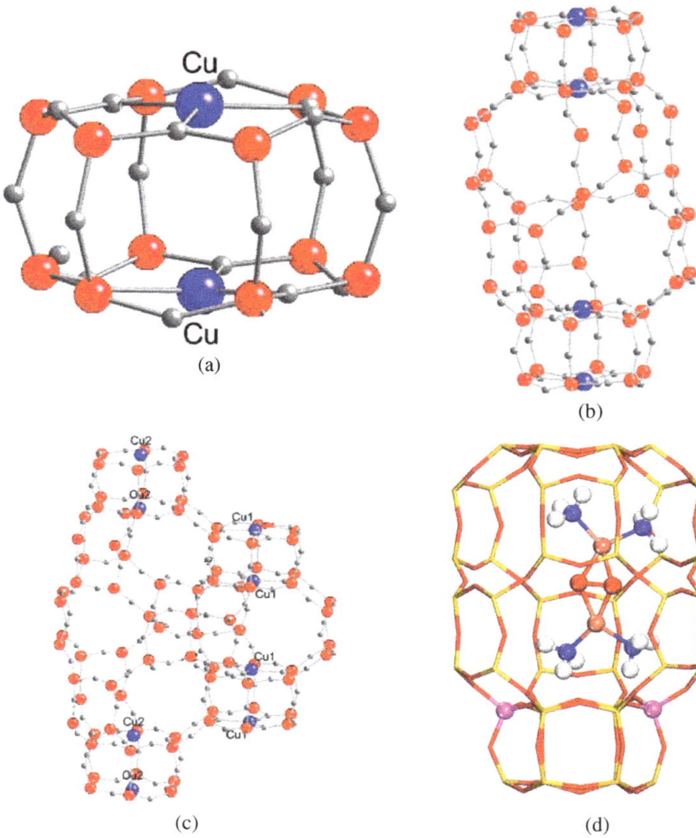

Figure 4.40 Location of copper ions in Cu-SSZ-13: (a) side view and (b) entire cage. (c) Locations of copper ions (Cu1 and Cu2) in Cu-SSZ-16 [232]. (d) Side view of the chabazite cages showing the Al position in the case with two atoms. It shows activated O_2 over the oxidized Cu complex pairs with two Al atoms situated in one six-membered ring. Atom color codes: Cu (orange), Si (yellow), Al (purple), O (red), N (blue), and H (white) [224].

intermediate structure with O_2 interacting with two Cu^+ cations coordinated also to ammonia is shown in Figure 4.40d.

In the Peden mechanism of Figure 4.39b, ammonia and NO adsorb on the Cu^{2+} cation. Lattice protons regenerate by reaction of water with NO^+. This gives intermediate ammonium nitrite that in a subsequent step decomposes to N_2. The system is then ready for reoxidation. The ammonium nitrite mechanism is supported by the spectroscopic experiments of Ref. [223].

The alternative proposal based on DFT computational studies by Janssens *et al.* [224], [234] is that N_2 is formed by the Fogel reaction similar to the vanadate catalyst. According to this proposal the HNO_2 intermediate reacts with ammonia to give water and H_2NNO that decomposes to N_2. The correspondence of both mechanisms is that N–H bond cleavage and reduction of Cu^{2+} compete with Cu^+ reoxidation.

The fast SCR reaction will reoxidize the single-site Cu^+ cation since it is reoxidized by NO_2.

A general mechanistic principle of reoxidation of redox systems is that a dual-site catalyst can shift to single-site catalysis when oxidation occurs not by molecular oxygen but by activated oxygen as in NO_2.

4.5 Summary; Elementary Reactions of Heterogeneous Selective Oxidation Catalysis

Oxidation catalysis is complex because of the many catalytic systems and great variation in reaction mechanisms. Metal catalysts such as Pt, Cu and Ag and mixed reducible metal oxides that have as main components high-valent Mo or V cations are discussed. Emphasis is on the relation between atomic structure and composition of the reaction center and the energetics of molecular X–H bond cleavage and oxygen insertion reactions.

Main selective oxidation reactions discussed concern oxidation of ammonia, methane, methanol, alkenes and alkanes. Activity and selectivity relate to cation oxidation state and topology of the respective reaction sites.

An important selectivity principle is whether reaction intermediate structure and reaction site topology allow for direct activation with surface metal atom or whether reaction of the alkane remains restricted to contact with the oxygen atoms that are coordinated by metal cations.

Selective reactions circumvent or limit direct contact with cations. Reactivity is determined by chemical bonding properties of lattice oxygen atoms. Bulk and supported metal oxide catalysts show important differences in activity and selectivity that relate to the unique reactivity of interphase oxygen atoms between reactive and supporting oxides.

This chapter treats mainly oxidation reactions catalyzed by unsupported or supported large particles. In Chapter 6 oxidation reactions catalyzed by single-site centers attached to high surface area reactive and non-reactive supports are discussed. Lewis acid-catalyzed reactions as well as Au-catalyzed reactions are presented.

Sections 4.5.1–4.5.6 contain a comprehensive overview of the elementary steps that one distinguishes in selective oxidation reactions. This overview is not limited to reaction steps on the metals and reducible metal oxides of this chapter but includes also elementary steps of the oxidation reactions discussed in Chapter 6.

4.5.1 *Selective Oxidation by Autocatalytic Radical Chain Reaction*

The autocatalytic radical chain reaction consists of reaction initiation, propagation and termination steps. In catalytic selective oxidation reactions, intermediate steps are catalyzed by reduction or oxidation of catalyst cations.

The reaction was introduced in Section 4.2.1.1. As described in detail in this section, oxidation radical chain reactions are initiated by reaction with catalyst surface oxygen atoms or with a cation. This generates a radical chain reaction with peroxides and hydroperoxides as described by Eqs. (4.2) and (4.3). The hydroperoxide intermediate plays an essential role [235]. In low-temperature liquid-phase redox systems their decomposition leads to oxygenates. In high-temperature gas-phase reactions, OH radicals and non-selective total combustion dominate.

In the following two examples the mechanisms of low- and high-temperature autocatalytic radical reactions are presented:

- An example of a low-temperature autocatalytic process is the liquid-phase oxidation of toluene to benzoic acid. The overall reaction is shown in Figure 4.41 [235].

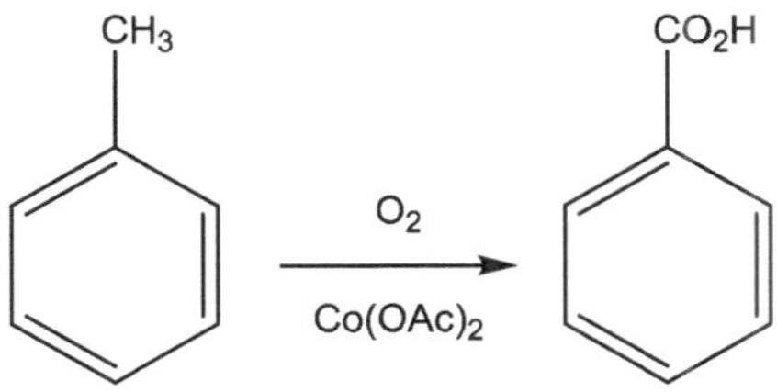

Figure 4.41 Oxidation of toluene to benzoic acid. Catalyst is $Co(OAc)_2$ in solvent HOAc; operation temperature is 445 K.

Elementary reaction steps of the toluene to aldehyde conversion reaction are given in Eqs. (4.10) [235]:

$$\begin{aligned}&ArCH_3 + CO^{3+} \rightarrow [ArCH_3]^{+}\cdot + CO^{2+}\\&[ArCH_3]^{+}\cdot \rightarrow ArCH_3\cdot + H^{+}\\&ArCH_3\cdot + O_2 \rightarrow ArCH_2O_2\cdot\\&ArCH_2O_2\cdot + CO^{2+} \rightarrow ArCH2O + HOCo^{2+}\end{aligned} \quad (4.10)$$

The benzaldehyde molecule is more reactive than toluene. In following autocatalytic reaction steps, it is oxidized to carboxylic acid.

- An example of a high-temperature reaction is the oxidative coupling of methane to ethene, introduced in Section 4.2.1. Via abstraction of a H atom by a surface radical site, methyl radicals are generated. These recombine in the gas phase. By subsequent hydrogen atom abstraction steps, ethene is formed. These and non-selective total combustion steps are due to gas-phase radical reactions with oxygen [9].

The catalytic cycle of radical and ethane formation as suggested by Lunsford for the Li/MgO catalyst is shown in Figure 4.42. The Li/MgO system has been identified early on as a promising candidate catalyst [236], [237] and has been the subject of extensive investigation.

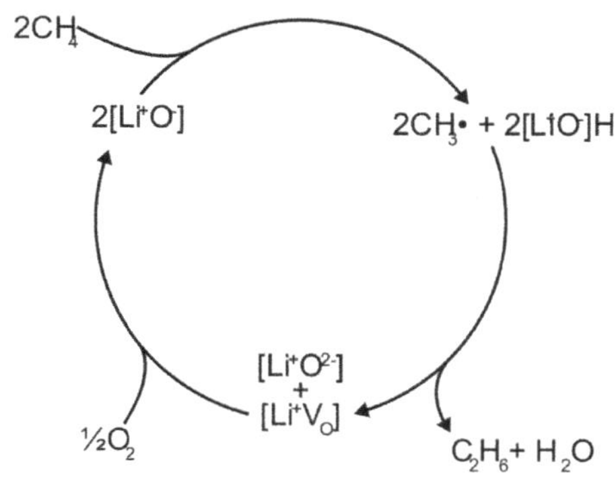

Figure 4.42 The Lunsford mechanism for the oxidative coupling of methane [238].

There are two proposals for the mechanism of methane activation by the Li/MgO system:

- Homolytic activation [239]
- Heterolytic activation [240]

According to the homolytic mechanism radical sites are created in MgO by substitution of two MgO units by a $[LiO]^-$ unit and $Li^+ V_o$ unit. V_o represents an oxygen vacancy. The $[LiO]^-$ site is a radical that reacts with methane C–H.

However, it was discovered that for this particular system there is a mismatch of activation energy of H atom abstraction from methane by the $[LiO]^-$ site and the overall reaction activation energy (12 kJ/mol versus 90 kJ/mol).

The alternative heterolytic mechanism activates methane by electron transfer to a non-dissociative co-adsorbed O_2 molecule as indicated in Eq. (4.11):

$$[Mg^{2+}O^{2-}]_{MgO} + CH_4 + O_2 \rightarrow \left(O_2\right)[HOMg^{2+}]^{+}_{MgO} + \bullet CH_3 \quad (4.11)$$

Intermediate superoxo O_2^- is stabilized by protonated magnesium oxide $(MgOH)^+$. The superoxo anion will reoxidize the surface through intermediate OOH formation. The computed activation energy of the mechanism compares well with experiment.

The heterolytic activation model received also experimental support. An important role of Li is to create a high surface area MgO catalyst with step-edge and corner sites. These coordinatively unsaturated sites create the highly reactive cationic sites that stabilize the O_2^- charge of the oxygen complex in Eq. (4.11). The computed activation energy compares well with that of experiment.

4.5.2 *Non-oxidative Radical Reactions*

Here the mechanism of non-oxidative methane activation is introduced. Comparison with this reaction is interesting since methyl radical formation is analogous.

A major difference with the oxidative process is that methane is not activated by adsorbed oxygen, but by reduction of the catalytically active metal cation.

Whereas in oxidative methane conversion total oxidation competes with ethene formation, in the non-oxidative high-temperature reaction carbonaceous residue formation limits selectivity.

Conversion of methane to higher organic molecules is an endothermic process. Decomposition into carbon and hydrogen has a lower free energy. Benzene formation becomes thermodynamically feasible at 800 K, and ethene and acetylene can be formed above 1000 K. Short contact times are necessary to obtain acceptable yield. At high temperature the methyl radicals, once generated, will give hydrocarbons through recombination and continued C–H abstraction. To suppress carbon deposition by recombination with reactor wall, a large reactor volume to reactor surface ratio is preferable [241].

The radical formation reaction has been studied in detail by Bao and coworkers at the Dalian Institute of Chemical Physics [242] with a uniquely synthesized catalyst that contains well-defined Fe single sites on a SiO_2 support. At 1350 K ethene as well as aromatics are produced with reasonable yield. The isolated Fe site transforms by reaction to a Si[C–Fe–C]Si site. Elementary reaction steps of methane activation are shown schematically in Figure 4.43.

The methane C–H bonds are activated by the iron carbide reaction site. Methyl coordinates to Fe and the hydrogen atom with the

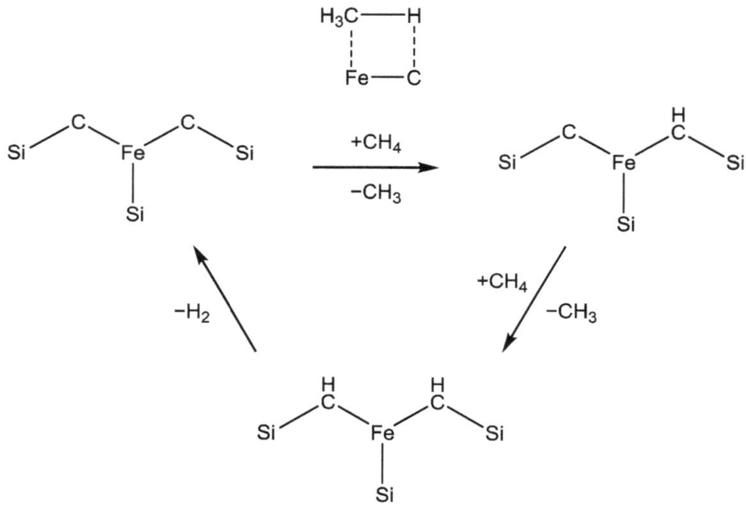

Figure 4.43 Activation of methane by FeC_2 complex [242].

carbon atoms. After two reaction steps with methane and desorption of methyl radicals, the hydrogen atoms recombine to give H_2. The mechanism is essentially heterolytic. The hydrogen atoms react with negatively charged carbon atoms as protons and cationic Fe binds the methyl anion. The Fe–CH_3 bond is weak and cleaves into radicals at high temperature.

A similar methane activation process, but for a different catalytic system, has been proposed by Wachs *et al.* [243]. It applies to the conversion of methane to aromatics by MoO_3 nanoclusters located in zeolite nanopores [244]. The Mo oxide nanocluster is converted into a carbide cluster and methane decomposition generates hydrogen atoms attached as C–H and CH_3 attached to Mo of the Mo/C cationic cluster. According to [245] this initiates a radical reaction within the zeolite nanopore (see Section 6.3.3).

4.5.3 *Alkane Activation by Reducible Oxide Atoms*

Here selective oxidation reactions of alkanes are considered that are activated by surface oxygen atoms of reducible metal oxides. The reaction is of radical character and there is no direct contact with the cations. At the reaction center the cation is shielded from reaction with reactant by surrounding oxygen atoms. Such non-radical bond activation is discussed at reaction sites where cation is not shielded from coordination to reactant.

Two kinds of radical C–H bond activation reactions are considered here: low-temperature oxidation of methane to methanol and high-temperature alkane dehydrogenation and oxygenation.

4.5.3.1 *The Rebound/Harpoon Mechanism of Methane to Methanol Oxidation*

Eq. (4.12) illustrates schematically the rebound mechanism of methane to methanol oxidation that is initiated by the radical harpoon reaction. It is a low-temperature reaction. This radical reaction is selective when there is a constraint on the mobility of free radical intermediates. A representative reaction center is $[FeO]^{2+}$, which can be part of a homogenous complex in the liquid phase, in an enzyme or coordinated to the zeolite framework [246], [250].

$$M^{n} = O + H - CH_3 \rightarrow M^{n-1} - OH + \bullet CH_3 \rightarrow M^{n-2} + HO - CH_3 \quad (4.12)$$

The oxygen atom abstracts a hydrogen atom from methane C–H. This generates a methyl radical that in a second step recombines directly with M–OH to give methanol [246], [251], [252]. In the radical rebound mechanism, there is no direct contact between metal cation and methane. This is different for the non-radical oxene mechanism of Section 4.5.5.

4.5.3.2 *Homolytic C–H Bond Activation*

Homolytic oxidative C–H bond activation of the alkane molecule is a reaction where hydrogen atom and alkyl radical both will coordinate with a surface oxygen atom. Such reactions happen with reactive metal oxide surfaces where the cations are screened from interaction with reactant by coordinative saturation with oxygen atoms, such as the basal plane of MoO_3 (see Figure 4.28a) or vanadium cations terminated by vanadyl V=O (see Figure 4.21c).

Figure 4.44 schematically illustrates the C–H bond activation and oxygen insertion reactions according to the homolytic reaction mechanism. This mechanism is representative for selective oxidation reactions as catalyzed by V- or Mo-based multicomponent

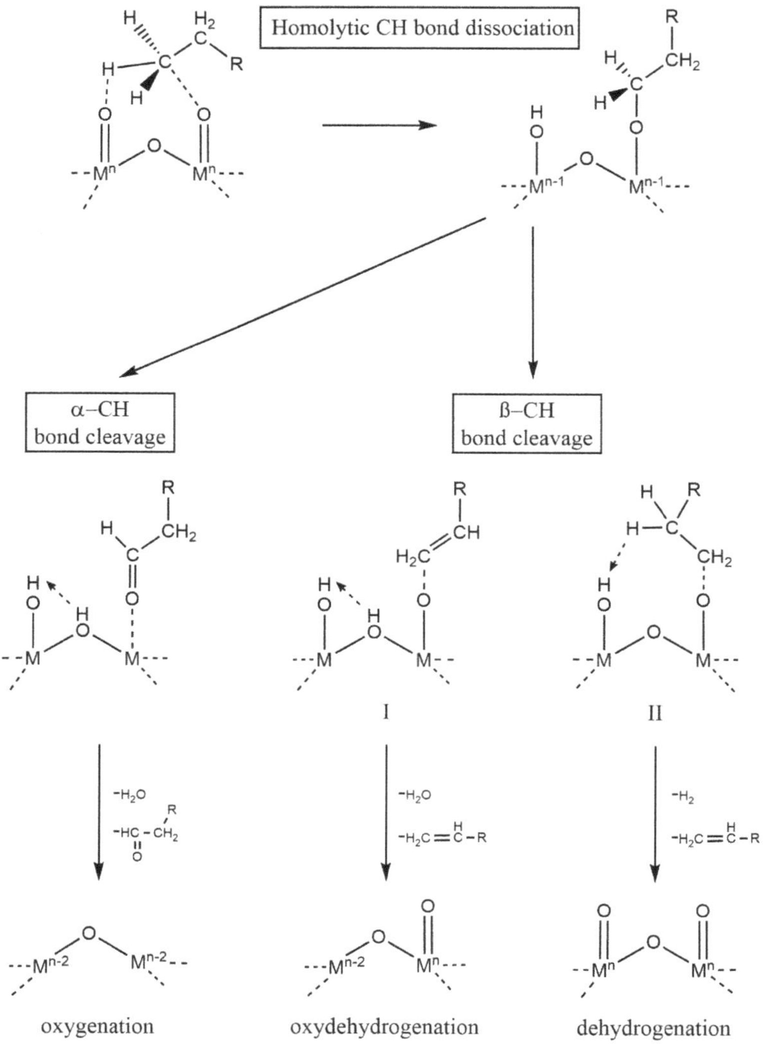

Figure 4.44 Schematics of the elementary reaction steps for radical oxidation of alkane by reducible metal oxide.

oxides that is more extensively discussed in Sections 4.4.3, 4.4.4.1, and 4.4.4.2.

Initial bond activation of C–H happens with formation of a surface O–H bond and an alkoxy intermediate. Two different C–H bond cleavage reactions can follow. The alkoxy α CH bond cleaves next to the C–O bond and an aldehyde or keto product results. This reaction can initiate non-selective total oxidation as for propane or selective oxygen insertion as for the butane to maleic anhydride reaction. Alternatively, a β C–H bond cleaves, and an alkene is product.

4.5.4 *Heterolytic Bond Activation Reactions*

Heterolytic bond cleavage happens on an oxycationic site. In the bond cleavage reaction of the RX–H molecule, the hydrogen atom adsorbs as a proton to the lattice or complex oxygen atom and the anionic RX^- adsorbs on a cation. This is schematically illustrated in Figure 4.45.

The cation acts as a Lewis acid. One distinguishes between redox and non-redox reactions. In non-redox reactions the site regenerates after reaction and hydrogen is consumed or produced.

In redox reactions two different sequences follow after bond cleavage: surface O–H hydroxyls recombine to give water and/or an oxygen atom inserts into a C–H bond with formation of hydroxyl, ketone or another oxygenate. The vacant site is regenerated by oxidation.

Non-redox heterolytic hydrogenation of alkene or dehydrogenation of alkane is discussed in Chapter 3 (Figure 3.5) and Chapter 6 (Section 6.3.1.3).

Compared to transition metals, transition metal oxides have reduced activity for alkane activation. An exception is PdO.

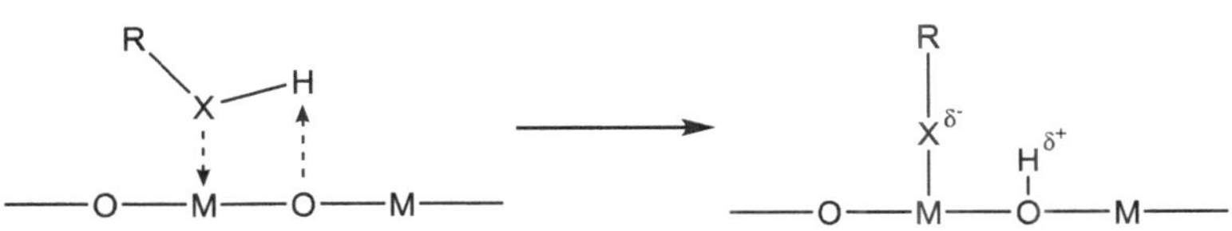

Figure 4.45 Schematic representation of heterolytic X–H bond cleavage reaction.

Methane is activated by PdO with a slightly lower activation energy than on the free metal [253], [254]. In this heterolytic reaction methane dissociates into CH_3 that adsorbs on the Pd cation and a hydrogen atom that binds to lattice oxygen. The oxygen atoms are reactive. PdO initiates gas-phase radical combustion reactions.

4.5.4.1 *Transition Metals*

Oxygen atoms co-adsorbed on a transition metal surface can act as co-reactants, promoters or site-blocking agents. C–H bond activation on the group 8 transition metals is a homolytic reaction with low activation energies (Chapter 3). As long as the surface state remains metallic, adsorbed oxygen atoms will inhibit alkane activation and oxidize surface reaction intermediates [255].

This is different for activation of the polar O–H bond in methanol or N–H bond of ammonia that are activated by co-adsorbed O atoms. Activation of ammonia by co-adsorbed oxygen is discussed in Section 4.3.1 (see Figure 4.6). The example of methanol activation is discussed here.

Figure 4.46 compares O–H dissociation of methanol with C–H bond dissociation of adsorbed methyl for the 4d and 5d transition metals. Results are from DFT calculations by Neurock *et al.* from Minnesota [256].

According to Figure 4.46 activation of the methanol O–H bond only has a low barrier when activated by O_{ad} or OH_{ads}. On the transition metals, including the noble metal Au, the adsorbed hydroxyl is most reactive, but on Cu and Ag activation by O_{ads} and OH_{ads} is comparable. The Ag results confirm the Wachs-Madix mechanism of the selective oxidation of methanol in Section 4.3.2. They suggest that initial O–H bond cleavage of methanol is activated by O_{ads}.

As is shown in Figure 4.46b activation of the C–H bond of adsorbed methyl is very different. Except for activation by Cu, Ag and Au, the free metal has the lower barrier and adsorbed oxygen or OH acts as a poison. With respect to the adsorbed methyl species, O_{ads} and OH_{ads} have comparable reactivities on the group IB metals.

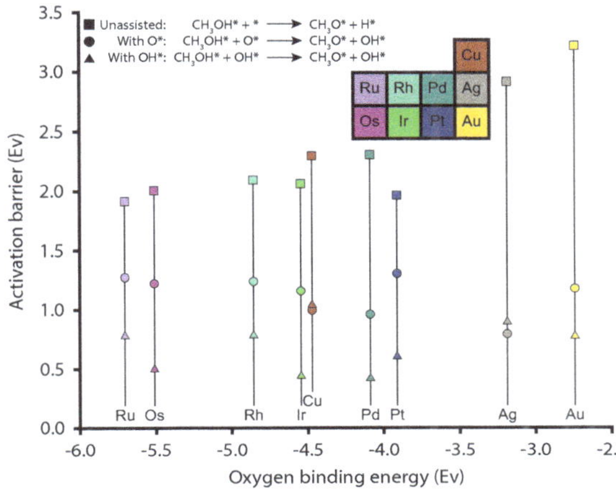

(a)

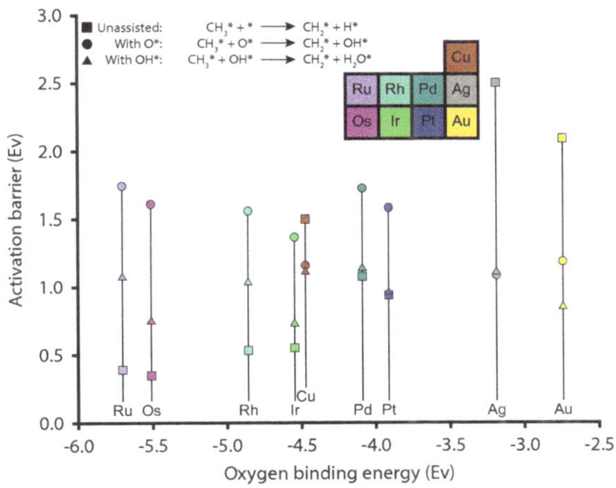

(b)

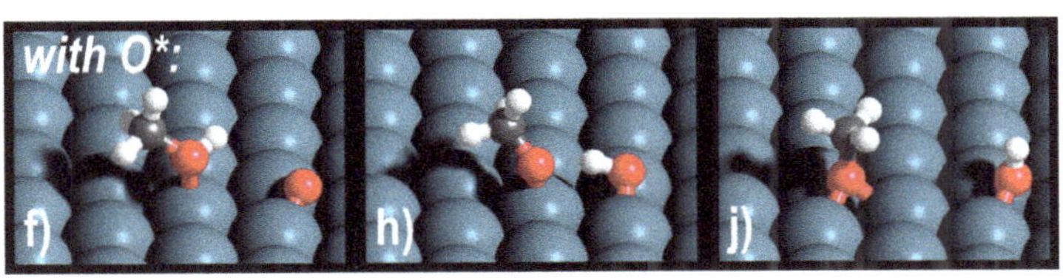

(c)

Figure 4.46 Heterolytic dissociation of methanol. (a) Activation barrier (referenced to methanol in the gas phase) for the O–H activation of methanol through direct, O*-assisted and OH*-assisted mechanisms for Group 8–11 metals. (b) A comparison of the activation barriers for the C–H activation of CH_3^* through direct O*-assisted and OH*-assisted reactions on Group 8–11 metals. (c) Structures of methanol reaction pathway for the activation of the O–H bond of methanol through O* abstraction mechanism on Pd(111) surface; (f) initial state, (h) transition state, (j) final state. Red is oxygen, white hydrogen, black carbon, blue Pd [256].

Figure 4.46c shows for the Pd(111) surface the reaction intermediates of heterolytic O–H bond cleavage of methanol. Methanol adsorbs by the oxygen atom to the transition metal and by reaction with adsorbed O converts to surface methoxy. In this elementary reaction step O_{ads} is converted into OH_{ads}

4.5.4.2 *Heterolytic Bond Dissociation by Oxides*

Alkane activation catalyzed by metal oxides is commonly a radical type of reaction as is discussed in Section 4.5.3. However, with methanol the O–H bond activation reaction is heterolytic. This reaction is discussed here. The H atom attaches to a lattice oxygen atom and methoxy adsorbs on a cation. The preferred oxides contain cations in high valence states and have high redox potential.

As mentioned in Section 4.4.3.2 MoO_3 is a component of the selective oxidation catalyst of methanol to formaldehyde. The coordination of the Mo cation largely determines the activation energy of the O–H bond cleavage reaction. This can be illustrated by the difference in activation energies when oxygen coordination of the Mo cation varies.

Figure 4.47 schematically illustrates the differences of the transition state structures for methanol activation by the basal MoO_3 surface, with the cation screened by six surrounding oxygen atoms and MoO_2, WHICH has the rutile structure. Half of the exposed Mo cations have an oxygen vacancy. The difference in Mo cation

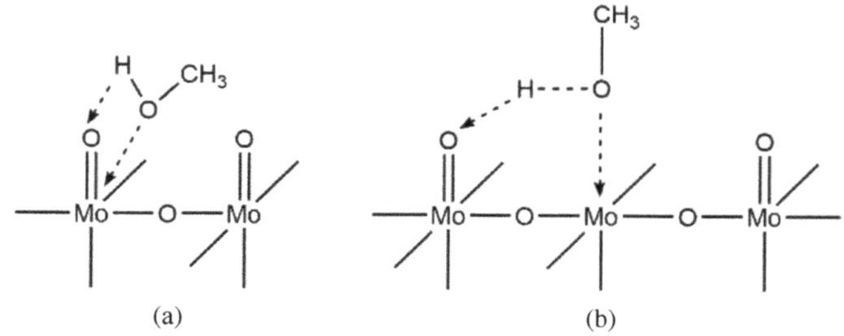

Figure 4.47 Schematic comparison of CH_3OH activation on MoO_3 and MoO_2. (a) Methanol activation on MoO_3 basal plane. (b) Activation of methanol on MoO_2 surface.

accessibility causes a difference of 100 kJ/mol in O–H bond activation energy [177].

Heterolytic activation reactions can also happen at the interphase of metal cluster and reducible oxide atom. This is the case for single-site catalytic systems. Such interphase reactions are extensively discussed in Chapter 6.

4.5.5 *The Oxene Oxygenation Reaction*

The heterolytic oxene mechanism applies to catalytic oxygenation of benzene by single Fe cationic sites in zeolites. This is part of the Panov reaction that is discussed in detail in Section 6.2.2.2 [11]. In this reaction N_2O is the oxidant.

Methane can be hydroxylated by a radical reaction such as the harpoon or rebound mechanism of oxygen (Section 4.5.3.1) as well as by heterolytic oxene insertion mechanism-related reactions. For methane heterolytic oxygen insertion has been demonstrated experimentally in gas-phase experiments using mass spectrometry and quantum-chemical studies [238], [257].

For reaction of methane with the $[FeO]^{2+}$ cationic complex, the oxene reaction pathway is schematically shown in Figure 4.48. The C–H bond of methane dissociates heterolytically, and CH_3 attaches to the metal cation and the hydrogen atom binds to the oxygen atom. In a second step CH_3 attaches to OH that is followed by formation of adsorbed methanol. In the process Fe^{4+} is reduced to Fe^{2+}.

On the other hand, in the hydroxylation reaction of benzene the cation oxygen atom activates the C–H bond by contact with a carbon atom of the aromatic ring. This is illustrated by Figure 4.49 [258].

The important difference between the hydroxylation of benzene and methane is that in benzene hydroxylation there is no intermediate C–H bond cleavage step.

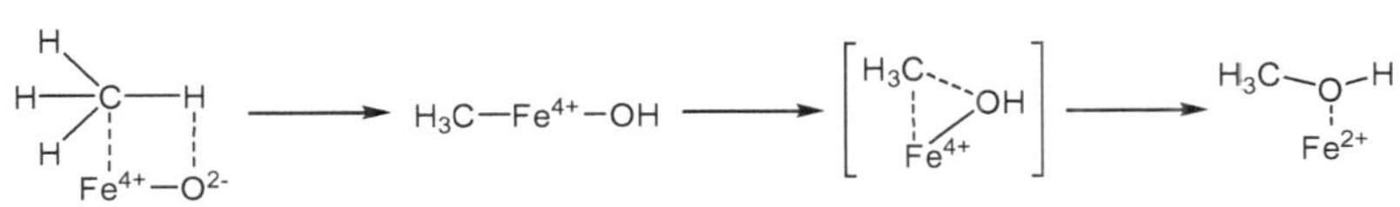

Figure 4.48 The oxene oxygenation reaction mechanism of methane.

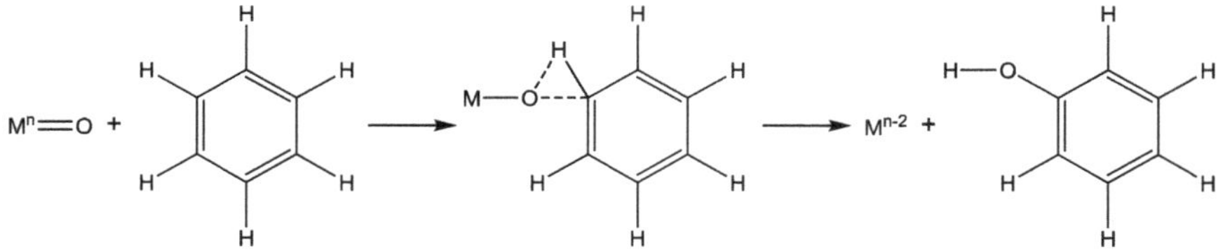

Figure 4.49 The heterolytic oxene reaction mechanism of benzene hydroxylation.

In Section 6.3.2.3 hydroxylation pathways for methane catalyzed by cationic Cu complexes in zeolites are discussed. The Panov catalytic system that consists of Fe cations located in zeolite nanopores can also produce methanol in low-temperature reactions [259], [260]. Isotope exchange experiments demonstrate the different hydroxylation reaction paths of methane versus benzene [261].

4.5.6 *Oxygen Insertion Into the Alkene π Bond; the Epoxidation Reaction*

Oxygenation reactions of alkenes by selective oxidation with molecular oxygen on metal and metal oxide catalysts are discussed respectively in Sections 4.3.3 and 4.4.3.1. Ethene gives an epoxide on Ag. Epoxidation of propene has low yield, but acrolein is the main product when Cu is the catalyst. The preferred catalyst for acrolein production from propene is the Bi_2O_3/MoO_3 catalytic system.

The difference in selectivity of ethene versus propene oxygenation relates to the high reactivity of the allylic C–H bond of propene. The mechanism of acrolein formation by the reducible oxide system is discussed in detail in Section 4.4.3.1. The reaction rate-controlling step is formation of the allyl in a radical-type reaction.

Here the mechanism of the epoxidation reaction is revisited.

The prototype reaction mechanism of oxygen insertion is the OMC proposed in 1977 by Sharpless [262] and studied computationally by Goddard [263], [264].

The stoichiometric reaction of the Sharpless epoxidation reaction with a Cr^{6+} or Mo^{6+} oxide complex is schematically given in Figure 4.50.

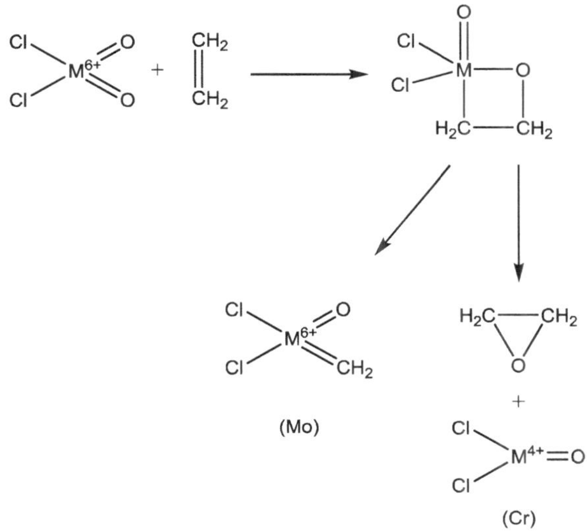

Figure 4.50 Epoxide formation versus carbene formation through oxametallic intermediate.

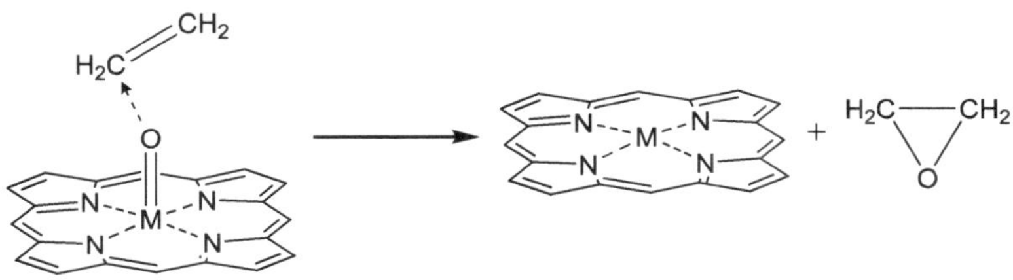

Figure 4.51 Direct oxygen transfer to alkene and epoxide formation by transition metal porphyrin complex (schematic).

This reaction occurs with complexes that contain d^0 transition metal cations such as Cr^{6+} and Mo^{6+}. The alkoxy OMC intermediate can decompose to the epoxide or the C=C bond can cleave and form a metal carbene complex. The epoxide will be formed with the complex with the lower redox potential.

The prototypical reaction of the non-cation-promoted direct epoxide formation mechanism is the epoxidation by Fe^{4+}, Mn^{5+} or Ru^{3+} porphyrin complexes or Schiff bases [265]–[268] as in the enzyme cytochrome P450. The intermediate of this reaction is sketched in Figure 4.51.

The OMC cannot form as intermediate because the planar ligands inhibit steric contact between reactant and cation site. The cations that are in high-valent states are highly reactive, and at low temperature will activate methane or alkanes according to the harpoon/rebound mechanism and give epoxide formation of alkenes at higher temperatures. The oxygen transfer reaction is of radical type. Reaction with the alkene π bond is asymmetric with respect to the C=C bond axis.

Reactive oxygenated complexes cannot be formed directly by activation of O_2, but oxygen atom addition can be done electrochemically, by intermediate H_2O_2 decomposition or can be generated by other activated O donors such as NaOCl.

The epoxidation reaction of ethene on Ag can occur via the OMC intermediates versus direct insertion as sketched in Figure 4.52.

As first suggested by Linic and Barteau [271] the OMC intermediate is formed on surfaces that are metallic. The molecule contacts a surface metal atom as well as the oxygen atoms. The intermediate can react to the product epoxide molecule by oxygen insertion. In a competing step acetaldehyde is formed. The latter reaction compares with the Wacker reaction (Section 4.2.2.2, Figure 4.2). The oxygen atom reacts with the C atom of ethene, and the α CH bond cleaves. The hydrogen atom then moves to the other carbon atom.

The direct epoxidation mechanism occurs on the oxidized surface as long as no oxygen vacancies are present. Then the

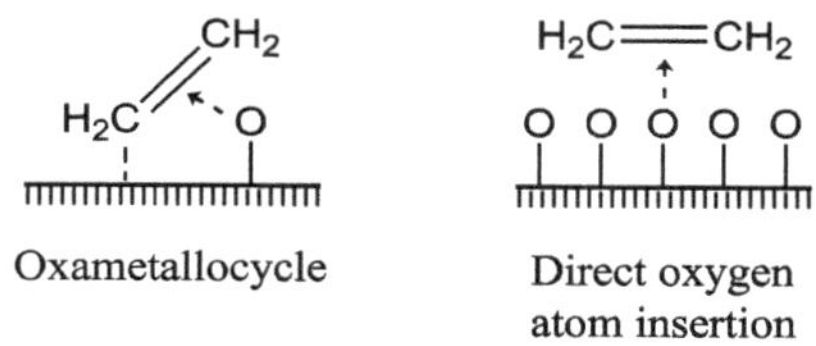

Figure 4.52 Schematic presentation of the oxametallocycle versus direct oxygen insertion into the π bond of ethene by Ag catalyst. In direct oxygen insertion the transition state is probably asymmetric [269]. According to [270] chemical bonding is radical-like.

epoxidation reaction has high selectivity. The oxygen atoms screen the silver cation from contact with reactant ethene. The acetaldehyde channel has a high barrier since C–H activation is not promoted by reaction with surface metal atom [269].

The higher alkene molecules with allylic C–H bonds can be oxidized selectively to the epoxide by reaction with hydrogen peroxide or hydroperoxide. In contact with hydroperoxide the allylic C–H bond is not activated. Epoxidation with the peroxide is a well-known reaction that is catalyzed in homogenous catalysis by high-valent d^0 transition metal cations such as Mo^{6+}, W^{6+}, V^{5+} and Ti^{4+} [272] that act as Lewis acids.

The analogous heterogeneous catalytic reaction is catalyzed by Ti immobilized on a siliceous support. This reaction is detailed in Section 6.3.1.1.

The mechanism of the epoxidation reaction with hydroperoxide is sketched in Figure 4.53.

Two reaction channels are shown. In both cases the initial state consists of an alkoxy intermediate MOR' (or MOH). As indicated on the left of Figure 4.53 the initial alkoxy species reacts with the hydroperoxide and leaves as the alcohol. Its ligand position is replaced by the peroxyl intermediate. In a subsequent step this

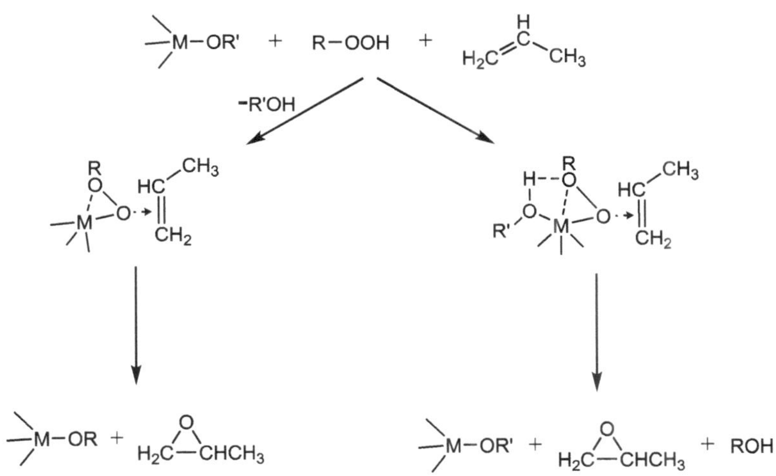

Figure 4.53 Mechanism of oxygen transfer from peroxyl metal complex [273].

reacts with alkene to the epoxide and the alkoxy intermediate is restored. R' is replaced by R.

In the alternative mechanism on the right of Figure 4.53 the hydroperoxide coordinates to the cation. The OOH proton forms a hydrogen bond with the spectator OR' alkoxy, which in a subsequent step protonates OR to H–OR. The other oxygen atom of the peroxide reacts with the alkene to the epoxide. Spectroscopy and DFT calculations of Thomas *et al.* [274] show that the latter mechanism is the most probable.

When a hydroperoxide is used as reactant the co-product of the reaction is an alcohol. In the case where hydrogen peroxide is used the co-product is water. As discussed in Section 6.3.1.1 the reaction with hydrogen peroxide is selective when catalyzed by Ti substituted in the lattice framework of zeolite.

4.5.7 *Reactivity Descriptors; the Dowden M-shaped Volcano Curve*

The strength of the adsorbate bond (M–O) defines the Sabatier maximum. In reducible metal oxides this relates to the redox potential, spin state and ionicity of the M–O bond.

In Section 4.4.2, the seven pillars of selective oxidation of Graselli are mentioned. These principles are formulated in terms of the structure of the catalyst surface and the reactivity of oxygen atoms. Here we will investigate the relation between the reactivity of oxygen atoms and chemical bonding features of the reducible metal oxides.

A kinetic tool to investigate the relation between catalyst performance and surface reactivity is the construction of volcano plots (see Section 2.4.4). When such a plot is made of the oxide cation position along a transition metal row of the periodic table, a correlation with the electronic structure of the cation is made. Such plots are shown here and will also be related with the reactivity of reducible oxide lattice oxygen.

Here volcano plot relations are present for the rate of C–H bond activation by reducible metal oxides in the absence of gas-phase oxygen and the potentials of electrocatalytic oxygen reduction reactions (ORR) and related oxygen evolution reactions (OER). The ORR is important for fuel cell catalysis and the OER for hydrogen evolution from water. The relative potentials measured at the same electric current are a measure of rate of the electrocatalytic reaction.

In 1972 Dennis Dowden, a leading industrial catalytic chemist at ICI in the 1960s and 1970s [275], discovered the twin-peaked M-shaped volcano curve of the reactivity of reducible oxides when their activity is plotted as a function of metal cation position along a row of the periodic table.

Dowden's M-shaped volcano curve is shown in Figure 4.54. Measured reaction rates of H_2–D_a exchange, a dehydrogenation reaction and a disproportionation reaction are plotted. The M-shaped volcano curve is very different from expected single maximum dependence that is found when such a volcano plot is constructed as a function of surface reactivity of the catalyst as probed by adsorbate bond energies (see Section 2.4.4). The M-shaped dependence will be seen to be related to the spin state of the cations.

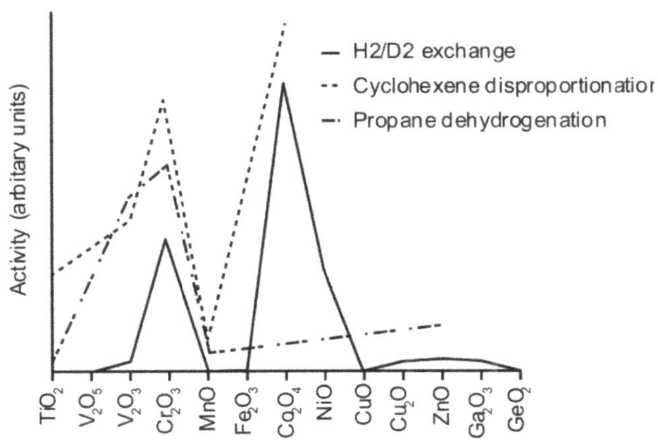

Figure 4.54 Activity pattern for H–H, C–H and C=C bond activation of reducible oxides as a function of metal cation position in the third row of the periodic table [278].

The M-shaped reactivity pattern is not only found for hydrogenation reactions, but also for oxidation reactions. Frank Stone of the University of Bath demonstrated similar M-shaped volcano curve dependence for N_2O decomposition or CO oxidation catalyzed by solid solutions of reducible transition metal cations in MgO [276], [277]. This dependence of reaction rate on position of third row transition metals is general.

Also, the potentials of electrocatalysis of OER or ORR show the M-shaped dependence when plotted as a function of d-valence electron number.

Here results of measurements with perovskite (the perovskite structure is shown in Figure 4.55a) electrodes of variable

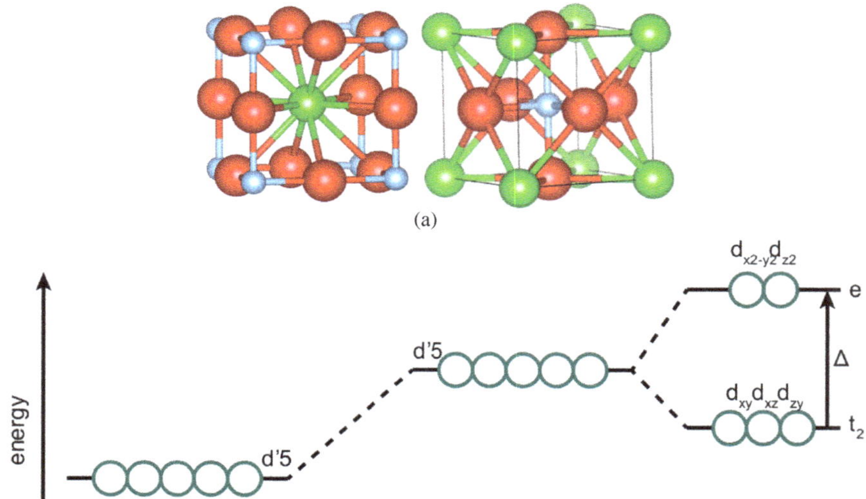

(b)

Figure 4.55 (a) The perovskite structure of $BaTiO_3$. Oxygen – red, barium – green, titanium – light blue. The Ba^{2+} cation is coordinated to 12 oxygen atoms, while the Ti^{4+} cation is octahedrally surrounded by 6 oxygen atoms. The reducible cations of the third row of the periodic table occupy the octahedral positions. (b) Ligand field splitting of d atomic orbitals of cation octahedrally surrounded by six oxygen atoms. The negative charge on the oxygen anions shifts the d-valence electron energies upwards. The d-valence electrons split into three bonding degenerate t_{2g} atomic orbitals and two degenerate anti-bonding e_g atomic orbitals.

composition with metal atoms of the third transition metal row of the periodic table by J. B. Goodenough *et al.* (Nobel award 2019) [279], [280], [285] are used to discuss the dependence of redox activity on cation electronic structure.

For the ORR in Figure 4.56a, the reduction potentials at finite current show the Dowden M-shaped reactivity pattern when plotted as a function of d-valence electron count of reducible perovskite cations. The larger the potential (that is measured with reference to hydrogen electrode), the larger the electrocatalytic rate.

Dowden already realized that the reactivity dependence he discovered relates to ligand field splitting and electron occupation of the d-atomic orbitals of the respective cations. Ligand field orbital splitting predicts differences in magnetic moments (electron spin state) as a function of electron distribution over bonding and anti-bonding orbitals.

The reducible metal cations of the third row of the periodic table occupy the oxygen octahedra of the perovskite structure (blue substitution site in Figure 4.55a). The ligand field splitting for a cation that has octahedral coordination is illustrated in Figure 4.55b. In the metal oxide the d-electrons are localized on the cation and not delocalized as in the metal [282]. The electrostatic field of the oxide will split the energies of the d-atomic orbitals that have no difference in the free cation. Orbital interactions and symmetry determine this d-valence electron energy splitting.

The 3d-valence electrons of a transition metal cation that is octahedrally surrounded by six oxygen atoms are distributed over three degenerate t_g and two degenerate e_g atomic orbitals (symbols indicate symmetry of orbitals).

Electron occupation of t_{2g} atomic orbitals gives a bonding contribution to the bond strength, while occupation of the e_g orbitals gives an anti-bonding contribution to the chemical bond energy. For the transition metal cations of the third row of the periodic table, the electron-electron repulsion that two electrons experience when positioned in the same atomic orbital is larger than the ligand field energy splitting Δ of Figure 4.55b.

The distribution of electrons as a function of d-valence electron count becomes the following: Ti^{2+} (t_{2g} = 2), V^{2+} (t_{2g} = 3), Cr^{2+} (t_{2g} = 3,

$e_g = 1$), Mn^{2+} ($t_{2g} = 3$, $e_g = 2$), Fe^{2+} ($t_{2g} = 4$, $e_g = 2$), Co^{2+} ($t_{2g} = 5$, $e_g = 2$), Ni^{2+} ($t_{2g} = 6$, $e_g = 2$). The interaction energy with neighbor oxygen atoms will be maximum for V^{2+} and Ni^{2+} and will be minimum for Mn^{2+}. Mn^{2+} has a high magnetic moment, because according to Hund's rule electron spins are oriented parallel when electrons occupy orbitals singly. The five valence electrons of Mn^{2+} in the perovskite lattice each occupy a different 3d orbital.

The electron distribution that corresponds to this high-spin state contrasts to that of the low-spin cation state where all bonding orbitals are doubly occupied before the anti-bonding orbitals fill. This happens for the electronic structure of the transition metals of the fourth or fifth row of the periodic table. For these metals the M-shaped volcano curve is absent and classical single maximum volcano curve dependence of reactivity results.

Due to ligand field splitting of the 3d transition metals the M^{2+}–O interaction energy will show an M-shaped dependence. The maxima in the catalytic experiments of Figure 4.54 are shifted compared to the electrochemical perovskite measurements of Figure 4.56a, because of the respective differences in cationic charge (the electron occupations are shifted by one).

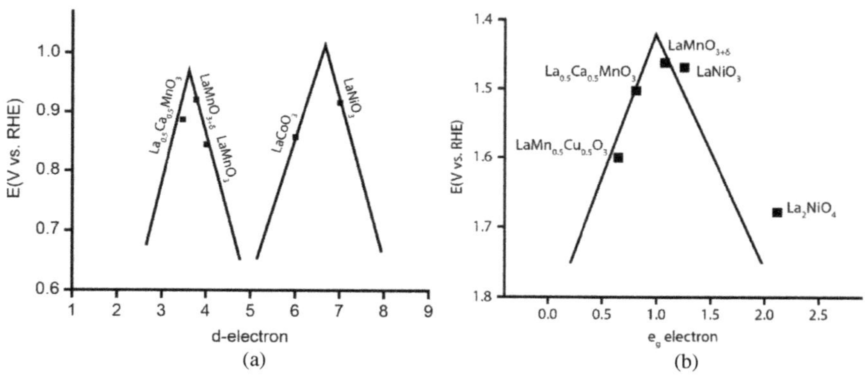

Figure 4.56 Oxygen reduction reaction activity of perovskite (ABO_3) catalysts. (a) Reduction potential as a function of reducible cation 3d-valence electron occupation. (b) The reduction potential as a function of electron occupation of cation anti-bonding e_g orbitals [283].

The twin-peaked feature in the potential plot of the ORR as a function of d-valence electron occupation of Figure 4.56a disappears when it is plotted against calculated electron occupation of the anti-bonding e_g orbitals of the third row transition metal cations [283]. Figure 4.56b shows that when the e_g orbital electron occupation is near one the ORR reaction rate is maximum. According to Goodenough *et al.*, the electron occupation of the e_g atomic orbitals is an electronic reactivity descriptor of the reducible oxide. This can be generalized to catalytic oxidation activity. However, this has been criticized by Nørskov [281] as is discussed below.

Similar volcano plots are constructed for the reverse OER reaction. In Figure 4.57a the OER potential is plotted as a function of the electron occupation of the cation e_g atomic orbitals. A very similar dependence as for the ORR is found. For the OER the lower the potential that generates current, the larger the electrode-catalyzed reaction rate. The DFT calculations of Figure 4.57b of Nørskov *et al.* [280] compare OER reactivity with adsorption energies of adsorbed O and OH. The free energy difference relates to the bond energy of

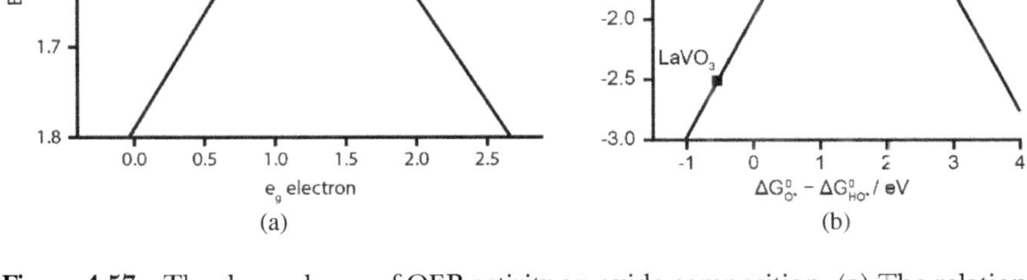

Figure 4.57 The dependence of OER activity on oxide composition. (a) The relation between the OER catalytic activity and the occupancy of the e_g atomic orbitals of the reducible perovskite transition metal [284]. (b) The OER overpotential as a function of M–O bond energy for substituted perovskites. Activity trends towards oxygen evolution plotted for perovskites. The negative theoretical overpotential is plotted against the standard free energy of the difference $\Delta G^0_{O*} - \Delta G^0_{HO*}$ [280].

surface oxygen. In this figure the calculated overpotential of the electrode reaction is plotted against the free energy difference of O_{ad} and OH_{ad}. As explained below, the closer the calculated overpotential is to zero, the better the electrocatalytic rate approximates the predicted value of the Sabatier volcano plot maximum. The perovskite compounds for which the calculated overpotential is near zero are close in composition to the compounds with maximum rate of the OER experiments of Figure 4.56.

Nørskov *et al.* [281] suggest that the M–O bond energy of intermediate oxygen adatoms relates linearly with the e_g anti-bonding electron occupation as well as t_{2g} bonding electron occupation. The electron distribution over both these orbitals determine the M–O bond strength. This is validated by early oxygen adsorption experiments for third row metal oxides performed by Kremenić *et al.* in 1985 [285]. When the oxygen adsorption energy is plotted as a function of metal cation position in their row of the periodic table, the M-shaped dependence is also found.

The assumed relation of Figure 4.57b between theoretically computed overpotential (the potential difference where current is measured and the equilibrium potential of reaction) and reaction rate is important to understand. Theory calculates the overpotential from equilibrium theory as is detailed below. The maximum rate prediction assumes that relative rates are dominated by reaction energy differences and that additional transition state energies are small.

Within equilibrium theory the optimum $\Delta G^0_{O*} - \Delta G^0_{HO*}$ value for maximum reaction rate has an elegant physical chemical interpretation. This will be illustrated for the OER reaction. The OER can be decomposed into four elementary reactions given in Eq. (4.13):

$$\begin{array}{ll} \text{a} & H_2O \rightarrow OH^* + H^+ + el \\ \text{b} & OH^* \rightarrow O^* + H^+ + el \\ \text{c} & H_2O + O^* \rightarrow OOH^* + H^+ + el \\ \text{d} & OOH^* \rightarrow O_2(g) + H^+ + el \end{array} \tag{4.13}$$

The overall decomposition energy of water into O_2 and H_2 is endothermic. This energy cost is overcome in a hydrolysis cell by applying a voltage. The Faraday relation gives the voltage ΔU that overcomes the reaction free energy $\Delta\mu$:

$$\Delta m = Z \cdot F \cdot \Delta U \tag{4.14}$$

Z is the number of electrons transferred, F the Faraday constant and ΔU the equilibrium electrochemical potential. For the OER, $\Delta U = 1.23$ V.

Only when the free energy cost of each of the steps of Eq. (4.14) is the same, the overpotential η is zero. The overpotential η is the additional electropotential over the thermodynamically required potential for the cell to have finite current [286].

Within equilibrium theory the maximum difference in energies of the reactions of Eq. (4.13) defines the overpotential. Calculations of adsorption energies define this difference and also the oxygen atom adsorption energy $E_{ads}(\mathrm{OH})$ that corresponds to the computed overpotential [286].

The free energy of the reactions of Eq. (4.13) can be calculated by equating the energy contribution of H^+ + el with ½ H_2.

From $E^*_{ads}(\mathrm{OH}) - E_{ads}(\mathrm{O})$ at the volcano curve maximum, the optimum value of $E^{opt}_{ads}(\mathrm{O})$ can be calculated. $E^*_{ads}(\mathrm{OH}) - E_{ads}(\mathrm{O})$ is given by Eq. (4.15):

$$E^*_{ads}(\mathrm{OH}) - E_{ads}(\mathrm{O}) = E\ (\text{O–H bond with respect to } ½\ \mathrm{H_2}) + E_{ads}(\mathrm{OH}) - E_{ads}(\mathrm{O}) \tag{4.15}$$

According to the Bond Order Principle (see Section 2.3.3) theory or scaling laws [287], [288]:

$$E_{ads}(\mathrm{OH}) \oplus \tfrac{1}{2} E_{ads}(\mathrm{O}).$$

As can be deduced from Figure 4.57b the volcano curve maximum occurs at the value of $E^*_{ads}(\mathrm{OH}) - E_{ads}(\mathrm{O}) = 1.7\mathrm{eV}$. This value differs by 0.47 eV from the thermodynamic value of 1.23 eV of the OER. The equilibrium theory argument of Eq. (4.13) equals this difference with the overpotential, which indeed is equal to the calculated minimum overpotential.

At zero overpotential the OER potential is 1.23 eV. Equilibrium theory can be used to predict the optimum value for $E_{ads}^{opt}(\mathrm{O})$ at this potential. One finds that this ideal value gives $E_{ads}^{opt}(\mathrm{O}) = 300$ kJ/mol (using an estimated value of 480 kJ/mol for the O–H bond dissociation energy).

This is close to calculated values of $E_{ads}(\mathrm{O})$ adsorbed on the Pt surface (see Section 2.3.3.3.1). Pt is used in practice as electrode material for this reaction [289], [290]. It is also close to $E_{ads}(\mathrm{O})$ of the Co_3O_4 electrode [291]–[294], which is extensively investigated for the OER.

The interpretation of Figure 4.56 is in accord with the Sabatier principle. When the M–O bond is too strong (left part of curve) the reaction rate is limited by the elementary rate of oxygen desorption, and when the M–O bond is too weak (right part of curve) the elementary reaction rate of water is rate limiting.

In summary, the strength of the surface M–O bond is an essential surface reactivity descriptor. Kinetic reactivity studies as a function of material should primarily relate catalyst performance with oxygen atom bond energies.

The M–O bond strength is a sensitive function of composition and structure of reducible oxides. The distribution of d-valence electrons over the bonding and anti-bonding orbitals between cation and neighboring oxygen atoms is an electronic reactivity descriptor. Here it is shown that for third row elements, in addition to the electron occupation of the d-valence electrons, the spin state of the cation may determine the M–O bond energy largely.

Metal oxides of different composition do not necessarily have M–O bond strengths that are also different. This is the reason for the twin volcano peak of catalytic rate when plotted as a function of cation position in the third row of the periodic table. Additional reasons for reactivity differences of the oxygen atoms are the polarity of the M–O bond that depends on size and charge of cations. This is discussed more extensively in Section 2.3.3.

Table 4.1 Oxidation reactions discussed in this chapter.

Ammonia to NO_x (Ostwald reaction)	4.3.1
Ammonia with NO_x to N_2	4.4.4.3
Benzene to maleic anhydride	4.2
Benzene to phenol	4.2, 4.5.6
Butane to acetic acid	4.2.2.1
Butane to maleic anhydride	4.2.2.4, 4.4.4.1
Carbon monoxide to hydrogen (water-gas shift)	4.2.2.3
Ethene to epoxide	4.2.2.3, 4.3.3.1
Ethene to aldehyde (Wacker)	4.2.2.2
Electrocatalytic hydrogen oxidation (ORR)	4.5.6
Methane to ethene (oxidation)	4.2, 4.5.2
Methane to ethene, benzene (non-oxidative)	4.5.3
Methane to synthesis gas (steam reforming)	4.2.2.3
Methanol to formaldehyde	4.2.2.3, 4.3.2, 4.4.3.2
Propane to acrylonitrile	4.4.3.1
Propane to propene	4.4.4.2
Propene to acrylonitrile	4.2.2.4, 4.4.3.1
Propene to acrolein	4.2.2.4, 4.3.3.2, 4.4.3.1
Propene to propene epoxide	4.3.3.2, 4.5.7
Toluene to benzoic acid	4.5.2
p-xylene to terephtalic	4.2.2.1

4.5.8 *Summary and List of Reactions*

This chapter provides mechanistic principles of selective oxidation reactions catalyzed by transition metals and reducible oxides. For most of the important selective oxidation reactions, catalytic cycles as well as elementary steps are discussed. The inorganic chemistry of the catalytic surface is an essential part of the oxidation reaction mechanism presentation. The mechanism sensitively depends on catalyst structure and composition.

After a summary of the chapter in Section 4.5, in Sections 4.5.1–4.5.6 a comprehensive review is given of the different elementary reactions that can be distinguished in selective heterogenous catalytic oxidation. The electronic structure of cations (catalyst reactivity descriptors) in relation to oxide reactivity is discussed in Section 4.5.7.

References

[1] L. B. Hunt, "Sir Humphry Davy on platinum, a footnote to the bicentenary celebrations," *Platin. Met. Rev.*, vol. 23, no. 1, pp. 29–31, 1979.

[2] G. B. Kauffman, "Johann Wolfgang Döbereiner's Feuerzeug," *Platin. Met. Rev*, vol. 43, no. 3, pp. 122–128, 1999.

[3] U. Bossel, *The Birth of the Fuel Cell, 1835-1845*. European Fuel Cell Forum, 2000.

[4] L. B. Hunt, "The ammonia oxidation process for nitric acid manufacture," *Platin. Met. Rev.*, vol. 2, no. 4, pp. 129–134, 1958.

[5] G. Chinchen, P. Davies, and R. J. Sampson, "The Historical Development of Catalytic Oxidation Processes," in *Catalysis*, J. R. Anderson and M. Boudart, Eds. Springer, 1987, pp. 1–67.

[6] B. K. Hodnett, *Heterogeneous Catalytic Oxidation: Fundametal and Technological Aspects of the Selective Oxidation of Organic Compounds*. John Wiley & Sons, 2000.

[7] G. E. Keller and M. M. Bhasin, "Synthesis of ethylene via oxidative coupling of methane I. Determination of active catalysts," *J. Catal.*, vol. 73, no. 1, pp. 9–19, 1982, doi: 10.1016/0021-9517(82)90075-6.

[8] U. Zavyalova, M. Holena, R. Schlögl, and M. Baerns, "Statistical analysis of past catalytic data on oxidative methane coupling for new insights into the composition of high-performance catalysts," *ChemCatChem*, vol. 3, no. 12, pp. 1935–1947, Dec. 2011, doi: 10.1002/CCTC.201100186.

[9] L. Luo, R. You, Y. Liu, J. Yang, Y. Zhu, W. Wen, Y. Pan, F. Qi, and W. Huang, "Gas-phase reaction network of Li/MgO-catalyzed oxidative coupling of methane and oxidative dehydrogenation of ethane," *ACS Catal.*, vol. 9, no. 3, pp. 2514–2520, Mar. 2019, doi: 10.1021/ACSCATAL.8B04728.

[10] A. Bielanski and J. Haber, *Oxygen in Catalysis*. Dekker Inc., 1991.

[11] A. K. Uriarte, M. A. Rodkin, M. J. Gross, A. S. Kharitonov, and G. I. Panov, "Direct hydroxylation of benzene to phenol by nitrous oxide," *Stud. Surf. Sci. Catal.*, vol. 110, pp. 857–864, Jan. 1997, doi: 10.1016/S0167-2991(97)81048-8.

[12] G. K. Boreskov, "The catalysis of isotopic exchange in molecular oxygen," *Adv. Catal.*, vol. 15, no. C, pp. 285–339, Jan. 1965, doi: 10.1016/S0360-0564(08)60556-8.

[13] G. I. Golodets, *Heterogeneous Catalytic Reactions Involving Molecular Oxygen*. Elsevier, 1983.

[14] M. Witko, E. Brocławik, and J. Haber, "Stereochemistry of the catalytic oxidation of benzene as revealed by quantum chemical calculations," *J. Mol. Catal.*, vol. 35, no. 2, pp. 179–189, May 1986, doi: 10.1016/0304-5102(86)87023-7.

[15] B. Grzybowska-Świerkosz, "Thirty years in selective oxidation on oxides: What have we learned?," *Top. Catal.*, vol. 11–12, no. 1–4, pp. 23–42, 2000, doi: 10.1023/a:1027256116098.

[16] R. A. Sheldon, "A History of Oxygen Activation: 1773–1993," in *The Activation of Dioxygen and Homogeneous Catalytic Oxidation*, D. H. R. Barton, A. E. Martell, and D. T. Sawyer, Eds. Springer, 1993, pp. 9–30.

[17] T. Ito, J. X. Wang, C. H. Lin, and J. H. Lunsford, "Oxidative dimerization of methane over a lithium-promoted magnesium oxide catalyst," *J. Am. Chem. Soc.*, vol. 107, no. 18, pp. 5062–5068, 1985, doi: 10.1021/JA00304A008.

[18] J. M. Thomas, *Design and Applications of Single-Site Heterogeneous Catalysts: Contributions to Green Chemistry, Clean Technology and Sustainability*. Imperial College Press, 2012.

[19] R. A. Sheldon and R. A. van Santen, *Catalytic Oxidation: Principles and Applications*. World Scientific, 1995.

[20] R. A. Sheldon and J. K. Kochi, *Metal-Catalyzed Oxidations of Organic Compounds: Mechanistic Principles and Synthetic Methodology Including Biochemical Processes*. Elsevier, 1981.

[21] J. Smidt, W. Hafner, R. Jira, J. Sedlmeier, R. Sieber, R. Rüttinger, and H. Kojer, "Katalytische Umsetzungen von Olefinen an Platinmetall-Verbindungen Das Consortium-Verfahren zur Herstellung von Acetaldehyd," *Angew. Chem.*, vol. 71, no. 5, pp. 176–182, Mar. 1959, doi: 10.1002/ANGE.19590710503.

[22] J. A. Keith and P. M. Henry, "The mechanism of the Wacker reaction: a tale of two hydroxypalladations," *Angew. Chem. Int. Ed.*, vol. 48, no. 48, pp. 9038–9049, Nov. 2009, doi: 10.1002/ANIE.2009 02194.

[23] A. B. Evnin, J. A. Rabo, and P. H. Kasai, "Heterogeneously catalyzed vapor-phase oxidation of ethylene to acetaldehyde," *J. Catal.*, vol. 30, no. 1, pp. 109–117, Jul. 1973, doi: 10.1016/0021-9517(73)90057-2.

[24] A. W. Stobbe-Kreemers, R. B. Dielis, M. Makkee, and J. J. F. Scholten, "Heteropolyanions as redox components in heterogeneous Wacker oxidation catalysts," *J. Catal.*, vol. 154, no. 2, pp. 175–186, Jul. 1995, doi: 10.1006/JCAT.1995.1158.

[25] J. G. Smith, "Frédéric Kuhlmann pioneer of platinum as an industrial catalyst," *Platin. Met. Rev.*, vol. 32, no. 2, p. 84, 1988.

[26] J. Pérez-Ramírez, F. Kapteijn, K. Schöffel, and J. A. Moulijn, "Formation and control of N2O in nitric acid production: Where do we stand today?," *Appl. Catal. B Environ.*, vol. 44, no. 2, pp. 117–151, Aug. 2003, doi: 10.1016/S0926-3373(03)00026-2.

[27] J. Wisniak, "The history of catalysis. From the beginning to Nobel Prizes," *Educ. Química*, vol. 21, no. 1, pp. 60–69, Jan. 2010, doi: 10.1016/S0187-893X(18)30074-0.

[28] U. Deka, I. Lezcano-Gonzalez, B. M. Weckhuysen, and A. M. Beale, "Local environment and nature of Cu active sites in zeolite-based catalysts for the selective catalytic reduction of NOx," *ACS Catal.*, vol. 3, no. 3, pp. 413–427, Mar. 2013, doi: 10.1021/CS300794S.

[29] J. Rostrup-Nielsen and L. J. Christiansen, *Concepts in Syngas Manufacture*. Imperial College Press, 2011.

[30] C. Murkin and J. Brightling, "Eighty years of steam reforming," *Johnson Matthey Technol. Rev.*, vol. 60, no. 4, pp. 263–269, 2016, doi: 10.1595/205651316X692923.

[31] G. Jones, J. Jakobsen, S. Shim, J. Kleis, M. Anderson, J. Rossmeisl, F. Abild-Pedersen, T. Ligaard, S. Helveg, and B. Hinnemann, "First principles calculations and experimental insight into methane steam reforming over transition metal catalysts," *J. Catal.*, vol. 259, no. 1, pp. 147–160, 2008, doi: 10.1016/j.jcat.2008.08.003.

[32] H. Adkins and W. R. Peterson, "The oxidation of methanol with air over iron, molybdenum, and iron-molybdenum oxides," *J. Am. Chem. Soc.*, vol. 53, no. 4, pp. 1512–1520, Apr. 1931, doi: 10.1021/JA01355A050.

[33] D. Sheldon, "Methanol production - A technical history," *Johnson Matthey Technol. Rev.*, vol. 61, no. 3., pp. 172–182, Jul. 01, 2017, doi: 10.1595/205651317X695622.

[34] M. Okumura, T. Fujitani, J. Huang, and T. Ishida, "A career in catalysis: Masatake Haruta," *ACS Catal.*, vol. 5, no. 8, pp. 4699–4707, Aug. 2015, doi: 10.1021/ACSCATAL.5B01122.

[35] J. C. Védrine, "Heterogeneous partial (amm)oxidation and oxidative dehydrogenation catalysis on mixed metal oxides," *Catalysts*, vol. 6, no. 2, p. 22, Jan. 2016, doi: 10.3390/CATAL6020022.

[36] "Sohio Acrylonitrile Process," *American Chemical Society*, May 16, 2022. https://www.acs.org/content/acs/en/education/whatischemistry/landmarks/acrylonitrile.html (accessed May 16, 2022).

[37] J. K. Bartley, N. F. Dummer, and G. J. Hutchings, "Vanadium phosphate catalysts," *Met. Oxide Catal.*, vol. 2, pp. 499–537, Mar. 2009, doi: 10.1002/9783527626113.CH12.

[38] F. Cavani, G. Centi, and P. Marion, "Catalytic Ammoxidation of Hydrocarbons on Mixed Oxides," in *Metal Oxide Catalysis*, S. David Jackson and J. S. J. Hargreaves, Eds. John Wiley & Sons, 2008, pp. 771–818.

[39] F. Cavani, N. Ballarini, and A. Cericola, "Oxidative dehydrogenation of ethane and propane: How far from commercial implementation?," *Catal. Today*, vol. 127, no. 1–4, pp. 113–131, Sep. 2007, doi: 10.1016/J.CATTOD.2007.05.009.

[40] H. Bosch and F. Janssen, "Preface," *Catal. Today*, vol. 2, no. 4, p. v, Mar. 1988, doi: 10.1016/0920-5861(88)80001-4.

[41] K. Tamaru and G. A. Mills, "Catalysts for control of exhaust emissions," *Catal. Today*, vol. 22, no. 2, pp. 349–360, Dec. 1994, doi: 10.1016/0920-5861(94)80109-6.

[42] L. Han, S. Cai, M. Gao, J. Hasegawa, P. Wang, J. Zhang, L. Shi, and D. Zhang, "Selective catalytic reduction of NOx with NH3 by using novel catalysts: state of the art and future prospects," *Chem. Rev.*, vol. 119, no. 19, pp. 10916–10976, Oct. 2019, doi: 10.1021/ACS.CHEMREV.9B00202.

[43] J. N. Armor, "A history of industrial catalysis," *Catal. Today*, vol. 163, no. 1, pp. 3–9, Apr. 2011, doi: 10.1016/J.CATTOD.2009.11.019.

[44] C. W. Nutt and S. Kapur, "Oxidation of ammonia on platinum," *Nature*, vol. 224, no. 5215, pp. 169–169, 1969, doi: 10.1038/224169a0.

[45] C. W. Nutt and S. Kapur, "Mechanism of oxidation of ammonia on platinum," *Nature*, vol. 220, no. 5168, pp. 697–698, 1968, doi: 10.1038/220697a0.

[46] T. Pignet and L. D. Schmidt, "Kinetics of NH3 oxidation on Pt, Rh, and Pd," *J. Catal.*, vol. 40, no. 2, pp. 212–225, Nov. 1975, doi: 10.1016/0021-9517(75)90249-3.

[47] E. V Rebrov, M. H. J. M. de Croon, and J. C. Schouten, "A kinetic study of ammonia oxidation on a Pt catalyst in the explosive region in a microstructured reactor/heat-exchanger," *Chem. Eng. Res. Des.*, vol. 81, no. 7, pp. 744–752, Aug. 2003, doi: 10.1205/026387603322302913.

[48] W. K. Offermans, A. P. J. Jansen, and R. A. van Santen, "Ammonia activation on platinum {{}111{}}: A density functional theory study," *Surf. Sci.*, vol. 600, no. 9, pp. 1714–1734, 2006, doi: 10.1016/j.susc.2006.01.031.

[49] G. Novell-Leruth, J. M. Ricart, and J. Pérez-Ramírez, "Pt(100)-catalyzed ammonia oxidation studied by DFT: Mechanism and microkinetics," *J. Phys. Chem. C*, vol. 112, no. 35, pp. 13554–13562, Sep. 2008, doi: 10.1021/JP802489Y.

[50] W. K. Offermans, A. P. J. Jansen, R. A. van Santen, G. Novell-Leruth, J. M. Ricart, and J. Perez-Ramirez, "Ammonia dissociation on Pt{100}, Pt{111}, and Pt{211}: A comparative density functional theory study," *J. Phys. Chem. C*, vol. 111, no. 47, pp. 17551–17557, Nov. 2007, doi: 10.1021/jp073083f.

[51] A. C. M. Van Den Broek, J. Van Grondelle, and R. A. van Santen, "Determination of surface coverage of catalysts: temperature

programmed experiments on platinum and iridium sponge catalysts after low temperature ammonia oxidation," *J. Catal.*, vol. 185, no. 2, pp. 297–306, Jul. 1999, doi: 10.1006/JCAT.1999.2506.

[52] R. J. Mukerji, A. S. Bolina, and W. A. Brown, "The influence of steps on the dissociation of NO on Pt surfaces: Temperature-programmed desorption studies of NO adsorption on Pt{211}," *J. Chem. Phys.*, vol. 119, no. 20, p. 10844, Nov. 2003, doi: 10.1063/1.1619940.

[53] Y. M. Fogel, B. T. Nadykto, V. F. Rybalko, and I. E. Shvachko, V. I. Korobchanskaya, "Mass-spectrometric investigation of the oxidation of ammonia on platinum," *Kinet. Catal.*, vol. 5, p. 431, 1964.

[54] K. Otto, M. Shelef, and J. T. Kummer, "Studies of surface reactions of nitric oxide by nitrogen-15 isotope labeling. I. Reaction between nitric oxide and ammonia over supported platinum at 200-250. deg.," *J. Phys. Chem.*, vol. 74, no. 13, pp. 2690–2698, 2002, doi: 10.1021/J100707A017.

[55] V. A. Kondratenko and M. Baerns, "Mechanistic insights into the formation of N2O and N2 in NO reduction by NH3 over a polycrystalline platinum catalyst," *Appl. Catal. B Environ.*, vol. 70, no. 1–4, pp. 111–118, Jan. 2007, doi: 10.1016/J.APCATB.2005.11.037.

[56] R. Imbihl, A. Scheibe, Y. F. Zeng, S. Günther, R. Kraehnert, V. A. Kondratenko, M. Baerns, W. K. Offermans, A. P. J. Jansen, and R. A. van Santen, "Catalytic ammonia oxidation on platinum: mechanism and catalyst restructuring at high and low pressure," *Phys. Chem. Chem. Phys.*, vol. 9, no. 27, pp. 3522–3540, Jul. 2007, doi: 10.1039/B700866J.

[57] R. Burch, S. T. Daniells, and P. Hu, "The mechanism of N2O formation via the (NO)2 dimer: A density functional theory study," *J. Chem. Phys.*, vol. 121, no. 6, p. 2737, Jul. 2004, doi: 10.1063/1.1767153.

[58] Q. Ge and M. Neurock, "Structure dependence of NO adsorption and dissociation on platinum surfaces," *J. Am. Chem. Soc.*, vol. 126, no. 5, pp. 1551–1559, Feb. 2004, doi: 10.1021/JA036575O.

[59] C. Popa, R. A. van Santen, and A. P. J. Jansen, "Density-functional theory study of NHx oxidation and reverse reactions on the Rh(111) surface," *J. Phys. Chem. C*, vol. 111, no. 27, pp. 9839–9852, Jul. 2007, doi: 10.1021/JP071072G.

[60] H. Ma and W. F. Schneider, "DFT and microkinetic comparison of Pt, Pd and Rh-catalyzed ammonia oxidation," *J. Catal.*, vol. 383, pp. 322–330, Mar. 2020, doi: 10.1016/J.JCAT.2020.01.029.

[61] L. Gang, B. G. Anderson, J. van Grondelle, and R. A. van Santen, "Intermediate species and reaction pathways for the oxidation of ammonia on powdered catalysts," *J. Catal.*, vol. 199, no. 1, pp. 107–114, Apr. 2001, doi: 10.1006/jcat.2000.3154.

[62] X. Bao, U. Wild, M. Muhler, B. Pettinger, R. Schlögl, and G. Ertl, "Coadsorption of nitric oxide and oxygen on the Ag(110) surface," *Surf. Sci.*, vol. 425, no. 2–3, pp. 224–232, Apr. 1999, doi: 10.1016/S0039-6028(99)00190-9.

[63] E. V. Kondratenko, V. A. Kondratenko, M. Richter, and R. Fricke, "Influence of O2 and H2 on NO reduction by NH3 over Ag/Al2O3: A transient isotopic approach," *J. Catal.*, vol. 239, no. 1, pp. 23–33, Apr. 2006, doi: 10.1016/J.JCAT.2006.01.008.

[64] A. Nagy and G. Mestl, "High temperature partial oxidation reactions over silver catalysts," *Appl. Catal. A Gen.*, vol. 188, no. 1–2, pp. 337–353, Nov. 1999, doi: 10.1016/S0926-860X(99)00246-X.

[65] G. I. N. Waterhouse, G. A. Bowmaker, and J. B. Metson, "Mechanism and active sites for the partial oxidation of methanol to formaldehyde over an electrolytic silver catalyst," *Appl. Catal. A Gen.*, vol. 265, no. 1, pp. 85–101, Jun. 2004, doi: 10.1016/J.APCATA.2004.01.016.

[66] I. E. Wachs and R. J. Madix, "The oxidation of methanol on a silver (110) catalyst," *Surf. Sci.*, vol. 76, no. 2, pp. 531–558, Sep. 1978, doi: 10.1016/0039-6028(78)90113-9.

[67] I. E. Wachs, "Extending surface science studies to industrial reaction conditions: Mechanism and kinetics of methanol oxidation over silver surfaces," *Surf. Sci.*, vol. 544, no. 1. pp. 1–4, Oct. 2003, doi: 10.1016/j.susc.2003.08.006.

[68] A. C. Van Veen, O. Hinrichsen, and M. Muhler, "Mechanistic studies on the oxidative dehydrogenation of methanol over polycrystalline silver using the temporal-analysis-of-products approach," *J. Catal.*, vol. 210, no. 1, pp. 53–66, Aug. 2002, doi: 10.1006/JCAT.2002.3682.

[69] X. Bao, M. Muhler, T. Schedel-Niedrig, and R. Schlögl, "Interaction of oxygen with silver at high temperature and atmospheric pressure: A spectroscopic and structural analysis of a strongly bound surface species," *Phys. Rev. B*, vol. 54, no. 3, p. 2249, Jul. 1996, doi: 10.1103/PhysRevB.54.2249.

[70] A. Andreasen, H. Lynggaard, C. Stegelmann, and P. Stoltze, "Simplified kinetic models of methanol oxidation on silver," *Appl. Catal. A Gen.*, vol. 289, no. 2, pp. 267–273, Aug. 2005, doi: 10.1016/j.apcata.2005.05.004.

[71] H. Aljama, J. S. Yoo, J. K. Nørskov, F. Abild-Pedersen, and F. Studt, "Methanol partial oxidation on Ag(111) from first principles," *ChemCatChem*, vol. 8, no. 23, pp. 3621–3625, Dec. 2016, doi: 10.1002/CCTC.201601053.

[72] H. H. Voge and C. R. Adams, "Catalytic oxidation of olefins," *Adv. Catal.*, vol. 17, no. C, pp. 151–221, Jan. 1967, doi: 10.1016/S0360-0564(08)60687-2.

[73] R. A. van Santen and H. P. C. E. Kuipers, "The mechanism of ethylene epoxidation," *Adv. Catal.*, vol. 35, no. C, pp. 265–321, Jan. 1987, doi: 10.1016/S0360-0564(08)60095-4.

[74] M. A. Barteau and R. J. Madix, "Low-pressure oxidation mechanism and reactivity of propylene on Ag(110) and relation to gas-phase acidity," *J. Am. Chem. Soc.*, vol. 105, no. 3, pp. 344–349, 1983, doi: 10.1021/JA00341A008.

[75] L. Cheng, C. Yin, F. Mehmood, B. Liu, J. Greeley, S. Lee, B. Lee, S. Seifert, R. E. Winans, D. Teschner, R. Schlögl, S. Vajda, and L. A. Curtiss, "Reaction mechanism for direct propylene epoxidation by alumina-supported silver aggregates: The role of the particle/support interface," *ACS Catal.*, vol. 4, no. 1, pp. 32–39, Jan. 2014, doi: 10.1021/CS4009368.

[76] M. T. Greiner, T. E. Jones, B. E. Johnson, T. C. R. Rocha, Z. J. Wang, M. Armbrüster, M. Willinger, A. Knop-Gericke, and R. Schlögl, "The oxidation of copper catalysts during ethylene epoxidation," *Phys. Chem. Chem. Phys.*, vol. 17, no. 38, pp. 25073–25089, Sep. 2015, doi: 10.1039/C5CP03722K.

[77] J. Teržan, M. Huš, B. Likozar, and P. Djinović, "Propylene epoxidation using molecular oxygen over copper-and silver-based catalysts: a review," *ACS Catal.*, vol. 10, no. 22., pp. 13415–13436, Nov. 2020, doi: 10.1021/acscatal.0c03340.

[78] H. Voge, "Mechanism of propylene oxidation over cuprous oxide," *J. Catal.*, vol. 2, no. 1, pp. 58–62, 1963, doi: 10.1016/0021-9517(63)90139-8.

[79] C. R. Adams and T. J. Jennings, "Investigation of the mechanism of catalytic oxidation of propylene to acrolein and acrylonitrile," *J. Catal.*, vol. 2, no. 1, pp. 63–68, Feb. 1963, doi: 10.1016/0021-9517(63)90140-4.

[80] W. M. H. H. Sachtler, "Mechanism of the catalytic oxidations of propene to acrolein and of isomeric butenes to butadiene," *Recl. des Trav. Chim. des Pays-Bas*, vol. 82, no. 3, pp. 243–245, Jan. 2010, doi: 10.1002/recl.19630820305.

[81] J. W. Harris, J. A. Herron, J. F. DeWilde, and A. Bhan, "Molecular characteristics governing chlorine deposition and removal on promoted Ag catalysts during ethylene epoxidation," *J. Catal.*, vol. 377, pp. 378–388, Sep. 2019, doi: 10.1016/J.JCAT.2019.07.043.

[82] T. Pu, H. Tian, M. E. Ford, S. Rangarajan, and I. E. Wachs, "Overview of selective oxidation of ethylene to ethylene oxide by Ag catalysts," *ACS Catal.*, pp. 10727–10750, 2019, doi: 10.1021/ACSCATAL.9B03443.

[83] G. H. Twigg, "The catalytic oxidation of ethylene," *Dep. Colloid Sci. Cambridge*, vol. 42, no. 81, p. 284, Jan. 1945, doi: 10.1039/TF9464200284.

[84] H. Kanoh, T. Nishimura, and A. Ayame, "Supported silver catalysts for the oxidation of ethylene: The effects of thermal treatment of alumina as catalyst support," *J. Catal.*, vol. 57, no. 3, pp. 372–379, 1979, doi: 10.1016/0021-9517(79)90003-4.

[85] J. H. Miller, A. Joshi, X. Li, and A. Bhan, "Catalytic degradation of ethylene oxide over Ag/α-Al2O3," *J. Catal.*, vol. 389, pp. 714–720, Sep. 2020, doi: 10.1016/J.JCAT.2020.07.008.

[86] E. L. Force and A. T. Bell, "The effect of dichloroethane moderation on the adsorbed species present during the oxidation of ethylene over silver," *J. Catal.*, vol. 44, no. 2, pp. 175–182, Aug. 1976, doi: 10.1016/0021-9517(76)90388-2.

[87] P. A. Kilty and W. M. H. Sachtler, "The mechanism of the selective oxidation of ethylene to ethylene oxide," *Catal. Rev.*, vol. 10, no. 1, pp. 1–16, Jan. 1974, doi: 10.1080/01614947408079624.

[88] W. M. H. Sachtler, C. Backx, and R. A. van Santen, "On the mechanism of ethylene epoxidation," *Catal. Rev.*, vol. 23, no. 1–2, pp. 127–149, Jan. 1981, doi: 10.1080/03602458108068072.

[89] J. W. Harris and A. Bhan, "Moderation of chlorine coverage and ethylene epoxidation kinetics via ethane oxychlorination over promoted Ag/α-Al2O3," *J. Catal.*, vol. 367, pp. 62–71, Nov. 2018, doi: 10.1016/J.JCAT.2018.08.021.

[90] C. T. Campbell and B. E. Koel, "Chlorine promotion of selective ethylene oxidation over Ag(110): Kinetics and mechanism," *J. Catal.*, vol. 92, no. 2, pp. 272–283, Apr. 1985, doi: 10.1016/0021-9517(85)90261-1.

[91] A. J. F. Van Hoof, E. A. R. Hermans, A. P. Van Bavel, H. Friedrich, and E. J. M. Hensen, "Structure sensitivity of silver-catalyzed ethylene epoxidation," *ACS Catal.*, vol. 9, no. 11, pp. 9829–9839, Nov. 2019, doi: 10.1021/ACSCATAL.9B02720.

[92] N. W. Cant and W. K. Hall, "Catalytic oxidation: VI. Oxidation of labeled olefins over silver," *J. Catal.*, vol. 52, no. 1, pp. 81–94, Mar. 1978, doi: 10.1016/0021-9517(78)90125-2.

[93] R. A. van Santen, J. Moolhuysen, and W. M. H. Sachtler, "The role of CH bond breaking in the epoxidation of ethylene," *J. Catal.*, vol. 65, no. 2, pp. 478–480, Oct. 1980, doi: 10.1016/0021-9517(80)90326-7.

[94] E. L. Force and A. T. Bell, "The relationship of adsorbed species observed by infrared spectroscopy to the mechanism of ethylene oxidation over silver," *J. Catal.*, vol. 40, no. 3, pp. 356–371, Dec. 1975, doi: 10.1016/0021-9517(75)90267-5.

[95] R. B. Grant and R. M. Lambert, "A single crystal study of the silver-catalysed selective oxidation and total oxidation of ethylene," *J. Catal.*, vol. 92, no. 2, pp. 364–375, Apr. 1985, doi: 10.1016/0021-9517(85)90270-2.

[96] R. A. van Santen and C. P. M. de Groot, "The mechanism of ethylene epoxidation," *J. Catal.*, vol. 98, no. 2, pp. 530–539, Apr. 1986, doi: 10.1016/0021-9517(86)90341-6.
[97] J. T. Gleaves, A. G. Sault, R. J. Madix, and J. R. Ebner, "Ethylene oxidation on silver powder: A tap reactor study," *J. Catal.*, vol. 121, no. 1, pp. 202–218, Jan. 1990, doi: 10.1016/0021-9517(90)90230-H.
[98] S. Böcklein, S. Günther, and J. Wintterlin, "High-pressure scanning tunneling microscopy of a silver surface during catalytic formation of ethylene oxide," *Angew. Chem. Int. Ed.*, vol. 52, no. 21, pp. 5518–5521, May 2013, doi: 10.1002/ANIE.201210209.
[99] T. C. R. Rocha, M. Hävecker, A. Knop-Gericke, and R. Schlögl, "Promoters in heterogeneous catalysis: The role of Cl on ethylene epoxidation over Ag," *J. Catal.*, vol. 312, pp. 12–16, 2014, doi: 10.1016/j.jcat.2014.01.002.
[100] S. Linic and M. A. Barteau, "Formation of a stable surface oxametallacycle that produces ethylene oxide," *J. Am. Chem. Soc.*, vol. 124, no. 2, pp. 310–317, Jan. 2002, doi: 10.1021/JA0118136.
[101] S. Linic, H. Piao, K. Adib, and M. A. Barteau, "Ethylene epoxidation on Ag: identification of the crucial surface intermediate by experimental and theoretical investigation of its electronic structure," *Angew. Chem.*, vol. 116, no. 22, pp. 2978–2981, May 2004, doi: 10.1002/ANGE.200353584.
[102] D. Torres, N. Lopez, F. Illas, and R. M. Lambert, "Why copper is intrinsically more selective than silver in alkene epoxidation: Ethylene oxidation on Cu(111) versus Ag(111)," *J. Am. Chem. Soc.*, vol. 127, no. 31, pp. 10774–10775, Aug. 2005, doi: 10.1021/JA043227T.
[103] R. M. Lambert, R. L. Cropley, A. Husain, and M. S. Tikhov, "Halogen-induced selectivity in heterogeneous epoxidation is an electronic effect—fluorine, chlorine, bromine and iodine in the Ag-catalysed selective oxidation of ethene," *Chem. Commun.*, vol. 3, no. 10, pp. 1184–1185, Apr. 2003, doi: 10.1039/B302620E.
[104] D. Torres, F. Illas, and R. M. Lambert, "Towards an understanding of promoter action in heterogeneously catalyzed ethene epoxidation: Why chlorine is the best halogen," *J. Catal.*, vol. 260, no. 2, pp. 380–383, 2008, doi: 10.1016/j.jcat.2008.10.011.
[105] M. Huš and A. Hellman, "Dipole effect on ethylene epoxidation: Influence of alkali metals and chlorine," *J. Catal.*, vol. 363, pp. 18–25, Jul. 2018, doi: 10.1016/J.JCAT.2018.04.008.
[106] M. Atkins, J. Couves, M. Hague, B. H. Sakakini, and K. C. Waugh, "On the role of Cs, Cl and subsurface O in promoting selectivity in Ag/α-Al2O3 catalysed oxidation of ethene to ethene epoxide," *J. Catal.*, vol. 235, no. 1, pp. 103–113, Oct. 2005, doi: 10.1016/J.JCAT.2005.07.019.

[107] H. Li, A. Cao, and J. K. Nørskov, "Understanding trends in ethylene epoxidation on group IB metals," *ACS Catal.*, vol. 11, no. 19, pp. 12052–12057, Oct. 2021, doi: 10.1021/ACSCATAL.1C03094.
[108] D. Chen, P.-L. Kang, and Z.-P. Liu, "Active site of catalytic ethene epoxidation: machine-learning global pathway sampling rules out the metal sites," *ACS Catal.*, vol. 11, no. 13, pp. 8317–8326, Jul. 2020, doi: 10.1021/ACSCATAL.1C02029.
[109] M. O. Özbek, I. Önal, and R. A. van Santen, "Ethylene epoxidation catalyzed by silver oxide," *ChemCatChem*, vol. 3, no. 1, pp. 150–153, Jan. 2011, doi: 10.1002/CCTC.201000249.
[110] M. O. Özbek, I. Önal, and R. A. van Santen, "Chlorine and caesium promotion of silver ethylene epoxidation catalysts," *ChemCatChem*, vol. 5, no. 2, pp. 443–451, Feb. 2013, doi: 10.1002/CCTC.201200690.
[111] R. A. van Santen, "The active site of promoted ethylene-epoxidation catalysts," in *Catalysis: Theory to Practice: Proceedings 9th international Congress on Catalysis*, vol. 3, M. J. Phillips and M. Ternan, Eds. Chemical Institute of Canada, 1988, pp. 1152–1158.
[112] H. R. Dettwiler, A. Baiker, and W. Richarz, "Kinetics of ethylene oxidation on a supported silver catalyst," *Helv. Chim. Acta*, vol. 62, no. 6, p. 1689, 1979, doi: 10.1002/aic.690170417.
[113] M. O. Ozbek, I. Onal, and R. A. van Santen, "Why silver is the unique catalyst for ethylene epoxidation," *J. Catal.*, vol. 284, no. 2, pp. 230–235, Dec. 2011, doi: 10.1016/j.jcat.2011.08.004.
[114] M. T. Greiner, T. E. Jones, A. Klyushin, A. Knop-Gericke, and R. Schlögl, "Ethylene epoxidation at the phase transition of copper oxides," *J. Am. Chem. Soc.*, vol. 139, no. 34, pp. 11825–11832, Aug. 2017, doi: 10.1021/jacs.7b05004.
[115] D. Torres, N. Lopez, F. Illas, and R. M. Lambert, "Low-basicity oxygen atoms: A key in the search for propylene epoxidation catalysts," *Angew. Chem. Int. Ed.*, vol. 46, no. 12, pp. 2055–2058, Mar. 2007, doi: 10.1002/anie.200603803.
[116] O. P. H. Vaughan, G. Kyriakou, N. MacLeod, M. Tikhov, and R. M. Lambert, "Copper as a selective catalyst for the epoxidation of propene," *J. Catal.*, vol. 236, no. 2, pp. 401–404, Dec. 2005, doi: 10.1016/J.JCAT.2005.10.019.
[117] J. Teržan, M. Huš, B. Likozar, and P. Djinović, "Propylene epoxidation using molecular oxygen over copper-and silver-based catalysts: a review," *ACS Catal.*, vol. 10, no. 22, pp. 13415–13436, Nov. 2020, doi: 10.1021/ACSCATAL.0C03340.
[118] I. Tezsevin, R. A. van Santen, and I. Onal, "A density functional theory study of propylene epoxidation mechanism on Ag2O(001) surface," *Phys. Chem. Chem. Phys.*, vol. 20, no. 41, pp. 26681–26687, Oct. 2018, doi: 10.1039/c8cp04210a.

[119] Q. Hua, T. Cao, X.-K. Gu, J. Lu, Z. Jiang, X. Pan, L. Luo, W.-X. Li, and W. Huang, "Crystal-plane-controlled selectivity of Cu2O catalysts in propylene oxidation with molecular oxygen," *Angew. Chem.*, vol. 126, no. 19, pp. 4956–4961, May 2014, doi: 10.1002/ANGE.201402374.

[120] J. Lu, J. J. Bravo-Suárez, M. Haruta, and S. T. Oyama, "Direct propylene epoxidation over modified Ag/CaCO3 catalysts," *Appl. Catal. A Gen.*, vol. 302, no. 2, pp. 283–295, Apr. 2006, doi: 10.1016/J.APCATA.2006.01.023.

[121] Y. Le, F. Mehmood, S. Lee, J. Greeley, B. Lee, S. Seifert, R. E. Winansl, W. Elám, R. J. Meyer, P. C. Redfern, D. Teschner, R. Schlö'Gl, M. J. Pellin, L. A. Curtiss, and S. Vajda, "Increased silver activity for direct propylene epoxidation via subnanometer size effects," *Science*, vol. 328, no. 5975, pp. 224–228, Apr. 2010, doi: 10.1126/SCIENCE.1185200.

[122] P. V. Geenen, H. J. Boss, and G. T. Pott, "A study of the vapor-phase epoxidation of propylene and ethylene on silver and silver-gold alloy catalysts," *J. Catal.*, vol. 77, no. 2, pp. 499–510, Oct. 1982, doi: 10.1016/0021-9517(82)90190-7.

[123] J. Llorca, M. Domínguez, C. Ledesma, R. J. Chimentão, F. Medina, J. Sueiras, I. Angurell, M. Seco, and O. Rossell, "Propene epoxidation over TiO2-supported Au–Cu alloy catalysts prepared from thiol-capped nanoparticles," *J. Catal.*, vol. 258, no. 1, pp. 187–198, Aug. 2008, doi: 10.1016/J.JCAT.2008.06.010.

[124] P. Mars and D. W. van Krevelen, "Oxidations carried out by means of vanadium oxide catalysts," *Chem. Eng. Sci.*, vol. 3, p. 41, Jan. 1954, doi: 10.1016/S0009-2509(54)80005-4.

[125] C. Doornkamp and V. Ponec, "The universal character of the Mars and Van Krevelen mechanism," *J. Mol. Catal. A Chem.*, vol. 162, no. 1–2, pp. 19–32, Nov. 2000, doi: 10.1016/S1381-1169(00)00319-8.

[126] G. W. Keulks, "The mechanism of oxygen atom incorporation into the products of propylene oxidation over bismuth molybdate," *J. Catal.*, vol. 19, no. 2, pp. 232–235, Nov. 1970, doi: 10.1016/0021-9517(70)90287-3.

[127] I. Matsuura and G. C. A. Schuit, "Adsorption equilibria and rates of reactions of adsorbed compounds on reduced and oxidized BiMo catalysts," *J. Catal.*, vol. 20, no. 1, pp. 19–39, Jan. 1971, doi: 10.1016/0021-9517(71)90004-2.

[128] R. K. Grasselli, "Site isolation and phase cooperation: Two important concepts in selective oxidation catalysis: A retrospective," *Catal. Today*, vol. 238, pp. 10–27, Dec. 2014, doi: 10.1016/j.cattod.2014.05.036.

[129] J. Haber, "Mechanism of Heterogeneous Catalytic Oxidation," in *Catalytic Oxidation: Principles and Applications*, R. A. Sheldon and R. A. van Santen, Eds. World Scientific, 1995, pp. 17–51.

[130] J. L. Callahan and R. K. Grasselli, "A selectivity factor in vapor-phase hydrocarbon oxidation catalysis," *AIChE J.*, vol. 9, no. 6, pp. 755–760, Nov. 1963, doi: 10.1002/AIC.690090610.

[131] M. J. Cheng and W. A. Goddard, "In silico design of highly selective Mo-V-Te-Nb-O mixed metal oxide catalysts for ammoxidation and oxidative dehydrogenation of propane and ethane," *J. Am. Chem. Soc.*, vol. 137, no. 41, pp. 13224–13227, Oct. 2015, doi: 10.1021/JACS.5B07073.

[132] C. C. McCain, G. Gough, and G. W. Godin, "Catalytic oxidation of propene," *Nature*, vol. 198, no. 4884, pp. 989–990, 1963, doi: 10.1038/198989a0.

[133] P. R. Huang, Y. He, C. Cao, and Z. H. Lu, "Impact of lattice distortion and electron doping on α-MoO3 electronic structure," *Sci. Rep.*, vol. 4, no. 1, pp. 1–7, Nov. 2014, doi: 10.1038/srep07131.

[134] A. Chakrabarti, K. Hermann, R. Druzinic, M. Witko, F. Wagner, and M. Petersen, "Geometric and electronic structure of vanadium pentoxide: A density functional bulk and surface study," *Phys. Rev. B*, vol. 59, no. 16, p. 10583, Apr. 1999, doi: 10.1103/PhysRevB.59.10583.

[135] J. M. Tatibouet, J. E. Germain, and J. C. Volta, "Structure-sensitive catalytic oxidation: Alcohols on graphite-supported molybdenum trioxide," *J. Catal.*, vol. 82, no. 1, pp. 240–244, Jul. 1983, doi: 10.1016/0021-9517(83)90136-7.

[136] Smokefoot, "File:CaWO4.tif," *Wikipedia*, Dec. 26, 2015. https://commons.wikimedia.org/wiki/File:CaWO4.tif (accessed Jul. 12, 2022).

[137] J. F. Brazdil, "Scheelite: a versatile structural template for selective alkene oxidation catalysts," *Catal. Sci. Technol.*, vol. 5, no. 7, pp. 3452–3458, Jun. 2015, doi: 10.1039/C5CY00387C.

[138] B. Grzybowska, J. Haber, and J. Janas, "Interaction of allyl iodide with molybdate catalysts for the selective oxidation of hydrocarbons," *J. Catal.*, vol. 49, no. 2, pp. 150–163, Aug. 1977, doi: 10.1016/0021-9517(77)90251-2.

[139] S. Pudar, J. Oxgaard, K. Chenoweth, A. C. T. Van Duin, and W. A. Goddard, "Mechanism of selective oxidation of propene to acrolein on bismuth molybdates from quantum mechanical calculations," *J. Phys. Chem. C*, vol. 111, no. 44, pp. 16405–16415, Nov. 2007, doi: 10.1021/JP074452A.

[140] J. Haber and B. Grzybowska, "Mechanism of the oxidation of olefins on mixed oxide catalysts," *J. Catal.*, vol. 28, no. 3, pp. 489–492, Mar. 1973, doi: 10.1016/0021-9517(73)90142-5.

[141] R. K. Grasselli, J. D. Burrington, D. J. Buttrey, P. DeSanto, C. G. Lugmair, A. F. Volpe, and T. Weingand, "Multifunctionality of active centers in (amm)oxidation catalysts: From Bi-Mo-Ox to

Mo-V-Nb-(Te, Sb)-Ox," *Top. Catal.*, vol. 23, no. 1–4, pp. 5–22, Aug. 2003, doi: 10.1023/A:1024859917786.

[142] A. Getsoian, Z. Zhai, and A. T. Bell, "Band-gap energy as a descriptor of catalytic activity for propene oxidation over mixed metal oxide catalysts," *J. Am. Chem. Soc.*, vol. 136, no. 39, pp. 13684–13697, Sep. 2014, doi: 10.1021/JA5051555.

[143] R. B. Licht and A. T. Bell, "A DFT investigation of the mechanism of propene ammoxidation over α-bismuth molybdate," *ACS Catal.*, vol. 7, no. 1, pp. 161–176, Jan. 2017, doi: 10.1021/ACSCATAL.6B 02523.

[144] R. B. Licht and A. T. Bell, "A DFT investigation of the mechanism of propene ammoxidation over α-bismuth molybdate," *ACS Catal.*, vol. 7, no. 1, pp. 161–176, Jan. 2017, doi: 10.1021/acscatal.6b02523.

[145] W. Jeitschko and IUCr, "Crystal structure of La2(MoO4)3, a new ordered defect Scheelite type," *Acta. Crystallogr. B*, vol. 29, no. 10, pp. 2074–2081, Oct. 1973, doi: 10.1107/S0567740873006138.

[146] A. W. Sleight and W. J. Linn, "Olefin oxidation over oxide catalysts with the scheelite structure*," *Ann. N. Y. Acad. Sci.*, vol. 272, no. 1, pp. 22–44, Apr. 1976, doi: 10.1111/J.1749-6632.1976.TB34222.X.

[147] J. D. Burrington, C. T. Kartisek, and R. K. Grasselli, "Mechanism of nitrogen insertion in ammoxidation catalysis," *J. Catal.*, vol. 81, no. 2, pp. 489–498, Jun. 1983, doi: 10.1016/0021-9517(83)90189-6.

[148] Y. H. Jang and W. A. Goddard, "Selective oxidation and ammoxidation of propene on bismuth molybdates, ab initio calculations," *Top. Catal.*, vol. 15, no. 2, pp. 273–289, 2001, doi: 10.1023/A:1016631100417.

[149] R. K. Grasselli, "Advances and future trends in selective oxidation and ammoxidation catalysis," *Catal. Today*, vol. 49, no. 1–3, pp. 141–153, Feb. 1999, doi: 10.1016/S0920-5861(98)00418-0.

[150] R. K. Grasselli, D. J. Buttrey, P. DeSanto, J. D. Burrington, C. G. Lugmair, A. F. Volpe, and T. Weingand, "Active centers in Mo-V-Nb-Te-Ox (amm)oxidation catalysts," *Catal. Today*, Jul. 2004, vol. 91–92, pp. 251–258, doi: 10.1016/j.cattod.2004.03.060.

[151] P. DeSanto, D. J. Buttrey, R. K. Grasselli, C. G. Lugmair, A. F. Volpe, B. H. Toby, and T. Vogt, "Structural characterization of the orthorhombic phase M1 in MoVNbTeO propane ammoxidation catalyst," *Top. Catal.*, vol. 23, no. 1–4, pp. 23–38, Aug. 2003, doi: 10.1023/A:1024812101856.

[152] P. DeSanto, D. J. Buttrey, R. K. Grasselli, C. G. Lugmair, A. F. Volpe, B. H. Toby, and T. Vogt, "Structural aspects of the M1 and M2 phases in MoVNbTeO propane ammoxidation catalysts," *Zeitschrift fur Krist.*, vol. 219, no. 3, pp. 152–165, Mar. 2004, doi: 10.1524/zkri.219.3.152.29091.

[153] J. N. Allison and W. A. Goddard, "Active sites on molybdate surfaces, mechanistic considerations for selective oxidation, and ammoxidation of propene," *ACS Symp. Ser.*, pp. 23–36, 1985, doi: 10.1021/BK-1985-0279.CH002.

[154] W. A. Goddard, L. Liu, J. E. Mueller, S. Pudar, and R. J. Nielsen, "Structures, mechanisms, and kinetics of ammoxidation and selective oxidation of propane over the M2 phase of MoVNbTeO catalysts," *Top. Catal.*, vol. 54, no. 10–12, pp. 659–668, Jul. 2011, doi: 10.1007/S11244-011-9688-8.

[155] P. A. Batist, A. H. W. M. Der Kinderen, Y. Leeuwenburgh, F. A. M. G. Metz, and G. C. A. Schuit, "The catalytic oxidation of 1-butene over bismuth molybdate catalysts: IV. Dependence of activity on the structures of the catalysts," *J. Catal.*, vol. 12, no. 1, pp. 45–60, Sep. 1968, doi: 10.1016/0021-9517(68)90072-9.

[156] I. Matsuura, R. Schut, and K. Hirakawa, "The surface structure of the active bismuth molybdate catalyst," *J. Catal.*, vol. 63, no. 1, pp. 152–166, May 1980, doi: 10.1016/0021-9517(80)90068-8.

[157] A. G. Mitchell, P. M. Lyne, K. F. Scott, and C. S. G. Phillips, "Catalytic and surface-structural investigations of bismuth molybdate selective-oxidation catalysts," *J. Chem. Soc. Faraday Trans. 1*, vol. 77, no. 10, pp. 2417–2427, Jan. 1981, doi: 10.1039/F19817702417.

[158] A. C. Sanfiz, T. W. Hansen, D. Teschner, P. Schnörch, F. Girgsdies, A. Trunschke, R. Schlögl, M. H. Looi, and S. B. A. Hamid, "Dynamics of the MoVTeNb oxide M1 phase in propane oxidation," *J. Phys. Chem. C*, vol. 114, no. 4, pp. 1912–1921, Feb. 2010, doi: 10.1021/JP909352U.

[159] W. E. Farneth, R. H. Staley, and A. W. Sleight, "Stoichiometry and structural effects in alcohol chemisorption/temperature-programmed desorption on MoO3," *J. Am. Chem. Soc.*, vol. 108, no. 9, pp. 2327–2332, 1986, doi: 10.1021/JA00269A031.

[160] W. E. Farneth, F. Ohuchi, R. H. Staley, U. Chowdhry, and A. W. Sleight, "Mechanism of partial oxidation of methanol over MoO3 as studied by temperature-programmed desorption," *J. Phys. Chem.*, vol. 89, no. 12, pp. 2493–2497, 1985, doi: 10.1021/J100258A014.

[161] M. Bowker, R. Holroyd, M. House, R. Bracey, C. Bamroongwongdee, M. Shannon, and A. Carley, "The selective oxidation of methanol on iron molybdate catalysts," *Top. Catal.*, May 2008, vol. 48, no. 1–4, pp. 158–165, doi: 10.1007/s11244-008-9058-3.

[162] C. Brookes, M. Bowker, and P. P. Wells, "Catalysts for the selective oxidation of methanol," *Catalyst*, vol. 6, no. 7, p. 92, Jun. 2016, doi: 10.3390/CATAL6070092.

[163] K. Routray, W. Zhou, C. J. Kiely, W. Grünert, and I. E. Wachs, "Origin of the synergistic interaction between MoO3 and iron

molybdate for the selective oxidation of methanol to formaldehyde," *J. Catal.*, vol. 275, no. 1, pp. 84–98, Sep. 2010, doi: 10.1016/J.JCAT.2010.07.023.

[164] E. Söderhjelm, M. P. House, N. Cruise, J. Holmberg, M. Bowker, J. O. Bovin, and A. Andersson, "On the synergy effect in MoO3-Fe2(MoO 4)3 catalysts for methanol oxidation to formaldehyde," *Top. Catal.*, vol. 50, no. 1–4, pp. 145–155, Nov. 2008, doi: 10.1007/S11244-008-9112-1.

[165] H. Hu and I. E. Wachs, "Catalytic properties of supported molybdenum oxide catalysts: In situ Raman and methanol oxidation studies," *J. Phys. Chem.*, vol. 99, no. 27, pp. 10911–10922, 1995, doi: 10.1021/j100027a035.

[166] S. D. Jackson and J. S. J. Hargreaves, *Metal Oxide Catalysis.* John Wiley & Sons, 2008.

[167] H. Hu and I. E. Wachs, "Surface structures of supported molybdenum oxide catalysts. Characterization by raman and Mo L3-edge XANES," *J. Phys. Chem.*, vol. 99, no. 27, pp. 10897–10910, 1995, doi: 10.1021/j100027a034.

[168] L. E. Briand, W. E. Farneth, and I. E. Wachs, "Quantitative determination of the number of active surface sites and the turnover frequencies for methanol oxidation over metal oxide catalysts: I. Fundamentals of the methanol chemisorption technique and application to monolayer supported molybdenum oxide catalysts," *Catal. Today*, vol. 62, no. 2–3, pp. 219–229, Nov. 2000, doi: 10.1016/S0920-5861(00)00423-5.

[169] L. J. Burcham, L. E. Briand, and I. E. Wachs, "Quantification of active sites for the determination of methanol oxidation turn-over frequencies using methanol chemisorption and in situ infrared techniques. 1: Supported metal oxide catalysts," *Langmuir*, vol. 17, no. 20, pp. 6164–6174, Oct. 2001, doi: 10.1021/LA010009U.

[170] T. Waters, R. A. J. O'Hair, and A. G. Wedd, "Catalytic gas phase oxidation of methanol to formaldehyde," *J. Am. Chem. Soc.*, vol. 125, no. 11, pp. 3384–3396, Mar. 2003, doi: 10.1021/JA028839X.

[171] M. Bowker, M. House, A. Alshehri, C. Brookes, E. K. Gibson, and P. P. Wells, "Selectivity determinants for dual function catalysts: applied to methanol selective oxidation on iron molybdate," *Catal. Struct. React.*, vol. 1, no. 2, pp. 95–100, Apr. 2015, doi: 10.1179/2055075815Y.0000000002.

[172] W. T. A. Harrison, U. Chowdhry, C. J. Machiels, A. W. Sleight, and A. K. Cheetham, "Preparation of ferric tungstate and its catalytic behavior toward methanol," *J. Solid State Chem.*, vol. 60, no. 1, pp. 101–106, Nov. 1985, doi: 10.1016/0022-4596(85)90169-0.

[173] I. E. Wachs and K. Routray, "Catalysis science of bulk mixed oxides," *ACS Catal.*, vol. 2, no. 6., pp. 1235–1246, Jun. 01, 2012, doi: 10.1021/cs2005482.

[174] W. E. Farneth, E. M. McCarron, A. W. Sleight, and R. H. Staley, "Comparison of the surface chemistry of two polymorphic forms of molybdenum trioxide," *Langmuir*, vol. 3, no. 2, pp. 217–223, Mar. 1987, doi: 10.1021/LA00074A013.

[175] J. C. Volta and J. L. Portefaix, "Structure sensitivity of mild oxidation reactions on oxide catalysts - a review," *Appl. Catal.*, vol. 18, no. 1, pp. 1–32, Sep. 1985, doi: 10.1016/S0166-9834(00)80296-1.

[176] J. N. Allison and W. A. Goddard, "Oxidative dehydrogenation of methanol to formaldehyde," *J. Catal.*, vol. 92, no. 1, pp. 127–135, Mar. 1985, doi: 10.1016/0021-9517(85)90242-8.

[177] M. Rellán-Piñeiro and N. López, "The active molybdenum oxide phase in the methanol oxidation to formaldehyde (Formox process): A DFT study," *ChemSusChem*, vol. 8, no. 13, pp. 2231–2239, Jul. 2015, doi: 10.1002/cssc.201500315.

[178] B. K. Hodnett and F. E. Cassidy, "Selective oxidation catalysts: An evaluation of the discriminating capacity of active sites on oxide catalysts with molecular oxygen as oxidant," *Cattech*, vol. 2, pp. 173–180, 1998.

[179] G. W. Coulston, S. R. Bare, H. Kung, K. Birkeland, G. K. Bethke, R. Harlow, N. Herron, and P. L. Lee, "The kinetic significance of V5+ in n-butane oxidation catalyzed by vanadium phosphates," *Science*, vol. 275, no. 5297, pp. 191–193, Jan. 1997, doi: 10.1126/SCIENCE.275.5297.191.

[180] V. A. Zazhigalov, A. I. Pyatnitskaya, I. V. Bacherikova, G. A. Komashko, G. Ladwig, and V. M. Belousov, "Dependence of the activity of vanadium-phosphorus catalysts on the V4+/V5+ ratio in oxidation of butane," *React. Kinet. Catal. Lett.*, vol. 23, no. 1, pp. 119–123, Aug. 2013, doi: 10.1007/BF02065675.

[181] F. Cavani, G. Centi, and F. Trifiro, "The chemistry of catalysts based on vanadium-phosphorus oxides: Note IV: Catalytic behaviour of catalysts prepared in organic medium in the oxidation of C4 fraction," *Appl. Catal.*, vol. 9, no. 2, pp. 191–202, Feb. 1984, doi: 10.1016/0166-9834(84)80063-9.

[182] F. Cavani, G. Centi, and F. Trifiro, "Study of n-butane oxidation to maleic anhydride in a tubular flow stacked-pellet reactor. Influence of phosphorus on the selectivity.," *Appl. Catal.*, vol. 15, no. 1, pp. 151–160, Mar. 1985, doi: 10.1016/S0166-9834(00)81495-5.

[183] M. J. Cheng, W. A. Goddard, and R. Fu, "The Reduction-Coupled Oxo Activation (ROA) mechanism responsible for the catalytic

selective activation and functionalization of n-butane to maleic anhydride by vanadium phosphate oxide," *Top. Catal.*, vol. 57, no. 14, pp. 1171–1187, Sep. 2014, doi: 10.1007/S11244-014-0284-6.

[184] W. C. O'Leary, W. A. Goddard, and M. J. Cheng, "Dual-phase mechanism for the catalytic conversion of n-butane to maleic anhydride by the vanadyl pyrophosphate heterogeneous catalyst," *J. Phys. Chem. C*, vol. 121, no. 43, pp. 24069–24076, Nov. 2017, doi: 10.1021/ACS.JPCC.7B07881.

[185] M. J. Cheng and W. A. Goddard, "The critical role of phosphate in vanadium phosphate oxide for the catalytic activation and functionalization of n-butane to maleic anhydride," *J. Am. Chem. Soc.*, vol. 135, no. 12, pp. 4600–4603, Mar. 2013, doi: 10.1021/JA3115746.

[186] M. A. Pepera, J. L. Callahan, M. J. Desmond, E. C. Milberger, P. R. Blum, and N. J. Bremer, "Fundamental study of the oxidation of butane over vanadyl pyrophosphate," *J. Am. Chem. Soc.*, vol. 107, no. 17, pp. 4883–4892, 1985, doi: 10.1021/JA00303A012.

[187] J. T. Gleaves, J. R. Ebner, and T. C. Kuechler, "Temporal analysis of products (TAP)—a unique catalyst evaluation system with submillisecond time resolution," *Catal. Rev.*, vol. 30, no. 1, pp. 49–116, Feb. 2006, doi: 10.1080/01614948808078616.

[188] H. Bluhm, M. Hävecker, E. Kleimenov, A. Knop-Gericke, A. Liskowski, R. Schlögl, and D. S. Su, "In situ surface analysis in selective oxidation catalysis: n-butane conversion over VPP," *Top. Catal.*, vol. 23, no. 1–4, pp. 99–107, Aug. 2003, doi: 10.1023/A:1024824404582.

[189] B. K. Hodnett and B. Delmon, "Factors influencing the activity and selectivity of vanadium-phosphorus oxide catalysts for n-butane oxidation to maleic anhydride," *Appl. Catal.*, vol. 15, no. 1, pp. 141–150, Mar. 1985, doi: 10.1016/S0166-9834(00)81494-3.

[190] M. Müller, M. Kutscherauer, S. Böcklein, G. D. Wehinger, T. Turek, and G. Mestl, "Modeling the selective oxidation of n-butane to maleic anhydride: From active site to industrial reactor," *Catal. Today*, vol. 387, pp. 82–106, Mar. 2022, doi: 10.1016/J.CATTOD.2021.04.009.

[191] G. Centi, G. Fornasari, and F. Trifirò, "On the mechanism of n-butane oxidation to maleic anhydride: Oxidation in oxygen-stoichiometry-controlled conditions," *J. Catal.*, vol. 89, no. 1, pp. 44–51, Sep. 1984, doi: 10.1016/0021-9517(84)90278-1.

[192] F. Cavani and F. Trifiro, "Catalyzing butane oxidation to make maleic anhydride," *Chemtech*, vol. 24, no. 4, Apr. 1994.

[193] Y. Zhang-Lin, M. Forissier, R. P. Sneeden, J. C. Védrine, and J. C. Volta, "On the mechanism of n-butane oxidation to maleic anhydride on VPO catalysts: I. A kinetics study on a VPO catalyst as compared to VPO reference phases," *J. Catal.*, vol. 145, no. 2, pp. 256–266, Feb. 1994, doi: 10.1006/JCAT.1994.1033.

[194] Y. Zhang-Lin, M. Forissier, J. C. Védrine, and J. C. Volta, "On the mechanism of n-butane oxidation to maleic anhydride on VPO catalysts .II. Study of the evolution of the VPO catalysts under n-butane, butadiene, and furan oxidation conditions," *J. Catal.*, vol. 145, no. 2, pp. 267–275, Feb. 1994, doi: 10.1006/JCAT.1994.1034.

[195] M. D. Wildberger, M. Maciejewski, J. D. Grunwaldt, T. Mallat, and A. Baiker, "Sol–gel bismuth–molybdenum–titanium mixed oxides: II. Oxidation of butadiene to furan," *Appl. Catal. A Gen.*, vol. 179, no. 1–2, pp. 189–202, Apr. 1999, doi: 10.1016/S0926-860X(98)00311-1.

[196] G. Centi and F. Trifirpò, "Furan production by oxygen insertion in the 1–4 position of butadiene on V-P-O-based catalysts," *J. Mol. Catal.*, vol. 35, no. 2, pp. 255–265, May 1986, doi: 10.1016/0304-5102(86)87031-6.

[197] M. Koichi, N. Taihei, M. Noritaka, O. Toshio, and M. Makoto, "The important step of the selective oxidation of n-butane over (VO)2P2O7," *Chem. Lett.*, vol. 17, no. 2, pp. 303–306, Mar. 2006, doi: 10.1246/CL.1988.303.

[198] M. Müller, M. Kutscherauer, S. Böcklein, G. Mestl, and T. Turek, "Improved kinetics of n-butane oxidation to maleic anhydride: the role of byproducts," *Ind. Eng. Chem. Res.*, vol. 60, no. 1, pp. 218–229, Jan. 2021, doi: 10.1021/ACS.IECR.0C05029.

[199] B. Chen and E. J. Munson, "Investigation of the mechanism of n-butane oxidation on vanadium phosphorus oxide catalysts: Evidence from isotopic labeling studies," *J. Am. Chem. Soc.*, vol. 124, no. 8, pp. 1638–1652, Feb. 2002, doi: 10.1021/JA010285V.

[200] J. Ziółkowski, E. Bordes, and P. Courtine, "Oxidation of butane and butene on the (100) face of (VO)2P2O7: A dynamic view in terms of the crystallochemical model of active sites," *J. Catal.*, vol. 122, no. 1, pp. 126–150, Mar. 1990, doi: 10.1016/0021-9517(90)90266-M.

[201] M. M. Bhasin, J. H. McCain, B. V. Vora, T. Imai, and P. R. Pujadó, "Dehydrogenation and oxydehydrogenation of paraffins to olefins," *Appl. Catal. A Gen.*, vol. 221, no. 1–2, pp. 397–419, Nov. 2001, doi: 10.1016/S0926-860X(01)00816-X.

[202] M. D. Argyle, K. Chen, A. T. Bell, and E. Iglesia, "Effect of catalyst structure on oxidative dehydrogenation of ethane and propane on alumina-supported vanadia," *J. Catal.*, vol. 208, no. 1, pp. 139–149, May 2002, doi: 10.1006/jcat.2002.3570.

[203] C. A. Carrero, R. Schloegl, I. E. Wachs, and R. Schomaecker, "Critical literature review of the kinetics for the oxidative dehydrogenation of propane over well-defined supported vanadium oxide catalysts," *ACS Catal.*, vol. 4, no. 10, pp. 3357–3380, Oct. 2014, doi: 10.1021/CS5003417.

[204] K. Alexopoulos, M. F. Reyniers, and G. B. Marin, "Reaction path analysis of propane selective oxidation over V2O5 and V2O5/

TiO2," *J. Catal.*, vol. 289, pp. 127–139, May 2012, doi: 10.1016/J.JCAT.2012.01.019.

[205] G. L. Dai, Z. H. Li, J. Lu, W. N. Wang, and K. N. Fan, "Deep oxidations in the oxidative dehydrogenation reaction of propane over V 2O 5(001): Periodic density functional theory study," *J. Phys. Chem. C*, vol. 116, no. 1, pp. 807–817, Jan. 2012, doi: 10.1021/JP208639T.

[206] K. Hermann, "Physical and Chemical Properties of Oxygen at Vanadium and Molybdenum Oxide Surfaces: Theoretical Case Studies," in *Computational Methods in Catalysis and Materials Science: An Introduction for Scientists and Engineers*, R. A. van Santen and P. Sautet, Eds. John Wiley & Sons, 2009, pp. 375–415.

[207] C. Xiong, S. Chen, P. Yang, S. Zha, Z. J. Zhao, and J. Gong, "Structure-performance relationships for propane dehydrogenation over aluminum supported vanadium oxide," *ACS Catal.*, vol. 9, no. 7, pp. 5816–5827, Jul. 2019, doi: 10.1021/ACSCATAL.8B04701.

[208] M. E. Harlin, V. M. Niemi, and A. O. I. Krause, "Alumina-supported vanadium oxide in the dehydrogenation of butanes," *J. Catal.*, vol. 195, no. 1, pp. 67–78, Oct. 2000, doi: 10.1006/JCAT.2000.2969.

[209] H. Tian, E. I. Ross, and I. E. Wachs, "Quantitative determination of the speciation of surface vanadium oxides and their catalytic activity," *J. Phys. Chem. B*, vol. 110, no. 19, pp. 9593–9600, May 2006, doi: 10.1021/JP055767Y.

[210] M. V. Martínez-Huerta, X. Gao, H. Tian, I. E. Wachs, J. L. G. Fierro, and M. A. Bañares, "Oxidative dehydrogenation of ethane to ethylene over alumina-supported vanadium oxide catalysts: Relationship between molecular structures and chemical reactivity," *Catal. Today*, vol. 118, no. 3–4, pp. 279–287, Dec. 2006, doi: 10.1016/J.CATTOD.2006.07.034.

[211] M. A. Bañares and I. E. Wachs, "Molecular structures of supported metal oxide catalysts under different environments," *J. Raman Spectrosc.*, vol. 33, no. 5, pp. 359–380, May 2002, doi: 10.1002/JRS.866.

[212] C. Kaidong, T. B. Alexis, and E. Iglesia, "Kinetics and mechanism of oxidative dehydrogenation of propane on vanadium, molybdenum, and tungsten oxides," *J. Phys. Chem. B*, vol. 104, no. 6, pp. 1292–1299, Feb. 2000, doi: 10.1021/JP9933875.

[213] B. Beck, M. Harth, N. G. Hamilton, C. Carrero, J. J. Uhlrich, A. Trunschke, S. Shaikhutdinov, H. Schubert, H. J. Freund, R. Schlögl, J. Sauer, and R. Schomäcker, "Partial oxidation of ethanol on vanadia catalysts on supporting oxides with different redox properties compared to propane," *J. Catal.*, vol. 296, pp. 120–131, Dec. 2012, doi: 10.1016/J.JCAT.2012.09.008.

[214] C. A. Carrero, C. J. Keturakis, A. Orrego, R. Schomäcker, and I. E. Wachs, "Anomalous reactivity of supported V2O5 nanoparticles for propane oxidative dehydrogenation: influence of the vanadium oxide precursor," *Dalton Trans.*, vol. 42, no. 35, pp. 12644–12653, Aug. 2013, doi: 10.1039/C3DT50611H.

[215] A. Brückner and E. Kondratenko, "Simultaneous operando EPR/UV–vis/laser–Raman spectroscopy — A powerful tool for monitoring transition metal oxide catalysts during reaction," *Catal. Today*, vol. 113, no. 1–2, pp. 16–24, Mar. 2006, doi: 10.1016/J.CATTOD.2005.11.006.

[216] J. Sauer and J. Döbler, "Structure and reactivity of V2O5: bulk solid, nanosized clusters, species supported on silica and alumina, cluster cations and anions," *Dalton Trans.*, no. 19, pp. 3116–3121, Sep. 2004, doi: 10.1039/B402873B.

[217] T. Blasco, A. Galli, J. M. López Nieto, and F. Trifiró, "Oxidative dehydrogenation of ethane and n-butane on VOx/Al2O3 catalysts," *J. Catal.*, vol. 169, no. 1, pp. 203–211, Jul. 1997, doi: 10.1006/JCAT.1997.1673.

[218] A. Khodakov, B. Olthof, A. T. Bell, and E. Iglesia, "Structure and catalytic properties of supported vanadium oxides: support effects on oxidative dehydrogenation reactions," *J. Catal.*, vol. 181, no. 2, pp. 205–216, Jan. 1999, doi: 10.1006/JCAT.1998.2295.

[219] N. Y. Topsøe, "Mechanism of the selective catalytic reduction of nitric oxide by ammonia elucidated by in situ on-line fourier transform infrared spectroscopy," *Science*, vol. 265, no. 5176, pp. 1217–1219, Aug. 1994, doi: 10.1126/SCIENCE.265.5176.1217.

[220] N. Y. Topsøe, J. A. Dumesic, and H. Topsøe, "Vanadia-titania catalysts for selective catalytic reduction of nitric-oxide by ammonia: I.I. Studies of active sites and formulation of catalytic cycles," *J. Catal.*, vol. 151, no. 1, pp. 241–252, Jan. 1995, doi: 10.1006/JCAT.1995.1025.

[221] L. Arnarson, H. Falsig, S. B. Rasmussen, J. V. Lauritsen, and P. G. Moses, "The reaction mechanism for the SCR process on monomer V5+ sites and the effect of modified Brønsted acidity," *Phys. Chem. Chem. Phys.*, vol. 18, no. 25, pp. 17071–17080, Jun. 2016, doi: 10.1039/C6CP02274J.

[222] G. He, Z. Lian, Y. Yu, Y. Yang, K. Liu, X. Shi, Z. Yan, W. Shan, and H. He, "Polymeric vanadyl species determine the low-temperature activity of V-based catalysts for the SCR of NOx with NH3," *Sci. Adv.*, vol. 4, no. 11, Nov. 2018, doi: 10.1126/SCIADV.AAU4637.

[223] T. Yu, T. Hao, D. Fan, J. Wang, M. Shen, and W. Li, "Recent NH3-SCR mechanism research over Cu/SAPO-34 catalyst," *J. Phys. Chem. C*,

vol. 118, no. 13, pp. 6565–6575, Apr. 2014, doi: 10.1021/JP4114199.

[224] L. Chen, T. V. W. Janssens, P. N. R. Vennestrøm, J. Jansson, M. Skoglundh, and H. Grönbeck, "A complete multisite reaction mechanism for low-temperature NH3-SCR over Cu-CHA," *ACS Catal.*, vol. 10, no. 10, pp. 5646–5656, May 2020, doi: 10.1021/ACSCATAL.0C00440.

[225] F. Gao, J. H. Kwak, J. Szanyi, and C. H. F. Peden, "Current understanding of Cu-exchanged chabazite molecular sieves for use as commercial diesel engine DeNOx catalysts," *Top. Catal.*, vol. 56, no. 15–17, pp. 1441–1459, Aug. 2013, doi: 10.1007/S11244-013-0145-8.

[226] U. S. Ozkan, Y. P. Cai, and M. W. Kumthekar, "Investigation of the reaction pathways in selective catalytic reduction of NO with NH3 over V2O5 catalysts: isotopic labeling studies using 18O2, 15NH3, 15NO, and 15N18O," *J. Catal.*, vol. 149, no. 2, pp. 390–403, Oct. 1994, doi: 10.1006/JCAT.1994.1306.

[227] M. Koebel, G. Madia, and M. Elsener, "Selective catalytic reduction of NO and NO2 at low temperatures," *Catal. Today*, vol. 73, no. 3–4, pp. 239–247, Apr. 2002, doi: 10.1016/S0920-5861(02)00006-8.

[228] A. Kato, S. Matsuda, T. Kamo, F. Nakajima, H. Kuroda, and T. Narita, "Reaction between NOx and NH3 on iron oxide-titanium oxide catalyst," *J. Phys. Chem.*, vol. 85, no. 26, pp. 4099–4102, 1981, doi: 10.1021/J150626A029.

[229] N. Y. Topsøe, H. Topsøe, and J. A. Dumesic, "Vanadia/titania catalysts for selective catalytic reduction (SCR) of nitric-oxide by ammonia: I. Combined temperature-programmed in-situ FTIR and on-line mass-spectroscopy studies," *J. Catal.*, vol. 151, no. 1, pp. 226–240, Jan. 1995, doi: 10.1006/JCAT.1995.1024.

[230] L. Arnarson, H. Falsig, S. B. Rasmussen, J. V. Lauritsen, and P. G. Moses, "A complete reaction mechanism for standard and fast selective catalytic reduction of nitrogen oxides on low coverage VOx/TiO2(0 0 1) catalysts," *J. Catal.*, vol. 346, pp. 188–197, Feb. 2017, doi: 10.1016/J.JCAT.2016.12.017.

[231] G. T. Went, L. J. Leu, R. R. Rosin, and A. T. Bell, "The effects of structure on the catalytic activity and selectivity of V2O5/TiO2 for the reduction of NO by NH3," *J. Catal.*, vol. 134, no. 2, pp. 492–505, Apr. 1992, doi: 10.1016/0021-9517(92)90337-H.

[232] D. W. Fickel and R. F. Lobo, "Copper coordination in Cu-SSZ-13 and Cu-SSZ-16 investigated by variable-temperature XRD," *J. Phys. Chem. C*, vol. 114, no. 3, pp. 1633–1640, Jan. 2010, doi: 10.1021/JP9105025.

[233] S. T. Korhonen, D. W. Fickel, R. F. Lobo, B. M. Weckhuysen, and A. M. Beale, "Isolated Cu2+ ions: active sites for selective catalytic reduction of NO," *Chem. Commun.*, vol. 47, no. 2, pp. 800–802, Dec. 2010, doi: 10.1039/C0CC04218H.

[234] T. V. W. Janssens, H. Falsig, L. F. Lundegaard, P. N. R. Vennestrøm, S. B. Rasmussen, P. G. Moses, F. Giordanino, E. Borfecchia, K. A. Lomachenko, C. Lamberti, S. Bordiga, A. Godiksen, S. Mossin, and P. Beato, "A consistent reaction scheme for the selective catalytic reduction of nitrogen oxides with ammonia," *ACS Catal.*, vol. 5, no. 5, pp. 2832–2845, May 2015, doi: 10.1021/CS501673G.

[235] R. A. SHELDON, "Liquid Phase Autoxidations," in *Catalytic Oxidation: Principles and Applications*, R. A. Sheldon and R. A. van Santen, Eds. World Scientific, 1995, pp. 151–174.

[236] T. Ito and J. H. Lunsford, "Synthesis of ethylene and ethane by partial oxidation of methane over lithium-doped magnesium oxide," *Nature*, vol. 314, no. 6013, pp. 721–722, 1985, doi: 10.1038/314721b0.

[237] J. X. Wang and J. H. Lunsford, "Characterization of [Li+O] centers in lithium-doped magnesium oxide catalysts," *J. Phys. Chem.*, vol. 90, no. 22, pp. 5883–5887, 2002, doi: 10.1021/J100280A084.

[238] H. Schwarz, "Chemistry with methane: concepts rather than recipes," *Angew. Chem. Int. Ed.*, vol. 50, no. 43, pp. 10096–10115, Oct. 2011, doi: 10.1002/ANIE.201006424.

[239] C. R. . Catlow, S. . French, A. . Sokol, and J. . Thomas, "Computational approaches to the determination of active site structures and reaction mechanisms in heterogeneous catalysts," *Philos. Trans. R. Soc. A*, vol. 363, no. 1829, pp. 913–936, Apr. 2005, doi: 10.1098/RSTA.2004.1529.

[240] K. Kwapien, J. Paier, J. Sauer, M. Geske, U. Zavyalova, R. Horn, P. Schwach, A. Trunschke, and R. Schlögl, "Sites for methane activation on lithium-doped magnesium oxide surfaces," *Angew. Chem. Int. Ed.*, vol. 53, no. 33, pp. 8774–8778, Aug. 2014, doi: 10.1002/anie.201310632.

[241] G. P. Van Der Zwet, P. A. J. M. Hendriks, and R. A. van Santen, "Pyrolysis of methane and the role of surface area," *Catal. Today*, vol. 4, no. 3–4, pp. 365–369, Feb. 1989, doi: 10.1016/0920-5861(89)85032-1.

[242] X. Guo, G. Fang, G. Li, H. Ma, H. Fan, L. Yu, C. Ma, X. Wu, D. Deng, M. Wei, D. Tan, R. Si, S. Zhang, J. Li, L. Sun, Z. Tang, X. Pan, and X. Bao, "Direct, nonoxidative conversion of methane to ethylene, aromatics, and hydrogen," *Science*, vol. 344, no. 6184, pp. 616–619, 2014, doi: 10.1126/SCIENCE.1253150.

[243] J. Gao, Y. Zheng, J.-M. Jehng, Y. Tang, I. E. Wachs, and S. G. Podkolzin, "Identification of molybdenum oxide nanostructures on zeolites for natural gas conversion," *Science*, vol. 348, no. 6235, pp. 686–690, 2015, doi: 10.1126/science.aaa7048.
[244] J. J. Spivey and Graham Hutchings, "Catalytic aromatization of methane," *Chem. Soc. Rev.*, vol. 43, no. 3, pp. 792–803, Jan. 2014, doi: 10.1039/C3CS60259A.
[245] N. Kosinov, A. S. G. Wijpkema, E. Uslamin, R. Rohling, F. J. A. G. Coumans, B. Mezari, A. Parastaev, A. S. Poryvaev, M. V. Fedin, E. A. Pidko, and E. J. M. Hensen, "Confined carbon mediating dehydroaromatization of methane over Mo/ZSM-5," *Angew. Chem. Int. Ed.*, vol. 57, no. 4, pp. 1016–1020, Jan. 2018, doi: 10.1002/ANIE.201711098.
[246] B. E. R. Snyder, M. L. Bols, H. M. Rhoda, D. Plessers, R. A. Schoonheydt, B. F. Sels, and E. I. Solomon, "Cage effects control the mechanism of methane hydroxylation in zeolites," *Science*, vol. 373, no. 6552, pp. 327–331, Jul. 2021, doi: 10.1126/SCIENCE.ABD5803.
[247] M. Ravi, M. Ranocchiari, and J. A. van Bokhoven, "The direct catalytic oxidation of methane to methanol—a critical assessment," *Angew. Chem. Int. Ed.*, vol. 56, no. 52, pp. 16464–16483, Dec. 2017, doi: 10.1002/ANIE.201702550.
[248] E. I. Solomon, A. Decker, and N. Lehnert, "Non-heme iron enzymes: Contrasts to heme catalysis," *Proc. Natl. Acad. Sci.*, vol. 100, no. 7, pp. 3589–3594, Apr. 2003, doi: 10.1073/PNAS.0336792100.
[249] P. E. M. Siegbahn and R. H. Crabtree, "Mechanism of C–H activation by diiron methane monooxygenases: quantum chemical studies," *J. Am. Chem. Soc.*, vol. 119, no. 13, pp. 3103–3113, Apr. 1997, doi: 10.1021/JA963939M.
[250] B. Ensing, F. Buda, M. C. M. Gribnau, and E. J. Baerends, "Methane-to-methanol oxidation by the hydrated iron(IV) oxo species in aqueous solution: a combined DFT and Car–Parrinello molecular dynamics study," *J. Am. Chem. Soc.*, vol. 126, no. 13, pp. 4355–4365, Apr. 2004, doi: 10.1021/JA038865A.
[251] Z.-J. Zhao, A. Kulkarni, L. Vilella, J. K. Nørskov, and F. Studt, "Theoretical insights into the selective oxidation of methane to methanol in copper-exchanged mordenite," *ACS Catal.*, vol. 6, no. 6, pp. 3760–3766, Jun. 2016, doi: 10.1021/ACSCATAL.6B00440.
[252] Á. Szécsényi, G. Li, J. Gascon, and E. A. Pidko, "Unraveling reaction networks behind the catalytic oxidation of methane with H2O2 over a mixed-metal MIL-53(Al,Fe) MOF catalyst," *Chem. Sci.*, vol. 9, no. 33, pp. 6765–6773, Aug. 2018, doi: 10.1039/C8SC02376J.

[253] Y.-H. C. Chin, C. Buda, M. Neurock, and E. Iglesia, "Consequences of metal-oxide interconversion for C-H bond activation during CH4 reactions on Pd catalysts.," *J. Am. Chem. Soc.*, vol. 135, no. 41, pp. 15425–15442, 2013, doi: 10.1021/ja405004m.

[254] A. Hellman, A. Resta, N. M. Martin, J. Gustafson, A. Trinchero, P.-A. Carlsson, O. Balmes, R. Felici, R. van Rijn, J. W. M. Frenken, J. N. Andersen, E. Lundgren, H. Grönbeck, R. van Rijn, J. W. M. Frenken, J. N. Andersen, E. Lundgren, and H. Grönbeck, "The active phase of palladium during methane oxidation," *J. Phys. Chem. Lett.*, vol. 3, no. 6, pp. 678–682, Mar. 2012, doi: 10.1021/JZ300069S.

[255] Y.-H. C. Chin, C. Buda, M. Neurock, and E. Iglesia, "Reactivity of chemisorbed oxygen atoms and their catalytic consequences during CH4–O2 catalysis on supported Pt clusters," *J. Am. Chem. Soc.*, vol. 133, no. 40, pp. 15958–15978, Oct. 2011, doi: 10.1021/ja202411v.

[256] D. Hibbitts and M. Neurock, "Promotional effects of chemisorbed oxygen and hydroxide in the activation of C–H and O–H bonds over transition metal surfaces," *Surf. Sci.*, vol. 650, pp. 210–220, Aug. 2016, doi: 10.1016/J.SUSC.2016.01.012.

[257] K. Yoshizawa, Y. Shiota, and T. Yamabe, "Abstraction of the hydrogen atom of methane by iron-oxo species: The concerted reaction path is energetically more favorable," *Organometallics*, vol. 17, no. 13, pp. 2825–2831, 1998, doi: 10.1021/om980067j.

[258] G. Li, E. A. Pidko, R. A. van Santen, Z. Feng, C. Li, and E. J. M. Hensen, "Stability and reactivity of active sites for direct benzene oxidation to phenol in Fe/ZSM-5: A comprehensive periodic DFT study," *J. Catal.*, vol. 284, no. 2, pp. 194–206, 2011, doi: 10.1016/j.jcat.2011.07.008.

[259] G. I. Panov, V. I. Sobolev, K. A. Dubkov, and A. S. Kharitonov, "Biomimetic oxidation on Fe complexes in zeolites," *Stud. Surf. Sci. Catal.*, vol. 101, pp. 493–502, Jan. 1996, doi: 10.1016/S0167-2991(96)80260-6.

[260] N. S. Ovanesyan, K. A. Dubkov, A. A. Pyalling, and A. A. Shteinman, "The Fe active sites in FeZSM-5 catalyst for selective oxidation of CH4 to CH3OH at room temperature," *J. Radioanal. Nucl. Chem.*, vol. 246, no. 1, pp. 149–152, 2000, doi: 10.1023/A:1006722207492.

[261] K. A. Dubkov, V. I. Sobolev, E. P. Talsi, M. A. Rodkin, N. H. Watkins, A. A. Shteinman, and G. I. Panov, "Kinetic isotope effects and mechanism of biomimetic oxidation of methane and benzene on FeZSM-5 zeolite," *J. Mol. Catal. A Chem.*, vol. 123, no. 2–3, pp. 155–161, Aug. 1997, doi: 10.1016/S1381-1169(97)00051-4.

[262] K. B. Sharpless, A. Y. Teranishi, and J. E. Bäckvall, "Chromyl chloride oxidations of olefins. Possible role of organometallic

intermediates in the oxidations of olefins by oxo transition metal species," *J. Am. Chem. Soc.*, vol. 99, no. 9, pp. 3120–3128, 1977, doi: 10.1021/JA00451A043.

[263] A. K. Rappé and W. A. Goddard, "Bivalent spectator oxo bonds in metathesis and epoxidation alkenes," *Nature*, vol. 285, no. 5763, pp. 311–312, 1980, doi: 10.1038/285311a0.

[264] A. K. Rappé and W. A. Goddard, "Hydrocarbon oxidation by high-valent group 6 oxides," *J. Am. Chem. Soc.*, vol. 104, no. 12, pp. 3287–3294, 1982, doi: 10.1021/JA00376A006.

[265] K.-P. Shing, B. Cao, Y. Liu, H. K. Lee, M.-D. Li, D. L. Phillips, X.-Y. Chang, and C.-M. Che, "Arylruthenium(III) porphyrin-catalyzed C–H oxidation and epoxidation at room temperature and [Ru V (Por)(O)(Ph)] intermediate by spectroscopic analysis and density functional theory calculations," *J. Am. Chem. Soc*, vol. 140, p. 20, 2018, doi: 10.1021/jacs.8b04470.

[266] A. Ishikawa and S. Sakaki, "Theoretical study of photoinduced epoxidation of olefins catalyzed by ruthenium porphyrin," *J. Phys. Chem. A*, vol. 115, p. 44, 2011, doi: 10.1021/jp200650q.

[267] J. T. Groves and T. E. Nemo, "Epoxidation reactions catalyzed by iron porphyrins. Oxygen transfer from iodosylbenzene," *J. Am. Chem. Soc.*, vol. 105, no. 18, pp. 5786–5791, 2002, doi: 10.1021/JA00356A015.

[268] W. Zhang, J. L. Loebach, S. R. Wilson, and E. N. Jacobsen, "Enantioselective epoxidation of unfunctionalized olefins catalyzed by salen manganese complexes," *J. Am. Chem. Soc.*, vol. 112, no. 7, pp. 2801–2803, 2002, doi: 10.1021/JA00163A052.

[269] M. O. Özbek and R. A. van Santen, "The mechanism of ethylene epoxidation catalysis," *Catal. Lett.*, vol. 143, no. 2, pp. 131–141, Jan. 2013, doi: 10.1007/S10562-012-0957-3.

[270] E. A. Carter and W. A. Goddard, "Chemisorption of oxygen, chlorine, hydrogen, hydroxide, and ethylene on silver clusters: A model for the olefin epoxidation reaction," *Surf. Sci.*, vol. 209, no. 1–2, pp. 243–289, Feb. 1989, doi: 10.1016/0039-6028(89)90071-X.

[271] S. Linic and M. A. Barteau, "Construction of a reaction coordinate and a microkinetic model for ethylene epoxidation on silver from DFT calculations and surface science experiments," *J. Catal.*, vol. 214, no. 2, pp. 200–212, Mar. 2003, doi: 10.1016/S0021-9517(02)00156-2.

[272] R. Ugo, *Aspects of Homogeneous Catalysis*. Springer, 1981.

[273] R. A. Sheldon, "Catalytic Oxidation and Fine Chemicals," in *Catalytic Oxidation: Principles and Applications*, R. A. Sheldon and R. A. van Santen, Eds. World Scientific, 1995, pp. 239–265.

[274] J. M. Thomas, C. R. A. Catlow, and G. Sankar, "Determining the structure of active sites, transition states and intermediates in heterogeneously catalysed reactions.," *Chem. Commun.*, no. 24, pp. 2921–2925, 2002, doi: 10.1039/b210679p.
[275] M. V. Twigg and M. S. Spencer, "Dennis dowden and the development of catalytic science," *Johnson Matthey Technol. Rev.*, vol. 58, no. 3, pp. 142–155, 2014, doi: 10.1595/147106714X682643.
[276] A. Cimino and F. S. Stone, "Oxide Solid Solutions," in *Handbook of Heterogeneous Catalysis*. John Wiley & Sons, 2008, doi: 10.1002/9783527610044.HETCAT0074.
[277] L. G. Tejuca, J. L. G. Fierro, and J. M. D. Tascón, "Structure and reactivity of perovskite-type oxides," *Adv. Catal.*, vol. 36, no. C, pp. 237–328, Jan. 1989, doi: 10.1016/S0360-0564(08)60019-X.
[278] D. A. Dowden, "Crystal and ligand field models of solid catalysts," *Catal. Rev.*, vol. 5, no. 1, pp. 1–32, Jan. 1972, doi: 10.1080/01614947208076863.
[279] J. Suntivich, H. A. Gasteiger, N. Yabuuchi, H. Nakanishi, J. B. Goodenough, and Y. Shao-Horn, "Design principles for oxygen-reduction activity on perovskite oxide catalysts for fuel cells and metal–air batteries," *Nat. Chem.*, vol. 3, no. 8, p. 647, Jul. 2011, doi: 10.1038/nchem.1093.
[280] I. C. Man, H.-Y. Su, F. Calle-Vallejo, H. A. Hansen, J. I. Martínez, N. G. Inoglu, J. Kitchin, T. F. Jaramillo, J. K. Nørskov, and J. Rossmeisl, "Universality in oxygen evolution electrocatalysis on oxide surfaces," *ChemCatChem*, vol. 3, no. 7, pp. 1159–1165, Jul. 2011, doi: 10.1002/cctc.201000397.
[281] A. Vojvodic and J. K. Nørskov, "Optimizing perovskites for the water-splitting reaction," *Science*, vol. 334, no. 6061, p. 1355, 2011, doi: 10.1126/science.1215081.
[282] T. A. Albright, J. K. Burdett, and M. H. Whangbo, *Orbital Interactions in Chemistry*, 2nd ed. John Wiley & Sons, 2013.
[283] J. Suntivich, H. A. Gasteiger, N. Yabuuchi, H. Nakanishi, J. B. Goodenough, and Y. Shao-Horn, "Design principles for oxygen-reduction activity on perovskite oxide catalysts for fuel cells and metal–air batteries," *Nat. Chem.*, vol. 3, no. 7, pp. 546–550, Jun. 2011, doi: 10.1038/nchem.1069.
[284] J. Suntivich, K. J. May, H. A. Gasteiger, J. B. Goodenough, and Y. Shao-Horn, "A perovskite oxide optimized for oxygen evolution catalysis from molecular orbital principles," *Science*, vol. 334, no. 6061, pp. 1383–1385, Dec. 2011, doi: 10.1126/SCIENCE.1212858.
[285] G. Kremenić, J. M. L. Nieto, J. M. D. Tascón, and L. G. Tejuca, "Chemisorption and catalysis on LaMO3 oxides," *J. Chem. Soc.*

Faraday Trans. 1, vol. 81, no. 4, pp. 939–949, Jan. 1985, doi: 10.1039/F19858100939.

[286] Á. Valdés, J. Brillet, M. Grätzel, H. Gudmundsdóttir, H. A. Hansen, H. Jónsson, P. Klüpfel, G.-J. Kroes, F. Le Formal, I. C. Man, R. S. Martins, J. K. Nørskov, J. Rossmeisl, K. Sivula, A. Vojvodic, and M. Zäch, "Solar hydrogen production with semiconductor metal oxides: new directions in experiment and theory.," *Phys. Chem. Chem. Phys.*, vol. 14, no. 1, pp. 49–70, Jan. 2012, doi: 10.1039/c1cp23212f.

[287] F. Abild-Pedersen, J. Greeley, F. Studt, J. Rossmeisl, T. R. Munter, P. G. Moses, E. Skúlason, T. Bligaard, and J. K. Nørskov, "Scaling properties of adsorption energies for hydrogen-containing molecules on transition-metal surfaces," *Phys. Rev. Lett.*, vol. 99, no. 1, pp. 16105–16109, Jul. 2007, doi: 10.1103/PHYSREVLETT.99.016105.

[288] R. A. van Santen, M. Neurock, and S. G. Shetty, "Reactivity theory of transition-metal surfaces: A brønsted-evans- polanyi linear activation energy-free-energy analysis," *Chem. Rev.*, vol. 110, no. 4, pp. 2005–2048, Apr. 2010, doi: 10.1021/CR9001808.

[289] H. G. Oswin and D. Phil, "Their function and applications in electrode structures," *Platin. Met. Rev.*, vol. 8, no. 2, p. 42, 1964.

[290] O. T. Holton and J. W. Stevenson, "The role of platinum in proton exchange membrane fuel cells," *Platin. Met. Rev.*, vol. 57, no. 4, pp. 259–271, Oct. 2013, doi: 10.1595/147106713X671222.

[291] S. Trasatti, *Electrodes of Conductive Metallic Oxides, Part B.* Elsevier, 1981.

[292] E. J. M. O'Sullivan and E. J. Calvo, "Chapter 3 Reactions at Metal Oxide Electrodes," *Compr. Chem. Kinet.*, vol. 27, no. C, pp. 247–360, Jan. 1988, doi: 10.1016/S0069-8040(08)70017-7.

[293] R. N. Singh, J. P. Pandey, N. K. Singh, B. Lal, P. Chartier, and J. F. Koenig, "Sol-gel derived spinel MxCo3–xO4 (M=Ni, Cu; 0≤x≤1) films and oxygen evolution," *Electrochim. Acta*, vol. 45, no. 12, pp. 1911–1919, Feb. 2000, doi: 10.1016/S0013-4686(99)00413-2.

[294] C. P. Plaisance and R. A. van Santen, "Structure sensitivity of the oxygen evolution reaction catalyzed by cobalt(II,III) oxide," *J. Am. Chem. Soc.*, vol. 137, no. 46, pp. 14660–14672, Nov. 2015, doi: 10.1021/JACS.5B07779.

Chapter 5

Solid Acid Catalysis

5.1 Introduction

Solid acid catalysts catalyze reactions at temperatures where liquid acids cannot be used.

The science of solid acid catalysis evolved closely with the development of new catalytic processes. The latter happened after the discoveries of the large oil reserves in the early part of the previous century and the expansion of automotive motion. Another driver for new processes was the need for aviation fuel in the Second World War. There was a rapid expansion of the oil refining and petrochemical industries.

Solid acids have become essential to many processes. In addition to the conversion of oil, they also are important in the transformation of natural gas or biomass to liquid fuels and chemicals.

The beginning of the petrochemical use of solid acid catalysts was the Houdry catalytic cracking process that became operational in 1937 in the USA [1]. The initial solid acids used were acidified clays or chlorinated alumina. A large impact to the refining processes came from the invention of synthetic zeolites. These are alumino-silicates with nanoporous structure that accommodate strongly acidic protons.

Acid-catalyzed reactions involve the formation of intermediate carbocations. This chemistry became well understood in the 1950s by the study of reactions catalyzed by H_2SO_4 and HF, which are applied at large scale in the production of airplane alkylate.

The mechanism of solid acid-catalyzed reactions is related to that of liquid acids.

An important advantage of solid acid catalysts is their use in high-temperature conversion reactions.

This chapter presents the inorganic chemistry of solid acid materials and the reaction mechanisms of major solid acid catalytic processes.

5.1.1 *Initial Developments*

Solid acid catalysts can be oxidic supports activated by acids. Mixed oxides have intrinsic Brønsted acidity.

In the catalytic cracking process crude oil is converted into hydrocarbons that have a size range and composition suitable for transportation fuel. The chemical process requires high temperature where liquid acids cannot be used. Early in the previous century [2], [3] $AlCl_3$ and clays treated with HCl or sulfuric acid had been explored for the upgrading of coal liquids. $AlCl_3$ became known at the end of the 19th century as a Friedel-Crafts reagent for the stoichiometric alkylation of aromatic compounds with chloro-hydrocarbons. In the 1940s and 1950s chlorinated alumina or clay was replaced initially by amorphous mixed oxide alumino-silicates and later by synthetic zeolite catalysts.

Synthetic mixed oxides with Brønsted acidic properties were invented in 1938 at United Oil Products (UOP) by Thomas [4]. They are amorphous alumino-silicates that sometimes also contain Mg or Zr. The materials have an acid strength that is strong enough for use as catalytic cracking.

Thomas also formulated the first theory of inorganic catalyst acidity; this is presented in Section 5.2. The amorphous mixed solid acid catalysts are the forerunner of crystalline synthetic zeolites that are introduced in Section 5.1.2.

Amorphous catalyst supports that are treated with liquid acids were and remain important for other refinery processes. An important example is the phosphoric acid/silica catalyst discovered in the 1930s by Ipatieff [5] for dehydration, oligomerization, and alkylation of alkenes. Ipatieff was a Russian scientist who, after leaving Russia in the 1930s and joined the then just established research company UOP in the USA, was responsible for the invention of several of the currently important catalytic refinery and petrochemical processes.

Processes that dissociate or isomerize hydrocarbons are based on endothermic and, thus, high-temperature reactions. Condensation reactions are exothermic, require low temperatures and liquid acid catalysis can be applied. The H_2SO_4- or HF-catalyzed isobutane-butene alkylation process, invented around 1940 by Pines [6], is such a process. In combination with catalytic cracking, this alkylation process was essential to provide aviation fuel to the Allies in the Second World War. The study of its reaction mechanism largely contributed to our present understanding of acid catalysis [7], [8].

The other important discovery of that period was the bifunctional acid catalyst promoted with Pt. Different from the catalytic cracking process, aromatics are produced for alkanes without excessive coke deposition.

As discussed earlier in Section 3.1.4 within the context of alloy catalysis, the discovery in the 1950s by Haensel not only gave catalysis an important new concept, but it also became of immense importance to the petroleum industry [9]. In 1949 the first platforming process that is based on this invention became operational. The process operates at high hydrogen pressure.

The Pt nanoparticles distributed at high dispersion over the acidic support serve to activate hydrogen and convert alkanes into alkenes, which readily react with catalyst support protons to form their isomers, lighter alkene molecules, or aromatics.

Deactivation of Pt nanoparticle contributes significantly to deactivation of the catalysts. Alloying of the Pt, for instance with Re [10], is used to prevent their deactivation. The large interest, industrially as well as academically, in alloy catalysis has its origin in this deactivation question (Section 3.1.4.2) [11], [12].

Hydrocracking and hydroisomerization processes [13], which work according to the same bifunctional catalyst principle, are also important refinery processes. Their reaction mechanism is discussed in detail in Section 5.4.3. The sections that follow have, as the major topic, shape-selective catalysis. The concept of restricted transition state catalysis and pre-transition state stabilization are introduced.

5.1.2 *The Discovery of Zeolite Catalysis*

The proton of the solid acid catalyst is Brønsted acidic. The internal surface O–H bond is weakened by Lewis acidic interaction with low-valent cation.

The accidental discovery of strong acidity of synthetic zeolites [14]–[16] by Rabo and colleagues at Union Carbide Research in 1957 is fundamental to the later innovation of the catalytic cracking process.

Zeolites are crystalline alumino-silicates that contain crystallographically defined microchannels and cavities. The dimensions of these channels and cavities are comparable to the size of organic molecules, and are activated by protons. This causes zeolite catalysts to be shape selective. Many different zeolite structures and compositions are known that are useful as solid acid catalysts.

The impact of the discovery of strong Brønsted acidity and the shape-selective properties of zeolites on the petrochemical industry has been enormous. It not only dramatically improved the efficiency of crude oil conversion to gasoline, but also largely decreased coke deposition compared to the amorphous alumino-silicates. The zeolite microporous channel structure suppresses formation of the large aromatic molecules (shape selectivity) that are coke formation precursors.

The catalytic cracking process innovation with the Brønsted acidic Zeolite Y catalyst is due to Plank and Rosinsky of Mobil Oil Company in 1964 [17], [18].

Because of the well-defined atomistic structure of the zeolite crystal the study of its inorganic chemistry was crucial to the theory of solid acidity. The now generally accepted structure of the zeolite acidic proton is close to that already suggested by Thomas in

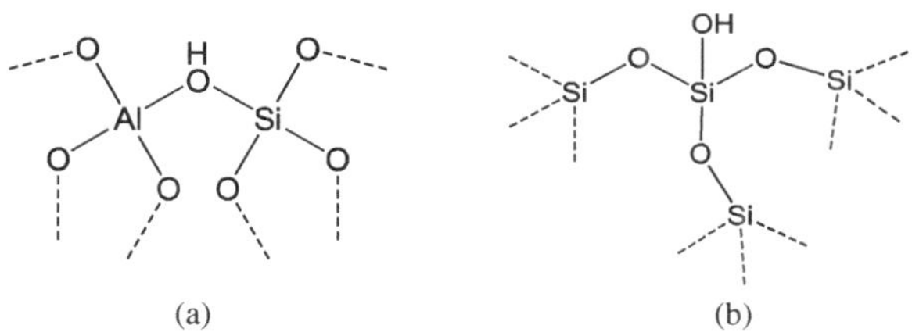

Figure 5.1 Comparison of the structure of zeolitic proton and surface silanol (schematic). (a) The zeolitic Brønsted site that bridges tetrahedrally coordinated Al^{3+} and Si^{4+}. (b) The surface silanol that is singly coordinated to a Si^{4+} cation.

1948 for acidic protons of amorphous alumino-silicates [4]. It was identified in 1965 by Hall and colleagues from the Carnegie Mellon Institute in Pittsburg [19]. They deduced its structure from infrared studies. It is sketched in Figure 5.1a. The lattice Si and Al cations have tetrahedral coordination with oxygen atoms that are singly connected. The network of such connected tetrahedra is the motif of the zeolite structure.

The acidic proton is attached to an oxygen atom that connects the Al- and Si-containing tetrahedra. In Figure 5.1b the structure is compared with that of silanol OH present at the silica surface. A major difference between the external surface of silica and the internal surface of the zeolite is that at the surface tetrahedral Si–O–Si bonds are broken, but the coordination of the cations of the zeolite lattice that determines its internal surface remains tetrahedral.

At the SiO_2 surface coordinatively unsaturated Si cations become four-coordinated by attachment of the OH silanol from reaction with water. The surface silanol group is weakly acidic. The zeolite proton is more acidic because it is activated by the Lewis acidic interaction with the Al^{3+} cation of the neighboring tetrahedron. In Section 5.3 zeolitic materials and their structure will be discussed in more detail.

The general rule that determines the acidity of a proton attached to a mixed oxide material depends on the cation coordination of the oxygen atom that also binds with the proton and the coordination and electronegativity of the cations. The latter relates to cation charge and size (Section 5.2).

Catalytic reactivity depends on proton acidity as well as adsorbate concentration near the protonic site. This is largely defined by the match of reaction intermediate size and shape with zeolite microcavity.

5.1.3 *Organic Carbocations*

Acid-catalyzed reactions produce intermediate carbocations. These activated intermediates initiate the catalytic reaction cycle.

In 1934 Whitmore proposed that a coordinatively unsaturated hydrocarbon by protonation forms an intermediate carbocation [20]. He supported the existence of carbocations with extensive studies of product patterns and use of deuterated systems.

Figure 5.2 illustrates three elementary reactions steps that are common to many acid-catalyzed systems and became understood around the middle of the previous century [4].

Figure 5.2 illustrates protonation and deprotonation of a propene molecule. A propyl carbocation is formed. This carbocation

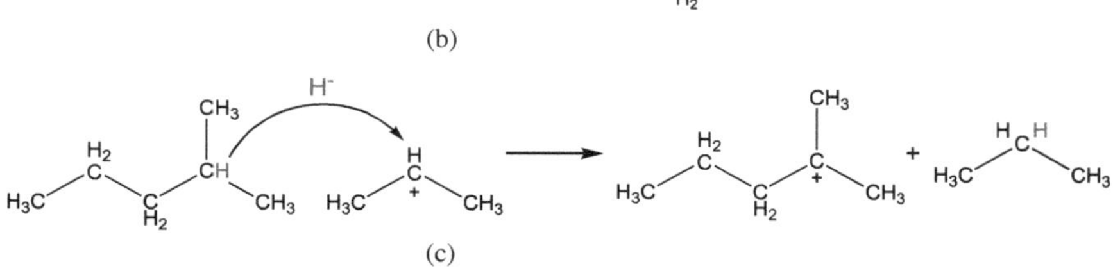

Figure 5.2 Carbenium ion reactions of alkenes or alkanes. (a) Protonation and deprotonation of propene and propyl carbenium ion. (b) Dimerization, the forward reaction of propene with propyl carbenium ion that gives a new C–C bond; the backward reaction is β C–C bond cleavage. (c) Hydride transfer between tertiary alkane carbon atom and propyl carbenium ion.

was initially called carbonium. Currently it is carbenium, because a positive charge becomes located on a threefold-coordinated carbon atom.

The positive charge makes the carbon atom reactive for dimerization, as illustrated by Figure 5.2b. The forward reaction is a reaction in which a C–C bond is formed, while the backward reaction, called β C–C bond cleavage, regenerates the carbenium ion and alkene. It is called β C–C cleavage because the C–C bond that is broken is at β location with respect to the positively charged C atom.

The valences of the alkane molecule are saturated, which makes it less reactive. It is activated by contact with a carbenium ion. Then the reacting carbenium ion can be converted into a saturated alkane (product), and the reacting alkane is converted into a carbenium ion that can undergo other carbenium ion-type transformations (see Figure 5.2c). A negatively charged hydride ion is transferred that converts the initial carbenium ion into alkane and reactant alkane into carbenium ion.

Reactions and intermediates as sketched in Figure 5.2 explain the products of catalytic cracking largely, but an important question remained. As Figure 5.2c illustrates, alkane molecules can be converted once carbenium ions are present due to reaction of protons with unsaturated hydrocarbons. But how will alkanes be activated when no unsaturated molecules are initially present in the reaction feed?

Around 1960 it was well known that at higher temperatures than needed for catalytic cracking, saturated hydrocarbons can convert into alkenes by a radical chain process. This happens in the thermal cracking process [21], at present still an important process that preceded the invention of catalytic cracking. The lower temperature of the catalytic cracking process (800 K versus 1050 K) made it unlikely that alkane activation is initiated by radical reactions that generate the necessary alkenes.

The answer to this question of alkane activation led to the discovery of protonated saturated hydrocarbons. These are non-classical carbocation intermediates that according to modern nomenclature are called carbonium ions. Examples of such

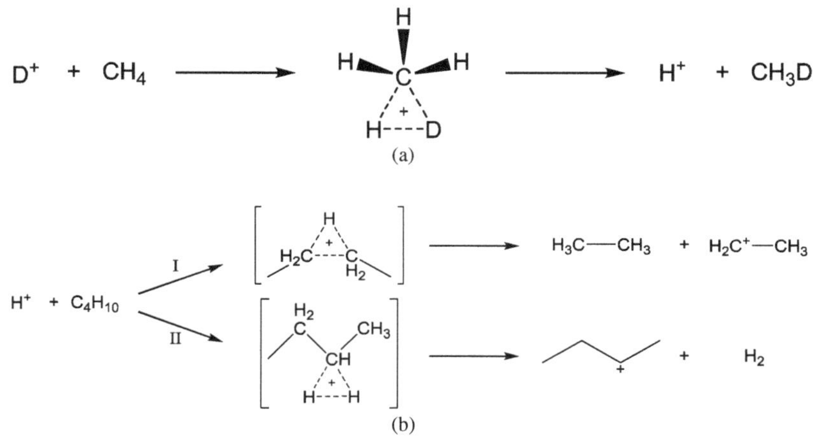

Figure 5.3 Non-classical organic carbocations (carbonium ions) and their reactions. (a) H–D exchange reaction of methane. (b) Carbonium ion decomposition of protonated butane. Reaction path I: Protonation of the C–C bond gives a shorter alkane and carbenium ion. Reaction path II cleaves a C–H bond, H_2 is product and a C_4 carbenium ion is formed.

intermediates and their reactions are sketched in Figure 5.3. As illustrated for protonated methane in Figure 5.3a the respective protonated hydrocarbon intermediates show a unique bonding feature: a two-electron (from the protonated σ bond) three-center delocalized bond. This bonding configuration is energetically unfavorable and hence the protonated alkane cations have short lifetimes. They are called non-classical because the carbon atom shares five chemical bonds instead of the classical maximum of four.

Observations of hydrogen evolution as well as deuterium exchange of alkane molecules catalyzed by solid acids suggest that alkane C–H bonds can be directly activated by a proton. As sketched in Figure 5.3b the reactive proton can act as a C–H as well as C–C bond cleavage catalyst [22], [23].

The nature of the reaction intermediates remained unresolved until Olah (Nobel prize 1994) and colleagues [24] discovered the carbonium ion intermediate in NMR studies of alkane activation by a superacid such as $HSbF_6$.

Also around the same time mass spectrometry investigations demonstrated the formation of non-classical carbocations in the gas phase [25].

The chemistry of carbonium and carbenium carbocations determines the kinetics of acid-catalyzed reactions, discussed in Section 5.4.

5.1.4 *The Catalytic Reaction Cycle; Shape-selective Catalysis*

The solid acid catalytic cycle is maintained by proton backdonation to the solid or by hydride transfer reactions.

The Zeolite Y catalyst initially used in the catalytic cracking process had a relatively high Al/Si framework ratio ($1 < \text{Al/Si} < 0.3$). Chemical treatment can reduce the Al/Si framework ratio, which results in an increase of proton reactivity [26]. This dependence of acid strength on Al/Si ratio provided an incentive to search for zeolitic materials with reduced Al/Si ratio.

Low Al/Si ratio materials were discovered by variation of zeolite synthesis conditions in 1969 by Argauer and Landolt [27], then employed by Mobil Oil Company. In the synthesis procedure they replaced the conventionally used inorganic base cations with particular organic base cations. The nanoporous structure of the zeolite lattice can be directed by the choice of the organic bases of particular shape [28]. Because of their size and positive charge, the use of organic base in zeolite synthesis mixtures reduces the zeolite framework Al/Si ratio.

With tetrapropylammonium cation as base Argauer and Landolt prepared a zeolite with nearly cylindrical nanopore structure that can have extremely low Al/Si ratio. Their zeolite material became known as H-ZSM-5.

Exploration of different organic base cations in synthesis with variation of composition led to a host of new zeolitic materials. The unique structure and acidity [29], [30] of H-ZSM-5 and other newly discovered zeolites [31] led to the exploration of many new

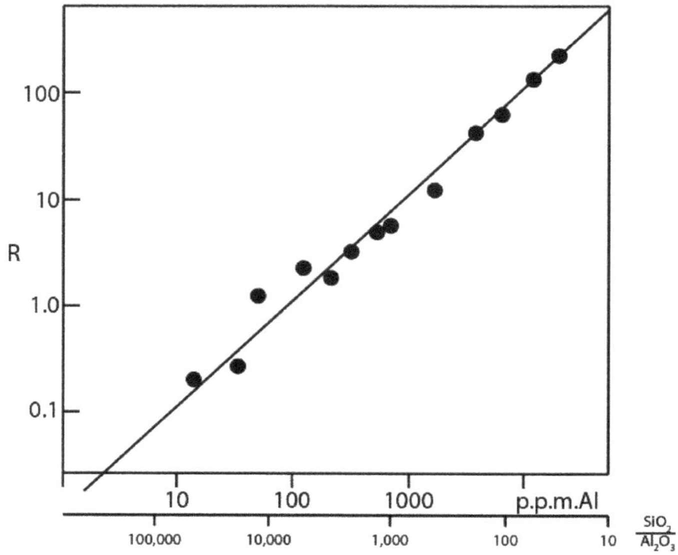

Figure 5.4 The hexane cracking activity denoted as R plotted against the aluminum framework content in H-ZSM-5 [35].

reactions, such as the reactions of methanol to olefins (MTO) and methanol to gasoline (MTG) (see Section 5.4.5) [32]–[34].

A fundamental observation is that proton strength becomes independent of Al/Si ratio when this ratio is less than 0.1.

This is illustrated by an important experiment from Mobil Oil researchers Haag and colleagues [35] in 1984. Figure 5.4 shows their measured rate of hexane conversion catalyzed by H-ZSM-5 as a function of Al/Si ratio that is varied over more than four orders of magnitude.

The reaction rate R of hexane cracking normalized per Al atom is seen to be independent of concentration. There is a one-to-one ratio of proton concentration and Al/Si framework concentration.

As will be discussed in Section 5.2.1.2 this independence of concentration holds as long as the Al-containing framework tetrahedra have no Al-substituted Si tetrahedra in next nearest neighbor sites [30]. This is the case as long as the Al/Si ratio is less than approximately 0.1. The detailed value depends on zeolite structure.

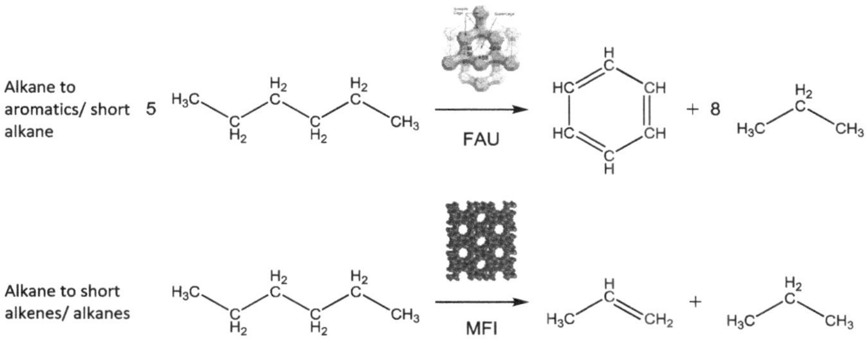

Figure 5.5 Shape selectivity of the alkane cracking reaction catalyzed by protonic zeolites. The zeolite with FAU structure such as Zeolite X or Y is compared with a zeolite with the MFI structure such as H-ZSM-5. The FAU structure has pores with a ring size of 12 framework tetrahedra, while the nanochannels of the MFI structure have a ring size of 10 tetrahedra [158].

The Haag experiment provides also proof that hexane cracking is a catalytic reaction which occurs by proton activation and is not a radical reaction.

The selectivity of the catalytic cracking reaction may vary when acidic zeolite catalysts of different structure are compared. Shape-selective catalysis of this reaction is illustrated by Figure 5.5. Catalytic cracking of alkanes by a protonic zeolite with large pores such as Zeolite X or Y gives aromatics and short alkanes as main product, but H-ZSM-5 with the smaller pores gives mainly short alkanes and alkenes.

This large difference in product selectivity arises because in the two systems the reaction intermediates of catalytic cracking are different. Whereas, as is explained below, in the wide-pore Zeolite X or Y a high concentration of carbenium ions dominate reactivity, in the narrow-pore H-ZSM-5 reaction routes with non-classical carbonium ions dominate.

Shape-selective catalysis can be caused by reactant or product diffusion limitations or chemical reaction selectivity [36], [37]. The difference of reactions dominated by carbonium or carbenium intermediates is an example of the latter. The main reason for this is that the narrow-pore zeolite suppresses bimolecular reactions. The zeolite nanopore size has to be large enough that bimolecular hydride transfer reactions (Figure 5.2c) can happen (Section 5.4.4.3). It results in a major alteration of the reaction network of the alkane cracking.

The mechanism of this reaction is schematically illustrated in Figure 5.6. As for many Brønsted acid-catalyzed reactions, the mechanism of catalytic cracking consists of several interconnected cycles. The reaction network can be understood as a combination of initiation, propagation, and deactivation cycles. Their relative contribution to the overall reaction depends strongly on zeolite catalyst structure and may also depend on reaction time.

Reaction is initiated by activation of the alkane and formation of an intermediate carbonium ion. This decomposes according to reactions of Figure 5.3 into a carbenium ion with formation of H_2 or a short alkane. Then two elementary reactions compete: the carbenium ion backdonates the proton, generating an alkene molecule, or a carbenium ion reacts with the reactant alkane by a hydride transfer process. The latter yields another carbenium ion (carbenium ion') and a short alkane, that can desorb as (co-)product. The first reaction of proton backdonation is monomolecular, while the second is bimolecular. The steady carbenium ion concentration that is maintained by the hydride transfer reaction initiates the propagation cycle.

Bimolecular reactions require larger cavity space and are only possible in wide-pore zeolites. The activation energies of the elementary reactions that involve carbenium ions are lower than carbonium ion-mediated reactions and hence, when physically possible, the propagation cycle takes over from the initiation cycle. This is the main reason that only wide-pore zeolites produce aromatics.

When hydride transfer is suppressed alkane activation can only occur by the high activation energy route via carbonium ion formation. β C–C bond cleavage then gives a lightweight alkane and alkenes as product, C–H bond activation gives dehydrogenation and the number of carbon atoms is maintained [38]. In the propagation cycle protons are not regenerated, but carbenium ions act as molecular organocationic catalysts. Many elementary reactions

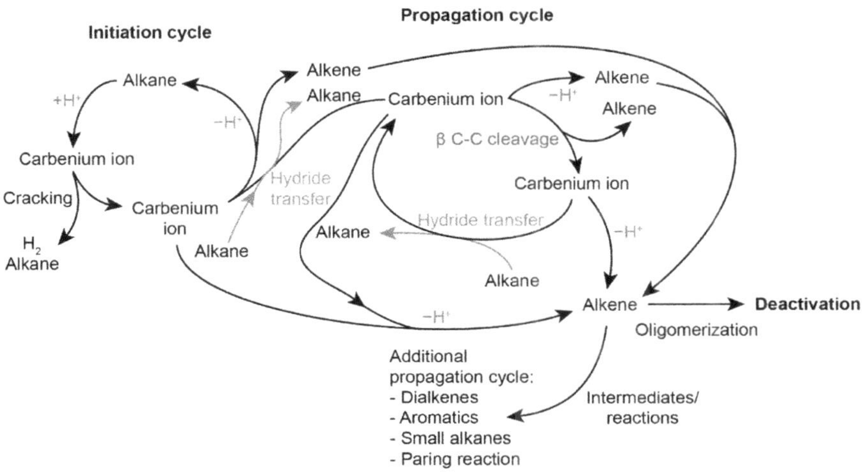

Figure 5.6 Catalytic cycles of the alkane cracking reaction [158]. Initiation and propagation cycles are indicated. Only part of propagation elementary reactions is sketched (see additional in text below the scheme). Colors indicate relevant elementary reaction steps.

participate. It creates a complex reaction network that leads to several products. Part of this reaction cycle of the process is sketched in Figure 5.6.

The mechanism of cyclization and aromatization reactions that are also part of the propagation cycle are discussed in detail in Section 5.4.3.1 [39]. Non-selective reactions of the alkenes lead to catalyst deactivation by formation of carbonaceous residue. Because of the size of the latter molecules the small nanopores of zeolite catalysts can suppress carbonaceous residue formation and hence catalyst deactivation is also shape selective.

Often in solid acid catalysis an unreactive coordinatively saturated molecule is to be converted. Conversion of an alkane or methanol is an example. Then time is needed to build up a high concentration of reactive intermediate carbenium ions for the propagation cycle to evolve. This implies that reaction proceeds through an initiation period that is followed by a longer quasi-steady state period. This period ends when the concentration of the non-selective deactivating compounds becomes too high.

The MTO and MTG [32], [33] reactions that are discussed in Section 5.4.5 are also shape selective. When small-pore zeolites are used suppression of part of the reaction cycle is responsible for shape selectivity.

It is possible to maintain the acid-catalyzed reaction, other than by hydride transfer or carbonium ion intermediate formation, with bifunctional solid acid catalysts. A bifunctional catalyst contains a transition metal such as Pt or Pd distributed on a solid acid support. The catalyst is used to isomerize or crack alkane molecules and is discussed in detail in Section 5.4.3. Here a short summary of the reaction mechanism is provided.

The reaction is executed in the presence of high-pressure hydrogen. The transition metal serves to activate C–H bonds as well as H_2 and establishes alkane-alkene equilibrium.

The protons activate the alkene intermediate. The carbenium ions formed upon protonation undergo the transformation reactions that ultimately lead to product intermediate alkenes.

The dehydrogenation of alkane to alkene is faster than hydride transfer between alkane and alkyl carbocation. After reaction the alkyl carbenium ion backdonates the proton. Hydrogenation of co-produced alkene is catalyzed by the transition metal to give product alkane. In these hydroisomerization and hydrocracking reactions the proton is regenerated in each cycle.

In bifunctional catalysis hydride transfer is circumvented and therefore shape selectivity is less significant than for the catalytic cracking reaction.

Alkane conversion is possible at relatively mild conditions but at the cost of high hydrogen pressure.

5.1.5 *Carbocations as Transition States*

Solid acid chemistry is activated complex chemistry.

At the end of the 20th century computational quantum chemistry reached a level of maturity that elementary reaction rate constants at surfaces could be calculated. Surface site models could be constructed that properly represent the geometry of proposed reaction sites. Calculation of an elementary rate constant implies simulation of ground state as well as transition state free energies of reaction intermediates and reaction paths.

Application to the protonation reaction catalyzed by solid acid catalysts generated important new insights. Acid catalysis generates carbonium and carbenium ions that are not necessarily part of ground states but are part of activated complexes or transition states. Solid acid chemistry is activated complex chemistry.

The alkane or alkene molecules initially physically adsorb on the solid acid surface and weakly interact with the protons. Carbenium- and carbonium-type intermediates are generated when the protonic O–H cleaves, and the proton is donated to the organic molecule.

In Figure 5.7 the energetics and reaction intermediates are given for these elementary steps from state-of-the-art quantum-chemical calculations.

Figure 5.7a shows the ground state and activation barrier energies for the hydrogen-deuterium exchange reaction of a zeolite proton with a propane molecule. The carbonium ion is part of the transition state. Whereas the intrinsic activation energy $\Delta H^{\ddagger}_{intr}$, that is measured with respect to the adsorbed state, does not vary with hydrocarbon chain length, the adsorption energy ΔH_{ads} of the hydrocarbon increases incrementally with the number of carbon atoms. This approximate insensitivity of intrinsic activation energy barrier with length of alkane molecule is also valid for other activation modes of the alkane. It has the important consequence that the variation of apparent activation energies $\Delta H^{\ddagger}_{app}$ of alkanes that differ in length does not relate to differences in proton activation energies but instead relates to respective adsorption energy differences. This is discussed in detail in Section 5.4.2.

Figures 5.7b$_1$ and b$_2$ show the potential energies of proton-activated C–C bond and C–H bond cleavage respectively. Before bond dissociation the transition state structures are carbonium ion-like. Bond dissociation also happens in the activated state and carbenium-like intermediates are generated. The activation energies for C–C and C–H bond cleavage of the alkanes are substantially higher

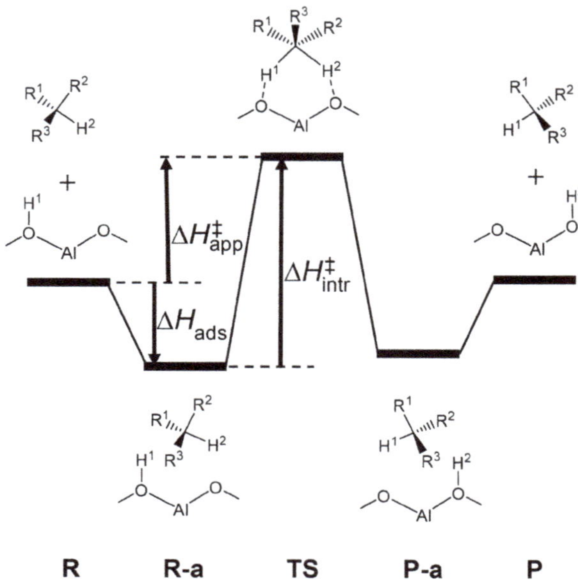

(a)

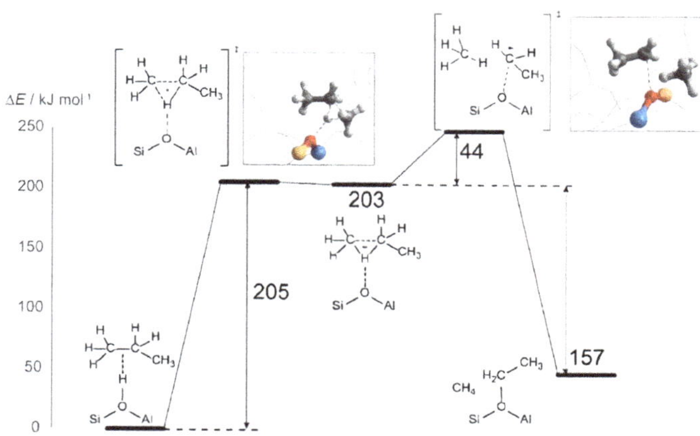

(b1)

Figure 5.7 Potential energies of proton-activated reactions by zeolite FAU (see Figure 5.18a) with low Al/Si framework composition. (a) Direct proton exchange mechanism. R^1, R^2, or R^3 are alkyl groups or hydrogen atoms. R-a is adsorbed alkane state, TS is carbonium ion-related transition state, P-a is product alkane. The oxygen positions at which the proton is located may be different in the initial and final structures. $\Delta H^{\ddagger}_{intr} = 125$ kJ/mol, ΔH_{ads} varies with length of hydrocarbon (ΔH_{ads} propane = 60 kJ/mol) [40]. (b1) The potential energy diagram of the propane C–C bond cleavage reaction. (b2) Proton-activated propane dehydrogenation to H_2 and propene (colored structures indicate zeolite reaction site fragment). (c) Ground state and transition state structures of protonation and dimerization of propene [41].

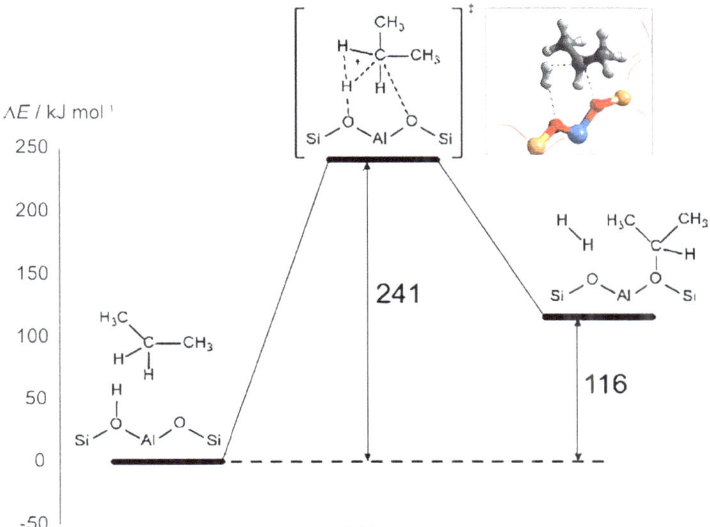

(b2)

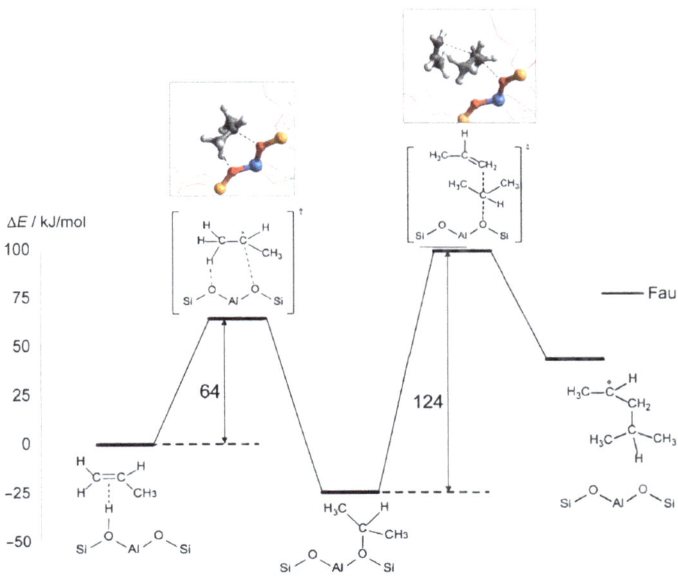

(c)

Figure 5.7 (*Continued*)

than those of hydrogen-deuterium exchange. After reaction the ground state of the carbenium ions is an alkoxy intermediate adsorbed to the zeolite wall.

Figure 5.7c shows the calculated activation energy of propene protonation and subsequent reaction with a second propene molecule. The transition state intermediates are carbenium ion-like. The ground state is an alkoxy intermediate.

The larger stability of carbenium ion intermediates compared to that of carbonium ion intermediates is reflected in the substantially lower activation energies of the protonation reaction of the alkene molecule.

5.2 Inorganic Chemistry of Solid Acidity

5.2.1 *The Hammett Function*

The protonation energy is determined by the heteropolar cleavage energy of the OH hydroxyl, the proton affinity of the reactant and attraction of the cation by the negative charge of the solid.

In this section the physical chemical relation of Brønsted acidity and material structure and composition is presented.

The reactivity of a surface proton depends on the protonation energy of an adsorbed molecule. When the base strength, or proton affinity, of the adsorbed molecule is large enough an equilibrium will establish between the non-protonated physically adsorbed molecule and the protonated molecular cation. When the proton affinity of the probe molecule is large enough it can overcome the deprotonation energy. Adsorption of probe molecules that change absorption spectrum upon protonation can be used to probe surface acidity.

This can be deduced from the Hammett function that measures proton affinity. Different indicator molecules are adsorbed on the surface and the pK value of the weakest base molecule is accepted as the Hammett function value of the respective proton [2].

The Hammett acidity function expression H_0 [42], [43] is defined by Eq. (5.1a):

$$H_0 = pK_{BH^+} + \log\left(\frac{[B]}{[BH^+]}\right) \tag{5.1a}$$

$$= -\log[H^+] \tag{5.1b}$$

$$pK_{BH^+} = -\log K_{BH^+} \tag{5.1c}$$

The Hammett function relates proton strength with the pK of probe molecule. This is the analogue of a pH measurement. The proton strength is defined by the equilibrium proton concentration (Eq. (5.1b)) when $\frac{[B]}{[BH^+]} = 1$. Indicator molecules that change color upon protonation are used to determine this equilibrium with the solid acid proton.

The Hammett function H_0 is a logarithmic function of the equilibrium constant K_{BH^+} and proportional to $-PA_{mol}/RT$. The proton affinity PA_{mol} is equal to the protonation energy of the molecule by a free proton.

The solid acid donates a proton to the probe molecule. The corresponding protonation energy E_p can be written as Eq. (5.2a):

$$E_p = -DPE + PA_{mol} + E_{el,stat} \tag{5.2a}$$

The protonation energy E_P is equal to the sum of three terms: the O–H deprotonation energy DPE, which is the energy cost of proton separation from the solid, the proton affinity PA_{mol} of the molecule that becomes protonated, and $E_{el,stat}$ the attractive electrostatic interaction energy of the protonated molecule with the negative charge on the solid.

When proton site and protonated indicator molecule are at equilibrium $E_p = 0$ and Eq. (5.2a) reduces to Eq. (5.2b):

$$DPE = PA_{mol} + E_{el,stat} \quad (E_p = 0) \tag{5.2b}$$

The deprotonation energy differs from PA_{mol} because of the electrostatic interaction between protonated indicator molecule and negatively charged solid. Since different indicators are used to determine H_0 this term will vary. It may cause a significant error in the determination of proton strength *DPE*.

Nonetheless the Hammett function H_0 is an often-used indicator of acidity. In the next section for mixed oxides it is seen that it correlates with electronegativity of the cationic atoms of which the solid acid material is composed. This is a global property that is averaged by the relative concentrations of the oxide cations.

This global relationship will not capture the substantial variation in measured proton strength of individual surface hydroxyls that is generally present on the solid oxide surface.

For this IR or NMR spectroscopic measurements are necessary.

There is an important difference between the Hammett acidity function and acidity as probed by the catalytic reactions.

The rate of a catalytic reaction is measured with respect to a reactant in the gas phase, whereas the Hammett acidity function relates to interaction of proton with adsorbed indicator molecule. Therefore, the free energy of adsorption is an essential part of the Hammett acidity function.

Also when the temperature of desorption of ammonia is used as a measure of acidity there will be a difference with the Hammett function expression, because the variation in adsorption energy of ammonia is an additional variable.

5.2.2 *The Acidity of Mixed Oxides*

Because charge is separated the deprotonation energy is larger for solid acids than liquid acids. The proton active site is electrostatically neutral.

In this subsection the Hammett function is used to probe correlation of solid acidity with the composition of mixed oxide materials.

Generally the acidity of the solid acid is found to be less than the reactivity of neat liquid acid or acid dissolved in water.

There is an important chemical difference between the protonation by the proton of the solid acid and that of an acid dissolved in water.

In water the protonating site is a positively charged H_3O^+. In contrast the solid acid site is neutral (see Figure 5.1a). Whereas in water proton transfer does not involve change of charge on the proton, in solid acids proton transfer implies charge separation. The positive charge develops when in contact with the reacting substrate.

In water or neat acid such as sulfuric acid the proton is present as H_3O^+ or $H_3SO_4^+$. In polar solvents such protonated molecules are due to self-protonation that is the result of molecular dissociation in cations and anions. The separation of charge is stabilized by the high dielectric constant of the liquids that is ≈ 80.

When solid acids are probed at gas-phase conditions such dielectric screening stabilization is absent. For comparison the dielectric constant of nanoporous zeolite is 4 [44]. The stability of protonated indicator molecule may be affected by additional stabilization due to nanopore size confinement [45].

The increased cost of charge separation at the solid surface compared to the liquid state is the reason for the lower reactivity of the protonic zeolites compared to neat acid solutions [46].

The Hammett function H_0 value of a highly siliceous protonic zeolite such as H-ZSM-5 is −10, while that of a high alumina-containing zeolite H–Y is −9. This is to be compared with H_0 values of neat liquid acids such as H_2SO_4 and HF that are respectively −12 and −15, and that of a superacid such as HF/SbF_5 which is as low as −23 [47]–[51].

The more negative the H_0 value, the more acidic the substance.

The acidity of a liquid acid decreases when the acid is more diluted in water. The Hammett function of 80% sulfuric acid in water is −8 compared to −12 for the neat acid. The reactivity in water is determined by the concentration of the less reactive H_3O^+ hydronium cation compared to that of $H_3OSO_4^+$ or H_2F^+. The reference energy to the proton affinity of indicator molecule is the self-ionization energy of water.

When surface acidity is probed differences in the Hammett function give differences in *DPE*: $-\Delta H_0 = \Delta(DPE + E_{el,stat})/RT$. The PA_{mol} of the indicator molecule that equilibrates with surface proton is smaller when protonation strength increases (or *DPE* decreases).

When differences in $E_{el,stat}$ are assumed to be small the difference in *DPE* between H-ZSM-5 and H-Y protons is estimated to be of the order of RT.

The spectroscopically determined *DPE* values that are discussed in Section 5.4.3 indicate that this underestimates the actual *DPE* difference at least by a factor of two.

The Hammett acidity function of solid oxides is sensitive to the degree of hydration of the solid. The relation with degree of hydration can be complex and is not very well understood.

For instance, when Nb_2O_5 is well hydroxylated [52], [53] its Hammett function is –8. This value is comparable to that of a solution of 80% sulfuric acid in water. When the oxide is heated the reactivity of Nb_2O_5 is maintained up to 350°C. Up to this temperature it is an active alcohol dehydration catalyst. At higher temperature the catalyst loses its reactivity due to water desorption.

Other solid acids have Hammett acidity function values similar to those of the neat liquid acids. At the condition of the determination of the Hammett function they are liquid-like.

For the heteropolyacids the Hammett acidity function indicates strong Brønsted acidity. The heteropolyacid consists of a Keggin unit that contains a central tetrahedrally coordinated cation such as P surrounded by 12 octahedrally coordinated W or Mo cations. The material has high acid strength due to the high charge of the tungsten and molybdenum cations. The heteropolyacid with W is the more acidic. For $H_3PW_{12}O_{40}.H_2O_x$ the reported Hammett function is comparable to that of sulfuric acid. For $Cs_{2.5}H_{0.5}PWO_{12}.H_2O_x$ a value of H_0 = −13.6 is reported [54]–[59]. This is due to self-protonation similar as in liquid acids. At higher temperature these materials catalyze at mild conditions dehydration as well as low-temperature alkane isomerization reactions. Also sulfonated zirconia catalysts have a large negative Hammett acidity function (H_0 = −14.5) [60] and catalyze alkane isomerization at the relatively low temperature of 600 K [61]. At low temperature the sulfate is in a liquid-like supported phase.

Figure 5.8 compares measured Hammett functions for a range of mixed oxides with equal molar ratio. The mixed oxides with compositions Si–Al, Si–Zr, Ti–Zr, and Ti–Si are the most acidic.

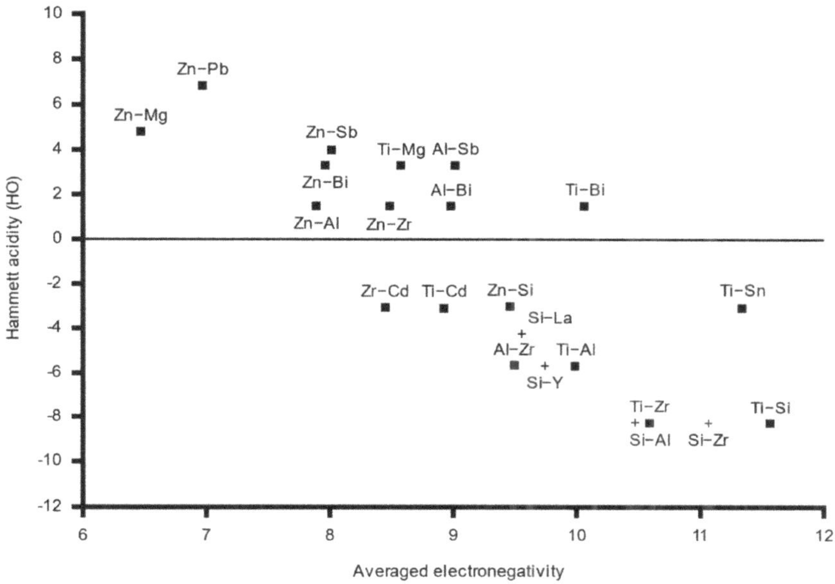

Figure 5.8 Highest acid strength (H_0) vs. averaged electronegativity of metal ions for binary oxides (molar ratio 1) [62].

The Hammett acidity function correlates with the averaged electronegativity [63] of the constituting cations. The electronegativity of an atom is defined by its tendency to attract shared electrons to form a chemical bond. This defines Lewis acid strength. Pauling [64] relates electronegativity with the polarity of a chemical bond. He defines it as the deviation of a heteroatomic bond strength from the average of two homolytic bonds. This polarity concept can be extended to inorganic compounds.

Electronegativity can also be defined as the average of the electron affinity and ionization potential of an atom, but this can depend on the environment of the atom. In 1961 a periodic table of electronegativity values for neutral atoms has been established that is currently widely used [65].

The electronegativity as a function of cation charge I can be deduced from the approximate expression Eq. (5.3) [66]:

$$\chi_i = \chi_0 + \sqrt{\Sigma_i I_i} \tag{5.3}$$

In this expression χ_0 is the electronegativity of the neutral atom, and I_i its ith ionization potential. The relation between strong acidity and electronegativity implies that the surface O–H bond strength becomes weaker the higher the electronegativity of the cations that bind to the oxygen atom connected with the proton. The larger this electronegativity, which relates to charge, radius, and redox properties of the cation, the more electron density on the oxygen is directed towards the cation. This depletion of electron density of the oxygen atom reduces its charge. In consequence the interaction with the positively charged proton will decrease.

In 1967 such an approximate bonding model had been suggested by Sanderson [67] to explain the acidity of the oxides. Based on a model where the charge on oxygen has a maximum of −1 it is proposed that when this charge is more negative than −0.5 the oxide is basic, but when the charge is in between −0.5 and −0.1 oxides in this category can be acidic, weakly basic as well as amphoteric. The acidic oxides dominate above the negative oxygen charge of −0.2.

Differences in charge can be measured with X-ray photoemission spectroscopy [68]. Sanderson mentions Re_2O, CrO_3, Tc_2O_7, and Mn_2O_7 as strongly acidic oxides, MoO_3 as medium acidic and WO_3 as weakly acidic. Na_2O is an example of a basic oxide.

A small size and high charge of the cation are conditions for strong acidity.

In the next section it is shown that at surfaces this is more complex. Surfaces can contain Brønsted acidic as well as Brønsted basic hydroxyls. Their chemical nature does not only depend on electronegativity of the cations that compose the oxide but also on oxygen atom coordination and local topology of the hydroxyl site.

The Hammett function value can correlate with catalytic reactivity for materials that differ largely in acidity. An example is given in Figure 5.9a that compares the temperature of maximum rate T_m of acid-catalyzed cracking of cumene for different acidic materials. T_m is determined in a thermally programmed reaction spectroscopy (TPSR) experiment [69].

Cumene is adsorbed at low temperature and the rate of benzene production is studied as a function of increasing temperature. Superacidic mixed oxides and zeolite catalysts are compared.

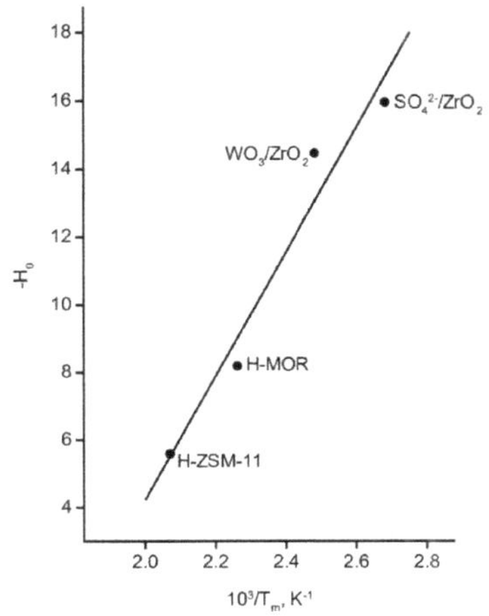

(a)

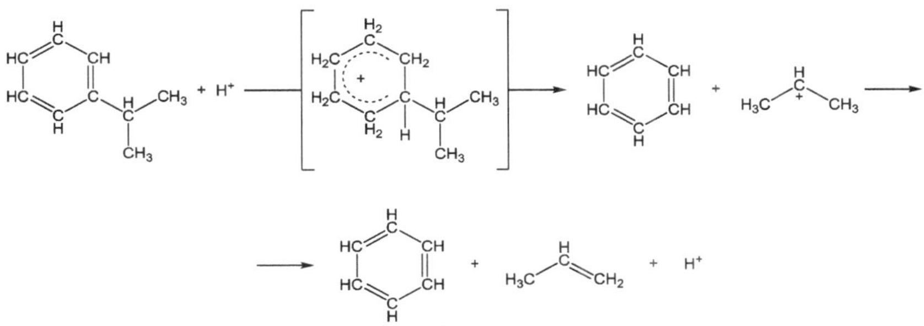

(b)

Figure 5.9 The cracking reaction of cumene catalyzed by solid acids. (a) Relationship between Hammett function of different solid acids and reciprocal cumene dealkylation temperature [70]. (b) The mechanism of the cumene cracking reaction.

The maximum T_m as a function of H_0 is plotted in Figure 5.9a. The reaction mechanism of cumene cracking is given in Figure 5.9b. The linear relation between H_0 and T_m^{-1} of Figure 5.9a implies that the activation energy of cumene cracking relates linearly with H_0 as in Eq. (5.4):

$$E_{\text{act}} = E_{\text{act}}(H_0 = 0) + \lambda H_0 \tag{5.4}$$

Eq. (5.4) is a kinetically important relation. It is a Brønsted-Evans-Polanyi relationship (Section 2.3.2) between activation energy and a thermodynamic energy indicator of acidity. The activation energy of a protonation reaction decreases when the deprotonation energy of the surface hydroxyl decreases.

The value deduced for λ is 0.15. It implies that the hydroxyl O–H bond only has to stretch by approximately 15% for the proton to transfer to the reactant.

The Hammett acidity function is a global descriptor of the surface acidity. It is not a very satisfactory descriptor since due to large inhomogeneity of the surfaces and adsorption features of the indicators H_0 values are sensitive to details of the solid surface structure and proton concentration.

A proper measure for solid acidity is the proton surface concentration and variation in corresponding *DPE*.

In the 1960s quantitative study of surface acidity by direct determination of *DPE* became possible with advances in vibrational spectroscopy [19], [71]. This provides an early foundation to atomic models of the surface hydroxyl as is discussed in the next section.

5.2.3 *The Definition of Deprotonation Energy (DPE)*

Infrared spectroscopy can be usefully applied to measure hydroxyl vibrational frequencies [72] that relate to the O–H bond strength. Additionally solid-state NMR spectroscopy [73] can probe bonding properties of surface hydroxyl groups by measurement of the charge of the hydroxyl proton.

Vibrational spectroscopy deduces proton strength from vibrational frequencies of surface hydroxyl, while solid-state NMR can be used to deduce the charge on the proton. Correlation of such surface data with catalytic reactivity of the proton is generally poor. This is partially due to the additional contribution to the catalytic rate of reactant adsorption energy variation with surface structure as discussed later in Section 5.4.3.3.

Measurement of the properties of the undisturbed proton gives information only on the chemical bond of O–H in the ground state, which correlates with the homolytic bond dissociation energy.

This poorly correlates with the deprotonation energy (*DPE*) that results from ionization of the O–H bond (see Figure 5.10).

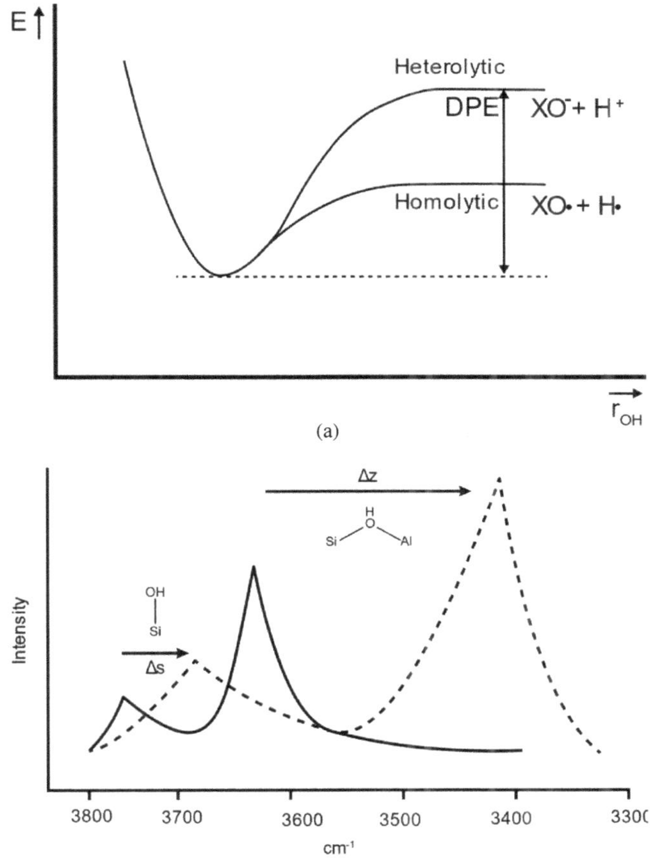

Figure 5.10 Potential energies and vibrational spectra of surface OH groups (schematic). (a) Potential energies of dissociation of the zeolite proton bond. Comparison of homolytic and heterolytic bond dissociation. (b) Comparison of shifts of vibrational O–H frequencies, Δs of silanol SiOH and Δz of zeolite Si(O-H) Al proton, when in contact with a probe molecule that binds through hydrogen bonding. The ratio of Δs/Δz is independent of adsorbate (— undisturbed, – hydrogen bonded).

The *DPE* consists of the sum of three energies as illustrated in Eq. (5.5):

$$DPE = E_{\mathrm{OH}}^{\mathrm{cov}} - E_{\mathrm{A,surface}} + IP_{\mathrm{proton}} \tag{5.5}$$

E_{OH}^{cov} is the homolytic bond cleavage energy of O–H into a hydrogen atom and a neutral surface site. The heterolytic bond dissociation energy is larger. Charge is separated and positive charge becomes located on the proton and negative charge on the surface. The ionization energy IP_{proton} is the energy cost to generate the proton and the electron affinity of the surface site $E_{\text{A,surface}}$ is the gain in energy due to electron attachment.

Differences of $E_{\text{A,surface}}$ largely determine differences in *DPE*. Accommodation of the negative charge on the surface depends on charge relaxation of the deprotonated surface. This is determined by ease of lattice polarization, which relates to the electron affinity of the cations that compose the oxide. It accounts for the relation of Hammett acidity function values H_0 and atom electronegativity of Figure 5.8.

As is illustrated in Figure 5.10a, with respect to homolytic dissociation of the O–H bond, the deprotonated state is an excited state. When the proton is in its equilibrium OH ground state, spectroscopies and calculations show that the charge on the proton is close to zero and the OH bond is only slightly polar [74], [75]. Ionicity of the O–H bond only develops when it interacts with a probe molecule.

The proton will transfer to the probe molecule when the proton affinity of the latter overcomes the zeolite framework deprotonation energy. Because of additional electrostatic stabilization (see Eq. (5.1)) the protonation energy *Ep* is less than *DPE*.

With weakly interacting adsorbates such as CO, N_2, and CH_4 the surface O–H bond remains intact [76]–[78] but the O–H bond weakens and the bond polarizes. A small positive charge appears on the H atom. As schematically indicated in Figure 5.10b the vibrational frequency shifts downward [76]–[78] and the infrared absorption intensity increases.

The shift Δv of the O–H frequency depends on the O–H bond strength and, as discussed below, can be correlated with its *DPE*, as shown by a comparison of the silanol frequency shift SiO–H at the silica surface with that of the zeolitic proton SiO–HAl. Their ground state frequencies are respectively 3750 and 3620 cm^{-1} for a zeolitic framework with a low Al/Si ratio. When CO adsorbs on the proton the silanol stretching frequency shifts downwards by 100 cm^{-1}, whereas the zeolite proton frequency shifts by 300 cm^{-1}. The O–H adsorption peak that belongs to the stronger bond (silanol with its higher frequency) has the smaller shift. It illustrates the generally observed result that the more acidic the proton the larger its shift when in contact with an adsorbed molecule.

For weakly interacting adsorbates the ratio Δν(solid acid Brønsted)/Δν(silanol) is a constant independent of the adsorption probe.

This property has been exploited by the Russian scientists Paukshtis and Yurchenko [79] to determine proton *DPE*. They discovered the empirical relation Eq. (5.6) that gives *DPE* and a function of Δν(solid acid Brønsted)/Δν(silanol):

$$DPE_{Brønsted} = DPE_{Silanol} - A\log\left[\frac{\Delta\nu_{Brønsted}}{\Delta\nu_{Silanol}}\right] A = \frac{1}{0.00226}\frac{kJ}{mol} \quad (5.6)$$

The *DPE* of silanol of 1390 kJ/mol can be determined by calibration with titration studies of acids in solution [80].

The Paukhstis-Yurchenko formula can be successfully applied to the study of proton acidity in zeolites. This is schematically illustrated in Figure 5.11 for the *DPE* of the protons in H-ZSM-5 as a function of Al/Si zeolite framework ratio [81].

The rate per proton of the catalytic cracking reaction of isooctane [30], *DPE*, and the OH frequency shift Δν(CO) induced by adsorbed CO are plotted as a function of the Al/Si zeolite framework ratio. There is an increase in reaction rate when Al/Si decreases from its maximum value of 1 until the Al/Si ratio reaches 0.1. Beyond this value the cracking rate is constant (see Figure 5.4).

It correlates with a lower value of *DPE* and a higher value of Δν(CO) when the Al/Si ratio is less than 0.1. For different zeolite structures the transition value of Al/Si where *DPE* becomes independent of framework concentration may slightly vary.

The *DPE* values of zeolitic protons vary by 50 kJ/mol. The chemical bonding cause of the dependence of *DPE* on framework Al/Si ratio is discussed in Section 5.3.3.

It is interesting to compare the covalent XO–H bond energy with the *DPE* as determined by the heterolytic bond energy. For the

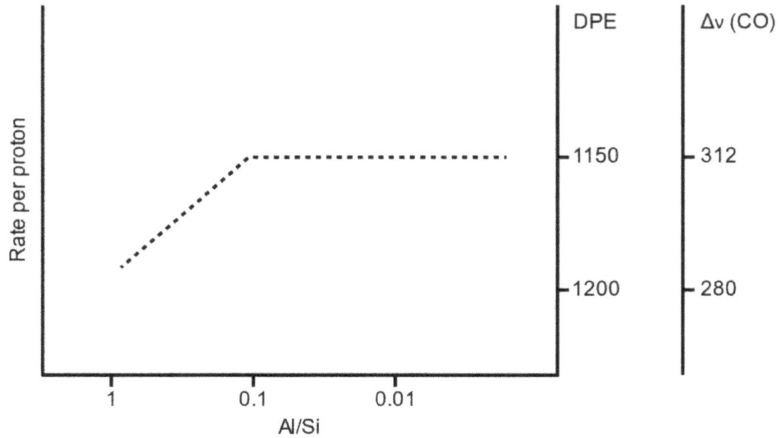

Figure 5.11 The catalytic cracking reaction rate of isooctane normalized per proton, measured *DPE* and vibrational shift Δν(CO) as a function of zeolite framework Al/Si ratio for H-ZSM-5 [30], [81].

surface silanol, computations determine for XO–H a covalent homolytic bond dissociation energy of 505 kJ/mol [82]. From the difference in vibrational frequencies one deduces that the covalent bond strength of the zeolite Si(O-H)Al is approximately 450 kJ/mol. The difference in the corresponding *DPE*s is much larger. The *DPE*s are respectively 1390 and 1150 kJ/mol.

This indicates that the difference in *DPE* of surface silanol and zeolite proton is dominated by the different stabilization of the negative surface charge.

An important additional conclusion derives from Figure 5.11. It indicates that proton reactivity does not correlate with the global average of the electronegativities of the constituting lattice framework cations but is a local property. It relates only to the cation composition near the protonic site. This defines the critical value of $Al/Si_{critical}$ where the O–H bond strength becomes independent of the Al/Si ratio.

5.2.4 *The Theory of Surface Brønsted Acidity*

Hydroxylated oxide surfaces contain Brønsted acid and Brønsted basic hydroxyls.

Here the reactivity of coordinatively unsaturated oxide surfaces is considered. These can be considered to be formed by cleavage of the corresponding bulk structures. In Section 5.3.3 acidity of zeolitic materials is discussed where the internal surface consists of coordinatively saturated atoms.

Measurement of the vibrational spectra of hydroxyls on high surface area oxides provides an atomistic understanding of surface acidity. For many materials these spectra indicate the presence of different hydroxyls. They can have Brønsted acidic as well as Brønsted basic reactivity.

In vacuum the coordinatively unsaturated oxygen anions or metal atom cations react as Lewis base or acid. On such a surface heterolytic dissociation of adsorbed water will create surface hydroxyls. A Brønsted basic OH will attach to the Lewis acid surface cation and a Brønsted acid proton to the oxygen surface anion. A spectroscopic measurement of such a system should at least give two values for the charge on O–H or its vibrational frequency.

On many materials more than two vibrational peaks are found. For the catalytically important γ-Al_2O_3 material Peri, then at American oil company, reported such spectra in the 1960s [71], [83]. State-of-the-art quantum-chemical methods that are discussed here provide a satisfactory interpretation of the relation between these spectra and surface site structure.

In the 1970s an important early insight came from suggestions from the Russian scientists Tsychanenko and Filiminov [84]. They suggested correctly that the differences in vibrational O–H frequencies of free hydroxyls relate to differences in coordination of the oxygen atoms of OH as illustrated in Figure 5.12a.

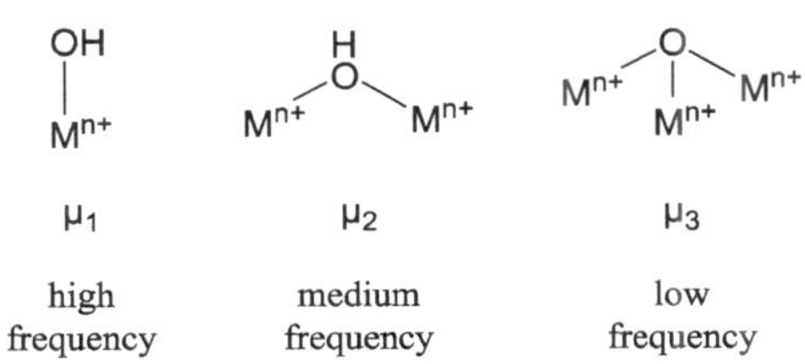

Figure 5.12 Schematic illustration of vibrational O–H frequencies as a function of OH coordination to metal cation.

OH attached to a single cation (μ_1) is a basic hydroxyl, whereas OH that is twofold (μ_2) or threefold (μ_3) coordinated has acidic properties.

The original infrared spectrum of Peri on the hydroxyls of γ-alumina [70] is shown in Figure 5.13a. In Figure 5.13b this is compared with results of quantum-chemical calculations published fifty years later by French scientists of IFP and the Ecole Normale Superieur [85].

Peri distinguished five vibrational adsorption peaks on γ-alumina. Calculations by Digne *et al.* [85] relate these spectral features to the structure of the hydroxylated surface. Their important discovery is that the complexity of the spectrum also derives from the different surfaces that compose the experimentally studied materials. As can be seen from Figure 5.13b adsorption frequencies from hydroxyls on different surfaces will show differences even when the local OH coordination is the same.

Comparison of the local structure of the hydroxyls validate the hypothesis of Tsychanenko and Filiminov that singly coordinated O–H has a high vibrational frequency and highly coordinated O–H has a low vibrational frequency. Frequency relates to OH bond strength. The higher coordination site has the lower vibrational frequency, which implies a weaker bond energy as one would deduce from Bond Order Conservation theory (Section 2.3.2.2). The corresponding O–H will be Brønsted acidic. Experimentally one can distinguish the different reactivities of the hydroxyls by ion exchange experiments with D_2O, D_2, or CD_4. These molecules will only undergo isotope exchange with the Brønsted acid sites.

The presence of different hydroxyls on oxide surfaces is quite general as can be seen from Figure 5.14. The number of vibrational frequencies, the ratios of intensities and their frequencies vary for different materials. The high frequency peaks correspond to Brønsted basic hydroxyls, the lower frequencies to Brønsted acidic or hydrogen-bonded hydroxyls.

Even on basic oxides such as MgO or CaO the signatures of Brønsted base (appr. 3750 cm^{-1}) and Brønsted acid (appr. 3620 cm^{-1}) are present. The low vibrational frequencies around

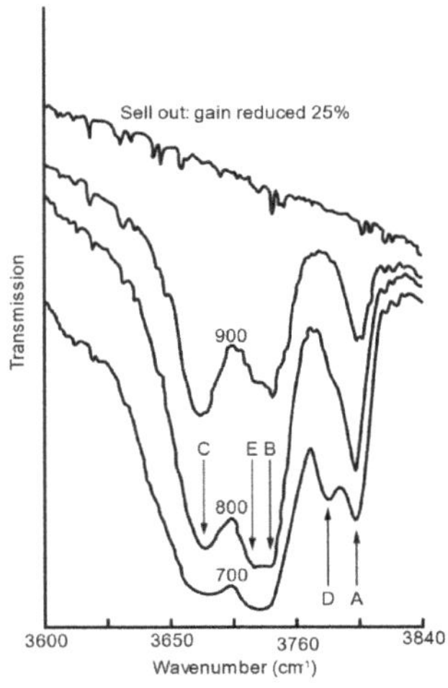

(a1)

Band	Wavenumber (cm^{-1}) Current
A	3800
B	3744
C	3700
D	3780
E	3733

(a2)

Site	Surface	ω_{cal}	ω_{exp}	
HO–μ_1–Al_{IV}	(110)	3842	3800–3785	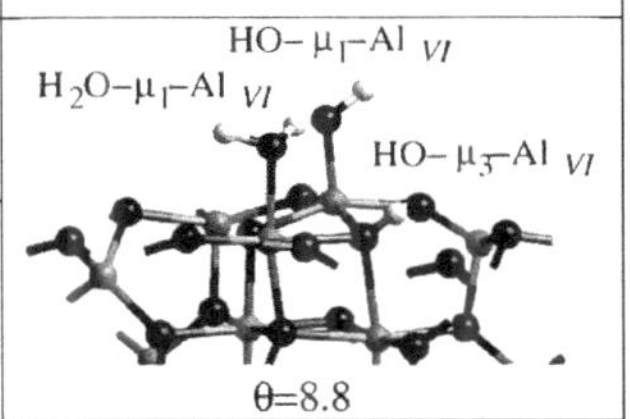
HO–μ_1–Al_{VI}	(100)	3777	3780–3760	
HO–μ_3–Al_{VI}	(111)	3752	3745–3740	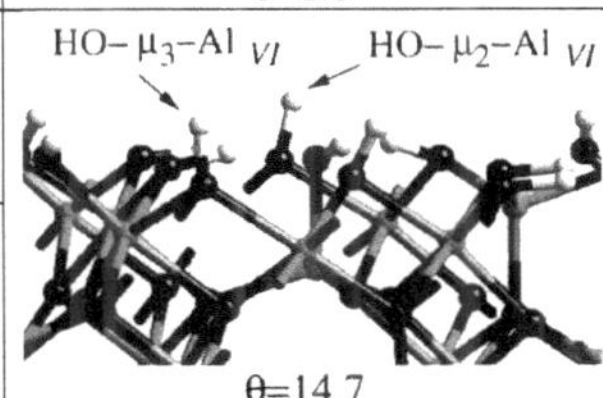
HO–μ_1–Al_V	(110)	3736	3735–3730	
HO–μ_2–Al_{VI}	(110)	3707	3710–3690	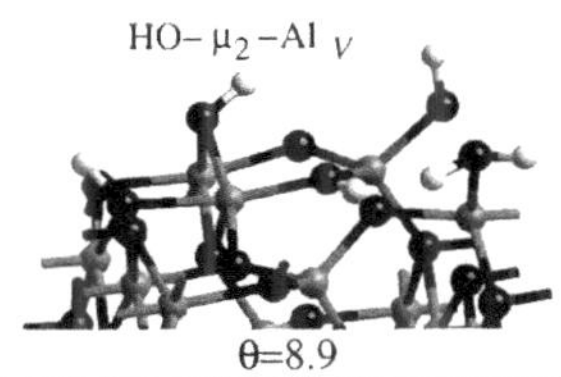
HO–μ_3–Al_{VI}	(101)	3589	3650–3590	

(b)

Figure 5.13 Experimental and computed OH frequencies of γ-alumina. (a1) Spectra of alumina dried at indicated temperatures (°C) [70], (a2) measured frequencies of surface OH groups [71]). (b) Results of DFT calculations. Local structures of hydroxylated (100), (110), and (111) surfaces and their calculated vibrational frequencies [85].

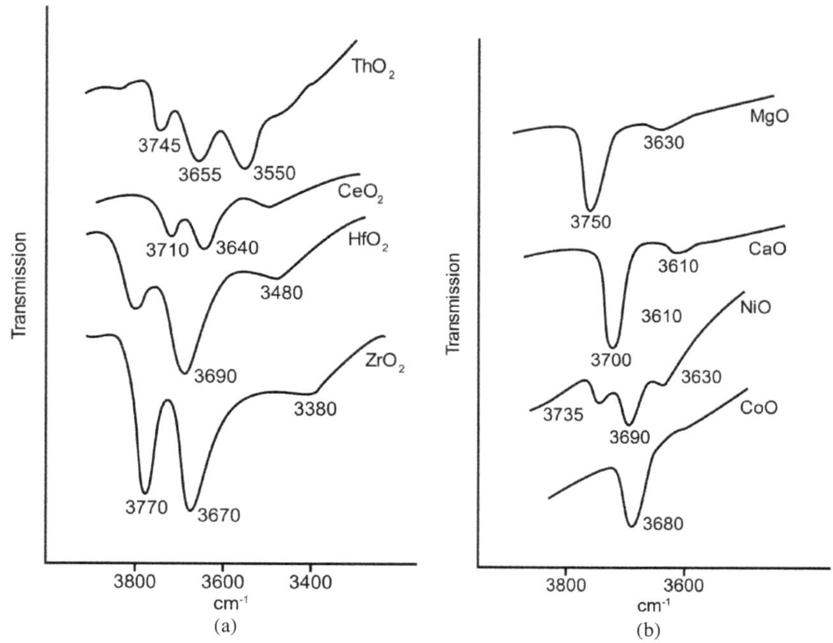

Figure 5.14 Infrared spectra of OH groups on the surface of some oxides after evacuation at 500°C. (a) Infrared spectra of surface hydroxyls on ThO_2, CeO_2, HfO_2, and ZrO_2 after evacuation at 500°C. (b) Infrared spectra of surface hydroxyls on MgO, CaO, NiO, and CoO after evacuation at 450, 500, 320 and 200°C, respectively [84].

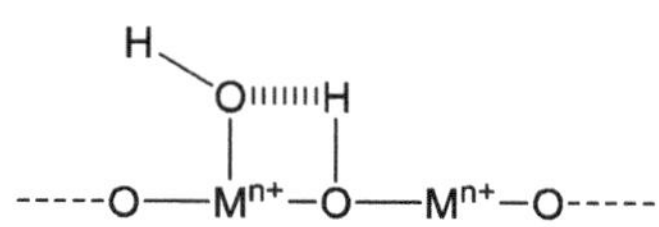

Figure 5.15 Hydrogen bonding between Brønsted basic hydroxyl and Brønsted acid surface proton.

3400 cm^{-1} are due to the hydrogen bonding interaction of basic O–H and acidic proton (see Figure 5.15).

A qualitative theory is available that predicts chemical reactivity of surface hydroxyls as a function of local coordination and oxide cation composition [86].

This theory is based on Pauling's electrostatic stability theory [87] and is an elaboration of early suggestions by C. L. Thomas [4]. He was the first to deduce differences in acid strength of mixed oxides from structural differences of protonic sites.

Essential to electrostatic bonding theory is knowledge of the structure of the oxides and the coordination of the cations and anions. Their respective formal charges are used. A model of the surface structure is deduced from bulk structure. It assumes there is no surface reconstruction (note: the quantum-chemical calculations of Figure 5.13b incorporate surface reconstruction that can follow hydroxylation). Also, the theory does not account for differences in electron affinity of the cations other than their charge. Cation size differences may have an indirect effect because it can lead to structural differences.

The electrostatic method is introduced here using the anatase structure of TiO_2 or corundum structure of α-Al_2O_3 as illustration. The ideal structures of the oxides differ in the way the octahedra are connected. In the anatase structure each oxygen atom shares three octahedra, while in the corundum the oxygen atoms share four octahedra.

Figure 5.16a illustrates for the bulk of the crystal the connectivity of three octahedra. At the surface metal atom cations and oxygen anions become coordinatively unsaturated. Figure 5.16b shows a

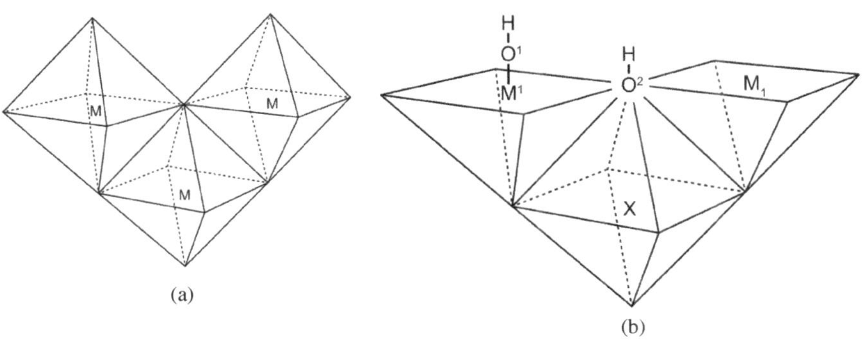

Figure 5.16 (a) Three octahedra tetrahedrally connected. (b) Hydroxylated surface site model.

surface model that is hydroxylated by dissociative adsorption of water.

When water dissociates the OH coordinates end-on to the metal cation. The proton adsorbs on a bridging anion. The Pauling electrostatic method can be used to deduce whether adsorbed hydroxyls are Brønsted acidic or basic.

We will introduce first the Pauling electrostatic method by analyzing the relative stability of a proposed structure. To decide how many octahedra the oxygen atoms will connect, or how many cations coordinate to an oxygen atom, Pauling introduced the notion of electron deficiency. The formal charge Q^- of the oxygen anion (−2) has to be compensated for by the electrostatic bond strength of surrounding cations. Pauling defined the electrostatic bond strength as the formal ion charge divided by cation coordination. The charge deficiency is defined as the sum of the electrostatic bond strengths minus the ion charge. When the charge deficiency does not differ by more than ±1/6 Pauling declares the structure stable.

The Pauling bond strength $S^{\pm}$ and charge deficiencies $\varepsilon_c^{\pm}$ are defined in Eq. (5.7):

$$S^{\pm} = \left|\frac{\text{formal ion charge}}{\text{number of nearest neighbour ions}}\right| \tag{5.7a}$$

$$\varepsilon_c^{+} = Q^{+} - \sum_i S_i^{-} \tag{5.7b}$$

$$\varepsilon_c^{-} = Q^{-} + \sum_i S_i^{+} \tag{5.7c}$$

Charge deficiencies at the surface are often larger than 1/6. This implies a large electrostatic imbalance at the location of that particular surface atom.

As a measure of the acid strength, ε_c^- is calculated on the oxygen atom that is attached to the proton. A larger positive value implies a large positively charged electrostatic field, which will repulse the proton. Then the proton behaves as a strong Brønsted acid. A large negative ε_c^- means electrostatic imbalance due to a negative electrostatic field. In this case the hydroxyl is a strong Brønsted base.

For the two metal oxides Table 5.1 lists the charge deficiencies of the oxygen atoms as a function of oxygen cation coordination. This changes when the oxygen atoms are at the oxide surface.

Since the stable situation requires zero charge excess one concludes that the oxygen atoms of bulk TiO_2 connect with three Ti^{4+} cations and the oxygen atoms of bulk Al_2O_3 connect with four Al^{3+} atoms. In TiO_2 half of the connecting octahedra are occupied by a cation, in Al_2O_3 two-thirds of the octahedra. In anatase TiO_2 four octahedra surround in tetrahedral geometry a central octahedron and the oxygen atoms are face-centered cubic packed. In α-Al_2O_3 (corundum structure) the octahedra are highly distorted. Around the oxygen atoms two Al cations are closer than the others. The structure is complex and oxygen atoms are hexagonal close-packed.

The increase in Brønsted acidity with increase of cation charge that coordinate OH shows for OH end-on adsorbed to the activating cation. The structure of the free silanol given in Figure 5.1b can be used as an example. The charge excess of silanol oxygen atom $\varepsilon_c^-(0) = 0$. There is no charge imbalance; the electrostatic model predicts that the hydroxyl is nor acid or base. Experimentally the silanol is found to be weakly acidic.

Table 5.1 $N_c(0)$ is the number of cations that coordinate with the bridging oxygen atom, S^+ is Pauling bond strength of cation bond, and $\varepsilon_c^-(0)$ is the charge deficiency of a bridging oxygen atom.

	$N_c(0)$	S^+	$\varepsilon_c^-(0)$
TiO_2	2	2/3	−2/3
	3	2/3	0
	4	2/3	+2/3
Al_2O_3	2	1/2	−1
	3	1/2	−1/2
	4	1/2	0

When in the structure of Figure 5.1b Si^{4+} is replaced by P^{5+} or S^{6+} the charge deficiencies $\varepsilon_c^-(0)$ are respectively +1/4 and +2/3. The charge deficiencies are positive and relatively large. Phosphate and sulfate react strongly Brønsted acidic. The OH acid strength increases the larger the charge on the cation that binds the hydroxyl and is at the center of the oxygen tetrahedron.

In agreement with this, experimentally phosphoric acid immobilized on silica behaves as a strong Brønsted acid but is less acidic than sulfuric acid.

For surface acidity this classical understanding of relation between proton strength and cation charge only holds for end-on (μ_1) adsorbed OH. In the following surface acidity is discussed as a function also of OH coordination number.

Figure 5.16b shows schematically a hydroxylated surface site at the anatase or corundum surface. Surface formation creates a surface metal atom that is fivefold-coordinated to oxygen atoms and an oxygen atom that is threefold- or twofold-coordinated to a metal cation. This depends on cation occupation of the octahedron below the surface atom.

Table 5.2 shows that singly coordinated μ_1 O–H is more basic when adsorbed to the cation with lower charge. The negative excess charge on O–H is larger. Whereas for the surface site of Al_2O_3 an acidic proton is only present when the octahedron below the connecting O_2 is occupied, for TiO_2 an acidic proton can only be stable when this octahedron is unoccupied (the charge excess of +1 implies

Table 5.2 Charge excesses ε_c on hydroxylated and non-hydroxylated surface sites. Comparison with different cation charges and occupation of octahedra. Atom numbers as in Figure 5.16b.

Hydroxylated		Non-hydroxylated			
$\varepsilon_c^-(0_1)$	$\varepsilon_c^-(0_2)$	$\varepsilon_c^+(m_1)$	$\varepsilon_c^-(0_2)$	Charge (m_1)	Charge X
−1/2	+1/2	+1/2	−1/2	3	3
−1/2	0	+1/2	−1	3	0
−1/3	+1	+2/3	0	4	4
−1/3	+1/3	+2/3	−2/3	4	0

the proton is unstable). On Al_2O_3 the acidic proton is more reactive compared to TiO_2 and is only stable when attached to a threefold-coordinated oxygen atom. On TiO_2 the acidic proton resides on an oxygen atom that is twofold coordinated.

Interestingly on the hydroxylated MgO(100) surface (in bulk MgO all octahedra are occupied and cation and anion have the same coordination), $S^{+}(\text{Mg}) = -S^{-}(\text{O}) = \frac{1}{3}$. The electrostatic model predicts the presence of two hydroxyls: one with $\varepsilon_c^{-}(\text{O}) = -\frac{2}{3}$, the other with $\varepsilon_c^{-}(\text{O}) = +\frac{2}{3}$. So even on a basic oxide as MgO acidic protons are present. Figure 5.14b confirms this prediction. The low frequency of the acidic proton is comparable to that of H-Y zeolite! The relative concentration may depend on the degree of MgO hydroxylation. In a high concentration of adsorbed water, the MgO will reconstruct and transform to $Mg(OH)_2$.

In catalysis a reactant that is liable for activation by acidic protons will selectively react with the acidic protons on a surface covered with hydroxyls of different reactivity. Even when the concentration of basic hydroxyls dominates, catalytically the oxide will be acidic, even when the concentration of acidic protons is low.

An example of such a selective acid surface site-sensitive reaction is the selective oxidation of ethene to give the epoxide. The ethene epoxidation reaction is discussed in Section 4.3.3.1. The catalyst consists of Ag particles dispersed on α-alumina. The acidic protons present on the α-alumina support isomerize the desired ethene epoxide product to the aldehyde. This decreases selectivity because of rapid combustion of the aldehyde by Ag.

An interesting conclusion of the comparison of the surface acid site of TiO_2 and Al_2O_3 is that the acid strength of the Brønsted acid hydroxyl is weaker on the TiO_2 than on the Al_2O_3 surface. Different from global predictions based on the correlation of acidity with average electronegativity for mixed oxides of Figure 5.8 the surface that contains the cation with the lower charge (lower electronegativity) has the higher proton acidity.

The Pauling electrostatic theory of ionic bonding is also a useful method to predict the acidity of mixed oxides. Here the example of mixed TiO_2/SiO_2, ZrO_2/SiO_2 oxides is discussed. Figure 5.17

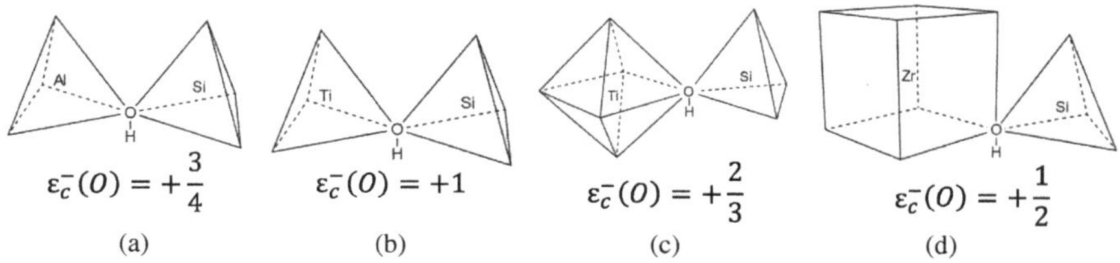

Figure 5.17 The cation coordination dependence of the electron deficiencies $\varepsilon_c^-(O)$ of protonic hydroxyl that bridges silicon tetrahedron and cations with different coordination. (a) Zeolite proton site, (b) tetrahedral Ti^{4+} (proton unstable), (c) octahedral Ti^{4+}, (d) cubic Zr^{4+}.

compares the oxygen electron deficiencies of the bridging hydroxyl of cations with different charge and oxygen coordination.

Figure 5.17 gives the charge excess $\varepsilon_c^-(O)$ of the oxygen atom that bridges the Si^{4+} cation with tetrahedral coordination and a second cation X^{n+}. In the case of the zeolite, X^{n+} is the Al^{3+} cation, also tetrahedrally coordinated. Its strong acidity (low *DPE*) is due to the high charge excess $\varepsilon_c^-(O) = +\frac{3}{4}$ on the protonated bridging oxygen atom. This is to be compared with $\varepsilon_c^-(O)$ when the oxygen atom bridges the Si^{4+} cation and the four-coordinated Ti^{4+}, six-coordinated Ti^{4+}, or the larger eight-coordinated Zr^{4+}. In the mixed oxides Ti^{4+}can be four- or six-coordinated.

When Ti^{4+} is four-coordinated there is no effective negative charge to compensate for the charge on the proton. The protonated site is unstable, and no acidity develops.

However, when Ti^{4+} has octahedral coordination, the electrostatic bond strength of the Ti–O bond decreases and a proton can be accommodated. Its charge excess becomes $\varepsilon_c^-(O) = +\frac{2}{3}$. The acid strength is slightly less than that of the Al/Si site. The Zr^{4+} cation that also has charge 4+ is larger than the Ti^{4+} cation. In the oxide the Zr cation may have eight coordination. In the mixed oxide the reduced electrostatic strength still creates an acidic proton: $\varepsilon_c^-(O) = +\frac{1}{2}$. This lower acidity agrees with expectation. The Si–Al, Si–Ti and Si–Zr mixed oxides belong to stronger solid acids (see Figure 5.8).

The electrostatic theory rationalizes differences in surface basicity and acidity. It illustrates the well-known chemistry that upon dissociative adsorption of water the stronger Lewis acid cation site gives the weaker Brønsted base, and the stronger Lewis base anion gives the weaker acid.

Only when the local coordination of the oxygen atom of the protonic site and the charges of the cations are known the proton reactivity can be estimated. The oxygen electron deficiency $\varepsilon_c^-(O)$ gives an indication of trends in surface O–H vibrational frequencies.

An important general conclusion is that instead of differences in cation electronegativity differences in surface topology can dominate differences in proton acidity.

5.3 Zeolite Catalysts, Their Structure and Acidity

Zeolites are nanoporous alumino-silicates. Their channel network and internal surface are defined by crystallography.

5.3.1 *Introduction*

Over the past fifty years the relation between the inorganic chemistry of zeolites and their catalytic functionality has become increasingly understood. Zeolite crystallites have a well-defined nanoporous structure that is accessible to reactant molecules. The dimensions of the zeolite nanopores are comparable to those of organic molecules, but there can be a mismatch of molecular size and form and zeolite cavity. Therefore the catalytic reactivity depends on the nanopore structure, and this structure is the cause of shape selectivity.

In mineralogy zeolites are known as alumino-silicates. They consist of a negatively charged alumino-silicate network that is stabilized by positively charged alkali or alkaline earth cations. The alumino-silicate network creates a microstructure of nanochannels and cavities in which the cations are located [88], [89].

There is a rich variety of synthetic inorganic materials with a structure that is related to the zeolite framework topology [90] but unknown for natural minerals. These materials with similar tetrahedral networks as in the zeolites are not necessarily alumino-silicates but may be alumino-phosphates [91]. In this book such materials will be called zeolitic.

By chemical treatment zeolitic materials can be converted into Brønsted acidic catalysts. They are robust and can be used at the high temperatures needed for many of the hydrocarbon conversion reactions [92].

The zeolite network is constructed according to an elementary building principle: it is a network of tetrahedra connected through their vertices. In alumino-silicates a tetravalent cation such as Si^{4+} or trivalent cation such as Al^{3+} is located at the center of each tetrahedron. The vertices of the tetrahedron are four oxygen anions. Currently there are 145 materials with different zeolite frameworks known and many more theoretical structures have been proposed based on their tetrahedral network [93], [94].

Two examples of catalytically important zeolite structures are given in Figure 5.18.

Figure 5.18a illustrates the FAU network of Zeolites X and Y that consists of double six-rings and sodalite cavities that are formed from rings that consist of four and six tetrahedra. The super cage is accessible through 12-ring windows to reacting molecules. Figure 5.18b_1 is a space-filling model that shows oxygen atoms (colored red) lining the zeolite microporous wall of the MFI network. The Brønsted acidic zeolite H-ZSM-5 has the MFI structure. Different from the FAU structure this low Al/Si framework is built from tetrahedra connected by a mixture of four, five, and six rings that form a three-dimensional nanoporous channel network with pore dimensions determined by rings of 10 tetrahedra, shown in Figure 5.18b_2. In this structure the cations, which compensate for the negative charge of the lattice framework, are only located along the channel walls. There are no additional cavities that can contain cations as in the FAU structure.

In the case that the tetrahedron contains a Si cation surrounded by four negatively charged oxygen atoms, the zeolite lattice framework will be electrostatically neutral. It has the SiO_2 stoichiometry similar to that of quartz, so it can be considered a polymorph. Such siliceous materials are catalytically unreactive.

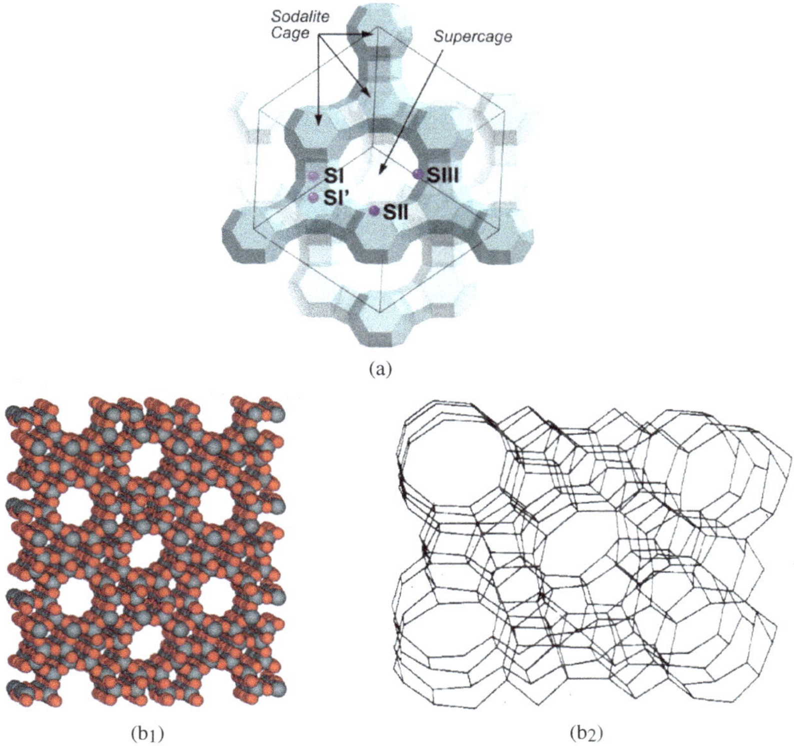

Figure 5.18 Framework structures of catalytically important zeolites. (a) The FAU structure of Zeolite X or Y [95]. (b) The MFI framework structure of H-ZSM-5 [29]; (b_1) space-filling model, (b_2) network model. The lines connect tetrahedral centers.

Substitution of the Si^{4+} cation by a cation of lower charge (often the Al^{3+} cation) creates a negative charge of the zeolite framework. In a stable structure this is compensated for by a cation, for instance an alkali or alkaline earth cation. In synthetic zeolites this can be also a structure-directing organic cation.

The maximum concentration of trivalent cations that substitute for Si^{4+} in the lattice framework of the zeolite is determined by $X^{3+}/Si^{4+} = 1$. This is the Löwenstein rule [96]. In the case that X is Al, according to this rule no two $[AlO_2]^-$ tetrahedra can share a connecting bridging O atom. At least one [SiO_2] tetrahedron has to be located between two [AlO_2]$^-$ tetrahedra.

The material can be converted into a solid acid by two different processes. Solid acid formation chemistry depends on the zeolite structure. The discovery in 1958 of the conversion of the Na^+-containing X and Y zeolites by ion exchange with Ca^{2+} into a solid acid is unique for the FAU structure with high Al/Si ratio [13]. The Na^+ cations are initially mainly located in the super cage. Two Na^+ cations can be exchanged by Ca^{2+} that locates in the double six-ring cage of the sodalite FAU structure. In this cage cations are inaccessible to reactant molecules.

The Ca^{2+} cation ionizes H_2O. A catalytically reactive proton attaches to a bridging O atom that is positioned between the Al^{3+}- and Si^{4+}-containing tetrahedra and the Ca^{2+} ion is converted into $Ca(OH)^+$.

The alternative and commonly used procedure is ion exchange of inorganic cations located in the zeolite channel by NH_4^+. Upon heating ammonia is released and a proton is left on an oxygen atom of the zeolite framework. The two solid acid-generating processes are illustrated in Figures 5.19a and b respectively.

The proton affinity of the solid acid generated by ion exchange of Ca^{2+} or cations such as Mg^{2+} or La^{3+} of Zeolite Y (Al/Si = 0.3) is comparable to that of the low Al/Si (Al/Si < 0.1) framework of the H-ZSM-5 catalyst [97], [98].

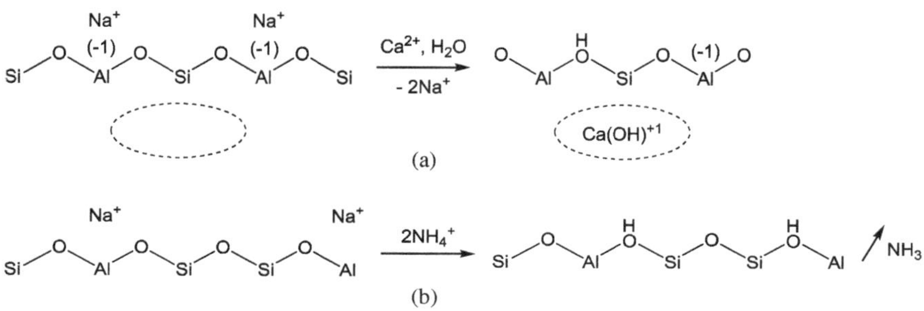

Figure 5.19 The generation of zeolite Brønsted acidity (schematic). (a) Ion exchange with Ca^{2+}, (b) ion exchange with NH_4^+.

It is discussed in the following subsections that the deprotonation energy *DPE* depends strongly on zeolite composition but is rather insensitive to zeolite structure. In contrast catalysis demonstrates large variation in catalytic functionality as a function of zeolite structure. This structural dependence of a catalytic reaction arises when protons interact with substrate molecules. It depends on match of size and shape of reactant molecule with zeolite nanocavity. This is the topic of Section 5.4.

A rich variation in the composition of the zeolite framework is possible. Differences in composition can importantly affect performance of the solid acid catalyst. Not only can the Si/Al atom zeolite framework ratio be varied over a large range, also the tetrahedral $[AlO_4]$ unit can be replaced by tetrahedral $[GaO_4]$ or $[FeO_4]$ units.

When in the SiO_2 framework two $[SiO_4]$ units are replaced by an $[AlO_4]$ unit and $[PO_4]$ unit, again a lattice is generated that is electrostatically neutral. When in the resulting $AlPO_4$ structure a P^{5+} cation is substituted by Si^{4+} an additional proton maintains charge neutrality. Such Brønsted acidic SAPO materials were discovered at Union Carbide in 1984 [91]. Because of their unique acidity and nanopore structure such synthetic silico-alumino-phosphate catalysts have been extensively explored. An important application is their use as catalyst for the conversion of methanol into short olefins (MTO process). This is an important step in the overall reaction of natural gas to base chemicals or liquid fuels [33], [99]–[101]. This and the related conversion reaction of methanol into aromatics is discussed in Section 5.4.5.

5.3.2 *The Structure Dependence of the Zeolite Deprotonation Energy*

The deprotonation energy of Brønsted acidic zeolites depends mainly on zeolite framework composition.

Whereas in the early episode of the discovery of zeolite acidity it was thought that the zeolite is to be considered a solid electrolyte with dissolved protons [102], the current view is quite different. The zeolite protons have strong covalent bonds with the zeolite framework. Protonic acidity is generated upon contact with a reacting molecule. After proton transfer to the reactant the positively charged carbocation is stabilized by interaction with the negatively charged zeolite framework.

The energy of zeolite framework deprotonation *DPE* consists of the difference between the energy of the O–H bond attached to the zeolite framework E_{OH}^{cov}, the energy of the free proton (i.e., the ionization energy of the hydrogen atom IP_{proton}), and the stabilization energy $E_{A,surface}$ of the negative charge left on the solid after heterolytic bond cleavage (see Eq. (5.5)).

Structure and composition differences will affect E_{OH}^{cov} and $E_{A,surface}$ differently. Differences in E_{OH}^{cov} are dominated by covalent bond interactions with the solid, whereas local electrostatics has a large effect on $E_{A,surface}$.

In Zeolite Y the presence of positively charged cations in zeolite wall cages has a large promoting effect on *DPE*. Calculations show that these additional electrostatic interactions may decrease *DPE* by 40 kJ/mol [103]. In addition to hydroxylated Ca and Mg, also oxycationic clusters of La^{3+} or Al^{3+} have this function [95]. The latter may be generated as extra framework cationic clusters by dealumination of the zeolite framework.

Different from expectation based on the ionic model of the zeolite lattice the charge of the zeolitic proton is close to neutral [73], [75], [104]. In contact with an adsorbate the O–H bond polarizes and charge develops on the proton (Figure 5.10). The near neutral charge of the free O–H proton is a reflection of the dominantly covalent nature of the respective O–H, Si–O, and Al–O bonds in the zeolite. Calculated charges on O, Al, and Si are typically half of the formal ion charges.

The zeolite lattice adapts to local changes [105]. Whereas distortion of the $[TO_4]$ tetrahedra has a large energy penalty, the bond bending energy of the Si–O–Si angle is less than 10 kJ/mol. This gives the zeolite lattice flexibility. It makes adaptation with little energy cost to slight local deformations possible. Structural relaxation happens when framework chemical bonds change upon deprotonation.

This relaxation effect depends on lattice position and may in principle make the reactivity of the proton dependent on zeolite lattice positions [106], [107]. Such relaxation is important because the structure of the [Al–(O–H)–Si] site is different from the deprotonated $[Al–O–Si]^-$ site. According to the Pauling covalent bonding model [87], when unconstrained, the bonds of three-coordinated O should be approximately sp^2 hybridized. Then the Al–O–Si angle is ideally 120°. This is in contrast with the unconstrained $[Al–O–Si]^-$ site, which is linear and the electrons of oxygen are sp hybridized.

When constrained by lattice strain the Al–O–Si angle will determine the covalent O–H bond energy. The smaller the Al–O–Si angle the stronger the O–H bond, because it increases the hydrogen 1s atomic orbital interaction with that of O 2s. This leads to differences in the vibrational frequencies of zeolite protons, which relate to the Si–O–Al bond angle of the proton site [Si–(O–H)–Al] at different lattice positions [108]–[110].

Siting of Al in the zeolite framework strongly depends on synthesis conditions and is difficult to determine. Therefore, modelling studies usually assume a statistical distribution of Al over different sites of the zeolite. In agreement with this assumption the experiment [35] shows that the proton-normalized rate of hexane cracking is independent of Al concentration in H-ZSM-5, although H-ZSM-5 zeolite has 12 distinguishable tetrahedral crystallographic positions (see Figure 5.4).

Computations on zeolites with low Al/Si concentration that average *DPE* over a statistical distribution of Al located at crystallographically different positions indeed find such an independence and confirm the independence of structure on proton reactivity [110]. In practice the differences become especially insignificant since proton mobility is high at the temperatures of catalysis.

This is illustrated in Figure 5.20 that give averaged calculated *DPEs* for different zeolite structures with low Al/Si ratio.

In the low Al/Si systems stabilization of the negative charge left on the reactive site upon deprotonation counteracts the changes in covalent bond strength of the protonated site.

The conclusion, that the reactivity of the proton in zeolites is independent of structure, implies that experimentally observed

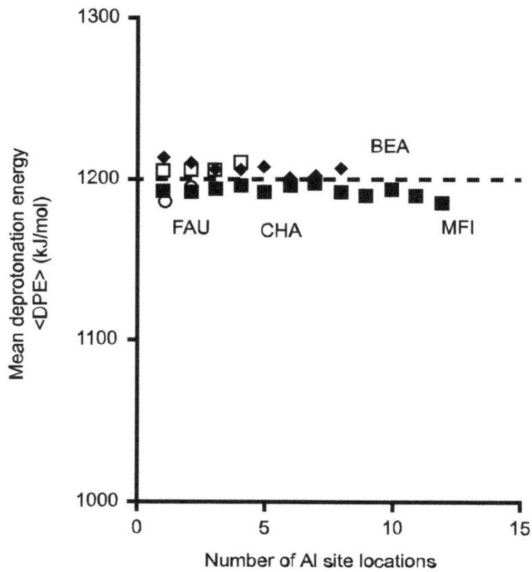

Figure 5.20 Comparison of DFT-calculated deprotonation energies for different zeolite structures. *DPEs* are averaged over the different site locations of the zeolite structures with low Al/Si concentration. They are seen to be independent of framework structure or proton location. MFI: ■, FAU: ○, FER: □, BEA: ◆ [110].

structure dependence of solid acid catalysis by siliceous zeolites has to have a different cause. In Section 5.4 it will be seen that this is dominated by the structure sensitivity of the activation of reactants and reaction intermediate to the dimensions of the zeolite nanopore. This is called the confinement effect.

5.3.3 *DPE as Function of Al/Si Framework Composition*

The averaged DPE is independent of Al/Si framework ratio as long as Al/Si < 0.1. DPE is a local property.

The dependence of the *DPE* on the concentration of the O–H zeolite groups is not an average global dependent property of framework Al/Si ratio but depends on the immediate environment of the zeolitic proton. Quantum-chemical calculations confirm this.

For a zeolite with the FER structure this is illustrated in Figure 5.21 [111]. The FER structure is related to the MFI structure, but different from MFI it contains tubular nanochannels that do not cross. Figure 5.21 shows DFT-computed *DPE* values as a function of global framework ratio as well as of local concentration topology. In the latter the Al concentration varies in the tetrahedra next to the the Si tetrahedron that is bonded by an oxygen atom to the proton. Differences in *DPE* only arise when Al substitution in next nearest neighbor positions with respect to proton is different.

The proton acidity decreases since *DPE* increases with 10–30 kJ/mol when the next nearest neighbor tetrahedral concentration of Al^{3+} increases. The proton acidity is maximum when the [Si–(O–H)–Al] site is only surrounded by silicon-containing tetrahedra and has a value of approximately 1200 kJ/mol.

Other important zeolitic materials are the SAPOs [31], [91]. Instead of the SiO_2 composition as reference, the SAPO material has

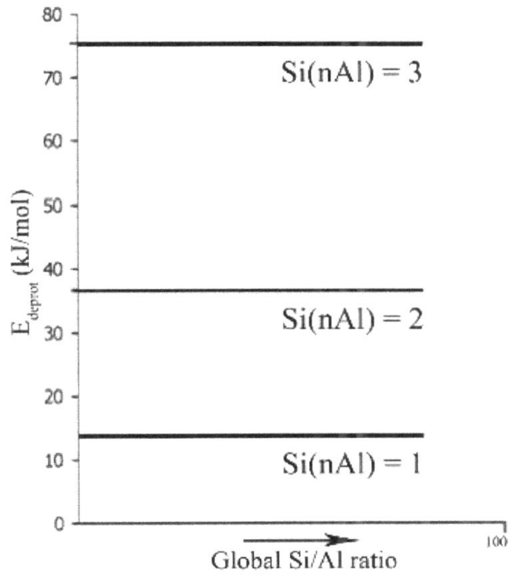

Figure 5.21 Schematic illustration of *DPE* dependence on global Si/Al ratio and on local Si(nAl) descriptor of the number of Al tetrahedra in next nearest neighbor sites. Results are from H-FER(8), H-FER(35), and H-FER(71) DFT model calculations [111].

the $AlPO_4$ stoichiometry as reference. The lattice framework building principle is the same. Charge neutrality is maintained by the alternating sequence of Al^{3+}- and P^{5+}-containing oxygen tetrahedra.

The material is converted into a solid acid via substitution of tetrahedrally coordinated P^{5+} by Si^{4+}. To maintain charge neutrality for each substitution a positively charged cation is to be introduced into the SAPO channel. Exchange of cations with NH_4^+ and subsequent NH_3 desorption gives the zeolitic proton [Al–OH–Si] site. The Si^{4+}-containing tetrahedron is connected to three additional Al^{3+}-containing tetrahedra and the Al^{3+}-containing tetrahedron is connected to three neighboring P^{5+}-containing tetrahedra. According to calculations [103] the acid strength of the proton that is part of the SAPO structure is weaker than that of the zeolite with alumino-silicate framework. The proton affinity of the [Al–OH–Si] site embedded in the $AlPO_4$ lattice increases by approximately 40 kJ/mol compared to that of the [Al–OH–Si] site embedded in a SiO_2 zeolitic lattice.

These changes of O–H bond energies due to variation of cation composition in the second coordination shell with respect to the hydroxyl cannot be described within the electrostatic oxide model of Section 5.2.4. This is limited to composition and structural change in the first coordination shells.

The increase in *DPE* with Al/Si ratio or by replacement of the SiO_2 lattice by the $AlPO_4$ lattice can be rationalized using the Bond Order Conservation rules (Section 2.3.3.3) of bond energy changes on surfaces with covalent bonds. The bonds in the SiO_2 or $AlPO_4$ structures are dominantly covalent. Bonding is more similar to that of organic molecules than in ionic sodium chloride.

According to Bond Order Conservation the valency of the atom is a constant and is equal to the sum of the bond orders of the bonds directed to it. Therefore, when one bond weakens other bonds to the atom will strengthen.

Application of the Bond Order Conservation principle to the change in OH bond strength of [Al–(O–H)–Si] embedded in the SiO_2 lattice with Al substituted next to the Si-containing tetrahedron is compared in Figures 5.22a and 5.22b. The O–H bond becomes

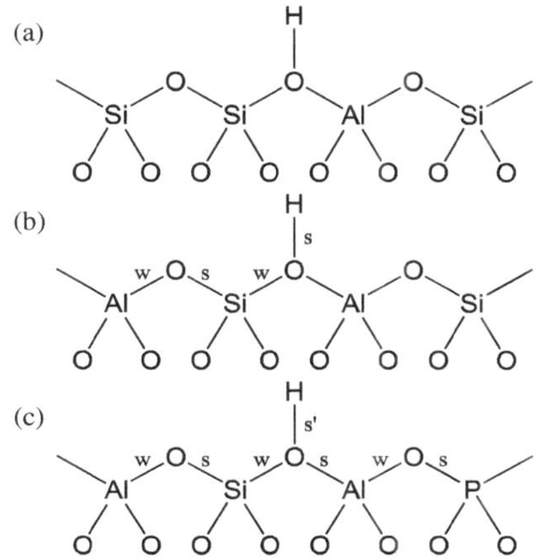

Figure 5.22 Schematic illustration of the Bond Order Conservation rule explanation of changes in OH bond strength when the embedding environment of the [Al–(O–H)–Si] site changes. (a) Isolated [Al–(O–H)–Si] in SiO_2 framework lattice, (b) [Al–(O–H)–Si] with Si surrounded by Al-containing tetrahedra, (c) [Al–(O–H)–Si] in $AlPO_4$ lattice.

stronger because the Al–O bond is weaker than the Si–O bond. Figures 5.22a and 5.22c compare [Al–(O–H)–Si] embedded in the SiO_2 lattice with that of the $AlPO_4$ lattice.

The change in O–H bond strength is predicted to be less because the weaker chemical bond of Al–O is partially compensated by the larger P–O bond strength. The latter is less influential because it weakens the already weaker Al–O bond compared to the Si–O bond with the zeolite proton. The calculations mentioned earlier show indeed that in the SAPO structure the strengthening of the O–H bond due to the weaker Al–O bond dominates.

DPE changes due to differences in local lattice composition are mainly due to differences in the covalent contribution to the O–H bond energy and follow changes in O–H vibrational frequencies [81], [112].

Covalent bond models and the electrostatic bond model give similar trends of the O–H bond strength because both count changes in bond valency and assume the valence of the atom to be a constant. It is the change in bond valency that determines the bond strength.

The usefulness of these models does not imply that it proves whether a bond is covalent or ionic. For this actual measurement of the charges on the respective atoms are necessary.

5.3.4 *DPE Variation Due to Al^{3+} Substitution by Fe^{3+} and Ga^{3+}*

The deprotonation energy has an inverse relation with the M–O bond energy.

Bond order conservation predicts also that the strength of the zeolite O–H bond will increase, and proton reactivity will decrease when Al^{3+} is replaced by a trivalent cation with a weaker M–O bond energy. Experiments that compare proton reactivity of zeolite alumino-silicate Al^{3+} substituted by Fe^{3+} and Ga^{3+} illustrate this decreased reactivity [113]–[115].

Also, proton reactivities scale inversely with O–H vibrational frequencies. It agrees with the trend prediction from the Bond Order Conservation rule.

A convenient probe of the protonation strength is the measurement of the heat of adsorption of ammonia or the rate of decomposition of adsorbed ammonium. As is explained in the next section this is a reliable probe as long as no comparison is made between zeolites with different structure [103], [110].

For zeolites with MFI structure Figure 5.23a relates O–H vibrational frequencies with the temperature of maximum decomposition rate of ammonium to ammonia of zeolites that have different cations in the tetrahedral lattice site. The higher this temperature the smaller the *DPE.* T_{max} correlates with measured vibrational frequencies, which increase in the order Al^{3+}, Fe^{3+}, and Ga^{3+}. This is the sequence expected from the corresponding increase in T–O bond (that are a measure of M–O bond strength) differences.

The different slopes of the increase in the rate of hexane conversion of Figure 5.23 with lattice cation concentration confirm the lower proton reactivities of Ga- and Fe-containing zeolites.

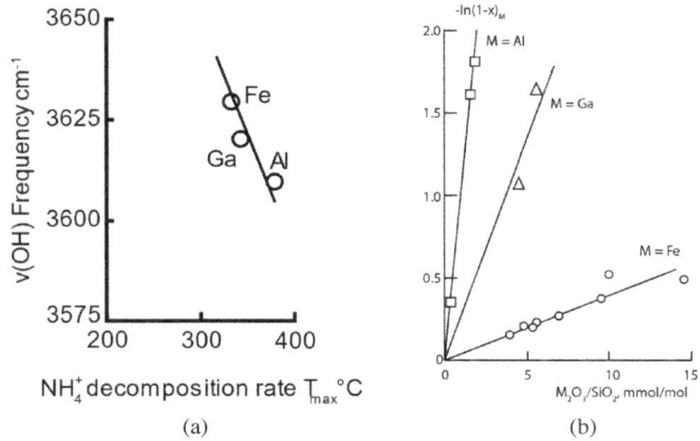

Figure 5.23 The effect of Al substitution by the two trivalent cations Ga^{3+} and Fe^{3+}. (a) T_{max} of NH_4^+ decomposition rate versus O–H vibrational frequency from FTIR spectra of isomorphous substituted ZSM-5 zeolites [113]. The respective DFT-calculated cation-oxygen bond lengths are: Al–O = 1.71 Å; Ga–O = 1.74 Å; Fe–O = 1.87 Å [116]. (b) Conversion of n-hexane over crystalline metal silicates (723 K, total pressure of 0.5 MPa, helium/n-hexane = 4/1 mol/mol). Effect of trivalent metal content of MFI silicate on reaction rate constant. Comparison of Al^{3+}-, Ga^{3+}-, and Fe^{3+}-substituted systems [114].

In conclusion:

- Alumino-silicate zeolites with low Al/Si framework ratio (Al/Si < 0.1) have proton sites with comparable (averaged) *DPE*. Proton affinity is independent of structure.
- Generally the *DPE* of zeolitic solid acid catalysts varies with zeolite framework concentration. This is the case when Al/Si > 0.1, when the protonic site is embedded in a different matrix (SiO_2 versus $AlPO_4$), or Al is substituted by a different trivalent cation. The *DPE* structure dependence is small.
- The *DPE* value of the [Al–(O–H)–Si] site that is dispersed with low concentration in SiO_2 matrix is, according to DFT calculations, 1200 kJ/mol (spectroscopy gives 1150 kJ/mol). *DPE* values are generally larger and can increase by 50–80 kJ/mol. Protons activated by hydroxylated oxycations occluded in microcavities (FAU) structure can have calculated *DPE* values lower than 1200 kJ/mol. The decrease can be as much as 40 kJ/mol.

5.4 Zeolite Catalysis, Structure Dependence and Shape Selectivity

Structure dependence of zeolite-catalyzed reactions depends largely on the adsorption energies of reactants or reaction intermediates. These are a sensitive function of match between the shape of the molecule and zeolite cavity.

5.4.1 *Introduction*

In this and following sections the reaction mechanisms of important solid acid-catalyzed reactions are presented. Main reactions discussed are hydrocarbon conversion and reactions of methanol. In Chapter 6 catalysis by Lewis acidic zeolitic systems is presented.

The two main types of sold acid catalysts that are discussed have only Brønsted acid catalytic reaction centers or are bifunctional catalysts. The latter contain an acidic catalytic site and a catalytic component such as a noble metal or metal sulfide that catalyzes hydrogenation and dehydrogenation reactions. Because of their nanoporous structure zeolite-catalyzed reactions are structure sensitive. The structure dependence reveals itself by differences in reaction rate and shape-selective catalysis.

The relation with structure sensitivity of zeolite-catalyzed reactions is an essential component of the mechanistic presentations.

There can be several causes of this structure sensitivity. Apart from limited access of the zeolite nanopore due to steric constraints, this structure dependence can in principle have two additional causes: elementary reaction rates can change or the concentration of reaction intermediates in the zeolite nanopore varies.

This can be illustrated by the rate expression of a monomolecular catalytic reaction, Eq. (5.8), that has as rate-limiting step the activation of reactant:

$$R = N_s k_r \theta \tag{5.8a}$$

$$\theta = \frac{K_{\text{ads}}[C]}{1 + K_{\text{ads}}[C]} \tag{5.8b}$$

According to Eq. (5.8a) the rate of the catalytic reaction R is proportional to N_S the number of reaction sites, k_r the elementary rate constant of reactant activation, and ϑ the reactant concentration at the reaction site.

The structure dependence can be due to variation of the elementary rate constant k_r, and/or ϑ, which depends on the adsorption equilibrium constant K_{ads} between free reactant and reactant adsorbed on the surface site. The latter is given by the Langmuir relation Eq. (5.8b) between surface coverage and reactant concentration [C] (Section 2.2.2.1).

Both k_r and ϑ may strongly depend on the match of adsorbate molecule dimension with zeolite microcavity. In zeolite science this is denoted as the confinement effect.

5.4.2 *Hydrocarbon Adsorption in Zeolites*

The adsorption energy of hydrocarbons relates to a match of hydrocarbon size and shape with zeolite nanopore.

In order to understand how hydrocarbons interact with the surface of the zeolite nanopore, the nature of chemical bonding in zeolites has to be revisited. In Section 5.3.2 it is discussed that chemical bonding in the zeolite lattice is dominantly covalent. Also the zeolite O–H bond is dominantly covalent and has a low dipole moment. The dielectric constant of the siliceous zeolite has a low value (≈ 2 [117]). Physically the dielectric constant relates to the polarizability of the large, slightly negatively charged oxygen atoms. The polarizability of Si or Al cations is small since these cations are substantially smaller in size.

When a hydrocarbon adsorbs into the microchannel of the zeolite it will have mainly contact with oxygen atoms of the zeolite wall, the Si and Al cations being hidden behind the oxygen atoms.

The attractive interaction between the hydrocarbon molecule and zeolite wall is determined by the dispersive van der Waals interactions between polarizable oxygen atoms and the hydrocarbon atoms. This dispersive interaction is additive. The larger the molecules and the more contact made with the oxygen atoms, the larger this interaction. In a wide-pore channel contact will be only with part of the oxygen atoms of the microchannel, but in a narrow-pore channel the molecule may become more surrounded with oxygen atoms and hence adsorption energy will be larger.

In Figure 5.24 this is illustrated for the change in adsorption energy of linear hydrocarbons as a function of zeolite nanopore size. The adsorption energy increases linearly with the length of the hydrocarbon. The adsorption energy decreases from its maximum in a narrow pore to a lower value when the nanopore diameter increases. The adsorption energies vary largely and can be of the same order of magnitude as activation energies of proton-activated elementary reaction constants.

The comparable dimension of hydrocarbon and nanochannel implies a very different diffusion than Knudsen diffusion in gases or liquids discussed in Section 2.2.2. In the latter case the rate of diffusion is determined by kinetic energy and collisions between

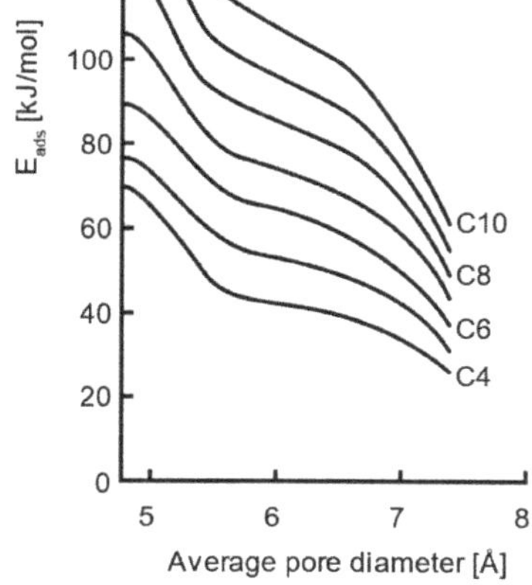

Figure 5.24 Adsorption energies E_{ads} of hydrocarbons with different chain length as a function of average nanopore size of different siliceous zeolite structures [118].

molecules. In contrast diffusion of molecules in the zeolite is dominated by collision of the molecule with zeolite wall atoms. In the absence of steric limitations this leads to so-called floating motion diffusion. This is orders of magnitude faster than Knudsen diffusion in the homogenous phase [119]–[121].

Whereas the adsorption energy of hexane in the hydrophobic siliceous zeolite nanopore can be of the order of 100 kJ/mol, its additional interaction with a proton is only of the order of 10 kJ/mol [122], [123].

The interaction energy is larger when an alkali cation substitutes for a proton. Near the inorganic cation there is a strong electrostatic field. The presence of such strong electrostatic interaction leads to strong polarization of hydrocarbon bonds that can be measured by NMR [124] and infrared spectroscopy [125], [126]. Zecchina *et al.* [77] determined the electrostatic field around the cations from the upwards shift of the vibrational frequency of CO or N_2 adsorbed to alkali cations. It depends on charge and radius of the cations. From an electric field strength of 6.3 V/nm for Na-ZSM-5 it decreases to 2.4 V/nm for Cs-ZSM-5.

In line with these differences in electrostatic field Zeolite Y gives a difference of 120 kJ/mol in heat of adsorption of ammonia on Na^+ versus 60 kJ/mol on Cs^+ [127]. Haag *et al.* compared the equilibrium adsorption constants of hexane to protonic and Cs^+ ion-exchanged ZSM-5. At a temperature of 250°C K_{ads} on the Cs-exchanged zeolite is increased by 10% [128]. The adsorption energy of a smaller molecule will of course change more since the relative fraction of molecular contact with the cation increases. Lewis acid catalysis by cations located in the zeolite micropore is the subject of Section 6.3.

When proton activation is reaction rate controlling and reaction rate is first order in hydrocarbon concentration, the kinetic consequence of the linear relation of adsorption energy with hydrocarbon length is that the apparent activation energy E_{act}^{app} linearly decreases with hydrocarbon chain length.

The zeolite proton activates the hydrocarbon locally and hence k_r may be expected to be independent of chain length. Then as described by Eq. (5.8), variation in rate with hydrocarbon chain

Table 5.3 Apparent and intrinsic activation energies of protolytic monomolecular alkane cracking and energies of adsorption as a function of n-alkane length. Reaction conditions 773 K, partial pressure 0.1–10 kbar [129].

Reactant	E_{ads} (kJ/mol)	E_{act}^{app} (kJ/mol)	E_{act}^{int} (kJ/mol)
Propane	43	155	198
n-Butane	62	135	197
n-Pentane	74	120	194
n-Hexane	92	105	197

length is only due to variation in surface coverage ϑ at the reaction center [128].

For conversion of linear alkanes by H-ZSM-5 this is illustrated by measurements of Lercher *et al.* [129]. At high reaction temperature and low partial pressure, the rate is first order in alkane concentration. It implies that surface coverage is low. Then the apparent activation energy of reaction can be written as (see Section 2.4):

$$E_{act}^{app} = E_{act}^{int} - |E_{ads}| \tag{5.9}$$

As shown in Table 5.3 the apparent activation energy E_{act}^{app} varies linearly with the adsorption energy. The intrinsic activation energy E_{act}^{int} of the elementary proton step that activates the alkane is independent of hydrocarbon chain length. The confinement effect that affects the conversion rate relates only to variation in adsorption energies of reactants.

5.4.3 *Bifunctional Catalytic Reactions*

The bifunctional acidic catalyst that also contains hydrogenation/dehydrogenation catalytic sites maintains the catalytic reaction cycle by replacing hydride transfer with alkane hydrogenation/dehydrogenation.

Here the hydrocracking of alkane or naphthene that gives lower-weight hydrocarbons, the related alkane hydroisomerization, and alkane aromatization are discussed [97], [98], [130]–[133]. Catalysts of these reactions are bifunctional and comprise a hydrogenation-active reaction site distributed on an acidic support. Reaction is executed in excess and high pressure of hydrogen. The hydrocracking reactions occur at the considerably lower temperature of 540 K compared to catalytic cracking (750 K).

In hydroisomerization an acidic zeolite, which operates at comparable temperatures to hydrocracking, that contains small amounts of a transition metal such as Pt or Pd as hydrogenation component is used as support.

Hydrocracking catalysts can also contain a metal sulfide as hydrogenation component in combination with a zeolite or amorphous alumino-silico catalyst. The alkane aromatization reaction [9], that is also discussed in this section, is catalyzed by an acidified alumina support promoted with Pt.

5.4.3.1 *The Mechanisms of Hydroisomerization, Hydrocracking, and Aromatization*

Bifunctional hydrocarbon conversion of saturated hydrocarbons is a stable reaction because the high pressure of hydrogen and presence of hydrogenation function maintain a low alkene concentration. Isomerization between cyclic hexyl and pentyl cations is a ready process important to alkane hydrocracking and aromatization. Isomerization of n-alkene proceeds by propyl cationic intermediates.

The catalysts for alkane hydroisomerization and hydrocracking are bifunctional Brønsted acidic zeolites. The bifunctional Brønsted acid catalyst of the aromatization reaction is an acidified alumina. The use of supports of different acidity relates to the difference in reaction conditions. Whereas exothermic hydroisomerization and hydrocracking reactions operate at 550 K, the endothermic aromatization reaction operates at 750 K.

Reactions are executed in excess hydrogen, which aims to maintain low alkene concentration so as to suppress catalyst deactivation by oligomerization. For instance in the hydroisomerization reaction the alkene/alkane ratio is $\approx 10^{-3}$.

As illustrated schematically in Figure 5.25 for hydroisomerization and hydrocracking the alkenes react with zeolite protons to carbocation intermediates that isomerize or undergo C–C bond cleavage. After deprotonation and subsequent hydrogenation products desorb as alkanes.

The catalyst has optimum functionality when the hydrogenation function establishes alkane-alkene equilibrium at a fast timescale compared to the timescale of proton activation [134]. This requires a minimum amount of noble metal in relation to proton concentration.

Some of the elementary reactions of the intermediate carbenium ions of Figure 5.25 are given in Figure 5.26.

Figure 5.26a indicates that isomerization of the n-alkyl cation proceeds through formation of a cyclopropyl cationic intermediate [136], [137]. This reaction mechanism, discovered by the Shell scientist Brouwer in the early 1970s, explains why monomolecular isomerization requires alkenes with at least five carbon atoms. Proton-activated monomolecular isomerization of n-butene is not possible, due to the difference in stability of primary and secondary versus tertiary organocations. The carbon atom with the positive charge that has two carbon neighbors (the secondary cation) is more stable by 30 kJ/mol than the primary end-on cation. The tertiary carbon atom with three neighboring carbon atoms is even more stable: 60 kJ/mol [138], [139] (see also Section 5.4.4.3).

In the case of isomerization of the n-butyl cation the corresponding cyclopropyl intermediate will only ring open as a primary cation. It has a high relative energy, which prevents its formation.

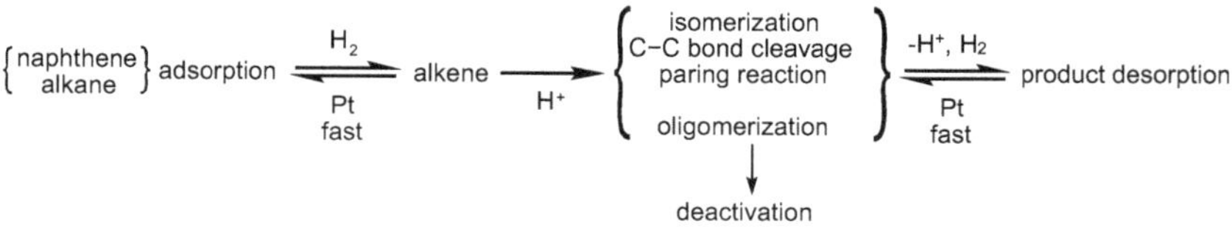

Figure 5.25 Bifunctional reaction scheme of hydroisomerization and hydrocracking. Proton actvation and desorption compete. The paring reaction applies to naphthene (R) activation, see Figure 5.26c.

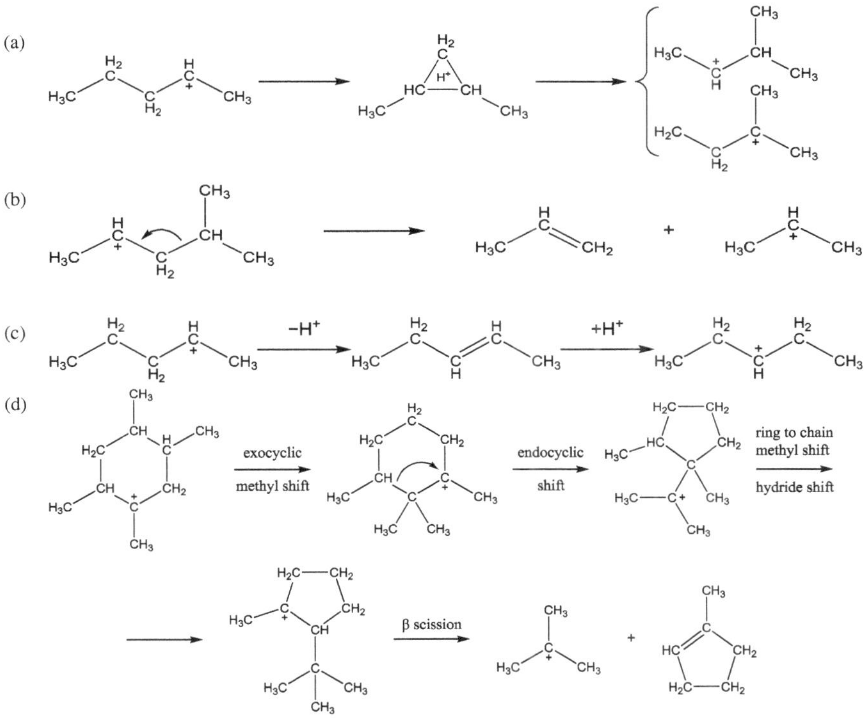

Figure 5.26 Elementary carbenium ion reactions of hydroisomerization and hydrocracking [135]. (a) Isomerization of n-alkyl to i-alkyl, (b) β C–C bond cleavage reaction to alkene and short carbenium cation, (c) proton shift as part of C=C bond shift reaction, (d) the cyclic alkyl cation paring reaction, at least three methyl substituents groups are to be present on the cyclohexyl cation.

In the case of protonated pentene, the cation intermediate is a secondary carbenium ion and isomerization becomes energetically possible.

The i-pentyl cation intermediate will not cleave C–C bonds. As can be deduced from Figure 5.26b that illustrates the β C–C bond cleavage mechanism, the initial carbenium ion would then be primary.

The hexyl cation will crack according to the β C–C bond scission reaction. In this case a propene molecule and secondary propyl cation are formed. When the hydrocarbon bond becomes longer double branching isomerization reactions become possible and the probability and rate of the C–C bond cleavage increases for longer alkane molecules. Then β C–C bond cleavage reactions become possible and the more stable tertiary carbenium ions are formed [140].

Figure 5.26c illustrates the proton-catalyzed C=C bond shift reaction by sequential deprotonation and protonation steps. The changing location of positive charge of the carbenium ion induces different reactivity especially for branched hydrocarbon intermediates.

In the proton-catalyzed reaction cycles of hydroisomerization as well as hydrocracking, the n-alkene to i-alkene isomerization precedes C–C bond cleavage. Consecutive β C–C bond cleavages occur dominantly in isomerized alkyl cations. β C–C bond cleavages only occur once tertiary carbenium ions are generated by the bond cleavage reaction. Since the longer hydrocarbons can undergo more isomerization, selective isomerization with maintained skeleton carbon number will decrease substantially for n-alkanes with chain length of seven and beyond [97]. Similarly, as in catalytic cracking discussed in the previous section, the reaction will increase with hydrocarbon length. However, the selectivity to isomerization then decreases [97].

Competive adsorption effects that relate to variation of adsorption energies of reactant alkanes also give rise to zeolite structure sensitivity. When catalyzed by Zeolite Y or Mordenite there is a difference in hydroisomerization catalysis of a mixture of C_5 and C_6 alkanes. Whereas in Zeolite Y with the larger nanocavities conversion of these alkanes is not different in a reaction mixture with these two molecules, in mordenite with the smaller channel dimension conversion of hexane is enhanced over pentane.

As for the cracking reaction the conversion rate of light alkanes is faster when the adsorption free energies of reactant increase. The hexane molecule has the higher adsorption energy and suppresses the adsorption of the shorter molecule that has the lower adsorption energy. In Zeolite Y the adsorption energy differences are smaller and are overcome by entropy differences (as we will see in Section 5.3.4.2) [141]–[144].

As illustrated in Figure 5.27a the hydrocracking pattern of linear alkanes is bell curved around a maximum concentration of hydrocarbons with half the number of carbon atoms of reactant alkane. Figure 5.27b shows that product distribution for a naphthene is double peaked. There is a maximum for isobutane and cyclic substituted cyclopentane molecules [132], [145]. These symmetric product distributions are found when the hydrogenation function is strong enough in relation to the acid function, so that the alkane-alkene equilibrium is maintained.

The bell-type curve of n-alkane cracking is the distribution of alkane fragments that results when the cleavage probability does not depend on position of the C–C bond. For example, the cracking of decane to C_5 fragments can be due to splitting of C_{10} in half, and also from the combinations C_4–C_6, C_3–C_7, and C_4–C_8, but the C_2 fragment can only be formed in combination with C_8.

The double-peaked product pattern of butylcyclohexane is due to the paring reaction, a reaction possible for cyclic alkanes and alkenes [147], [187]. The paring reaction of Figure 5.26d is unique for cyclic alkanes [135], [146]. For the hydrocracking reaction this mechanism was proposed in 1962 by Egan *et al.* from the California Research Corporation (Chevron).

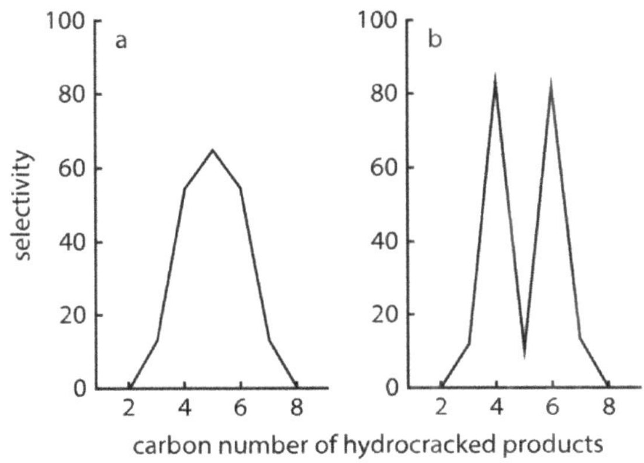

Figure 5.27 Carbon number distributions of the hydrocracked products from n-decane (bell-type curve) and butylcyclohexane (M-type curve) on bifunctional zeolite catalysts. (a) n-decane Pt/Ca–Y and (b) butylcyclohexane Pd/H–Y [145].

The paring reaction requires intermediate cyclohexyl cations that have at least three alkyl substituents. The cyclohexyl cation undergoes a transition to a cyclopentenyl cation by internal bond shift. Subsequent isomerization reactions give the isobutyl cation and methyl cyclopentene as product intermediates, which deprotonate and hydrogenate in following steps. There is no C_5 product.

The mechanism of the aromatization of n-alkanes is fundamental to the platforming process; it is presented in Figure 5.28.

The metal function ideally establishes alkane-alkene equilibrium. At high hydrogen pressure the low alkene concentration suppresses deactivating oligomerization reactions. The ring closure is catalyzed by protonation of hexadiene. This gives a methyl cyclopentenyl cation. In a following isomerization step the five-ring is transformed to the six-ring cyclohexyl cation. This isomerization is the reverse of the six- to five-ring transformation of the paring reaction of Figure 5.24d. Subsequent deprotonation and dehydrogenation give the aromatic molecule. At the high temperature of reaction, the transformation from the five- to six-ring cation is driven by benzene product formation.

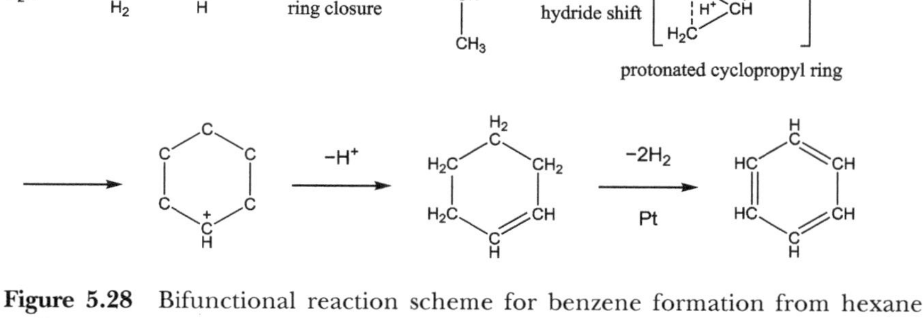

Figure 5.28 Bifunctional reaction scheme for benzene formation from hexane [147]. The transition metal activates C–H bonds and H_2. The acidic proton catalyzes ring closure and hydrocarbon skeleton isomerization.

For solid acids the activation energy of alkenyl isomerization E_{iso}^{int} varies between 120 and 140 kJ/mol [137], [148]. This is the activation energy of n-alkyl cation isomerization with respect to its adsorbed n-alkoxy state. For the larger alkanes this activation energy can become comparable to their energies of adsorption. For zeolites with medium-sized nanopores, when the hydrocarbon length exceeds C_{10} the adsorption energy can become substantially larger than 100 kJ/mol. At reaction condition for the longer hydrocarbons the nanochannels of the zeolite then become fully occupied. The reaction rate of product molecule desorption may become reaction rate controlling.

5.4.3.2 *The Kinetics of the Hydroisomerization and Hydrocracking Reaction; Inverse Shape Selectivity*

When product desorption is rate controlling shape selectivity is due to differences in adsorption equilibria of reaction intermediates in the zeolite nanopore and therefore independent of elementary reaction rate constants of proton activation.

In the hydroisomerization and hydrocracking reactions the occupation of the nanopore is dominated by concentation of alkanes. This nanopore occupation is high for large hydrocarbons that have adsorption energies which exceed activation energies of proton-activated elementary reaction rate constants.

Due to transport limitations equilibration of reaction intermediates in the nanopore will be fast and product distribution is not determined anymore by elementary reaction rates of proton activation, but by relative concentrations of intermediates in the nanopore. This is discussed in the second part of this section.

For alkane hydroisomerization, when the elementary reaction rate of proton activation k_r is reaction rate controlling then the reaction rate R is proportional to the concentration of alkene at the proton reaction site as in Eq. (5.10a):

$$\frac{R}{N_s} = k_r \cdot \theta(C^=) \tag{5.10a}$$

$$= k_r \frac{K_{ads}^{=}[C^{=}]}{1 + K_{ads}[CH] + K_{ads}^{=}[C^{=}]}; \tag{5.10b}$$

$$[C^{=}] = \frac{K_{eq}[CH]}{[H_2]} \tag{5.10c}$$

Due to the high relative concentration of alkane versus that of alkene there is competition between alkane and alkene adsorption. The adsorption energy of longer hydrocarbons to the protonic site is not very different from that of a small hydrocarbon since a substantial part of the adsorbed alkoxy cation interacts with the siliceous part of the zeolite. This is the reason for the two adsorption equilibrium contributions in the denominator of Eq. (5.10b). K_{ads} and $K_{ads}^{=}$ are the respective adsorption equilibrium constants of alkane and alkene. In Eq. (5.10c) K_{eq} is the equilibrium constant of the alkane-alkene equilibrium.

In the case that the alkane and alkene adsorption concentrations are low the reaction rate depends on the ratio $[alkane/H_2]^x$, with the order of x between 1 and 0. In the limit of high alkane concentration the reaction rate becomes zero order in alkane and negative order in H_2. These are the conventionally found dependencies on concentration [149], [150].

However, beyond a particular hydrocarbon size or chain length the nanopores get such a high occupation that desorption or diffusion of product becomes reaction rate controlling. Then there is no equilibrium between molecules that occupy the nanopore and gas-phase concentration. Product distribution becomes determined by the equilibrium distribution in the nanopore of adsorbed reaction intermediates.

For hydrocracking of long-chain n-C_{16} alkane molecules Zones at Chevron discovered an inverse shape selectivity effect. For zeolites with the smaller nanopore he found that the more branched molecule is the preferred product molecule [151], when he used the dimethylbutane/hexane ratio as a probe of shape selectivity. It is an inverse shape-selective effect since it runs counter to expectation for zeolites.

Simulations by Smit *et al.* showed that this inverse shape selectivity relates to an adsorption entropy effect that happens at high nanopore concentration [152], [153]. Statistical mechanical grand-canonical equilibrium simulations of the adsorption of binary mixtures of dimethylbutane and n-hexane are shown in Figure 5.29. Computed dimethylbutane/hexane ratios are plotted as a function of the cavity size of different zeolites. Comparison with experimental data shows reasonable agreement. Importantly this agreement is only found when interaction between adsorbed molecules is explicitly taken into account. In a micropore of low occupation, where molecules do not interact, differences in the Henry coefficients (that ignore interaction between adsorbates, green curve in Figure 5.29) of the adsorption equilibrium constants between

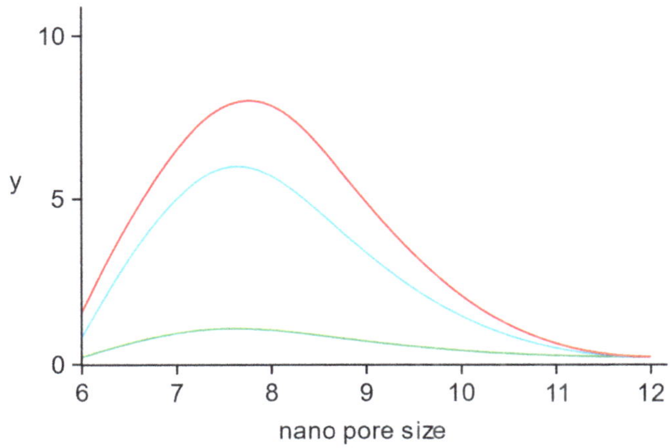

Figure 5.29 Simulated hydrocracking equilibrium product distributions of n-C_{16} hydrocarbon conversion [153]. Schematic representation of normalized DMB/n-C_6 yield ratios (y) for various zeolite structures as a function of nanopore dimension. The DMB/n-C_6 ratios are normalized by setting the value for the FAU-type zeolite, which has the larger nanopore size at one. The experimental ratios (red) were determined from n-C_{16} hydrocarbon conversion experiments [151], the calculated ratios were taken from simulated adsorption isotherms of 2,2-DMB/n-C_6 (blue), or from Henry coefficient calculations (green). The adsorption concentrations derive from calculations of binary mixtures of DMB and hexane at pressures and temperatures of the experiment. The numbers in parentheses are the average pore sizes [Å]. DMB = dimethylbutane, n-C_6 = n-hexane.

dimethylbutane and hexane do not depend on zeolite structure. The interaction energies of both molecules with the zeolite wall are comparable.

The difference in adsorption equilibria originates from non-ideal mixing behavior of adsorbed intermediates. Pores occupied with high concentration of adsorbate can accommodate smaller than larger molecules. Bulkier branched molecules have a smaller diameter than linear molecules that become stretched in a small pore and hence have a larger effective size. This favors selectivity of reaction towards formation of the branched products. The effect increases with decreasing micropore size. Selectivity is optimum for micropores of intermediate size [153].

5.4.3.3 *Reaction Rate as a Function of Zeolite Structure; the Catalytic Hammett Acidity Function*

Global catalytic acidity relates with proton reactivity when corrected for differences in reactant adsorption energies. Due to entropy-related compensation reactions activated at intermediate temperature prefer medium-sized nanopores.

In Section 5.4.2 it has been discussed that differences in the monomolecular catalytic cracking rates of n-alkanes of different chain length relate to differences of the adsorption energies of the hydrocarbons (see Table 5.3).

Here the dependence of reaction rate on confinement is discussed as a function of zeolite structure. The question is investigated as to whether differences in proton reactivity or hydrocarbon adsorption energies dominate zeolite structure dependence. For zeolites with comparable Al/Si framework the reaction rate of the protolysis of propane and hydroisomerization of pentane are compared. A Hammett-type expression of catalytic acidity is presented, analogous to the Hammett function of Brønsted acidity.

When catalyzed by different zeolites differences in rate of the monomolecular cracking reaction of light alkanes are also dominated by variation in adsorption free energies. Gounder and Iglesia [154] studied the cracking reaction of propane catalyzed by three zeolites of different nanopore structure and dimension with low Al/Si ratios. The reaction conditions were selected such that cracking is only monomolecular and reaction rate is first order in propane.

Table 5.4 presents the respective adsorption energies and apparent and intrinsic activation energies.

The intrinsic propane activation energy is independent of zeolite structure. This follows when the apparent activation energy is corrected by the adsorption energy of propane. The intrinsic activation energy differences of C–C bond cleavage and dehydrogenation are close to values computed quantum-chemically (see Figure 5.7). Their difference relates to differences in respective C–H (411 kJ/mol) and C–C (346 kJ/mol) bond cleavage energies.

The reaction rate differences are controlled by differences in ϑ, the concentration of propane adsorbed at the reaction center. Differences in nanopore curvature of the zeolites account for the differences in propane adsorption energies E_{ads}.

For propane the differences in adsorption entropies in relation to nanocavity size are minor. For larger alkanes changes in adsorption free energy as well as adsorption entropy have to be accounted for [128]. Entropy and activation energy tend to counteract. In a narrower pore the adsorption energy increases but the mobility of the hydrocarbon will decrease. This is called the compensation effect [155], [156]. Which of the two changes dominate will depend on temperature. At higher temperature entropy differences tend to dominate, while at lower temperature adsorption energy differences

Table 5.4 Comparison of E_{act}^{app}, E_{act}^{int}, and E_{ads} (kJ/mol) in the monomolecular propane cracking reaction [154]. H-MFI has crossing channels with 10 tetrahedral ring diameters, H-FER has one-dimensional channels of 10 rings, and H-MOR has one-dimensional 12-ring channels. T = 748 K.

		E_{act}^{app}		E_{act}^{int}	
Zeolite	E_{ads}	cracking	dehydrogenation	cracking	dehydrogenation
H-MFI	-45	158	200	203	245
H-FER	-49	157	195	206	244
H-MOR	-41	163	193	204	234 (±10)

dominate. This is illustrated by kinetic measurements of the bifunctional hydroisomerization reaction of n-hexane to i-hexene [157].

As is discussed in Section 5.4.3.2 the reaction rate-controlling step is proton-activated isomerization of the reaction intermediate n-hexene to i-hexene.

Turnover frequencies (TOF = reaction rate normalized per proton), measured adsorption equilibia constants, and deduced elementary reaction rate constants k_{iso} are presented in Table 5.5. A comparison is made of the TOF measured for the 12-ring mordenite catalyst versus the rate of H-ZSM-22 with 10-ring channels. Both structures have one-dimensional channels. One notes that differences in TOF are substantially larger than the difference in elementary reaction rate constants. It relates to the difference in K_{ads}. At the particular temperature of this reaction K_{ads} is larger for the wide-pore zeolite notwithstanding the lower heat of adsorption of hexane. The gain in entropy for the wider pore overcomes the loss in adsorption energy.

Whereas from a catalytic point of view one would be tempted to assign the differences in catalytic rates of the zeolites to differences in acidity, the intrinsic reactivity of the protons is similar. Differences in reactivity rather relate to structural variation that varies adsorption equilibrium constants. Reaction rate varies due to alteration in nanopore occupation.

For monomolecular kinetics the reaction rate expression Eq. (5.8) can be used to deduce an expression analogous to the

Table 5.5 Turnover frequencies (TOF, reaction rate normalized per proton) of the hydroisomerization reaction, adsorption equilibrium constants of hexane K_{ads} (T = 550 K; p_{H_2}/p_{nC_6} = 14), and elementary reaction rate constant for isomerization k_{iso} [157]. H-MOR has one-dimensional 12-ring channels (with small 8-ring pockets), and H–ZSM-22 (TON, FER) has one-dimensional 10-ring channels.

Zeolite	TOF (s^{-1})	K_{ads} (Pa^{-1})	K_{iso} (s^{-1})
H-MOR	$1.1 \cdot 10^{-2}$	$3.3 \cdot 10^{-4}$	$2.7 \cdot 10^{-2}$
H-ZSM-22	$1.6 \cdot 10^{-3}$	$1.4 \cdot 10^{-5}$	$1.7 \cdot 10^{-2}$

equilibrium Hammett acidity function (Eq. (5.2)) for the kinetic catalytic acidity (Cat_A) [158]. According to the Eyring transition reaction rate equation (Section 2.3.2.3) the transition state can be considered quasi-equilibrated with the reactant ground state.

This quasi-equilibrium assumption gives a relation for the transition state site coverage θ_{T_r} and gas-phase reactant concentration $[C_r]$:

$$\theta_{T_r} / \left(1 - \theta_{T_r}\right) = K_{\text{quasi}} \left[C_r\right] \tag{5.11a}$$

In this expression K_{quasi} follows the apparent reaction rate constant expression $k_r K_{\text{ads}}$:

$$K_{\text{quasi}} = \frac{h k_r K_{\text{ads}}}{kT} \tag{5.11b}$$

The $\frac{h}{KT}$ term in Eq. (5.11b) removes the frequency dependence of k_r (Eq. (2.8)).

Analogous to the definition of the Hammett acidity function one defines the catalytic acidity function Cat_A:

$$Cat_A\left(\theta_{T_r}\right) = -\ln\left[C_r\right] = \ln K_{\text{quasi}} - \ln \theta_{T_r} / \left(1 - \theta_{T_r}\right) \tag{5.12a}$$

$Cat_A(\theta_{T_r} = \frac{1}{2})$ is the analogue of the Hammett acidity function when used to define proton acidity:

$$Cat_A\left(\theta_{T_r} = \frac{1}{2}\right) = \ln\left(\frac{h}{kT}\right) k_r K_{\text{ads}} = -\frac{\left(G_{\text{act}}^{\text{int}} + G_{\text{ads}}\right)}{kT} = -\frac{G_{\text{act}}^{\text{app}}}{kT} \tag{5.12b}$$

$Cat_A(\theta_{T_r} = \frac{1}{2})$ relates to the apparent free energy of the reaction. Cat_A becomes less negative when the reactivity of the catalyst increases.

The apparent free activation energy $G_{\text{act}}^{\text{app}}$ relates through $G_{\text{act}}^{\text{int}}$ with the deprotonation energy *DPE* and with the free energy of adsorption G_{ads}.

Eq. (5.12b) is the mathematical expression that shows why catalytic differences in reaction rate do not necessarily correlate with *DPE*. It has to be corrected for with additional contributions that relate to the adsorption energies of reactant and reaction intermediates.

This is also the reason that adsorption energy measurements with acid probe molecules such as ammonia are not predictive for differences in proton reactivity when used for protonic zeolites of different structures [103], [157]. The measurements are made with respect to the gas phase. An important contribution to the protonation energy of ammonia is the van der Waals interaction with the zeolite cavity that is a function of zeolite curvature. Only when the zeolite cavity remains the same will the temperature for the programmed desorption of ammonia provide a reasonable estimate.

5.4.4 *Shape-selective Elementary Reactions*

5.4.4.1 *Restricted Transition State Selectivity*

Restricted transition state selectivity evolves when zeolite cavity size or cavity interconnections limit or constrain bimolecular reaction events. It may suppress hydride transfer or other substituent transfer reactions.

In this and following subsections shape selectivity of proton-activated reactions is introduced.

Shape-selective zeolite catalysis has similarity to the classical lock and key model of enzyme catalysis of Emiel Fischer [159], [160], but the geometric fit is never as precise as in enzyme catalysis. An important difference is that the organic enzyme catalyst is flexible and adapts to reactant geometry. In modern enzyme catalysis the rigid lock and key model has been modified into the induced fit model [161], [162] between reactant and catalyst cavity.

The limited flexibility of the zeolite lattice will restrain such adaptation. The fit between reactant and cavity should not be too well. Otherwise the product would not desorb.

There is a major additional difference between proton activation in the zeolite and in the enzyme. In the latter case the reactant can be activated synchronously by several interactions with the enzyme molecule [163]. In proton catalysis activation is a single-site reaction.

One of the first papers on shape-selective zeolite catalysis that appeared in 1960 indicates a role of diffusion. This was suggested by Weisz, the director of Socony-Mobil Oil Research, who played a large role in zeolite refining process innovation. He and his colleagues showed that whereas the rate of dehydration of 1-butanol and isobutanol is comparable when catalyzed by amorphous alumino-silicate, conversion of 1-butanol readily occurs over nanoporous Zeolite X (FAU structure) but isobutanol is not converted. This is due to the mismatch of the larger isobutanol with the smaller-sized nanocavity of Ca-exchanged Zeolite X [36], [164], [165].

Another example is the inverse cracking reactivity of a mixture of n- and i-alkanes. When catalyzed by amorphous alumino-silicate i-alkanes will more readily convert than n-alkanes, but when zeolite is used the reactivity of the branched molecule is suppressed. The mismatch of shape of molecule with zeolite cavity prevents its entry into the nanopores.

Also product diffusion limitation may cause shape-selective catalysis. An example of the latter is provided by the structure sensitivity of methanol conversion. Figure 5.30 indicates different selectivity of methanol when catalyzed by H-SAPO-34, which has the chabazite (CHA) structure, compared to catalysis by H-ZSM-5 with MFI structure.

The CHA structure consists of large super cavities (6.7 × 10 Å) formed by stacked rows of double six-rings. The super cavities are connected through 8-ring openings. The MFI structure of H-ZSM-5 consists of a three-dimensional network of 10-ring channels.

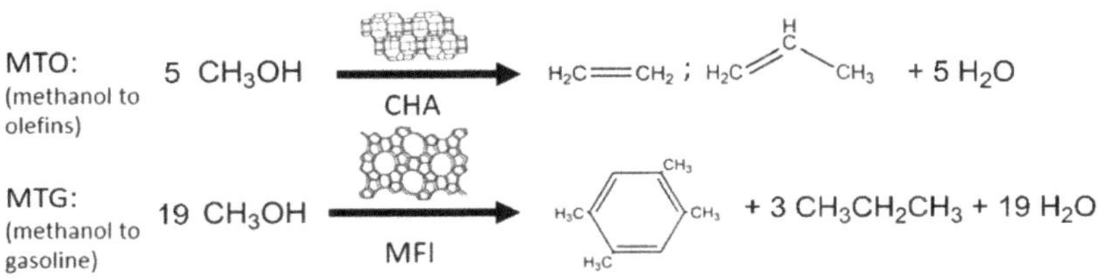

Figure 5.30 Shape-selective conversion of methanol [158]. The MTO reaction is catalyzed by chabazite (CHA) structure with 8-ring connections between cavities and the MTG reaction by MFI structure that has 10-ring diameter channels.

Brønsted acid-catalyzed methanol dehydration produces a wide spectrum of product molecules. It varies between substituted aromatics and ethene or propene. The substituted benzene product molecule will not diffuse through the 8-ring openings of CHA, in contrast to ethene or propene. On the other hand, methylated benzene molecules diffuse readily through the H-ZSM-5 channels, resulting in aromatics and alkanes as major products [166], [167]. Mechanistic aspects and stereochemistry of the MTO and MTG reactions are discussed in Section 5.4.5.

Csisery, then at Chevron Research Company, formulated in 1984 the concept of restricted transition state selectivity [37]. In the 1970s Csisery discovered shape selectivity of the trans-alkylation reaction of substituted benzene molecules. This is a bimolecular reaction that scrambles alkyl substituents around the benzene ring. Protonation of the benzene ring activates the alkyl substituent that transfers as carbenium cation to another aromatic molecule.

In the one-dimensional pore of mordenite there is steric bias to the formation of para-alkylated products. As is discussed in detail in Section 5.4.4.4 ortho- or meta-substituted product formation is hindered by spatial constraint of these more spatially demanding transition states.

Restricted transition state selectivity is also the reason that in H-ZSM-5 catalysis there is a bias for monomolecular alkane cracking. The bimolecular hydride transfer reactions between carbenium ions and alkanes are suppressed by limited reaction space.

5.4.4.2 *Protonation of Isobutene; Curvature Effects*

Transition state energies for primary carbenium ions are higher than those of secondary and tertiary carbenium ion intermediates. Differences in activation energies of protonation reactions are due to confinement result from variation in electrostatic screening and steric repulsion.

Fundamental to many of the processes presented in this chapter is the protonation reaction of an olefin. Here the dependence on nanopore structure and size for the protonation of isobutene is discussed. Important reference experiments for the reactivity of isobutene are classic deuterium isotope exchange experiments from the 1960s [168]. These studies in liquid H_2SO_4 demonstrated that isobutene will only be protonated on its primary carbon atom. These mechanistic studies were done within the context of the isobutane-alkene alkylation reaction, discussed in Section 5.4.6.2.

When the alkene molecule adsorbs on the solid acid a hydrogen-bonded π-adsorbed complex initially forms with the proton. When protonated the alkene molecule converts into a carbenium ion that, unless prevented by repulsive steric interactions with the zeolite cavity, adsorbs as an alkoxy intermediate to the proton vacant site.

In Figure 5.31 quantum-chemical DFT calculations of the protonation energies of isobutene adsorbed in different zeolites are presented. Protonation occurs via primary or tertiary carbenium ion-like transition states.

The two protonation reaction pathways of isobutene are schematically illustrated in Figure 5.31a. The carbenium-related structure of the respective transition states is indicated.

When the primary carbon atom is protonated, the positive charge will reside on the tertiary carbon atom. When vice versa the tertiary carbon atom is protonated, the positive charge is located on the primary carbon atom.

The transition state energies with respect to the π-adsorbed state of isobutene are shown in Figure 5.31b. Due to the destabilization of the primary cation with respect to the tertiary cation [169] there is a difference of 80 kJ/mol in the respective transition state energies. Consistent with the liquid H_2SO_4 experiments protonation of the primary carbon atoms of isobutene is the favored pathway.

Whereas the activation energies of primary carbenium ion formation vary between 20–40 kJ/mol when zeolite nanopore cavity size changes the differences are less for tertiary carbenium ion formation. Only on MOR with the least curved nanopore walls the tertiary alkoxy species is stable. The instability on the other surfaces that have higher curvature is due to steric repulsion. The bulky isobutyl cation cannot approach the zeolite wall as closely as the n-butyl cation.

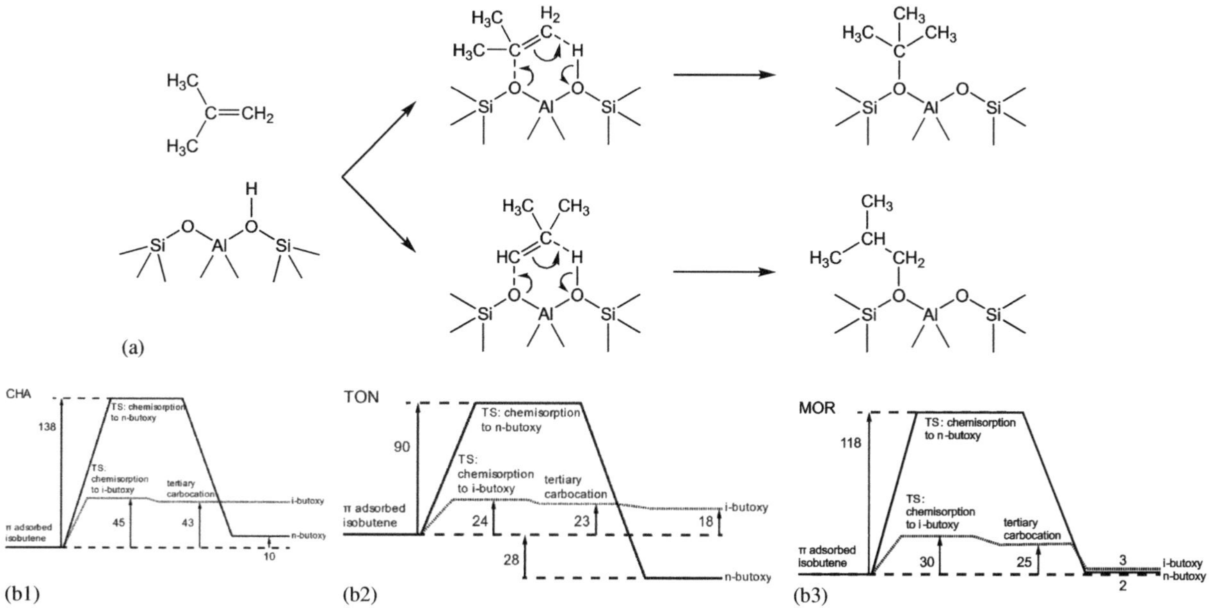

Figure 5.31 The protonation of isobutene [170]. (a) Reaction paths of protonation of isobutene. (b) Potential energy diagrams of isobutene pronation in different acidic zeolites of low Al/Si framework ratio obtained from the periodic DFT calculations. Respective cavity sizes are: (b1) CHA (6.7 × 10 Å), (b2) TON (5.7 × 4.6 Å), (b3) MOR (6.5 × 7.0 Å). Energies in kJ/mol.

For the zeolites with higher curvatures only the primary carbenium ions form stable alkoxy species. The tertiary carbenium ions are free from the surface and quasi-stable intermediates. Calculations of the free energies that include entropy indicate that the isobutyl cation can already be considered a freely moving cation at a temperature of 120 K [171].

The activation energies decrease when pore size decreases. Wide-pore CHA has the highest barrier, followed by mordenite with the larger cylindrical nanopore and TON with the smaller cylindrical nanopore. When zeolite cavity size decreases the match between carbocation size and cavity improves. The charge on the positively charged carbenium ion becomes increasingly stabilized by polarization of the oxygen atoms of the zeolite wall [169].

In Section 5.4.3.3 it has been discussed that for zeolites of different structure adsorption entropy differences counteract differences in adsorption energies. This is also the case for the activation free energies of proton activation. The relatively small differences in activation energies of secondary and tertiary carbenium ion formation are counteracted by differences in activation entropies that result from mobility constraints. This causes the elementary reaction rates of proton-activated alkene isomerization to be relatively independent of cavity size as observed in Table 5.5.

5.4.4.3 *Pre-transition State Stabilization; Methanol Alkylation of Toluene*

Restricted transition state selectivity relates primarily to differences in ground state energies of the pre-transition state complex.

Here, as a model of restricted transition state selectivity, the DFT-simulated reaction paths of the alkylation of toluene with methanol indicate that differences in regioselectivity have mainly a geometric cause. The structure of the zeolite cavity determines the pre-arrangement of the reacting molecules that leads to differences in pre-transition state stabilization.

Alkylation of toluene can give ortho-, meta-, or para-xylene. Para-xylene is a base chemical essential for the production of polyethylene terephthalate (PET). In the liquid phase alkylation occurs with little preference for ortho- versus para-xylene. Shape-selective catalysis is possible when reaction is executed by zeolites with confined space. Shape-selective catalysis of para-xylene production is found for mordenite, which has tubular one-dimensional nanochannels [172]–[174].

Figure 5.32a illustrates the successive steps of the reaction. Toluene and methanol have to adsorb in close vicinity. Then for para-xylene formation the methyl group of toluene has to orient para with respect to the methyl group of methanol. This reorganization of reacting molecules gives a pre-transition complex. The polar methanol is initially protonated. Activation of methanol leads to addition of CH_3^+ to the toluene molecule. Para-xylene is formed upon backdonation of the proton from the protonated molecule.

Figure 5.32b gives the computed energies of the different stages of the reaction. The differences in energy of the three relevant transition complexes are given. The respective pre-transition state configurations have shapes close to that of the respective product molecules.

Steric constraints arise from misfit of some of the bent pre-transition state orientations of the reaction intermediates with the linear mordenite channel. The linear para-xylene molecule fits well, but the bent ortho- as well meta-xylene molecules have less favorable interactions.

The pre-transition state intermediate of the para-xylene is more stable by 16 or 17 kJ/mol than that of the ortho- and meta-intermediates.

The difference in intrinsic activation energies with respect to the transition state energies of the para- and ortho-methyl addition reactions is only 1 kJ/mol. For the toluene meta position the activation energy is 7 kJ/mol higher.

The similarity in transition state energies of ortho and para addition and the higher barrier for meta addition agrees with expectation based on classical physical organic chemistry [176].

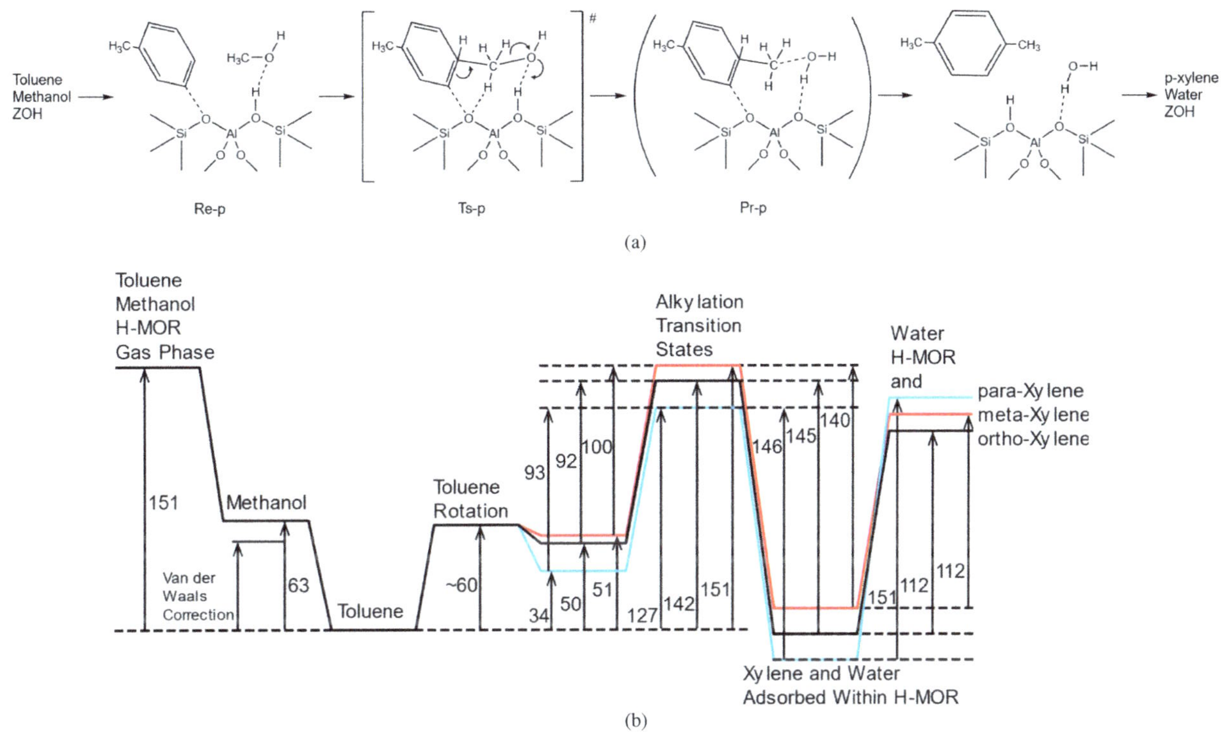

Figure 5.32 Simulation of shape-selective toluene alkylation with methanol catalyzed by H-mordenite [175]. (a) Structures of reaction intermediates of alkylation of toluene with methanol that gives p-xylene. Re-p: coadsorbed toluene and methanol, Ts-p: transition state, within brackets protonated xylene, Pr-p: adsorbed products xylene and water. (b) Dispersive energy-corrected DFT-calculated potential energies of the elementary reaction steps of the alkylation of toluene with methanol. A comparison is made of reaction paths that lead to the formation of p-xylene (blue), m-xylene (red), or o-xylene (black) (all energy values in kJ/mol).

The apparent activation energy for para-xylene formation is 15 kJ/mol lower than that of ortho-xylene formation and 24 kJ/mol lower compared to meta-xylene formation.

Restricted transition state selectivity has its root in the difference of the adsorption energies of sterically different pre-transition state complexes. It is due to different van der Waals stabilization of reactant molecules before bond rearrangement reactions.

5.4.5 *Zeolite-catalyzed Dehydration of Methanol to Alkenes, Alkanes, and Aromatics*

The hydrocarbon pool concept is consistent with catalysis by intermediate organocationic complexes. In the paring reaction a substituted cyclopentadienyl cation is formed that catalyzes short alkene formation.

The conversion of methanol to aromatics and olefins is an important invention from Mobil Research in the 1970s [32]. It led to the MTG process with a plant in New Zealand that was constructed in 1986. It then supplied that country with 1/3 of its transportation fuel.

The MTG process is part of the overall route to convert natural gas to gasoline. In this sense it is a competitor of the Fischer-Tropsch process that is based on the conversion of synthesis gas (see Section 3.2.2.1).

The catalyst of choice for the MTG process is the H-ZSM-5 catalyst. This catalyst is also a useful catalyst to convert biomass to liquid hydrocarbon fuel [177]. As with many discoveries of new reactions, the discovery of the MTG process was also serendipitous.

By selecting different zeolitic materials the reaction can be adapted to selectively produce olefins. This is the MTO process [178], [179].

Catalyst inventions at Union Carbide of the zeolitic SAPO material led to the discovery of shape-selective solid acid catalysis by the H-SAPO-34 structure for ethene and propene production. In China as well as in Norway the MTO process with this and related catalysts became extensively explored.

Because of the lattice structure with narrow nanopores that enclose large cavities the H-SAPO-34 catalyst, which has CHA structure, rapidly deactivates. A low-pressure fluidized bed reactor design is used, similar to the catalytic cracking process (see Section 7.2.2), that enables efficient temperature control and continuous regeneration.

In the 1990s Norsk Hydro (now INEOS) together with UOP built a first demonstration plant.

Research on the MTO reaction had also already started in the early 1980s at the Dalian Institute of Chemical Physics by Liu and colleagues [180]. In China the first commercial MTO plant was brought onstream in 2010.

The product pattern of the reaction is illustrated in Figure 5.33, that as a function of contact time (conversion) shows the products of methanol conversion catalyzed by H-ZSM-5 [32].

The methanol reaction to alkenes or aromatics is a dehydration reaction that gives "CH_2" as intermediate with water as co-product. At low conversion dimethyl ether is the main product. When conversion increases initially olefins and water are formed and finally aromatics. Alkene and aromatics are produced after an initiation

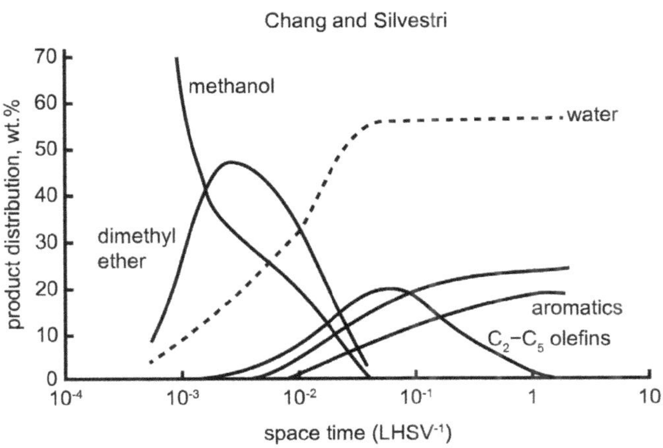

Figure 5.33 Conversion of methanol and products of the methanol to hydrocarbon reaction catalyzed by H-ZSM-5 (371°C) [32].

period. This is followed by a period of quasi-steady state productivity and finally catalyst deactivation.

The reaction network of this solid acid-catalyzed reaction is highly complex. Over time it has gradually become understood at the molecular level [167], [181]–[186].

A schematic representation of the reaction mechanism is presented in Figure 5.34.

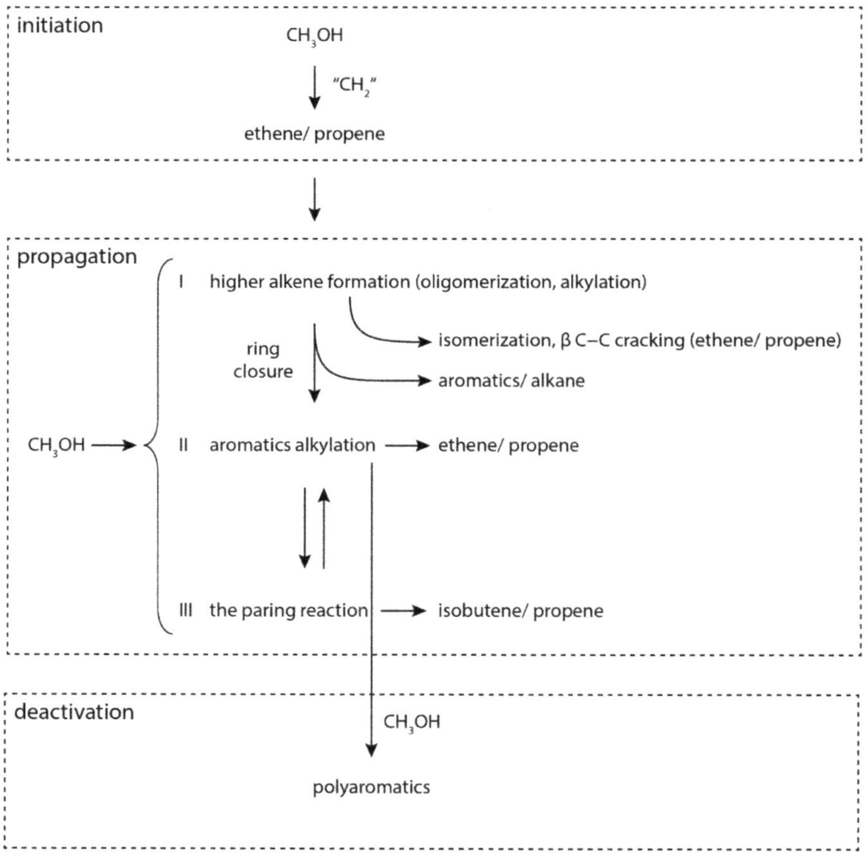

Figure 5.34 The three essential reactions of the MTO process. The reaction of the methanol to hydrocarbon dehydration process; once ethene and propene are formed in the propagation stage, initiation is replaced by autocatalytic reactions [191].

Research on the mechanism of the methanol conversion reaction has largely benefitted from studies on zeolitic H-SAPO-34 and related catalysts.

In 1993 the Norwegian scientists Dahl and Kolboe [187], [188] discovered the Hydrocarbon Pool mechanism. Using isotope labelling experiments, they discovered that part of the carbon atoms of the reactant remain contained in the catalyst. It implies that a reservoir of carbon atoms is present. Later NMR and IR studies revealed that the hydrocarbon pool consists of substituted benzenium and cyclopentadienyl cations [189], [190]. At steady state these cations are organocatalytic molecules which catalyze C–C bond formation reactions that give short alkenes.

In the initiation period the organocatalytic molecules are formed that later maintain the propagation cycle. These carbocations undergo subsequent alkylation reactions with methanol. Once such unsaturated molecules have formed an autocatalytic reaction network gets underway [191] that, next to formation of aromatic products, regenerates the organocatalytic molecules.

There is a multitude of proposals on the mechanism of the initiation reaction [183]. The proposals concern different paths to intermediate "CH_2" formation that can form C–C bonds.

Initially the $O(CH_3)_3^+$ oxonium cation was proposed [192] as intermediate. The ylide $(CH_2)O(CH_3)_2$, formed by deprotonation, is thought to be alkylated (by reaction with surface methoxy) to give $(CH_3CH_2)O(CH_3)_2^+$.

More recently formaldehyde ($H_2C{=}O$) has been suggested as intermediate. A hydride transfer reaction of methanol with surface methoxy gives methane and formaldehyde. Formaldehyde can decompose to CO that in a subsequent reaction inserts into surface methoxy. Acetic acid is produced by hydrolysis with water. Also, formaldehyde may react directly with surface methoxy to give the aldehyde. Subsequent recombination and decarboxylation reactions will give ethene [193]–[198].

A common feature of all proposals of initial C–C bond formation is that the apparent activation energy is high compared to the subsequent alkylation reactions of methanol, which happen in the quasi-steady state episode. This is the reason for the relatively long initiation period of the reaction. This initiation period can be shortened by addition of alcohol or ethene to the methanol feed at the beginning of the reaction.

Once the reaction is at steady state methanol is consumed in three cycles that interact in a complex way. The three reaction networks are sketched in the propagation part of Figure 5.34.

In reaction network I there is alkene oligomerization and alkylation with methanol. Isomerization and β C–C cracking are important reactions. Ethene and propene are main products.

Ring closure reactions generate aromatics. These reactions are possible once alkadienyl or alkatrienyl cation intermediates are formed. The mechanism of aromatics formation in the methanol reactions is similar to that in the catalytic cracking reaction (Figure 5.6). Alkadiene or alkatriene molecules result from hydride transfer reactions as illustrated in Figure 5.35a. The allylic C–H bond at β position with respect to the C=C bond of the alkene is very reactive. The carbenium cation that accepts the hydride converts into an alkane. Short alkanes are a co-product of aromatics formation.

In acid catalysis the alkene molecule hydrogenates by accepting a proton from the zeolite and hydride ion from another reactive unsaturated hydrocarbon.

As is illustrated in Figure 5.35b protonation of hexatriene initiates the ring closure reaction to methylcyclopentadiene. Subsequent hydride transfer and isomerization reactions with an additional deprotonation step leads to benzene.

Route I produces short alkenes including ethene as well as aromatics. Route II consists of methanol alkylation reactions of the aromatic molecules. Propene is formed through the side-chain mechanism of Figure 5.36a [199], [200]. The side-chain reaction is a reaction of multi-methylated benzene. In the process the benzene ring itself does not transform. This happens in route III. Aromatics alkylation transforms the cyclic six-ring aromatic molecule into a substituted cyclopentadienyl cation [181], [201]. The paring reaction gives propene and isobutene as products. This is shown in

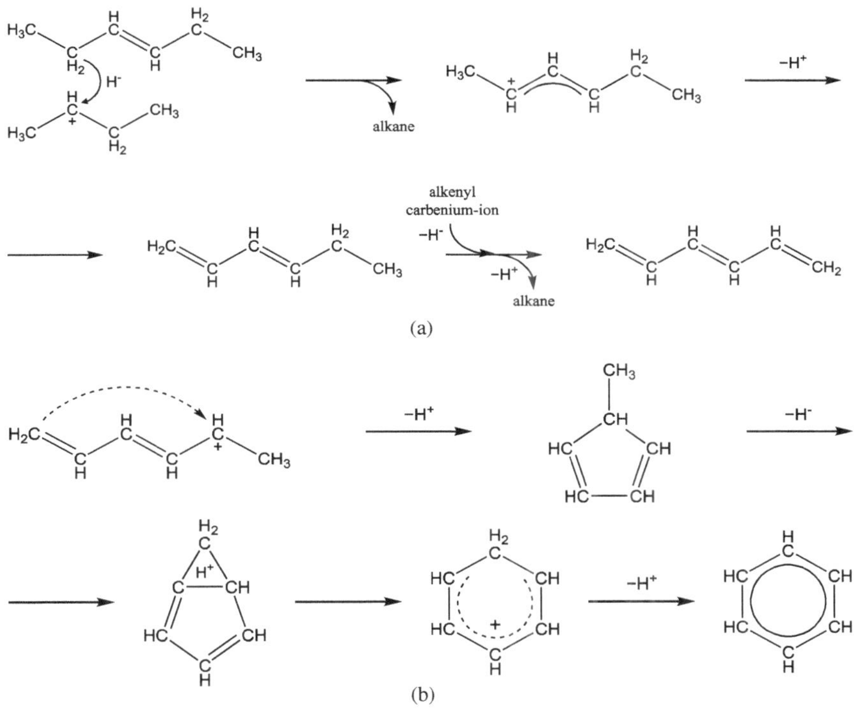

Figure 5.35 The ring closure reaction of hexatriene. (a) The hydride transfer reaction that gives hexatriene. (b) The ring closure reaction through intermediate cyclopentadienyl formation (compare Figure 5.28 ring closure via cyclopentenyl intermediate).

Figure 5.36b. The benzenium cation plays a key catalytic role in route II, while the substituted cyclopentadienyl is the catalytic cation of route III.

The paring reaction of the MTO process (Figure 5.36b) is related to that of hydrocracking of naphthenes in Figure 5.26d. The paring reaction in the MTO process does not happen with the cyclopentenyl cation, as is currently generally accepted for hydrocracking, but with the unsaturated cyclopentadienyl cation.

The paring mechanism with the cyclopentadienyl cation intermediate is due to Sullivan *et al.* [201], who were the first to suggest such a reaction in 1961.

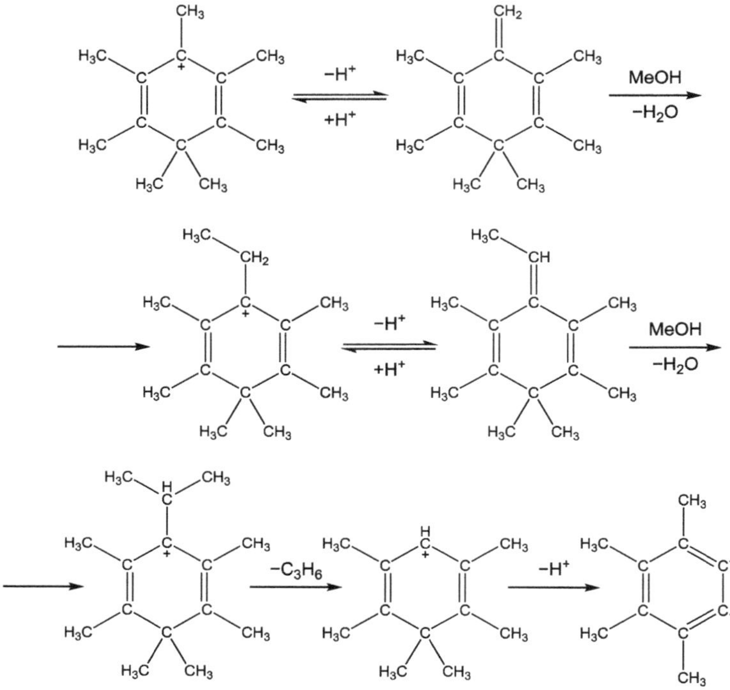

(a)

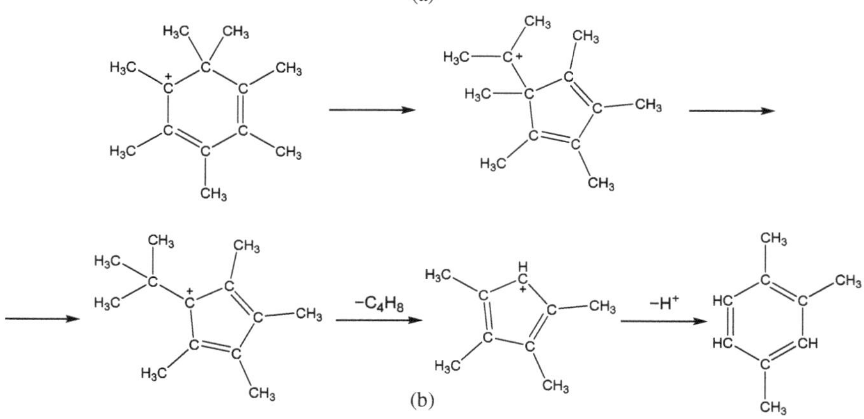

(b)

Figure 5.36 Alkylation reaction that gives propene and isobutane [202]. (a) Propene formation by reaction of methanol with protonated hexamethylbenzene. (b) The paring reaction that gives isobutene as product.

The three routes of Figure 5.34 form a complex network of interconnected elementary reactions since they share common reaction intermediates. It is consistent with the dual cycle reaction mechanism concept [203], [204] that explains propene and isobutene formation from the aromatic-based II,III routes, and ethene aromatics and alkane formation from route I. The main deactivation reaction is formation of polyaromatics.

The differences in shape selectivity of H-ZSM-5, H-SAPO-34, and H-ZSM-22 (TON) are due to diffusional limitations and transition state selectivity [182]. It reflects in different preferences of reaction cycles I, II or III.

The one-dimensional 10-ring system H-ZSM-22 produces selectively propene oligomers and aromatics formation is suppressed. Route I is mainly operational.

H-SAPO-34 with narrow cavity connections has a high selectivity to ethene and propene. Here the main reaction path is route II.

H-ZSM-5 with its three-dimensional pore system and larger nanopore space at the channel cross-sections is the preferred catalyst for aromatics formation. Also a substantial amount of isobutane is produced. It implies that for this zeolite the main reaction paths are aromatics formation in route I, followed by routes II and III.

Restricted transition state selectivity can be due to increased reaction intermediate stabilization or by steric inhibition [205]. This can be demonstrated by DFT calculations on zeolitic catalysts with cavities of different size and shape, but all connected though 8-ring nanopores as in the CHA structure [206].

Experimental and computational studies of the methanol conversion reaction are shown in Figure 5.37. The three zeolite structures that are compared are H-SAPO-35, H-SAPO-34, and DNL-6. The structures of the different zeolitic systems are shown in Figure 5.37b. Sizes of cavities increase in the same order.

Figure 5.37a shows that the smaller cavity material H-SAPO-35 has low reactivity and produces selectively ethene. The materials with larger cavities have higher reactivity and produce selectively propene and butene.

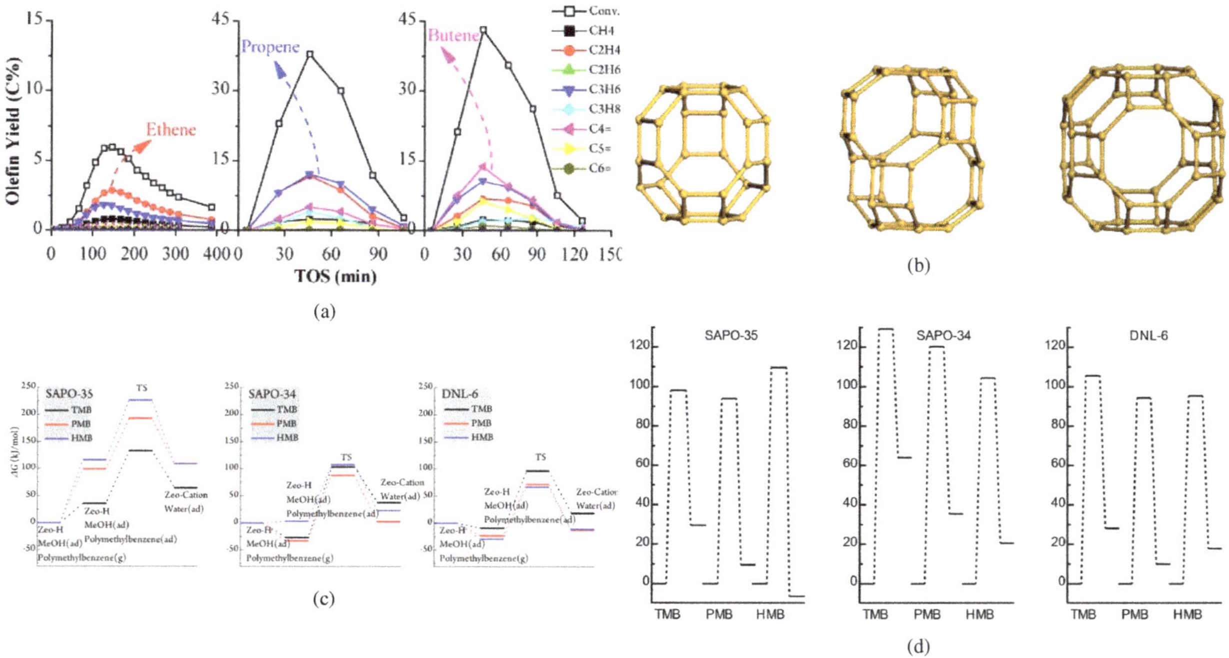

Figure 5.37 Comparison of methanol to alkene catalysis by H-SAPO-35, H-SAPO-34, and H-DNL-6 [206]. (a) Conversion and olefin yields over H-SAPO-35 (left), H-SAPO-34 (middle), and H-DNL-6 (right) in the MTO reaction at 275°C. (b) Cavity structures of 8MR molecular sieves: H-SAPO-35 (left, *lev* cavity, 6.3 × 7.3 Å), H-SAPO-34 (middle, *cha* cavity, 6.7 × 10 Å), and H-DNL-6 (right, α cavity, 11.4 × 11.4 Å). (c) Computed free-energy profiles of the methylation of tetraMB (TMB), pentaMB (PMB), and hexaMB (HMB) over H-SAPO-35, H-SAPO-34, and H-DNL-6 at 275°C. TS: transition state. (d) Computed intrinsic free energy barriers (275°C) of the methylation of TMB, PMB, and HMB over H-SAPO-35, H-SAPO-34, and H-DNL-6. In the respective figures (c) and (d) activation energies are with reference to individual methanol and polyMB.

Spectroscopy indicates that in the experimental systems the cavities are occupied by different intermediates. Polymethyl benzene cations (polyMB+) dominate the larger cavity and polymethyl cyclopentadienyl cations (polyMCP+) occupy the smaller cavity.

Steric constraint by the cavity induces product formation by the paring reaction that is the least cavity space demanding. The side-chain alkylation can occur in the larger cavities. This reaction is catalyzed by the larger polyMB+ cation and selectively occurs in DNL-6.

The combined effect of attractive and repulsive interactions that determines the transition state free energies follows from Figure 5.37c. It compares methylation of benzene molecules with different degrees of substitution.

The cavity size is large enough for the side-chain alkylation (Figure 5.26a) to occur in the larger cavities. This reaction is catalyzed by the larger polyMB+ cation and has a low barrier in H-DNL-6. The largest hexamethylated benzene (HMB) molecule has the lower transition state energy in this large-cavity zeolite. Attractive interactions dominate. Activation energies for different polyMB+ cations vary little, but the activation energy is slightly higher for the alkylation of smaller TMB.

In the small cavity of H-SAPO-35 the situation is different. In this cavity repulsive interactions dominate. All three polyMBs are too large. It provides the driving force for the cyclohexadienyl molecule to convert to the smaller cyclopentadienyl cation (not shown). Products of the paring reaction are expected. As Figure 5.37d shows intrinsic activation barriers again do not differ significantly. For the larger molecule the reaction barrier is slightly higher.

The cavity of H-SAPO-34 has an intermediate size. The two smaller polyMBs have dominantly attractive interactions, but the larger molecule experiences repulsive interactions. It causes the intrinsic activation barrier of the larger reacting system to be lower. This system is already strained in the ground state. The intrinsic activation energy of the smaller unstrained system has the higher barrier.

In summary, the mechanism of methanol conversion to alkenes and aromatics is complex. This methanol dehydration reaction illustrates the important mechanistic idea that at steady state catalytic organocations may take over the role of direct catalytic proton activation. These organocatalytic molecules are substituted cyclohexadienyl or cyclopentadienyl cations.

Methanol to olefins and aromatics catalysis is autocatalytic. Alkene or benzene molecules are not only products but also precursors to the organocatalytic molecules.

The interplay of attractive and repulsive interactions of reacting molecules in a zeolite cavity determines differences in alkylation activation energies. Shape selectivity depends on cavity size, which determines energetics of intermediates, as well as on dimensions of cavity pore connections. The latter constrains size or shape of product molecules by diffusive limitation.

In the next two sections the kinetics of bimolecular reactions is presented. An interesting consideration is that in catalysis such reactions are often first order in reactant. This is a consequence of catalysis by carbocations that are formed *in situ*.

5.4.6 *Kinetics of Bimolecular Solid Acid-catalyzed Reactions*

In bimolecular reactions protonic sites are replaced by carbenium ions. The latter are the catalytic reaction centers that regenerate.

In solid acid catalysis of hydrocarbons, when a reaction is bimolecular one of the components will usually become a protonated carbenium ion intermediate. When the reaction rate-controlling step is a combination of the carbenium ion with a reactant molecule, bimolecular reactions can behave quasi-monomolecular. These reactions are interesting since the rate-controlling step is not reaction with a proton but with an organocatalytic intermediate.

5.4.6.1 *Bimolecular Reaction Kinetics of the Dimerization of Alkene*

Bimolecular reactions of alkenes have often first-order kinetics.

Here for the dimerization of propene the change of second-order to first-order kinetics at high pressure is discussed.

This is demonstrated for the oligomerization reaction of small olefins in experiments by Iglesia *et al.* from UC Berkeley [207]. Figure 5.38a shows that the dimerization reactions of ethene, propene, and isobutene are first order in alkene pressure over a large range. The large molecule has larger reaction rate, because of its higher adsorption energy.

The zeolite used in this experiment has the TON structure, which has cylindrical nanopores consisting of rings of 10 tetrahedra. The low-pressure experiment of Figure 5.38b shows the initially second-order dependence for conversion of propene.

Figure 5.38c shows for the low-pressure experiment the rapid saturation of respective alkoxide surface concentrations with increase of pressure as observed by *in situ* infrared vibrational spectroscopy [208].

As the quantum-chemical calculations of ethene oligomerization of Figure 5.7 illustrate, initially one molecule is π-adsorbed to the proton. Upon protonation this is converted into the alkoxy state.

In the oligomerization reaction a second molecule reacts with the alkoxy intermediate. At the condition where the dimerization rate is first order, the protonated alkyl cation is the MARI (Major Abundant Reactive Intermediate). In the rate-controlling step the alkyl cation reacts as alkoxide with the alkene.

The transition from second- to first-order kinetics evolution follows from Eqs. (5.13) for the oligomerization reaction rate:

$$\frac{R}{N_{\mathrm{c}}} = k_{int}\theta_{C^=}\theta_{\mathrm{alk}};\theta_{\mathrm{alk}} = \frac{K_{\mathrm{ads}}^{\mathrm{p}}\left[C^=\right]}{1+K_{\mathrm{ads}}^{\mathrm{p}}\left[C^=\right]} \tag{5.13a}$$

$$= k_{int}K_{\mathrm{ads}}^{\mathrm{c}}K_{\mathrm{ads}}^{\mathrm{p}}\left[C^=\right]^2;\left(K_{\mathrm{ads}}^{\mathrm{c}}\left[C^=\right] \ll 1;K_{\mathrm{ads}}^{\mathrm{p}}\left[C^=\right] \ll 1\right) \tag{5.13b}$$

$$= k_{int}\,K_{\mathrm{ads}}^{\mathrm{c}}\left[C^=\right] \quad \left(K_{\mathrm{ads}}^{\mathrm{c}}\left[C^=\right] \ll 1;\; K_{\mathrm{ads}}^{\mathrm{p}}\left[C^=\right] \gg 1\right) \tag{5.13c}$$

One distinguishes the protonated alkene that adsorbs as alkoxy θ_{alk} and surface coverage θ_C= of non-protonated alkene adsorbed to the siliceous nanopore wall. θ_C= is defined by the adsorption equilibrium constant $K_{\mathrm{ads}}^{\mathrm{c}}$ of physically adsorbed alkene. $K_{\mathrm{ads}}^{\mathrm{p}}$ is the equilibrium constant of gas-phase alkene with alkoxide intermediate: $K_{\mathrm{ads}}^{\mathrm{c}} \ll K_{\mathrm{ads}}^{\mathrm{p}}$.

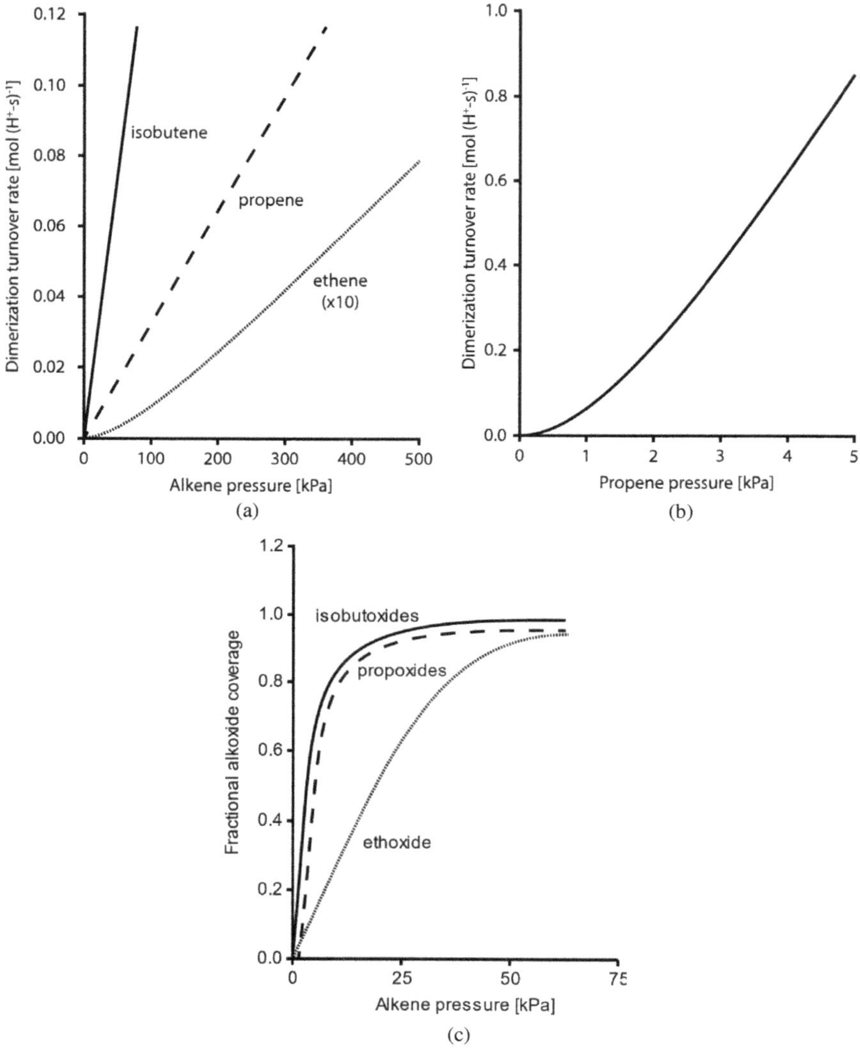

Figure 5.38 Dimerization kinetics of small alkenes [208]. (a) Alkene dimerization turnover rates on TON for C_2H_4, C_3H_6, and i-C_4H_8 as a function of alkene pressure. (b) Propene dimerization turnover rates on TON at low pressures (503 K). (c) Fractional alkoxide coverage measured from the decrease in the intensity of the 3600 cm^{-1} OH band in the infrared spectra of TON (Si/Al = 40) at 503 K as a function of alkene pressure.

As Eq. (5.13b) illustrates, in the low-pressure regime the reaction rate is second order.

In the higher-pressure first-order regime θ_{alk} saturates and is equal to one. The structure dependence of the reaction rate is determined by the apparent elementary rate constant $k_{int}K^{c}_{ads}$ (Eq. (5.13c)). As long as the intrinsic rate constant k_{int} only depends on contact of the two molecules, the confinement-dependent rate constant will vary with the van der Waals dispersive interaction of reactant with zeolite cavity.

For the oligomerization reaction of propene this is confirmed by the experimental results presented in Figure 5.39 [207]. The propene dimerization reaction is compared for a variety of solid acids. Figure 5.39a gives a plot of calculated deprotonation energies *DPE* against measured reaction constants $k_{int}K^{c}_{ads}$ of the first-order reaction of dimerization of propene. The rate of reaction decreases with

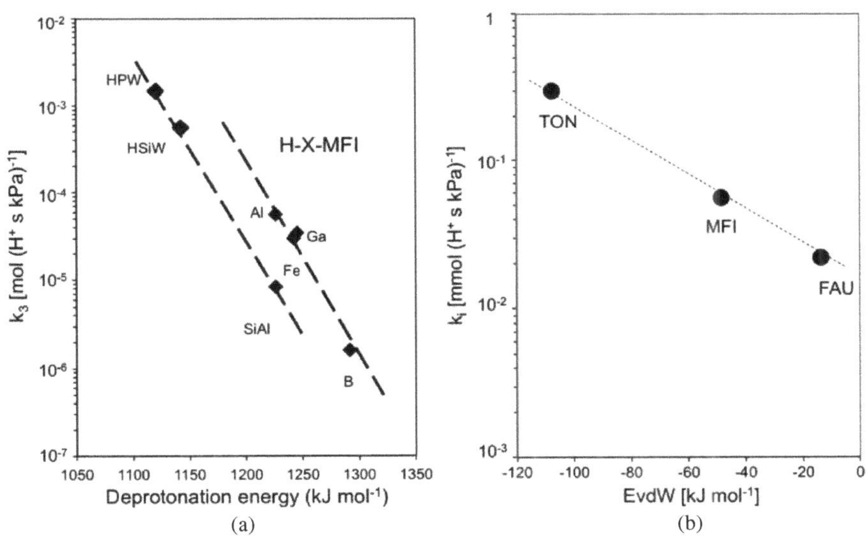

Figure 5.39 Dimerization kinetics of propene [207]. (a) Normalized first-order rate constant for propene dimerization (k_3) as a function of deprotonation energy for both non-confined and MFI-confined protons (503 K). (b) First-order rate constants for propene dimerization ($k_i = 10^3\ k_3$) as a function of E_{vdW}. This is the adsorption energy of propene to the siliceous wall of the zeolite cavity (503 K).

increase of *DPE*. This logarithmic dependence of the apparent reaction rate constant on *DPE* for solid acids with comparable adsorption energies demonstrates the independence of k_{int} on adsorption energy.

The reason for the lower reactivity of the catalysts with the open surface in Figure 5.39a is their lower propene adsorption energy compared to that of the nanoporous systems. The plot of the reaction rate as a function of the van der Waals interaction energy E_{vdW} of propene for different zeolites, Figure 5.39b, shows that in the first-order regime the logarithm of the apparent reaction rate constant is proportional to the adsorption energy of physically adsorbed propene. The *DPE* of the respective protons in the different zeolites is the same. This structure dependence relates to the increase of K^{c}_{ads} with the physical adsorption energy of propene.

In terms of kinetics the reaction rate is determined by the MARI. In the first-order regime the MARI is the alkoxy intermediate. Reaction rate is not proportional to the number of free protons, but is proportional to the MARI concentration.

The reaction mechanism is of the Eley-Rideal type (Section 2.4.2). The dependence of reaction rate on zeolite structure is due to the differences in stability of the Eley-Rideal intermediate, which is the pre-transition state complex. The intrinsic reaction rate constant k_{int} of the pre-transition state complex does not depend on adsorption energy but only on *DPE*. The pre-transition state complex of the adsorbed propene molecule rearranges to propyl alkoxy, and a C–C bond is formed. The activation energy of k_{int} decreases when *DPE* increases.

5.4.6.2 *The Alkylation of Isobutane and Alkene*

The MARI of the alkylation of isobutane and alkene is the isobutyl cation. Therefore alkylation is first order in alkene. The low concentration of the alkyl cation causes the reaction rate of alkene oligomerization to be second order in alkene.

The alkylation of isobutane with propene or n-butene that produces branched C_6 or C_7 hydrocarbons is an important refinery process for producing high-octane liquid fuel. The process, invented in the 1930s, is catalyzed by neat sulfuric acid or hydrogen fluoride. Study of its reaction mechanism led to an early understanding of carbocation catalysis [7], [8], [209].

Over the years an ongoing research effort to replace the hazardous liquid acids by solid zeolitic acids has taken place [210]–[213]. Successful operation of a zeolite-based process is possible and has been demonstrated for the ALKYclean process developed by Albemarle, ABB Lummus, and Neste Oil [214].

A major issue is the fast deactivation of the zeolite solid acid catalyst that is used in the reaction. Alkene oligomerization leads rapidly to formation of deactivating carbonaceous residue. This reaction competes with alkylation, which also is a reaction with alkene. As will be seen the choice of reactor is important. Long life requires the use of a continuously stirred tank reactor (CSTR) instead of a tubular plug flow reactor (PFR) [215].

To minimize alkene concentration the reaction is to be executed in excess isobutane at reaction condition where alkene is 100% converted. In the CSTR, the low concentration of alkene is uniformly distributed in the reactor. Even at 100% conversion the unidirectional reactant flow in the PFR causes an alkene concentration gradient in the reactor that is detrimental to its deactivation time (reduction of lifetime by two orders of magnitude).

Zeolites such as X or Y (FAU structure) with wide nanopores that allow for rapid diffusion of the bulky alkylate molecules are used, promoted by La to increase acid strength [216].

In the following the mechanism of the solid acid-catalyzed alkylation of propene with isobutane is introduced. It is shown that for this bimolecular alkylation reaction, at reaction condition kinetics is also first order in alkene. The concept of MARI again is essential to deduce global kinetics from microkinetics.

As for other catalytic reactions one has to distinguish different catalyst lifetime regions: initiation, quasi-steady state, and deactivation.

The mechanism for the reaction of propene with isobutane is schematically illustrated in Figure 5.40a [7], [212]. At quasi-steady state the propagation cycle takes over from the initiation reactions.

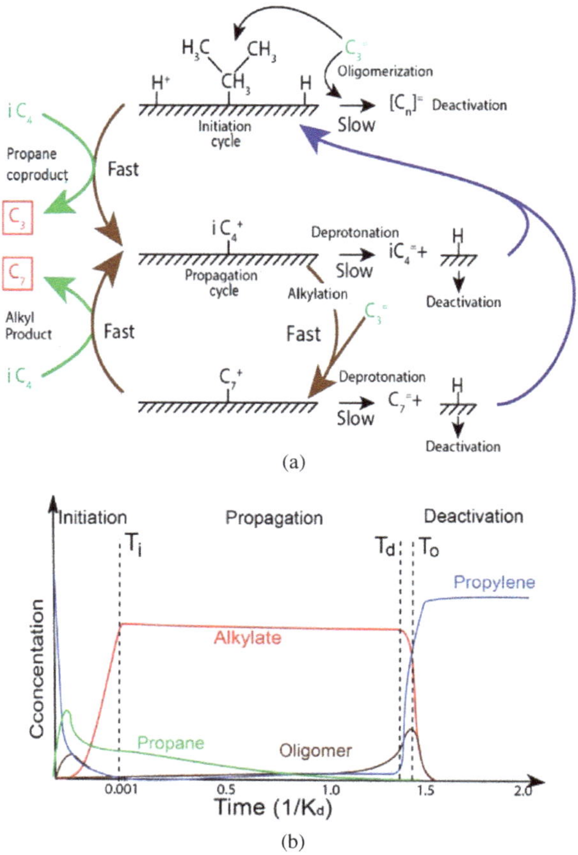

Figure 5.40 The alkylation of isobutane with propene. (a) Scheme of the alkylation reaction cycle including deactivation paths [217]. Desirable elementary rate constant relations are indicated for high selectivity of alkylate and slow deactivation rate. (b) Time evolution of reactant, product, and reaction intermediate concentrations [218]. $1/K_d$ is dimensionless time, T_i initiation time, T_d deactivation time, and T_o time of maximum oligomer production.

Figure 5.40b illustrates the reactant and product time dependence of the reaction. Initially propane is the main product but after a short time this is taken over by alkylate production. In the quasi-steady state region catalytic reactivity decreases and deactivation sets in when reactivity has reduced too much, and propene conversion is not 100% anymore.

At the low temperature of this exothermic alkylation reaction (slightly below or above room temperature) the saturated isobutane molecule will not react, but alkene will rapidly protonate and adsorb as propyl alkoxy intermediate.

As indicated in Figure 5.40 a reaction is initiated by conversion of isobutane into isobutyl cation in a hydride transfer step to a propyl cation. This gives propane as co-product.

Deuterium exchange experiments of the 1950s by Shell researchers in the USA [168] demonstrated the operation of this hydride transfer step in neat H_2SO_4. Consistent with this only hydrogen atoms of primary carbon atoms of isobutane exchange with deuterium atoms.

This exchange happens via intermediate isobutyl cation formation by initial loss of the hydrogen atom attached to the tertiary carbon atom as hydride ion by reaction with a proton of sulfuric acid. Through intermediate formation of isobutene this carbenium ion will exchange protons or deuterium. Isobutane is recovered by hydride transfer of isobutyl ion with another isobutane molecule.

In the alkylation reaction the propyl cation accepts the hydride ion from isobutane. The isobutyl cation is the organocatalytic molecule of the reaction.

Once a finite concentration of isobutyl cation builds up, at time T_i the propagation cycle starts.

In an Eley-Rideal-type reaction propene reacts with the alkyl cation. This pre-transition state complex transforms to adsorbed C_7^+ by formation of a C–C bond between isobutyl cation and adsorbed propene. Hydride transfer from another isobutane molecule to C_7^+ regenerates the isobutyl cation and gives product alkylate molecule C_7.

In case no deprotonation reactions of C_7^+ or iC_4^+ happen, the reaction cycle that alternates between the C_7^+ and iC_4^+ intermediates continues as long as reactant molecules remain available. There is no deactivation and alkene is consumed and converted to alkylate by the iC_4^+ cation that acts as an organocatalyst. Also in this reaction proton catalysis is replaced by organocatalysis.

However, when the respective cations deprotonate and alkenes are produced, the initiation cycle restarts. In this cycle protonated propene (or product alkene) will not only react with isobutane but also alkene oligomerization reactions will occur that cause catalyst deactivation. This second-order reaction of alkene will remain suppressed as long as reactor alkene concentration is low, but at time T_d it increases rapidly when the propene concentration increases.

Alkene oligomerization and related consecutive reactions may lead to stable (protonated) aromatic or cyclopentadienyl molecules that are more basic than the deprotonated zeolite lattice (see Figure 5.36b) that consumes protons [181], [219]. Carbonaceous residue molecules may also block reaction sites.

The protonated olefinic and aromatic deactivating compounds can be readily hydrogenated. In the Alkyclean process the solid acid catalyst is promoted with a small amount of Pt. Alternating reaction cycles are used. When catalyst deactivation starts, reactant is switched to hydrogen. Deactivating molecules are removed by hydrogenation and reaction is resumed.

In the quasi-steady state regime of selective alkylation, the MARI is the isobutyl cation. The low concentration of propene makes it possible that the C–C bond formation with the isobutyl cation competes with hydride transfer.

The quasi-steady state global kinetic equations of the alkylation reaction can be deduced from microkinetic equations [218] and are presented in simplified form in Eqs. (5.14).

$$-\frac{d}{dt}\left[C_3^{=},t\right]=\Phi\left(\left[C_3^{=},t\right]-\left[C_3^{=}[O]\right]\right)+k_1\left[C_3^{=},t\right]\theta\left(iC_4^{+},t\right) \quad (5.14a)$$
$$+k_2\left[C_3^{=},t\right]^2\theta\left(iC_4^{+},t\right)$$

$$\frac{d}{dt}\left[C_i^{=}(t)\right]=f\left(\left[C_3^{=}\right],\sum_i C_i^{=}(t),\theta\left(iC_4^{+}\right)\right)-\Phi\left[C_i^{=}(t)\right] \quad (5.14b)$$

$$-\frac{d}{dt}\theta\left(iC_4^{+},t\right)=F\left(\left[C_3^{=},t\right]\right)\left(\sum_i k_i\left[C_i^{=}(t)\right]\right)\theta\left(iC_4^{+},t\right) \quad (5.14c)$$

Φ^{-1} is the CSTR contact time of reactant with catalyst.

Global kinetic expressions depend explicitly on reactant and product concentration in the gas phase and reaction rates are proportional to the concentration of surface reactant sites.

In the quasi-steady state regime this is the concentration of $iC_4^+; N_c\theta(iC_4^+, t)$. The reaction is quasi-steady state since the catalytic center concentration iC_4^+ varies with time.

The reaction rate is fast compared to the deactivation rate. The concentration $\theta(iC_4^+)$ is initially close to one. To deduce Eqs. (5.14) from microkinetics the concentrations of all other surface intermediates are related to $\theta(iC_4^+)$ using the assumption that iC_4^+ is MARI. Eq. (5.14a) gives the consumption rate of propene and Eq. (5.14c) the rate of change of $\theta(iC_4^+, t)$. This rate of change does not only depend on propene concentration but also on the concentrations of other reaction intermediates that can be calculated from expressions of type Eq. (5.14b). Because the catalyst deactivation rate is slow, the steady state equations give calculated concentrations of reaction intermediates at a particular time that can be used in Eq. (5.14c) to calculate the deactivation rate.

For simplicity and clarity Eqs. (5.14) are only presented in implicit form. The corresponding explicit expressions can be found in [218]. Figure 5.40b is constructed from the solution to Eqs. (5.14).

Propene is consumed in a reaction that gives alkylate, with rate constant k_1, and by oligomerization, with rate constant k_2. k_1 and k_2 are the effective rate constants that like rate constants k_i and function F of Eq. (5.14c) depend in a complex way on elementary rate constants of the reaction network.

The alkylation reaction is first order in propene concentration. It proceeds by the Eley-Rideal complex of propene with isobutyl alkoxy. Different from the oligomerization reaction discussed in the previous section, in the alkylation reaction the deactivating oligomerization rate is quadratic in propene concentration. This is because surface coverage $\theta_{C_3^=}$ is extremely low as long as the propagation cycle dominates the reaction.

Figure 5.40a indicates that for high alkylation selectivity the elementary rate constant of hydride transfer from isobutane to carbenium cation has to be fast compared to elementary deprotonation rate constants that produces undesirable alkene. Since it will stabilize the carbenium ion with respect to deprotonation, a catalyst of high proton affinity (low *DPE*) is needed to prolong catalyst life [217], [220]. Also operation in excess isobutane biases the hydride transfer reaction versus oligomerization.

The concentration parameter that defines catalyst deactivation is low concentration of alkene. This is common to all solid-state-catalyzed reactions. Oligomerization rates have a low barrier and hence will often compete with desirable reactions. In bifunctional catalytic conversion of alkanes hydrogenation of intermediate alkene extends catalyst lifetime.

Also shape-selective catalysis may decrease the catalyst deactivation rate because bulky, deactivating carbonaceous molecules cannot be accommodated in smaller zeolite cavities.

In alkylation catalysis bulky branched C_7 or C_8 are products and deactivation is initiated by substituted aromatics or cyclopentadienyl cations. Because the respective molecular sizes are comparable the requirement of large nanopores for the bimolecular alkylation conflicts with the demand of small nanopores to suppress deactivation reactions.

5.5 Summary and List of Reactions

The mechanisms of reactions catalyzed by solid and liquid acids have many similarities. Study of the mechanism of homogenous acid-catalyzed reactions has generated the physical organic basis of the mechanism of proton-activated reactions that is presented in this chapter.

Unique features of solid acid catalysis are:

- the very different energetics of proton transfer compared to that of liquid phase acid catalysis
- the important role of adsorption and diffusion
- the confinement effect and shape selectivity, which are due to nanocavity occlusion (specific for zeolites).

Additionally, the robustness of inorganic solid acid materials makes them useful in high-temperature reactions where homogeneous acid catalysis cannot be used. Examples are the endothermic catalytic cracking and alcohol dehydration reactions.

An important practical issue is rapid deactivation of the catalyst. The successful use of nanoporous zeolites in catalytic cracking is based on suppression of coke formation by size constraint of the zeolite nanopores.

The solid acid material can also support additional catalytic components such as transition metals or those that activate hydrogen. They catalyze hydroisomerization and hydrocracking reactions at mild conditions. In this case the hydrogenation function of the catalyst reduces the rate of deactivation.

Two kinds of carbocations are important intermediates of proton-activated reactions. A proton can be added to a saturated molecule as the alkane molecule to create an unstable carbonium ion. The proton can also be added to an unsaturated molecule such as an alkene. This gives the more stable carbenium ion. The carbonium ion is called a non-classical organic cation, since one or two molecular carbon atoms become five-coordinated. This is the reason for its instability. The carbenium ion is more stable than the carbonium ion because its protonated carbon atom is three-coordinated and sp^2 hybridized.

In solid acid catalyst systems the carbonium and carbenium ions are activated complexes and often part of transition states. The solid acid catalytic transformation is on top of the potential energy hill chemistry.

The cycle of proton-activated catalysis is propagated in two ways. In one mechanism there is a sequence of protonation of reactant molecule and carbocation transformation reactions. Upon product formation the proton is backdonated to the solid. Activation of alkane through a carbonium ion intermediate is followed by secondary C–C bond cleavage reactions that give a short alkane and a carbenium ion. An alkene product is formed upon proton backdonation.

Alternatively, a second mechanistic route becomes possible once intermediate carbenium ions are formed by reaction. The carbenium carbocation intermediate accepts a hydride ion from a saturated reactant hydrocarbon. This produces again a saturated but different hydrocarbon and a carbenium ion. This commonly occurring elementary reaction is called hydride transfer. It activates alkanes by reaction with carbenium ions. Since hydride transfer has a low activation energy it substitutes for alkane protonation and formation of the highly activated carbonium ion. The later formation rate has a high activation energy.

Shape-selective catalysis discriminates between the two reaction paths. When zeolites with small nanopores are used that can only accommodate one single molecule, the alkane can only be activated by the carbonium ion mechanism. Products are short alkanes and alkene molecules.

With solid acidic materials which have wider nanopores that can accommodate two molecules, the low-activation-energy hydride transfer can take over. The intermediate carbenium ion is converted into alkane and reactant alkane is converted into carbenium ion. In this case short alkanes are produced.

Reactions that proceed by hydride transfer can have a complex product pattern since not only alkanes but also alkenes can be hydride-donating agents. The latter provides a route for aromatics formation.

The mechanism that discriminates between bimolecular and monomolecular reactions is called transition state selectivity.

In bifunctional catalysis the hydride transfer reaction and carbonium ions are circumvented. Reaction happens in excess hydrogen. A promoting transition metal catalyzes alkane dehydrogenation and alkene hydrogenation. Reactant alkane dehydrogenates to alkene, which rapidly reacts with a proton. The carbenium ion undergoes transformation reactions and a product carbenium ion backdonates proton to solid and is hydrogenated to alkane product. The high hydrogen pressure prevents deactivating oligomerization reactions.

Composition and structure of a solid acid catalyst strongly affects its proton reactivity. There are two classes of structurally solid acids:

In one class of materials the surface of the metal oxide is created by cleavage of bulk chemical bonds. The surface is reactive and upon contact with water becomes partially hydroxylated.

Adsorbed hydroxyls can have Brønsted acid or Brønsted base reactivity. The reactivity of the hydroxyls depends on the coordination of the proton, structure of the solid, and solid composition. A hydroxyl coordinated to a single surface cation is usually Brønsted basic. The charge and size of the cation determine its reactivity. Brønsted acidic protons are coordinated to oxygen atoms that bridge surface cations. Proton donation activity increases when the cations that coordinate the oxygen atoms are small and have high positive charge. The Pauling electrostatic model is useful to estimate differences in base and acid reactivity.

The second class of solid acids contains protons that are attached to the internal surface of a nanoporous material. This is the case for zeolitic materials, in which the internal surface of the zeolite framework surrounds the nanocavities. The atoms that constitute the zeolite framework are coordinatively saturated. The acidic zeolite proton is attached to an oxygen atom that bridges two cations. These cations are four-coordinated to surrounding lattice oxygen atoms.

One cation is to have a charge of 4+, the other 3+. Usually one cation is Si^{4+}, the other Al^{3+} but other substitutions are possible.

In zeolites the proton strength is dominated by lattice composition but is rather insensitive to framework structure. The proton strength is independent of composition as long as the framework Al/Si ratio is less than 0.1.

The proton strength can be measured with Hammett indicator molecules. These molecules change color by protonation. The pK of coloring relates with the strength of the proton. It measures an average proton strength. Individual proton strengths can be measured from vibrational shifts of the O–H bond by contact with weakly interacting probe molecules.

Zeolite catalysis is sensitive with respect to nanopore dimensions and the structure of its three-dimensional nanopore network. This is not due to variation in proton strength but relates to the sensitivity of physical adsorption of reactants or reaction intermediates to match the size and shape of the nanopores.

Whereas physical adsorption energy is only a few kJ/mol per atom-atom contact, the adsorption energies of large molecules can become comparable to that of proton activation energies.

Table 5.6 The mechanism and kinetics of the following reactions discussed in this chapter.

Alkene oligomerization	Transition state (Figure 5.7c), kinetics (Figures 5.38, 5.39a)
Alkylation of toluene with methanol to xylene	Pre-transition state (Figure 5.32, Section 5.4.4.4)
Alkylation of alkene with isobutane	Mechanism and deactivation (Figure 5.40, Section 5.4.6.2)
Catalytic cracking alkanes	Shape selectivity (Figure 5.5), catalytic reaction cycle (Figure 5.6), transition state energies (C–C, C–H bond cleavage) (Figure 5.7b)
Cumene dealkylation	Relation with Hammett acidity function and mechanism (Figure 5.9)
Hexane to benzene	Bifunctional reaction mechanims (Figure 5.28)
Hydroisomerization of hexane	Kinetics (Table 5.5); reaction scheme (Figure 5.25), kinetic equations (Section 5.4.3.3)
Hydrocracking of n-decane, butyl hexane, product distributions	n-C_{16} simulated product distributions (Figure 5.29)
Methanol to aromatics and alkenes	Mechanism and shape selectivity (Figures 5.30, 5.33, 5.34, Section 5.4.5)
Monomolecular hexane cracking	Relation with proton concentration (Figure 5.4), relation with zeolite composition (Figure 5.23b)
Monomolecular alkane cracking	Apparent activation energies (Tables 5.3, 5.4)

Short versions of parts of Sections 5.2, 5.3 and 5.4 are also present in *Comprehensive Inorganic Chemistry III, Part 6: Heterogeneous Inorganic Catalysis*, Chapter 13, R. A. van Santen, "The Zeolite Protonic Site"; Chapter 14, R. A. van Santen, "Catalytic Chemistry of Proton Activation," Elsevier, 2023.

For larger molecules there is multipoint contact of molecular (mainly hydrogen) atoms and zeolite lattice oxygen atoms. The oxygen atoms are large compared to the small cations and, because of the larger charge, they are more polarizable. This is the electronic interaction that dominates the van der Waals interaction.

The adsorption energy of reactants and intermediates strongly depends on match of hydrocarbon dimensions and zeolite nanopore dimensions.

As long as the rate of product desorption is fast and temperature of reaction is high, the apparent activation energy for a monomolecular reaction is $E_{app} = E_{int} - E_{ads}$ (see Section 2.4.3). E_{int} is the activation energy of the elementary proton activation rate constant and E_{ads} the energy of adsorption of the reactant. This expression illustrates that change in structure of the zeolite has a large effect on catalytic reactivity not because of a change in proton reactivity, but because of a difference in adsorption energy. There is a linear Brønsted-Evans-Polanyi type of relation between the change in E_{app} and E_{ads}: $\Delta E_{app} = -\Delta E_{ads}$.

In summary, the structure dependence in zeolite catalysis relates to transition state selectivity, change in physical adsorption, and diffusional inhibition. The latter happens when molecules or reaction intermediates are too large to rapidly diffuse through the zeolite nanopores. The main parameter that influences proton reactivity is composition of the mixed oxides.

References

[1] "Houdry Process for Catalytic Cracking — National Historic Chemical Landmark — American Chemical Society." https://www.acs.org/content/acs/en/education/whatischemistry/landmarks/houdry.html (accessed Jan. 10, 2022).

[2] K. Tanabe, *Solid Acids and Bases: Their Catalytic Properties*, 1st ed. Academic Press, 1970.

[3] C. F. Thiele, "Absorption clays and oils," *Pet. Age*, vol. 7, pp. 45–47, 1920.

[4] C. L. Thomas, "Chemistry of cracking catalysts," *Ind. Eng. Chem.*, vol. 41, no. 11, pp. 2564–2573, Nov. 1934, doi: 10.1021/IE50479A042.

[5] V. Ipatieff, B. B. Corson, and G. Egloff, "Polymerization, a new source of gasoline," *Ind. Eng. Chem.*, vol. 27, no. 9, pp. 1077–1081, Sep. 1934, doi: 10.1021/IE50309A027.

[6] A. Corma, "12th International Congress on Catalysis: proceedings of the 12th ICC, Granada, Spain, July 9-14, 2000," p. 3948, 2000.

[7] L. Schmerling, "The mechanism of the alkylation of paraffins," *J. Am. Chem. Soc.*, vol. 67, no. 10, pp. 1778–1783, Oct. 1945, doi: 10.1021/ja01226a048.

[8] L. Schmerling, "The mechanism of the alkylation of paraffins. II. Alkylation of isobutane with propene, 1-butene and 2-butene," *J.*

Am. Chem. Soc., vol. 68, no. 2, pp. 275–281, Feb. 1946, doi: 10.1021/ja01206a038.

[9] V. Haensel and H. S. Bloch, "Duofunctional platinum catalysts in the petroleum industry," *Platin. Met. Rev*, vol. 8, no. 1, pp. 2–8, 1964.

[10] J. C. Rasser, W. H. Beindorff, and J. J. F. Scholten, "Characterization and performance of platinum-iridium reforming catalysts," *J. Catal.*, vol. 59, no. 2, pp. 211–222, Oct. 1979, doi: 10.1016/S0021-9517(79)80026-3.

[11] J. H. Sinfelt, "Catalysis by alloys and bimetallic clusters," *Acc. Chem. Res.*, vol. 10, no. 1, pp. 15–20, Jan. 1977, doi: 10.1021/ar50109a003.

[12] W. M. H. Sachtler, "Chemisorption complexes on alloy surfaces," vol. 14, no. 1, pp. 193–210, Jan. 2006, doi: 10.1080/03602457608073411.

[13] F. G. Ciapetta and J. Hunter, "Isomerization of saturated hydrocarbons in presence of hydrogenation-cracking catalysts: normal hexane," *Ind. Eng. Chem.*, vol. 45, no. 1, pp. 147–155, Jan. 1953, doi: 10.1021/IE50517A046.

[14] J. A. Rabo and M. W. Schoonover, "Early discoveries in zeolite chemistry and catalysis at Union Carbide, and follow-up in industrial catalysis," *Appl. Catal. A Gen.*, vol. 222, no. 1–2, pp. 261–275, Dec. 2001, doi: 10.1016/S0926-860X(01)00840-7.

[15] J. A. Rabo, P. E. Pickert, D. N. Stamires, and J. E. Boyle, "Molecular sieve catalyst in hydrocarbon reactions," in *Actes du Deuxieme Congres Internationale de Catalyse*, 1960, pp. 2055–2074.

[16] J. A. Rabo, P. E. Pickert, E. Dempsey, and V. Schomacker, "Zeolite cation with strong electrostatic fields as carponiogenic cataltic centers," in *Proceedings of the 3rd International Congress on Catalysis*, 1964, pp. 714–728.

[17] C. J. Plank, E. J. Rosinski, and W. P. Hawthorne, "Acidic crystalline aluminosilicates. New superactive, superselective cracking catalysts," *Ind. Eng. Chem. Prod. Res. Dev. 3*, vol. 3, no. 3, pp. 165–169, 2002, doi: 10.1021/I360011A001.

[18] C. J. Plank, "The invention of zeolite cracking catalysts — a personal viewpoint," in *Heterogeneous Catalysis: Selected American Histories*, B. H. Davis and W. P. Hettinger, Eds. American Chemical Society, 1983, pp. 253–271.

[19] J. B. Uytterhoeven, L. G. Christner, and W. K. Hall, "Studies of the hydrogen held by solids. VIII. The decationated zeolites," *J. Phys. Chem.*, vol. 69, no. 6, pp. 2117–2126, 1965, doi: 10.1021/j100890a052.

[20] F. C. Whitmore, "The common basis of intramolecular rearrangements," *J. Am. Chem. Soc.*, vol. 54, no. 8, pp. 3274–3283, Aug. 1932, doi: 10.1021/JA01347A037.

[21] J. G. Speight, "Thermal Cracking of Petroleum," in *Natural and Laboratory-Simulated Thermal Geochemical Processes,* R. Ikan, Ed. Springer, 2003, pp. 31–52.
[22] P. Backus, J. F. Stauffer Madison, W. M. J Johnson, H. Calvert, E. Dulaney, J. J. Stefaniak, F. B. Gailey, and F. G. Jarvis, "The mechanism of paraffin isomerization," *J. Am. Chem. Soc.*, vol. 68, no. 1, p. 153, Jan. 1946, doi: 10.1021/JA01205A519.
[23] G. Parravano, E. F. Hammel, and H. S. Taylor, "The exchange reaction between methane and deuteromethanes on silica — alumina cracking catalysts," *J. Am. Chem. Soc.*, vol. 70, no. 6, pp. 2269–2270, Jun. 1948, doi: 10.1021/JA01186A083.
[24] P. V. R. Schleyer, W. E. Watts, R. C. Fort, M. B. Comisarow, and G. A. Olah, "Stable carbonium ions. X.1 Direct nuclear magnetic resonance observation of the 2-norbornyl cation," *J. Am. Chem. Soc.*, vol. 86, no. 24, pp. 5679–5680, Dec. 1964, doi: 10.1021/JA01078A056.
[25] F. H. Field Yol and F. H. Field, "Chemical ionization mass spectrometry," *Acc. Chem. Res.*, vol. 1, no. 2, pp. 42–49, Jan. 1968, doi: 10.1021/AR50002A002.
[26] D. Barthomeuf, "Description of a characteristic infrared frequency for acidic hydroxy-groups in zeolites," *J. Chem. Soc. Chem. Commun.*, no. 21, pp. 743–744, 1977, doi: 10.1039/C39770000743.
[27] R. J. Argauer, "Crystalline zeolite zsm-5 and method of preparing the same," 3702886, Oct. 10, 1972.
[28] R. M. Barrer, J. W. Baynham, F. W. Bultitude, and W. M. Bleier, "Hydrothermal chemistry of silicates. Part XIV. Zeolite crystallisation in presence of mixed bases," *J. Chem. Soc. A*, no. 0, pp. 1470–1475, Jan. 1970, doi: 10.1039/J19700001470.
[29] D. H. Olson, G. T. Kokotailo, S. L. Lawton, and W. M. Meier, "Crystal structure and structure-related properties of ZSM-5," *J. Phys. Chem.*, vol. 85, no. 15, pp. 2238–2243, 1981, doi: 10.1021/j150615a020.
[30] D. Barthomeuf, "Zeolite acidity dependence on structure and chemical environment. Correlations with catalysis," *Mater. Chem. Phys.*, vol. 17, no. 1–2, pp. 49–71, Apr. 1987, doi: 10.1016/0254-0584(87)90048-4.
[31] E. M. Flanigen, "Zeolites and Molecular Sieves an Historical Perspective," in *Studies in Surface Science and Catalysis,* vol. 58, no. C, Elsevier, 1991, pp. 13–34.
[32] C. D. Chang and A. J. Silvestri, "The conversion of methanol and other o-compounds to hydrocarbons over zeolite catalysts," *J. Catal.*, vol. 259, no. 2, pp. 249–259, May 1977, doi: 10.1016/0021-9517(77)90172-5.

[33] J. Liang, H. Li, S. Zhao, W. Guo, R. Wang, and M. Ying, "Characteristics and performance of SAPO-34 catalyst for methanol-to-olefin conversion," *Appl. Catal.*, vol. 64, no. C, pp. 31–40, 1990, doi: 10.1016/S0166-9834(00)81551-1.

[34] S. A. Gembicki, "New solid acid based breakthrough technologies," *Stud. Surf. Sci. Catal.*, vol. 130, pp. 147–156, Jan. 2000, doi: 10.1016/S0167-2991(00)80951-9.

[35] W. O. Haag, R. M. Lago, and P. B. Weisz, "The active site of acidic aluminosilicate catalysts," *Nature*, vol. 309. pp. 589–591, 1984, doi: 10.1038/309589a0.

[36] P. B. Weisz, "Molecular shape-selective catalysis," *Pure Appl. Chem.*, vol. 52, no. 9, pp. 2091–2103, Jan. 1980, doi: 10.1351/PAC198052092091.

[37] S. M. Csicsery, "Shape-selective catalysis in zeolites," *Zeolites*, vol. 4, no. 3, pp. 202–213, Jul. 1984, doi: 10.1016/0144-2449(84)90024-1.

[38] H. Krannila, W. O. Haag, and B. C. Gates, "Monomolecular and bimolecular mechanisms of paraffin cracking: n-butane cracking catalyzed by HZSM-5," *J. Catal.*, vol. 135, no. 1, pp. 115–124, May 1992, doi: 10.1016/0021-9517(92)90273-K.

[39] J. A. Martens and P. A. Jacobs, "Introduction to acid catalysis with zeolites in hydrocarbon reactions," in *Studies in Surface Science and Catalysis*, vol. 137, H. van Bekkum, E. M. Flanigen, P. A. Jacobs, and J. C. Jansen, Eds. Elsevier, 2001, pp. 633–671.

[40] M. Rybicki and J. Sauer, "Ab initio prediction of proton exchange barriers for alkanes at brønsted sites of zeolite H-MFI," *J. Am. Chem. Soc.*, vol. 140, no. 51, pp. 18151–18161, Dec. 2018, doi: 10.1021/JACS.8B11228.

[41] R. A. van Santen and C. Liu, "Theory of Zeolite Catalysis: An Introductory Account," in *Modelling and Simulation in the Science of Micro- and Meso-Porous Materials*, C. R. A. Catlow, V. van Speybroeck, and R. A. van Santen, Eds. Elsevier, 2018, pp. 151–188.

[42] L. P. Hammett, *Physical Organic Chemistry*, vol. 29, no. 7. John Wiley & Sons, 1940.

[43] D. Hammett and T. Journal, "The acid strength of surfaces," *J. Am. Chem. Soc.*, vol. 72, no. 3, pp. 1164–1168, Mar. 2002, doi: 10.1021/JA01159A025.

[44] Z. Li, Y. Zhang, Y. Zhang, and Z. Li, "Monodisperse silica-coated polyvinylpyrrolidone/NaYF4 nanocrystals with multicolor upconversion fluorescence emission," *Angew. Chem.*, vol. 118, no. 46, pp. 7896–7899, Nov. 2006, doi: 10.1002/ANGE.200602975.

[45] W. Chen, X. Yi, L. Huang, W. Liu, G. Li, D. Acharya, X. Sun, and A. Zheng, "Can Hammett indicators accurately measure the acidity

of zeolite catalysts with confined space? Insights into the mechanism of coloration," *Catal. Sci. Technol.*, vol. 9, no. 18, pp. 5045–5057, 2019, doi: 10.1039/c9cy01392j.

[46] V. N. Solkan, I. V. Kuz'min, and V. B. Kazanskii, "Quantum-chemical study of alkyl carbenium ions in 100% sulfuric acid," *Kinet. Catal.*, vol. 42, no. 3, pp. 411–417, 2001, doi: 10.1023/A:1010477718602.

[47] S. Sitthisa, T. Sooknoi, Y. Ma, P. B. Balbuena, and D. E. Resasco, "Kinetics and mechanism of hydrogenation of furfural on Cu/SiO2 catalysts," *J. Catal.*, vol. 277, no. 1, pp. 1–13, Jan. 2011, doi: 10.1016/J.JCAT.2010.10.005.

[48] K. Arata, "Solid superacids," *Adv. Catal.*, vol. 37, no. C, pp. 165–211, Jan. 1990, doi: 10.1016/S0360-0564(08)60365-X.

[49] H. Matsuhashi, H. Motoi, and K. Arata, "Determination of acid strength of solid superacids by temperature programmed desorption using pyridine," *Catal. Lett. 1994 263*, vol. 26, no. 3, pp. 325–328, Sep. 1994, doi: 10.1007/BF00810605.

[50] M. J. Jorgenson and D. R. Hartter, "A critical re-evaluation of the Hammett acidity function at moderate and high acid concentrations of sulfuric acid. New H 0 values based solely on a set of primary aniline indicators," *J. Am. Chem. Soc.*, vol. 85, no. 7, pp. 878–883, Apr. 1963, doi: 10.1021/ja00890a009.

[51] G. A. Olah, G. K. Surya Prakash, J. Sommer, and A. Molnar, *Superacid Chemistry*. John Wiley & Sons, 2009.

[52] S. H. Chai, H. P. Wang, Y. Liang, and B. Q. Xu, "Sustainable production of acrolein: Gas-phase dehydration of glycerol over Nb2O5 catalyst," *J. Catal.*, vol. 250, no. 2, pp. 342–349, Sep. 2007, doi: 10.1016/J.JCAT.2007.06.016.

[53] M. Luisa Marin, G. L. Hallett-Tapley, S. Impellizzeri, C. Fasciani, S. Simoncelli, J. C. Netto-Ferreira, and J. C. Scaiano, "Synthesis, acid properties and catalysis by niobium oxide nanostructured materials," *Catal. Sci. Technol.*, vol. 4, no. 9, pp. 3044–3052, Aug. 2014, doi: 10.1039/C4CY00238E.

[54] T. Okuhara, T. Nishimura, and M. Misono, "Novel microporous solid 'Superacids': CsxH3–xPW12O40 (2≤x≤3)," *Stud. Surf. Sci. Catal.*, vol. 101, pp. 581–590, Jan. 1996, doi: 10.1016/S0167-2991(96)80269-2.

[55] B. B. Bardin, S. V. Bordawekar, M. Neurock, and R. J. Davis, "Acidity of Keggin-type heteropolycompounds evaluated by catalytic probe reactions, sorption microcalorimetry, and density functional quantum chemical calculations," *J. Phys. Chem. B*, vol. 102, no. 52, pp. 10817–10825, 1998, doi: 10.1021/JP982345Y.

[56] I. V Kozhevnikov, "Advances in catalysis by heteropolyacids," *Russ. Chem. Rev.*, vol. 56, no. 9, pp. 811–825, Sep. 1987, doi: 10.1070/RC1987V056N09ABEH003304.
[57] X. Song and A. Sayari, "Sulfated zirconia-based strong solid-acid catalysts: recent progress," *Catal. Rev.*, vol. 38, no. 3, pp. 329–412, 2006, doi: 10.1080/01614949608006462.
[58] M. Hino and K. Arata, "Synthesis of solid superacid catalyst with acid strength of H0⩽−16.04," *J. Chem. Soc. Chem. Commun.*, no. 18, pp. 851–852, Jan. 1980, doi: 10.1039/C39800000851.
[59] N. Mizuno and M. Misono, "Heteropolyacid catalysts," *Curr. Opin. Solid State Mater. Sci.*, vol. 2, no. 1, pp. 84–89, Feb. 1997, doi: 10.1016/S1359-0286(97)80109-X.
[60] F. R. Chen, G. Coudurier, J. F. Joly, and J. C. Vedrine, "Superacid and catalytic properties of sulfated zirconia," *J. Catal.*, vol. 143, no. 2, pp. 616–626, Oct. 1993, doi: 10.1006/JCAT.1993.1304.
[61] V. Adeeva, H. Y. Liu, B. Q. Xu, and W. M. H. Sachtler, "Alkane isomerization over sulfated zirconia and other solid acids," *Top. Catal.*, vol. 6, no. 1, pp. 61–76, 1998, doi: 10.1023/A:1019114406219.
[62] S. Katsue, K. Tadamitsu, K. Jun, S. Takashi, and T. Kozo, "Acidic properties of binary metal oxides," *Bull. Chem. Soc. Jpn.*, vol. 46, no. 10, pp. 2985–2988, Mar. 1973, doi: 10.1246/BCSJ.46.2985.
[63] K. I. Tanaka and A. Ozaki, "Acid-base properties and catalytic activity of solid surfaces," *J. Catal.*, vol. 8, no. 1, pp. 1–7, May 1967, doi: 10.1016/0021-9517(67)90274-6.
[64] Linus Pauling, *The Nature of the Chemical Bond: An Introduction to Modern Structural Chemistry*. Cornell University Press, 1960.
[65] A. L. Allred, "Electronegativity values from thermochemical data," *J. Inorg. Nucl. Chem.*, vol. 17, no. 3–4, pp. 215–221, Jun. 1961, doi: 10.1016/0022-1902(61)80142-5.
[66] M. Misono, E. Ochiai, Y. Saito, and Y. Yoneda, "A new dual parameter scale for the strength of lewis acids and bases with the evaluation of their softness," *J. Inorg. Nucl. Chem.*, vol. 29, no. 11, pp. 2685–2691, Nov. 1967, doi: 10.1016/0022-1902(67)80006-X.
[67] R. T. Sanderson, *Inorganic Chemistry*. Reinhold Publishing Corporation, 1967.
[68] L. Q. Wu, Y. C. Li, S. Q. Li, Z. Z. Li, G. D. Tang, W. H. Qi, L. C. Xue, X. S. Ge, and L. L. Ding, "Method for estimating ionicities of oxides using O1s photoelectron spectra," *AIP Adv.*, vol. 5, no. 9, p. 097210, Sep. 2015, doi: 10.1063/1.4931996.
[69] J. M. Jehng, I. E. Wachs, G. S. Patience, and Y. M. Dai, "Experimental methods in chemical engineering: Temperature programmed

surface reaction spectroscopy — TPSR," *Can. J. Chem. Eng.*, vol. 99, no. 2, pp. 423–434, Feb. 2021, doi: 10.1002/CJCE.23913.

[70] V. V. Brei, K. N. Khomenko, and D. V. Shystka, "A correlation between Hammett acidity of catalysts and their activity in temperature-programmed reaction of cumene dealkylation," *React. Kinet. Catal. Lett.*, vol. 81, no. 1, pp. 197–202, 2004, doi: 10.1023/B:REAC.0000016535.01531.95.

[71] J. B. Peri, "Infrared and gravimetric study of the surface hydration of γ-alumina," *J. Phys. Chem.*, vol. 69, no. 1, pp. 211–219, 1965, doi: 10.1021/J100885A032.

[72] J. A. Lercher, C. Gründling, and G. Eder-Mirth, "Infrared studies of the surface acidity of oxides and zeolites using adsorbed probe molecules," *Catal. Today*, vol. 27, no. 3–4, pp. 353–376, Feb. 1996, doi: 10.1016/0920-5861(95)00248-0.

[73] M. Hunger, "Multinuclear solid-state NMR studies of acidic and non-acidic hydroxyl protons in zeolites," *Solid State Nucl. Magn. Reson.*, vol. 6, no. 1, pp. 1–29, 1996, doi: 10.1016/0926-2040(95)01201-X.

[74] E. M. Flanigen, J. C. Jansen, and H. van. Bekkum, *Introduction to Zeolite Science and Practice.* Elsevier, 1991.

[75] R. A. van Santen and G. J. Kramer, "Reactivity theory of zeolitic broensted acidic sites," *Chem. Rev.*, vol. 95, no. 3, pp. 637–660, 1995, doi: 10.1021/cr00035a008.

[76] A. Zecchina, R. Buzzoni, S. Bordiga, F. Geobaldo, D. Scarano, G. Ricchiardi, and G. Spoto, "Host-guest interactions in zeolite cavities," *Stud. Surf. Sci. Catal.*, vol. 97, pp. 213–222, 1995, doi: 10.1016/S0167-2991(06)81892-6.

[77] A. Zecchina and C. Otero Areán, "Diatomic molecular probes for mid-IR studies of zeolites," *Chem. Soc. Rev.*, vol. 25, no. 3, pp. 187–197, 1996, doi: 10.1039/CS9962500187.

[78] S. Bordiga, C. Lamberti, F. Bonino, A. Travert, and F. Thibault-Starzyk, "Probing zeolites by vibrational spectroscopies," *Chem. Soc. Rev.*, vol. 44, no. 20, pp. 7262–7341, 2015, doi: 10.1039/c5cs00396b.

[79] E. A. Paukshtis and E. N. Yurchenko, "Study of the acid–base properties of heterogeneous catalysts by infrared spectroscopy," *Russ. Chem. Rev.*, vol. 52, no. 3, pp. 242–258, Mar. 1983, doi: 10.1070/rc1983v052n03abeh002812.

[80] P. G. Rouxhet and R. E. Sempels, "Hydrogen bond strengths and acidities of hydroxyl groups on silica–alumina surfaces and in molecules in solution," *J. Chem. Soc. Faraday Trans. 1*, vol. 70, no. 0, pp. 2021–2032, Jan. 1974, doi: 10.1039/F19747002021.

[81] L. Kubelková, S. Beran, and J. A. Lercher, "Determination of proton affinity of zeolites and zeolite-like solids by low-temperature adsorption of carbon monoxide," *Zeolites*, vol. 9, no. 6, pp. 539–543, Nov. 1989, doi: 10.1016/0144-2449(89)90052-3.

[82] D. J. Lucas, L. A. Curtiss, and J. A. Pople, "Theoretical study of the silicon–oxygen hydrides SiOHn (n=0–4) and SiOH+n (n=0–5): Dissociation energies, ionization energies, enthalpies of formation, and proton affinities," *J. Chem. Phys.*, vol. 99, no. 9, p. 6697, Aug. 1998, doi: 10.1063/1.465837.

[83] J. B. Peri and R. B. Hannan, "Surface hydroxyl groups on γ-alumina," *J. Phys. Chem.*, vol. 64, no. 10, pp. 1526–1530, 1960, doi: 10.1021/J100839A044.

[84] A. A. Tsyganenko and V. N. Filimonov, "Infrared spectra of surface hydroxyl groups and crystalline structure of oxides," *J. Mol. Struct.*, vol. 19, no. 2, pp. 579–589, Dec. 1973, doi: 10.1016/0022-2860(73)85136-1.

[85] M. Digne, P. Sautet, P. Raybaud, P. Euzen, and H. Toulhoat, "Use of DFT to achieve a rational understanding of acid–basic properties of γ-alumina surfaces," *J. Catal.*, vol. 226, no. 1, pp. 54–68, Aug. 2004, doi: 10.1016/J.JCAT.2004.04.020.

[86] T. Kataoka and J. A. Dumesic, "Acidity of unsupported and silica-supported vanadia, molybdena, and titania as studied by pyridine adsorption," *J. Catal.*, vol. 112, no. 1, pp. 66–79, Jul. 1988, doi: 10.1016/0021-9517(88)90121-2.

[87] L. Pauling, *Nature of the Chemical Bond*. Cornell University Press, 1939.

[88] D. S. Coombs *et al.*, "Recommended nomenclature for zeolite minerals: report of the subcommittee on zeolites of the International Mineralogical Association, Commission on new Minerals and Mineral names," *Can. Mineral.*, vol. 35, no. 6, pp. 1571–1606, 1997.

[89] F. Liebau, *Structural Chemistry of Silicates*. Springer, 1985.

[90] D. W. Breck, *Zeolite Molecular Sieves: Structure, Chemistry, and Use*. John Wiley & Sons, 1973.

[91] B. M. Lok, C. A. Messina, R. L. Patton, R. T. Gajek, T. R. Cannan, and E. M. Flanigen, "Silicoaluminophosphate molecular sieves: another new class of microporous crystalline inorganic solids," *J. Am. Chem. Soc.*, vol. 106, no. 20, pp. 6092–6093, 1984, doi: 10.1021/ja00332a063.

[92] J. D. Sherman, "Synthetic zeolites and other microporous oxide molecular sieves," *Proc. Natl. Acad. Sci.*, vol. 96, no. 7, pp. 3471–3478, Mar. 1999, doi: 10.1073/pnas.96.7.3471.

[93] C. Baerlocher, L. B. McCusker, and D. H. Olson, *Atlas of Zeolite Framework Types*. Elsevier, 2007.

[94] M. M. J. Treacy, I. Rivin, E. Balkovsky, K. H. Randall, and M. D. Foster, "Enumeration of periodic tetrahedral frameworks. II. Polynodal graphs," *Microporous Mesoporous Mater.*, vol. 74, no. 1–3, pp. 121–132, 2004, doi: 10.1016/j.micromeso.2004.06.013.

[95] C. Liu, G. Li, E. J. M. Hensen, and E. A. Pidko, "Nature and catalytic role of extraframework aluminum in faujasite zeolite: a theoretical perspective," *ACS Catal.*, vol. 5, no. 11, pp. 7024–7033, Nov. 2015, doi: 10.1021/acscatal.5b02268.

[96] W. Lowenstein, "The distribution of aluminum in the tetra-hedra of silicates and aluminates," *Am. Mineral.*, vol. 39, no. 1–2, pp. 92–96, 1954.

[97] J. Weitkamp, "Isomerization of long-chain n-alkanes on a Pt/CaY zeolite catalyst," *Ind. Eng. Chem. Prod. Res. Dev.*, vol. 21, no. 4, pp. 550–558, Dec. 1982, doi: 10.1021/i300008a008.

[98] J. Weitkamp, P. A. Jacobs, and J. A. Martens, "Isomerization and hydrocracking of C9 through C16 n-alkanes on Pt/HZSM-5 zeolite," *Appl. Catal.*, vol. 8, no. 1, pp. 123–141, Oct. 1983, doi: 10.1016/0166-9834(83)80058-X.

[99] J. P. Lourenco, M. F. Ribeiro, F. R. Ribeiro, J. Rocha, Z. Gabelica, N. Dumont, and E. G. Derouane, "Study of catalytic properties of SAPO-40," *Stud. Surf. Sci. Catal.*, vol. 84, no. C, pp. 867–874, 1994, doi: 10.1016/S0167-2991(08)64092-6.

[100] U. Olsbye, S. Svelle, K. P. Lillerud, Z. H. Wei, Y. Y. Chen, J. F. Li, J. G. Wang, and W. B. Fan, "The formation and degradation of active species during methanol conversion over protonated zeotype catalysts," *Chem. Soc. Rev.*, vol. 44, no. 20, pp. 7155–7176, 2015, doi: 10.1039/c5cs00304k.

[101] P. Tian, Y. Wei, M. Ye, and Z. Liu, "Methanol to olefins (MTO): from fundamentals to commercialization," *ACS Catal.*, vol. 5, no. 3, pp. 1922–1938, Mar. 2015, doi: 10.1021/ACSCATAL.5B00007.

[102] J. A. Rabo, R. D. Bezmam, and L. M. Poutsma, "Unifying principles of zeolite catalysis," *Acta Phys Chem. Szeged*, vol. 24, p. 39, 1978.

[103] C. Liu, I. Tranca, R. A. van Santen, E. J. M. Hensen, and E. A. Pidko, "Scaling relations for acidity and reactivity of zeolites," *J. Phys. Chem. C*, vol. 121, no. 42, pp. 23520–23530, Oct. 2017, doi: 10.1021/ACS.JPCC.7B08176.

[104] M. Sierka, U. Eichler, J. Datka, and J. Sauer, "Heterogeneity of Brønsted acidic sites in faujasite type zeolites due to aluminum content and framework structure," *J. Phys. Chem. B*, vol. 102, no. 33, pp. 6397–6404, 1998, doi: 10.1021/jp983843+.

[105] R. A. van Santen, "Solid acid catalysis; part I, the zeolite protonic site," *Comprehensive Inorganic Chemistry III,* 3rd ed, J. Reedijk and K. R. Poeppelmeier, Eds. Elsevier, 2021, pp. 271–292.

[106] U. Eichler, M. Brändle, and J. Sauer, "Predicting absolute and site specific acidities for zeolite catalysts by a combined quantum mechanics/interatomic potential function approach," *J. Phys. Chem. B*, vol. 101, no. 48, pp. 10035–10050, Nov. 1997, doi: 10.1021/jp971779a.

[107] M. Brändle and J. Sauer, "Acidity differences between inorganic solids induced by their framework structure. A combined quantum mechanics/molecular mechanics ab initio study on zeolites," *J. Am. Chem. Soc.*, vol. 120, no. 7, pp. 1556–1570, 1998, doi: 10.1021/ja9729037.

[108] P. J. O'Malley and J. Dwyer, "An ab initio quantum chemical investigation on the effect of the magnitude of the T-O-T angle on the Brønsted acid characteristics of zeolites," *J. Phys. Chem.*, vol. 92, no. 10, pp. 3005–3007, 1988, doi: 10.1021/j100321a059.

[109] K. P. Schröder, J. Sauer, M. Leslie, C. Richard, A. Catlow, and J. M. Thomas, "Bridging hydrodyl groups in zeolitic catalysts: a computer simulation of their structure, vibrational properties and acidity in protonated faujasites (HY zeolites)," *Chem. Phys. Lett.*, vol. 188, no. 3–4, pp. 320–325, 1992, doi: 10.1016/0009-2614(92)90030-Q.

[110] A. J. Jones and E. Iglesia, "The strength of brønsted acid sites in microporous aluminosilicates," *ACS Catal.*, vol. 5, no. 10, pp. 5741–5755, Oct. 2015, doi: 10.1021/acscatal.5b01133.

[111] L. Grajciar, C. O. Areán, A. Pulido, and P. Nachtigall, "Periodic DFT investigation of the effect of aluminium content on the properties of the acid zeolite H-FER," *Phys. Chem. Chem. Phys.*, vol. 12, no. 7, pp. 1497–1506, 2010, doi: 10.1039/b917969k.

[112] D. Barthomeuf, "Description of a characteristic infrared frequency for acidic hydroxy-groups in zeolites," *J. Chem. Soc. Chem. Commun.*, no. 21, pp. 743–744, Jan. 1977, doi: 10.1039/C39770000743.

[113] C. T. W. Chu and C. D. Chang, "Isomorphous substitution in zeolite frameworks. 1. Acidity of surface hydroxyls in [B]-, [Fe]-, [Ga]-, and [Al]-ZSM-5," *J. Phys. Chem.*, vol. 89, no. 9, pp. 1569–1571, 1985, doi: 10.1021/j100255a005.

[114] M. F. M. Post, T. Huizinga, C. A. Emeis, J. M. Nanne, and W. H. J. Stork, "An infrared and catalytic study of isomorphous substitution in pentasil zeolites," *Stud. Surf. Sci. Catal.*, vol. 46, no. C, pp. 365–375, Jan. 1989, doi: 10.1016/S0167-2991(08)60993-3.

[115] A. J. Jones, R. T. Carr, S. I. Zones, and E. Iglesia, "Acid strength and solvation in catalysis by MFI zeolites and effects of the identity,

concentration and location of framework heteroatoms," *J. Catal.*, vol. 312, pp. 58–68, 2014, doi: 10.1016/j.jcat.2014.01.007.

[116] S. P. Yuan, J. G. Wang, Y. W. Li, and H. Jiao, "Brønsted acidity of isomorphously substituted ZSM-5 by B, Al, Ga, and Fe. Density functional investigations," *J. Phys. Chem. A*, vol. 106, no. 35, pp. 8167–8172, Sep. 2002, doi: 10.1021/jp025792t.

[117] Z. Li, M. C. Johnson, M. Sun, E. T. Ryan, D. J. Earl, W. Maichen, J. I. Martin, S. Li, C. M. Lew, J. Wang, M. W. Deem, M. E. Davis, and Y. Yan, "Mechanical and dielectric properties of pure-silica-zeolite low-k materials," *Angew. Chem.*, vol. 118, no. 38, pp. 6477–6480, Sep. 2006, doi: 10.1002/ange.200602036.

[118] J. F. Haw and T. Xu, "NMR studies of solid acidity," *Adv. Catal.*, vol. 42, no. C, pp. 115–180, Jan. 1998, doi: 10.1016/S0360-0564(08)60628-8.

[119] E. G. Derouane, J. M. Andre, and A. A. Lucas, "Surface curvature effects in physisorption and catalysis by microporous solids and molecular sieves," *J. Catal.*, vol. 110, no. 1, pp. 58–73, Mar. 1988, doi: 10.1016/0021-9517(88)90297-7.

[120] W. O. Haag, R. M. Lago, and P. B. Weisz, "Transport and reactivity of hydrocarbon molecules in a shape-selective zeolite," *Faraday Discuss. Chem. Soc.*, vol. 72, no. 0, pp. 317–330, Jan. 1981, doi: 10.1039/DC9817200317.

[121] J. Kärcher, *Sitzungsberichte der Sächsischen Akademie der Wissenschaften zu Leipzig, Mathematisch-Naturwissenchaftliche Klasse*, vol. Band 128 Heft 6. 2003.

[122] F. Eder and J. A. Lercher, "On the role of the pore size and tortuosity for sorption of alkanes in molecular sieves," *J. Phys. Chem. B*, vol. 101, no. 8, pp. 1273–1278, Feb. 1997, doi: 10.1021/JP961816I.

[123] Y. H. Yeh, R. J. Gorte, S. Rangarajan, and M. Mavrikakis, "Adsorption of small alkanes on ZSM-5 zeolites: influence of Brønsted sites," *J. Phys. Chem. C*, vol. 120, no. 22, pp. 12132–12138, Jun. 2016, doi: 10.1021/ACS.JPCC.6B03855.

[124] C. Sievers, A. Onda, R. Olindo, and J. A. Lercher, "Adsorption and polarization of branched alkanes on H-LaX," *J. Phys. Chem. C*, vol. 111, no. 14, pp. 5454–5464, 2007, doi: 10.1021/jp067312u.

[125] L. Chen, L. Lin, Z. Xu, T. Zhang, Q. Xin, P. Ying, G. Li, and C. Li, "Fourier transform-infrared investigation of adsorption of methane and carbon monoxide on HZSM-5 and Mo/HZSM-5 zeolites at low temperature," *J. Catal.*, vol. 161, no. 1, pp. 107–114, 1996, doi: 10.1006/jcat.1996.0167.

[126] A. Y. Khodakov, L. M. Kustov, V. B. Kazansky, and C. Williams, "Infrared spectroscopic study of the interaction of cations in zeolites with simple molecular probes. Part 3. — Adsorption and polarization of methane and ethane on cationic forms of high-silica zeolites," *J. Chem. Soc. Faraday Trans.*, vol. 89, no. 9, pp. 1393–1395, Jan. 1993, doi: 10.1039/FT9938901393.

[127] A. Auroux, P. Artizzu, I. Ferino, R. Monaci, E. Rombi, and V. Solinas, "Conversion of 4-methylpentan-2-ol over alkali–metal ion-exchanged X and Y zeolites: a microcalorimetric and catalytic investigation," *Microporous Mater.*, vol. 11, no. 3–4, pp. 117–126, Sep. 1997, doi: 10.1016/S0927-6513(97)00030-8.

[128] W. O. Haag, "Catalysis by zeolites – science and technology," *Stud. Surf. Sci. Catal.*, vol. 84, no. C, pp. 1375–1394, Jan. 1994, doi: 10.1016/S0167-2991(08)63680-0.

[129] T. F. T. F. Narbeshuber, H. Vinek, and J. A. J. A. Lercher, "Monomolecular conversion of light alkanes over H-ZSM-5," *J. Catal.*, vol. 157, no. 2, pp. 388–395, Dec. 1995, doi: 10.1006/jcat.1995.1304.

[130] P. B. Weisz, "Stepwise reaction via intermediates on separate catalytic centers [3]," *Science*, vol. 123, no. 3203, pp. 887–888, May 1956, doi: 10.1126/SCIENCE.123.3203.887.

[131] J. C. Platteeuw, H. De Ruiter, D. Van Zoonen, and H. W. Kouwenhoven, "Hydroisomerization of olefins," *Ind. Eng. Chem. Prod. Res. Dev.*, vol. 6, no. 1, pp. 76–80, Mar. 1967, doi: 10.1021/I360021A014.

[132] H. L. Coonradt and W. E. Garwood, "Mechanism of hydrocracking. Reactions of paraffins and olefins," *Ind. Eng. Chem. Process Des. Dev.*, vol. 3, no. 1, pp. 38–45, Jan. 1964, doi: 10.1021/I260009A010.

[133] G. F. Asselin, H. S. Bloch, G. R. Donaldson, V. Haensel, and E. L. Pollitzer, "Isomerization of paraffins," *Preprints*, vol. 17, no. 3, 1972, doi: 10.1021/BA-1973-0121.CH048.

[134] P. B. Weisz and E. W. Swegler, "Stepwise reaction on separate catalytic centers: Isomerization of saturated hydrocarbons," *Science*, vol. 126, no. 3262, pp. 31–32, Jul. 1957, doi: 10.1126/SCIENCE.126.3262.31.

[135] W. Souverijns, R. Parton, J. A. Martens, G. F. Froment, and P. A. Jacobs, "Mechanism of the paring reaction of naphtenes," *Catal. Lett.*, vol. 37, no. 3, pp. 207–212, 1996, doi: 10.1007/BF00807755.

[136] D. M. Brouwer and H. Hogeveen, "Electrophilic Substitutions at Alkanes and in Alkylcarbonium Ions," in *Progress in Physical Organic Chemistry*, vol. 9, A. Streitwieser Jr. and R. W. Taft, Eds. John Wiley & Sons, 2007, pp. 179–240.

[137] T. Demuth, X. Rozanska, L. Benco, J. Hafner, R. A. van Santen, and H. Toulhoat, "Catalytic isomerization of 2-pentene in H-ZSM-22 — A DFT investigation," *J. Catal.*, vol. 214, no. 1, pp. 68–77, Feb. 2003, doi: 10.1016/S0021-9517(02)00074-X.
[138] J. L. Franklin, "Calculation of the heats of formation of gaseous free radicals and ions," *J. Chem. Phys.*, vol. 21, no. 11, p. 2029, 1953, doi: 10.1063/1.1698737.
[139] J. L. Franklin and F. H. Field, "The energies of strained carbonium ions," *J. Chem. Phys.*, vol. 21, no. 3, p. 550, 1953, doi: 10.1063/1.1698944.
[140] M. N. Mazar, S. Al-Hashimi, M. Cococcioni, and A. Bhan, "β-scission of olefins on acidic zeolites: A periodic PBE-D study in H-ZSM-5," *J. Phys. Chem. C*, vol. 117, no. 45, pp. 23609–23620, Nov. 2013, doi: 10.1021/JP403504N.
[141] J. A. Rabo, P. E. Pickert, and R. L. Mays, "Pentane and hexane isomerization," *Ind. Eng. Chem.*, vol. 53, no. 9, pp. 733–736, Sep. 1961, doi: 10.1021/IE50621A029.
[142] J. Spivey, "Hydroisomerization of Pentane and Hexane Mixtures on Zeolite Catalysts," *LSU Historical Dissertations and Theses 3537*, Jan. 01, 1980. https://digitalcommons.lsu.edu/gradschool_disstheses/3537 (accessed Jan. 16, 2022).
[143] J. J. Spivey and P. A. Bryant, "Hydroisomerization of n-C5 and n-C6 mixtures on zeolite catalysts," *Ind. Eng. Chem. Process Des. Dev.*, vol. 21, pp. 750–760, 1982, doi: 10.1021/i200019a034.
[144] A. Chica, A. Corma, and P. J. Miguel, "Isomerization of C5–C7 n-alkanes on unidirectional large pore zeolites: activity, selectivity and adsorption features," *Catal. Today*, vol. 65, no. 2–4, pp. 101–110, Feb. 2001, doi: 10.1016/S0920-5861(00)00578-2.
[145] J. Weitkamp, "Catalytic hydrocracking-mechanisms and versatility of the process," *ChemCatChem*, vol. 4, no. 3, pp. 292–306, 2012, doi: 10.1002/cctc.201100315.
[146] C. J. Egan, G. E. Langlois, R. J. White, and J. Egan, "Selective hydrocracking of C9 - to C12 -alkylcyclohexanes on acidic catalysts. Evidence for the paring reaction," *J. Am. Chem. Soc.*, vol. 84, pp. 1204–1212, 1962, doi: 10.1021/ja00866a028.
[147] P. G. Menon and Z. Paá, "Some aspects of the mechanisms of catalytic reforming reactions," *Ind. Eng. Chem. Res.*, vol. 36, no. 8, pp. 3282–3291, 1997, doi: 10.1021/IE960606P.
[148] A. Van De Runstraat, J. Van Grondelle, and R. A. van Santen, "On the temperature dependence of the arrhenius activation energy for hydroisomerization catalyzed by Pt/mordenite," *J. Catal.*, vol. 167, no. 2, pp. 460–463, Apr. 1997, doi: 10.1006/JCAT.1997.1603.

[149] M. Steijns and G. F. Froment, "Hydroisomerization and hydrocracking. 3. Kinetic analysis of rate data for n-decane and n-dodecane," *Ind. Eng. Chem. Prod. Res. Dev.*, vol. 20, no. 4, pp. 660–668, Dec. 2002, doi: 10.1021/I300004A014.

[150] G. F. Froment, "Kinetics of the hydroisomerization and hydrocracking of paraffins on a platinum containing bifunctional Y-zeolite," *Catal. Today*, vol. 1, no. 4, pp. 455–473, Aug. 1987, doi: 10.1016/0920-5861(87)80009-3.

[151] D. S. Santilli, T. V Harris, and S. I. Zones, "Inverse shape selectivity in molecular sieves: Observations, modelling, and predictions," *Microporous Mater.*, vol. 1, no. 5, pp. 329–341, Sep. 1993, doi: 10.1016/0927-6513(93)80049-Z.

[152] B. Smit and T. L. M. Maesen, "Molecular simulations of zeolites: adsorption, diffusion, and shape selectivity," *Chem. Rev.*, vol. 108, no. 10, pp. 4125–4184, Oct. 2008, doi: 10.1021/cr8002642.

[153] M. Schenk, S. Calero, T. L. M. Maesen, L. L. Van Benthem, M. G. Verbeek, and B. Smit, "Understanding zeolite catalysis: Inverse shape selectivity revised," *Angew. Chem. Int. Ed.*, vol. 41, pp. 2499–2502, 2002, doi: 10.1002/1521-3773(20020715)41:14<2499::AID-ANIE2499>3.0.CO;2-T.

[154] R. Gounder and E. Iglesia, "Catalytic consequences of spatial constraints and acid site location for monomolecular alkane activation on zeolites," *J. Am. Chem. Soc.*, vol. 131, no. 5, pp. 1958–1971, Feb. 2009, doi: 10.1021/JA808292C.

[155] A. Bhan, R. Gounder, J. Macht, and E. Iglesia, "Entropy considerations in monomolecular cracking of alkanes on acidic zeolites," *J. Catal.*, vol. 253, no. 1, pp. 221–224, Jan. 2008, doi: 10.1016/j.jcat.2007.11.003.

[156] R. Gounder and E. Iglesia, "The roles of entropy and enthalpy in stabilizing ion-pairs at transition states in zeolite acid catalysis," *Acc. Chem. Res.*, vol. 45, no. 2, pp. 229–238, Feb. 2012, doi: 10.1021/ar200138n.

[157] R. A. van Santen and F. J. M. M. De Gauw, "Acidity in zeolite catalysis," *Stud. Surf. Sci. Catal.*, vol. 130 A, no. 2, pp. 127–135, 2000, doi: 10.1016/s0167-2991(00)80949-0.

[158] R. A. van Santen, "Solid acid catalysis. Part II, Catalytic chemistry of proton activation," in *Comprehensive Inorganic Chemistry III*, 3rd ed, J. Reedijk and K. R. Poeppelmeier, Eds. Elsevier, 2021, pp. 293–310.

[159] E. Fischer, "Ueber die optischen Isomeren des Traubenzuckers, der Gluconsäure und der Zuckersäure," *Berichte der Dtsch. Chem. Gesellschaft*, vol. 23, no. 2, pp. 2611–2624, Jul. 1890, doi: 10.1002/CBER.189002302157.

[160] E. Fischer, "Einfluss der Configuration auf die Wirkung der Enzyme," *Berichte der Dtsch. Chem. Gesellschaft*, vol. 27, no. 3, pp. 2985–2993, Oct. 1894, doi: 10.1002/CBER.18940270364.

[161] D. E. Koshland, Jr., "Application of a theory of enzyme specificity to protein synthesis," *Proc. Natl. Acad. Sci.*, vol. 44, no. 2, p. 98, Feb. 1958, doi: 10.1073/PNAS.44.2.98.

[162] D. E. Koshland, Jr., "The key–lock theory and the induced fit theory," *Angew. Chem. Int. Ed.*, vol. 33, no. 23–24, pp. 2375–2378, Jan. 1995, doi: 10.1002/ANIE.199423751.

[163] L. Stryer, R. Tasker, and C. Rhodes, *Biochemistry*, 4th ed. W.H. Freeman and Company, 1995.

[164] P. B. Weisz and V. J. Frilette, "Intracrystalline and molecular-shape-selective catalysis by zeolite salts," *J. Phys. Chem.*, vol. 64, no. 3, p. 382, 1960, doi: 10.1021/J100832A513.

[165] V. J. Frilette, P. B. Weisz, and R. L. Golden, "Catalysis by crystalline aluminosilicates I. Cracking of hydrocarbon types over sodium and calcium 'X' zeolites," *J. Catal.*, vol. 1, no. 4, pp. 301–306, Aug. 1962, doi: 10.1016/0021-9517(62)90057-X.

[166] C. D. Chang, "The New Zealand gas-to-gasoline plant: An engineering tour de force," *Catal. Today*, vol. 13, no. 1, pp. 103–111, Mar. 1992, doi: 10.1016/0920-5861(92)80190-X.

[167] U. Olsbye *et al.*, "Conversion of methanol to hydrocarbons: How zeolite cavity and pore size controls product selectivity," *Angew. Chem. Int. Ed.*, vol. 51, no. 24, pp. 5810–5831, Jun. 2012, doi: 10.1002/anie.201103657.

[168] J. W. Otvos, D. P. Stevenson, C. D. Wagner, and O. Beeck, "The behavior of isobutane in concentrated sulfuric acid 1," *J. Am. Chem. Soc.*, vol. 73, pp. 5741–5746, 1951, doi: 10.1021/ja01156a075.

[169] X. Rozanska, R. A. van Santen, T. Demuth, F. Hutschka, and J. Hafner, "A periodic DFT Study of isobutene chemisorption in proton-exchanged zeolites: dependence of reactivity on the zeolite framework structure," *J. Phys. Chem. B*, vol. 107, no. 6, pp. 1309–1315, Feb. 2003, doi: 10.1021/jp021646b.

[170] X. Rozanska and R. A. van Santen, "Reaction mechanisms in protonic zeolites," *Computer Modelling of Microporous Materials*, C.R.A. Catlow, R.A. van Santen and B. Smit, Eds. Elsevier, 2004, pp. 165–200.

[171] C. Tuma and J. Sauer, "Protonated isobutene in zeolites: tert-butyl cation or alkoxide?," *Angew. Chem. Int. Ed.*, vol. 44, no. 30, pp. 4769–4771, Jul. 2005, doi: 10.1002/anie.200501002.

[172] W. W. Kaeding, C. Chu, L. B. Young, B. Weinstein, and S. A. Butter, "Selective alkylation of toluene with methanol to produce

para-Xylene," *J. Catal.*, vol. 67, no. 1, pp. 159–174, Jan. 1981, doi: 10.1016/0021-9517(81)90269-4.

[173] T. C. Tsai, S. Bin Liu, and I. Wang, "Disproportionation and transalkylation of alkylbenzenes over zeolite catalysts," *Appl. Catal. A Gen.*, vol. 181, no. 2, pp. 355–398, May 1999, doi: 10.1016/S0926-860X(98)00396-2.

[174] J. A. Martens, J. Perez-Pariente, E. Sastre, A. Corma, and P. A. Jacobs, "Isomerization and disproportionation of m-xylene: selectivities induced by the void structure of the zeolite framework," *Appl. Catal.*, vol. 45, no. 1, pp. 85–101, Jan. 1988, doi: 10.1016/S0166-9834(00)82395-7.

[175] A. M. Vos, X. Rozanska, R. A. Schoonheydt, R. A. van Santen, F. Hutschka, and J. Hafner, "A theoretical study of the alkylation reaction of toluene with methanol catalyzed by acidic mordenite," *J. Am. Chem. Soc.*, vol. 123, no. 12, pp. 2799–2809, 2001, doi: 10.1021/ja001981i.

[176] L. Hammett, *Physical Organic Chemistry: Reaction Rates, Equilibria and Mechanisms.* McGraw Hill, 1940.

[177] P. B. Weisz, W. O. Haag, and P. G. Rodewald, "Catalytic production of high-grade fuel (gasoline) from biomass compounds by shape-selective catalysis," *Science*, vol. 206, no. 4414, pp. 57–58, Oct. 1979, doi: 10.1126/SCIENCE.206.4414.57.

[178] F. J. Keil, "Methanol-to-hydrocarbons: process technology," *Microporous Mesoporous Mater.*, vol. 29, no. 1–2, pp. 49–66, Jun. 1999, doi: 10.1016/S1387-1811(98)00320-5.

[179] M. Stöcker, "Methanol-to-hydrocarbons: catalytic materials and their behavior," *Microporous Mesoporous Mater.*, vol. 29, no. 1–2, pp. 3–48, 1999, doi: 10.1016/S1387-1811(98)00319-9.

[180] P. Tian, Y. Wei, M. Ye, and Z. Liu, "Methanol to olefins (MTO): From fundamentals to commercialization," *ACS Catal.*, vol. 5, no. 3, pp. 1922–1938, Mar. 2015, doi: 10.1021/ACSCATAL.5B00007.

[181] J. F. Haw, W. Song, D. M. Marcus, and J. B. Nicholas, "The mechanism of methanol to hydrocarbon catalysis," *Acc. Chem. Res.*, vol. 36, no. 5, pp. 317–326, May 2003, doi: 10.1021/ar020006o.

[182] J. Li, Y. Wei, G. Liu, Y. Qi, P. Tian, B. Li, Y. He, and Z. Liu, "Comparative study of MTO conversion over SAPO-34, H-ZSM-5 and H-ZSM-22: Correlating catalytic performance and reaction mechanism to zeolite topology," *Catal. Today*, vol. 171, no. 1, pp. 221–228, Aug. 2011, doi: 10.1016/J.CATTOD.2011.02.027.

[183] T. Li, T. Shoinkhorova, J. Gascon, and J. Ruiz-Martinez, "Aromatics production via methanol-mediated transformation routes," *ACS*

Catal., vol. 11, no. 13, pp. 7780–7819, Jul. 2021, doi: 10.1021/ACSCATAL.1C01422.
[184] W. Zhang, J. Chen, S. Xu, Y. Chu, Y. Wei, Y. Zhi, J. Huang, A. Zheng, X. Wu, X. Meng, F. Xiao, F. Deng, and Z. Liu, "Methanol to olefins reaction over cavity-type zeolite: cavity controls the critical intermediates and product selectivity," *ACS Catal.*, vol. 8, no. 12, pp. 10950–10963, Dec. 2018, doi: 10.1021/ACSCATAL.8B02164.
[185] D. M. McCann, D. Lesthaeghe, P. W. Kletnieks, D. R. Guenther, M. J. Hayman, V. Van Speybroeck, M. Waroquier, and J. F. Haw, "A complete catalytic cycle for supramolecular methanol-to-olefins conversion by linking theory with experiment," *Angew. Chem.*, vol. 120, no. 28, pp. 5257–5260, Jun. 2008, doi: 10.1002/ange.200705453.
[186] D. M. McCann, D. Lesthaeghe, P. W. Kletnieks, D. R. Guenther, M. J. Hayman, V. Van Speybroeck, M. Waroquier, and J. F. Haw, "A complete catalytic cycle for supramolecular methanol-to-olefins conversion by linking theory with experiment," *Angew. Chem. Int. Ed.*, vol. 47, no. 28, pp. 5179–5182, Jun. 2008, doi: 10.1002/ANIE.200705453.
[187] I. M. Dahl and S. Kolboe, "On the reaction mechanism for hydrocarbon formation from methanol over SAPO-34: 2. Isotopic labeling studies of the co-reaction of propene and methanol," *J. Catal.*, vol. 161, no. 1, pp. 304–309, Jun. 1996, doi: 10.1006/JCAT.1996.0188.
[188] I. M. Dahl and S. Kolboe, "On the reaction mechanism for hydrocarbon formation from methanol over SAPO-34: I. Isotopic labeling studies of the co-reaction of ethene and methanol," *J. Catal.*, vol. 149, no. 2, pp. 458–464, Oct. 1994, doi: 10.1006/JCAT.1994.1312.
[189] B. Arstad and S. Kolboe, "The reactivity of molecules trapped within the SAPO-34 cavities in the methanol-to-hydrocarbons reaction," *J. Am. Chem. Soc.*, vol. 123, no. 33, pp. 8137–8138, 2001, doi: 10.1021/JA010668T.
[190] J. F. Haw, J. B. Nicholas, W. Song, F. Deng, Z. Wang, T. Xu, and C. S. Heneghan, "Roles for cyclopentenyl cations in the synthesis of hydrocarbons from methanol on zeolite catalyst HZSM-5," *J. Am. Chem. Soc.*, vol. 122, no. 19, pp. 4763–4775, May 2000, doi: 10.1021/JA994103X.
[191] N. Y. Chen and W. J. Reagan, "Evidence of autocatalysis in methanol to hydrocarbon reactions over zeolite catalysts," *J. Catal.*, vol. 59, no. 1, pp. 123–129, Aug. 1979, doi: 10.1016/S0021-9517(79)80050-0.
[192] G. A. Olah, H. Doggweiler, J. D. Felberg, S. Frohlich, M. J. Grdina, R. Karpeles, T. Keumi, S. Inaba, W. M. Ip, K. Lammertsma, G.

Salem, and D. Tabor, "Onium ylide chemistry. 1. Bifunctional acid-base-catalyzed conversion of heterosubstituted methanes into ethylene and derived hydrocarbons. The onium ylide mechanism of the C1 to C2 conversion," *J. Am. Chem. Soc.*, vol. 106, pp. 2143–2149, 1984, doi: 10.1021/ja00319a039.

[193] L. Kubelková, J. Nováková, and P. Jírů, "Reaction of small amounts of methanol on HZSM-5, Hy and modified Y zeolites," *Stud. Surf. Sci. Catal.*, vol. 18, no. C, pp. 217–224, Jan. 1984, doi: 10.1016/S0167-2991(09)61157-5.

[194] G. J. Hutchings, F. Gottschalk, M. V. M. Hall, and R. Hunter, "Hydrocarbon formation from methylating agents over the zeolite catalyst ZSM-5. Comments on the mechanism of carbon?carbon bond and methane formation," *J. Chem. Soc. Faraday Trans. 1*, vol. 83. p. 571, 1987, doi: 10.1039/f19878300571.

[195] Y. Liu, F. M. Kirchberger, S. Müller, M. Eder, M. Tonigold, M. Sanchez-Sanchez, and J. A. Lercher, "Critical role of formaldehyde during methanol conversion to hydrocarbons," *Nat. Commun.*, vol. 10, no. 1, pp. 1–9, Apr. 2019, doi: 10.1038/s41467-019-09449-7.

[196] Y. Liu, S. Müller, D. Berger, J. Jelic, K. Reuter, M. Tonigold, M. Sanchez-Sanchez, and J. A. Lercher, "Formation mechanism of the first carbon-carbon bond and the first olefin in the methanol conversion into hydrocarbons," *Angew. Chem. Int. Ed.*, vol. 55, no. 19, pp. 5723–5726, May 2016, doi: 10.1002/ANIE.201511678.

[197] L. Yang, T. Yan, C. Wang, W. Dai, G. Wu, M. Hunger, W. Fan, Z. Xie, N. Guan, and L. Li, "Role of acetaldehyde in the roadmap from initial carbon-carbon bonds to hydrocarbons during methanol conversion," *ACS Catal.*, vol. 9, no. 7, pp. 6491–6501, Jul. 2019, doi: 10.1021/ACSCATAL.9B00641.

[198] A. D. Chowdhury, A. L. Paioni, K. Houben, G. T. Whiting, M. Baldus, and B. M. Weckhuysen, "Bridging the gap between the direct and hydrocarbon pool mechanisms of the methanol-to-hydrocarbons process," *Angew. Chem. Int. Ed.*, vol. 57, no. 27, pp. 8095–8099, Jul. 2018, doi: 10.1002/ANIE.201803279.

[199] T. Mole, J. A. Whiteside, and D. Seddon, "Aromatic co-catalysis of methanol conversion over zeolite catalysts," *J. Catal.*, vol. 82, no. 2, pp. 261–266, Aug. 1983, doi: 10.1016/0021-9517(83)90192-6.

[200] T. Mole, G. Brett, and D. Seddon, "Conversion of methanol to hydrocarbons over ZSM-5 zeolite: An examination of the role of aromatic hydrocarbons using 13carbon- and deuterium-labeled feeds," *J. Catal.*, vol. 84, no. 2, pp. 435–445, Dec. 1983, doi: 10.1016/0021-9517(83)90014-3.

[201] R. F. Sullivan, C. J. Egan, G. E. Langlois, and R. P. Sieg, "A new reaction that occurs in the hydrocracking of certain aromatic hydrocarbons," *J. Am. Chem. Soc.*, vol. 83, no. 5, pp. 1156–1160, Mar. 1961, doi: 10.1021/JA01466A036.

[202] A. Sassi, M. A. Wildman, H. J. Ahn, P. Prasad, J. B. Nicholas, and J. F. Haw, "Methylbenzene chemistry on zeolite hbeta: multiple insights into methanol-to-olefin catalysis," *J. Phys. Chem. B*, vol. 106, no. 9, pp. 2294–2303, Mar. 2002, doi: 10.1021/JP013392K.

[203] S. Svelle, F. Joensen, J. Nerlov, U. Olsbye, K. P. Lillerud, S. Kolboe, and M. Bjørgen, "Conversion of methanol into hydrocarbons over zeolite H-ZSM-5: Ethene formation is mechanistically separated from the formation of higher alkenes," *J. Am. Chem. Soc.*, vol. 128, no. 46, pp. 14770–14771, Nov. 2006, doi: 10.1021/JA065810A.

[204] M. Bjørgen, S. Svelle, F. Joensen, J. Nerlov, S. Kolboe, F. Bonino, L. Palumbo, S. Bordiga, and U. Olsbye, "Conversion of methanol to hydrocarbons over zeolite H-ZSM-5: On the origin of the olefinic species," *J. Catal.*, vol. 249, no. 2, pp. 195–207, Jul. 2007, doi: 10.1016/J.JCAT.2007.04.006.

[205] D. Lesthaeghe, B. De Sterck, V. Van Speybroeck, G. B. Marin, and M. Waroquier, "Zeolite shape-selectivity in the gem-methylation of aromatic hydrocarbons," *Angew. Chem. Int. Ed.*, vol. 46, no. 8, pp. 1311–1314, 2007, doi: 10.1002/anie.200604309.

[206] J. Li, Y. Wei, J. Chen, S. Xu, P. Tian, X. Yang, B. Li, J. Wang, and Z. Liu, "Cavity controls the selectivity: Insights of confinement effects on MTO reaction," *ACS Catal.*, vol. 5, no. 2, pp. 661–665, Feb. 2015, doi: 10.1021/CS501669K.

[207] M. L. Sarazen, E. Doskocil, and E. Iglesia, "Catalysis on solid acids: Mechanism and catalyst descriptors in oligomerization reactions of light alkenes," *J. Catal.*, vol. 344, pp. 553–570, 2016, doi: 10.1016/j.jcat.2016.10.010.

[208] M. L. Sarazen and E. Iglesia, "Stability of bound species during alkene reactions on solid acids," *Proc. Natl. Acad. Sci.*, vol. 114, no. 20, pp. E3900–E3908, May 2017, doi: 10.1073/PNAS.1619557114.

[209] L. Schmerling, "Reactions of hydrocarbons," *Ind. Eng. Chem.*, vol. 45, no. 7, pp. 1447–1455, Jul. 2002, doi: 10.1021/IE50523A030.

[210] A. Corma and A. Martinez, "Chemistry, catalysts, and processes for isoparaffin–olefin alkylation: actual situation and future trends," *Catal. Rev.*, vol. 35. pp. 483–570, 1993, doi: 10.1080/01614949308013916.

[211] S. I. Hommeltoft, "Isobutane alkylation: Recent developments and future perspectives," *Appl. Catal. A Gen.*, vol. 221, no. 1–2, pp. 421–428, Nov. 2001, doi: 10.1016/S0926-860X(01)00817-1.

[212] A. Feller and J. A. Lercher, "Chemistry and technology of isobutane/alkene alkylation catalyzed by liquid and solid acids," *Adv. Catal.*, vol. 48, pp. 229–295, Jan. 2004, doi: 10.1016/S0360-0564(04)48003-1.

[213] S. Singhal, S. Agarwal, S. Arora, N. Singhal, and A. Kumar, "Solid acids: potential catalysts for alkene–isoalkane alkylation," *Catal. Sci. Technol.*, vol. 7, no. 24, pp. 5810–5819, Dec. 2017, doi: 10.1039/C7CY01554B.

[214] V. D'Amico, J. Gieseman, H. Nousiainen, E. van Broekhoven, and E. van Rooijen, "Consider new methods to debottleneck clean alkylate production: solid-acid technology enables an inherently safer approach to produce high-octane blending components," *Hydrocarb. Process.*, vol. 85, no. 2, pp. 65–70, Feb. 2006.

[215] K. P. de Jong, C. M. A. M. Mesters, D. G. R. Peferoen, P. T. M. van Brugge, and C. de Groot, "Paraffin alkylation using zeolite catalysts in a slurry reactor: Chemical engineering principles to extend catalyst lifetime," *Chem. Eng. Sci.*, vol. 51, no. 10, pp. 2053–2060, 1996, doi: 10.1016/0009-2509(96)00062-0.

[216] F. Schüßler, E. A. Pidko, R. Kolvenbach, C. Sievers, E. J. M. Hensen, R. A. van Santen, and J. A. Lercher, "Nature and location of cationic lanthanum species in high alumina containing faujasite type zeolites," *J. Phys. Chem. C*, vol. 115, no. 44, pp. 21763–21776, Nov. 2011, doi: 10.1021/JP205771E.

[217] A. Sengar, R. A. van Santen, E. Steur, J. A. M. Kuipers, and J. Padding, "Deactivation kinetics of solid acid catalyst with laterally interacting protons," *ACS Catal.*, vol. 8, no. 10, pp. 9016–9033, Oct. 2018, doi: 10.1021/ACSCATAL.8B01511.

[218] A. Sengar, R. A. van Santen, and J. A. M. Kuipers, "Deactivation kinetics of the catalytic alkylation reaction," *ACS Catal.*, vol. 10, no. 13, pp. 6988–7006, Jul. 2020, doi: 10.1021/ACSCATAL.0C00932.

[219] J. Pater, F. Cardona, C. Canaff, N. S. Gnep, G. Szabo, and M. Guisnet, "Alkylation of isobutane with 2-butene over a HFAU zeolite. Composition of coke and deactivating effect," *Ind. Eng. Chem. Res.*, vol. 38, no. 10, pp. 3822–3829, 1999, doi: 10.1021/IE9902232.

[220] C. Liu, R. A. van Santen, A. Poursaeidesfahani, T. J. H. H. Vlugt, E. A. Pidko, and E. J. M. M. Hensen, "Hydride transfer versus deprotonation kinetics in the isobutane–propene alkylation reaction: a computational study," *ACS Catal.*, vol. 7, no. 12, pp. 8613–8627, Dec. 2017, doi: 10.1021/acscatal.7b02877.

Chapter 6

Molecular Heterogenous Catalytic Reactions

6.1 Introduction

This chapter introduces molecular heterogeneous catalysis, which originated in the 1950s and 1960s. It is part of the second golden age of catalysis discoveries.

Molecular heterogeneous catalysts, which are supported single-atom or molecular cluster catalysts, were explored due to the spectacular mid-century advances of coordination and organometallic chemistry. By then important homogeneous catalytic processes had been implemented based on this new chemistry [1].

A successful early example is the hydroformylation reaction that produces an aldehyde or ketone from reaction of alkene with CO and H_2. This homogeneous reaction was discovered by Roelen at Ruhrchemie in 1938 [2] and is catalyzed by the coordination complex $HCo(CO)_4$. Discoveries of organometallic chemistry in the 1960s were recognized in 1973 with the Nobel prizes of Wilkinson and Fischer. The newly discovered materials led to the invention of important new reactions.

In the following sections material inventions and corresponding reaction mechanisms are presented that shape molecular heterogeneous catalysis. A rich variety of single-atom and single-site catalysts that have unique reactivity will be introduced. Access to sophisticated characterization tools as well as computational techniques makes for impressively detailed mechanistic understanding of reactivity in relation to structure and composition of the catalytic centers [3].

Unique reactivity stems especially from catalysts with single-atom sites analogous to metalorganic complexes. In the course of time it was realized that the nature of the support materials is not only important to provide anchoring sites for the molecular complexes, but also to create synergetic interactions. Shape- and stereoselective catalysis were explored with zeolitic materials where the framework is substituted with catalytically reactive cations [4]–[7], [8], [9].

The following sections of this chapter are organized according to related catalysis or catalyst. Here a short overview of the different topics is presented. The chapter opens with the Nobel prize-winning discoveries of the metathesis reaction and propene polymerization. Also the early invention from 1964 of disproportionation catalysis at Phillips Petroleum, that precedes metathesis catalysis, will receive attention in Section 6.2. This important discovery was based on a coordination complex catalyst distributed on a high surface area support. Earlier in 1953 the catalytic polymerization of ethene had been discovered in the USA (also at Phillips Petroleum) as well as in Mülheim, Germany, while the catalytic polymerization of propene was discovered in 1954 in Milan, Italy. In all cases the reactions are catalyzed by a coordination complex. The mechanisms are well understood. The important difference is that polymerization of ethene and propene occurs through alkyl group coordinated to a reactive cation, whereas the metathesis reaction is catalyzed by a carbene complex.

In the following section the rich chemistry of Lewis acid catalytic systems is presented. Lewis acidic non-redox and redox reactions are discussed. Non-redox Lewis acid selective oxidation catalysts with single- or dual-site Ti reaction centers have been discovered that epoxidize propene with hydrogen peroxide or hydroperoxide. These are major discoveries also because the environmentally harmful chlorohydrin process is replaced by processes that do not produce chloride waste. Propene oxide is currently produced by these reactions in large-scale chemical plants of Shell Chemical and Dow/BASF.

An early Lewis acidic catalyst that catalyzes heterolytic C–H bond activation of alkanes is Cr_2O_3 supported on Al_2O_3 (Section 3.1.2). Similarly, single-site Ga or Zn oxycationic complexes located in the nanopores of bifunctional Brønsted acidic zeolite catalysts also activate heterolytic alkane C–H bond cleavage. These bifunctional catalysts are useful in the conversion of short alkanes such as ethane and propane to aromatics. Different from hydroisomerization catalysts (Section 5.4.3), they operate without hydrogen. This reaction is part of the UOP-BP Cyclar process invented at BP in 1985.

Also discussed is the related bifunctional reaction of methane to benzene that is catalyzed by a bifunctional ZSM-5 zeolite catalyst activated by FeO_x or MoO_x, which convert into catalytically reactive carbide clusters.

Zeolitic catalysts substituted with non-reducible Lewis acid cations are also selective catalysts for glucose isomerization and dehydration reactions. The Lewis cations activate polar groups. The key mechanistic step of these interesting reactions is hydride transfer that leads to dehydrogenation or hydrogenation between keto and hydroxyl substituents.

Single-site redox Lewis acid catalysts are explored in selective oxidation reactions that insert oxygen into alkane or benzene C–H bonds. The Russian scientist Panov discovered in 1988 the reaction of N_2O with benzene that gives phenol. It is catalyzed by single-site or dual-site Fe oxycationic clusters that are located in the nanopores of ZSM-5 zeolites. This reaction is of interest since it provides a useful outlet to N_2O that is a co-product in nylon production. N_2O is a greenhouse gas.

Direct oxidation of methane to methanol remains one of the great challenges in heterogenous catalysis. Catalytic systems are presented that contain oxycationic clusters of Fe or Cu in zeolite nanopores, which catalyze this reaction in a process where methane oxidation and catalyst reoxidation alternate.

Zeolitic catalysts with reducible cations located in its lattice framework have been explored in the 1990s for selective oxidation by Thomas from University of Cambridge. These radical reactions are quenched in the zeolite nanopore. For this reason they are structure sensitive with respect to the size and topology of the zeolite nanopores. Selective alkane oxidation reactions such as the oxidation of cyclohexane to adipic acid, which enable the production of nylon without N_2O co-production, are discussed.

Single-atom catalysis promoted by embedding (solid solution) of reactive metal atoms in a reducible oxide matrix such as CeO_2 is the subject of Section 6.4. These systems were initially discovered as automotive exhaust emission reduction catalysts. A revolutionary discovery is the unique reactivity of Au single-atom catalysts dispersed on reducible supports. The advantage of Au is its weak interaction with adsorbates. Therefore, it is not readily poisoned and remains reactive at low temperatures. Amongst others, application to the oxidation of furfural to 2,5-furandicarboxylic acid is discussed. This biomass molecule is an intermediate for biodegradable polymers.

6.2 Disproportionation and Polymerization Catalysis

Disproportionation or metathesis catalysts are metal-carbene complexes, while Ziegler-Natta polymerization catalysts are metal alkyl complexes.

The single-site catalyst that started the age of molecular heterogeneous catalysis is the silica-supported chromate catalyst for low-pressure polymerization of ethene [10]. It was discovered in 1953 by Banks and Hogan at Phillips Petroleum company in Oklahoma. The structure and proposed mechanism of the Cr-catalyzed ethene polymerization reaction is shown in Figure 6.1 [16], [17]. Ethene activates the reduced catalyst by a radical reaction. The Cr^{3+} cation is connected through two Si–O–Cr bonds to the SiO_2 surface. In the polymerization reaction a growing alkyl chain develops by subsequent insertion of ethene. The polymer product desorbs by a β C–H bond cleavage step. The reaction propagates by formation of an initial alkyl from reaction of ethene with the Cr–H intermediate.

Around the same time as the invention of the Phillips catalyst, Ziegler at Mühlheim serendipitously discovered that $TiCl_4$/

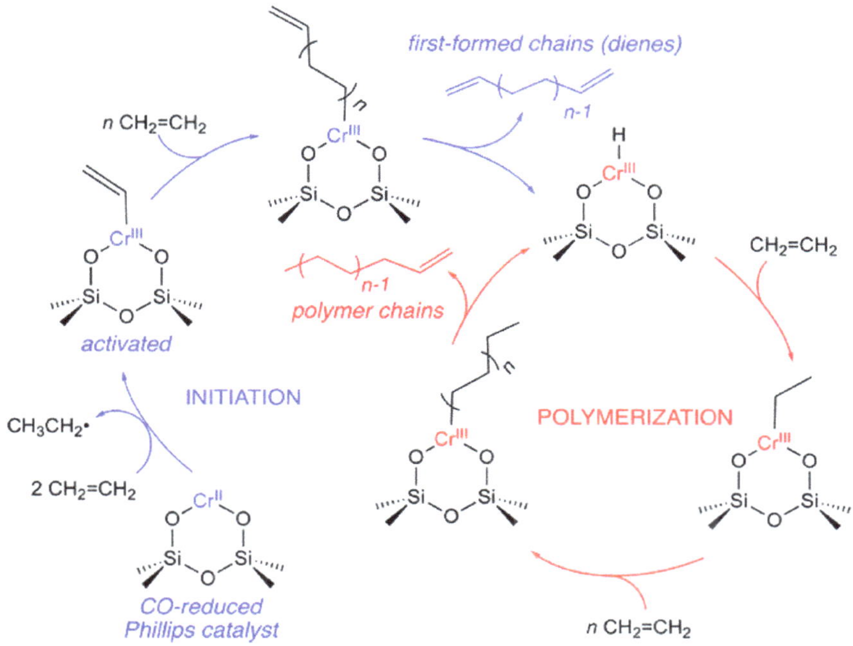

Figure 6.1 The reaction cycle of the ethene polymerization catalyst [11].

$Al(C_2H_5)_3$ could also polymerize ethene to polyethene [12], [13]. Originally Ziegler had tried with aluminum alkyl compounds, but all he got was oligomers. But when one of his co-workers tried to polymerize ethene with $Al(C_2H_5)_3$ in an autoclave that had been used before for a different reaction with a transition metal compound, he discovered that polyethene had been formed. Thereafter many other transition metal salts were tried in combination with aluminum alkyls and the use of transition metal compounds in the polymerization of ethene was patented. When Natta at the Technical University of Milan heard of this invention, he wondered if other alkenes wouldn't polymerize as well. This proved to be possible and the polymerization of alkenes containing side groups was patented. Furthermore, Natta realized that the polymerization of propene and higher alkenes should lead to stereoregular polymers, because the carbon atom bearing the alkyl group in the polyalkene chain is connected to four inequivalent groups [14]–[16]. When for instance in polypropene all methyl groups have the same orientation, isotactic polypropene would be formed, when the methyl groups would alternate in orientation syndiotactic polypropene would be formed, and when the methyl group orientation would be random atactic polypropene would be formed. The physical properties of these polymers are different. Whereas atactic polypropene has low melting point and rubbery properties, isotactic and syndiotactic polypropene have higher melting points, crystalline properties, and they can be used in applications that require elevated temperatures, such as the sterilization of plastic containers (cups for butter and margarine). For these inventions, Ziegler and Natta received the Nobel prize in 1963. A supported heterogenous $TiCl_3$/$MgCl_2$ catalyst was introduced in industry in 1970. At the end of the last century homogeneous catalysts based on late transition metals were discovered. They have led to the commercialization of specialty polymers.

The Ziegler-Natta catalyst is one of the first catalytic systems that were probed by early quantum-chemical models and calculations. For this catalyst Cossee proposed in 1964 [17] the currently accepted insertion mechanism of ethene and propene polymerization using a semi-empirical ligand field quantum-chemical model. Ligand field theory had then just been developed [18], [19] and describes the relation between the electronic structure of the transition metal cations and the relative stability of the coordination complex.

The Cossee-Arlman model is illustrated in Figure 6.2. Propene inserts into the growing alkyl chain, which is coordinated to the Ti^{4+} cation. Similar to the Cr catalyst, after β C–H bond cleavage the polymer desorbs as long-chain alkene and leaves a hydrogen atom attached to Ti. Reaction of propene with the hydrogen atom reinitiates the reaction by formation of adsorbed alkyl intermediate.

With the advance of *ab initio* first-principle quantum-chemical methods Clementi *et al.* [21] at IBM in 1978 did the first of such calculations to probe the Cossee-Arlman polymerization model. The Clementi calculations are also the first first-principle calculations

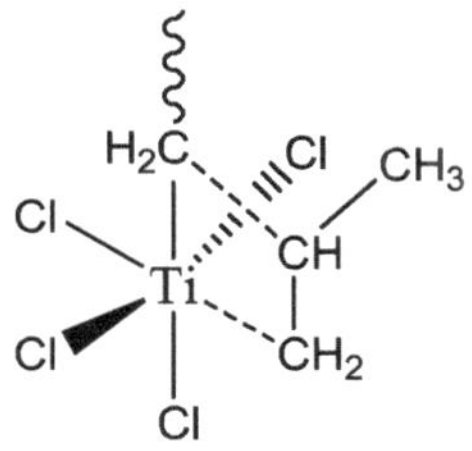

Figure 6.2 The chlorine vacant titanium chloride site that illustrates the insertion of the propene molecule into the alkyl chain according to the Cossee-Arlman mechanism of propene polymerization [20].

applied to a catalytic reaction. Later advanced DFT quantum-chemical calculations that support the Cossee-Arlman mechanism were performed by Ziegler *et al.* in 2003 [22].

Another single-site catalyst was discovered by Banks and Bailey in 1964 at Phillips Petroleum Company [23]. This Mo catalyst catalyzes the disproportionation of propene to ethene and butene. To prepare the catalysts a $Mo(CO)_6$ coordination complex is impregnated on silica and subsequently calcined.

In the same period of the discovery of the Cossee-Arlman mechanism and the formulations of its quantum-chemical founding, Mango and Schachtschneider [24] proposed a quantum-chemical model of the disproportionation reaction. At that time Cossee and Arlman as well as Mango and Schachtschneider were researchers in Shell.

The Mango and Schachtschneider proposal is based on the quantum-chemical Woodward-Hoffmann symmetry rule of alkene cycloaddition reaction, which was just published in those days [25].

In the disproportionation reaction two propene molecules react and produce an ethene and butene molecule. The Mango-Schachtschneider proposal is schematically sketched in Figure 6.3. The two alkene molecules adsorb symmetrically in the same way with the transition metal cation. It is suggested that in a follow-up reaction a cyclobutane complex is formed. In the next step the C–C bonds of the reacting molecules cleave and alkene molecules desorb with newly formed C–C bonds.

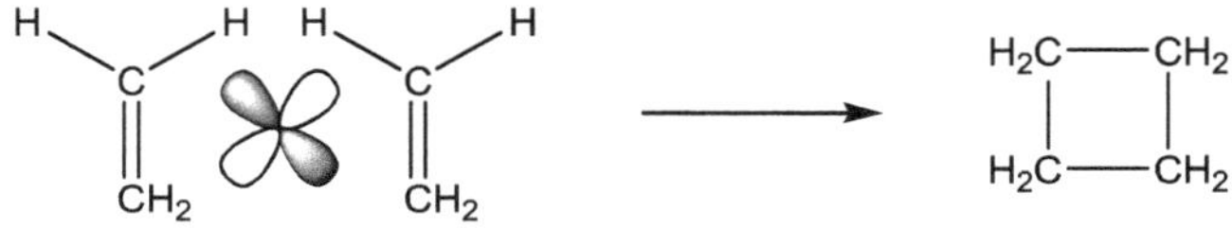

Figure 6.3 Schematic representation of the disproportionation C–C bond formation step of the Mango-Schachtschneider mechanistic proposal. The reaction intermediate cyclobutane forms from two ethene molecules. This step becomes Woodward-Hoffmann symmetry allowed by symmetry breaking through interaction with empty d-valence atomic orbitals of the transition metal cation [24].

According to the Woodward-Hoffmann orbital symmetry rule (R. B. Woodward is a synthetic organic chemist and Nobel award winner 1965, R. Hoffmann is a quantum chemist and Nobel laureate 1981), formation of a cyclobutane ring by reaction of two alkene molecules in the absence of catalyst will not happen. There is mismatch of orbital symmetries of the reacting alkene molecules. The reaction is orbital symmetry forbidden.

As illustrated in Figure 6.3 the asymmetry of the empty transition metal d-valence atomic orbital located in between the reacting molecules introduces asymmetry into the system. This breaks the Woodward-Hoffmann symmetry rule.

The Mango-Schachtschneider mechanistic proposal was rejected by later experiments. The pathway of the disproportionation reaction which later was named metathesis is very different from initially thought.

In 1971 Herisson and Chauvin from the Institut Français du Pétrol correctly proposed a carbene mechanism for the disproportionation/metathesis reaction. The now generally accepted Chauvin metathesis mechanism has been deduced from the redistribution of carbon atoms in the product patterns of mixtures of alkenes [26].

Once the reaction mechanism was established new homogenous molecular complexes with W, Ta, and Re [27] were discovered that became important to ring-opening monomer polymerization (ROMP) of cyclic olefins. For their contribution to metathesis catalysis Nobel Prizes in Chemistry were awarded in 2005 to Chauvin, Grubbs, and Schrock [28].

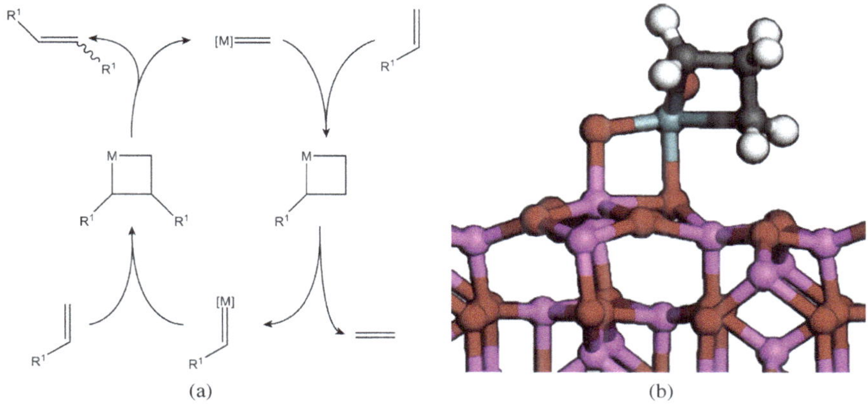

Figure 6.4 (a) The Chauvin mechanism of the metathesis reaction [29]. (b) The four-center metallocycle of carbene with ethene of Mo cation immobilized on a SiO_2 surface [30].

As indicated in Figure 6.4a, the Chauvin mechanism starts with a reaction of a metal $M{=}CH_2$ carbene complex with alkene. The metal atom becomes part of a four-ring metallocycle complex of three carbon atoms and the metal atom (Figure 6.4b). Cleavage of the weakened C=C bond of the reacting alkene gives a new carbene and product alkene molecule. The catalytic cycle then resumes with the carbene fragment of the reactant left. The Chauvin carbene complex also initiates polymerization of cyclic alkenes by ring opening and recombination of ring-opened molecules. This is the ROMP reaction.

Figure 6.4b shows the calculated structure of the metallocycle intermediate that is part of a Mo single-site complex attached to a SiO_2 surface. DFT calculations by Hanzlik and Sautet [30] show that only a unique configuration with a specific number and angles of Mo–O–Si bonds give low activation energy of the overall bond rearrangement reaction. The active Mo metallocycle complex contains a free dangling M=O double bond. This is a spectator oxygen bond that itself does not react with reactant. It activates the reaction of alkene with $Mo{=}CH_2$ to give the metallocycle intermediate [31], [32]. Experiment shows that only a small fraction of Mo immobilized on the SiO_2 surface is active for the reaction.

In summary, in the Cossee-Arlman mechanism alkene inserts into the growing alkyl hydrocarbon chain. The alkene polymerization reaction is catalyzed by a transition metal of the first row in the periodic table. The reaction site atoms are cationic, nearly or completely ionized and in a low valence state (3+ or 4+).

In the Chauvin mechanism the reactive complex contains a carbene species (CH_2). The metathesis reaction is catalyzed by single-atom transition metal cations in their highest 5+, 6+, or 7+ valence states.

6.3 Lewis Acid Single-site Heterogenous Catalysts

The selective oxidation of propene to propene epoxide with hydrogen peroxide or hydroperoxide catalyzed by Ti^{4+} cations is a successful example of catalysis by a non-reducible Lewis acidic cation. Ti^{4+} can be supported on an amorphous support such as SiO_2 or can be part of the zeolite framework (Section 6.3.1.1).

New catalysis emerges when analogous Lewis acid catalysts with Sn^{4+} in its framework are used in non-oxidation reactions. This is illustrated by catalysis of glucose isomerization (Section 6.3.1.2). In glucose conversion catalysis, a key step is hydride transfer from alcohol to ketone. Heterolytic O–H bond cleavage is a synergistic reaction of Lewis acid cation and Lewis basic lattice oxygen. Additionally, the Lewis acid cation activates the keto group. Such reactions can be part of ring-opening and ring closure reactions of carbohydrates.

The heterolytic C–H bond activation of small alkanes is catalyzed by Ga^{3+} or Zn^{2+} oxycationic clusters (Section 6.3.1.3). This reaction is used in the conversion of short alkanes to aromatics, which is catalyzed by bifunctional solid acid catalysts activated by Ga oxycation clusters. Related anaerobic conversion of methane by bifunctional redox catalysts is presented in Section 6.3.3.

Reducible single cations can be substituted in the framework of zeolite catalysts or can be also part of oxycationic clusters located in the zeolite nanopore. In Section 6.3.2 such catalysts useful for selective oxidation reactions are presented. Hydroxylation of benzene to phenol and methane to methanol are discussed. Selective oxidation of alkanes with molecular oxygen catalyzed by zeolitic systems are also presented. Selectivity of radical reactions is tuned by nanopore connectivity and size.

6.3.1 *Catalysis by Non-reducible Lewis Acid Cations*

6.3.1.1 *Selective Oxygen Atom Insertion into Propene and Cyclohexanone*

Lewis acid single-site Ti catalysts selectively epoxidize propene to propene epoxide with H_2O_2 or hydroperoxide.

Non-selective catalytic oxidation of propene with molecular oxygen to propene epoxide is known. This low selectivity to propene epoxide is due to the presence of allylic hydrogen in propene. This makes propene liable to total oxidation or selective formation of acrolein (Sections 4.3.3.2, 4.4.3.1) [33], [34]. Selective epoxidation of propene is possible when instead of molecular oxygen a peroxide is used as reactant.

In these reactions an oxygen atom inserts into the olefinic π bond. A related reaction with hydrogen peroxide is the oxidation of cyclohexanone into a cyclic ester lactone. In this reaction the oxygen atom is introduced into the σ C–C bond next to the keto group.

First epoxidation catalysis is discussed, followed by cyclohexanone oxidation.

Since the early 1920s it is known that alkenes can be selectively epoxidized by peracetic acid [35]. This principle is basic to two heterogeneous catalytic processes for the selective oxidation of propene to propylene oxide. One process uses an organic hydroperoxide molecule in an aprotic medium, in the other hydrogen peroxide is oxidant dissolved in water. Water is detrimental to the catalyst selectivity. Oxidation with hydrogen peroxide can be done with Ti substituted in the framework of a siliceous zeolite. This creates locally an apolar environment.

When hydroperoxide is used a co-product of the epoxide is an alcohol, while with hydrogen peroxide water is the co-product. Homogeneous catalytic propene epoxidation by complexes containing Mo^{6+}, W^{6+}, V^{5+}, or Ti^{4+} [36] were industrially applied around 1970. The heterogeneous process with a single-site Ti catalyst immobilized on silica developed by Shell became operational in 1979. In the SMPO (Styrene Monomer Propene Oxide) process propene is converted to propene epoxide by reaction with ethylbenzene hydroperoxide. This produces the epoxide and ethylbenzene alcohol. In a subsequent reaction the latter is dehydrated to styrene, that is of use as monomer for styrene-butadiene copolymerization.

Instead of ethylbenzene hydroperoxide, isobutyl hydroperoxide can also be used as oxidant. Both peroxides are produced in autocatalytic oxidation reactions from respectively ethylbenzene or isobutane [37].

The preparation of the Ti catalyst by reaction of an organometallic complex, in this case the $Ti(alkoxy)_4$ complex, with a silica surface is an early example of the by now well-developed surface organometallic chemistry approach for the preparation of well-defined molecular complexes attached to oxidic surfaces [4], [5], [38]. The Ti complex attached to the silica surface is calcined. By this process it becomes anchored through two or three Si–O–Ti bonds to the silica surface.

The mechanism of the epoxidation reaction as deduced from EXAFS measurements and DFT calculations is illustrated by Figure 6.5 [39].

Ti^{4+} accommodates the ROO^- peroxy species after proton transfer from hydroperoxyl to Ti-bound alkoxy. The proton makes a hydrogen bridge with Ti-attached ROO^-. The oxygen atom of the peroxy species attached to the Ti ion is transferred to propene. Because of its bond to Ti and the hydrogen promotion the oxygen atom that inserts into the alkene π bond is highly electrophilic. After reaction the alcoholate remains adsorbed to Ti, and reacts with another ROOH molecule which initiates the next cycle.

This mechanism of Ti catalyst grafted to the surface of a porous silica catalyst is very similar to that suggested earlier in 1981 for the

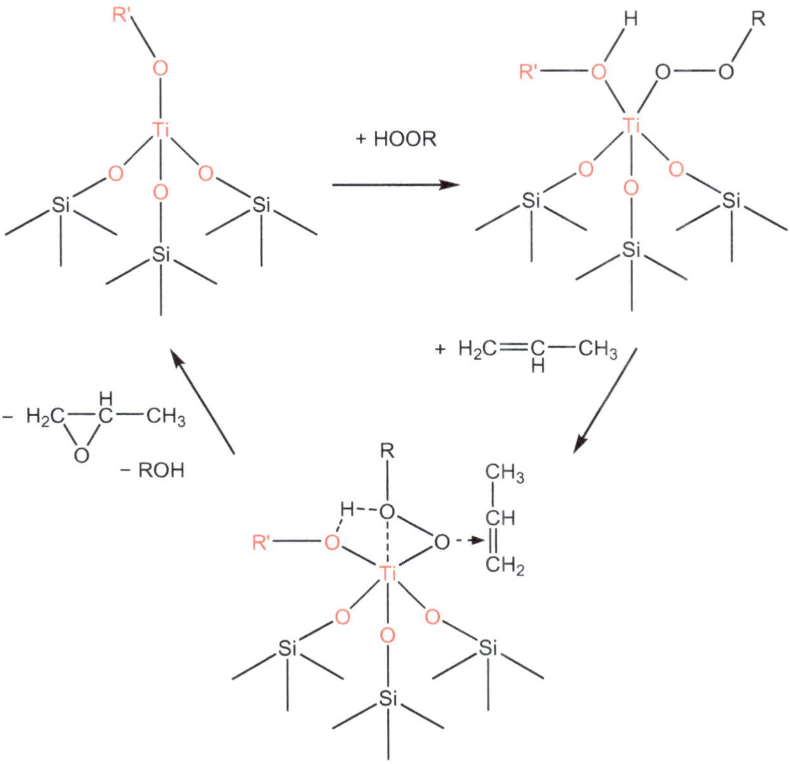

Figure 6.5 Schematic representation of propene epoxidation by hydroperoxide with Ti/SiO_2 catalyst; proton changes position between peroxide atoms [39].

homogeneous enantiomeric epoxidation of alkenes. Stereoselective catalysis by an organometallic Ti complex with chiral tartrate ligands was discovered by the American chemist Sharpless (Nobel award 2001) [40]–[42].

In 1983 at Enichem company in Italy the epoxidation of propene with H_2O_2, also catalyzed by a heterogenous catalyst, was discovered. It is catalyzed by a siliceous zeolite called silicalite that has the MFI structure (see Figure 5.1) in which a small number of Si cations is substituted by Ti. This substitution maintains charge neutrality of the framework. Silicalite was synthesized in the 1970s as part of the high silica framework zeolite synthesis program at Mobil Oil [45]. An introduction to these zeolitic materials, their synthesis, and use as solid acids can be found in Chapter 5. The catalyst is applied in the Dow-BASF epoxidation process since 2008.

The catalyst of this process, named TS-1 by the inventors, was originally thought to be a single-atom site of Ti [43], [44]. The siliceous environment of the silicalite nanopores makes the Ti site hydrophobic. Water is not adsorbed, which prevents hydrolysis of propylene oxide to glycol.

Recently it was discovered by Gordon *et al.* that the catalytic site is not a single Ti atom but actually contains two Ti atoms [46]. They showed this using $^{18}O_2$ labelling in combination with advanced spectroscopic studies and DFT simulations [46]. As schematically illustrated in Figure 6.6 the reactive site consists of two Ti atoms that are located in neighboring tetrahedra. When reacted with hydrogen peroxide the single oxygen atom bridge is replaced by two OOH hydroperoxide substituents that bridge the two Ti cations. The coordination of the Ti atoms becomes pentagonal. Reaction with propene is analogous to the epoxidation mechanism of grafted Ti shown in Figure 6.5. The hydrogen atom of the peroxide forms a hydrogen bond with an oxygen atom next to Ti of the zeolite framework.

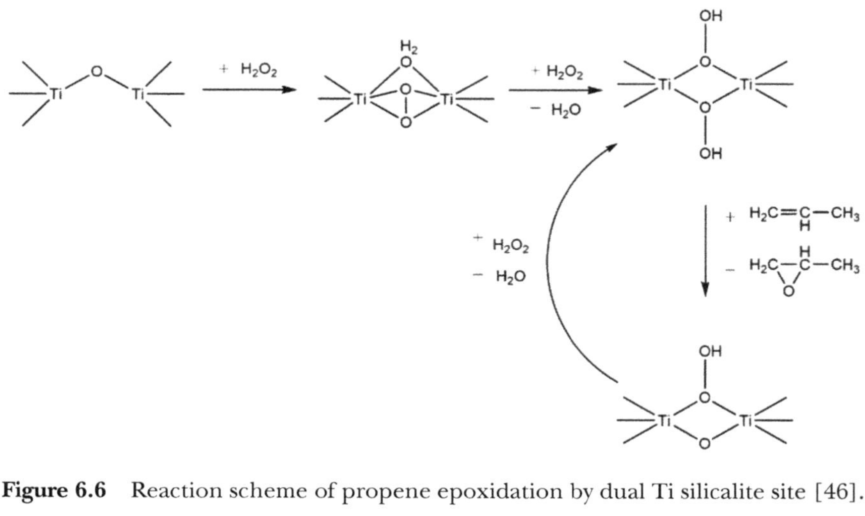

Figure 6.6 Reaction scheme of propene epoxidation by dual Ti silicalite site [46].

The concentration of Ti in the material is very low. It cannot be excluded that in parallel also single-atom Ti sites are reactive. The question of whether a single-atom site or a dual or higher atom nanocluster is the selective site is a general key question to be addressed in many related systems.

Similar to the single-atom model site there is also an organometallic dimer complex, the di-μ-oxotitanium(salene) molecule that reacts analogously to the dual-site Ti epoxidation mechanism [47].

The epoxidation of propene with hydrogen peroxide or hydroperoxides can be described as a reaction catalyzed by Ti^{4+} that acts as a Lewis acidic site. The valency of the Ti cation does not change. It provides an anchor site to the intermediate hydroperoxide substituents and promotes the electrophilic insertion of the oxygen atom into the alkene π bond.

Corma *et al.* from Valencia Institute of Technology [48] demonstrated the general use of Lewis acid-catalyzed reactions with H_2O_2 by their discovery of selective oxidation of ketones. They discovered that methylcyclohexanone converts selectively to the corresponding lactone catalyzed by wide nanoporous zeolite β, with Sn incorporated in its framework. Lactone, a cyclic carboxylic ester, is an important fine chemical and intermediate for polyester plastics. The reaction is shown in Figure 6.7.

The Lewis acidic Sn cation activates the carbonyl that attaches through the oxygen atom to the Sn cation. The OOH^- reacts and inserts the oxygen into the cyclohexyl ring. Zeolites with Ti or Zr are less selective. The Sn cation has been shown to have the stronger interaction with the carbonyl group of the ketone [49].

The Sn catalyst is not only useful in selective oxidation reactions but has been found to also catalyze hydride transfer between alcohols and ketones as discussed in the next section.

6.3.1.2 *Lewis Acid-catalyzed Hydride Transfer Reactions in Polar Molecules; Carbohydrate Conversion Catalysis*

Hydrogen atom transfer between ketone and alcohol is promoted by Lewis acid cations.

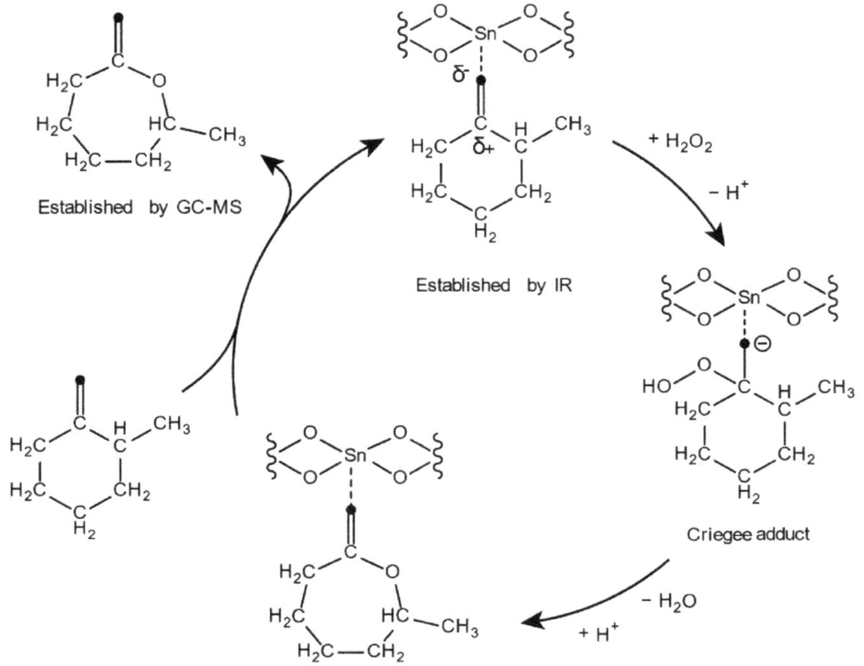

Figure 6.7 Selective oxidation of 1-methylcyclohexanone to lactone catalyzed by Lewis acidic Sn^{4+} in zeolite β [48]. The ketone coordinates to the Lewis acid center and, thereby, the carbonyl group is activated. Then the hydrogen peroxide attacks the more electrophilic carbonyl carbon atom. After the rearrangement step, the lactone product is replaced by a new substrate molecule.

The discovery of hydride transfer reactions of Sn-promoted zeolite β catalysts, which are important in the carbohydrate conversion reactions that are discussed here, are preceded by the discovery by van Bekkum *et al.* from Delft Technical University [50], [51] that zeolite β substituted by Ti selectively activates hydrogen transfer by polar molecules through their hydroxyl or keto group. They demonstrated also shape-selective catalysis of the hydride transfer reaction between bulky alcohol and ketone molecules.

The dehydrogenation of a secondary alcohol by a ketone is a classic organic reaction promoted by metal alkoxides. It is called the

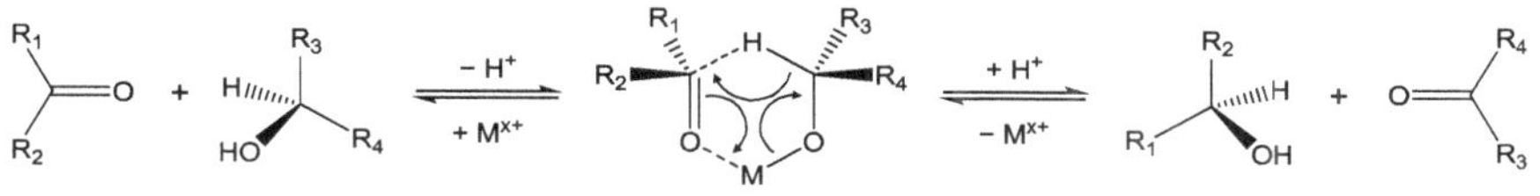

Figure 6.8 The Meerwein-Ponndorf-Verley-Oppenauer reaction [50].

Meerwein-Ponndorf-Verley-Oppenauer reaction. It is illustrated in Figure 6.8.

In the reaction a hydride ion is transferred between the respective carbon atoms and at the same time a proton between the respective oxygen atoms. One can describe it as a heterolytic dehydrogenation reaction. It is analogous to the hydride transfer between a carbenium cation and an alkane in acid catalysis (see Figure 5.2c).

Sn-promoted Lewis acid catalysts are also excellent catalysts for the isomerization of glucose to fructose [52] and its consecutive conversion to lactic acid [53].

Whereas Brønsted acidic zeolites will catalyze dehydration of fructose molecules to hydroxymethylfurfural and related molecules, Lewis acidic cations promote isomerization reactions from glucose to fructose and can produce lactic acid as product [54].

The zeolite-catalyzed isomerization of glucose to fructose is a potential replacement for the conventional glucose isomerase enzyme reaction [55], [56]. Figure 6.9 presents the mechanism of the glucose-to-fructose isomerization that is catalyzed by Sn-substituted zeolite β as modelled by a computational DFT study [57].

The reaction is initiated by heterolytic bond cleavage of a glucose O–H group. This is followed by backdonation of the proton to the ether O atom, which activates C–O bond cleavage and opening of the hexose ring. In the next step an internal hydride transfer relocates a hydrogen atom to the end-on C atom. In this rearrangement process a CO– and C=O group interchange. The positive charge of Sn^{4+} stabilizes negative charges on the oxygen atoms that change their respective charges in the process.

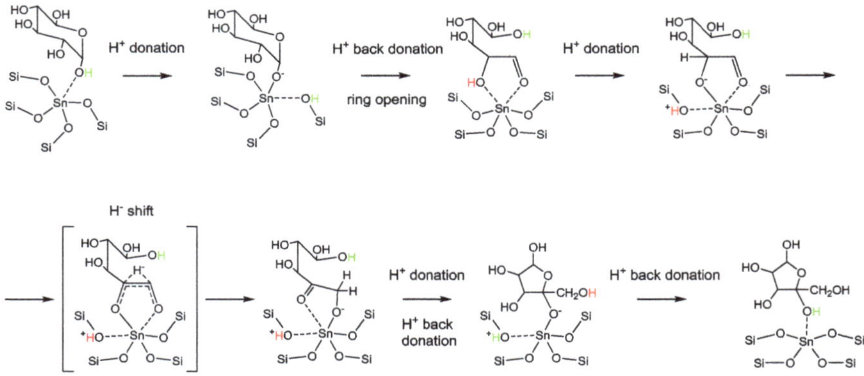

Figure 6.9 Mechanism of glucose isomerization to fructose catalyzed by a Sn lattice site in zeolite β [57].

The rearrangements of the H atom and charge are followed by the ring closure reaction that forms the fructose pentose ring. In the process a proton is donated to the zeolite and also a proton is backdonated to C–O⁻. In Figure 6.9 the respective protons are indicated with different colors. The reaction is concluded with a final proton backdonation step to ionized fructose. The hydride transfer is the key step of this reaction. The parallel transformation of alcohol to ketone relates to steps of the Meerwein-Ponndorf-Verley-Oppenauer reaction.

Metal cation-activated isomerization of sugar xylose isomerase shows analogous hydride transfer reaction steps [58]. The inorganic system biomimicks the enzyme. The negatively charged zeolite framework accommodates proton exchange reactions. The large Sn cation stabilizes negative charge of intermediate anionic organic intermediates.

6.3.1.3 *Heterolytic C–H Bond Activation by Ga and Zn Cations*

The dehydrogenation of alkane by single cation sites is a concerted reaction that involves β C–H bond cleavage.

Not only transition metals such as Pt can be used in bifunctional catalytic reactions for alkane activation. Zeolites activated by Lewis acid cations of Zn or Ga are also useful as hydrocarbon conversion catalysts. They are of use in reactions where hydrogen is detrimental. In the absence of hydrogen Pt nanoparticles rapidly deactivate due to non-selective carbonaceous residue-forming reactions [59].

Bifunctional acidic zeolites exploit the reactivity of Zn or Ga cations for ethane and propane oligomerization to alkenes or aromatics that cannot be executed in the presence of hydrogen [60]. Oligomerization of short alkanes is a desirable reaction to convert natural gas components to liquid fuel. The Zn or Ga cations catalyze dehydrogenation of ethane or propane. The zeolite protons catalyze the consecutive alkene oligomerization and aromatization reactions.

In the 1970s the oligomerization of short alkanes was discovered by researchers at British Petroleum Company [61]. At present it is part of the BP-UOP Cyclar process announced in 1991. The endothermic reaction requires a temperature of 780 K. Catalyst recycling by high-temperature coke removal is part of the overall process. The catalyst is H-ZSM-5 promoted with Ga. Elements other than Ga can also be used. Ga is the preferred H-ZSM-5 promoter, since Zn is too volatile to be sustained through the catalyst recycling process [62].

When Ga^{3+} is substituted for Al^{3+} in the framework of the zeolite, the catalyst will only react as a solid acid catalyst. Catalysis will be similar to catalytic cracking discussed in Chapter 5. Only when Ga becomes located in extra framework positions the alkane oligomerization reaction is catalyzed [63]. Ga is present as single cation or as part of an oxy(hydroxy)cation. An important difference with proton-catalyzed dehydrogenation, where C–C bond cleavage is a competitive reaction (see Section 5.4.2), is that the C–C bond cleavage is not catalyzed by Ga or Zn oxycationic complexes.

A major kinetic question is whether in the activation reaction of alkane the zeolite proton and Lewis cation act independently. This question has not yet been completely resolved [60]. This is not surprising since as will be seen the mechanism of C–H activation involves generation and consumption of zeolite protons.

The mechanism of the dehydrogenation reaction of the alkane shows important differences when catalyzed by an oxycationic complex or a single cation. This is sketched in Figure 6.10.

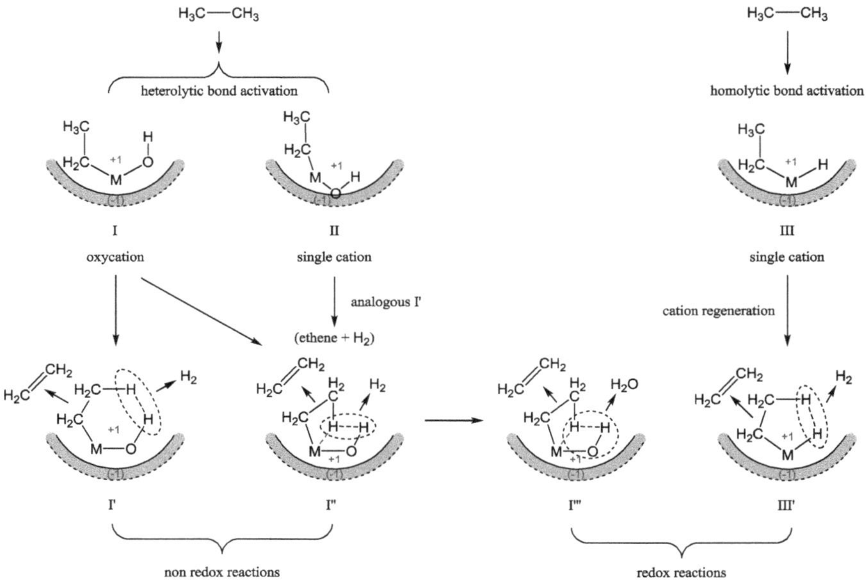

Figure 6.10 Dehydrogenation of alkane catalyzed by oxycation versus single cation. (a) Non-redox reactions with oxycationic complex I: charge M is 3+; complex II: charge M is 1+. As complexes I′ and II′ illustrate, alkene formation can occur concerted (I′) or in consecutive steps (I″). (b) Redox reactions: C–H bond cleavage is oxidative addition reaction (charge M: 1+ → 3+), hydrogen atom recombination is reductive elimination (charge M: 3+ → 1+).

The oxycation has a positive charge and compensates the negative charge on the zeolite wall. The formal charge on the metal atom in the oxycation is 3+. This is different when the atom is present as a single cation; its charge then is 1+. Heterolytic C–H bond cleavage leads to attachment of the alkyl group to the metal cation and proton to the oxygen atom. In the case of the oxycation (complex I, Figure 6.10) the oxygen atom is the extra framework oxygen atom that is part of the oxycationic complex. For the single cation (complex II of Figure 6.10) this is a zeolite lattice oxygen atom. Generally, the extra framework oxygen atom is more reactive and C–H bond activation occurs more readily on the oxycationic complex.

The two reactions of the alkane of complexes I and II are non-redox reactions and charge on metal cation does not change.

With the single cation, in addition to heterolytic C–H bond cleavage, a redox reaction can also occur. Then the C–H bond cleavage is an oxidative addition and the formal charge on the metal cation increases to 3+. When the reaction is completed with formation of alkene and H_2, in a reductive elimination step the charge on the metal cation is reduced back to 1+.

As discussed in Section 3.1.2, heterolytic bond cleavage reactions of alkanes are also known for the bulk oxide surfaces of Cr_2O_3 and ZnO. Compared to the oxycationic clusters the oxygen atoms are less reactive and the activation energy of C–H bond cleavage is higher.

Once a C–H bond is broken alkene formation can happen in a concerted reaction (I′) or in consecutive steps (II′). In the former β C–H bond cleavage happens by H–H bond formation with the hydrogen atom attached to the oxygen atom. In the latter case ethene desorbs leaving behind a hydrogen atom on the metal atom that in a subsequent step recombines to form the hydrogen molecule.

Also, as is illustrated by complex I′, the two hydrogen atoms can react with the oxygen atom of the oxycation and the complex reduces to a single cation.

A quantum-chemical calculation with GaO^+ in the nanopore of ZSM-5 finds that the overall activation energy for the concerted dehydrogenation of ethane equals 155 kJ/mol. The concerted reaction is favored by 40 kJ/mol over the overall activation energy of the reaction that happens by consecutive steps. Deoxygenation according to complex III′ also has a 40 kJ/mol lower barrier than reaction through complex I′ [64].

The reduction of a binuclear Ga oxycation in reaction with propane has been confirmed experimentally by Hensen *et al.* [65], [66]. The initial high conversion of the reaction catalyzed by the Ga oxycation decreases fivefold in time. This is because oxycation oxygen is reduced and the single Ga cation that is left is substantially less reactive than the oxycation. At steady state Ga will be present as a single cation.

For the gallium cation the heterolytic C–H bond dissocation involving a Lewis basic lattice oxygen atom that gives complex II is preferred over formation of complex III by homolytic oxidative addition.

Calculations of ethane activation give an activation energy of 374 kJ/mol for oxidative addition versus 210 kJ/mol for heterolytic C–H bond activation [67], [68]. Note the significantly higher activation energies of Ga^+ compared to the earlier mentioned activation by GaO^+.

At steady state the dominant state of Ga is most likely a mixture of Ga^+ and $[GaH_2]^+$ intermediates [66], [69], [70].

The reactivity difference of single cation versus oxycation for non-reducible Zn^{2+} is different than for Ga^{3+} or Ga^+. Calculations of C–H bond activation indicate that for Zn-promoted catalysts the oxy cationic complex $ZnOZn^{2+}$ has the higher barrier. Heterolytic activation by Zn^{2+} cations is favored [71].

The decomposition of $[GaH_2]^+$ to Ga^+ has a high barrier for the same reason that homolytic dissociation of alkane C–H for addition to Ga^+ also has a high activation energy. In Ga^+ the highest occupied valence atomic orbital has s symmetry. The C–H or H–H bond is also β symmetric. Electron transfer between occupied and non-occupied orbitals initiates chemical reaction. According to the Woodward-Hoffmann rule recombination is symmetry forbidden. Because of mismatched symmetries interaction between occupied orbital of one fragment with empty unoccupied orbital of another fragment is impossible.

Heterolytic reactions as in structures I′, II′, or III′ that react hydrogen with alkyl intermediates circumvent this selection rule [72].

The activation energies of proton-activated alkane dehydrogenation and those activated by single cations of Ga or Zn are of the same order of magnitude. However, importantly the Ga and Zn cations only promote C–H bond cleavage and, differently from protons, do not activate C–C bonds.

The protons of the bifunctional Ga-promoted H-ZSM-5 catalyst are necessary for the alkene oligomerization and ring closure reactions. The Ga cation promotes alkane to alkene dehydrogenation. It also plays a role in consecutive reactions. In acid catalysis hydride transfer steps are necessary for aromatics formation. However, the small pores of ZSM-5 limit the probability of the bimolecular hydride transfer reaction (Figure 5.5). The Ga or Zn cations take over the dehydrogenation function and enhance propane conversion and rate of aromatization [61].

6.3.2 *Single-site Redox Catalysis; Selective Oxidation*

Here shape-selective oxidation reactions are introduced for catalytic redox systems with reaction centers located in the zeolite framework or nanopore.

Single-site redox catalysts with reactive cations located in the framework of the zeolite are discussed in Section 6.3.2.1. These catalysts can be used for selective oxidation with molecular oxygen that are autocatalytic radical reactions.

Whereas autocatalytic oxidation is intrinsically non-selective, when such reactions happen in the zeolite nanopore selectivity can be modified by exploitation of zeolite shape.

This concept has generated rich catalytic oxidation chemistry [6], [7]. Amongst others selective oxidation of alkanes to carboxylic acids is presented. Selective catalysis is due to selective reaction of hydroperoxide intermediates.

Selective formation of phenol from benzene and methanol from methane are catalyzed by redox cation complexes located in the zeolite nanopore. These reactions are tuned by the type of oxycation cluster and by choice of zeolite structure [73]. Elementary steps are single oxygen atom reactions with molecular hydrocarbon fragments. The hydroxylation reactions are discussed in Sections 6.3.2.2 and 6.3.2.3.

6.3.2.1 *Redox-selective Oxidation by Zeolite Compounds*

Radical reactions quench in the zeolite nanopore.

The rich variety in composition and structure of substituted zeolitic $AlPO_4$ systems makes them attractive to exploration in selective oxidation and related reactions. These are autocatalytic oxidation reactions with molecular oxygen. Reaction of substrate with intermediate hydroperoxides that are generated *in situ* give selective catalysis.

At the end of the last century single-site heterogeneous catalysts with redox cations substituted in an AlPO4 framework for fine chemical production were developed and investigated by J. M. Thomas from University of Cambridge [38], [74], [75]. According to Thomas single-site heterogenous catalysts are catalysts in which the active centers are spatially isolated from one another and uniformly distributed through the solid such that each site has the same interaction with incoming reactants. Examples are the Co/Mn-AlPO-18 and Fe-AlPO-31 (the number denotes a particular structure) materials designed for the selective oxidation of alkanes by molecular oxygen [76]–[78]. $AlPO_4$ structures are selected that upon substitution with redox cations regioselectively oxidize cyclohexane or n-hexane.

Thomas *et al.* studied selective oxidation of alkanes to dicarboxylic acids. A desirable product molecule is adipic acid (hexanedioic acid) with, at each end of the linear alkane, a carboxylic acid group. Direct oxidation of alkane to adipic acid is potentially an environmentally beneficial reaction that could replace the current stoichiometric process with nitric acid, which has the greenhouse gas N_2O as co-product. Adipic acid is a monomer important in nylon production.

Reaction of oxygen with alkane initiates autooxidation. According to [79] the unique selectivity that Thomas and his colleagues discovered relates to the effect of nanopore size on the relative rates of radical termination. In the gas phase, termination of tertiary peroxides is much faster than that of primary peroxides. In the zeolitic material restricted transition state shape selectivity limits encounters of oxyradical and bulky reactants (see Section 5.4.4).

The radical chain reaction proceeds by initiation, propagation, and termination as indicated in Eqs. (6.1):

(a) initiation

$$RH + O_2 \rightarrow R\cdot + HOO\cdot$$

(b) propagation

$$ROO\cdot + RH \rightarrow ROOH + R\cdot \text{ peroxide formation}$$

(c) redox reactions

$$ROOH + M^{3+} \rightarrow ROO\cdot + H^+ + M^{2+} \text{ oxidation}$$
$$ROOH + M^{2+} + H^+ \rightarrow RO\cdot + H_2O + M^{3+} \text{ reduction}$$

(d) termination

$$2\,R(H)OO\cdot \rightarrow ROH + RCHO + O_2 \tag{6.1}$$

The initiation step generates OOR radicals from reaction of O_2 with the organic molecule. In propagation steps these radicals give hydroperoxides (Eq. (6.1b)). Reaction with redox cations leads to oxygenated hydrocarbons (Eq. (6.1c)). The ROO radical can be regenerated in a reduction reaction or ROOH decomposes into RO by oxidation with the metal cation. A zeolite proton is generated or consumed in the process. In the redox reaction change of the cation charge is possible, accommodated by alternation of non-protonated charge-neutral site with protonated site.

Radical chain termination (Eq. (6.1d)) happens by recombination of two hydroperoxides. Alcohol, ketone, or aldehyde are products. Suppression of this bimolecular reaction by restricted transition state selectivity biases propagation. Shape selectivity favors primary end-on carbon atom oxidation. This siting of the hexane molecule in a M_xAl_1-xPO_4 site is illustrated in Figure 6.11.

The $M_xAl_{1-x}PO_4$ material with smallest nanopore dimension shows the highest selectivity for terminal oxidation of n-alkanes. For the small nanopore system inter-nanopore connections only allow access of the terminal carbon atom to the oxidation reaction center.

With cyclohexane as reactant the smaller AlPO-31 structure, in this case substituted with Fe cations, is the preferred zeolite for adipic acid production. The Fe-$AlPO_4$ material with wider pore selectively produces cyclohexanol and cyclohexanone. These are products one expects when shape selectivity does not apply [81]. For the smaller nanopore material the adipic acid is produced due to

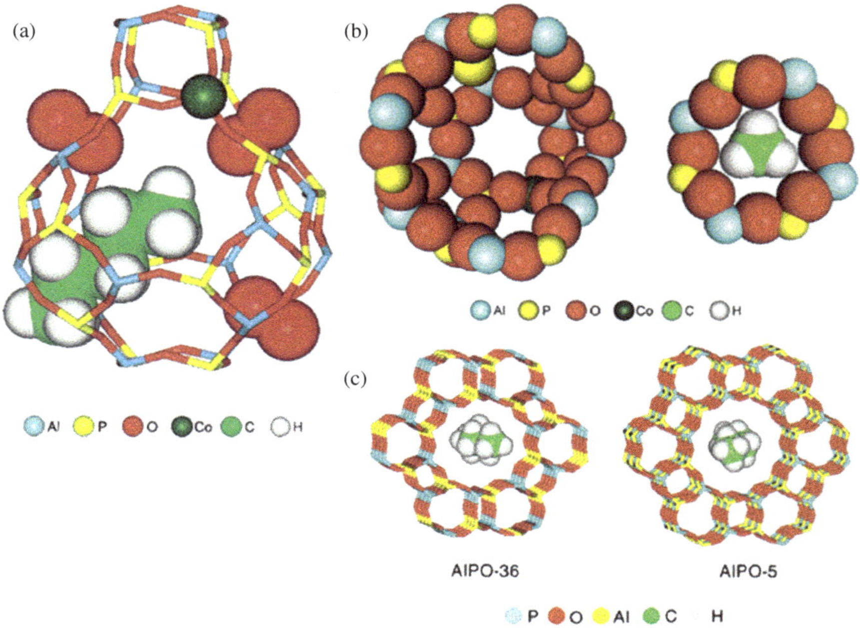

Figure 6.11 (a) Skeletal outline of a single chabazite cage through which molecules of dioxygen (red lobes) permeate freely. Alkanes, on the other hand, can enter the chabazite cage only by an *end-on* approach (see bottom left). (b) Views of (left) the interior of the chabazite cage in CoAlPO-18 that is lined with a Co^{3+} ion, substitutionally replacing a framework Al^{3+} ion; and (right) the terminal methyl group of a linear alkane (n-hexane) fitting snugly into the aperture, the extremities of the van der Waals radii of the methyl group very nearly touching those of the oxygen atoms of the framework. (c) Representation of the n-hexane molecule inside the larger-pore AlPO-36 and AlPO-5 structures. [80].

consecutive reactions because size restriction increases the residence time of oxygenated molecules.

Another illustration of the use of *in situ* generated peroxides to synthesize a desirable product molecule is the invention by Thomas *et al.* [82] of direct synthesis of ε-caprolactam with molecular oxygen. ε-caprolactam is also a monomer for nylon manufacture.

It is another example of the green chemistry program of the 1990s to develop alternatives to stoichiometric organic synthesis processes that produce harmful waste molecules as co-product. In

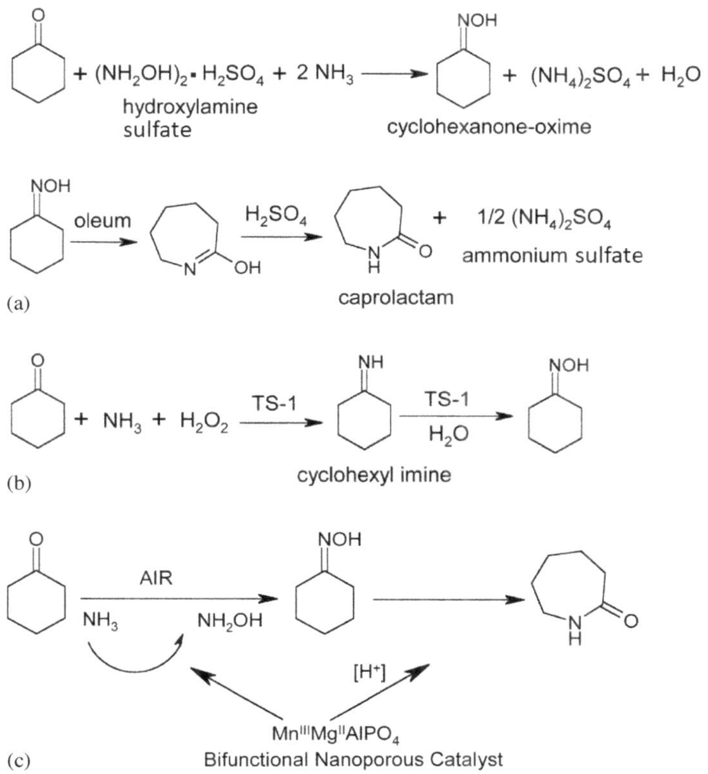

Figure 6.12 Selective oxidation of cyclohexanone to caprolactam [83]. (a) The classical oxidation process with ammonium salt waste. (b) Cyclohexyl imine reaction with ammonia and hydrogen peroxide catalyzed by Ti-silicalite. (c) The Thomas reaction with oxygen catalyzed by bifunctional nanoporous catalyst.

the case of ε-caprolactam current large-scale processes use sulfuric acid and produce ammonium sulfate as waste.

Figure 6.12a illustrates the present process.

Cyclohexanone reacts with hydroxylamine sulfate and ammonia to give cyclohexanone oxime. In a consecutive reaction with neat sulfuric acid cyclohexanone oxime is converted to ε-caprolactam.

The Ti-silicalite catalyst introduced in Section 6.3.1.1 for the selective epoxidation of propene with hydrogen peroxide is also a selective catalyst for conversion of cyclohexanone with ammonia and hydrogen peroxide into cyclohexanone oxime. Thomas *et al.*

designed a wide-pore bifunctional catalyst that converts cyclohexanone with ammonia and oxygen in one reaction to ε-caprolactam. To generate intermediate peroxides in the autocatalytic radical process part of reactants are consumed in sacrificial reactions.

The optimum functioning catalyst has large nanopores with composition $Mn^{3+}Mg^{2+}AlPO$-5. The Mn and Mg cations both substitute for Al^{3+}. Substitution of Al^{3+} by Mg^{2+} introduces a Brønsted acidic proton to the system. The two reactions of Figure 6.12c are executed in gas phase at mild temperature.

The use of Thomas catalysts is preferably in gas phase, since in polar solvent the reactive cations will leach, and the catalyst deactivates.

6.3.2.2 *The Panov Benzene Hydroxylation Reaction*

Catalytic hydroxylation of benzene is an insertion reaction.

The Panov reaction, in which benzene is oxidized with N_2O to phenol, is catalyzed by a low Fe content ZSM-5 zeolite catalyst [84]. The reaction is potentially attractive since it provides an outlet for N_2O, which is an environmentally undesirable greenhouse gas. N_2O is a stoichiometric co-product in the production of adipic acid monomer [85]. In the previous section a potential green reaction is discussed that produces adipic acid by direct oxidation of hexane without N_2O as co-product.

The N_2O reaction with benzene to produce phenol was originally discovered by Japanese chemists in 1983 [86].

The zeolite H-ZSM-5 gives the highest activity. Subsequent research led in 2000 to the announcement by Solutia (formerly Monsanto) on commercial implementation of this hydroxylation reaction. The superior catalyst used by Solutia had been developed jointly with the Panov group from the Boreskov Institute in Novosibirsk [87]. A major event in the development of this catalyst was the discovery that Fe present as an impurity in the H-ZSM-5 catalyst is responsible for its catalytic activity. Fe^{3+} cations that substitute for Al^{3+} in the framework of the zeolite are oxidized during reaction, leave lattice positions and become located as Fe oxycationic clusters in the zeolite nanopores. The reaction is a redox reaction catalyzed by the Fe oxycations. This mechanism is in conflict with initial proposals that the reaction is catalyzed by zeolite protons.

An extensive research effort by several groups concluded that there is one uniquely reactive Fe cation site that activates oxygen of N_2O selectively into the benzene C–H bond [88], [89]. The selective site of the Panov catalyst is a single extra framework Fe^{2+} cation that is charge compensated by the negative charges of two neighboring Al^{3+}-containing framework tetrahedra.

The Fe^{2+} cation converts into $[FeO]^{2+}$ when reacted with N_2O. The selective elementary step that produces phenol is a concerted insertion step of O into the C–H bond of benzene, which proceeds via intermediate oxene. The currently accepted mechanism of the reaction is illustrated in Figure 6.13.

Figure 6.13a schematically illustrates the catalytic cycle, while Figures 6.13b1 and 6.13b2 compare DFT-calculated potential energy changes of the respective elementary reactions for Fe^{2+} and $[FeO]^+$ adsorbed into the nanochannel of zeolite ZSM-5. The activation energy of N_2O is higher on the Fe^{3+} cation than Fe^{2+}. This relates to the smaller oxygen atom adsorption energy with the Fe^{3+} cation, on which already an oxygen atom is attached.

The insertion of the oxygen atom into the benzene C–H bond via an oxene-type transition state has lower activation energy for the Fe^{3+} cation. The desorption energy of phenol has the higher activation energy, is comparable for the two systems, and is rate controlling.

A major difference between the two systems is that a competitive reaction with adsorbed phenol on the FeO^+ complex happens that is inhibited for the Fe^{2+} cation. The oxygen atom of $[FeO]^+$ is strongly Lewis basic. Product phenol readily cleaves its O–H bond in a heterolytic reaction and the phenoxy species adsorbs on the Fe^{3+} cation. Once phenoxy is formed deactivation reactions initiate, because of oligomerization of phenol [91], [92]. A phenoxy-forming heterolytic O–H bond cleavage reaction does not happen for phenol adsorbed on a single cation because of the much lower Lewis basicity of the

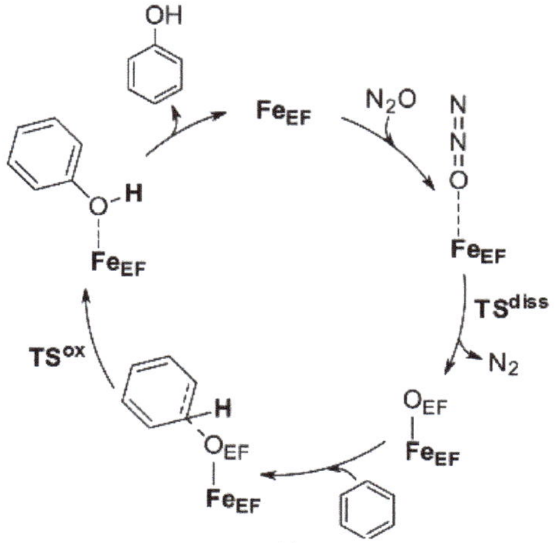

(a)

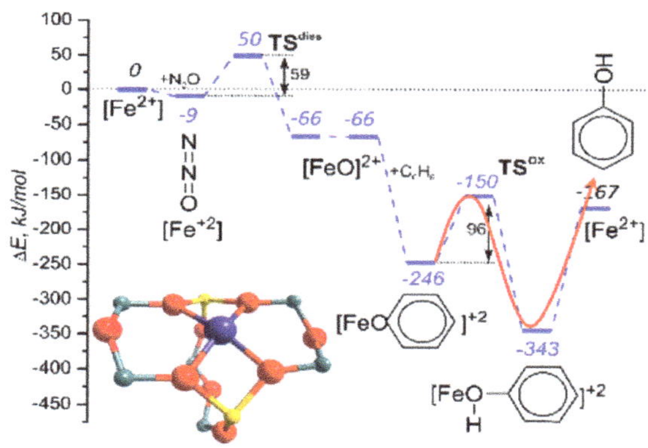

(b1)

Figure 6.13 The hydroxylation of benzene by N_2O [90]. (a) The catalytic cycle of benzene to phenol via oxidation with N_2O by extra framework iron (Fe_{EF}) sites. (b) The respective calculated potential energies of the selective oxidation of benzene. Comparison of activation by Fe^{2+} (b1) and FeO^{+} (b2) cation (dark blue, Fe; yellow Al; red O; light blue Si).

zeolite lattice oxygen atoms. Differences in the rate of catalyst deactivation determine the preferable Panov catalyst. In agreement with computational suggestions the $[FeO]^{2+}$ site has been experimentally identified as the site of selective oxidation by Snyder *et al.* [93].

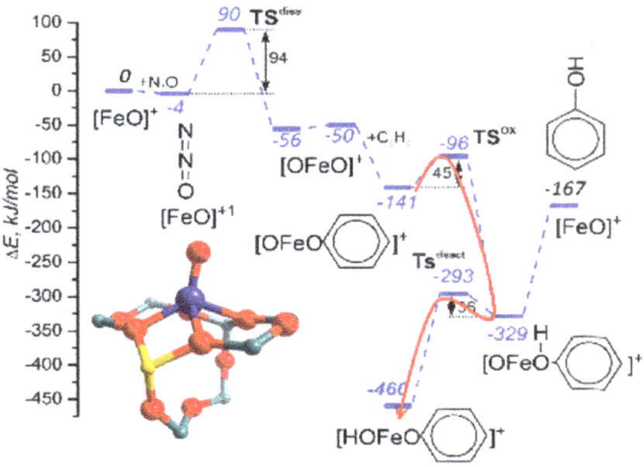

(b2)

Figure 6.13 (*Continued*)

6.3.2.3 *Methane to Methanol Oxidation*

Mimicking enzyme catalysis; the harpooning mechanism. Shape-selective catalysis by frustrated radical diffusion.

Direct oxidation of methane to methanol is a highly desirable reaction. It would replace current methanol synthesis processes from syngas. The methane to methanol oxidation reaction has low selectivity because methanol has a higher reactivity than methane [94], [95]. The reaction is therefore preferably executed by selecting a procedure where methanol and oxidant have no direct or indirect contact [96]. This is the case in the biocatalytic oxidation of methane to methanol by the monooxygenase enzyme, where the active oxygenated intermediate is *in situ* generated by an electrochemical reaction or by oxidation of the catalytically reactive metal oxide complex [97], [98].

Most man-made systems that are discussed in this section use two-step procedures, where in one step the catalyst is oxidized and in a second step, in the absence of gas-phase oxidant, the reaction between catalyst and methane occurs. The exceptional case is one-step oxidation of methane with hydrogen in water [99].

Already in 1997 Panov *et al.* [100] investigated the question of whether the catalyst that catalyzes hydroxylation of benzene with N_2O will also catalyze conversion of methane to methanol. They observed that methane generates methanol at low temperature when it reacts with a catalyst that had been pre-treated at high temperature with N_2O. Methanol remains strongly adsorbed on the oxycationic reaction center. Methanol could be identified by an extraction process with a mixture of water and additional methanol. When methanol is desorbed in gas phase it decomposes. H/D isotope effect measurements showed a large effect for methane and a smaller effect for phenol. It agrees with the mechanism of methanol formation by complete C–H bond rupture of methane, whereas the C–H bond remains intact in the transition state of benzene hydroxylation [101].

Using magnetic circular dichroism spectroscopy of the Panov catalyst, Snyder *et al.* in 2016 [93] identified the structure of the reactive Fe cation. They demonstrated that the single $[FeO]^{2+}$ cation not only hydroxylates benzene but also methane.

Selective methane activation by the oxycationic center can be initiated by a homolytic radical reaction or by heterolytic C–H bond activation. The radical reaction corresponds to the rebound or harpoon mechanism [93], [102]–[104].

6.3.2.3.1 The Rebound/Harpoon Mechanism of Methane Hydroxylation

The rebound/harpooning radical mechanism mimics enzyme catalytic chemistry. Shape-selective catalysis by frustrated radical diffusion.

The rebound hydroxylation mechanism of alkanes was proposed in the 1970s by Groves *et al.*, who studied Fe^{2+} complexes such as cytochrome P450, the active component of the heme-iron enzyme [109]. Since then the mechanism has become widely accepted and is also found to apply for the zeolitic systems discussed here. These systems biomimic the enzyme-catalyzed reactions.

Cu oxycationic complexes in zeolites have been extensively explored for the methane hydroxylation reaction. There is a variety of dicopper cationic complexes known. The mechanism of hydroxylation depends on the radical character of reactive oxygen atoms.

The reactivity of the Cu oxycations to generate CH_3 radicals relates to the radical character of the oxycation O atoms. This is minimum for $[CuOCu]^{2+}$, where the oxygen atom has a formal charge of −2. It increases for $[Cu_2O_2]^{2+}$ in which the oxygen atoms are part of O_2^{2-}, and further to $[Cu_3O_3]^{2+}$ with a formal oxygen charge of −2/3. The actual composition of the zeolite strongly depends on preparation conditions and zeolite matrix.

For Cu in ZSM-5 and mordenite the $[Cu_2O_2]^{2+}$ or $[CuOCu]^{2+}$ cluster has been suggested to activate the C–H bond of methane. The $[CuOCu]^{2+}$ complex is the most active [102], [110], [111]. When it is oxidized at high temperature the $[Cu_3O_3]^{2+}$ complex is formed. This also is an active methane hydroxylation catalyst [103], [112].

A schematic representation of the radical rebound mechanism is given in Figure 6.14 for a model of the $[Cu_3O_3]^{2+}$ cation. The initial step is abstraction of the hydrogen atom from the C–H bond of methane and a CH_3 radical is generated. In a second step this radical intermediate reacts with cluster bound O–H. The product methanol desorbs in a reductive elimination step. The Cu complex can be regenerated by oxidation with molecular oxygen.

A confirmation of the rebound mechanism for methane hydroxylation by the Fe cationic systems is shape selectivity of this reaction, as was demonstrated in a joint research effort by researchers from Leuven and Stanford [113]. The selectivity of methanol production by Fe^{2+} in zeolite β (with 12-ring channels) and chabazite (with 8-ring connectivity between cavities) was compared. These systems are analogous to the Panov catalyst. The reaction was executed in

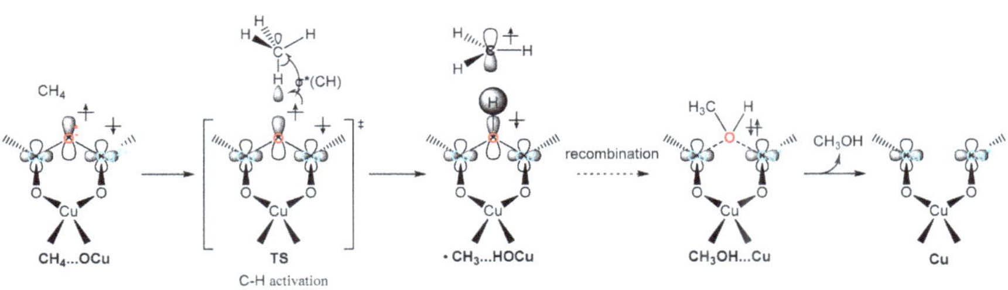

Figure 6.14 The rebound mechanism [103]. Schematic representation of orbital interactions involved in C–H bond activation by the reactive $O\cdot^-$ radical in Cu-oxo clusters. Green: Cu d-orbital, red: O p-orbital.

two steps. In a high-temperature step the Fe cation was oxidized by decomposition of N_2O. At low temperature the $[FeO]^{2+}$ complex was reacted with methane. The state of the Fe complex was followed using Mossbauer spectroscopy. A large difference in methanol yield was measured between wide-pore zeolite β and chabazite that has narrow eight-ring connections between its cavities. In zeolite β a significant fraction of the freely moving methyl radical fragments will not react with $(FeOH)^{2+}$ to give methanol but with still present non-reacted free FeO^{2+} cations. The latter reaction with the methyl radical fragment leads to total combustion and selectivity for methanol formation is low. In contrast, in chabazite the narrow pore connections limit movement of CH_3. Its probability to react with $(FeOH)^{2+}$ is increased and selectivity of reaction is higher.

6.3.2.3.2 The Heterolytic Pathway of Methane Oxidation

Heterolytic C–H bond activation of methane is promoted by reaction with water.

The non-radical character of the $[CuOCu]^{2+}$ complex is the cause of its liability for heterolytic activation of methane. Water promotes reaction and protonation of zeolite lattice oxygen atoms. This has been demonstrated by van Bokhoven *et al.* from ETH Zürich for catalysis by the di-copper cationic cluster located in the nanopore of mordenite. In the experiment a reaction cycle of methane activation and water extraction was used. They used *in situ* infrared, XANES, and mass spectrometry to observe reaction intermediates of heterolytic C–H bond dissociation of methane. In a two-step reaction a Cu-methoxy species and zeolitic proton is generated [106], [108]. This rejects an alternative suggestion of Li *et al.* [103] who instead proposed formation of methoxide attached to the zeolite framework oxygen atom and Cu cluster O–H formation.

The other important observation by van Bokhoven *et al.* is that reaction can be executed anaerobically. Methanol is produced in high yield in alternating cycles of methane (9 bar) and water in He (1 bar) at 473 K. Hydrogen was co-product. Isotope experiments with ^{18}O-labelled water demonstrated incorporation of the oxygen atom of water into the methanol molecule. Water reoxidizes Cu^+ to Cu^{2+}.

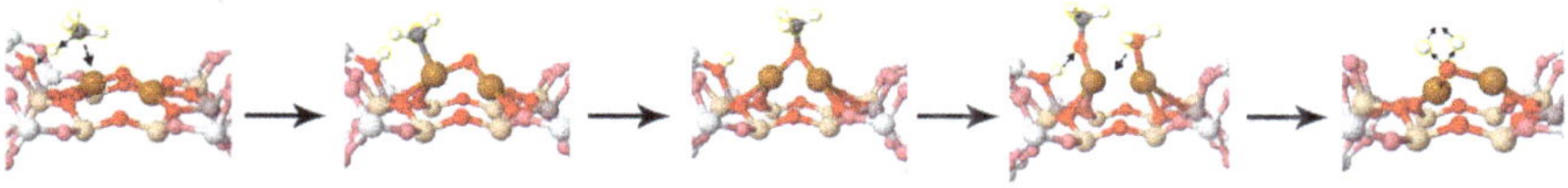

Figure 6.15 The succession of reaction intermediates for the oxidation of methane to methanol promoted in water by a di-copper cationic cluster in mordenite [106].

The reaction of methane and water to methanol and hydrogen is endothermic. Reaction is possible since it is two-step. Subsequent reactions with methane and water are performed at different conditions. In the overall reaction methanol is formed with 97% selectivity. The succession of elementary steps is illustrated in Figure 6.15.

As this figure illustrates the initial reaction of methane with the $[CuOCu]^{2+}$ oxycation is heterolytic and a zeolite proton and Cu-methyl intermediate are formed. This Cu-methyl intermediate is not stable and converts to the more stable Cu-methoxy intermediate. In a next concerted step adsorbed H_2O and the zeolite proton react with the methoxy intermediate. Methanol is formed and the oxycation is reoxidized after desorption of H_2.

The two reaction paths for methane hydroxylation are the homogenous radical rebound mechanism and the heterolytic C–H bond activation. The latter has also been demonstrated to happen by reaction with $[FeO]^{2+}$ in gas-phase mass spectrometric experiments (see also Section 4.5.6), but is a spin-forbidden high-activation reaction [104]. The heterolytic pathway requires low-spin complexes.

The main mechanistic difference between the hydroxylation of benzene and methane is that in the latter the C–H bond cleaves before C–OH bond formation. In benzene hydroxylation, the C–H bond only cleaves in the oxene complex. Its formation is assisted by the stabilizing interaction with the aromatic ring π electrons.

6.3.2.3.3 The Hydrogen Peroxide Reaction with Methane

Hydrogen peroxide decomposition initiates radical reactions. OH intermediates give methanol.

Hydrogen peroxide in water selectively oxidizes methane to methanol by a radical reaction. Hutchings et *al.* at the University of Cardiff [99], [114] discovered this reaction catalyzed by Fe-containing ZSM-5. This reaction is not unique to the zeolite ZSM-5; it is also catalyzed by the $[FeOFe]^{2+}$ cation in the nanoporous metal organic framework MIL-53 [115]. A dramatic improvement in selectivity of methanol was found when the Fe catalyst was promoted by Cu. The main contribution of Cu is to reduce excess OH radical formation that causes consecutive oxidation of methanol.

Figure 6.16 schematically shows the suggested mechanism of the hydrogen peroxide reaction with methane to methanol catalyzed by the di-iron oxide cation.

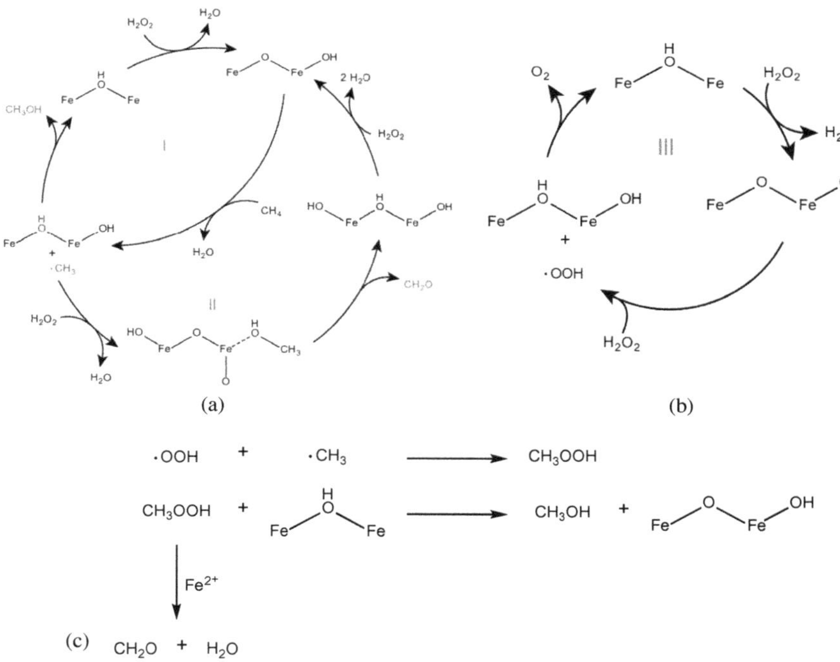

Figure 6.16 The oxidation of methane to methanol with hydrogen peroxide catalyzed by the di-Fe oxycation. (a) The catalytic cycles I and II of the oxidation of CH_4 to CH_3OH and CH_2O. (b) The decomposition of H_2O_2 to O_2 (cycle III), and generation of OOH radicals. (c) The recombination of CH_3 and OOH radicals, formation of methanol and formaldehyde from methylperoxide.

The cycle is initiated by reaction of methane with the bridging O atom of the di-Fe oxycation with formation of a CH_3 radical. Hydrogen peroxide reacts with OH of the $[FeO–HFe]^{2+}$. This gives adsorbed (radical) OH and H_2O. This OH species recombines with the CH_3 radical to give methanol (cycle I, Figure 6.16a). It can also react with another hydrogen peroxide to give the OOH radical and O_2 evolution (cycle III, Figure 6.16b) or oxidize methanol to formaldehyde (cycle II, Figure 6.16a). The latter will initiate total oxidation.

The OOH radicals can generate additional OH radicals but can also combine with a CH_3 radical to CH_3OOH, which was observed by Hutchings *et al.* It can decompose to methanol according to Figure 5.16c with formaldehyde as co-product.

According to [115] the decomposition rate of hydrogen peroxide catalyzed by the iron oxycation is fast compared to the rate of methane oxidation. The Fe-Cu oxycation suppresses this decomposition reaction.

6.3.3 *Methane to Aromatics Catalysis; The Methane Dehydro-aromatization Reaction*

Anaerobic methane to aromatics catalysis is a high-temperature radical reaction. It is initiated by reaction of methane with a reducible cation.

Methane oligomerization to higher hydrocarbons is an endothermic dehydrogenation reaction. According to thermodynamics at 800 K aromatics can be formed, above 1000 K ethene, and ethyne at 1200 K. At these temperatures product-forming reactions are gas-phase radical reactions. The main challenge in non-oxidative selective methane conversion is to suppress deactivating coke formation because this is thermodynamically most favorable.

There are two approaches to suppress coke formation. Short residence time of the reactant will reduce consecutive reactions and hence coke formation. When zeolitic systems are used shape-selective catalysis will suppress formation of bulky coke-forming intermediates.

The rate of high-temperature gas-phase methane conversion is greatly enhanced by coating a quartz reactor wall with the amorphous single-site Fe/SiO_2 catalyst discovered for this reaction in 2004 by Bao *et al.* from Dalian Research Institute in China [116]. Methane is converted to ethene and aromatics at 1300 K. The reaction is selective when operated with millisecond residence time. The radical mechanism relates to that of non-catalytic classical industrial processes for ethyne production from methane in arc plasmas such as the Ashland/ISP Marl, which operates on a comparable timescale [123]. When the quartz reactor is coated with the Bao Fe/SiO_2 catalyst it has a tenfold increase in reaction rate, due to increased rate of CH_3 radical formation catalyzed by the isolated Fe sites. Coke selectivity is 25%. This is high compared to coke selectivity of the shape-selective catalysts. However, deposition of coke does not suppress reaction rate, but gradually fills the reactor, which leads finally to deactivation. The coke production rate relates to the surface/volume ratio of the reactor [130].

The active Fe/SiO_2 is a unique single-site catalyst. Fe is part of a carbide FeC_2 cluster. Fe is coordinated to two carbon atoms as schematically shown in Figure 4.43. Methane activation is heterolytic. C–H bond cleavage gives H attached to a C atom around Fe and CH_3 coordinates to Fe. In subsequent steps a hydrogen molecule is formed by recombination of the H atoms and the CH_3 radical initiates oligiomerization and aromatics-forming radical reactions.

Shape-selective zeolite catalysts produce aromatics at the lower temperature of 1000 K. They are promoted by Mo and are only selective for benzene when they have ten-ring channel cross-sections. Zeolites with nanopores of larger ring size rapidly deactivate by coke formation and those with smaller nanopores cannot form aromatics due to space restriction [127].

The later widely investigated Mo-H-ZSM-5 catalyst for the methane to aromatics reaction was discovered in 1993 by Wang *et al.*, also at Dalian Research Institute [125]. Zeolite catalysts promoted with Fe also catalyze this reaction [117]. Reviews of these reaction systems can be found in [118], [119].

The catalyst as prepared contains single metal atom Mo(VI) oxo centers that reduce upon reaction with methane [120]. The catalytically selective Mo cation sites are located in the nanopores of the zeolite. In quasi-steady state condition the Mo oxo center transforms into a Mo(VI) oxo carbon complex that activates methane heterolytically [121], similar to the Bao carbide Fe/SiO_2 system. Coke formation is suppressed inside the zeolite by the limited size of the nanopore. Part of the Mo cations agglomerate as Mo_2C species, which will move externally of the zeolite nanopores. Whereas protons are not strictly necessary for the reaction, catalysts that contain protons have enhanced reactivity. This is ascribed to improved anchoring of the carbided Mo species inside the zeolite nanopores.

The reaction proceeds through three stages: activation of Mo oxide cluster compounds by reduction with methane, formation of MoC_x species, and autocatalytic formation of benzene [122]–[124]. The latter is reminiscent of the gas-phase reactions where deposited coke also activates radical intermediate formation.

As Wang *et al.* mention in their original paper [125] the idea to select a bifunctional acidic zeolite for the methane activation reactions stems from an early Russian paper by Minachev *et al.* [125]. They suggested that methane can be converted to aromatics by the Zn or Ga bifunctional zeolite catalysts used in the Cyclar reaction for the aromatization of ethane.

Instead of the radical mechanism it has been originally proposed that the Mo-H-ZSM-5 catalyst operates as a bifunctional catalyst. The Mo-containing reaction center generates ethene. In a consecutive reaction ethene is converted to aromatics by reactions catalyzed by the zeolite protons [126].

Figure 6.17 schematically illustrates a mechanistic proposal supported by DFT calculations [127] of a reaction path for ethene from methane. The key intermediate is a CH_2 carbene species attached to the metal cation (see Section 6.2 for the related disproportionation reaction). Transition metal cations in high valence states are known to stabilize such carbene intermediates.

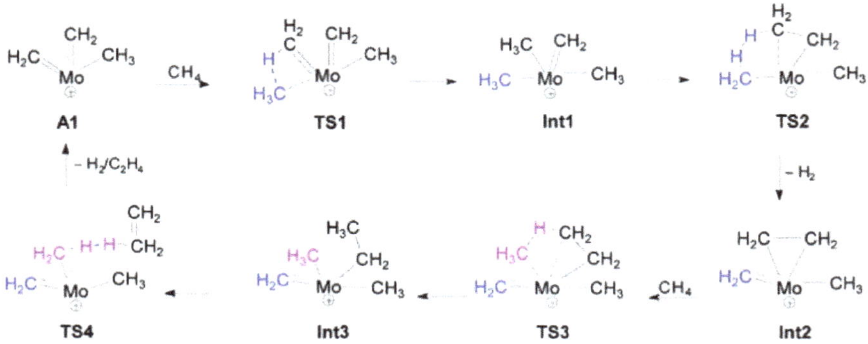

Figure 6.17 Reaction pathway for methane coupling to ethene over a single Mo cation [127].

Methane reaction with CH_2 gives two adsorbed CH_3 intermediates. These species dehydrogenate to adsorbed ethene. Additional reactions with methane regenerate the initial complex. In the mechanistic proposal of Figure 6.17, after ethene formation a third carbon-containing ligand remains as part of the reaction complex.

The experimental observation that reactive Mo sites are carbide clusters does not agree with the carbene mechanism of ethene formation of Figure 6.17. The activity of carbided Mo or Fe clusters for C–H bond cleavage is consistent with the reactivity of iron and molybdenum carbides for hydrocarbon activation [128], [129].

6.3.4 *Summary; Clusters in Zeolites*

The discovery that single- or multi-metal clusters occluded in the nanopores of zeolites give new and interesting catalytic reactivity has led to their extensive exploration. In previous sections catalysis by non-redox and redox systems is introduced. A variety of synthesis approaches of such systems exists. Coordination complex synthesis, activation of metal organic compounds, as well as classical impregnation methods are used [8].

An early example of the practical application of metal clusters in zeolites is the chabazite catalyst with Cu oxide clusters that is used as diesel exhaust treatment catalyst (see Figure 4.40). As discussed in previous sections of this chapter, such redox oxycationic clusters are also active hydroxylation catalysts of methane. In addition to Cu clusters Fe oxycationic clusters are also active as selective oxidation catalysts. Anaerobic methane and ethane aromatization reactions have been discussed with Ga, Zn, and Mo. The mechanisms of these reactions are discussed in detail in the earlier sections.

Salient practical features of the relation between nanocluster composition and catalytic reactivity are:

— Often a distribution of different nanoclusters is present in the nanopores of the zeolite;
— The reactivity differences between single-atom site and multi-metal atom nanocluster can be large;
— Which structure or composition is optimum may depend strongly on reaction and reaction conditions;
— The distribution of metal clusters and their composition may change with time onstream. Ultimately often one nanocluster structure and composition dominate reactivity. This can be a minority of the overall nanocluster distribution present in the zeolite nanopore;
— The reactivity of clusters deposited externally or internally of the zeolite particle is usually very different. Cluster particles deposited externally of the zeolite often catalyze undesirable non-selective and deactivating reactions.

6.4 Single-atom/Reducible Support Catalysts; Au Catalysis

6.4.1 *Single-atom Catalysis*

Molecular bond dissociation by transition metal cation requires reducible oxide vacancy. Activity is high because, different from a metal at low temperature, a cation is not poisoned by reactant.

Reactive metal cations dispersed on reducible supports such as CeO_2 are uniquely reactive. Reactant molecules are activated by a two-point contact with transition metal cation and support oxygen vacant site. Here mechanisms of reactions catalyzed by these single-site systems are introduced.

Reducible support catalysis with CeO_2 was discovered as part of automotive gasoline exhaust catalysts, which were widely implemented in automotive vehicles after 1981.

In the exhaust gas the catalyst converts NO_x to N_2 by reaction with CO, H_2, or a hydrocarbon. This reduction requires the absence of excess oxygen in the exhaust. The air-to-fuel ratio has to be equal to one. Therefore the exhaust composition is regulated with oxygen sensors, which enables operation at near neutral conditions of oxidation-reduction stoichiometry.

The catalyst consists of one or several of the transition metals Pt, Pd, and Rh distributed on CeO_2/ZrO_2 reducible support [130] and is sensitive to lead or sulfur poisoning. Non-leaded fuel is mandatory in the USA since 1975 and sulfur levels in gasoline are reduced by increasingly severe legislation.

Pd mainly functions as oxidation catalyst for CO and hydrocarbons. Rh is an active NO_x reduction catalyst that reacts with hydrogen, CO, or hydrocarbon. Pt serves both functions.

In the exhaust treatment catalyst the function of the reducible mixed CeO_2 support oxide is to balance oxygen stoichiometry in the exhaust gas mixture. It stores oxygen when excess oxygen is present and provides oxygen at reducing conditions. It took some time before it was recognized that CeO_2 also has another unique catalytic role.

Proper understanding of the composition and structure of these systems is only recent. Main contributions came from High Resolution Electron Microscopy that can visualize monoatomically dispersed transition metal atoms on the reducible supports [131], [132].

The transition metal atoms are present as isolated cations or as small (oxidic) metal cation clusters dispersed on the reducible support [9]. In the single-site catalysts, the metal atoms are present as cations and substitute for the Ce cations. Charge imbalance introduced by difference in charge of the transition metal cation with the Ce cation is accommodated through oxygen vacancy formation of the oxide.

The mechanism of a reaction catalyzed by the single-site/reducible oxide catalyst is illustrated here for the reduction of NO with CO. Single metal atoms cannot dissociate the N=O bond, which is a necessary step to generate the N atoms that upon recombination give N_2. Due to the synergetic interaction of a single metal atom with the oxygen surface vacant site of CeO_2, N=O bond dissociation becomes possible [133]–[135]. The oxygen vacant site of the reducible oxide assists in N=O bond dissociation because then the Ce cation can accommodate the oxygen atom that results from N=O dissociation.

The mechanism of NO reduction with CO to give N_2 catalyzed by a Pd cation incorporated in the CeO_2 surface is shown in Figure 6.18 that is based on DFT calculations [136].

Reaction initiates by reduction of CeO_2 with CO. This makes possible the adsorption of two NO molecules on a reaction center that composes of a Pd cation and CeO_2 surface vacant site. Intermediate N_2O is formed, and an oxygen atom adsorbs on the vacant site of CeO_2. In a subsequent step the oxygen atom of N_2O reacts with a second CeO_2 vacant site and an N_2 molecule is generated. The surface vacancies are regenerated by successive reaction with CO and CeO_2 oxygen atoms to give CO_2.

The uniqueness of the CeO_2 support relates to its low energy of surface oxygen atom vacancy formation [137], about 200 kJ/mol

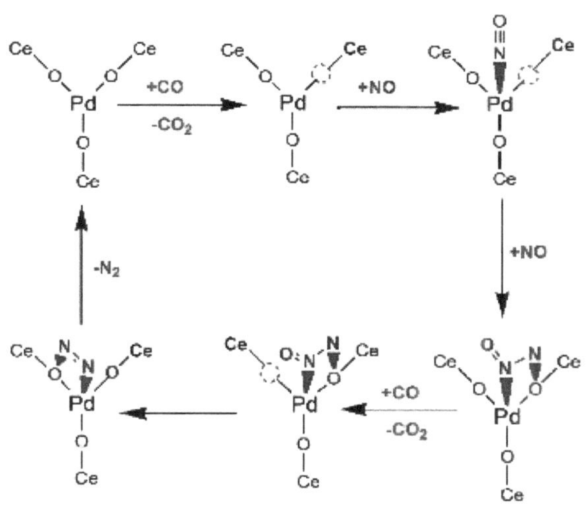

Figure 6.18 Schematic representation of catalytic cycle derived for the 2 NO + 2 CO → N_2 + 2 CO_2 reaction on a Pd/CeO_2(111) model catalyst [136].

[138]. This is substantially lower than the surface oxygen vacancy formation energies of selective oxidation catalysts such as MoO_3 or V_2O_5 that are of the order of 600 kJ/mol [139].

A spectacular aspect of these systems is their extremely high reactivity. Single-atom catalysts of Pt adsorbed on CeO_2 oxidize CO to CO_2 or catalyze the water-gas shift reaction of CO with H_2O at a temperature as low as room temperature. In addition to catalysts supported by CeO_2 other reducible supports such as Fe_2O_3 or TiO_2 can also be used.

The reactivity at low temperature is due to the metal atom state as a cation. Cations such as Pt^{2+} binds CO weakly, whereas CO has a strong chemisorption bond with the reduced metal atom (representative numbers are 180 kJ/mol for CO adsorption to Pt surface versus 80 kJ/mol for CO adsorbed to the respective cation).

The temperature of reaction is controlled by the temperature of CO desorption. Instead of room temperature for the cation, for a reduced transition metal this temperature is 450–500 K. Catalysis by metal cation instead of metal atom is also the reason for the low-temperature CO oxidation reaction when catalyzed by small Rh_2O_3 particles dispersed on CeO_2 support [140].

The low temperature of CO oxidation makes the catalyst suitable as PReferential OXidation (PROX) catalyst. The PROX reaction is useful in fuel cell operation, where CO poisoning of the Pt electrode has to be prevented. CO is often present in hydrogen as an impurity. The PROX catalyst selectively oxidizes CO in the presence of hydrogen [141], [142].

Whereas adsorbed CO will readily react with reactive O from the CeO_2 support to give CO_2, dissociation of O_2 is not possible on a single metal atom. Jointly with the surface vacancy low activation barrier dissociation of O_2 becomes possible that regenerates the dual reaction site [143].

6.4.2 *Au Catalysis*

Au catalyzes molecular dissociation when activated by Lewis acidic support cations or water.

The discovery by Japanese scientists in 1987 of low-temperature CO oxidation by Au dispersed on Fe_2O_3 or other reducible supports is a milestone for several reasons [144].

Au had long been considered to be chemically inert. This is one reason that this discovery of Au catalysis was unexpected, along with the understanding that its reactivity relates to synergetic interaction with reducible oxide or Lewis acidic supports. Single Au^{3+} cations are also active CO oxidation catalysts when part of a solid solution such as La_2O_3 [145].

The discovery in 1987 had been preceded by earlier unrelated observations of Au reactivity by Bond *et al.* [146] that Au atoms (probably as partially chlorinated Au^{3+}, the catalyst had been prepared by impregnation with $AuCl_3$) dispersed on a silica or alumina support have modest activity as alkene hydrogenation catalyst. Around the same time Hutchings discovered that $AuCl_3$ is an active ethyne hydrochlorination catalyst [147]. Especially since the discoveries of Haruta *et al.* Au particle catalysis has widely been explored. It is a promising catalyst for several other important reactions [148]–[151].

The causes of the unique reactivity of Au are now reasonably well understood. Au can be active as a small particle or a single cation. The dual interaction with a reactive support is usually essential, but there are also systems where the support does not play a role. However, the presence of water is then essential. For low-temperature oxidation Haruta *et al.* discovered in 1991 this reactivity for hydrated larger Au particles. It is initiated by reaction of O_2 with H_2O [152].

A comparison of the activation of O_2 according to the dual site mechanism and by water is illustrated by a revealing experiment of Haruta *et al.* [153], who made this comparison for the CO oxidation reaction.

Figure 6.19 illustrates the substantially higher reactivity of an Au particle supported by the reducible oxide TiO_2 than by the non-reducible oxide Al_2O_3 or SiO_2. The dual site mechanism is only operational on the TiO_2-supported catalyst. The mechanism of this activation is discussed in detail in Section 6.4.2.1.

The reactivity of the Au/Al_2O_3 catalysts increases at least two orders of magnitude by water addition and approaches that of the TiO_2-supported Au catalyst.

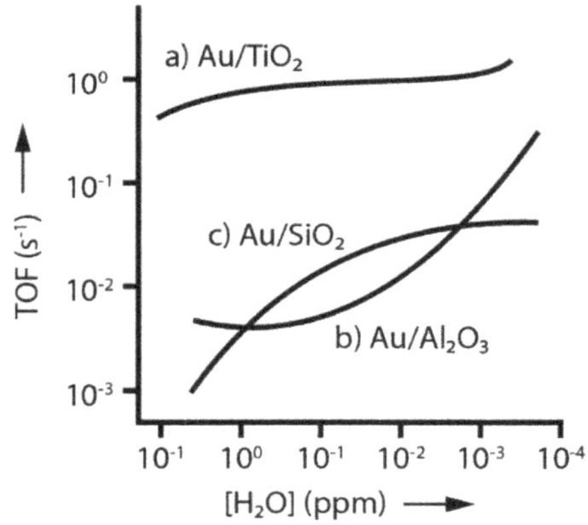

Figure 6.19 Turnover frequencies per surface gold atom at 273 K for CO oxidation catalyzed by (a) Au/TiO_2, (b) Au/Al_2O_3 and (c) Au/SiO_2 as a function of moisture concentration [153].

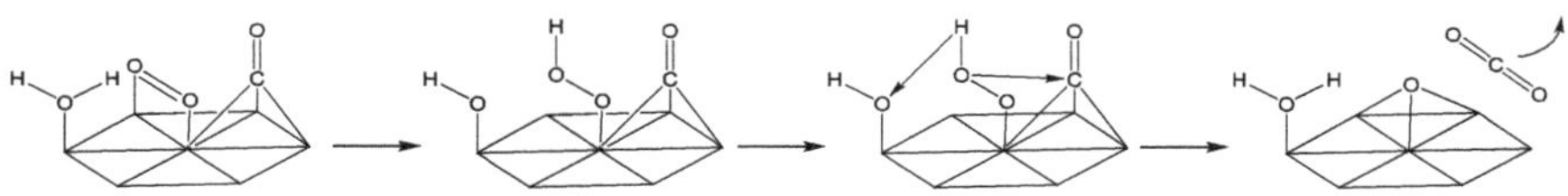

Figure 6.20 Schematic illustration of H_2O-promoted CO oxidation on Au surface [154].

A suggested mechanism of O_2 activation based on quantum-chemical calculations and H/D exchange rate measurements [154]–[156] is illustrated in Figure 6.20.

This schematic shows initiation via hydrogen atom abstraction by molecular oxygen from adsorbed water. The resulting adsorbed OOH intermediate oxidizes adsorbed CO. One oxygen atom is transferred to adsorbed CO, while the other proton is backdonated to water. In Section 6.4.2.3 an alternative mechanism is presented where the O–OH bond cleaves before the oxygen atom reacts with substrate and the oxidation-active agent is the OH radical [153].

The promoting role of water is not necessarily only mechanistic. As has been pointed out by Kung *et al.* of Northwestern University [157] catalysts may deactivate by carbonate formation at the dual sites. Water will decompose the carbonate and restore reactivity.

One reason for its unique low-temperature reactivity is the very inertness of Au. The adsorption energy of molecules on Au is small, therefore the surface will not be readily poisoned by reactants that otherwise strongly adsorb on the more reactive transition metals. This inertness of Au is used in transition metal alloys to change selectivity of hydrocarbon conversion reactions. The chemistry of catalytic sites changes by the presence of non-reactive Au by reduction of the ensemble size of the transition metal atoms (Section 3.1.4.2). An example from oxidation catalysis is the Pd/Au catalysts for the selective production of hydrogen peroxide from H_2 and O_2. The catalyst is selective because isolated Pd atoms will not dissociate O_2 [158], [159].

6.4.2.1 *CO Oxidation and the Water-gas Shift Reaction; The Dual Site Mechanism*

Dissociation of O_2 and water occurs on dual sites at the interphase between Au particle and Lewis acid surface cations. Au is not poisoned by CO. This makes low-temperature reactions possible.

Already more than half a century ago the phenomenon known as Strong-Metal-Support-Interaction (SMSI) was identified [160]–[162]. SMSI refers to catalytic reactivity of a reactive metal particle when in contact with a reducible support. At a molecular level the cause of SMSI remained poorly understood for a long time. Here simulations of the reactivity of the interphase of an Au particle and TiO_2 are presented, which illustrate SMSI events.

For CO oxidation the atomic details of the elementary events that happen at the interphase of Au particle and TiO_2 support have been recently elucidated [163].

Yates and Neurock, then at Virginia University, used infrared kinetic measurements in combination with quantum-chemical calculations to study this low-temperature reaction at 120 K.

As illustrated in Figure 6.21 O_2 adsorbs on a dual Au/Ti site and is activated by co-adsorbed CO. The molecular O_2 cleaves and CO_2 forms by reaction of an oxygen atom with CO. The Au edge atoms assist O_2 to dissociate but are not poisoned by CO as would happen at low temperature with more reactive transition metals. The coordinative unsaturation of Au edge atoms in combination with the oxophilicity of Ti^{4+} makes dissociation of O_2 possible.

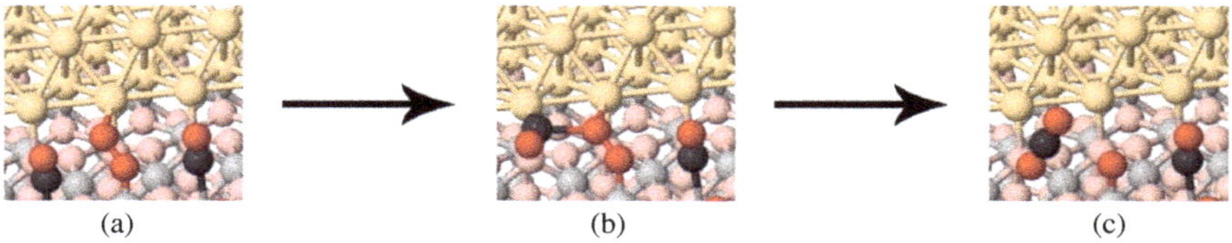

Figure 6.21 CO oxidation at the interphase of an Au particle on TiO_2 [163]. (a) Activation of O_2 by Au edge atom, CO adsorbs on TiO_2. (b) and (c) show CO_2 formation by the successive steps of adsorbed CO with Au-attached O_2 and finally O_2 dissociation.

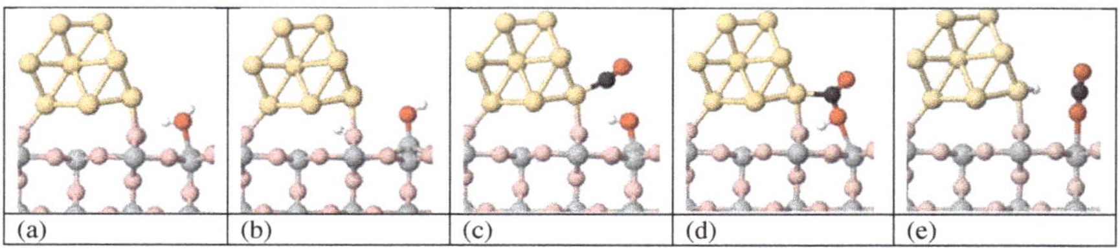

Figure 6.22 Successive stages of reaction of H_2O with CO at an Au/TiO_2 interphase [167]. (a), (b) Dissociation of H_2O at the Au/TiO_2 interface. (c), (d) Reaction of CO and OH, CO preferentially binds to the Au sites whereas OH binds to the Ti^{4+} cation. (e) Deprotonation of *COOH at the Au/TiO_2 interface. The Au, Ti, and O lattice atoms are shown in yellow, grey, and pink, respectively, whereas the O and H atoms from water are shown in red and white, respectively.

An analogous reaction mechanism is that of the low-temperature water-gas shift reaction catalyzed by the Au/TiO_2 catalyst. The water-gas shift reaction produces H_2 from reaction of CO with H_2O. A low temperature for this exothermic reaction is desirable. Whereas the classical process with $Cu/Zn/Al_2O_3$ catalysts operates at 600 K [164], Au/TiO_2 catalyzes this reaction already at 400 K [165], [166].

Simulated stages of the reaction are illustrated in Figure 6.22. Reaction happens at the edge of the Au particle near a surface cation site. Upon H_2O bond dissociation a proton attaches to the oxygen atom at the interphase of Au particle and support, and OH^- binds to the surface Ti^{4+} cation. In a follow-up step the support OH reacts with CO to give intermediate $COOH^-$. CO_2 is liberated by reaction of hydride with the Au particle. The H_2 molecule desorbs upon combination of the H atoms. The reaction of CO and OH is the slow step.

Au supported on TiO_2 is more reactive than on Al_2O_3. This difference in reactivity relates to the weaker Ti–OH bond strength compared to that of Al–OH. Also Cu catalysts have higher activity when nanosized Cu particles are deposited on TiO_2 [164]. The water-gas shift mechanism is similar to that of the Au catalyst. The temperature of reaction catalyzed by Cu is higher because at lower temperature Cu sites are poisoned by co-adsorbed spectator formate or carbonate species [165], [166].

6.4.2.2 *Alcohol Oxidation in Gas Phase*

The dual site of gas-phase alcohol oxidation is the Au nanoparticle edge and a Lewis basic oxygen of the reducible oxide.

Selective oxidation of alcohol to the aldehyde with O_2 is catalyzed by an Au particle distributed on a reducible support. Reaction is initiated by deprotonation of the alcohol O–H group by a Lewis basic oxygen atom.

Here the mechanism of the gas-phase reaction is discussed for an Au particle on a $MgCuCr_2O_4$ spinel oxide support [168], [169]. The spinel-supported Au nanoparticle catalyst is uniquely stable in comparison to support on MoO_3 [170] or TiO_2 [171], [172]. The reaction is executed at a temperature of 530 K.

Figure 6.23 schematically illustrates the activation of ethanol as suggested by quantum-chemical DFT calculations [169].

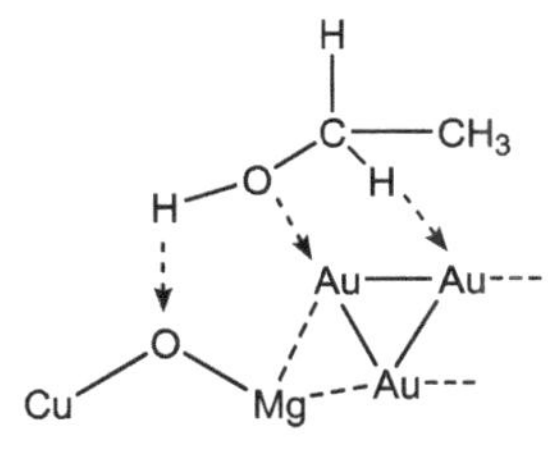

Figure 6.23 Schematic representation of the activation of ethanol by an Au nanoparticle on $MgCuCr_2O_4$ [169].

Reactive oxygen atoms of the $MgCuCr_2O_4$ spinel bridge Mg and Cu ions. The hydroxyl of the alcohol molecule that is adsorbed to the edge of the Au particle reacts with a support oxygen atom. The proton becomes attached to this Lewis base oxygen atom. The alkoxy intermediate generated by alcohol deprotonation is stabilized by adsorption on an Au edge atom. The α C–H bond of the adsorbed alkoxy readily dissociates to give the aldehyde and an H atom adsorbed on Au. The catalytic cycle closes by recombination of the H atom with surface OH to give water and regeneration of Cu-bound oxygen atoms by O_2 dissociation.

A dual site of transition metal particle with Brønsted acid proton catalyzes dihydroxylation. This system is discussed in Section 3.4.2 for hydrodeoxygenation of phenol to benzene [172]. The reactive site is the edge of a Ru particle with a proton attached to TiO_2 (see Figure 3.42).

6.4.2.3 *Alcohol Oxidation in the Water Phase*

Au-catalyzed oxidation of ethanol or glycerol in water by molecular oxygen requires high pH. Oxidation of the alcohol occurs by water-derived hydroxyls as demonstrated by isotope exchange studies with $^{18}O_2$ and $H_2{}^{18}O$ and the oxygen isotope distribution of the carboxylic acid. The reaction mechanism by Davis and Neurock of Virginia University and Minnesota University, respectively, in 2010 is presented here [173]–[175].

Au catalyzes the reaction of molecular oxygen with water that generates the OOH and OH radicals in a step similar to the reaction illustrated in Figure 6.20. This step initiates the oxidation reaction. The succession of elementary steps of the overall oxidation reaction is given by Eq. (6.2). Eq. (6.2a) show the oxidation of the alcohol with a hydroxyl radical to carboxylic acid. These reactions formally reduce the metal and generate adsorbed H atoms. The reactions of oxygen and water represented by Eq. (6.2b) produce OH radicals. This is formulated as consumption of electrons and generation of OH^-. H_2O_2 is a co-product of the reaction that generates adsorbed OH radicals by recombination of OOH.

$$
\begin{aligned}
&RCH_2OH_{|} + OH^-_{|} + * &&\rightarrow RCH_2O_{ads} + H_2O + el \\
&RCH_2O_{ads} + * &&\rightarrow RCHO_{ads} + H_{ads} \\
&RCHO_{ads} + OH^-_{|} &&\rightarrow RCHOOH_{ads} + el \\
&RCHOOH_{ads} &&\rightarrow RCOOH_{|} + H_{ads} \\
&RCH_2OH_{|} + 2\ OH^-_{|} + 2* &&\rightarrow RCOOH_{|} + 2\ H_{ads} + H_2O + 2\ el \quad (6.2a)
\end{aligned}
$$

$$
\begin{aligned}
&2\ O_{2,ads} + 2\ H_2O_{ads} &&\rightarrow 2\ OOH_{ads} + 2\ OH_{ads} \\
&2\ OH_{ads} + 2\ H_{ads} &&\rightarrow 2\ H_2O_{ads} \\
&2\ OOH_{ads} + 2\ H_2O_{ads} &&\rightarrow 2\ H_2O_{2,|} + 2\ OH_{ads} \\
&2\ OH_{ads} + 2\ el &&\rightarrow 2\ OH^-_{|} \qquad (6.2b)
\end{aligned}
$$

The oxidation of alcohol by adsorbed OH_{ads} and liquid-phase OH^- to carboxylic acid proceeds in two steps. The alcohol O–H bond cleaves by reaction with OH^- in the liquid phase. This gives an alcoholate that adsorbs as alkoxy intermediate on the Au surface (the metallic Au surface will not dissociate the alcohol O–H bond). In a consecutive reaction the α C–H bond of alcoholate cleaves by reaction with adsorbed OH. This gives adsorbed aldehyde as intermediate. In a second step the aldehyde reacts with OH_{ads} or liquid phase OH^- to give the acetal intermediate. The carboxylic acid is formed when the second α C–H bond cleaves by reaction with adsorbed OH.

The dominant role of OH intermediates in this low-temperature liquid-phase reaction is reminiscent of the dominant role of OH intermediates in gas-phase radical autocatalytic reactions (Chapter 4, Eqs. (4.2)).

The common denominator of the two systems is that peroxyl reaction intermediates by O–O bond cleavage give reactive OH, whereas the catalyst cannot cleave the O=O bond directly.

The need for adsorbed hydroxide to activate both the C–H and O–H bonds on Au explains the experimentally observed increase in catalytic activity of the noble metal at high pH.

On Au hydrogen peroxide is co-produced with carboxylic acid. It is co-generated in the reaction of two OOH radicals that gives OH radicals. On more reactive Pt the hydrogen peroxide decomposes. In water the reactivity of Au and Pt is comparable because the more reactive Pt surface is deactivated by strongly adsorbing hydroxyl.

The rate of oxidation of the aldehyde to carboxylic acid is faster than that of alcohol to aldehyde. In the oxidation of 5-hydroxymethylfurfural this is the cause of its rapid deactivation. In the next section a chemical procedure is introduced that prevents deactivation.

6.4.2.4 *Selective Oxidation of 5-hydroxymethylfurfural (HMF) to 2,5-furandicarboxylic Acid (FDCA)*

The oxidation of aldehyde is fast compared to that of the alcohol. To oxidize HMF selectively a protective group has to be used.

Biomass molecules are, amongst others, of potential interest as monomer in the production of biodegradable polymers. The product FDCA of the oxidation of HMF is of interest, since it is a monomer of polyethylene furanoate (PEF). PEF has similar properties as polyethylene terephthalate (PET), which is a commonly used high-performance polyester. The Dutch Avantium Company is building a plant for PEF production that will be operational in 2023 [183]. HMF is produced by dehydration from fructose, which derives from glucose by isomerization catalysis [184]. The mechanism of this reaction catalyzed by a zeolite is illustrated in Figure 6.9.

Several routes exist to oxidize HMF to FDCA [178], [179], [187]. Selective oxidation of HMF in the liquid phase requires high pH. The reaction is catalyzed by Au or Pt nanoparticles often distributed on a carbon support. The reaction network of the selective oxidation of HMF to FDCA is shown in Figure 6.24 [188], [191]. Less reactive –CH_2OH has to be oxidized in the presence of more reactive –CHO. As described in the previous section, whereas molecular oxygen is the oxidant, the oxygen atoms inserted into FDCA are derived from water.

Whereas Au is the preferred catalyst for oxidation of the aldehyde substituent, more reactive Pt or Pd is needed to oxidize the alcohol group at comparable conditions. However, the transition metals Pt or Pd will also activate C–H bonds.

The aerobic oxidation of HMF in high concentration is also hampered by rapid site reactions of reactive –CH_2OH and –CHO

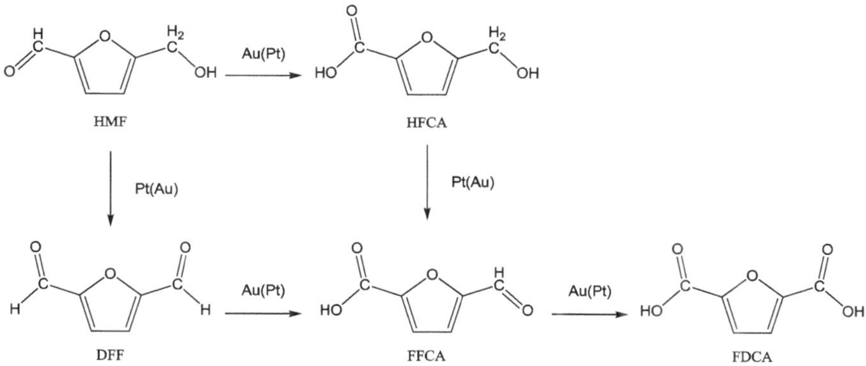

Figure 6.24 Reaction scheme of the oxidation of HMF to FDCA at high pH; Au is preferred over Pt [180]. FDCA: 2,5-furan dicarboxylic acid, HMF: 5-hydroxymethyl fufural, HFCA: 2-hydroxymethyl furan carboxylic acid, FFCA: 5-formyl-2-furan carboxylic acid, DFF: diformyl furan.

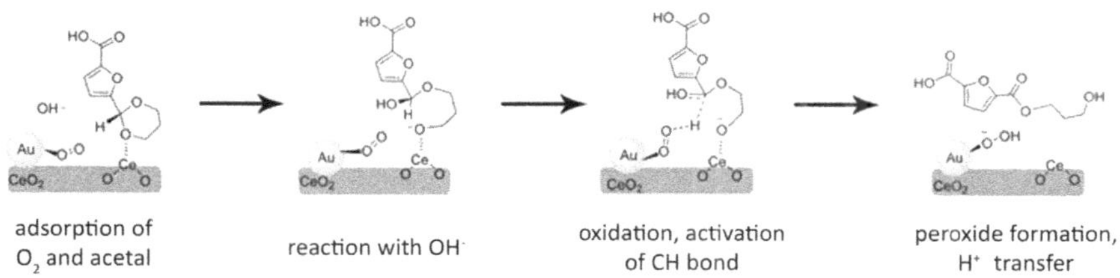

Figure 6.25 Oxidation of the FFCA acetal into FDCA by the Au/CeO_2 catalyst [181].

that lead to oligomerization reactions. This rapidly deactivates the system.

An approach that prevents these oligomerization reactions is to deactivate the furfural substituents with readily removable reactants. This can be done by reaction of HMF with 1,3-propanediol to the corresponding acetal as proposed by Nakayima *et al.* from Hokkaido University [181]. Oxidation is catalyzed by an Au/CeO_2 catalyst. The acetal is readily synthesized with a Brønsted acidic Amberlyst-15 resin catalyst. In a subsequent step the Au/CeO_2 catalyst converts the HMF cyclic acetal to intermediate FFCA acetal. This oxidizes with high yield to the FDCA ether. The ether hydrolyses and liberates FDCA. The reaction sequence from cyclic acetal to FDCA ether is shown in Figure 6.25.

The FFCA acetal interacts with a vacant site of the CeO_2 surface and O_2 attaches to the Au particles. OH^- from the solution reacts with the acetal C–O bond. Ring opening is assisted by interaction with the Lewis acidic Ce cation. Molecular oxygen adsorbed to the Au particle oxidizes H–C–OH to the C=O of carboxylate. OOH^- generates by reaction of O_2 with the hydride ion. Additional proton transfer gives the FDCA ether. OH^- and co-product peroxide is produced by reaction of HOO^- with H_2O.

Similar to alcohol oxidation discussed in Section 6.4.2.3, in the overall oxidation process an oxygen atom of water is incorporated into the HMF.

This reaction illustrates the dual-site mechanistic concept that is central to this section. The dual site of the Au/reducible oxide surface as exemplified by Au/CeO_2 provides access to surface vacant Lewis acid sites. The latter stabilize polar groups. This activates molecular bond dissociation reactions. Interaction with the cation stabilizes one dissociation fragment. The other fragment is stabilized by Au. Additionally, Au catalyzes reaction at low temperature because its low reactivity inhibits deactivating adsorption of reactants or products.

6.5 Summary and List of Reactions

Molecular heterogeneous catalytic systems are discussed in this chapter. The reactions are mainly organized according to different catalyst type. The catalytic systems share as common feature the fact that they are molecular or atomic complexes immobilized on a variety of reducible and non-reducible supports. The catalytic centers may have redox as well as non-redox properties.

Table 6.1 List of catalytic reactions of this chapter.

1. Polymerization and oligomerization reactions of alkenes:	
Metathesis (disproportionation) reaction of alkene	Mo, reaction cycle and intermediate, Section 6.2 (Figure 6.4)
Polymerization of ethene	Cr/SiO_2, reaction cycle, Section 6.2 (Figure 6.1)

Table 6.1 (*Continued*)

Polymerization of propene	Ti/$MgCl_2$, reaction intermediate, Section 6.2 (Figure 6.2)
2. Selective oxidation reactions	
Alcohol oxidation with O_2	Water-phase Au/TiO_2, gas-phase Au/$MgCuCr_2O_4$, Section 6.4.2 (Figures 6.22, 6.23)
Benzene to phenol with N_2O	Fe/zeolite, mechanism, Section 6.2.2.2 (Figure 6.13)
CO oxidation with O_2, water-gas shift	Pt, CeO_2, Rh_2O_3, Au/TiO_2, mechanism, Sections 6.4.1, 6.4.2 (Figure 6.19)
Cyclohexanone with hydrogen peroxide to lactone	Sn/zeolite, mechanism, Section 6.3.1.1 (Figure 6.7)
Cyclohexane, n-hexane to adipic acid with O_2	redox zeolites, Section 6.3.2.1 (Figure 6.11)
Epoxidation of propene with hydroperoxide	Ti, mechanism, Section 6.3.1.1 (Figure 6.5)
Epoxidation of propene with hydrogen peroxide	Ti/zeolite, mechanism, Section 6.3.1.1 (Figure 6.6)
α-caprolactam, direct synthesis	redox zeolites, mechanism, Section 6.3.2.1 (Figure 6.12)
5-Hydroxymethylfurfural (HMF) to 2,5-furandicarboxylic acid (FDCA) with O_2	Au, Pt, mechanism, Section 6.4.3.3 (Figure 6.24)
Methane to methanol with O_2	Cu, mechanism, Section 6.3.2.3.1.2 (Figure 6.14, 6.15)
Methane to methanol with hydrogen peroxide	Fe, mechanism, Section 6.3.2.3.1.3 (Figure 6.16)
3. Dehydrogenation (anaerobic) reactions	
Alkane to alkene and aromatics	Ga, Zn, zeolite(H^+), mechanism, Section 6.3.1.3 (Figure 6.10)
Dehydrogenation by ketone	Ti, mechanism, Section 6.3.1.2 (Figure 6.8)
Methane to aromatics	Fe/Si, Mo/zeolite(H^+), mechanism, Section 6.3.2.3.2 (Figure 6.17)

(*Continued*)

Table 6.1 (*Continued*)

4. Isomerization	
Glucose to fructose isomerization	Sn/zeolite, mechanism, Section 6.3.1.2 (Figure 6.9)
5. NO reduction	
NO reduction with CO	Pt, Pd/CeO_2, mechanism, Section 6.4.1 (Figure 6.18)
6 Chlorination	
Ethyne hydrochlorination	$AuCl_3$, Section 6.4.2

**Short versions of parts of these sections are also present in Comprehensive Inorganic Chemistry III, Part 6: Heterogeneous Inorganic Catalysis, Chapter 1, R. A. van Santen and E. J. M. Hensen, "Introduction and Short History of Single Site Catalysis," Elsevier, 2023.*

References

[1] R. Whyman, "Piet W. N. M. Van Leeuwen. Homogeneous catalysis — understanding the art. Kluwer, Dordrecht, 2004, 407 pp; (UK). ISBN 1-4020-1999-8," *Appl. Organomet. Chem.*, vol. 19, no. 8, pp. 994–994, Aug. 2005, doi: 10.1002/AOC.939.

[2] G. D. Frey, "75 years of oxo synthesis — The success story of a discovery at the OXEA Site Ruhrchemie," *J. Organomet. Chem.*, vol. 754. pp. 5–7, Mar. 2014, doi: 10.1016/j.jorganchem.2013.12.023.

[3] B. C. Gates, "Atomically dispersed supported metal catalysts: seeing is believing," *Trends Chem.*, vol. 1, no. 1, pp. 99–110, Apr. 2019, doi: 10.1016/J.TRECHM.2019.01.004.

[4] C. Copéret, M. Chabanas, R. Petroff Saint-Arroman, and J. M. Basset, "Homogeneous and heterogeneous catalysis: bridging the gap through surface organometallic chemistry," *Angew. Chem. Int. Ed.*, vol. 42, no. 2, pp. 156–181, Jan. 2003, doi: 10.1002/ANIE.200390072.

[5] J. Guzman and B. C. Gates, "Supported molecular catalysts: metal complexes and clusters on oxides and zeolites," *Dalton Trans.*, vol. 3, no. 17, pp. 3303–3318, Aug. 2003, doi: 10.1039/B303285J.

[6] J. M. Thomas, "Uniform heterogeneous catalysts: the role of solid-state chemistry in their development and design," *Angew. Chem. Int. Ed.*, vol. 27, no. 12, pp. 1673–1691, Dec. 1988, doi: 10.1002/ANIE.198816731.

[7] J. M. Thomas, *Design and Applications of Single-Site Heterogeneous Catalysts: Contributions to Green Chemistry, Clean Technology and Sustainability.* Imperial College Press, 2012.

[8] B. C. Gates, "Metal clusters in zeolites: nearly molecular catalysts for hydrocarbon conversion," *Stud. Surf. Sci. Catal.*, vol. 100, no. C, pp. 49–63, Jan. 1996, doi: 10.1016/S0167-2991(96)80007-3.

[9] M. Flytzani-Stephanopoulos and B. C. Gates, "Atomically dispersed supported metal catalysts," *Annu. Rev. Chem. Biomol. Eng.*, vol. 3, pp. 545–574, Jul. 2012, doi: 10.1146/ANNUREV-CHEMBIOENG-062011-080939.

[10] J. P. Hogan, "Ethylene polymerization catalysis over chromium oxide," *J. Polym. Sci. A*, vol. 8, no. 9, pp. 2637–2652, Sep. 1970, doi: 10.1002/POL.1970.150080929.

[11] C. Brown, A. Lita, Y. Tao, N. Peek, M. Crosswhite, M. Mileham, J. Krzystek, R. Achey, R. Fu, J. K. Bindra, M. Polinski, Y. Wang, L. J. van de Burgt, D. Jeffcoat, J. Salvatore Profeta, A. E. Stiegman, and S. L. Scott, "Mechanism of initiation in the phillips ethylene polymerization catalyst: ethylene activation by Cr(II) and the structure of the resulting active site," *ACS Catal.*, vol. 7, no. 11, pp. 7442–7455, Nov. 2017, doi: 10.1021/ACSCATAL.7B02677.

[12] K. Ziegler, E. Holzkamp, H. Breil, and H. Martin, "Polymerisation von Äthylen und anderen Olefinen," *Angew. Chem.*, vol. 67, no. 16, pp. 426–426, Aug. 1955, doi: 10.1002/ANGE.19550671610.

[13] K. Ziegler, E. Holzkamp, H. Breil, and H. Martin, "Das M{ü}lheimer Normaldruck-Poly{ä}thylen-Verfahren," *Angew. Chem.*, vol. 67, no. 19–20, pp. 541–547, 1955, doi: 10.1002/ange.19550671902.

[14] G. Natta, "Une nouvelle classe de polymeres d'α-olefines ayant une régularité de structure exceptionnelle," *J. Polym. Sci.*, vol. 16, no. 82, pp. 143–154, Apr. 1955, doi: 10.1002/POL.1955.120168205.

[15] G. Natta, P. Pino, P. Corradini, F. Danusso, E. Mantica, G. Mazzanti, and G. Moraglio, "Crystalline high polymers of α-olefins," *J. Am. Chem. Soc.*, vol. 77, no. 6, pp. 1708–1710, 2002, doi: 10.1021/JA01611A109.

[16] G. Natta, P. Pino, and G. Mazzanti, "No Title," *Chim. Ind.*, vol. 37, p. 927, 1955.

[17] P. Cossee, "Ziegler-Natta catalysis I. Mechanism of polymerization of α-olefins with Ziegler-Natta catalysts," *J. Catal.*, vol. 3, no. 1, pp. 80–88, Feb. 1964, doi: 10.1016/0021-9517(64)90095-8.

[18] J. Chatt and L. A. Duncanson, "586. Olefin co-ordination compounds. Part III. Infra-red spectra and structure: attempted preparation of acetylene complexes," *J. Chem. Soc.*, no. 0, pp. 2939–2947, Jan. 1953, doi: 10.1039/JR9530002939.

[19] C. J. Ballhausen and B. J. Ballhausen, *Introduction to Ligand Field Theory*. McGraw-Hill, 1962.

[20] E. J. Arlman and P. Cossee, "Ziegler-Natta catalysis III. Stereospecific polymerization of propene with the catalyst system TiCl30/AlEt3," *J. Catal.*, vol. 3, no. 1, pp. 99–104, Feb. 1964, doi: 10.1016/0021-9517(64)90097-1.

[21] O. Novaro, E. Blaisten-Barojas, E. Clementi, G. Giunchi, and M. E. Ruiz-Vizcaya, "Theoretical study on a reaction pathway of Ziegler–Natta-type catalysis," *J. Chem. Phys.*, vol. 68, no. 5, pp. 2337–2351, Mar. 1978, doi: 10.1063/1.436004.

[22] M. Seth and T. Ziegler, "Polymerization properties of a heterogeneous Ziegler–Natta catalyst modified by a base: a theoretical study," *Macromolecules*, vol. 36, no. 17, pp. 6613–6623, Aug. 2003, doi: 10.1021/MA0301247.

[23] R. L. Banks and G. C. Bailey, "Olefin disproportionation. a new catalytic process," *Ind. Eng. Chem. Prod. Res. Dev.*, vol. 3, no. 3, pp. 170–173, Sep. 1964, doi: 10.1021/I360011A002.

[24] F. D. Mango and J. H. Schachtschneider, "Molecular orbital symmetry conservation in transition metal catalyzed transformations," *J. Am. Chem. Soc.*, vol. 89, no. 10, pp. 2484–2486, May 1967, doi: 10.1021/JA00986A044.

[25] R. Hoffmann and R. B. Woodward, "Orbital symmetries and endo-exo relationships in concerted cycloaddition reactions," *J. Am. Chem. Soc.*, vol. 87, no. 19, pp. 4388–4389, 1965, doi: 10.1021/JA00947A033.

[26] P. Jean-Louis Hérisson and Y. Chauvin, "Catalyse de transformation des oléfines par les complexes du tungstène. II. Télomérisation des oléfines cycliques en présence d'oléfines acycliques," *Die Makromol. Chemie*, vol. 141, no. 1, pp. 161–176, Feb. 1971, doi: 10.1002/macp.1971.021410112.

[27] J. R. Anderson and M. Boudart, Eds., *Catalysis: Science and Technology.* Springer, 1996.

[28] C. P. Casey, "2005 Nobel Prize in Chemistry. Development of the olefin metathesis method in organic synthesis," *J. Chem. Educ.*, vol. 83, no. 2, pp. 192–195, 2006, doi: 10.1021/ED083P192.

[29] Y. Chauvin, "Olefin metathesis: the early days (Nobel Lecture)," *Angew. Chem. Int. Ed.*, vol. 45, no. 23, pp. 3740–3747, Jun. 2006, doi: 10.1002/anie.200601234.

[30] J. Handzlik and P. Sautet, "Active sites of olefin metathesis on molybdena-alumina system: A periodic DFT study," *J. Catal.*, vol. 256, no. 1, pp. 1–14, May 2008, doi: 10.1016/J.JCAT.2008.02.016.

[31] A. K. Rappé and W. A. Goddard, "Bivalent spectator oxo bonds in metathesis and epoxidation alkenes," *Nature*, vol. 285, no. 5763, pp. 311–312, 1980, doi: 10.1038/285311a0.

[32] A. K. Rappé and W. A. Goddard, "Olefin metathesis — a mechanistic study of high-valent Group VI catalysts," *J. Am. Chem. Soc.*, vol. 104, no. 2, pp. 448–456, 2002, doi: 10.1021/JA00366A013.

[33] M. Akimoto, K. Ichikawa, and E. Echigoya, "Kinetic and adsorption studies on vapor-phase catalytic oxidation of olefins over silver," *J. Catal.*, vol. 76, no. 2, pp. 333–344, Aug. 1982, doi: 10.1016/0021-9517(82)90264-0.

[34] W. M. H. H. Sachtler, "Mechanism of the catalytic oxidations of propene to acrolein and of isomeric butenes to butadiene," *Recl. des Trav. Chim. des Pays-Bas*, vol. 82, no. 3, pp. 243–245, Jan. 2010, doi: 10.1002/recl.19630820305.

[35] T. W. Findley, D. Swern, and J. T. Scanlan, "Epoxidation of unsaturated fatty materials with peracetic acid in glacial acetic acid solution," *J. Am. Chem. Soc.*, vol. 67, no. 3, pp. 412–414, Mar. 1945, doi: 10.1021/JA01219A018.

[36] R. Landau, G. A. Sullivan, and D. Brown, "Propylene oxide by the co-product processes," *Chem. Tech.*, vol. 9, p. 602, 1979.

[37] K. Weissermel and H. J. Arpe, "Oxidation Products Ethylene," in *Industrial Organic Chemistry*, 4th ed. John Wiley & Sons, 2003, pp. 145–192.

[38] J. M. Thomas, *Design and Applications of Single-Site Heterogeneous Catalysts: Contributions to Green Chemistry, Clean Technology and Sustainability*. Imperial College Press, 2012.

[39] C. R. Catlow, S. French, A. Sokol, and J. Thomas, "Computational approaches to the determination of active site structures and reaction mechanisms in heterogeneous catalysts," *Philos. Trans. R. Soc. A*, vol. 363, no. 1829, pp. 913–936, Apr. 2005, doi: 10.1098/RSTA.2004.1529.

[40] M. G. Finn and K. B. Sharpless, "Mechanism of asymmetric epoxidation. 2. Catalyst structure," *J. Am. Chem. Soc.*, vol. 113, no. 1, pp. 113–126, 1991, doi: 10.1021/JA00001A019.

[41] T. Katsuki and K. B. Sharpless, "The first practical method for asymmetric epoxidation," *J. Am. Chem. Soc.*, vol. 102, no. 18, pp. 5974–5976, 1980, doi: 10.1021/JA00538A077.

[42] E. N. Jacobsen, "Transition Metal-catalyzed Oxidations: Asymmetric Epoxidation," in *Comprehensive Organometallic Chemistry II*, E. W. Abel, F. G. A. Stone, and G. Wilkinson, Eds. Elsevier, 1995, pp. 1097–1135.

[43] S. Bordiga, S. Coluccia, C. Lamberti, L. Marchese, A. Zecchina, F. Boscherini, F. Buffa, F. Genoni, and G. Leofanti, "XAFS study of Ti-silicalite: structure of framework Ti(IV) in the presence and absence of reactive molecules (H2O, NH3) and comparison with

ultraviolet-visible and IR results," *J. Phys. Chem.*, vol. 98, no. 15, pp. 4125–4132, 2002, doi: 10.1021/J100066A036.

[44] S. Bordiga, F. Bonino, A. Damin, and C. Lamberti, "Reactivity of Ti(IV) species hosted in TS-1 towards H2O2–H2O solutions investigated by ab initio cluster and periodic approaches combined with experimental XANES and EXAFS data: a review and new highlights," *Phys. Chem. Chem. Phys.*, vol. 9, no. 35, pp. 4854–4878, Aug. 2007, doi: 10.1039/B706637F.

[45] E. M. Flanigen, "Zeolites and molecular sieves an historical perspective," in *Studies in Surface Science and Catalysis*, vol. 58. Elsevier, 1991, pp. 13–34.

[46] C. P. Gordon, H. Engler, A. S. Tragl, M. Plodinec, T. Lunkenbein, A. Berkessel, J. H. Teles, A.-N. Parvulescu, and C. Copéret, "Efficient epoxidation over dinuclear sites in titanium silicalite-1," *Nature*, vol. 586, no. 7831, pp. 708–713, Oct. 2020, doi: 10.1038/s41586-020-2826-3.

[47] Y. Sawada, K. Matsumoto, and T. Katsuki, "Titanium-catalyzed asymmetric epoxidation of non-activated olefins with hydrogen peroxide," *Angew. Chem. Int. Ed.*, vol. 46, no. 24, pp. 4559–4561, Jun. 2007, doi: 10.1002/ANIE.200700949.

[48] A. Corma, L. T. Nemeth, M. Renz, and S. Valencia, "Sn-zeolite beta as a heterogeneous chemoselective catalyst for Baeyer-Villiger oxidations," *Nature*, vol. 412, no. 6845, pp. 423–425, 2001, doi: 10.1038/35086546.

[49] A. Corma, M. E. Domine, L. Nemeth, and S. Valencia, "Al-free Sn-Beta zeolite as a catalyst for the selective reduction of carbonyl compounds (Meerwein-Ponndorf-Verley Reaction)," *J. Am. Chem. Soc.*, vol. 124, no. 13, pp. 3194–3195, Apr. 2002, doi: 10.1021/JA012297M.

[50] J. C. van der Waal, P. J. Kunkeler, K. Tan, and H. van Bekkum, "Zeolite titanium beta: a selective catalyst for the gas-phase Meerwein–Ponndorf–Verley, and Oppenauer reactions," *J. Catal.*, vol. 173, no. 1, pp. 74–83, Jan. 1998, doi: 10.1006/JCAT.1997.1901.

[51] E. J. Creyghton, S. D. Ganeshie, R. S. Downing, and H. Van Bekkum, "Stereoselective reduction of 4-tert-butylcyclohexanone to cis-4-tert-butylcyclohexanol catalysed by zeolite BEA," *J. Chem. Soc. Chem. Commun.*, vol. 18, no. 18, pp. 1859–1860, Jan. 1995, doi: 10.1039/C39950001859.

[52] M. Moliner, Y. Román-Leshkov, and M. E. Davis, "Tin-containing zeolites are highly active catalysts for the isomerization of glucose in water," *Proc. Natl. Acad. Sci.*, vol. 107, no. 14, pp. 6164–6168, Apr. 2010, doi: 10.1073/PNAS.1002358107.

[53] M. S. Holm, S. Saravanamurugan, and E. Taarning, "Conversion of sugars to lactic acid derivatives using heterogeneous zeotype catalysts," *Science*, vol. 328, no. 5978, pp. 602–605, Apr. 2010, doi: 10.1126/SCIENCE.1183990.

[54] X. Zhang, K. Wilson, and A. F. Lee, "Heterogeneously catalyzed hydrothermal processing of C5-C6 sugars," *Chem. Rev.*, vol. 116, no. 19, pp. 12328–12368, Oct. 2016, doi: 10.1021/ACS.CHEMREV.6B00311.

[55] A. Achari, S. E. Marshall, H. Muirhead, R. H. Palmieri, and E. A. Noltmann, "Glucose — 6 — phosphate isomerase," *Philos. Trans. R. Soc. London. B*, vol. 293, no. 1063, pp. 145–157, Jun. 1981, doi: 10.1098/RSTB.1981.0068.

[56] S. H. Bhosale, M. B. Rao, and V. V Deshpande, "Molecular and industrial aspects of glucose isomerase," *Microbiol. Rev.*, vol. 60, no. 2, pp. 280–300, Jun. 1996, doi: 10.1128/MR.60.2.280-300.1996.

[57] G. Yang, E. A. Pidko, and E. J. M. Hensen, "The mechanism of glucose isomerization to fructose over Sn-BEA zeolite: A periodic density functional theory study," *ChemSusChem*, vol. 6, no. 9, pp. 1688–1696, 2013, doi: 10.1002/cssc.201300342.

[58] T. D. Fenn, D. Ringe, and G. A. Petsko, "Xylose isomerase in substrate and inhibitor Michaelis states: atomic resolution studies of a metal-mediated hydride shift," *Biochemistry*, vol. 43, no. 21, pp. 6464–6474, Jun. 2004, doi: 10.1021/bi049812o.

[59] P. C. Doolan and P. R. Pujado, "Make aromatics from LPG," *Hydrocarb. Process*, vol. 68(9), pp. 72–76, 1989.

[60] A. Bhan and W. N. Delgass, "Propane aromatization over HZSM-5 and Ga/HZSM-5 catalysts," *Catal. Rev.*, vol. 50, no. 1, pp. 19–151, Mar. 2008, doi: 10.1080/01614940701804745.

[61] J. A. Biscardi and E. Iglesia, "Structure and function of metal cations in light alkane reactions catalyzed by modified H-ZSM5," *Catal. Today*, vol. 31, no. 3–4, pp. 207–231, Dec. 1996, doi: 10.1016/S0920-5861(96)00028-4.

[62] Y. Xiang, H. Wang, J. Cheng, and J. Matsubu, "Progress and prospects in catalytic ethane aromatization," *Catal. Sci. Technol.*, vol. 8, no. 6, pp. 1500–1516, Mar. 2018, doi: 10.1039/C7CY01878A.

[63] C. R. Bayense, A. J. H. P. van der Pol, and J. H. C. van Hooff, "Aromatization of propane over MFI-gallosilicates," *Appl. Catal.*, vol. 72, no. 1, pp. 81–98, May 1991, doi: 10.1016/0166-9834(91)85030-Y.

[64] E. A. Pidko, E. J. M. Hensen, and R. A. van Santen, "Dehydrogenation of light alkanes over isolated gallyl ions in Ga/ZSM-5 zeolites," *J. Phys. Chem. C*, vol. 111, no. 35, pp. 13068–13075, Sep. 2007, doi: 10.1021/JP072110Z.

[65] E. J. M. Hensen, E. A. Pidko, N. Rane, and R. A. van Santen, "Water-promoted hydrocarbon activation catalyzed by binuclear gallium sites in ZSM-5 zeolite," *Angew. Chem. Int. Ed.*, vol. 46, no. 38, pp. 7273–7276, 2007, doi: 10.1002/anie.200702463.

[66] N. Rane, A. R. Overweg, V. B. Kazansky, R. A. van Santen, and E. J. M. Hensen, "Characterization and reactivity of Ga+ and GaO+ cations in zeolite ZSM-5," *J. Catal.*, vol. 239, no. 2, pp. 478–485, Apr. 2006, doi: 10.1016/J.JCAT.2006.03.004.

[67] E. A. Pidko, V. B. Kazansky, E. J. M. Hensen, and R. A. van Santen, "A comprehensive density functional theory study of ethane dehydrogenation over reduced extra-framework gallium species in ZSM-5 zeolite," *J. Catal.*, vol. 240, no. 1, pp. 73–84, May 2006, doi: 10.1016/J.JCAT.2006.03.011.

[68] Frash and R. A. van Santen, "Activation of small alkanes in Ga-exchanged zeolites: A quantum chemical study of ethane dehydrogenation," *J. Phys. Chem. A*, vol. 104, no. 11, pp. 2468–2475, Mar. 2000, doi: 10.1021/JP993414R.

[69] N. M. Phadke, E. Mansoor, M. Bondil, M. Head-Gordon, and A. T. Bell, "Mechanism and kinetics of propane dehydrogenation and cracking over Ga/H-MFI prepared via vapor-phase exchange of H-MFI with GaCl3," *J. Am. Chem. Soc.*, vol. 141, no. 4, pp. 1614–1627, Jan. 2019, doi: 10.1021/JACS.8B11443.

[70] M. W. Schreiber, C. P. Plaisance, M. Baumgärtl, K. Reuter, A. Jentys, R. Bermejo-Deval, and J. A. Lercher, "Lewis-Brønsted acid pairs in Ga/H-ZSM-5 to catalyze dehydrogenation of light alkanes," *J. Am. Chem. Soc.*, vol. 140, no. 14, pp. 4849–4859, Apr. 2018, doi: 10.1021/JACS.7B12901.

[71] E. A. Pidko and R. A. van Santen, "Activation of light alkanes over zinc species stabilized in ZSM-5 zeolite: A comprehensive DFT study," *J. Phys. Chem. C*, vol. 111, no. 6, pp. 2643–2655, 2007, doi: 10.1021/JP065911V.

[72] N. O. Gonzales, A. K. Chakraborty, and A. T. Bell, "A density functional theory study of hydrogen recombination and hydrogen-deuterium exchange on Ga/H-ZSM-5," *Top. Catal.*, vol. 9, no. 3, pp. 207–213, 1999, doi: 10.1023/A:1019118823908.

[73] B. E. R. Snyder, M. L. Bols, H. M. Rhoda, D. Plessers, R. A. Schoonheydt, B. F. Sels, and E. I. Solomon, "Cage effects control the mechanism of methane hydroxylation in zeolites," *Science*, vol. 373, no. 6552, pp. 327–331, Jul. 2021, doi: 10.1126/SCIENCE.ABD5803.

[74] A. Corma and H. García, "Lewis acids as catalysts in oxidation reactions: From homogeneous to heterogeneous systems," *Chem. Rev.*, vol. 102, no. 10, pp. 3837–3892, Oct. 2002, doi: 10.1021/CR010333U.

[75] B. Kraushaar-Czarnetzki, W. G. M. Hoogervorst, and W. H. J. Stork, "Oxidation of saturated hydrocarbons involving CoAPO molecular sieves as oxidants and as catalysts," *Stud. Surf. Sci. Catal.*, vol. 84, no. C, pp. 1869–1876, Jan. 1994, doi: 10.1016/S0167-2991(08)63743-X.

[76] J. M. Thomas, R. Raja, G. Sankar, and R. G. Bell, "Molecular-sieve catalysts for the selective oxidation of linear alkanes by molecular oxygen," *Nature*, vol. 398, no. 6724, pp. 227–230, Mar. 1999, doi: 10.1038/18417.

[77] L. Gómez-Hortigüela, F. Corà, and C. R. A. Catlow, "Modeling hydrocarbon oxidation mechanisms catalyzed by microporous materials," *Model. Simul. Sci. Micro- Meso-Porous Mater.*, pp. 265–295, 2018, doi: 10.1002/1521-3773(20000703)39:13<2313::AID-ANIE2313>3.0.CO;2-Z.

[78] M. Dugal, G. Sankar, R. Raja, and J. M. Thomas, "Designing a heterogeneous catalyst for the production of adipic acid by aerial oxidation of cyclohexane," *Angew. Chem. Int. Ed.*, vol. 39, p. 2310, 2000.

[79] J. A. Labinger, "Selective alkane oxidation: hot and cold approaches to a hot problem," *J. Mol. Catal. A Chem.*, vol. 220, no. 1, pp. 27–35, Sep. 2004, doi: 10.1016/J.MOLCATA.2004.03.051.

[80] J. M. Thomas, R. Raja, G. Sankar, and R. G. Bell, "Molecular sieve catalysts for the regioselective and shape-selective oxyfunctionalization of alkanes in air," *Acc. Chem. Res.*, vol. 34, no. 3, pp. 191–200, 2001, doi: 10.1021/ar970020e.

[81] J. M. Thomas, R. Raja, G. Sankar, and R. G. Bell, "Molecular sieve catalysts for the regioselective and shape-selective oxyfunctionalization of alkanes in air," *Acc. Chem. Res.*, vol. 34, no. 3, pp. 191–200, 2001, doi: 10.1021/ar970020e.

[82] J. M. Thomas and R. Raja, "Design of a 'green' one-step catalytic production of epsilon-caprolactam (precursor of nylon-6).," *Proc. Natl. Acad. Sci.*, vol. 102, no. 39, pp. 13732–13736, 2005, doi: 10.1073/pnas.0506907102.

[83] J. M. Thomas and R. Raja, "Design of a 'green' one-step catalytic production of ε-caprolactam (precursor of nylon-6)," *Proc. Natl. Acad. Sci.*, vol. 102, no. 39, pp. 13732–13736, Sep. 2005, doi: 10.1073/PNAS.0506907102.

[84] G. I. Panov, "Advances in oxidation catalysis; oxidation of benzene to phenol by nutrous oxide," *CATTECH*, vol. 4, no. 1, pp. 18–31, 2000, doi: 10.1023/A:1011991110517.

[85] M. H. Thiemens and W. C. Trogler, "Nylon production: An unknown source of atmospheric nitrous oxide," *Science*, vol. 4996, no. 251, pp. 932–934, 1991, doi: 10.1126/SCIENCE.251.4996.932.

[86] M. Iwamoto, J. Hirata, K. Matsukami, and S. Kagawa, "Catalytic oxidation by oxide radical ions. 1. One-step hydroxylation of benzene to phenol over Group 5 and 6 oxides supported on silica gel," *J. Phys. Chem.*, vol. 87, no. 6, pp. 903–905, 2002, doi: 10.1021/J100229A001.
[87] A. K. Uriarte, M. A. Rodkin, M. J. Gross, A. S. Kharitonov, and G. I. Panov, "Direct hydroxylation of benzene to phenol by nitrous oxide," *Stud. Surf. Sci. Catal.*, vol. 110, pp. 857–864, Jan. 1997, doi: 10.1016/S0167-2991(97)81048-8.
[88] J. Jia, K. S. Pillai, and W. M. H. Sachtler, "One-step oxidation of benzene to phenol with nitrous oxide over Fe/MFI catalysts," *J. Catal.*, vol. 221, no. 1, pp. 119–126, Jan. 2004, doi: 10.1016/J.JCAT.2003.08.010.
[89] E. J. M. Hensen, Q. Zhu, R. A. J. Janssen, P. C. M. M. Magusin, P. J. Kooyman, and R. A. van Santen, "Selective oxidation of benzene to phenol with nitrous oxide over MFI zeolites: 1. On the role of iron and aluminum," *J. Catal.*, vol. 233, no. 1, pp. 123–135, Jul. 2005, doi: 10.1016/J.JCAT.2005.04.009.
[90] G. Li, E. A. Pidko, R. A. van Santen, Z. Feng, C. Li, and E. J. M. Hensen, "Stability and reactivity of active sites for direct benzene oxidation to phenol in Fe/ZSM-5: A comprehensive periodic DFT study," *J. Catal.*, vol. 284, no. 2, pp. 194–206, 2011, doi: 10.1016/j.jcat.2011.07.008.
[91] L. Meng, X. Zhu, and E. J. M. Hensen, "Stable Fe/ZSM-5 nanosheet zeolite catalysts for the oxidation of benzene to phenol," *ACS Catal.*, vol. 7, no. 4, pp. 2709–2719, Apr. 2017, doi: 10.1021/ACSCATAL.6B03512.
[92] D. P. Ivanov, V. I. Sobolev, and G. I. Panov, "Deactivation by coking and regeneration of zeolite catalysts for benzene-to-phenol oxidation," *Appl. Catal. A Gen.*, vol. 241, no. 1–2, pp. 113–121, Feb. 2003, doi: 10.1016/S0926-860X(02)00462-3.
[93] B. E. R. Snyder, P. Vanelderen, M. L. Bols, S. D. Hallaert, L. H. Böttger, L. Ungur, K. Pierloot, R. A. Schoonheydt, B. F. Sels, and E. I. Solomon, "The active site of low-temperature methane hydroxylation in iron-containing zeolites," *Nature*, vol. 536, no. 7616, pp. 317–321, Aug. 2016, doi: 10.1038/nature19059.
[94] J. H. Lunsford, "Catalytic conversion of methane to more useful chemicals and fuels: a challenge for the 21st century," *Catal. Today*, vol. 63, no. 2–4, pp. 165–174, Dec. 2000, doi: 10.1016/S0920-5861(00)00456-9.

[95] B. L. Farrell, V. O. Igenegbai, and S. Linic, "A viewpoint on direct methane conversion to ethane and ethylene using oxidative coupling on solid catalysts," *ACS Catal.*, vol. 6, no. 7, pp. 4340–4346, Jul. 2016, doi: 10.1021/ACSCATAL.6B01087.

[96] A. A. Latimer, A. Kakekhani, A. R. Kulkarni, and J. K. Nørskov, "Direct methane to methanol: the selectivity-conversion limit and design strategies," *ACS Catal.*, vol. 8, no. 8, pp. 6894–6907, Aug. 2018, doi: 10.1021/ACSCATAL.8B00220.

[97] R. Banerjee, Y. Proshlyakov, J. D. Lipscomb, and D. A. Proshlyakov, "Structure of the key species in the enzymatic oxidation of methane to methanol," *Nature*, vol. 518, no. 7539, pp. 431–434, Jan. 2015, doi: 10.1038/nature14160.

[98] D. Park and J. Lee, "Biological conversion of methane to methanol," *Korean J. Chem. Eng.*, vol. 30, no. 5, pp. 977–987, Apr. 2013, doi: 10.1007/S11814-013-0060-5.

[99] S. J. Freakley, N. Dimitratos, D. J. Willock, S. H. Taylor, C. J. Kiely, and G. J. Hutchings, "Methane oxidation to methanol in water," *Acc. Chem. Res.*, vol. 54, no. 11, pp. 2614–2623, Jun. 2021, doi: 10.1021/ACS.ACCOUNTS.1C00129.

[100] G. I. Panov, V. I. Sobolev, K. A. Dubkov, V. N. Parmon, N. S. Ovanesyan, A. E. Shilov, and A. A. Shteinman, "Iron complexes in zeolites as a new model of methane monooxygenase," *React. Kinet. Catal. Lett.*, vol. 61, no. 2, pp. 251–258, 1997, doi: 10.1007/BF02478380.

[101] G. I. Panov, A. K. Uriarte, M. A. Rodkin, and V. I. Sobolev, "Generation of active oxygen species on solid surfaces. Opportunity for novel oxidation technologies over zeolites," *Catal. Today*, vol. 41, no. 4, pp. 365–385, Jun. 1998, doi: 10.1016/S0920-5861(98)00026-1.

[102] M. H. Groothaert, P. J. Smeets, B. F. Sels, P. A. Jacobs, and R. A. Schoonheydt, "Selective oxidation of methane by the bis(μ-oxo) dicopper core stabilized on ZSM-5 and mordenite zeolites," *J. Am. Chem. Soc.*, vol. 127, no. 5, pp. 1394–1395, Feb. 2005, doi: 10.1021/JA047158U.

[103] G. Li, P. Vassilev, M. Sanchez-Sanchez, J. A. Lercher, E. J. M. Hensen, and E. A. Pidko, "Stability and reactivity of copper oxo-clusters in ZSM-5 zeolite for selective methane oxidation to methanol," *J. Catal.*, vol. 338, pp. 305–312, Jun. 2016, doi: 10.1016/J.JCAT.2016.03.014.

[104] H. Schwarz, "Chemistry with methane: concepts rather than recipes," *Angew. Chem. Int. Ed.*, vol. 50, no. 43, pp. 10096–10115, Oct. 2011, doi: 10.1002/ANIE.201006424.

[105] K. Kervinen, P. C. A. Bruijnincx, A. M. Beale, J. G. Mesu, G. Van Koten, R. J. M. Klein Gebbink, and B. M. Weckhuysen, "Zeolite framework stabilized copper complex inspired by the 2-His-1-carboxylate facial triad motif yielding oxidation catalysts," *J. Am. Chem. Soc.*, vol. 128, no. 10, pp. 3208–3217, Mar. 2006, doi: 10.1021/JA0567992.

[106] V. L. Sushkevich, D. Palagin, M. Ranocchiari, and J. A. Van Bokhoven, "Selective anaerobic oxidation of methane enables direct synthesis of methanol," *Science*, vol. 356, no. 6337, pp. 523–527, May 2017, doi: 10.1126/SCIENCE.AAM9035.

[107] R. Balasubramanian, S. M. Smith, S. Rawat, L. A. Yatsunyk, T. L. Stemmler, and A. C. Rosenzweig, "Oxidation of methane by a biological dicopper centre," *Nature*, vol. 465, no. 7294, pp. 115–119, Apr. 2010, doi: 10.1038/nature08992.

[108] P. Tomkins, M. Ranocchiari, and J. A. Van Bokhoven, "Direct conversion of methane to methanol under mild conditions over Cu-zeolites and beyond," *Acc. Chem. Res.*, vol. 50, no. 2, pp. 418–425, Feb. 2017, doi: 10.1021/ACS.ACCOUNTS.6B00534.

[109] X. Huang and J. T. Groves, "Beyond ferryl-mediated hydroxylation: 40 years of the rebound mechanism and C–H activation," *J. Biol. Inorg. Chem.*, vol. 22, no. 2, pp. 185–207, Dec. 2016, doi: 10.1007/S00775-016-1414-3.

[110] J. S. Woertink, P. J. Smeets, M. H. Groothaert, M. A. Vance, B. F. Sels, R. A. Schoonheydt, and E. I. Solomon, "A [Cu2O]2+ core in Cu-ZSM-5, the active site in the oxidation of methane to methanol.," *Proc. Natl. Acad. Sci.*, vol. 106, no. 45, pp. 18908–18913, 2009, doi: 10.1073/pnas.0910461106.

[111] Z.-J. Zhao, A. Kulkarni, L. Vilella, J. K. Nørskov, and F. Studt, "Theoretical insights into the selective oxidation of methane to methanol in copper-exchanged mordenite," *ACS Catal.*, vol. 6, no. 6, pp. 3760–3766, Jun. 2016, doi: 10.1021/ACSCATAL.6B00440.

[112] S. Grundner, M. A. C. Markovits, G. Li, M. Tromp, E. A. Pidko, E. J. M. Hensen, A. Jentys, M. Sanchez-Sanchez, and J. A. Lercher, "Single-site trinuclear copper oxygen clusters in mordenite for selective conversion of methane to methanol," *Nat. Commun.*, vol. 6, p. 7546, Jun. 2015, doi: 10.1038/ncomms8546.

[113] B. E. R. Snyder, M. L. Bols, H. M. Rhoda, D. Plessers, R. A. Schoonheydt, B. F. Sels, and E. I. Solomon, "Cage effects control the

mechanism of methane hydroxylation in zeolites," *Science*, vol. 373, no. 6552, pp. 327–331, Jul. 2021, doi: 10.1126/SCIENCE.ABD5803.

[114] C. Hammond *et al.*, "Direct catalytic conversion of methane to methanol in an aqueous medium by using copper-promoted Fe-ZSM-5," *Angew. Chem.*, vol. 124, no. 21, pp. 5219–5223, May 2012, doi: 10.1002/ANGE.201108706.

[115] Á. Szécsényi, G. Li, J. Gascon, and E. A. Pidko, "Unraveling reaction networks behind the catalytic oxidation of methane with H2O2 over a mixed-metal MIL-53(Al,Fe) MOF catalyst," *Chem. Sci.*, vol. 9, no. 33, pp. 6765–6773, Aug. 2018, doi: 10.1039/C8SC02376J.

[116] S. C. Oh, E. Schulman, J. Zhang, J. Fan, Y. Pan, J. Meng, and D. Liu, "Direct non-oxidative methane conversion in a millisecond catalytic wall reactor," *Angew. Chem.*, vol. 131, no. 21, pp. 7157–7160, May 2019, doi: 10.1002/ANGE.201903000.

[117] Y. Lai and G. Veser, "The nature of the selective species in Fe-HZSM-5 for non-oxidative methane dehydroaromatization," *Catal. Sci. Technol.*, vol. 6, no. 14, pp. 5440–5452, Jul. 2016, doi: 10.1039/C5CY02258D.

[118] N. Kosinov and E. J. M. Hensen, "Reactivity, selectivity, and stability of zeolite-based catalysts for methane dehydroaromatization," *Adv. Mater.*, vol. 32, no. 44, p. 2002565, Nov. 2020, doi: 10.1002/ADMA.202002565.

[119] P. Schwach, X. Pan, and X. Bao, "Direct conversion of methane to value-added chemicals over heterogeneous catalysts: challenges and prospects," *Chem. Rev.*, vol. 117, no. 13, pp. 8497–8520, Jul. 2017, doi: 10.1021/ACS.CHEMREV.6B00715.

[120] S. Liu, L. Wang, R. Ohnishi, and M. Ichikawa, "Bifunctional catalysis of Mo/HZSM-5 in the dehydroaromatization of methane with CO/CO2 to benzene and naphthalene," *Kinet. Catal.*, vol. 41, no. 1, pp. 132–144, 2000, doi: 10.1007/BF02756152.

[121] J. Gao, Y. Zheng, J.-M. Jehng, Y. Tang, I. E. Wachs, and S. G. Podkolzin, "Identification of molybdenum oxide nanostructures on zeolites for natural gas conversion," *Science*, vol. 348, no. 6235, pp. 686–690, 2015, doi: 10.1126/science.aaa7048.

[122] N. Kosinov, A. S. G. Wijpkema, E. Uslamin, R. Rohling, F. J. A. G. Coumans, B. Mezari, A. Parastaev, A. S. Poryvaev, M. V. Fedin, E. A. Pidko, and E. J. M. Hensen, "Confined carbon mediating dehydroaromatization of methane over Mo/ZSM-5," *Angew. Chem. Int. Ed.*, vol. 57, no. 4, pp. 1016–1020, Jan. 2018, doi: 10.1002/ANIE.201711098.

[123] M. Agote-Arán, R. E. Fletcher, M. Briceno, A. B. Kroner, I. V. Sazanovich, B. Slater, M. E. Rivas, A. W. J. Smith, P. Collier,

I. Lezcano-González, and A. M. Beale, "Implications of the molybdenum coordination environment in MFI Zeolites on methane dehydroaromatisation performance," *ChemCatChem*, vol. 12, no. 1, pp. 294–304, Jan. 2020, doi: 10.1002/CCTC.201901166.

[124] E. A. Uslamin, H. Saito, Y. Sekine, E. J. M. Hensen, and N. Kosinov, "Different mechanisms of ethane aromatization over Mo/ZSM-5 and Ga/ZSM-5 catalysts," *Catal. Today*, vol. 369, pp. 184–192, Jun. 2021, doi: 10.1016/J.CATTOD.2020.04.021.

[125] V. I. Yakerson, T. V. Vasina, L. I. Lafer, V. P. Sytnyk, G. L. Dykh, A. V. Mokhov, O. V. Bragin, and K. M. Minachev, "The properties of zinc and gallium containing pentasils — The catalysts for the aromatization of lower alkanes," *Catal. Lett. 1989 34*, vol. 3, no. 4, pp. 339–345, Jul. 1989, doi: 10.1007/BF00766072.

[126] L. Y. Chen, L. Lin, Z. Xu, X. Li, and T. Zhang, "Dehydro-oligomerization of methane to ethylene and aromatics over molybdenum/HZSM-5 catalyst," *J. Catal.*, vol. 157, no. 1, pp. 190–200, Nov. 1995, doi: 10.1006/JCAT.1995.1279.

[127] D. Zhou, S. Zuo, and S. Xing, "Methane dehydrogenation and coupling to ethylene over a Mo/HZSM-5 catalyst: a density functional theory study," *J. Phys. Chem. C*, vol. 116, no. 6, pp. 4060–4070, Feb. 2012, doi: 10.1021/JP209037C.

[128] R. B. Levy and M. Boudart, "Platinum-like behavior of tungsten carbide in surface catalysis," *Science*, vol. 181, no. 4099, pp. 547–549, Aug. 1973, doi: 10.1126/SCIENCE.181.4099.547.

[129] M. Boudart, S. T. Oyama, and L. Leclercq, "Molybdenum carbide, oxycarbide and nitride as catalysts in the activation of CO, N-N, C-C and H-H Bonds," *Stud. Surf. Sci. Catal.*, vol. 7, pp. 578–590, Jan. 1981, doi: 10.1016/S0167-2991(09)60300-1.

[130] K. Tamaru and G. A. Mills, "Catalysts for control of exhaust emissions," *Catal. Today*, vol. 22, no. 2, pp. 349–360, Dec. 1994, doi: 10.1016/0920-5861(94)80109-6.

[131] J. Jones, H. Xiong, A. T. DeLaRiva, E. J. Peterson, H. Pham, S. R. Challa, G. Qi, S. Oh, M. H. Wiebenga, X. I. P. Hernández, Y. Wang, and A. K. Datye, "Thermally stable single-atom platinum-on-ceria catalysts via atom trapping," *Science*, vol. 353, no. 6295, pp. 150–154, Jul. 2016, doi: 10.1126/SCIENCE.AAF8800.

[132] L. Nie, D. Mei, H. Xiong, B. Peng, Z. Ren, X. I. P. Hernandez, A. DeLaRiva, M. Wang, M. H. Engelhard, L. Kovarik, A. K. Datye, and Y. Wang, "Activation of surface lattice oxygen in single-atom Pt/CeO2 for low-temperature CO oxidation," *Science*, vol. 358, no. 6369, pp. 1419–1423, Dec. 2017, doi: 10.1126/SCIENCE.AAO2109.

[133] M. S. Hegde, G. Madras, and K. C. Patil, "Noble metal ionic catalysts," *Acc. Chem. Res.*, vol. 42, no. 6, pp. 704–712, Jun. 2009, doi: 10.1021/AR800209S.

[134] Weiyu Song, A. P. J. Jansen, and E. J. M. Hensen, "A computational study of the influence of the ceria surface termination on the mechanism of CO oxidation of isolated Rh atoms," *Faraday Discuss.*, vol. 162, no. 0, pp. 281–292, Jul. 2013, doi: 10.1039/C3FD20129E.

[135] L. S. Kibis, D. A. Svintsitskiy, E. A. Derevyannikova, T. Y. Kardash, E. M. Slavinskaya, O. A. Stonkus, V. A. Svetlichnyi, and A. I. Boronin, "From highly dispersed Rh3+ to nanoclusters and nanoparticles: Probing the low-temperature NO+CO activity of Rh-doped CeO2 catalysts," *Appl. Surf. Sci.*, vol. 493, pp. 1055–1066, Nov. 2019, doi: 10.1016/J.APSUSC.2019.07.043.

[136] W.-C. Ding, X.-K. Gu, H.-Y. Su, and W.-X. Li, "Single Pd atom embedded in CeO2(111) for NO reduction with CO: a first-principles study," *J. Phys. Chem. C*, vol. 118, no. 23, pp. 12216–12223, Jun. 2014, doi: 10.1021/JP503745C.

[137] A. Trovarelli, *Catalysis by Ceria and Related Materials.* Imperial College Press, 2002.

[138] M. Nolan, S. Grigoleit, D. C. Sayle, S. C. Parker, and G. W. Watson, "Density functional theory studies of the structure and electronic structure of pure and defective low index surfaces of ceriax," *Surf. Sci.*, vol. 576, no. 1–3, pp. 217–229, Feb. 2005, doi: 10.1016/J.SUSC.2004.12.016.

[139] K. Hermann, "Physical and Chemical Properties of Oxygen at Vanadium and Molybdenum Oxide Surfaces: Theoretical Case Studies," in *Computational Methods in Catalysis and Materials Science: An Introduction for Scientists and Engineers*, R. A. van Santen and P. Sautet, Eds. John Wiley & Sons, 2009, pp. 375–415.

[140] D. A. J. M. Ligthart, R. A. van Santen, and E. J. M. Hensen, "Supported rhodium oxide nanoparticles as highly active CO oxidation catalysts," *Angew. Chem. Int. Ed.*, vol. 50, no. 23, pp. 5306–5310, May 2011, doi: 10.1002/anie.201100190.

[141] O. Korotkikh and R. Farrauto, "Selective catalytic oxidation of CO in H2: fuel cell applications," *Catal. Today*, vol. 62, no. 2–3, pp. 249–254, Nov. 2000, doi: 10.1016/S0920-5861(00)00426-0.

[142] K. Liu, A. Wang, and T. Zhang, "Recent advances in preferential oxidation of CO reaction over platinum group metal catalysts," *ACS Catal.*, vol. 2, no. 6, pp. 1165–1178, Jun. 2012, doi: 10.1021/CS200418W.

[143] G. Spezzati, A. D. Benavidez, A. T. DeLaRiva, Y. Su, J. P. Hofmann, S. Asahina, E. J. Olivier, J. H. Neethling, J. T. Miller, A. K Datye, and

E. J. M. Hensen, “CO oxidation by Pd supported on CeO2(100) and CeO2(111) facets,” *Appl. Catal. B Environ.*, vol. 243, pp. 36–46, Apr. 2019, doi: 10.1016/J.APCATB.2018.10.015.

[144] M. Haruta, T. Kobayashi, H. Sano, and N. Yamada, “Novel gold catalysts for the oxidation of carbon monoxide at a temperature far below 0 °C,” *Chem. Lett.*, vol. 16, no. 2, pp. 405–408, Mar. 1987, doi: 10.1246/CL.1987.405.

[145] J. C. Fierro-Gonzalez, V. A. Bhirud, and B. C. Gates, “A highly active catalyst for CO oxidation at 298 K: mononuclear AuIII complexes anchored to La2O3 nanoparticles,” *Chem. Commun.*, vol. 8, no. 42, pp. 5275–5277, Oct. 2005, doi: 10.1039/B509629D.

[146] G. C. Bond and P. A. Sermon, “Gold catalysts for olefin hydrogenation,” *Gold Bull.*, vol. 6, no. 4, pp. 102–105, Dec. 1973, doi: 10.1007/BF03215018.

[147] G. J. Hutchings, “Vapor phase hydrochlorination of acetylene: Correlation of catalytic activity of supported metal chloride catalysts,” *J. Catal.*, vol. 96, no. 1, pp. 292–295, Nov. 1985, doi: 10.1016/0021-9517(85)90383-5.

[148] D. Thompson, “New advances in gold catalysis part I,” *Gold Bull.*, vol. 31, no. 4, pp. 111–118, 1998, doi: 10.1007/BF03214775.

[149] D. Thompson, “New advances in gold catalysis Part II,” *Gold Bull.*, vol. 32, no. 1, pp. 12–19, 1999, doi: 10.1007/BF03214784.

[150] G. J. Hutchings, “Heterogeneous gold catalysis,” *ACS Cent. Sci.*, vol. 4, no. 9, pp. 1095–1101, Sep. 2018, doi: 10.1021/ACSCENTSCI.8B00306.

[151] S. Cattaneo, M. Stucchi, A. Villa, and L. Prati, “Gold catalysts for the selective oxidation of biomass-derived products,” *ChemCatChem*, vol. 11, no. 1, pp. 309–323, Jan. 2019, doi: 10.1002/CCTC.201801243.

[152] M. Haruta, T. Takase, and T. Kobayashi, “Low-temperature combustion catalysts mainly for CO oxidation,” *J. Japan Pet. Inst.*, vol. 37, no. 5, pp. 480–491, Sep. 1994, doi: 10.1627/JPI1958.37.480.

[153] M. Daté, M. Okumura, S. Tsubota, and M. Haruta, “Vital role of moisture in the catalytic activity of supported gold nanoparticles,” *Angew. Chem.*, vol. 116, no. 16, pp. 2181–2184, Apr. 2004, doi: 10.1002/ANGE.200453796.

[154] A. Bongiorno and U. Landman, “Water-enhanced catalysis of CO oxidation on free and supported gold nanoclusters,” *Phys. Rev. Lett.*, vol. 95, no. 10, p. 106102, Sep. 2005, doi: 10.1103/PHYSREVLETT.95.106102.

[155] J. Saavedra, H. A. Doan, C. J. Pursell, L. C. Grabow, and B. D. Chandler, “The critical role of water at the gold-titania interface in

catalytic CO oxidation," *Science*, vol. 345, no. 6204, pp. 1599–1602, Sep. 2014, doi: 10.1126/SCIENCE.1256018.

[156] G. M. Mullen and C. B. Mullins, "Waters place in Au catalysis," *Science*, vol. 345, no. 6204, pp. 1564–1565, Sep. 2014, doi: 10.1126/SCIENCE.1259527.

[157] C. K. Costello, J. H. Yang, H. Y. Law, Y. Wang, J. N. Lin, L. D. Marks, M. C. Kung, and H. H. Kung, "On the potential role of hydroxyl groups in CO oxidation over Au/Al2O3," *Appl. Catal. A Gen.*, vol. 243, no. 1, pp. 15–24, Mar. 2003, doi: 10.1016/S0926-860X(02)00533-1.

[158] J. K. Edwards, S. J. Freakley, A. F. Carley, C. J. Kiely, and G. J. Hutchings, "Strategies for designing supported gold-palladium bimetallic catalysts for the direct synthesis of hydrogen peroxide," *Acc. Chem. Res.*, vol. 47, no. 3, pp. 845–854, Mar. 2014, doi: 10.1021/AR400177C.

[159] J. Pritchard, M. Piccinini, R. Tiruvalam, Q. He, N. Dimitratos, J. A. Lopez-Sanchez, D. J. Morgan, A. F. Carley, J. K. Edwards, C. J. Kiely, and G. J. Hutchings, "Effect of heat treatment on Au–Pd catalysts synthesized by sol immobilisation for the direct synthesis of hydrogen peroxide and benzyl alcohol oxidation," *Catal. Sci. Technol.*, vol. 3, no. 2, pp. 308–317, Jan. 2013, doi: 10.1039/C2CY20234D.

[160] S. J. Tauster, S. C. Fung, and R. L. Garten, "Strong metal-support interactions. Group 8 noble metals supported on titanium dioxide," *J. Am. Chem. Soc.*, vol. 100, no. 1, pp. 170–175, 1978, doi: 10.1021/JA00469A029.

[161] V. Ponec, "On some real and apparent carrier effects in catalysis by metals," *Stud. Surf. Sci. Catal.*, vol. 11, no. C, pp. 63–75, Jan. 1982, doi: 10.1016/S0167-2991(09)61378-1.

[162] D. E. Resasco and G. L. Haller, "A model of metal-oxide support interaction for Rh on TiO2," *J. Catal.*, vol. 82, no. 2, pp. 279–288, Aug. 1983, doi: 10.1016/0021-9517(83)90194-X.

[163] I. X. Green, W. Tang, M. Neurock, and J. T. Yates, "Spectroscopic observation of dual catalytic sites during oxidation of CO on a Au/TiO 2 catalyst," *Science*, vol. 333, no. 6043, pp. 736–739, Aug. 2011, doi: 10.1126/SCIENCE.1207272.

[164] J. A. Rodriguez, P. Liu, X. Wang, W. Wen, J. Hanson, J. Hrbek, M. Pérez, and J. Evans, "Water-gas shift activity of Cu surfaces and Cu nanoparticles supported on metal oxides," *Catal. Today*, vol. 143, no. 1–2, pp. 45–50, May 2009, doi: 10.1016/J.CATTOD.2008.08.022.

[165] D. Tibiletti, A. Goguet, F. C. Meunier, J. P. Breen, and R. Burch, "On the importance of steady-state isotopic techniques for the investigation of the mechanism of the reverse water-gas-shift reaction," *Chem.*

Commun., vol. 4, no. 14, pp. 1636–1637, Jul. 2004, doi: 10.1039/B403438D.
[166] R. Burch, "Gold catalysts for pure hydrogen production in the water–gas shift reaction: activity, structure and reaction mechanism," *Phys. Chem. Chem. Phys.*, vol. 8, no. 47, pp. 5483–5500, Nov. 2006, doi: 10.1039/B607837K.
[167] M. Shekhar, W. S. Lee, M. C. Akatay, L. Maciel, W. Tang, J. T. Miller, E. A. Stach, M. Neurock, W. N. Delgass, and F. H. Ribeiro, "Water-gas shift reaction over supported Au nanoparticles," *J. Catal.*, vol. 405, pp. 475–488, Jan. 2022, doi: 10.1016/J.JCAT.2021.12.021.
[168] P. Liu and E. J. M. Hensen, "Highly efficient and robust Au/MgCuCr2O4 catalyst for gas-phase oxidation of ethanol to acetaldehyde," *J. Am. Chem. Soc.*, vol. 135, no. 38, pp. 14032–14035, Sep. 2013, doi: 10.1021/JA406820F.
[169] W. Song, P. Liu, and E. J. M. Hensen, "A mechanism of gas-phase alcohol oxidation at the interface of Au nanoparticles and a MgCuCr2O4 spinel support," *Catal. Sci. Technol.*, vol. 4, no. 9, pp. 2997–3003, Aug. 2014, doi: 10.1039/C4CY00462K.
[170] T. Takei, N. Iguchi, and M. Haruta, "Support effect in the gas phase oxidation of ethanol over nanoparticulate gold catalysts," *New J. Chem.*, vol. 35, no. 10, pp. 2227–2233, Sep. 2011, doi: 10.1039/C1NJ20297A.
[171] O. A. Simakova, V. I. Sobolev, K. Y. Koltunov, B. Campo, A. R. Leino, K. Kordás, and D. Y. Murzin, "'Double-Peak' catalytic activity of nanosized gold supported on titania in gas-phase selective oxidation of ethanol," *ChemCatChem*, vol. 2, no. 12, pp. 1535–1538, Dec. 2010, doi: 10.1002/CCTC.201000298.
[172] V. I. Sobolev, K. Y. Koltunov, O. A. Simakova, A. R. Leino, and D. Y. Murzin, "Low temperature gas-phase oxidation of ethanol over Au/TiO2," *Appl. Catal. A Gen.*, vol. 433–434, pp. 88–95, Aug. 2012, doi: 10.1016/J.APCATA.2012.05.003.
[173] W. C. Ketchie, Y. L. Fang, M. S. Wong, M. Murayama, and R. J. Davis, "Influence of gold particle size on the aqueous-phase oxidation of carbon monoxide and glycerol," *J. Catal.*, vol. 250, no. 1, pp. 94–101, Aug. 2007, doi: 10.1016/J.JCAT.2007.06.001.
[174] M. A. Sanchez-Castillo, C. Couto, W. B. Kim, and J. A. Dumesic, "Gold-nanotube membranes for the oxidation of CO at gas–water interfaces," *Angew. Chem.*, vol. 116, no. 9, pp. 1160–1162, Feb. 2004, doi: 10.1002/ANGE.200353238.
[175] B. N. Zope, D. D. Hibbitts, M. Neurock, and R. J. Davis, "Reactivity of the gold/water interface during selective oxidation catalysis,"

Science, vol. 330, no. 6000, pp. 74–78, Oct. 2010, doi: 10.1126/SCIENCE.1195055.

[176] K. Loos, R. Zhang, I. Pereira, B. Agostinho, H. Hu, D. Maniar, N. Sbirrazzuoli, A. J. D. Silvestre, N. Guigo, and A. F. Sousa, "A perspective on PEF synthesis, properties, and end-life," *Front. Chem.*, vol. 8, p. 585, Jul. 2020, doi: 10.3389/FCHEM.2020.00585.

[177] V. M. Chernyshev, O. A. Kravchenko, and V. P. Ananikov, "Conversion of plant biomass to furan derivatives and sustainable access to the new generation of polymers, functional materials and fuels," *Russ. Chem. Rev.*, vol. 86, no. 5, pp. 357–387, May 2017, doi: 10.1070/RCR4700.

[178] A. D. K. Deshan, L. Atanda, L. Moghaddam, D. W. Rackemann, J. Beltramini, and W. O. S. Doherty, "Heterogeneous catalytic conversion of sugars into 2,5-furandicarboxylic acid," *Front. Chem.*, vol. 8, p. 659, Jul. 2020, doi: 10.3389/FCHEM.2020.00659/BIBTEX.

[179] M. Tan, L. Ma, M. S. U. Rehman, M. A. Ahmed, M. Sajid, X. Xu, Y. Sun, P. Cui, and J. Xu, "Screening of acidic and alkaline pretreatments for walnut shell and corn stover biorefining using two way heterogeneity evaluation," *Renew. Energy*, vol. 132, pp. 950–958, Mar. 2019, doi: 10.1016/J.RENENE.2018.07.131.

[180] S. E. Davis, L. R. Houk, E. C. Tamargo, A. K. Datye, and R. J. Davis, "Oxidation of 5-hydroxymethylfurfural over supported Pt, Pd and Au catalysts," *Catal. Today*, vol. 160, no. 1, pp. 55–60, Feb. 2011, doi: 10.1016/J.CATTOD.2010.06.004.

[181] M. Kim, Y. Su, A. Fukuoka, E. J. M. Hensen, and K. Nakajima, "Aerobic oxidation of 5-(hydroxymethyl)furfural cyclic acetal enables selective furan-2,5-dicarboxylic acid formation with CeO2-supported gold catalyst," *Angew. Chem. Int. Ed.*, vol. 57, no. 27, pp. 8235–8239, Jul. 2018, doi: 10.1002/ANIE.201805457.

Chapter 7

The Catalytic Enterprise

7.1 Introduction

What is the contribution of mechanistic knowledge to catalyst design?

The mechanisms of catalytic reactions are the subject of this book. The network of elementary reactions determines the outcome of the catalytic event, and its dynamics is a function of catalyst properties and reaction conditions. Therefore, reaction mechanism is the key to unraveling the relation between catalyst performance and composition. Such relations aim to assist the design of improved catalysts or discovery of new reactions.

Here the question is addressed as to whether the current status of mechanistic understanding warrants this expectation.

The content and material presented in the previous chapters provide an opportunity to investigate the question.

Interestingly the very formulation of this question raises the issue of the role of science. Should not science be the intellectual activity that discovers the natural laws of the material world without any purpose other than to reveal the truth?

What, then, is the position of the science of catalysis?

As has been recounted in the first chapter and also illustrated by the history of several of the catalytic processes elsewhere, innovation in catalysis is driven by technological challenges posed by society and by technical capabilities to address them. These technical capabilities, as the discovery of catalyst materials or reactions, also develop in the course of process invention.

Catalysis science and catalytic technology are closely intertwined. It is this interconnection that creates the impact of catalytic chemistry on the development of major industrial processes.

Essential to the understanding of the dynamics of catalytic technology is also insight on the status of catalysis science. What are its contributions and, possibly more importantly, what are its limitations? A different formulation of this question is whether progress in scientific understanding has altered the practice of catalyst invention or improvement.

Is catalysis science a predictive science? There is a large, at least empirical, understanding of existing catalytic materials that can be used in a predictive sense. However, can catalytic reactivity be predicted?

As described in previous chapters for many reactions, there is still a debate as to which reaction network is the true one. The different mechanistic proposals for the same reaction are often caused by a different view on the state of the reactive surface.

The holy grail for a definitive understanding of catalyst performance structure is possibly within reach, but not yet reached.

What has been accomplished for many systems is an increasing understanding how mechanism and catalyst composition cohere. This is a great theoretical advance.

It then raises the experimental question of how to determine the actual structure of the working catalyst surface. As discussed in Section 7.3 this is a hard question to answer.

Why is a conclusive relationship between catalyst performance and composition so difficult to obtain? Catalyst performance is a global macroscopic property that is expressed as a production rate and depends on reactant concentration and reaction condition.

The chemical relations that define global performance depend on molecular surface chemistry that is only indirectly addressed in global experiments.

The length and timescales of global catalytic kinetics and those of the bond cleavage and bond formation reactions of adsorbates activated by the catalyst surface differ by many orders of magnitude. It took a major part of the previous century to understand this relationship.

This understanding was accompanied by appreciation for the limitation of our capability to predict catalyst performance. Knowledge that was lacking is becoming understood. This is a major scientific advance.

Science progresses by the formulation of research questions that can be experimentally addressed. This furthers understanding of the fundamental processes that determine a physical phenomenon.

In the following section views on what drives catalytic innovation and what the role of basic science is are discussed. Philosophers of science and technology have developed a framework to understand this interaction. Three case studies of catalytic discovery demonstrate the rich variety of ways by which this interaction is realized. Catalysis is a technology that is well positioned to address technical chemical challenges posed by demands of society.

Section 7.3 highlights the status of catalysis science and the contribution of scientific discovery to progress in catalyst invention and application.

Specifically, the role of mechanistic understanding is discussed.

To determine the relationship between catalyst performance and composition, knowledge of the actual structure of the working catalyst surface is essential.

It is argued that catalysis is a well-founded and maturing science. Indeed, many aspects of catalytic reactivity are well understood. The still open research questions of current catalysis science are presented. The challenge remains to understand the systems in its full complexity.

Powerful supporting design tools have been developed, but catalyst design ultimately will require tinkering of catalyst formulations.

7.2 Catalytic Science-Technology Dynamics

7.2.1 *Science Philosophy Views*

Science and technology are synergetic and enforce each other in two-way processes.

There are three science philosophical models: Stokes science quadrant, Casimir's science technology spiral, and Arthur's science and technology symbiosis theory.

The first question to address is that of science culture. What motivates scientific research?

Stokes [1] wrote in 1997 an important report to the USA government on this question. He distinguished three different investigation cultures: pure basic research, use-inspired basic research, and pure applied research that he clarified in the quadrant reproduced in Figure 7.1.

The Stokes report made a large impact, because its view on the scientific enterprise opposed the then generally accepted science philosophy within government agencies and industrial corporations, that fundamental science and applied science are separate cultures.

This separation of fundamental (free) and mission-driven research (goal oriented) happened after the Second World War. It was due to another important report, "Science: the endless frontier," formulated by Vannevar Bush in 1945 [2].

Vannevar Bush argued that free basic science would deliver novel discoveries, that in the hands of applied scientists (for security, preferentially outside universities) will deliver useful practical devices.

The goal of the report was political. After the Second World War, scientists that during the war had become very involved in military

		Considerations of use?	
		No	Yes
Quest for fundamental understanding?	Yes	Pure basic research (Bohr)	Use-inspired basic research (Pasteur)
	No		Pure applied research (Edison)

Figure 7.1 Quadrant model of scientific research [1].

applications became very concerned about the use of military technologies and in particular the atomic bomb. The aim of the disconnection of the scientists into the two cultures of fundamental and applied research was to remove responsibility of the former for the latter applications. Vannevar Bush was an engineer of MIT and became a high-ranking USA government official. He had a leading role in the USA industrial-military complex system.

Since that period many identify university research with basic research and industry with applied research.

However, the history of catalyst invention shows that the interplay of basic and applied research is at the root of many of the truly innovative discoveries. The interaction between the two research cultures is not unidirectional.

Below some examples will be given. At universities as well as at industrial research laboratories the cultures should overlap at least partially. Implementation and development of catalytic processes at industrial scale require an infrastructure that cannot be realized at most universities.

Forty years after the Vannevar Bush report the merit of the Stokes report is that it states the premise of separation of fundamental and applied science is incorrect and inhibits technical innovation.

As illustrated in Figure 7.1 Stokes distinguished three scientific cultures: pure basic research, pure applied research, and use-inspired basic research.

Stokes defined pure basic research as directed towards a question within science. It is not obvious whether the solution has practical use. The question is what the causal relations fundamental to a phenomenon are. Stokes gave as an example of basic research the discoveries by Bohr (Nobel Prize 1922) in quantum mechanics. He founded the electronic theory of matter as consisting of discrete energy levels.

The explanation of the atomic spectra was his concern. His discovery, which rejects classical mechanics, provided the basis for quantum physical methods to calculate such spectra.

Quantum theory became a tool to make quantitative predictions.

Stokes defines pure applied research as the other extreme. Edison's inventions, for instance the light bulb, are identified with this research profile. In Edison's case the invention was an art, the result of tinkering. Science does not necessarily play a role. Interestingly, as catalysis history shows, serendipitous discoveries may become recognized also as breakthrough fundamental science advances. An example is the discovery by Ziegler of catalytic polymerization of ethene mentioned in Section 6.2.

The third quadrant of Figure 7.1 is called use-inspired basic research. The French scientist Pasteur is often considered to symbolize this science culture. The topics of research relate to practical technical activities, but the aim of the research is not to directly address them. The aim of the research is to understand the working of useful technical systems. Indirectly it may resolve a practical problem.

Pasteur's discoveries of immunization and pasteurization are the result of such an approach.

Serendipitous discovery by scientists active in "use-inspired basic research" can be the inspiration for breakthrough innovations. Their relevance is recognized by the scientist with an eye for it.

Haber's invention of ammonia synthesis fits in the quadrants of "use-inspired basic research" as well as "pure applied research." He had a scientific conflict with Nernst to identify the thermodynamics of the ammonia synthesis process and used his data to discover the ammonia synthesis process. In the design of the process there was a large knowledge component. However, the invention of the catalyst was strictly empirical.

Nernst belongs more strictly to the category of "use-inspired basic research". As he explains in his Nobel lecture of 1920 his primary motivation was understanding the galvanic cell.

Haber's work is in the full tradition of 19th-century chemistry where research of scientists was mainly motivated to find technical solutions to problems in society. At the same time important fundamental discoveries were made. The discovery of Davy of the mine lamp went together with the discovery of the catalytic oxidation activity of platinum.

After the Vannevar Bush report university research was declared as fundamental, and applied research was to belong in the mission-driven institutes. In the 1950s research management in industry took over this vision.

In the 1950s and 1960s research laboratories in major industrial corporations that were largely free to select their research topics were set up. In these departments, such as at Shell Research, "use-inspired basic research" became the research culture. The Cossee model of the coordination complex of polymerization (Section 6.2) finds its origin in such industrial "use-inspired basic research." Also applied research successes are due to this opportunity of "free research" inspired by the problems of their industry, such as the invention of ethene polymerization and disproportionation catalysis at Phillips petroleum company (Section 6.2).

A well-known example where the division between basic and applied research breaks down is the Nobel Prize of Langmuir, who in the first part of the previous century was research scientist at General Electric.

He contributed with many inventions that were often based on his parallel basic research. His work is basic to catalysis and discussed in Chapter 2.

In the 1990s Thomas (Section 6.3.2.1) directed a research program at the University of Cambridge to discover new zeolitic catalysts for organic chemical reactions. His laboratory was equipped with state-of-the-art spectroscopic and computational facilities. This research is also at the interphase of "use-inspired basic research" and "pure applied research."

One can conclude that catalysis science belongs to the two categories in the right column of Figure 7.1. Its research topics are inspired by the question of the working mechanism of the catalyst, which is posed within the context of the design of new or improved catalysts.

The first question of the "working mechanism" is the "what" question of knowledge, the second question is the "how" question of design [3].

The solution to the "how" question usually is discovered with significant tinkering.

Mechanistic knowledge of a reaction is an answer to the "what" question. The answer to the question of to what extent mechanistic knowledge can contribute to catalyst design has many colors of gray. As will be seen in this and following sections, there are technical as well as non-technical contributions.

In "use-inspired basic research" and "pure basic research" the research strategy is Cartesian reductionistic. The system to be investigated is simplified in order to become amenable to a well-defined experiment. It aims to answer a hypothesis. The system of investigation will emulate only partially the applied system.

The ever-important question is the relevance of the reductionistic answer to the practical system. This is an inductionistic issue since the practical system is different from the simplified system for which an answer had been found. The answer is not necessarily valid in a more general context. Through time this inductionistic question caused and still causes much debate in catalysis.

A key example of the reductionistic approach to heterogenous catalysis is the surface science method (see Section 2.3.2). The complex catalytic system is reduced to a single crystal surface and reactivity is probed at model conditions.

Can one formulate in a general way the contribution of basic research to the science enterprise? One contribution is the understanding of a phenomenon formulated in mathematical equations. The solution of these equations provides a tool for prediction. The contribution of the reductionistic approach to science is invaluable. Great discoveries relate to it. In catalysis the equations of chemical thermodynamics and chemical kinetic approaches are examples of modelling tools provided by basic science. However, since the theory is the outcome of a reductionistic approach the question that has to be posed is how well it relates to the wholeness of the phenomenon.

Is the whole properly understood as more than the sum of its parts? This question is very relevant to the science of the catalytic system.

Within this context the limitation of kinetic modelling and mechanistic models to catalytic description has to be questioned. This is extensively discussed in Section 7.3.2.

The tools developed can be mathematical equations but also new instrumental and, more recently, computational techniques. Specific spectroscopic or kinetic techniques were discovered to characterize the catalyst surface. Also, the discovery of new chemistry such as the catalytically active molecular coordination complexes belongs to this category.

Basic science refines the way we interrogate nature. It enlarges the unknown. The way forward is to formulate a hypothesis that can be experimentally tested. Scientific discoveries that refer to one specific situation may also be valid in a new, previously not investigated, situation. This induction process generates new questions and design of experiments to discover an answer. This process of question and answer is the fundamental scientific process but is also highly fruitful within the design context of applied science. It is the reason for breakthrough scientific discoveries in applied science.

Still today the practice of catalyst design uses the intuitive approach as Dowden did at ICI in the 1960s and 1970s. Hutchings from Cardiff University describes the process as follows [4]: the idea is to dissect a desired reaction into imaginary elementary steps and then to search for catalytic systems that are known to activate such steps. A new catalyst can be attempted that is a complex mixture of the component materials [5]. Obviously, the better the reaction mechanism and the relation between activation of adsorbates and surface reactivity are understood, the more refined the design of the catalyst material. Significant tinkering will still be necessary to produce a working system.

An example in catalysis where the known mechanism of one reaction led to the hypothesis of a new reaction is the oligomerization of methane to higher alkanes [6], [7].

In Section 3.2.2.1.1 the mechanism of the Fischer-Tropsch reaction is discussed. In this reaction higher alkane molecules are formed from the reaction of CO with hydrogen. In the mechanism an essential step is the dissociative adsorption of CO, which generates intermediate CH_x species.

This idea is used in the methane oligomerization reaction. Methane is decomposed by reaction with a transition metal and CH_x surface intermediates are formed. Growing hydrocarbon intermediates are then formed by combination of the surface CH_x intermediates, and in a final step, alkane molecules are produced by hydrogenation of the growing hydrocarbon intermediates. Such steps are similar to those known from the Fischer-Tropsch reaction.

Mechanistic discovery can also affect investment decisions in applied industrial research. The question may be raised as to whether present performance of a particular catalytic system can be further improved.

This is an issue for the epoxidation of ethylene.

As discussed in Section 4.3.3.1, in the 1970s there was a debate about whether molecular oxygen or atomic oxygen is the selective intermediate that gives the epoxide. The molecular oxygen mechanism predicts a maximum selectivity of 86% whereas the atomic oxygen mechanism in principle makes 100% selectivity possible.

This was resolved once the atomic oxygen mechanism was experimentally supported. At a time when the catalysts gave selectivity around 86% this knowledge of the oxygen atom mechanism was crucial. It indicates that commercial performance can be improved further.

Interesting in this context is also the discovery of chlorine promotion of the silver-catalyzed epoxidation reaction. This discovery was made serendipitously in France in the 1940s. Air polluted with chlorinated molecules from other factories appeared to be the cause of the exceptional selectivity of the plant concerned.

It was discovered that chlorine adsorbed on the Ag catalyst is the cause of the exceptional selectivity. In turn the study of this chlorine promotion became an important fundamental tool to probe the mechanism of the reaction.

The issue that Stokes addressed is whether it is correct to assume that applied research harvests the fruits of pure basic research and

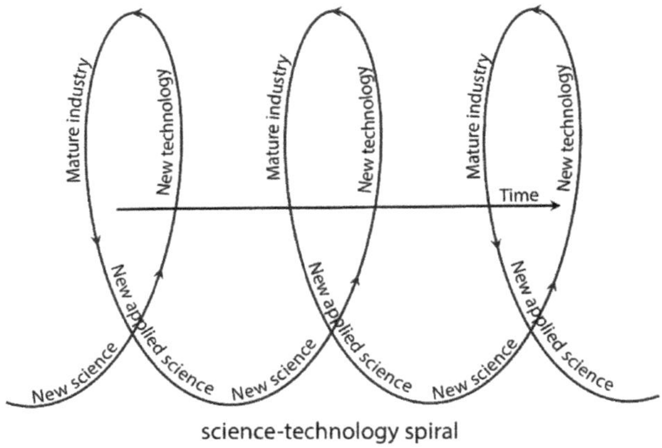

Figure 7.2 The science-technology spiral of Casimir.

that these linearly and sequentially relate. He argues that the process is not unidirectional. This was alluded to earlier and as shown below the advance of catalysis science and its technologies provides a clear example.

One can use the science technology spiral of Figure 7.2 as suggested by Hendrik Casimir, theoretical physicist and former director of Philips Research in Eindhoven, The Netherlands, to illustrate the interplay of scientific discovery and technological rise [8].

At a particular moment in time the present state of science and technology provides the technical basis for new discoveries and inventions as a response to perceived technological needs or challenges of society. The new technologies that arise are usually partially understood and generate "use-inspired basic research." Industries grow and mature and are succeeded by new industries. The origin of new industries will be new discoveries and inventions that follow from science and technologies developed in a previous period.

In Philips company an example is the development of the radio industry that followed the initial development of the glass light bulb industry. The common denominator is the vacuum bulb. The vacuum transistor is a complex sister of the light bulb with essential add-on features.

These are part of great industries because they solve great human needs. The light bulb solves the issue of darkness and provides ready and affordable light, while the vacuum transistor and with it the radio contribute to human needs for communication.

Changes in catalytic technology often result from changes in raw material supply. In catalytic technology, the conversion of oil into liquid fuel with zeolite catalysts used in the cracking of crude oil was succeeded by the conversion of natural gas with adapted zeolitic materials thirty years later. This is another example where new industries arose by additional scientific discoveries that build on earlier developed technology. The particular case of zeolite-based technologies is discussed in part of the second case study of Section 7.2.2.

Another example is maleic acid anhydride production. Maleic anhydride is an important chemical that is used to make polyester resins and coatings. In the early part of the previous century, it was produced from benzene (from coal liquid or oil) by oxidation catalyzed by vanadium oxide. At present this process is replaced by the use of butane (from natural gas) as feedstock. Again, the catalyst is related and contains vanadium as main component, but it is a new vanadium phosphate material.

The economist W. B. Arthur [9] proposed an evolutionary view on how technology evolves based on theories of how complex systems evolve and adapt. He defines technology as usage of phenomena for a purpose and supports the view of the non-unidirectional process of interacting agents.

The difference with random biological evolution is that dynamics is human manipulated. The ultimate technology that develops may not have been in the minds of researchers that are part of the science innovation network. Goals may be redefined halfway. There may be many unknowns and unexpected problems may arise that ask for novel solutions.

The opportunity for an alternative suddenly within reach may arise. A similarity with biology is that technology may behave opportunistically. It tends to enforce itself by new combinations of technologies. The new technology may be very different from technologies in which these components were originally used.

In the chemical industry the complexity way of thinking of Arthur's reflects in the increasing reaction engineering complexity of processes and the multiple ways that different processes enforce one another.

The presence of one catalytic process makes another catalytic process possible. In the fertilizer industry hydrogen production (by steam gasification and water-gas shift reaction) technology, ammonia synthesis reaction, and Ostwald ammonia oxidation are to be combined into one industrial process.

A process where homogeneous and heterogeneous catalytic reactions are combined is the Shell Higher Olefins Process that is detailed in Section 1.3.3.

The combination of processes that have interdependence or that also can partially replace each other can lead to a stable industrial ecology with minimum emission of waste or optimal energy use. The modern integrated refinery or industrial chemical complex is an example. The undesirable co-product of one process can be a desirable product of the other. Heat produced in one process is of use in the other. It is a model of a cyclic economy with minimal waste.

In this broader context catalysis is a support technology. It supports chemical processes of an industrial chemical site. In turn the refinery supports automotive motion or airplanes by producing the liquid fuels of the right quality.

The view of Arthur relates to the answer of science historian Joel Mokyr [10] to the question: what was the cause of the exponential growth of technology (the beginning of the Industrial Revolution) at the beginning of the 19th century? It is due to the accumulation of practical and, as he called it, systematic knowledge. An example is the invention of the steam engine, that came underpinned by ("use-oriented basic science") the formulation of thermodynamics. Technical innovation is the fruit of the fertilizing interaction of scientific knowledge and technical skill.

This also happened in several stages of catalytic process invention. Scientific understanding (tools and techniques) accumulated jointly with practical knowledge of inorganic chemistry. This and the expanding economies around the beginning and the middle of the 20th century were the drivers of the two golden episodes of catalysis as described in Chapter 1.

7.2.2 *Three Case Studies from Catalysis*

> *The catalytic innovation process is opportunistic. It is fueled by catalyst discovery.*

Desulfurization of liquid fuels is discussed in combination with the introduction of automotive exhaust catalyst as an example of evolutionary process discovery and implementation. The catalytic cracking reaction is an example where science discovery and technological invention enforce each other. As a third example the relation will be outlined between the Roelen discovery of hydroformylation and the discovery of propylene polymerization. Unexpected new catalytic chemistry leads to molecular inorganic chemistry, founding a new episode of novel catalytic discoveries.

7.2.2.1 *Sulfur Reduction and Automotive Exhaust Reduction*

Here for catalyst development the suggestion of Arthur of evolutionary dynamic technological innovation is illustrated by the role of sulfur reduction in gasoline. It is an example of how government regulation of exhaust emissions was one of the major drivers for new catalyst development processes (the USA Clean Air Act of 1990). The initial driver for the reduction of sulfur content of liquid fuels was the need to reduce acid rain damage in forests.

The reduction of the sulfur content in fuels started around the same time that the lead content of fuel became regulated as well (the USA Safe Drinking Water and Toxic Enforcement Act of 1986). This reduction of lead required adaptation of fuel content (xylene, toluene, branched hydrocarbons) that initiated additional development of new refinery processes such as catalytic alkylation and transalkylation (see Section 5.4.4.2).

The need to reduce sulfur content led to major industrial and academic research activities and brought with it novel generations of adapted and effective hydrodesulfurization catalysts. The mechanism of the hydrodesulfurization reaction is discussed in Section 3.4.2. The major improvement came by variation of hydrodesulfurization catalyst composition and structure. In parallel in industry and universities there was a large presence of "use-inspired basic research." The major contribution of the latter was knowledge of performance-composition relation, improved catalyst preparation and tool development to analyze catalyst composition and structure.

But unexpectedly sulfur reduction also became important for the implementation of automotive exhaust catalysis. Otherwise, the catalyst would be poisoned by the sulfur compounds in the car exhaust as these catalysts contain noble metals such as Pt and Pd that become readily poisoned by sulfate and lead oxides. It established a large automotive exhaust catalyst industry [11].

For the proper functioning of the exhaust catalyst an additional requirement is the ratio of the exhaust oxidant (air) versus reductant. The catalyst will only operate when this ratio remains between particular limits. In the exhaust pipe catalyst sensors have to be present to measure exhaust gas emission concentration as well as a computer system that feeds back this information to the engine to regulate fuel injection. Novel for catalysis was the invention of Ford company in the 1970s to use for the noble metals a reducible support such as CeO_2 that manages fluctuation in exhaust oxygen content [12].

A combination of several processes partially developed in parallel and others in succession ultimately led to the catalytic exhaust converter. The science-technology-society process dynamics was not predetermined but evolved in an evolutionary way.

Improved, altered catalysts as well as new catalytic systems were part of it. The exhaust catalyst was realized due to governmental regulation and reduction of lead and sulfur content by oil companies. There was a joint effort by catalyst manufacturers and car manufacturers to design appropriate catalysts and control of engine operation.

7.2.2.2 *Zeolite-Catalyzed Processes*

The second example of evolutionary technology development deals initially with catalyst and process innovation in an existing process. This had major societal impact and was followed by important and unexpected new catalytic technologies. This detailed history of solid acid catalyst discovery and process invention illustrates also the contagious nature of the following success story.

The catalytic cracking reaction of crude oil to liquid fuels currently produces more than half of the world's transportation fuels. The present fluid catalytic cracking process took forty years to be realized. As described below many actors in oil companies and universities contributed to its discovery [13]. The mechanism of the reaction, catalyzed by solid acids, is discussed in Section 5.2.

The Friedel-Crafts alkylation catalyst $AlCl_3$ that had been discovered in the second part of the 19th century inspired scientists in the beginning of the 20th century to explore such and related catalysts more generally as solid acid catalysts. Already in 1913 McAfee explored the cracking of crude oils by such catalysts.

In parallel in 1902 the Russian scientist Vladimir Ipatieff explored [14] related solid acid catalysts for dehydration reactions. In the 1930s Ipatieff moved to the USA and became director of the United Oil Products (UOP) organization. Under his leadership several important refinery processes that exploit solid acidic catalysts were developed.

The first commercial catalytic cracking process is the Houdry process that since 1936 is in operation in the USA. Eugene Houdry was a French inventor who already in 1927 had developed a three-stage process for the production of motor fuel based on oil derived from lignite.

He moved to the USA where Sun Oil corporation had become interested in developing a catalytic cracking process based on crude oil. The catalyst was a chlorinated clay material with alumina-silicate as major component (in 1990 Houdry was recognized for his inventions in the USA National Inventors Hall of Fame).

A major difficulty of the process was its rapid deactivation. The catalyst was covered with carbonaceous residue and coke within 10 minutes. As a consequence, the reactor had to be operated in a swing mode. Short production by a flow of oil was alternated with catalyst regeneration in a flow of steam and air.

In 1941 the process was changed into a continuous process by implementation of moving bed technology, in which the catalyst was moved between reactors.

The next major invention in reactor technology was based on suggestions by Lewis and Gilliland from MIT in Cambridge, USA (they are considered founding fathers of USA chemical engineering) for the design of the fluid catalyst bed reactor.

The catalyst will behave as a fluid when it is reduced to small finely divided particles in contact with vaporized oil. Then the reactor can be designed as a recirculation system. In one part of the reactor system oil is added and product distilled off. In a second part of the reactor at different conditions coke is burned off and catalyst is regenerated. This design became a great success.

The first commercial fluid catalytic cracking facility was at Standard Oil company (now Exxon) and began production in 1942. New production methods had to be invented to produce the solid acid catalyst particles that survive the harsh conditions of the fluid flow cracking process. Synthetic amorphous alumino-silicates earlier discovered in 1936 at UOP were applied [15], [16].

A major economic drawback of the process at that time was the low yield of product fuel based on input of crude oil. A non-insignificant fraction of crude oil was converted to coke to be burned off in the regeneration part of the process.

The next major step in the final design of the process was new catalyst discovery that ultimately resolved the low yield of production. This part of the story starts in Great Britain where Richard Barrer (originally from New Zealand) started to synthesize nanoporous zeolites (see Section 5.3) that at the time mainly were used in separation technology [17], [18].

As a professor at Imperial College in London Barrer was also active as a consultant to Union Carbide where in the 1950s a large synthetic zeolite research effort had started and recruited several of Barrer's students.

Over several years this research group generated a variety of the zeolitic materials that are currently applied in major industrial processes [13].

As partially told in Chapter 5 the accidental discovery by the Union Carbide scientists Rabo *et al.* that the synthetic zeolites could be converted into acidic zeolites generated interest to apply them as solid acid catalysts [19].

The Union Carbide chemist Edith Flanigan synthesized zeolite Y that became part of the new generation of cracking catalysts. At Mobil Oil the discovery was made that this catalyst has highly superior performance than those used so far in this process. In 1964 Mobil Oil started the first fluid catalytic cracking plant with zeolite Y as main component. The process was very successful and became widely applied at refineries. The highly reduced coke deposition and gas production meant a more selective use of crude oil and brought large economic advantages. The inventors Plank and Rosinsky at Mobil Oil are also memorialized in the USA National Inventors Hall of Fame [20].

The reduced coke deposition is due to the nanoporous structure of the zeolite that inhibits formation of large aromatic coke-forming intermediates. There is a relation between size and shape of reaction intermediates and the dimensions of the zeolite pore. The concept of shape-selective catalysis by inorganic materials was new to catalysis science. Its discovery is a great fundamental contribution that has its origin in applied catalysis (see Section 5.4).

Research continued at Mobil Oil and Union Carbide. It generated an extensive range of new zeolitic catalysts that became applied in previously unknown reactions. The two examples to be mentioned are the low Al/Si content Mobil Oil H-ZSM-5 synthetic zeolite discovery, useful for converting methanol to aromatics (an important step in the conversion of natural gas to liquid fuel, the MTG process of 1986, see Section 5.4.5), and the alumino-phosphate zeolitic catalysts developed at Union Carbide. The latter was applied in a related process that converts methanol to olefins (MTO process, China 2010, see Section 5.4.5).

This history illustrates the tortuous evolutionary path of catalytic process invention, with an outcome that initially could not be foreseen. Who could have imagined in the 1940s that the zeolite sieving agents would be applied fifty years later in a process that converts natural gas to liquid fuels or chemicals? The development of the catalytic cracking process is also an illustration of the cross-fertilization of reactor engineering design and chemical catalyst innovation.

7.2.2.3 *Heterogenous Coordination Complex Catalysts*

The third example of catalyst discovery and process invention deals mainly with homogenous reactions.

It serves to illustrate that discovery of the mechanism of a reaction can lead to design of a new catalyst.

This happened in metathesis catalysis (Section 6.2). This was successful because such catalysts are molecular complexes that can be synthesized in full molecular detail.

One of the first catalytic reactions catalyzed by a molecular inorganic complex was the oxo reaction discovered in 1938 by Roelen at Ruhr Chemie (see Figure 3.18). This reaction produces an aldehyde by reaction of CO and hydrogen with an olefin. Aldehyde was discovered by accident in the liquid phase of the product mixture of the heterogenous catalytic Fischer-Tropsch process that converts CO and hydrogen into liquid hydrocarbons. In the Fischer-Tropsch process, a homogenous catalyst is formed due to leaching of the Co of the Fischer-Tropsch catalyst into the oil phase of the reaction. The catalyst became identified as the molecular $HCo(CO)_4$ complex.

The mechanism of the reaction has been unraveled in 1961 by Hecke and Breslow [23].

The oxo reaction was a great applied discovery and became an important industrial process [21], [22].

The oxygenate product is, amongst others, important for the detergent industry. It is another example of an important technology in one field of application (in this case liquid fuels production) that connects to another industry (in this case the detergent industry).

In 1968 Wilkinson *et al.* from Imperial College in London discovered that organometallic Rh compounds complexed with phosphine ligands have superior activity [24], [25].

Compared to the carbonyl complex, Co or Rh complexed with phosphine ligands has a 1000-fold higher activity. In the 1950s the oxo reaction was renamed as the hydroformylation reaction.

The competitive reaction of oxygenate formation from alkene by selective oxidation was discovered in 1957, again in Germany, by Wacker company (Figure 4.2). The reaction is catalyzed by a Pd coordination complex.

These developments were part of the rapidly developing field of organometallic chemistry in the 1960s and 1970s. Its significance was recognized in the Nobel awards to Wilkinson and Fischer in 1973 (see also Figure 3.2).

Another important reaction catalyzed by a carbonyl complex was discovered at Philips Petroleum in 1964 (see Section 6.2). The disproportionation reaction of two propylene molecules into ethylene and butylene is catalyzed by an alumina catalyst impregnated with $Mo(CO)_6$, which at the time had become a well-known coordination complex.

This disproportionation reaction is industrially important (this reaction is discussed in Section 1.3.2 in the context of the Shell SHOP process). The mechanism of the reaction, by that time renamed as metathesis, was discovered in 1971 by Yves Chauvin from the Institut Français du Pétrole (IFP). The debate on the reaction mechanism mainly with Robert Grubbs was concluded in 1987 by a joint paper on living polymerization reactions.

Their interest was ring-opening catalysis of cyclic alkenes (Chauvin, Grubbs, and Richard Schrock were jointly awarded the Nobel Prize in 2002). Schrock was for a short time employed by Dupont company (where polymerization of cyclic alkenes was known as coordination polymerization in 1960) and became a professor at MIT. Grubbs was a professor at California Institute of Technology.

The important contribution of the Chauvin mechanism is that the key catalyst intermediate was identified as $M=CH_2$, a carbene complex. Complexes containing the carbene motive can be synthesized and turn out to be highly active and industrially useful. Such catalysts are called Schrock carbenes, because he had been developing them since the 1970s.

This short summary of the chemical events that led to the Schrock-Grubbs metathesis catalysts provides an illustration of how insight in reaction mechanism leads to novel or improved catalysis.

As for the other two case studies, it shows the importance of scientific and technical communication and the transfer of one technology to another that generates unexpected possibilities.

In the third case study the discovery of a Co complex in the oil of the Fischer-Tropsch reaction in Germany connects with ring-opening polymerization catalysis forty years later in the USA. The design of the latter reaction owns to mechanistic knowledge discovered in France in the 1980s.

Technology develops evolutionarily in a seemingly random process. Technical invention and scientific discovery reinforce each other. The interference of researchers with an eye for practical opportunity or scientific discovery is the key to societal impact.

7.3 Catalysis Science

7.3.1 *Introduction*

Catalysis science aims to predict the catalyst performance for a particular reaction from composition and structure of the catalyst. Knowledge of reaction mechanism is a necessity to establish this relation. One way to evaluate the status of catalysis science is to ask for the significance of reaction mechanistic knowledge in the design of improved or new catalytic systems. The answer to this question relates to the basic issue of how well the working principles of a catalyst are understood.

As a prelude to this discussion, it is good to realize that the word "mechanism" is not always clear. One can speak of the mechanism of the working catalytic system. This refers to the complex interactions of surface organic chemistry, inorganic chemistry of the catalyst surface, and reaction network.

A mechanism is usually considered as the network of elementary reactions that are input to a kinetic model. Here an important difference is to be made between reaction mechanism used in a chemical engineering context and that of the chemist.

For the chemical engineer the reaction mechanism is a model that provides the best fit to kinetic experiments. Elementary reaction rate constants are considered model parameters that are adapted by fitting with experiment. The reaction mechanism is the best available kinetic model.

To the chemist the reaction mechanism provides the link between catalytic site chemistry and catalyst performance. Conventionally, based on kinetic data and measured catalyst structure and composition, a correlation between catalyst performance and composition is sought for. However, due to lack of molecular reactivity information, causal relations are not established. In the history of catalysis science this is the cause of extensive discussions and contradictory opinions on mechanisms of heterogeneous catalytic reactions.

Even with today's sophisticated experimental instrumentation and advances in computational capabilities these scientific debates continue. This is a healthy sign of creativity in science.

7.3.2 *Reaction Mechanism and Catalyst Design*

7.3.2.1 *The Working Catalyst*

A new view on catalytic action: the dynamic formation of selective reaction sites in local equilibrium with catalyst and reactant environment. A major

> *reaction activity insight: the temperature of reaction is often determined by the need to reduce surface site blocking, rather than by activation energies of elementary steps. Surface site blocking results from equilibration with reaction medium molecules.*

Here views are presented that indicate that there are difficulties for full understanding of the working catalytic system. This is relevant to studies of reaction mechanisms that aim for structure-function relationships.

The views of two catalytic scientists are presented. It is argued that kinetic models also have to incorporate surface inorganic chemical transformations. Molecular information on surface chemical reactivity is indispensable.

One view as expressed by Geoffrey Bond, a highly respected scientist who had an industrial career at Johnson and Matthey and became professor at Brunel University in the UK [26], was:

> "For practical as well as philosophical reasons the complete specification of the mechanism of a surface-catalyzed reaction is probably an unattainable goal. The process of refining mechanistic statements has been likened to peeling an onion; the removal of each layer reveals another of even greater subtlety and complexity, and the middle is inaccessible."

With this statement Bond refers to the mathematical process of the derivation of a reaction mechanistic model from global kinetics. A difficulty arises when the reaction steps in a mechanistic model are identified with molecular elementary steps. In global kinetics reaction rate constants are lumped parameters that implicitly depend on true elementary rate constants, which relate with surface chemistry. The latter is the one we desire to know but have to be disentangled from the lumped equations.

The philosophical reasons he refers to are this kinetic complexity and, importantly also, reductionism. The kinetic model is the result of a reductionistic process. Assumptions have been made on what to include or what to ignore.

Another statement of Bond is:

> "Comprehensive measurement of kinetics is the only reliable way of accessing the transition state, and any mechanism derived from other information must be regarded as tentative until confirmed by the kinetics, and any that is not in harmony with them is unacceptable."

The required consistency is correct but whether transition states can be accessed is, as argued below, questionable.

It is good to realize that Bond's statements refer to mainstream catalysis science as practiced for the main part of the previous century. The nature of surface intermediates could only be deduced indirectly. To obtain this information isotope-labelled kinetic experiments were extensively used. Proper understanding between surface state and electronic structure as probed by spectroscopy still had yet to develop.

What is needed to close the gap between kinetic model prediction and the chemistry of reacting catalyst? With respect to this question the view of another respected catalytic scientist Robert Schlögl, from the Max Planck Institute in Berlin, is of interest [27].

Whereas Bond represents the chemical engineering and kinetics approach to catalysis, Schlögl is a scholar in materials science. His laboratory has a strong tradition in the use of advanced spectroscopies and atomistic structure analysis of complex materials with a focus on heterogeneous catalysts. Schlögl makes the important point that usually the state of the catalyst surface is quite different from what one tends to deduce from its bulk structure.

His arguments can be illustrated with the chemistry of the complex metal oxide catalysts of Section 4.4. There it is discussed that the surface composition or structure of complex metal oxide catalytic systems such as $Fe_2(MO_3)_2$ or $(VO)_2P_2O_7$ become dramatically different from the original bulk structure when exposed to reactant. For instance, the $(VO)_2P_2O_7$ catalyst is a dynamic system in which different phase structures alternate. The surface of $Fe_2(MO_3)_2$ is mainly composed of MoO_3.

Such changes of the inorganic chemistry of the surface layer in catalytic systems is general. A new surface phase of a few atomic layers develops in a local equilibrium with gas or liquid phase. The inorganic atoms in the surface layer are in a dynamic disordered environment that can be even liquid-like.

For the ethylene epoxidation catalyst of Section 4.3.3.1 it has been suggested that the Ag/Cs/Cl/O surface layer is liquid-like and is a eutectic [28].

Surface reconstruction of transition metal surfaces induced by reaction is also well documented and discussed in Section 3.3.

Schlögl gave the following definition of the working catalytic system:

> "A heterogeneous catalyst is a functional material that continuously creates active sites with its reactants under reaction conditions. These sites change the rates of chemical reactions of the reactants localized on them without changing the thermodynamic equilibrium between the materials."

This formulation of catalytic action has important implications on our understanding of the catalytic system. In essence it gives a catalytic action description that intrinsically relates to a time. Macroscopically in the life of the working catalyst three episodes can be distinguished. When the fresh catalyst is contacted with reactant, reactivity changes until a stationary state of catalyst performance is reached. After some time the catalyst starts to deactivate. The three episodes are: initiation, stationary state and deactivation.

The Schlögl definition implies that reaction mechanism should include the inorganic transformation of the catalyst or at least a prediction of the state of the system in the stationary state.

Fundamental to commonly used kinetics is the assumption that the nature of the surface is unchanged by reaction. Schlögl defines this as the standard kinetic model. It is stationary and assumes a particular surface structure. For this surface elementary reaction rate constants are calculated or experimentally deduced, which are input to the kinetic model [29].

Already for the standard kinetic model, and in line with Bond's objections, this raises an additional issue. In addition to the kinetically major abundant reactive intermediate (MARI) that defines the reaction rate-controlling step, different molecules can be adsorbed on the surface. They can be product molecules that are in equilibrium with the gas phase, or strongly adsorbed reactant molecules that have to desorb to create vacant sites for molecules to dissociate. The concentration of these adsorbates may dominate. These adsorbates are often non-reactive spectator molecules (major adsorbed spectator molecules, MASI). Spectroscopically it may be difficult to disentangle the MARI, that may have low concentration, from the MASI that can dominate the surface at reaction condition (this high concentration of strong adsorbates may induce surface reconstructions).

The kinetic relevance of the presence of MASI is that it blocks reaction sites. These spectator molecules have to desorb to make it possible for reactant molecules to become activated by the catalyst surface. This implies that temperature of reaction is not necessarily determined by activation energies of the reaction but by the desorption temperature of the MASI. This is the main reason that Bond's suggestion that kinetics can determine transition states cannot be generally accepted.

This temperature selection to prevent parasitic consumption of sites is the cause of the high temperature of many catalytic oxidation reactions. Water or CO_2 has to desorb so as to make reactive surface sites available. For the Fischer-Tropsch reaction the temperature of CO desorption largely determines the temperature of reaction (Section 3.2.2.1.4). In Section 6.4.1 single-site metal cation/reducible metal oxide catalysts are discussed that are reactive at low temperature because reactant does not poison the cationic metal cations. When reduced transition metal particles are used, strong adsorption energies cause the catalysts to be only active at significantly higher temperature.

Schlögl discusses the commercially important dehydrogenation reaction of ethylbenzene to styrene, which is deactivated by coke, as an example where temperature of reaction is not determined by reaction rate of surface intermediates but by inorganic chemical transformations that maintain the catalyst at steady state. The high temperature of 870 K is not needed for the desired ethylbenzene dehydrogenation reaction but serves to desorb the coke that is removed *in situ* by steam gasification with water. The latter reaction is at equilibrium. Reaction is executed in tenfold excess of water. The catalyst is a potassium-promoted iron oxide catalyst. The metal oxide component catalyzes the dehydrogenation reaction, and potassium catalyzes the steam gasification.

An extended standard kinetic model of the reaction is proposed that also includes inorganic surface metal oxide transformations [30].

The surface transformation reaction that has to be included is that of Fe_3O_4 to Fe_2O_3. Carbon is deposited mainly on the Fe_3O_4 phase, which has low reactivity. Fe_2O_3 is the reactive phase that deactivates by reduction with hydrogen and has to be regenerated by oxygen and steam.

The high simulation methods have become available for an extension of the standard kinetic model. Simulations can be done to deduce the state of the surface in contact with a reactive ambient atmosphere. This state can in principle be deduced when one assumes equilibrium with the adsorbing gases. A commonly used method is the surface thermodynamics approach developed by Scheffler *et al.* [31]. Molecular dynamics simulation techniques are also being developed that predict surface phase transitions [32]–[35].

The physical cause of surface transformation processes is high coverage with adsorbates that strongly interact with the surface. It usually leads to formation of surface islands with different phases. The boundary between these phases may have unique reactivity.

Under particular reaction conditions different surface phases alternate which leads to oscillatory instead of stationary kinetic behavior. Such alternation can be the result of an oxidized versus reduced surface phase. Even when reaction is at quasi-steady state and stationary, the catalyst surface may contain a mixture of such

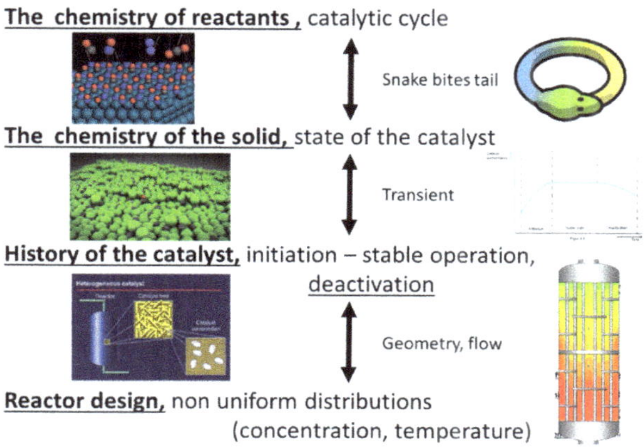

Figure 7.3 The four main features of a working catalyst: the catalytic reaction cycle and surface state of reaction center; the three stages of catalyst reaction initiation, quasi-steady state and deactivation; the non-uniform distribution of concentrations in the reactor.

phases that are distributed randomly and their local concentration changes with time [36], [37].

Figure 7.3 illustrates four major aspects of the working catalyst, which highlight the experimental complexity of the system.

The catalyst changes in contact with reaction medium. Conversion of reactant relates to surface state and state of surface in turn depends on reaction medium composition. The prediction of the state of the catalyst surface is a self-reference problem. This self-reference is symbolized by the "snake bites tail" figure.

To solve this self-reference problem of state and structure of the working catalyst that determines reaction mechanism, where surface state itself depends on reaction conditions, remains one of the holy grails of catalysis science.

The inorganic chemistry of the catalyst determines which surface state develops. Major surface reconstruction may happen that is part of the complex inorganic chemistry of the catalyst. Different surface phases may alternate in time.

Chemical changes are usually a function of reaction time. When exposed to reaction medium the fresh catalyst will change. This is the initiation period of the catalyst. This is a transient state that ends when the catalyst reaches its steady state. Small changes may continue that finally will lead to catalyst deactivation.

The different stages of initiation, steady state, and catalyst deactivation affect the reproducibility of catalytic measurements. Reproducible experimentation with catalyst powders or particles is a well-known difficulty. This becomes understandable when one realizes the different time and length scales which are involved in the catalytic event.

The reactor, catalyst shape, and packing have a large effect on catalyst performance because of inhomogeneity in reaction medium distribution and possibly also in temperature. This will cause catalyst initiation and deactivation times to be also non-uniform in the catalyst bed and affect average performance of catalytic reaction.

In a flow reactor the distribution of reactants will be inhomogeneous because at the entrance of the reactor conversion it is high and it decreases along the reactor bed. This makes the state of catalyst surface different along the catalyst bed. Also due to diffusion the packing, size, and shape of the catalyst particles will make the surface state non-uniform.

For supported particles there is a difference between exterior and interior and also the distribution of catalytic particles may be non-uniform in size.

These factors will cause the practical system to have a memory of how catalytic reaction has been started.

To catalyst performance the history of how the steady state has been reached may essentially determine catalyst performance.

7.3.2.2 *The Simulation of Surface Reactivity*

Surface chemical binding theory and surface reactivity are joined once transition states of surface elementary reactions could be accessed.

To relate catalyst function with catalyst inorganic chemistry, elementary reaction rate constants have to be known as a function of catalyst surface composition and structure. For the most part of the previous century proper understanding of surface reactivity was lacking. This largely contributed to the speculative nature of suggested reaction models. Here some of the shortcomings and oversimplifications of early approaches are discussed.

Key to understanding the state of adsorbed intermediates and their reactivity properly is to know their structure at the atomic level.

Today it is understood that adsorbates can coordinate to a single surface atom or have twofold or threefold coordination (see Figure 2.10). This difference in coordination will cause a difference in adsorption energy that is unrelated to the intrinsic reactivity of the surface atoms.

Kinetic models start to distinguish such energetic and chemically relevant differences when the surface site is not indicated anymore as a * or [] but is described in atomic detail as a coordination complex.

In the classical Langmuir model atomic detail of the surface is absent. The surface consists of unspecified site positions. When a molecule adsorbs or dissociates an assumption is made as to how many vacant sites are consumed; it is not specified as to how many adatoms are coordinated. This is aggregated into the single site. Within this Langmuirian site aggregation model the only way to describe differences in adsorption modes is by defining differences in binding affinity. This is schematically illustrated in Figure 7.4.

Differences in binding are only related to differences in the chemical bonding properties of the adsorbate. Surface structural information is not incorporated.

In line with this, Bond emphasizes that his main interest is to understand differences in reactivity of molecules. Differences in

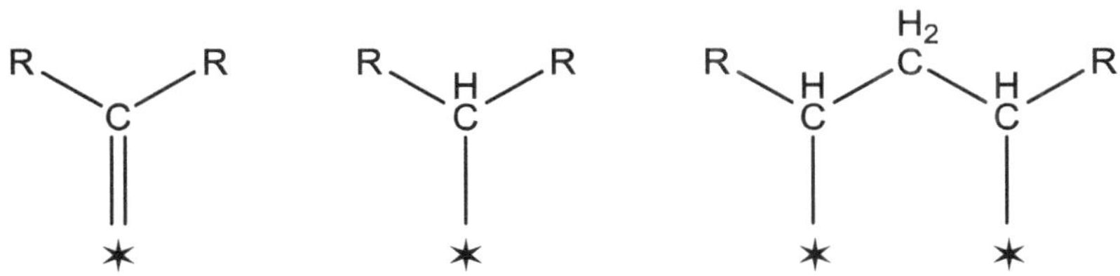

Figure 7.4 Binding modes of adsorbates in early mechanistic models.

reactivity of surfaces are described as variation of bonding of reaction intermediates.

Only with the advance of surface science and computational catalysis proper atomistic understanding of coordination modes of adsorbates became possible. It is described in Section 2.3.

Modern microkinetic simulations can now incorporate the coordination differences of adsorbates and also the structures of transition states. It is only at the end of the 1990s that modern reactivity theory was formulated. This is due to proper understanding of the nature of the surface chemical bond and its relation with the electronic structure of materials [38], [39].

It is illuminating to realize how, compared to earlier theories, current thinking differs on how electronic structure and chemical reactivity relate.

Two quotes from the history paper by van Veen are of interest [40].

He noted that in the 1930s William Frankenburger, a colleague of Alwin Mittasch (the inventor of the ammonia synthesis catalyst), suggested that the richness of the line emission spectrum of an element indicates its readiness for electron exchange. For the Ag ethylene epoxidation catalyst he suggested that a match of the radiation spectra of catalyst and reactant predicts chemical bonding properties. These are early speculative suggestions that there is a relation between the electronic structure of the solid and surface chemical reactivity.

In the middle of the previous century electronic structure theories of solid-state materials became well developed. It was suggested that there is relation between electron density at the Fermi level and catalytic reactivity. The electron density at the Fermi level was a well-understood property in solid-state physics. It relates with conductivity and magnetism of materials. In Section 2.3 the rejection of this view by modern theories of chemical bonding is described.

Even in the 1980s one of the first kinetic models based on quantum-chemical electronic structure theory was still based on the assumption of a correlation with the electron density of states at the Fermi level. These microkinetic simulations of ammonia synthesis by Stolze and Nørskov, then at Topsøe laboratories in Denmark, in 1987 [41] are based on the mechanistic model of the ammonia synthesis reaction as proposed by Ertl (Section 3.2.1) [42].

Nørskov, who is also professor at Lyngby University, can be considered one of the founding fathers of computational catalysis.

His electronic structure calulations were based on approximate DFT effective medium theory. They were suitable for the calculation of adsorption energies. Transition state energies that are necessary for prediction of elementary reaction rate constants could not be calculated and had to be estimated [43].

The techniques to calculate transition state energies became available only twenty years later.

The simulations compare reactivity of transition metals along a row of the periodic table. When elements are compared from the left to the right of the periodic table the number of d-valence electrons increases. Nørskov and Stolze correctly predicted a uniform decrease in adsorption energy with an increase in d-valence electron band occupation. However, they assumed incorrectly that the transition state configuration of the dissociating nitrogen molecule is a situation of weak interaction with the surface. For such a weakly interacting state physical theory predicts that interaction energy follows the electron density of state at the Fermi level.

The kinetic equations were also solved assuming that N_2 dissociation is reaction rate controlling. When the rate of ammonia production is plotted as a function of d-valence electron count of transtion metal the volcano-like dependence of Figure 7.5 is found.

As expected the reaction rate closely follows the approximate bell-shaped dependence of the partial density of states of the d-valence electron band as a function of Fermi level energy of the transition metals (Figure 2.18d).

This conclusion, however, is not substantiated by modern catalysis science. Twenty years later computation of transition state energies demonstrated that activation energies relate linearly with adsorption energies of surface atoms (the Brønsted-Evans-Polanyi relations, Section 2.3.2.3) [44]–[46]. In contrast with earlier assumptions, in the transition state there is a strong interaction with the

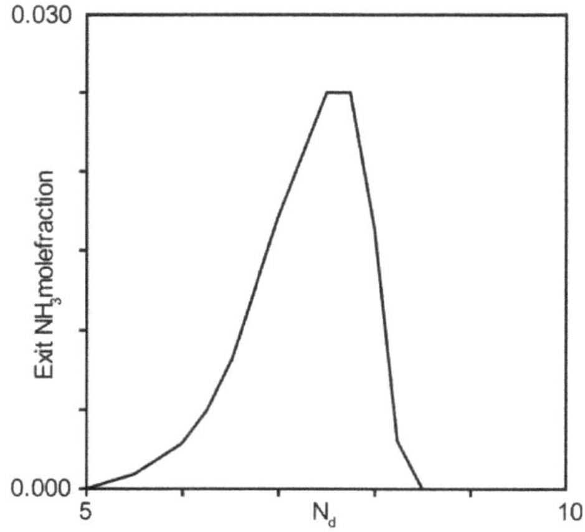

Figure 7.5 Calculated rate of ammonia production from N_2 and H_2 per unit catalyst surface area as a function of N_d, the d-valence electron occupation of the catalyst transition metal atom [41].

catalyst surface. Volcano-type dependence on surface reactivity does not reflect differences in chemical bonding with the surface. They have a kinetic origin that is explained by the Sabatier principle (see Section 2.4.4). The corrected volcano curve dependence of the ammonia synthesis reaction, based on DFT-calculated transition states also by Nørskov *et al.*, was published fourteen years after the first simulations (see Figure 2.32) [47]).

The simulations represent a major advance in catalytic kinetics. Whereas the Sabatier principle was already known since the 1960s as the cause of the volcano curves, its quantitative construction with a direct relation to surface structure and composition had not been realized before.

The early Stolze-Nørskov simulation is instructive. Their simulation accidentally matches with experiment. It illustrates how fallacious an observed relation between global kinetic property and microscopic reactivity descriptor may be; in this case, due to poor understanding of surface chemical bonding. It also illustrates that even at the end of the previous century there was a poor understanding of bonding principles in relation to chemical reactivity.

Microkinetic simulations such as the Nørskov studies presented above are of great use to assist catalyst design. In the mechanistic chapters of the book results of many related simulations are used to understand structure-function relations. This is notwithstanding the fact that such simulations are based on the standard kinetics model, which has significant limitations as explained in the previous section.

7.3.3 *Catalysis, a Predictive Science?*

From correlation to causality.

Science is about the desire to understand the "what" of a particular phenomenon. In catalysis this question concerns the relation between catalyst composition and structure and its perfomance. This is different from the design question: this concerns the "how" to improve the performance of a catalyst for a known reaction or to design a catalyst for an unknown reaction. This ultimately leads to the question of how well we can predict the performance of a particular catalytic material.

In the previous sections the deepened understanding of the inorganic and physical chemistry of the working catalyst has been sketched. One cannot escape the impression that the more we learn, the more questions remain. This is one of the ironic marks of progress in science. On the other hand increased understanding leads to sharpened insights that help to formulate research questions or hypotheses relevant for further investigations. The better the phenomenon is understood, the closer the question is to its answer. This is the ultimate contribution of catalysis science to catalyst design.

The empirical correlation of catalyst performance with its material properties has been the main route to progress and discovery. Refinement in experimentation and computation have transformed catalytic kinetics into a deterministic modelling tool based on an understanding of the causal relations between surface reactivity, elementary rate constants, and reaction mechanism.

This implies a deep understanding of the chemistry of catalytic reactivity for many of the important catalytic systems. Knowledge on the level of elementary reaction steps brings with it the possibility of transferability between different systems.

Also, due to computational catalysis there is a large database of elementary reaction rate constants available for an artificial intelligence program to search for new catalysts of a particular reaction.

Artificial Intelligence provides a very different mathematical approach to the search for new catalysts or their improvement compared to mechanistic kinetic modelling. A successful example of a related combinatorial problem is a recent advance on the prediction of the structure of proteins. The AlphaFold algorithm from the Artificial Intelligence company DeepMind in the UK works with a computer network built around 128 machine learning processors and is trained with the data of all 170,000 approximately known protein structures [48]. It successfully predicted the folding of proteins where first-principle molecular dynamics studies failed. Examples of related machine learning approaches in heterogeneous catalysis are given in [49]–[52]. Also in modelling of catalytic reactivity this is an important additional tool to discover the configuration and composition of catalyst particles that reconstruct by reaction, and for the design of mechanistic models of unknown reactions [53]. An interesting example is the exploration of millions of Fe complexes for the conversion of methane to methanol and the identification of an unexpected new reactive complex [54]. On a molecular level the mechanism of catalytic transformations is described in terms of sequences of chemical bond cleavage and formation reactions, which can happen in a multitude of combinations. The richness of this chemistry is due to the great variety of catalytic systems. This is documented and organized in the mechanistic chapters of this book.

The four themes of major progress are:

- A proper theory of the surface chemical bond is now available. This is due to the discovery of quantum mechanics and started with electronic structure theories of the solid state. Major concepts were developed by Pauling, Mott, and Hoffmann [55]–[57].

Currently first-principle calculations of the electronic structure provide surface bond energies and structures for any material that one desires to study as catalyst. This implies that for any proposed reaction intermediate its structrure and adsortion energy can be calculated as well as activation free energies for bond breaking and formation [39], [58].

– The composition of most catalysts is complex. Often catalysts are multicomponent systems with parts present as molecular clusters or nanoparticles distributed over high surface area porous materials.

Spectroscopies have become available that are able to determine composition as well as structure up to the nano or even atomistic level. Operando spectroscopies have been developed that probe the surface state at reaction conditions. Electronic structure calculations of proposed structures are important to assist identification of structure and composition of reaction centers. Simulations to predict the complex composition of the reconstructing surface have become available.

– The catalytic reaction is a cycle of elementary steps in which the reaction center is regenerated. The reaction network is a complex dynamic system, with positive and negative feedback relations. Often several different catalytic cycles that are catalyzed by different reaction centers in multicomponent systems combine.

The catalytic center will have a different composition in the initiation stage than in the quasi-steady state regime. Major inorganic restructuring of the catalytic surface may happen. The quasi-steady state regime will have chemical memory of how this kinetic state was reached from the initial start-up of the reaction. Catalyst deactivation can be due to slow reactions of undesired co-products or instabilities of the inorganic components that have long-time constants.

The scientific reductionistic approach to describe catalyst reactivity is useful to understand isolated steps, but the functioning of the catalytic system requires holistic understanding. For this it is also important to understand that the catalytic reaction is a multiscale phenomenon, where longer time and space events depend on shorter time- and length-scale events. Each time and length scale has different dynamics determined by the shorter time and length events. For instance, whereas microkinetic equations explicitly depend on surface molecular information and concentration of surface reaction intermediates, global kinetic equations are expressed as a function of reactant and product concentration. The lumped kinetic parameters of global kinetics implicitly depend on reaction intermediate concentrations, and will be only valid within limited temperature and concentration regions.

– Catalyst improvement or invention happens by discovery of new catalytic materials. The science of catalysis assists design by prediction of optimum structure and composition of the catalytic center based on atomic models.

A major challenge is to predict the stability or changes to the catalyst surface at reaction conditions.

This will allow us to incorporate self-organization of the surface into the design of the catalyst.

Most catalyst synthesis involves solution-phase chemistry. Understanding the details of the interaction of solutes with inorganic surfaces remains an important subject of study [59].

7.3.4 *Future Perspective*

Today's technological challenges are many and questions to catalysis science remain.

Can one expect a third golden age to heterogeneous catalysis? Of course, there is no way to make predictions on the future, but previous sections may provide some ideas.

The science of catalysis clearly belongs in the Pasteur square of Stone's quadrant (Figure 7.1). The research is fundamental in that it is directed towards understanding but investigates systems that may technically be useful. Major drivers to catalytic innovation are scientific discoveries, technical opportunities, and practical art.

Initially, catalysis had its emphasis on reactor engineering and physical characterization of the materials. Preparation of catalytic materials became a well-developed art. In the second stage application of spectroscopic tools brought heterogeneous catalysis into the realm of molecular chemistry.

The present state of heterogeneous catalysis is powerful. The catalytic system can be probed in its full complexity. It proffers the catalytic scientist a deep understanding of structure-function relationships. The interplay of the inorganic chemistry and reactive medium is complex and specific for each system. To understand this dynamics is one of the great present challenges.

Advances in materials science have dramatically affected catalyst synthesis, through new solid-state as well as molecular inorganic chemistry. New catalytic reactions are realized with the discovery of new materials.

In the eyes of the philosopher catalysis has become a mature science, with the promise of important discoveries yet to be made. The science is not fully predictive, but it contains an impressive body of tools, empirical knowledge, and understanding. Because of advances with respect to catalyst design it knows better what not to do and which options are useful.

Because of the political climate in the beginning of the 20th century there was a large awareness in society of the need for artificial nitrate production as well as processes that convert coal to liquid fuel. Ample financial and industrial resources became available to build the new processes.

Society at present is in a comparable situation. For more than twenty years there has been ample awareness of the challenges posed by climate change due to temperature increase. This temperature increase is thought to be mainly caused by emission of CO_2. Technology is needed to reduce these emissions, that to a large part are due to the combustion of fossil fuels. This calls for more energy-efficient processes and replacement with other resources. Many governments take measures through subsidies or tax to generate financial resources to implement renewable technologies.

For catalysis this provides great opportunities. Not only will there be a need to optimize efficiency of processes, but it also implies a search for alternative energy conversion routes.

Windmills, solar panels, and nuclear energy will aim to substitute for electricity generated from fossil fuel. Additionally, a widely supported strategy is to substitute fuel-driven engines by electrical motors. An important example is the promotion of the electric car. It implies an expansion of electricity as an energy carrier.

But electricity storage and transport are still prohibitively expensive. Conversion of electricity into hydrogen or other energy carriers is a widely discussed option. This gives a large boost to electrocatalytic research, since improvement of energy efficiency of fuel cells and electrolysis of water is highly desired.

The ultimate process that would help to solve the CO_2 emission problem is artificial photosynthesis [60].

This would combine CO_2 capture from air with production of methane or other hydrocarbons [64]. Such an artificial process should be more efficient than biological photosynthesis. This is, so far, not yet within reach. Partial reaction steps have been successfully investigated. Hydrogen production by photocatalysis via water splitting can be done using semiconductor catalysts [61]–[63].

A source of carbon for the chemical industry apart from fossil fuels or CO_2 is biomass. Catalysis research to convert lipids [65] into liquid fuels has been quite successful to convert fats or oils from seeds. Also, processes for the conversion of carbohydrates such as starch to fuels and chemicals have been developed [66]. However, the raw materials are crops, a food source. Hence there is a need for other processes that are based on non-food-competing resources. Instead of starch the chemistry of complex lignocellulose conversion is investigated.

Lignocellulose has to be broken down into cellulose, hemicellulose, and lignin. Each constituent requires different treatment processes. This has led to the concept of the biorefinery, an integrated complex of different interrelated bioprocesses for the production of chemicals [67]. An example of an interesting environmentally sustainable application is the production of a polymer such as polyethylene furanoate that competes with polyethylene terephthalate but is now made from biodegradable biomass [68], [69].

An area of interest is the conversion of the aromatic lignin into a product other than pulp. This is a topic of extensive investigation that has led to the discovery of new chemistry where heterogeneous catalysts find important application. Biomaterial-based phenol production that is catalyzed by solid acid zeolites has been demonstrated [70]–[72].

Biomass can provide the carbon-containing molecules needed for essential chemicals.

These new systems are the subject of future catalysis science. The catalyst material will be probed with increasing resolution in time and space. Instrumentation will develop improved and novel capabilities.

A major scientific question is the identification of the unique reaction centers, which are often present at low concentration. This state relates to the inorganic chemistry of the material in which it is embedded and composition of the reaction medium.

The holy grail is to predict the optimum composition and state of the catalyst for maximum activity, selectivity, and stability. Theory needs to capture and incorporate in kinetics the full dynamics of the catalytic system, which has several different time scales. There is the dynamics of transformations in the transition states, as well as the slower dynamics of the surface transformations.

There is also the slower overall reaction rate and the slower deactivation processes.

Advances in the determination of transition state free energies make discrimination between mechanistic proposals possible. But so far there is no *a priori* method that predicts the mechanism of a particular reaction and catalyst. Kinetic models still depend on input from chemical intuition based on the extensive knowledge of reaction mechanisms as described in this book.

The technological and scientific challenges, the arsenal of advanced techniques that can probe the catalyst structure-performance relations, and the rich portfolio of catalytically reactive materials promise a future with important new discoveries.

References

[1] D. Stokes, *Pasteur's Quadrant: Basic Science and Technological Innovation.* Brookings Institution, 1997.

[2] G. P. Zacchary, *Endless Frontier.* MIT Press, 1999.

[3] A. C. Grayling, *The Frontiers of Knowledge: What We Know About Science, History and The Mind.* Penguin, 2021.

[4] G. J. Hutchings, "Heterogeneous catalysts—discovery and design," *J. Mater. Chem.*, vol. 19, no. 9, pp. 1222–1235, Feb. 2009, doi: 10.1039/B812300B.

[5] J. S. J. Hargreaves, G. J. Hutchings, R. W. Joyner, and S. H. Taylor, "Methane partial oxidation to methanol over Ga_2O_3 based catalysts: use of the CH_4/D_2 exchange reaction as a design tool," *Chem. Commun.*, no. 4, p. 523, Jan. 1996, doi: 10.1039/cc9960000523.

[6] M. Belgued, P. Pareja, A. Amariglio, and H. Amariglio, "Conversion of methane into higher hydrocarbons on platinum," *Nature*, vol. 352, no. 6338, pp. 789–790, 1991, doi: 10.1038/352789a0.

[7] T. Koerts and R. A. van Santen, "A low temperature reaction sequence for methane conversion," *J. Chem. Soc. Chem. Commun.*, no. 18, pp. 1281–1283, Jan. 1991, doi: 10.1039/C39910001281.

[8] A. Sarlemijn and H. Brugt Gerhard Casimir, *Tussen academie en industrie. Casimirs visie op wetenschap en researchmanagement.* Amsterdam, 1984.

[9] B. Tonn, "The Nature of Technology: What It Is and How It Evolves, W.B. Arthur, Free Press, New York (2009), 246 pp.," *Futures*, vol. 42, no. 9, pp. 1032–1033, Nov. 2010, doi: 10.1016/J.FUTURES.2010.08.015.

[10] J. Mokyr, *The Gifts of Athena : Historical Origins of the Knowledge Economy.* Princeton University Press, 2005.

[11] C. Morgan, "40 years of cleaner air: The evolution of the autocatalyst," *Johnson Matthey Technol. Rev.*, vol. 58, no. 4, pp. 217–220, 2014, doi: 10.1595/205651314X684726.

[12] K. C. Taylor, "Nitric oxide catalysis in automotive exhaust systems," *Catal. Rev.*, vol. 35, no. 4, pp. 457–481, Dec. 1993, doi: 10.1080/01614949308013915.

[13] N. Nichols, "History of the HPI: The 1960s: Synthetic oils, zeolite catalysts, LLDPE, OPEC and and creation of the PLC," *Hydrocarb. Process.*, vol. May, no. 100th anniversary, 2022.

[14] W. Ipatiew, "Pyrogenetische contactreactionen organischer verbindungen," *Berichte der Dtsch. Chem. Gesellschaft*, vol. 35, no. 1, pp. 1047–1057, Jan. 1902, doi: 10.1002/CBER.190203501173.

[15] C. L. Thomas, "Chemistry of cracking catalysts," *Ind. Eng. Chem.*, vol. 41, no. 11, pp. 2564–2573, Nov. 1934, doi: 10.1021/IE50479A042.

[16] V. P. Doronin and T. P. Sorokina, "Chemical design of cracking catalysts," *Russ. J. Gen. Chem. 2007 7712*, vol. 77, no. 12, pp. 2224–2231, Dec. 2007, doi: 10.1134/S1070363207120274.

[17] L. V. C. Rees, "Richard Maling Barrer. 16 June 191012 September 1996," *Biogr. Mem. Fellows R. Soc.*, vol. 44, pp. 37–49, Jan. 1998, doi: 10.1098/RSBM.1998.0003.

[18] D. E. W. Vaughan, "From Zeolites to Porous MOF Materials — The 40th Anniversary of International Zeolite Conference: Contributions of R. M. Barrer to zeolite synthesis," in *Studies in Surface Science and Catalysis*, R. Xu, J. Chen, and Z. Gao, Eds. Elsevier, 2007, pp. 87–95.

[19] J. A. Rabo and M. W. Schoonover, "Early discoveries in zeolite chemistry and catalysis at Union Carbide, and follow-up in industrial catalysis," *Appl. Catal. A Gen.*, vol. 222, no. 1–2, pp. 261–275, Dec. 2001, doi: 10.1016/S0926-860X(01)00840-7.

[20] C. J. Plank, "The invention of zeolite cracking catalysts," *ACS Symp. Ser.*, vol. 222, no. 4, pp. 253–271, 1983.

[21] R. Franke, D. Selent, and A. Börner, "Applied hydroformylation," *Chem. Rev.*, vol. 112, no. 11, pp. 5675–5732, Nov. 2012, doi: 10.1021/CR3001803.

[22] A. M. Trzeciak, "Hydroformylation," in *Comprehensive Inorganic Chemistry II*, 2nd ed., J. Reedijk and K. Poeppelmeier, Eds. Elsevier, 2013, pp. 25–46.

[23] R. F. Heck and D. S. Breslow, "The reaction of cobalt hydrotetracarbonyl with olefins," *J. Am. Chem. Soc.*, vol. 83, no. 19, pp. 4023–4027, Oct. 1961, doi: 10.1021/ja01480a017.

[24] D. Evans, J. A. Osborn, and G. Wilkinson, "Hydroformylation of alkenes by use of rhodium complex catalysts," *J. Chem. Soc. A*, vol. 62, no. 0, pp. 3133–3142, Jan. 1968, doi: 10.1039/J19680003133.

[25] D. Evans, G. Yagupsky, and G. Wilkinson, "The reaction of hydridoc arbonyltris(triphenylphosphine)rhodium with carbon monoxide, and of the reaction products, hydridodicarbonylbis(triphenylphosp hine)rhodium and dimeric species, with hydrogen," *J. Chem. Soc. A*, no. 0, pp. 2660–2665, Jan. 1968, doi: 10.1039/J19680002660.

[26] G. C. Bond, "The use of kinetics in evaluating mechanisms in heterogeneous catalysis," vol. 50, no. 4, pp. 532–567, 2009, doi: 10.1080/01614940802480338.

[27] R. Schlögl, "Heterogeneous catalysis," *Angew. Chem. Int. Ed.*, vol. 54, no. 11, pp. 3465–3520, Mar. 2015, doi: 10.1002/ANIE.201410738.

[28] R. A. van Santen, "The active site of promoted ethylene-epoxidation catalysts," in *Catalysis: Theory to Practice: Proceedings 9th International Congress on Catalysis*, vol. 3, M. J. Phillips and M. Ternan, Eds. Chemical Institute of Canada, 1988, pp. 1152–1158.

[29] G. B. Marin, G. S. Yablonsky, and D. Constales, *Kinetics of Chemical Reactions: Decoding Complexity*, 2nd ed. John Wiley & Sons, 2019.

[30] A. Schüle, O. Shekhah, W. Ranke, R. Schlögl, and G. Kolios, "Microkinetic modelling of the dehydrogenation of ethylbenzene to styrene over unpromoted iron oxides," *J. Catal.*, vol. 231, no. 1, pp. 172–180, Apr. 2005, doi: 10.1016/J.JCAT.2005.01.013.

[31] C. Stampfl, H. J. Kreuzer, S. H. Payne, H. Pfnür, and M. Scheffler, "First-principles theory of surface thermodynamics and kinetics," *Phys. Rev. Lett.*, vol. 83, no. 15, pp. 2993–2996, Oct. 1999, doi: 10.1103/PhysRevLett.83.2993.

[32] Z. Zhang, B. Zandkarimi, and A. N. Alexandrova, "Ensembles of metastable states govern heterogeneous catalysis on dynamic interfaces," *Acc. Chem. Res.*, vol. 53, no. 2, pp. 447–458, Feb. 2020, doi: 10.1021/ACS.ACCOUNTS.9B00531.

[33] J. S. Lim, J. Vandermause, M. A. Van Spronsen, A. Musaelian, Y. Xie, L. Sun, C. R. O'Connor, T. Egle, N. Molinari, J. Florian, K. Duanmu, R. J. Madix, P. Sautet, C. M. Friend, and B. Kozinsky, "Evolution of metastable structures at bimetallic surfaces from microscopy and machine-learning molecular dynamics," *J. Am. Chem. Soc.*, vol. 142, no. 37, pp. 15907–15916, Sep. 2020, doi: 10.1021/JACS.0C06401.

[34] R. Réocreux, Girel, P. Clabaut, A. Tuel, M. Besson, A. Chaumonnot, A. Cabiac, P. Sautet, and C. Michel, "Reactivity of shape-controlled crystals and metadynamics simulations locate the weak spots of alumina in water," *Nat. Commun.*, vol. 10, no. 1, pp. 1–8, Jul. 2019, doi: 10.1038/s41467-019-10981-9.

[35] S. Vandenhaute, S. M. J. Rogge, and V. Van Speybroeck, "Large-scale molecular dynamics simulations reveal new insights into the phase transition mechanisms in MIL-53(Al)," *Front. Chem.*, vol. 9, p. 699, Aug. 2021, doi: 10.3389/FCHEM.2021.718920.

[36] R. J. Gelten, A. P. J. Jansen, R. A. van Santen, J. J. Lukkien, J. P. L. Segers, and P. A. J. Hilbers, "Monte Carlo simulations of a surface reaction model showing spatio-temporal pattern formations and

oscillations," *J. Chem. Phys.*, vol. 108, no. 14, p. 5921, 1998, doi: 10.1063/1.476003.

[37] G. Ertl, *Reactions at Solid Surfaces.* John Wiley & Sons, 2009.

[38] J. K. Nørskov, F. Studt, F. Abild-Pedersen, and T. Bligaard, *Fundamental Concepts in Heterogeneous Catalysis.* John Wiley & Sons, 2014.

[39] R. A. van Santen, *Modern Heterogeneous Catalysis: An Introduction.* John Wiley & Sons, 2017.

[40] J. A. R. van Veen, "What's new? On the development of sulphidic HT catalysts before the molecular aspects," *Catal. Today*, vol. 292, pp. 2–25, Sep. 2017, doi: 10.1016/J.CATTOD.2016.09.027.

[41] J. K. Nørskov and P. Stoltze, "Theoretical aspects of surface reactions," *Surf. Sci.*, vol. 189–190, no. C, pp. 91–105, Oct. 1987, doi: 10.1016/S0039-6028(87)80419-3.

[42] G. Ertl, "Surface science and catalysis—studies on the mechanism of ammonia synthesis: The P. H. Emmett Award Address," *Sci. Eng.*, vol. 21, no. 2, pp. 201–223, Jan. 1980, doi: 10.1080/036024580080 67533.

[43] J. K. Nørskov, "Covalent effects in the effective-medium theory of chemical binding: Hydrogen heats of solution in the 3d metals," *Phys. Rev. B*, vol. 26, no. 6, p. 2875, Sep. 1982, doi: 10.1103/PhysRevB.26.2875.

[44] S. Dahl, A. Logadottir, C. J. H. Jacobsen, and J. K. Nørskov, "Electronic factors in catalysis: the volcano curve and the effect of promotion in catalytic ammonia synthesis," *Appl. Catal. A Gen.*, vol. 222, no. 1–2, pp. 19–29, Dec. 2001, doi: 10.1016/S0926-860X(01)00826-2.

[45] T. Bligaard, J. K. Nørskov, S. Dahl, J. Matthiesen, C. H. Christensen, and J. Sehested, "The Brønsted–Evans–Polanyi relation and the volcano curve in heterogeneous catalysis," *J. Catal.*, vol. 224, no. 1, pp. 206–217, May 2004, doi: 10.1016/j.jcat.2004.02.034.

[46] K. Honkala, A. Hellman, I. N. Remediakis, A. Logadottir, A. Carlsson, S. Dahl, C. H. Christensen, and J. K. Nørskov, "Ammonia synthesis from first-principles calculations.," *Science*, vol. 307, no. 5709, pp. 555–558, Jan. 2005, doi: 10.1126/science.1106435.

[47] C. J. H. Jacobsen, S. Dahl, B. G. S. Clausen, S. Bahn, A. Logadottir, and J. K. Nørskov, "Catalyst design by interpolation in the periodic table: bimetallic ammonia synthesis catalysts," *J. Am. Chem. Soc.*, vol. 123, no. 34, pp. 8404–8405, 2001, doi: 10.1021/JA010963D.

[48] R. F. Service, "'The game has changed.' AI triumphs at protein folding: In milestone, software predictions finally match structures calculated from experimental data," *Science*, vol. 370, no. 6521, pp. 1144–1145, Dec. 2020, doi: 10.1126/SCIENCE.370.6521.1144.

[49] J. G. Freeze, H. R. Kelly, and V. S. Batista, "Search for catalysts by inverse design: artificial intelligence, mountain climbers, and alchemists," *Chem. Rev.*, vol. 119, no. 11, pp. 6595–6612, Jun. 2019, doi: 10.1021/ACS.CHEMREV.8B00759.

[50] W. Yang, T. T. Fidelis, and W.-H. Sun, "Machine learning in catalysis, from proposal to practicing," *ACS Omega*, vol. 5, no. 1, pp. 83–88, Jan. 2019, doi: 10.1021/ACSOMEGA.9B03673.

[51] B. R. Goldsmith, J. Esterhuizen, J.-X. Liu, C. J. Bartel, and C. Sutton, "Machine learning for heterogeneous catalyst design and discovery," *AIChE J.*, vol. 64, no. 7, pp. 2311–2323, Jul. 2018, doi: 10.1002/AIC.16198.

[52] J. R. Kitchin, "Machine learning in catalysis," *Nat. Catal.*, vol. 1, no. 4, pp. 230–232, Apr. 2018, doi: 10.1038/s41929-018-0056-y.

[53] L. Grajciar, C. J. Heard, A. A. Bondarenko, M. V. Polynski, J. Meeprasert, E. A. Pidko, and P. Nachtigall, "Towards operando computational modeling in heterogeneous catalysis," *Chem. Soc. Rev.*, vol. 47, no. 22, pp. 8307–8348, Nov. 2018, doi: 10.1039/C8CS00398J.

[54] A. Nandy, C. Duan, C. Goffinet, and H. J. Kulik, "New strategies for direct methane-to-methanol conversion from active learning exploration of 16 million catalysts," *J. Am. Chem. Soc.*, May 2022, doi: 10.1021/JACSAU.2C00176.

[55] L. Pauling, *The Nature of the Chemical Bond and the Structure of Molecules and Crystals: An Introduction to Mode.* Cornell University Press, 1960.

[56] N. F. Mott and L. Friedman, "Metal-insulator transitions in VO_2, Ti_2O_3 and Ti2-x V x O3," vol. 30, no. 2, pp. 389–402, 1990, doi: 10.1080/14786439808206565.

[57] R. Hoffmann, *Solids and Surfaces: A Chemist's View of Bonding in Extended Structures.* John Wiley & Sons, 1988.

[58] J. K. Nørskov, F. Studt, F. Abild-Pedersen, and T. Bligaard, *Fundamental Concepts in Heterogeneous Catalysis.* John Wiley & Sons, 2014.

[59] K. P. de Jong, *Synthesis of Solid Catalysts.* John Wiley & Sons, 2009.

[60] D. G. Nocera, "The artificial leaf," *Acc. Chem. Res.*, vol. 45, no. 5, pp. 767–776, May 2012, doi: 10.1021/AR2003013.

[61] H. Nishiyama, T. Yamada, M. Nakabayashi, Y. Maehara, M. Yamaguchi, Y. Kuromiya, H. Tokudome, S. Akiyama, T. Watanabe, R. Narushima, S. Okunaka, N. Shibata, T. Takata, T. Hisatomi, and K. Domen, "Photocatalytic solar hydrogen production from water on a 100 m2-scale," *Nature,* vol. 598 pp. 304–307, Aug. 2021, doi: 10.1038/s41586-021-03907-3.

[62] T. Hisatomi and K. Domen, "Reaction systems for solar hydrogen production via water splitting with particulate semiconductor photocatalysts," *Nat. Catal.*, vol. 2, no. 5, pp. 387–399, Mar. 2019, doi: 10.1038/s41929-019-0242-6.

[63] J. Rongé, T. Bosserez, D. Martel, C. Nervi, L. Boarino, F. Taulelle, G. Decher, S. Bordiga, and J. A. Martens, "Monolithic cells for solar fuels," *Chem. Soc. Rev.*, vol. 43, no. 23, pp. 7963–7981, Nov. 2014, doi: 10.1039/C3CS60424A.

[64] B. Smit, J. A. Reimer, C. M. Oldenburg, and I. C. Bourg, *Introduction to Carbon Capture and Sequestration.* Imperial College Press, 2014.

[65] M. Mittelbach, "Fuels from oils and fats: Recent developments and perspectives," *Eur. J. Lipid Sci. Technol.*, vol. 117, no. 11, pp. 1832–1846, Nov. 2015, doi: 10.1002/EJLT.201500125.

[66] D. M. Alonso, J. Q. Bond, and J. A. Dumesic, "Catalytic conversion of biomass to biofuels," *Green Chem.*, vol. 12, no. 9, pp. 1493–1513, Sep. 2010, doi: 10.1039/C004654J.

[67] Y. Liao *et al.*, "A sustainable wood biorefinery for low-carbon footprint chemicals production," vol. 367, no. 6484, pp. 1385–1390, Mar. 2020.

[68] V. M. Chernyshev, O. A. Kravchenko, and V. P. Ananikov, "Conversion of plant biomass to furan derivatives and sustainable access to the new generation of polymers, functional materials and fuels," *Russ. Chem. Rev.*, vol. 86, no. 5, pp. 357–387, May 2017.

[69] N. O. Balayeva, N. Zheng, R. Dillert, and D. W. Bahnemann, "Visible-light-mediated photocatalytic aerobic dehydrogenation of n-heterocycles by surface-grafted TiO_2 and 4-amino-TEMPO," *ACS Catal.*, vol. 9, no. 12, pp. 10694–10704, 2019, doi: 10.1021/ACSCATAL.9B03322.

[70] A. J. Ragauskas *et al.*, "Lignin valorization: improving lignin processing in the biorefinery," *Science*, vol. 344, no. 6185, 2014, doi: 10.1126/SCIENCE.1246843.

[71] X. Huang, O. M. M. Gonzalez, J. Zhu, T. I. Korányi, M. D. Boot, and E. J. M. Hensen, "Reductive fractionation of woody biomass into lignin monomers and cellulose by tandem metal triflate and Pd/C catalysis," *Green Chem.*, vol. 19, no. 1, pp. 175–187, Jan. 2017, doi: 10.1039/C6GC02962K.

[72] J. Yan, Q. Meng, X. Shen, B. Chen, Y. Sun, J. Xiang, H. Liu, and B. Han, "Selective valorization of lignin to phenol by direct transformation of Csp2–Csp3 and C–O bonds," *Sci. Adv.*, vol. 6, no. 45, Nov. 2020, doi: 10.1126/SCIADV.ABD1951.

Index

CATALYTIC SCIENCE SERIES

(Continued from page ii)

Vol. 12 *Catalysis by Ceria and Related Materials (Second Edition)*
edited by A. Trovarelli and P. Fornasiero

Vol. 11 *Supported Metals in Catalysis (Second Edition)*
by J. A. Anderson

Vol. 10 *Concepts in Syngas Manufacture*
by J. Rostrup-Nielsen and L. J. Christiansen

Vol. 9 *Deactivation and Regeneration of Zeolite Catalysts*
edited by M. Guisnet and F. R. Ribeiro

Vol. 8 *Petrochemical Economics: Technology Selection in a Carbon Constrained World*
by D. Seddon

Vol. 7 *Combinatorial Development of Solid Catalytic Materials: Design of High-Throughput Experiments, Data Analysis, Data Mining*
edited by M. Baerns and M. Holeňa

Vol. 6 *Catalysis by Gold*
edited by G. C. Bond, C. Louis and D. T. Thompson

Vol. 5 *Supported Metals in Catalysis*
edited by J. A. Anderson and M. F. García

Vol. 4 *Isotopes in Heterogeneous Catalysis*
edited by Justin S. J. Hargreaves, S. David Jackson and Geoff Webb

Vol. 3 *Zeolites for Cleaner Technologies*
edited by Michel Guisnet and Jean-Pierre Gilson

Vol. 2 *Catalysis by Ceria and Related Materials*
edited by Alessandro Trovarelli

Vol. 1 *Environmental Catalysis*
edited by F. J. J. G. Janssen and R. A. van Santen

Other Titles by the Author

Theoretical Heterogeneous Catalysis
Fundamental Aspects of Heterogeneous Catalysis Studies by Particle Beams
Elementary Reaction Steps in Heterogeneous Catalysis
Chemical Kinetics and Catalysis
Catalytic Oxidation: Principles and Applications
Transition Metal Sulfides
Catalysis: An Integrated Approach
Environmental Catalysis
Computer Modelling of Microporous Materials
Molecular Heterogeneous Catalysis: A Conceptual and Computational Approach
2030: Technology That Will Change the World
Catalysis: From Principles to Applications
Computational Methods in Catalysis and Materials Science: An Introduction for Scientists and Engineers
Modern Heterogeneous Catalysis: An Introduction
Modelling and Simulation in the Science of Micro- and Meso-Porous Materials
Complexity Science: An Introduction
Catalysis for Renewables: From Feedstock to Energy Production

Milton Keynes UK
Ingram Content Group UK Ltd.
UKHW021114190924
448415UK00041B/2

9 781800 614000